Lecture Notes on Data Engineering and Communications Technologies

Volume 90

Series Editor

Fatos Xhafa, Technical University of Catalonia, Barcelona, Spain

The aim of the book series is to present cutting edge engineering approaches to data technologies and communications. It will publish latest advances on the engineering task of building and deploying distributed, scalable and reliable data infrastructures and communication systems.

The series will have a prominent applied focus on data technologies and communications with aim to promote the bridging from fundamental research on data science and networking to data engineering and communications that lead to industry products, business knowledge and standardisation.

Indexed by SCOPUS, INSPEC, EI Compendex.

All books published in the series are submitted for consideration in Web of Science.

More information about this series at https://link.springer.com/bookseries/15362

Deepak Gupta · Zdzislaw Polkowski ·
Ashish Khanna · Siddhartha Bhattacharyya ·
Oscar Castillo

Editors

Proceedings of Data Analytics and Management

ICDAM 2021, Volume 1

 Springer

Editors
Deepak Gupta
Maharaja Agrasen Institute of Technology
New Delhi, India

Zdzislaw Polkowski
Jan Wyzykowski University
Polkowice, Poland

Ashish Khanna
Maharaja Agrasen Institute of Technology
New Delhi, India

Siddhartha Bhattacharyya
Rajnagar Mahavidyalaya
Birbhum, West Bengal, India

Oscar Castillo
Tijuana Institute of Technology
Tijuana, Mexico

ISSN 2367-4512 ISSN 2367-4520 (electronic)
Lecture Notes on Data Engineering and Communications Technologies
ISBN 978-981-16-6291-1 ISBN 978-981-16-6289-8 (eBook)
https://doi.org/10.1007/978-981-16-6289-8

This Springer imprint is published by the registered company Springer Nature Singapore Pte Ltd.
The registered company address is: 152 Beach Road, #21-01/04 Gateway East, Singapore 189721,
Singapore

Dr. Deepak Gupta would like to dedicate this book to his father Sh. R. K. Gupta, his mother Smt. Geeta Gupta for their constant encouragement, his family members including his wife, brothers, sisters, kids, and to his students close to his heart.

Dr. Zdzislaw Polkowski would like to dedicate this book to his Wife, Daughter, and Parents.

Dr. Ashish Khanna would like to dedicate this book to his mentors Dr. A. K. Singh and Dr. Abhishek Swaroop for their constant encouragement and guidance and his family members including his mother, wife, and kids. He would also like to dedicate this work to his (Late) father Sh. R. C. Khanna with folded hands for his constant blessings.

Prof. (Dr.) Siddhartha Bhattacharyya would like to dedicate this book to Dr. Sujit Pal, Joint Director of Public Instructions, Department of Higher Education, Government of West Bengal.

ICDAM 2021 Steering Committee Members

Patrons

Dr. Tadeusz Kierzyk, Prof. UJW, Rector of Jan Wyzykowski University, Polkowice, Poland
Shri. Rakesh Tayal, Panipat Institute of Engineering and Technology, India
Shri. Hariom Tayal, Panipat Institute of Engineering and Technology, India
Shri. Shubham Tayal, Panipat Institute of Engineering and Technology, India
Shri. Suresh Tayal, Panipat Institute of Engineering and Technology, India
Dr. Ashok Gupta, IIS Deemed to be University, Jaipur
Shri. B. S. Yadav, Hon'ble Chancellor-IES University, Bhopal
Dr. Sunita Singh, Hon'ble Pro Chancellor, IES University, Bhopal
Shri. Devansh Singh, CEO, IES University, Bhopal

General Chairs

Prof. Dr. Janusz Kacprzyk, Polish Academy of Sciences, Systems Research Institute, Poland
Prof. Dr. Cesare Alippi, Polytechnic University of Milan, Italy
Prof. Dr. Siddhartha Bhattacharyya, CHRIST (Deemed to be University), Bangalore, India
Prof. Dr. Shakti Kumar, Panipat Institute of Engineering and Technology, India
Prof. T. N. Mathur, IIS Deemed to be University, Jaipur
Dr. Jyotiram Sawale, Registrar, IES University, Bhopal

Honorary Chairs

Prof. Dr. Aboul Ella Hassanien, Cairo University, Egypt
Prof. Dr. Vaclav Snasel, Rector, VSB-Technical University of Ostrava, Czech Republic
Dr. Raakhi Gupta, IIS Deemed to be University, Jaipur
Dr. Naveen Chandra, Vice Chancellor, IES University, Bhopal

Conference Chairs

Dr. Zdzislaw Polkowski, Prof. UJW, Jan Wyzykowski University, Polkowice, Poland
Prof. Dr. Joel J. P. C. Rodrigues, Universidade Estadual do Piau Teresina, Brazil
Prof. Dr. Abhishek Swaroop, Bhagwan Parshuram Institute of Technology, Delhi, India
Prof. Dr. Anil K. Ahlawat, KIET Group of Institutes, Ghaziabad, India
Prof. Dr. Vijay Athavale, Panipat Institute of Engineering and Technology, India
Dr. Suresh Chand Gupta, Panipat Institute of Engineering and Technology, India
Prof. Vijay Singh Rathore, IIS Deemed to be University, Jaipur
Dr. O. P. Modi, Principal, IES Institute of Technology and Management, IES University, Bhopal
Dr. G. K. Pandey, Principal, IES College of Technology, Bhopal

Technical Program Chairs

Dr. Stanislaw Piesiak, Prof. UJW, Jan Wyzykowski University, Polkowice, Poland
Dr. Jan Walczak, Jan Wyzykowski University, Polkowice, Poland
Dr. Anna Wojciechowicz, Jan Wyzykowski University, Lubin, Poland
Dr. Anju Bhandari, Panipat Institute of Engineering and Technology, India
Dr. Dinesh Verma, Panipat Institute of Engineering and Technology, India
Dr. Arti Jain, Jaypee Institute of Information Technology (JIIT)
Dr. Sushil Kumar Singh, Seoul National University of Science and Technology, Seoul, South Korea
Dr. Pallavi Bhatnagar, IES College of Technology, Bhopal
Dr. Pramod Kumar Patel, IES College of Technology, Bhopal
Dr. Anil Kumar Yadav, IES College of Technology, Bhopal
Dr. Nikhat Raza Khan, IES College of Technology, Bhopal
Dr. Rajesh Kumar Nema, IES College of Technology, Bhopal
Dr. Jitendra Mathur, IES College of Technology, Bhopal

Prof. Khushbu Kriplani, IES College of Technology, Bhopal
Prof. Sonu Lal, IES College of Technology, Bhopal
Prof. Jamvant Omkar, IES College of Technology, Bhopal

Technical Program Co-chairs

Prof. Dr. Victor Hugo C. de Albuquerque, Universidade de Fortaleza, Brazil
Dr. Gulshan Shrivastava, Sharda University, Greater Noida, India
Dr. Akhilesh Kumar Mishra, Panipat Institute of Engineering and Technology, India
Dr. Mukesh Chawla, Panipat Institute of Engineering and Technology, India
Rattendeep Aneja, Panipat Institute of Engineering and Technology, India

Conveners

Dr. Ashish Khanna, Maharaja Agrasen Institute of Technology (GGSIPU), New Delhi, India
Dr. Deepak Gupta, Maharaja Agrasen Institute of Technology (GGSIPU), New Delhi, India
Dr. Pradeep Kumar Mallick, KIIT Deemed to be University Bhubaneswar, Odisha, India
Dr. Akash Kumar Bhoi, Sikkim Manipal University, India
Dr. Bhawna Singla, Panipat Institute of Engineering and Technology, India
Prof. K. S. Sharma, IIS Deemed to be University, Jaipur

Publication Chairs

Dr. Jerzy Widerski, Prof. UJW, V-Rector of Jan Wyzykowski University, Polkowice, Poland
Dr. Vicente García Díaz, University of Oviedo, Spain
Akanksha, Panipat Institute of Engineering and Technology, India
Sandeep Jaglan, Panipat Institute of Engineering and Technology, India
Shakti Arora, Panipat Institute of Engineering and Technology, India

Publicity Chairs

Dr. Pawel Gren, Prof. UJW, V-Rector of Jan Wyzykowski University, Polkowice, Poland

Dr. Aditya Khamparia, Lovely Professional University, Punjab, India
Saurab Gupta, Panipat Institute of Engineering and Technology, India
Tarun Miglani, Panipat Institute of Engineering and Technology, India
Ms. Ginni, Panipat Institute of Engineering and Technology, India

Co-conveners

Dr. Deepak Wadhwa, Panipat Institute of Engineering and Technology, India
Mr. Moolchand Sharma, Maharaja Agrasen Institute of Technology, India
Dr. Vaishali Mehta, Panipat Institute of Engineering and Technology, India
Dr. Anubha Jain, IIS Deemed to be University, Jaipur

Advisory Committee

Advisory Committee

Dr. Tadeusz Kierzyk, Prof. UJW, Rector of Jan Wyzykowski University, Polkowice, Poland

Prof. Vincenzo Piuri, University of Milan, Italy

Prof. Aboul Ella Hassanien, Cairo University, Egypt

Prof. Marcin Paprzycki, Polish Academy of Science, Poland

Prof. Valentina Emilia Balas, Aurel Vlaicu University of Arad, Romania

Prof. Marius Balas, Aurel Vlaicu University of Arad, Romania

Prof. Mohamed Salim Bouhlel, Sfax University, Tunisia

Prof. Cenap Ozel, King Abdulaziz University, Saudi Arabia

Prof. Ashiq Anjum, University of Derby, Bristol, UK

Prof. Mischa Dohler, King's College London, UK

Prof. David Camacho, Universidad Autonoma de Madrid, Spain

Prof. Parmanand, Dean, Galgotias University, Uttar Pradesh, India

Prof. Maryna Yena, Medical University of Kiev, Ukraine

Prof. Giorgos Karagiannidis, Aristotle University of Thessaloniki, Greece

Prof. Tanuja Srivastava, Department of Mathematics, IIT Roorkee

Dr. D. Jude Hemanth, Karunya University, Coimbatore

Prof. Tiziana Catarci, Sapienza University, Rome, Italy

Prof. Salvatore Gaglio, University Degli Studi di Palermo, Italy

Prof. Bozidar Klicek, University of Zagreb, Croatia

Prof. A. K. Singh, NIT Kurukshetra, India

Prof. Anil Kumar, KIET Group of Institutes, India

Prof. Chang-Shing Lee, National University of Tainan, Taiwan

Dr. Paolo Bellavista, Alma Mater Studiorum–Università di Bologna

Prof. Sanjay Misra, Covenant University, Nigeria

Prof. Benatiallah Ali, Adrar University, Algeria

Prof. Suresh Chandra Satapathy, PVPSIT, Vijayawada, India

Prof. Marylene Saldon-Eder, Mindanao University of Science and Technology

Prof. Özlem Onay, Anadolu University, Eskisehir, Turkey

Prof. Kei Eguchi, Department of Information Electronics, Fukuoka Institute of Technology

Prof. Zoltan Horvath, Kasetsart University
Dr. A. K. M. Matiul Alam Vancouver British Columbia, Canada
Prof. Joong Hoon Jay Kim, Korea University
Prof. Sheng-Lung Peng, National Dong Swa University, Taiwan
Dr. Dusanka Boskovic, University of Sarajevo, Sarajevo
Dr. Periklis Chat Zimisios, Alexander TEI of Thessaloniki, Greece
Dr. Nhu Gia Nguyen, Duy Tan University, Vietnam
Dr. Ahmed Faheem Zobaa, Brunel University, London
Prof. Ladjel Bellatreche, Poitiers University, France
Prof. Victor C. M. Leung, The University of British Columbia, Canada
Prof. Huseyin Irmak, Cankiri Karatekin University, Turkey
Dr. Alex Norta, Tallinn University of Technology, Estonia
Prof. Amit Prakash Singh, GGSIPU, Delhi, India
Prof. Abhishek Swaroop, Bhagwan Parshuram Institute of Technology, Delhi
Prof. Christos Douligeris, University of Piraeus, Greece
Dr. Brett Edward Trusko, President and CEO (IAOIP) and Assistant Professor, Texas
A&M University, Texas
Prof. Joel J. P. C. Rodrigues, National Institute of Telecommunications (Inatel),
Brazil; Instituto de Telecomunicações, Portugal
Prof. Victor Hugo C. de Albuquerque, University of Fortaleza (UNIFOR), Brazi
Dr. Atta ur Rehman Khan, King Saud University, Riyadh
Dr. João Manuel R. S. Tavares, FEUP—DEMec
Prof. Ku Ruhana Ku Mahamud, School of Computing, College of Arts and Sciences,
Universiti Utara Malaysia, Malaysia
Prof. Ghasem D. Najafpour, Babol Noshirvani University of Technology, Iran
Prof. Sanjeevikumar Padmanaban, Aalborg University, Denmark
Prof. Frede Blaabjerg, President (IEEE Power Electronics Society), Aalborg Univer-
sity, Denmark
Prof. Jens Bo Holm Nielson, Aalborg University, Denmark
Dr. Abu Yousuf, University Malaysia Pahang Gambang, Malaysia
Dr. Ahmed A. Elngar, Faculty of Computers and Information, Beni-Suef University,
Egypt
Prof. Dijana Oreski, Faculty of Organization and Informatics, University of Zagreb,
Varazdin, Croatia
Prof. Prasad K. Bhaskaran, Ocean Engineering and Naval Architecture, IIT
Kharagpur
Dr. Yousaf Bin Zikria, Yeungnam University, South Korea
Dr. Sanjay Sood, C-DAC, Mohali
Prof. Ajay Rana, Senior Vice President and Advisor—Amity Education Group,
Amity University, Noida, India
Dr. Florin Popentiu Vladicescu, University Politehnica of Bucharest, Romania
Dr. Pawel Gren, Prof. UJW, Jan Wyzykowski University, Polkowice, Poland
Prof. Joanna Jozefowska, Pro-Rector for Research (etc.) of Poznan University of
Technology
Prof. Gerhard-Wilhelm Weber, Poznan University of Technology, Poland

Preface

We hereby are delighted to announce that Jan Wyzykowski University, Polkowice, Poland, Panipat Institute of Engineering and Technology, IIS University, and IES University, India, have hosted the eagerly awaited and much coveted International Conference on Data Analytics and Management (ICDAM 2021). The second version of the conference was able to attract a diverse range of engineering practitioners, academicians, scholars, and industry delegates, with the reception of abstracts including more than 2600 authors from different parts of the world. The committee of professionals dedicated toward the conference is striving to achieve a high-quality technical program with tracks on data analytics, data management, big data, computational intelligence, and communication networks. All the tracks chosen in the conference are interrelated and are very famous among present-day research community. Therefore, a lot of research is happening in the above-mentioned tracks and their related sub-areas. More than 650 full-length papers have been received, among which the contributions are focused on theoretical, computer simulation-based research, and laboratory scale experiments. Among these manuscripts, 131 papers have been included in Springer proceedings after a thorough two-stage review and editing process. All the manuscripts submitted to ICDAM 2021 were peer-reviewed by at least two independent reviewers, who were provided with a detailed review proforma. The comments from the reviewers were communicated to the authors, who incorporated the suggestions in their revised manuscripts. The recommendations from two reviewers were taken into consideration while selecting a manuscript for inclusion in the proceedings. The exhaustiveness of the review process is evident, given the large number of articles received addressing a wide range of research areas. The stringent review process ensured that each published manuscript met the rigorous academic and scientific standards. It is an exalting experience to finally see these elite contributions materialize into the two book volumes as ICDAM proceedings by Springer entitled *Proceedings of Data Analytics and Management*.

ICDAM 2021 invited three keynote speakers, who are eminent researchers in the field of computer science and engineering, from different parts of the world. In addition to the plenary sessions on each day of the conference, twelve concurrent technical sessions are held every day to assure the oral presentation of around

131 accepted papers. Keynote speakers and session chair(s) for each of the concurrent sessions have been leading researchers from the thematic area of the session. The delegates were provided with a book of extended abstracts to quickly browse through the contents, participate in the presentations, and provide access to a broad audience of the audience. The research part of the conference was organized in a total of 40 special sessions. These special sessions provided the opportunity for researchers conducting research in specific areas to present their results in a more focused environment.

An international conference of such magnitude and release of the ICDAM 2021 proceedings by Springer has been the remarkable outcome of the untiring efforts of the entire organizing team. The success of an event undoubtedly involves the painstaking efforts of several contributors at different stages, dictated by their devotion and sincerity. Fortunately, since the beginning of its journey, ICDAM 2021 has received support and contributions from every corner. We thank them all who have wished the best for ICDAM 2021 and contributed by any means toward its success. The edited proceedings volumes by Springer would not have been possible without the perseverance of all the steering, advisory, and technical program committee members.

All the contributing authors owe thanks from the organizers of ICDAM 2021 for their interest and exceptional articles. We would also like to thank the authors of the papers for adhering to the time schedule and for incorporating the review comments. We wish to extend our heartfelt acknowledgment to the authors, peer reviewers, committee members, and production staff whose diligent work put shape to the ICDAM 2021 proceedings. We especially want to thank our dedicated team of peer reviewers who volunteered for the arduous and tedious step of quality checking and critique on the submitted manuscripts. We wish to thank our faculty colleague Mr. Moolchand Sharma for extending their enormous assistance during the conference. The time spent by them and the midnight oil burnt are greatly appreciated, for which we will ever remain indebted. The management, faculties, administrative, and support staff of the college have always been extending their services whenever needed, for which we remain thankful to them.

Lastly, we would like to thank Springer for accepting our proposal for publishing the ICDAM 2021 conference proceedings. Help received from Mr. Aninda Bose, the acquisition senior editor, in the process has been very useful.

Ashish Khanna
Deepak Gupta
Organizers, ICDAM 2021
New Delhi, India

Contents

About the Editors

Dr. Deepak Gupta received a B.Tech. degree in 2006 from the Guru Gobind Singh Indraprastha University, India. He received M.E. degree in 2010 from Delhi Technological University, India and Ph.D. degree in 2017 from Dr. A. P. J. Abdul Kalam Technical University, India. He has completed his Post-Doc from Inatel, Brazil. With 13 years of rich expertise in teaching and two years in the industry; he focuses on rational and practical learning. He has contributed massive literature in the fields of Intelligent Data Analysis, BioMedical Engineering, Artificial Intelligence, and Soft Computing. He has served as Editor-in-Chief, Guest Editor, Associate Editor in SCI and various other reputed journals (IEEE, Elsevier, Springer, and Wiley). He has actively been an organizing end of various reputed International conferences. He has authored/edited 43 books with National/International level publishers (IEEE, Elsevier, Springer, Wiley, Katson). He has published 162 scientific research publications in reputed International Journals and Conferences including 83 SCI Indexed Journals of IEEE, Elsevier, Springer, Wiley and many more.

Dr. Zdzislaw Polkowski is an Adjunct Professor at Faculty of Technical Sciences at the Jan Wyzykowski University, Poland. He is also the Rector's Representative for International Cooperation and Erasmus Programme and former dean of the Technical Sciences Faculty during the period of 2009–2012. His area of research includes Management Information Systems, Business informatics, IT in business and administration, IT security, Small Medium Enterprises, CC, IoT, Big Data, Business Intelligence, and Block chain. He has published around 60 research articles. He has served the research community in the capacity of author, professor, reviewer, keynote speaker, and co-editor. He has attended several international conferences in the various parts of the world. He is also playing the role principal investigator.

Dr. Ashish Khanna has expertise in Teaching, Entrepreneurship, and Research and Development of 16 years. He received his Ph.D. degree from National Institute of Technology, Kurukshetra in March 2017. He has completed his M.Tech. and B.Tech. from GGSIPU, Delhi. He has completed his PDF from Internet of Things Lab at Inatel, Brazil. He has around 100 research papers along with book chapters including more than 40 papers in SCI indexed Journals with cumulative impact factor of above 100 to his credit. Additionally, He has authored, edited and editing 19 books. Furthermore, he has served the research field as a Keynote Speaker/Session Chair/Reviewer/TPC member/Guest Editor and many more positions in various conferences and journals. His research interest includes image processing, Distributed Systems and its variants, and Machine learning. He is currently working at the CSE, Maharaja Agrasen Institute of Technology, Delhi. He is convener and organizer of ICICC Springer conference series.

Dr. Siddhartha Bhattacharyya (FRSA, FIET (UK), FIEI, FIETE, LFOSI, SMIEEE, SMACM, SMIETI, LMCSI, LMISTE) is currently the Principal of Rajnagar Mahavidyalaya, Birbhum, India. Prior to this, he was a Professor at CHRIST (Deemed to be University), Bangalore, India. He also served as the Principal of RCC Institute of Information Technology, Kolkata, India. He has served VSB Technical University of Ostrava, Czech Republic as a Senior Research Scientist. He is the recipient of several coveted national and international awards. He received the Honorary Doctorate Award (D.Litt.) from The University of South America and the SEARCC International Digital Award ICT Educator of the Year in 2017. He was appointed as the ACM Distinguished Speaker for the tenure 2018–2020. He has been appointed as the IEEE Computer Society Distinguished Visitor for the tenure 2021–2023. He is a co-author of 6 books and the co-editor of 75 books and has more than 300 research publications in international journals and conference proceedings to his credit.

Dr. Oscar Castillo holds the Doctor in Science degree (Doctor Habilitatus) in Computer Science from the Polish Academy of Sciences (with the Dissertation *Soft Computing and Fractal Theory for Intelligent Manufacturing*). He is a Professor of Computer Science in the Graduate Division, Tijuana Institute of Technology, Tijuana, Mexico. Currently, he is President of HAFSA (Hispanic American Fuzzy Systems Association) and Past President of IFSA (International Fuzzy Systems Association). Prof. Castillo is also Chair of the Mexican Chapter of the Computational Intelligence Society (IEEE). His research interests are in Type-2 Fuzzy Logic, Fuzzy Control, Neuro-Fuzzy and Genetic-Fuzzy hybrid approaches. He has published over 300 journal papers, 10 authored books, 40 edited books, 200 papers in conference proceedings, and more than 300 chapters in edited books, in total 865 publications according to Scopus (H index = 60), and more than 1000 publications according to Research Gate (H index = 72 in Google Scholar).

Psychological Stress and Mental Health Among Seafarers

Graziano Pallotta, Gopi Battineni, Giulio Nittari, and Francesco Amenta

Abstract **Objective**: This work aims to determine whether the mental health of seafarers is a significant problem, by providing actual epidemiological information, to identify the factors that are supporting or undermining the mental health of seafarers and to propose solutions and practices aimed at improving the health of this class of workers. **Methods**: This study is an epidemiological investigation of mental and behavioral disorders among seafarers onboard commercial ships without a physician. The aim is to propose solutions to improve the quality of life in this difficult working environment. We examined 38.477 requests of assistance from patients embarked on ships assisted by the CIRM from 2011 to 2019. All the diagnosed diseases have been categorized based on the ICD-10 classification system by the WHO. **Results**: From 2012 to 2020, 376 cases of "mental and behavioral disorders" were officially diagnosed. The most common form of mental disorder was anxious syndrome (119 cases), followed by depressive disorder (103 cases), insomnia (51 cases), panic attacks (35 cases), etc. Over the 9 years analyzed, a total of 37 suicide and 4 attempted suicide

Authors' contributions. GP: wrote the manuscript, organized the research, and conducted the statistical analysis, data collection, and interpretation. GR, GB: contributed to statistic and data interpretation. FA: supervision and final draft revision, and GN: conducted statistical analysis, data collection, and interpretation. All authors read and approved the final manuscript.

G. Pallotta · G. Battineni (✉) · G. Nittari · F. Amenta
Telemedicine and Telepharmacy Centre, School of Medicinal and Health Products Sciences, University of Camerino, Camerino, Italy
e-mail: gopi.battineni@unicam.it

G. Pallotta
e-mail: graziano.pallotta@unicam.it

G. Nittari
e-mail: giulio.nittari@unicam.it

F. Amenta
e-mail: francesco.amenta@unicam.it

F. Amenta
Research Department, International Radio Medical Centre (C.I.R.M.), Rome, Italy

cases occurred. The number of mental and behavioral disorders diagnosed on board—as well as the number of suicides—shows that the seafarers' mental problems are a concrete and serious issue.

Keywords Mental health · Seafarers · Telemedicine · Psychological stress · Epidemiology

1 Introduction

The specific and characteristic isolation of the naval environment can put a strain on the mental health of the crews. Being in the middle of the ocean, performing complex and sometimes dangerous tasks can significantly increase the stress of these workers, with important repercussions on the health of seafarers [1, 2]. Working always in the same spaces, isolated from the rest of the world, but at the same time always in contact with the same people, the same faces, feeling the same noises and smells, few hours' sleep, can cause a psychological deterioration that may lead to chronic disorders and even severe pathologies [3–5].

The logistics of the work onboard ships requires a rigorous division of roles in the various sectors (bridge, engine, etc.), which inevitably leads to a physical and human separation between seafarers—all of this exacerbates relationships, emphasizing the negative aspects [6]. It is no coincidence that the World Labor Organization (ILO) created a study group on the working conditions of maritime crews in 1993 [7], nor that the International Maritime Organization (IMO) first and the various national and community legislators after have launched precise protection rules for seafarers [8, 9].

On the other hand, it is a fact that research has often focused mainly on accidents and injuries or physical pathologies [9–11], sometimes neglecting psychological discomfort, which profoundly affects health, defined by the WHO as "a complete state of physical, social, and mental well-being" and, therefore, not just the absence of diseases or infirmities [12]. After all, if technology has ensured contacts with the mainland and with families via e-mail and the Internet, on the other hand, life onboard keeps on being, today as centuries ago, a highly penalizing dimension.

Several studies agree that the main cause of mental health problems of seafarers is the long absence from home [13]. A life inscribed in a dimension dominated by the roll of the ship, where the perception of existence is punctuated by work shifts, finds it difficult to deal with an ordinary life, albeit much desired. Nostalgia for home, that "pain for the desire to return," as the Greek etymology attests, from Homer onward, substantiates the individual even when he has returned to his family because he has been completely impregnated and altered by it.

By listening to many experiences, another data emerged to pay utmost attention to—the "transition" period, that is, the phase relating to the first days after returning home, a real period of re-adaptation, similar to what the seafarer will suffer before departure and in the early days spent on board. This is why, uneasiness and tension are expressed in couples and families, while children, especially if they are young, see their fathers as "strangers," elements extraneous to their life, which becomes "normal" when they leave [14].

This work aims to determine whether the mental health of seafarers is a significant problem, by providing actual epidemiological information—furthermore, to propose solutions and practices aimed at improving the health of this class of workers. The research questions that we aim to answer by this work as follows.

RQ1: Do mental health disorders represent a health issue for seafarers and shipping companies?

RQ2: What factors can improve or worsen the mental health of seafarers?

RQ3: What initiatives can be applied to improve life on board?

2 Methods

This work is an epidemiological investigation carried out on 38,477 requests for assistance from seafarers embarked on commercial vessels without health personnel on board and assisted by the International Radio Medical Center (C.I.R.M.), in the 9 years 2012–2020, focusing on the mental health of seafarers.

As in these vessels, there is no medical personnel onboard—health care is provided through telehealth techniques. For each patient assisted by the C.I.R.M., an electronic medical report was established and called "electronic health record" (EHR), which has been updated following every contact with the ship [15]. These records represent the basis of the investigations carried out in this study. All the diagnosed diseases have been categorized accordingly to the International Statistical Classification of Diseases and Related Health Problems, 10th Revision (ICD-10) by the World Health Organization (WHO) [16].

Among the different ICD-10 categories, we took into consideration the diagnoses included in the ICD-10 Class "V," called "mental and behavioral disorders." All data were analyzed with standard statistic methods. Microsoft Excel was the software used for information processing and result analysis. Data are expressed in the text as means ± SD.

Data were anonymized before being used for research purpose. The survey is a part of the project called health protection and safety on board ships (acronym: HEALTHY SHIP). It is a project of disease prevention and health protection onboard sailing ships through information campaigns on the major health risks for seafarers and on their approved by ethical committee at International Radio Medical Center (C.I.R.M.) [17].

3 Results

The results of the epidemiological analysis are summarized in Tables 1 and 2, and Fig. 1 presents mental disorder distribution. The diagnoses were performed by competent physicians of the C.I.R.M., adequately trained, via telemedicine platforms capable of communicating with the patient by phone or video call. Available for

Table 1 Specific diagnoses and total cases included in Class "V" of the ICD-10 system, sorted by year

Disorder	2012	2013	2014	2015	2016	2017	2018	2019	2020	Total
Anxious syndrome	16	14	14	15	13	8	12	15	12	119
Behavioral disorder	1	1	1	1	2	1	2	2	2	13
Confusion state	1	0	1	2	0	2	1	1	1	9
Depressive disorder	8	14	7	10	16	12	16	4	16	103
Generalized malaise	0	0	1	2	1	0	0	1	1	6
Insomnia	1	6	4	6	7	11	6	2	8	51
Panic attack	2	3	2	9	4	4	3	3	5	35
Psychomotor agitation	2	1	0	1	2	1	3	1	2	13
Psychotic syndrome	3	3	1	2	3	2	2	1	4	21
Spatiotemporal disorientation	1	0	0	1	1	0	1	1	1	6
Total	35	42	31	49	49	41	46	31	52	376

Table 2 A number of suicides and suicide attempts reported in the 9 years 2012– 2020

Year	Suicide attempts	Suicides	Total (suicides + attempts)
2012	0	4	4
2013	1	2	3
2014	0	5	5
2015	0	3	3
2016	1	3	4
2017	0	4	4
2018	1	4	5
2019	1	6	7
2020	0	6	6
Total	4	37	41

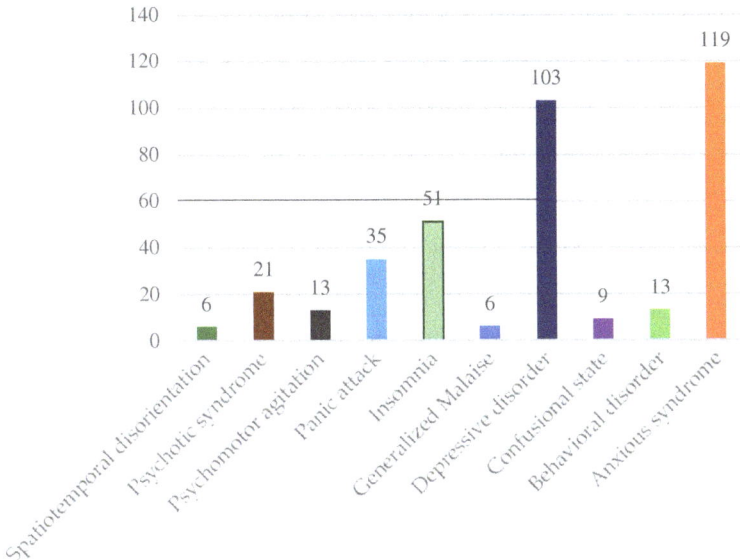

Fig. 1 Total cases of mental disorders recorded in 2012–2020, distinguished according to their specific diagnosis

doctors, the patient's electronic health record (EHR), which indicates all the known medical information of the subject, as well as their complete recorded medical history.

The diagnoses were confirmed following the guidelines of the ICD-10 system by the WHO and by the Diagnostic and Statistical Manual of Mental Disorders (DSM-5) published by the American Psychiatric Association.

From 2012 to 2020, 376 cases of "mental and behavioral disorders" were officially diagnosed with a slight increase in cases over the years (refer Fig. 1), hand in hand with an increased number of patients treated in total. Between 2011–2019, the most common form of mental disorder was anxious syndrome (119), followed by depressive disorder (103), next are insomnia (51), panic attacks (35), psychotic syndrome (21), psychomotor agitation (13 cases), behavioral disorder (13), confessional state (9), spatiotemporal disorientation (6), generalized malaise (6). Over the 9 years analyzed, a total of 37 suicide and 4 attempted suicide cases occurred (Table 2). These are worrying numbers that may be strictly related to the mental disorders diagnosed and mentioned above. The number of suicides over the years appears fairly constant, without a visible tendency to increase or decrease. Also, worrying is the fact that the number of successful suicides is significantly higher than suicide attempts (37 vs. 4).

The main mental health problems onboard diagnosed from 2012 to 2020 are anxious syndrome (119) and depressive disorder (103). Anxiety disorders are a set of psychiatric disorders characterized by unjustified fear and anxiety, often associated with psychosomatic manifestations that create considerable discomfort for the subject. Commonly, these are chronic disorders present in a latent form in the individual, and that occurs as a result of particular psychophysical stress, or following a traumatic event. They are characterized by a strong anxious and phobic component, in most cases unjustified. The naval environment may trigger these pathological conditions, as suggested by the high number of cases diagnosed on board.

A depressive disorder is a complex psychiatric pathological condition characterized by episodes of depressed mood, low self-esteem, loss of interest in normal life activities, etc. It is a disabling disease capable of compromising the patient's social, physical, and working life. It normally starts manifesting itself at 20 years of age, with a peak in subjects between their 30s and 40s [18]. For both anxiety and depressive disorders, drug therapy associated with targeted psychotherapy is essential [19]. Although the causes of depression are still being studied, it is now widely believed that both biological and genetic factors, as well as environmental and psychological factors, can determine the onset of depressive disorder [20].

4 Discussion

Epidemiological information relating to the mental health of seafarers has been rather scarce so far. This is mainly due to a historical lack of interest in the naval environment epidemiology and also to a greater emphasis on physical pathologies and injuries [9, 10].

After all, determining seafarers' mental health is quite difficult. First, we are talking about remote workers, located far from the land and difficult to reach. They are selected following specific medical examinations. This determines the immediate elimination of the most fragile and sick subjects, creating a population of workers we could define as "selected" and, therefore, theoretically "stronger." Therefore, even from a merely statistical point of view, it is difficult to compare maritime workers with the rest of the population, or with other groups.

Among the environmental and social factors that would seem to play a greater role in the etiology of depression, we have childhood trauma, family problems, bereavements, divorces, serious health problems, poverty, unemployment, social isolation, etc. [21] Work mobbing, bullying, and prolonged work stress are also recognized as possible causes [22]. There are situations that, according to the most recent literature, occur quite frequently on board [23]. Indeed, physical and/or verbal offenses, humiliation, bullying, persistent criticism, etc. often occur onboard ships [24]. The maritime authorities must consider these social issues like these can have harmful consequences for the health of seafarers. Aggressive, violent behavior (both physically and verbally) should be reported, and the perpetrators were identified and severely punished. Repeated physical and verbal abuse phenomena must be predicted and punished under the regulations in force.

The number of suicides may appear as low after a first observation of the data (Table 2), and this number takes a different perspective when compared to the number of confirmed deaths per year. For example, in 2020, the year with the highest number of confirmed deaths (26), 6 suicides occurred (23% of all deaths). In 2018, 4 of the 27 deaths recorded were suicides (15% of all deaths). The same applies to 2016 where 3 suicides occurred out of 23 total deaths (13% of all deaths). This means that suicides account for an important part of all the deaths that occur on board. Although the number of mental illnesses diagnosed on board is low, it would be useful to understand why cases of suicide on board are such a significant constant in all the years analyzed. Data about repatriation could provide interesting information—however, it emerged that many seafarers prefer to suffer in silence, fearing to be diagnosed with a psychic and/or mood disorder, which may lead to repatriation and a possible compromise of their future job opportunities [23]. This may lead to a serious underestimation of diagnoses of mental illness and stress on board. Monitoring and support programs aimed at the psychophysical health of the maritime patient are, therefore, crucial.

According to the most recent literature, several factors can compromise the mental health of seafarers. First of all, the distance from home, followed by the condition of isolation, loneliness [25–27], few leave, too long work contracts [28], poor-quality food [23], bad social relationships on board, and [28] fear of losing their job [29, 30], fear of losing their family [26, 28, 31–33]. The researchers are also quite sure that the different tasks onboard significantly affect the mental stress to which seafarers are subjected. From this point of view, the most fragile category would seem to be the officers, engineers, and engine crew [4, 25, 31, 34]. The social relationships that can be established in the naval environment can have their impact as well-strong, and important friendships can improve life on board. On the contrary, enmities, bullying, fights, and bad relationships can significantly compromise an already difficult co-existence [23]. The mental health problems of seafarers are not limited to their period on board. The problems ashore can be just as serious. Separation and divorces are very frequent among seafarers [35]. The situation is complicated for those couples with children, where the maritime member is often not to see their child grow up, from whom he is then seen as a sort of distant relative [23]. Often, in the few months on the ground, the seafarer is forced to face problems such as family issues, illnesses, economic problems, bereavements in the family, bills, political and social situations.

The work done by ITF Seafarers Trust and Yale University has led to the identification of needs that seafarers have, by recognizing what they think would be good to make them happier and reduce stress [23]. Since naval isolation and distance from home are the main causes of depression and mood disorders on board, there is a need to enhance the possibilities of communication and contact with the mainland. Communicating with families and friends on the ground must be simpler and more immediate. At the same time, there is a need for something that breaks the daily routine and diversifies the working days, so that no two days are alike. Recreational activities, social events, the opportunity to practice sports, better quality food are the main needs on board. Themed events, group activities, customizable and thematic lunch/dinner menus are some of the seafarers' wishes to break the monotony of a work environment which can be insidious to their psyche. These are needs that can be met with a minimum of organization and method, but which may also have the great potential to improve the general health of this class of workers.

5 Conclusions

In terms of numbers the specific cases of maritime patients affected by mental disorders appear low when compared with other classes of pathologies, there is evidence that these health problems may actually be more widespread and often hidden by the seafarer himself. Equally worrying is the number of suicides that occur on board every year, covering an important number of all recorded deaths. Although epidemiological information does not show a large diffusion of mental disorders on board, from an elaboration of these and the analysis of the literature, the authors hypothesize that these data may be underestimated. The mental problems on board are concrete and capable of seriously compromising the seafarer's health, work, and life in general.

6 Recommendations

The problem has so far been acknowledged only by a small group of shipping companies, which are finally starting to seriously consider the problem of mental health, proposing strategies aimed at improving co-existence onboard commercial vessels. The scientific research will aim to better identify the needs of the seafarers to propose reactive solutions and guarantee mental well-being on board. Authorities and shipping companies need to take patient health more seriously and establish monitoring and mental support programs. The mental health of these people must become of interest to both naval authorities and societies.

Conflicts of interest The authors declare that they have no competing interests.

Funding Part of this work was supported by a grant of the ITF Trust (No. 1276/2018) for the epidemiological analysis part.

References

1. Sandal GM, Leon GR, Palinkas L (2006) Human challenges in polar and space environments. Rev Environ Sci Biotechnol. https://doi.org/10.1007/s11157-006-9000-8
2. Palinkas LA (2003) The psychology of isolated and confined environments: understanding human behavior in Antarctica. Am Psychol. https://doi.org/10.1037/0003-066X.58.5.353
3. Doyle N et al (2016) Resilience and well-being amongst seafarers: cross-sectional study of crew across 51 ships. Int Arch Occup Environ Health. https://doi.org/10.1007/s00420-015-1063-9
4. Carter T (2005) Working at sea and psychosocial health problems. Report of an International Maritime Health Association. Travel Med Infect Dis. https://doi.org/10.1016/j.tmaid.2004.09.005.

5. Hystad SW, Eid J (2016) Sleep and fatigue among seafarers: the role of environmental stressors, duration at sea and psychological capital. Saf Health Work. https://doi.org/10.1016/j.shaw.2016.05.006
6. Forsyth CJ, Bankston WB (1984) The social psychological consequence of a life at sea: a causal model. Marit Policy Manag. https://doi.org/10.1080/03088838400000005
7. "Maritime Labor Convention (2006) International Labour Organization (ILO)
8. Nittari G, Pallotta G, Di Canio M, Traini E, Amenta F (2018) Benzodiazepine prescriptions on merchant ships without a doctor on board: Analysis from medical records of Centro Internazionale Radio Medico (CIRM). Int Marit Health. https://doi.org/10.5603/IMH.2018.0005
9. Pallotta G, Di Canio M, Scuri S, Amenta F, Nittari G (2019) First surveillance of malaria among seafarers: evaluation of incidence and identification of risk areas. Acta Biomed. https://doi.org/10.23750/abm.v90i3.8612
10. Nilsson R, Nordlinder R, Högstedt B, Karlsson A, Järvholm B (1997) Symptoms, lung and liver function, blood counts, and genotoxic effects in coastal tanker crews. Int Arch Occup Environ Health. https://doi.org/10.1007/s004200050166
11. Nittari G et al (2019) Comparative analysis of the medicinal compounds of the ship's 'medicine chests' in European Union maritime countries. Need for improvement and harmonization. Int Marit Health. https://doi.org/10.5603/IMH.2019.0023
12. WHO (2020) WHO frequently asked questions
13. Battineni G et al (2020) Cloud-based framework to mitigate the impact of COVID-19 on seafarers' mental health. Int Marit Health 71(3):213–214. https://doi.org/10.5603/IMH.2020.0038
14. Finkelman JM (1994) A large database study of the factors associated with work- induced fatigue. Hum Factors. https://doi.org/10.1177/001872089403600205
15. Battineni G, Sagaro GG, Chintalapudi N, Amenta F (2020) Conceptual framework and designing for a Seafarers' Health Observatory (SHO) based on the Centro Internazionale Radio Medico (C.I.R.M.) Data Repository. Sci World J 2020. https://doi.org/10.1155/2020/8816517
16. WHO (2011) International statistical classification of diseases and related health problems— 10th revision. World Health Organization, vol 2, no. 5, pp 1–252, 2011, ISBN: 978 92 4 549165
17. Amenta F, Carotenuto A, Grappasonni I (2011) Healthy Ship: an innovative project for healthcare improvement and health promotion on board ships. Luxemburg (2011)
18. American Psychiatric Association (2012) Major depressive disorder and the bereavement exclusion. American Psychiatric Publications
19. Kirsch I, Deacon BJ, Huedo-Medina TB, Scoboria A, Moore TJ, Johnson BT (2008) Initial severity and antidepressant benefits: a meta-analysis of data submitted to the food and drug administration. PLoS Med. https://doi.org/10.1371/journal.pmed.0050045
20. Rutter M, Harrington R, Quinton D, Pickles A (1994) Adult outcome of conduct disorder in childhood: Implications for concepts and definitions of patterns of psychopathology. In: Adolescent problem behaviors: issues and research
21. Weich S, Lewis G (1998) Poverty, unemployment, and common mental disorders: Population based cohort study. Br Med J. https://doi.org/10.1136/bmj.317.7151.115
22. Kendler KS, Hettema JM, Butera F, Gardner CO, Prescott CA (2003) Life event dimensions of loss, humiliation, entrapment, and danger in the prediction of onsets of major depression and generalized anxiety. Arch Gen Psychiatry. https://doi.org/10.1001/archpsyc.60.8.789
23. "Seafarer Mental Health Study," ITF Seafarers' Trust & Yale University (2019)
24. International Chamber of Shipping (2016) Guidance on eliminating shipboard harassment and bullying
25. Mellbye A, Carter T (2017) Seafarers' depression and suicide. Int Marit Health. https://doi.org/10.5603/IMH.2017.0020
26. Jepsen JR, Zhao Z, van Leeuwen WMA (2015) Seafarer fatigue: a review of risk factors, consequences for seafarers' health and safety and options for mitigation. Int Marit Health. https://doi.org/10.5603/IMH.2015.0024

27. Battineni G, Amenta F (2020) Designing of an expert system for the management of Seafarer's health. Digit Heal 6(1):205520762097624. https://doi.org/10.1177/2055207620976244
28. Iversen RTB (2012) The mental health of seafarers. Int Marit Health 63(2):78–89
29. Swift O (2015) Social isolation of seafarers. ISWAN
30. Jezewska M, Iversen RTB, Leszczyńska I (2013) MENHOB–mental health on board, In: 12th International symposium on maritime health brest, France, 6 June 2013
31. Carotenuto A, Molino I, Fasanaro AM, Amenta F (2012) Psychological stress in seafarers: a review. Int Maritime Health
32. Leszczyńska I, Jezewska M, Jaremin B (2008) Work-related stress at sea. Possibilities of research and measures of stress. Int Marit Health 59(1–4):93–102
33. Oldenburg M, Jensen HJ, Latza U, Baur X (2009) Seafaring stressors aboard merchant and passenger ships. Int J Public Health. https://doi.org/10.1007/s00038-009-7067-z
34. Lefkowitz RY, Slade MD, Redlich CA (2019) Rates and occupational characteristics of international seafarers with mental illness. Occup Med (Chic. Ill). https://doi.org/10.1093/occmed/kqz069.
35. "Divorce Rate Among Cruise Ship Crew Members Rising," Crew Center

Real-Time Face Mask Detection and Analysis System

Juhi Singh, Lakshy Gupta, Rahul Kushwaha, Tanya Varshney, and Akshat Chauhan

Abstract Due to COVID-19 situation, we need to wear face masks in public places. Reports say that wearing face mask at public places and at workspace reduces the transmission of virus as the SARS-CoV-2 spreads through atmosphere among people, at gathering in any environment. In this paper, a real-time face mask detection system is presented which will detect mask presence on the face using TensorFlow. We are using MobileNetV2 model to provide a greater accuracy in determining the mask presence. Accuracy obtained is 99%. Older systems do not provide a proper working system. A face mask detector has been designed with computer vision using Python, OpenCV, Keras, and TensorFlow. Video surveillance input can be given directly, and our primary purpose is to identify to check people are wearing masks on daily basis or not wearing masks and prepare a weekly and monthly report based on this observation and display the data on an interactive web application. System provides option to see the historical records, thereby reducing transmission.

Keywords Deep learning · Computer vision · OpenCV · MobileNetV2 · TensorFlow · Keras · Face detection · Face mask detection · Real-time mask detection · Python · Django · CNN · Architecture · Caffemodel · Report · Artificial intelligence · Linear bottlenecks · Average pooling · Methodology · Surveillance

1 Introduction

We all are witnessing this after COVID-19 era where wearing mask is a must, and the trend of wearing mask is increasing around the globe. Prior to COVID-19, people generally wore mask to avoid pollution, but after the community spread of corona virus, it is mandatory to wear mask to avoid virus transmission. Contamination occurs through gathering of a large number of people; thus, there is a need for a surveillance system for ensuring this at the entry/exit points of various places. This problem can be tackled using deep learning approach. Machine learning and artificial intelligence

J. Singh (✉) · L. Gupta · R. Kushwaha · T. Varshney · A. Chauhan
Department of Computer Science and Engineering, ASET Delhi, New Delhi, India
e-mail: jsingh7@amity.edu

© The Author(s), under exclusive license to Springer Nature Singapore Pte Ltd. 2022   11
D. Gupta et al. (eds.), *Proceedings of Data Analytics and Management*,
Lecture Notes on Data Engineering and Communications Technologies 90,
https://doi.org/10.1007/978-981-16-6289-8_2

played a vital role for any such solution. In this document, we have applied convolutional neural network approach (CNN) to extract meaningful features from images and differentiate among them by using certain learning parameters. In convolutional neural network, there are two primary processes. First process being convolution and the latter being pooling. Multiple filters can be applied on an image to extract relevant features using convolution using features map. Pooling operation, also known as subsampling, reduces parameters of feature maps from the previous operation. This face mask detection model can be easily combined with digital cameras to impede the virus transmission in real time. The model is developed by the integration of deep learning and machine learning algorithms. Using the MobileNetV2 for object detection, we achieved the highest accuracy. We used up least time in this system by using this approach.

The organization of the remaining paper is given as follows. Section 2 presents the description of tools. Sections 3 and 4 describe related and proposed works. Section 5 presents methodology. The paper is concluded in Sect. 6.

2 Description of Tools

2.1 OpenCV

OpenCV is used to provide faster acceleration in computer vision. This library enables us to use various important methods to perform image operations accurately.

2.2 TensorFlow

TensorFlow, an open-source software, provides libraries to train models with higher accuracy in less time. It provides us the good and flexible working framework for development.

3 Related Work

Adrian Rosebrock [1] presents an algorithm to determine the face mask detection with MobileNetV2. The algorithm is able to identify whether a person is wearing face mask or not wearing a mask on images using facial landmarks, i.e., eyes, nose, mouth, and jawline. As MobileNetV2 can detect object, video stream faces can also be detected using this algorithm in real time.

According to [2] study, it gives a system for face detection in video using improved version of AdaBoost classifier having Haar cascade classifier, and classifier and support vector machine (SVM) depending upon the facial expression, pose, and light effects, it gives almost accurate results for face detection on video stream.

According to [3] study, a basic approach is established to detect masks on peoples' faces. A deep learning model based on TensorFlow and Keras is proposed to solve the problem, while preprocessing of data would be done using OpenCV.

Meenpal [4] aims to propose a binary face classifier which can find any face present in the frame without considering of its alignment.

Siegfried [5] compares the MobileNetV2, ResNet50V2, Caffemodel, and Xception image classification models and evaluates them based on accuracy and iteration time of each epoch.

Study [6] presents the systematic approaches used in the past to tackle pandemic situations. The study compares the usage of masks and its effectiveness in various types of diseases.

Study [7] gives an overall view on variations in cultural and societal paradigms of mask presence in different demographical areas. It also presents the role of policymakers as a compelling character for people to use appropriate masks.

In study [8], a one-stage detector is proposed. It is having a pyramid network to fuse high-level information with multiple map features and a novel context attention framework to gain a focal point on face detection.

According to [9] study, two-stage architecture is proposed for detecting masked and unmasked faces. The two stages are ROI extraction and resizing and batching.

4 Proposed Work

In this study, we proposed a real-time mask detection and analysis system using deep learning. Due to current COVID-19 situation, it is very necessary to take precautions, and sometimes, we need to use proper methods to detect the issues. Our system provides a full-fledged working surveillance for the mask detection and gives the results who are more vulnerable, so that admin can take necessary steps to reduce the contamination. Our system generates a report consisting of historical data. This data consists of various parameters. Crowded area requires fast and proper surveillance, so that contamination can be reduced. Our system recognizes the persons who are not wearing masks along with the parameters day, time, date. For face mask recognition, we have used MobileNetV2 [10] model giving a very good accuracy. System is running as web interface developed using Django framework. System can be installed in malls, markets, theaters, parks, etc. The dataset is trained on masked and unmasked images of 1376 images. Our system provides a robust way to tackle the COVID-19 situation in an advanced manner. Data generated can be used to predict the outcome before it happens. Contamination can be reduced at much higher level. System will enable admins to take necessary steps to prevent the spread of disease (Fig. 1).

Fig. 1 Proposed work flow

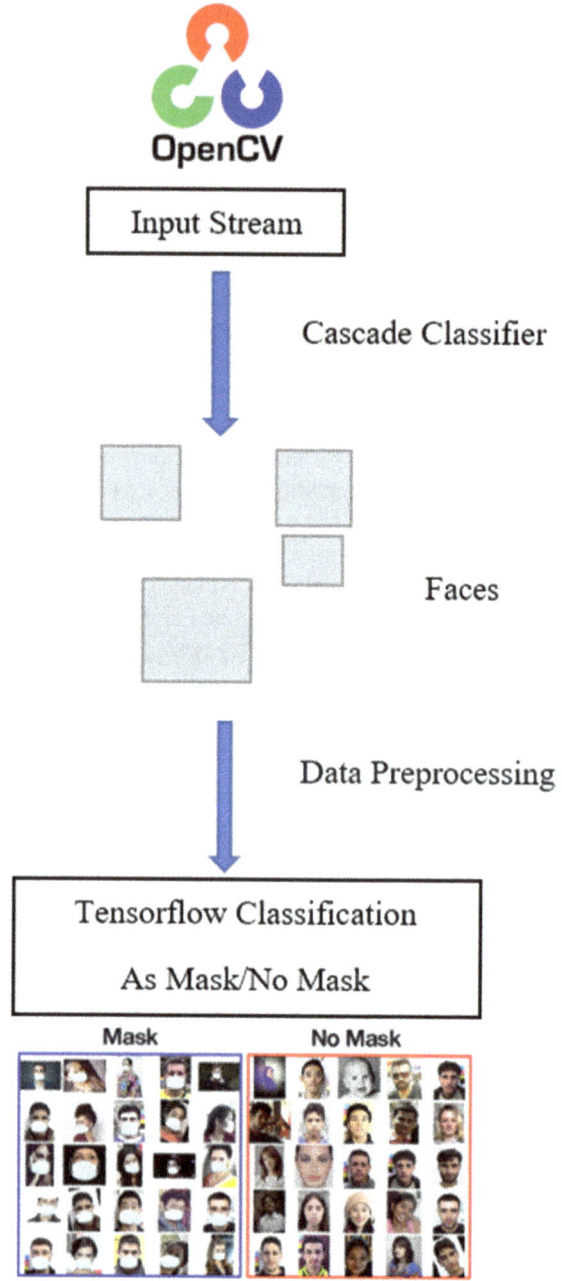

5 Methodology

5.1 Model Architecture

In this work, we used the training model MobileNetV2. MobileNetV2 is a CNN architecture. This model will eventually give two new features upon MobileNetV1 model.

(1) Having linear bottlenecks to reduce the data flow
(2) Shortcut connections between bottlenecks

Four sequential layers on top of MobileNetV2 model were developed. Those are:

(1) Average pooling
(2) Linear layer with ReLu activation function
(3) Dropout layer
(4) Linear layer with softmax activation function

After training, the model is integrated with OpenCV to perform real-time mask detection. Training is carried using TensorFlow and Keras libraries. We used the MobileNetV2 model based on the baseline results as shown in Paper [11]. Table 1 shows the generic MobileNet backbones and respective performance (Tables 2 and 3).

Model is trained using res10_300 × 300 Caffemodel. Due to the ability of Caffe's high learning and processing speed, deep learning models can be trained in less time.

Column t represents the expansion rate of the channels, c shows number of input channels, and n represents the frequency of block repetition. This trained model gives the accuracy of 99% on the test dataset as shown in Paper [12] (Fig. 2).

Total parameters are 2,257,984.

Trainable parameters are 2,223,872.

Non-trainable parameters are 34,112.

Table 1 Generic MobileNet backbones and their respective performance on WIDER FACE	Backbone	Hard	Overall	n_{params}
	$Mnet_{\alpha=0.25}$	87.11	90.90	8.94×10^5
	$MnetV2_{\alpha=0.25}$	87.27	91.03	7.08×10^5
	MnetV3-Small	86.52	89.37	1.46×10^6
	$Mnet_{\alpha=1.0}$	89.10	93.10	4.57×10^6
	$MnetV2_{\alpha=1.0}$	89.87	93.28	3.62×10^6
	MnetV3-Large	88.23	92.12	4.33×10^6

Table 2 Architecture

Given input	Having operator	t	c	n	s
1×1	ReLU	–	–	–	–
$224^2 \times 3$	Conv2d	–	32	1	2
$112^2 \times 32$	Bottleneck	1	16	1	1
$112^2 \times 16$	Bottleneck	6	24	2	2
$56^2 \times 24$	Bottleneck	6	32	3	2
$28^2 \times 32$	Bottleneck	6	64	6	2
$14^2 \times 64$	Bottleneck	6	96	3	1
$14^2 \times 96$	Bottleneck	6	160	6	2
$7^2 \times 160$	Bottleneck	6	320	1	1
$7^2 \times 320$	Conv2d 1×1	–	1280	1	1
$7^2 \times 1280$	Averagepool 7×7	–	1	1	–
$1 \times 1 \times 1280$	Conv2d 1×1	–	k	–	–

Table 3 Django model

Attribute	Datatype
with_Mask	IntegerField
without_Mask	IntegerField
date_time	CharField
day	CharField
age	CharField

Fig. 2 Accuracy/loss curve

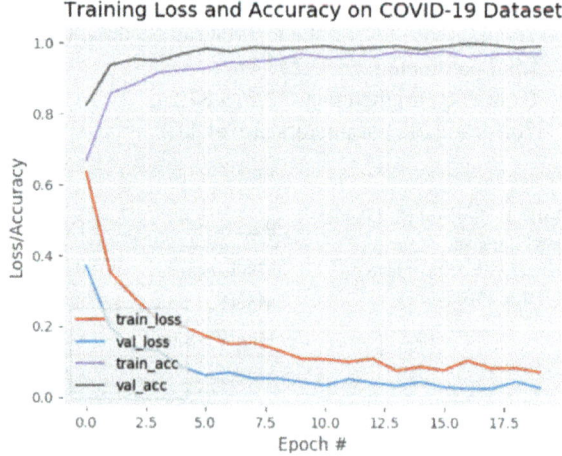

Training Loss and Accuracy on COVID-19 Dataset

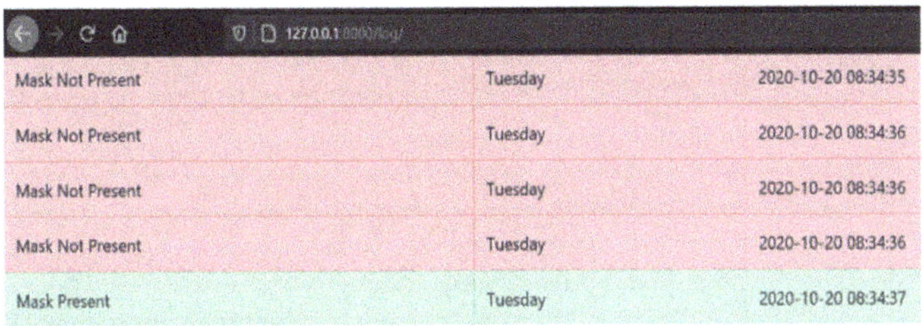

Mask Not Present	Tuesday	2020-10-20 08:34:35
Mask Not Present	Tuesday	2020-10-20 08:34:36
Mask Not Present	Tuesday	2020-10-20 08:34:36
Mask Not Present	Tuesday	2020-10-20 08:34:36
Mask Present	Tuesday	2020-10-20 08:34:37

Fig. 3 Log output

5.2 Approach

Input stream is given through the surveillance cameras located at the entrance gate of area. The system requires proper lightning. The classifier is having many simpler classifiers, so these are used to find the region of interests (ROI). The dataset consists of 1,376 images. The mask is artificially added to all images to make it real-world application. Every time, when a new face is detected, its information will be recorded for proper surveillance along with other parameter including day of week, time, and date (Fig. 3).

5.3 Result Analysis and Discussion

This model will easily detect high moving faces accurately and determines the output. This model was trained using dataset which consists of 686 images without mask and 690 with mask images (Fig. 4).

Fig. 4 Output

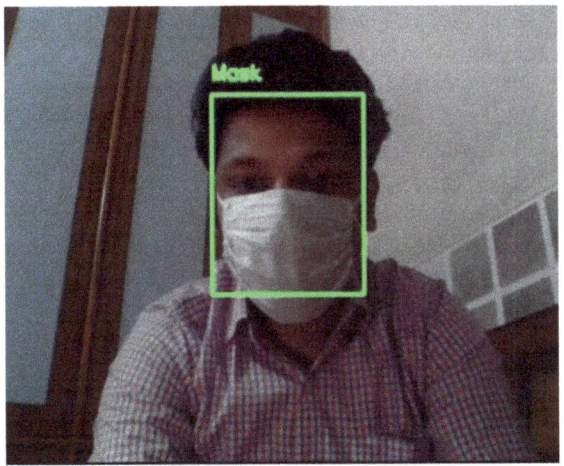

Fig. 5 Accuracy versus latency [13]

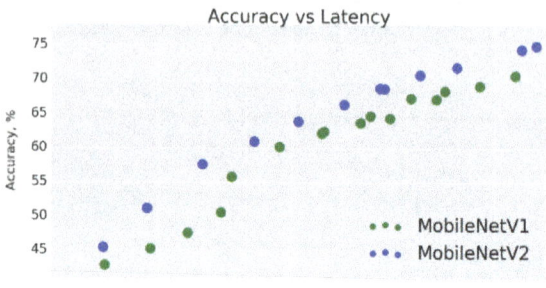

Fig. 6 Analytics results

System is developed which maintains the record of persons. Improving upon the previous research, we have also provided a user interface to view and manage the results for the algorithm.

Django database structure contains the parameters:

We used MobileNetV2 in comparison with MobileNetV1 due to its higher accuracy and with approximately same latency (Fig. 5).

Using OpenCV and TensorFlow, we have generated the analytics using the records. Concepts are taken from books [14]. Our system can also be used as biometric attendance, facial features extraction, and security access control [15] (Fig. 6).

6 Conclusion

Due to COVID-19 situation, there was a need of proper surveillance system, particularly in crowded area. Our system provides much faster. Accuracy came out to be

99.016%. After using MobileNetV2, its accuracy is much higher than before. This is an upgradation of MobileNetV1 model. System generates a weekly/monthly report consisting of various parameters which eventually be useful for the admin to take necessary steps. From the trends, one can understand who are following norms or not, thereby decreasing the spread of disease. System is more efficient and reliable.

References

1. Adrian R (2020) COVID-19: face mask detector with OpenCV, Keras/TensorFlow, and deep learning [Online]. Available: https://www.pyimagesearch.com/2020/05/04/covid-19-face-mask-detector-with-opencv-keras-tensorflow-and-deep-learning/
2. Ahmad F, Najam A, Ahmed Z (2013) Image-based face detection and recognition: state of the art. IJCSI Int J Comput Sci Issues 9(6) (2013)
3. Nerpagar T, Junnare S, Raut J, Shah A, Patil P (2020) Face mask recognition using machine learning. IJIRT 7(7)
4. Meenpal T, Balakrishnan A, Verma A (2019) Facial mask detection using semantic segmentation. In: 4th International conference on computing, communications and security (ICCCS)
5. Siegfried I (2020) Comparative study of deep learning methods in detection face mask utilization. Independent Res
6. Greenhalgh T, Schmid M, Czypionka T, Bassler D, Gruer L (2020) Face masks for the public during the covid-19 crisis. BMJ 369
7. Feng S, Shen C, Xia N, Song W, Fan M, Crowlling B (2020) Rational use of face masks in the covid-19 pandemic. Lancet Resp Med
8. Jiang M, Fan X, Yan H (2020) Retina FaceMask: a face mask detector
9. Chavda A, Dsouza J, Badgujar S, Damani A (2020) Multi-stage CNN architecture for face mask detection
10. Vinitha V, Velantina V (2020) Covid-19 facemask detection with deep learning and computer vision. Int Res J Eng Technol (IRJET) 07:3131
11. Samuel WF, Pavit N, Justin AC, Ankush G (2019) Face detection with feature pyramids and landmarks, p 06
12. Smaranjit G, Suhrid D (2020) Automatic facial mask detection using deep learning. Mukt Shabd J IX(VII):1121
13. Mark S, Andrew H (2018) MobileNetV2: the next generation of on-device computer vision networks [Online]. Available: https://ai.googleblog.com/2018/04/mobilenetv2-next-generation-of-on.html. 03 April 2018
14. Garcia GB (2015) Learning image processing with OpenCV, PACKT
15. Kumar A, Kaur A, Kumar M (2019) Face detection techniques: a review. Artif Intell Rev 52:927–948

Performance Evaluation of Brahmagupta-Bhaskara Equation Based Algorithm Using OpenMP

Veena Narayanan, R. Kavitha, and R. Srikanth

Abstract The development of a user-friendly cryptographic algorithm passes through various processes like effort analysis, duration analysis and productivity. Also, the performance of such applications can be improved with the aid of OpenMP tools. In the present study, the inital values of the various attributes like effort, duration and productivity of Brahmagupta–Bhaskara equation-based cryptographic algorithm by using basic constructive cost model (COCOMO) equations are discussed. Further, the experimental speed-up values and efficiency in parallel mode are evaluated and compared with serial mode in this study. Also, the COCOMO values in OpenMP for encryption and decryption are evaluated and compared with serial mode to estimate the effort and cost of the application to be developed for the processing of this new cryptographic algorithm.

Keywords Brahmagupta–Bhaskara equation · Parallel computing · Constructive cost model equations · Cryptographic algorithms

1 Introduction

Even though a variety of computer applications exist for solving large problems, some real-world problems are there where a vast amount of data is required to execute the problem and makes it difficult to manage using serial computing. In such situations, parallel computing techniques give a hand to program developers for making better work of hardware within the required time-bound. The widely

V. Narayanan (✉) · R. Srikanth
Department of Mathematics, SASTRA Deemed university, Thanjavur, Tamil Nadu 613401, India
e-mail: veenanarayanan@sastra.ac.in

R. Srikanth
e-mail: srikanth@maths.sastra.edu

R. Kavitha
School of computing, SASTRA Deemed university, Thanjavur, Tamil Nadu 613401, India
e-mail: kavitha_r@cse.sastra.edu

© The Author(s), under exclusive license to Springer Nature Singapore Pte Ltd. 2022   21
D. Gupta et al. (eds.), *Proceedings of Data Analytics and Management*,
Lecture Notes on Data Engineering and Communications Technologies 90,
https://doi.org/10.1007/978-981-16-6289-8_3

used parallel programming models have shared memory (without threads), threads, distributed memory, etc. The application programming interface (API), Open Multiprocessing (OpenMP) assist shared memory parallel programming written in C/C++ and Fortran.

The constructive cost model (COCOMO) is one of the most commonly used software estimation models in the world derived by Barry Boehm in 1981. According to their requirements, one can choose any of the three different COCOMO, say, Organic model, Semideatched model and Embedded model. The crucial attributes that decide the quality of any software project are effort and schedule that are also outcomes of COCOMO. COCOMO models do not consider customer skills and hardware issues, while it mostly depends on time factors. Many of the developers treat this as a disadvantage of the COCOMO model. A lot of research has been done worldwide about COCOMO model and its applications in various fields [1–3].

The Diophantine equation is a polynomial equation in two or more unknown variables that seeks only integer solutions. The Brahmagupta-Bhaskara (BB) equation is a quadratic Diophanitne equation of the form $NX^2 + D = Y^2$ where D is any interger and N is a positive integer whose square root is irrational. Murthy and Swamy developed an add-on algorithm [4] based on this BB equation for the further enhancement of already existing symmetric key algorithms. There are a lot of studies that have been carried out worldwide about the parallelization of the most commonly used cryptographic algorithm called RSA [5–7].

In the present work, the authors enroot a C++ code in OpenMP and predict the initial effort and cost for developing a software project for the BB based algorithm using the basic COCOMO model. In Sect. 2, all the relevant works related to the present work is depicted. Section 3, the speed-up factor as well as efficiency of the BB based algorithm is evaluated and in Sect. 4, the initial effort, schedule and cost of the software to be developed for BB based algorithm is evaluated using basic COCOMO model both in serial and parallel computing.

2   Related Works

Evaluating perfomance of various cryptographic algorithms is necessary in the modern era as securing information in various applications is necessary. Priyadarshini et al. performed a comprehensive evaluation of various cryptographic algorithms [8]. In 2018, Alzubi et al. invented a new method [9] for combining an ensemble of classifiers and achieved a significant improvement in classification accuracy over the product. They developed a new theoretical framework for a linear combining method and evaluated its accuracy by comparing with the existing methods. Later, Gheisari et al. developed a new optimal mathematical model [10] for the prediction of the degree of stakeholder satisfaction. Also, Jain et al. [11] newly designed the real-time video processing system using Field Programmable Gate Arrays.

Table 1 System configurations for the study

S. No.	Configuration	Description
1	Processor	Intel(R)Core(TM)i5-8250U@1.60GHz 1.80GHz
2	RAM	8.00GB
3	System type	64 bit OS, x64-based processor
4	Environment	Windows 10
5	Platform	GCC infrastructure
6	Compiler	TDM-GCC 4.9.2 64 bit Release

In 2020, Alzubi et al. introduced a new algorithm for dynamic programming based ensemble design algorithm [12], and its performance is compared with classical ensemble model. They observed that DPED outperforms on all data sets in terms of both accuracy and ensemble size. Movassagh et al. trained the neural networks using meta-heuristic approaches [13] and then resulted in the reduction of prediction error. Recently, Alzubi proposed a novel authentication technique for medical IoT systems [14] and the security level got enhanced by 7% .

3 Evaluation of Speed-up Factor and Efficiency

In this section, we evaluate the speed-up factor and the efficiency of BB based algorithm using OpenMP interface, which is compared with the serial programming mode. The experimental findings are discussed with an input of a prime number which is used to generate the number of bits required in the binary form of plaintext. Table 1 represents the experimental setup, and Table 2 comapres the execution time in serial and parallel computing.

It is to be noted that the execution time will vary depending upon the size of the input prime number. Table 2 illustrates the execution time differences between serial and parallel modes during the encryption.

The speed-up factor is a tool to express actual benefit of answering a computationally large problem in parallel. It is defined as

$$S_k = \frac{T_s}{T_k} \tag{1}$$

where T_s and T_k are the corresponding execution time (in seconds) with single processor and k processors. In this study, we evaluate the execution time, speed-up and efficiency (the ratio of speed-up factor to the number of threads) of encryption and decryption for different thread counts in a single physical CPU node and is tabulated in Tables 3 and 4.

Table 2 The CPU execution time (in seconds) for serial and parallel mode during encryption

Prime number	Serial mode	Parallel mode
5	14	11
11	18	14
53	22	18
103	22	20
199	25	17
257	30	20
701	23	19
997	27	22

Table 3 The execution time (CPU time) for the encyption in parallel mode

Threads	Execution time (in seconds)	Speed-up factor	Efficiency
1	24	1	1
2	22	1.09	0.54
3	20	1.2	0.4

Table 4 The execution time (CPU time) for the decryption in parallel mode

Threads	Execution time (in seconds)	Speed-up factor	Efficiency
1	14	1	1
2	13	1.15	0.58
3	12	1.25	0.42

From Tables 3 and 4 it is clear that the execution time decreases as the number of the threads increases, whereas the efficiency decreases. This depicts the effective reduction of run time through parallelization.

4 Cost and Effort Evaluation Using COCOMO

In this section, we provide an estimate for the effort and cost of the new application to be developed to implement the BB based algorithm using organic COCOMO equations.

The basic COCOMO model equations for organic model software project are

$$A = \alpha \times (L)^{\beta} \tag{2}$$

$$R = \gamma \times (A)^{\delta} \tag{3}$$

$$P = A/R \tag{4}$$

where A is the effort applied in persons-months, R is the development time in chronological months, L is the estimated number of lines of code for the project (expressed in thousands) and P is the number of persons. The coefficients α and γ and the exponents β and δ for different project types are in Table 5.

The process of calculating the effort and duration time of the application takes place by applying the COCOMO equation once the number of lines of codes in the source file is counted. As the number of lines of codes increases, it is clear that the effort and duration increases accordingly after applying the COCOMO equation. Table 6 shows the estimated effort, duration time and number of persons for a diiferent line of code for developing the source code.

The size of the BB based algorithm in C++ application is 113 lines of code for encryption and 122 lines of code for decryption. So from Table 1, the values of the different parameters are $\alpha = 2.4$, $\beta = 1.05$, $\gamma = 2.5$ and $\delta = 0.38$. Thus, the corresponding COCOMO model equations will become

$$A = 2.4 \times (L)^{1.05} \tag{5}$$

$$R = 2.5 \times (A)^{0.38} \tag{6}$$

Table 5 Coefficient and exponent values [2]

Software project	α	β	γ	δ
Organic	2.4	1.05	2.5	0.38
Semi-detached	3.0	1.12	2.5	0.35
Embedded	3.6	1.20	2.5	0.32

Table 6 Estimated effort, duration and number of persons for a different size of lines of code

Lines of code	Effort (person-month)	Duration (month)	Number of persons
50	0.10	1.04	0.09
150	0.33	1.64	0.20
250	0.56	2.00	0.28
550	1.28	2.75	0.47
1200	2.91	3.75	0.78
1400	3.42	3.99	0.86
1800	4.45	4.41	1.01
2000	4.96	4.59	1.08
4000	10.28	6.06	1.69
5000	13.00	6.62	1.96

Table 7 Comparison of Encryption code with serial and parallel computing

Attribute	Serial	Parallel
Lines of code	92	113
Effort (E)	0.1959	0.24
Duration (D)	1.34	1.45
No. of persons	0.1456	0.16
Productivity	0.4696	0.47

Table 8 Comparison of Decryption code with serial and parallel computing

Attribute	Serial	Parallel
Lines of code	102	122
Effort (E)	0.2183	0.26
Duration (D)	1.40	1.49
No. of persons	0.1559	0.17
Productivity	0.467	0.47

Thus for encoding the effort $A = 0.24$ person-month, duration time $R = 1.45$ months, the number of persons $P = 0.16$ and the productivity $= L/A = 0.47$, whereas for the decryption $A = 0.26$, $R = 1.49$, $P = 0.17$ and productivity $= 0.47$. This is tabulated and is compared with the corresponding serial program. Tables 7 and 8 illustrates the comparison of COCOMO values with the serial and parallel methods.

Figures 1 and 2 show the estimated effort, duration, productivity and number of persons needed using COCOMO equations for developing the software for newly proposed algorithm. In this proposed algorithm, both encryption and decryption are parallelized. If decryption is not parallelized, the corresponding lines of code will also

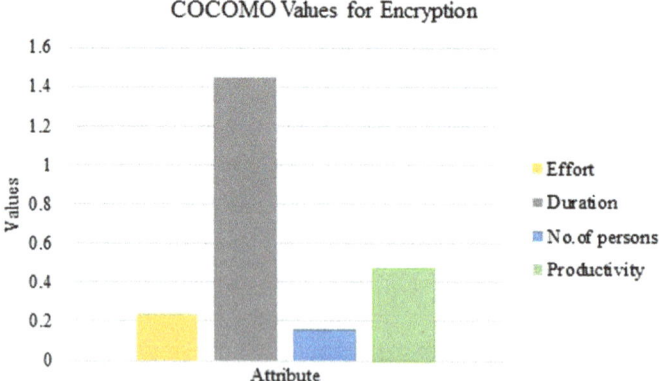

Fig. 1 COCOMO values for encryption

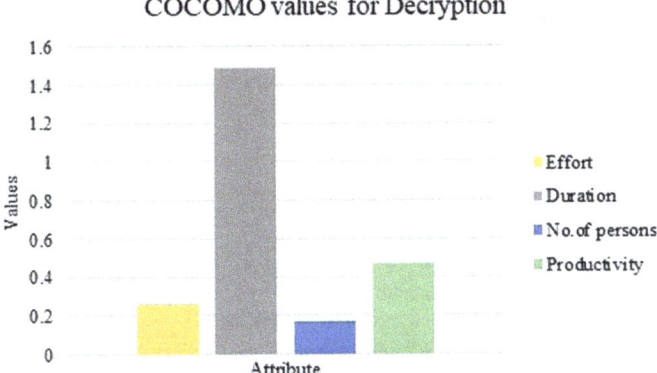

Fig. 2 COCOMO values for decryption

decrease and thus reduce the COCOMO values. The COCOMO based equations are applied for parallel mode having 3 threads. If number of threads increases, number of lines of code also increases and thus the effort and duration will automatically increase eventhough the execution time will decrease. But for encryption and decryption processes, one will expect less time for execution and thus we have to go for parallel method. Thus, for software implementation, parallel method will demand more effort and cost compared to serial method. Also, in this study, we described the effort and cost required to develop a software project for this algorithm both in serial and parallel computing mode.

5 Conclusion

The expediency of a cryptographic algorithm relies on the choice and obstructions of integrated circuit (IC) technology. The experimental results in the present work showed that observed speed-up values increase as the number of processors increases and ensures improving a cryptographic algorithm based on BB equation in OpenMP by reducing the time complexity. The fundamental COCOMO model values showed that the effort and cost are slightly higher in parallel mode than in the conventional method. Eventhough, the COCOMO values are high compared to the traditional mode, and parallel computing provided less execution time for encryption and decryption. Thus, parallel computing-based programming in OpenMP saves a lot of energy and time in applications that demand massive data.

References

1. Boehm VW, Valredi R (2008) Achievements and challenges in COCOMO-based software resource estimation. IEEE softw 25:74–83
2. Mahmood Y, Abdulqader A (to Appear) A platform for porting IPv4 applications to IPv6. Int J comput Digit Syst
3. Pal G, Kumar M, Barala K (2015) A review paper on cocomo model. J Comput Sci Eng 1:83–87
4. Murthy NR, Swamy NS (2006) Cryptographic applications of Brahmagupa-Bhaskara equation. IEEE Trans Circ Syst I Reg Pape 53:1565–1571
5. Ayub MA, Onik ZA, Smith S (2019) Parallelized RSA algorithm: an analysis with performance evaluation using OpenMP library in high performance computing environment. In: 22nd International conference of computer and information technology
6. Fadhil HM, Younis MI (2014) Parallelizing RSA algorithm on multicore CPU and GPU. Int J Comput Appl 87:15–22
7. Fan W, Chen X, Li X (2010) Parallelization of RSA algorithm Based on compute unified device architecture. In: Ninth international conference on grid and cloud computing
8. Patil P, Narayankar P, Narayan DG, Meena SM (2016) A comprehensive evaluation of cryptographic algorithms: DES, 3DES, AES, RSA and blowfish. Procedia Comput Sci 78:617–624
9. Alzubi OA, Alzubi JA, Tedmori S, Rashaideh H, Almomani O (2018) Consensus-based combining method for classifier ensembles. Int Arab J Inf Tech 15:76–86
10. Gheisari M, Panwar D, Tomar P, Harsh H, Zhang X, Solanki A, Nayyar A, Alzubi JA (2019) An Optimization model for software quality prediction with case study analysis using MATLAB. IEEE Access 7:85123–85138
11. Jain DK, Jacob S, Alzubi J, Menon V (2019) An efficient and adaptable multimedia system for converting PAL to VGA in real–time video processing. JRTIP 12:2113–2125
12. Alzubi OA, Alzubi JA, Alweshah M, Qiqieh I, Al-Shami S, Ramachandran M (2020) An optimal pruning algorithm of classifier ensembles: dynamic programming approach. Neural Comput Appl 32:16091–16107
13. Movassagh AA, Alzubi JA, Gheisari M, Rahimi M, Mohan SK, Abbasi AA, Nabipour N (2021) Artificial neural networks training algorithm integrating invasive weed optimization with differential evolutionary model. J Ambient Intell Humaniz Comput. https://doi.org/10.1007/s12652-020-02623-6
14. Alzubi JA (2021) Blockchain-based Lamport Merkle digital signature: authentication tool in IoT healthcare. Comput Commun 170:200–208

Gait Recognition Biometric System

Rahul Gupta, Naman Gupta, Tushar Gupta, Aditya Srivastava, Ritu Gupta, and Abhilasha Singh

Abstract As, we enter the digital era, a reliable security system is the need of the hour. Human gait is a unique feature to identify and authenticate a person. Gait has many attractive properties as a biometric, and there are many factors which can confuse the biometric systems. This paper provides the overview of gait analysis, data and factors used in gait and motion analysis, assessment methods, existing gait recognition technologies, how it works and recognizes the gait of a human being, development of a failsafe full proof biometrics using gait analysis, its application in human identification, its advantages over other biometric systems and its limitations which are the concerned topics of research nowadays. This paper proposes a model to efficiently record the selected features of a person's gait using multiple machine learning algorithms like SVM, NN, K-NN from the video feed and recognize/authenticate the registered personnel via a video input. This paper also refers Hanavan's model toward the approach for the recognition of a person's gait. The paper explores various hardware/software requirements such as PyCharm IDE along with additional packages like OpenCV, keras, and Tkinter for the proposed model and analyzes the scope and future prospects of the model.

Keywords Human gait · Biometric · Gait analysis · Machine learning (ML) · Support vector machine (SVM) · K-nearest neighbors (K-NN) · Neural networks (NN) · Hanavan's model · OpenCV · Keras

1 Introduction

Gait can be defined as a coordinated and cyclic combination of body movements of a human that result in its locomotion. The coordination of movements is in such a way that they occur with a specific pattern for a gait to occur. As it is cyclic as well as

R. Gupta · N. Gupta · T. Gupta (✉) · A. Srivastava · A. Singh
Department of IT, Amity University, Noida, Uttar Pradesh 201313, India

R. Gupta
Bhagwan Parshuram Institute of Technology, IP University, New Delhi, India

© The Author(s), under exclusive license to Springer Nature Singapore Pte Ltd. 2022
D. Gupta et al. (eds.), *Proceedings of Data Analytics and Management*,
Lecture Notes on Data Engineering and Communications Technologies 90,
https://doi.org/10.1007/978-981-16-6289-8_4

coordinated in nature, this motion makes it a unique phenomenon [1]. Our purpose is to use this property, i.e., gait analysis to construct a gait recognition security system. The two main reasons for this are:

- First, several systems require a recognition system on which they can rely to recognize or identify an individual and confirm their identity. The main objective of such security biometric systems is to assure that only the authorized persons can access the offered services. Human beings have an ability to recognize a person from a distance simply by recognition of the way of walking of that person. This experience combined with the interests in biometrics security systems has led the research for the development of a gait recognition security system for biometric identification.
- Another reason for the implementation of such a system is that as you know in today's era of wide variety with vast advancements in technology enables many individuals to bypass the security systems such as fingerprint scanner, facial recognition systems, etc.

1.1 Overview

Traditional methods like knowledge based which means that one knows and it maintains the secrecy. Example, passwords, PINs, etc. Object-based authentication is characterized by possession. However, generally token-based approach is used with the combination of knowledge-based approach. Example: a bank card with a PIN code. But in these traditional approaches, passwords, PIN codes, bank cards or tokens can be lost, mistaken, forgotten or stolen. Usability is also one of a disadvantage. And managing numerous passwords and PINs, and remembering and retaining strong passwords are very difficult. Biometrics gives a natural authenticate solution for management of identities by using automated or semi-automated systems to identify individuals [2]. As gait biometric has many attractive properties. Collection of images and portraying the individual's gait, it can easily be done in public areas by using simple instrumentation, and it does not require any physical interaction or the cooperation and even the awareness of an individual who is under observation [3]. There are a lot of confounding properties of gait in biometrics. Unlike face, fingerprints, DNA, we do not know how much the gait of an individual is unique. Furthermore, there are many factors which causes changes in the gait, which includes injury, fatigue, exhaustion, footwear, etc. [4].

This research paper explores the proposed concept for gait analysis as an efficient biometric for personnel authentication, as an extra layer of security over the existing biometric authentication techniques such as fingerprint, facial recognition, etc.

1.2 Gait Biometric Recognition

Gait biometric has been exhibited as an alternative identifier in recognition systems. Single measure is not enough to recognize the human gait, many measures and a complete set of dynamics are required for a recognition of human gait. Several measures can be used to recognize the gait of a human, like visual approaches as shown in Fig. 1 which take pictures from different angles from a distance, it involves cameras, and sensor approaches, it collects the data of gait when in contact with the individual.

As discussed earlier, the gait of an individual is a cyclic or periodic motion, in which each cycle consists of two strides: (i) Left foot forward (ii) Right foot forward.

There are two aspects of gait biometrics, i.e., shape and dynamics. Shape is defined as the configuration, shape or structure of a person as it performs various gait phases. Dynamics is said to be the transition rate between the phases. So, basically research in the field of gait biometrics and its development has to consider both the shape and dynamics of humans. Developing a gait recognition biometrics is not a difficult task but the main challenge is to resolve the problem of variations in the gait motions due to footwear, uneven walking surfaces, different clothing, carrying an object, etc. As we lack in understanding and modeling of effects of factors mentioned above on shape and dynamics of gait, it relies on its datasets. Dataset can expose the problems. Technically, gait recognition biometric can be classified into two types: (i) Holistic approach (also called Model Free approach) (ii) Model-based approach as shown in Fig. 2.

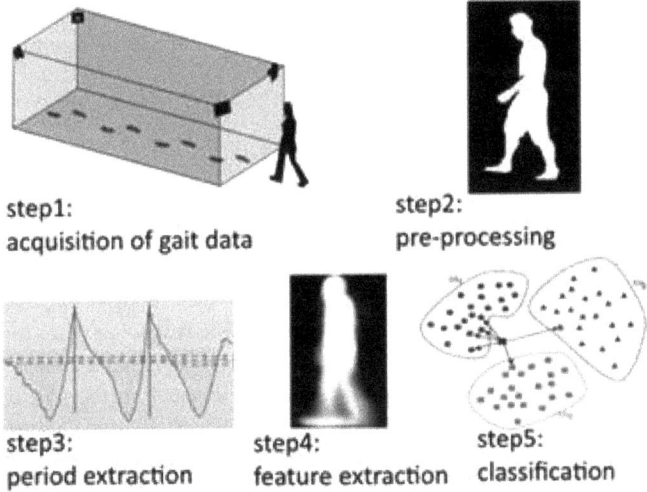

step1:
acquisition of gait data

step2:
pre-processing

step3:
period extraction

step4:
feature extraction

step5:
classification

Fig. 1 Steps in gait recognition [5]

Fig. 2 Classification of
authentication systems

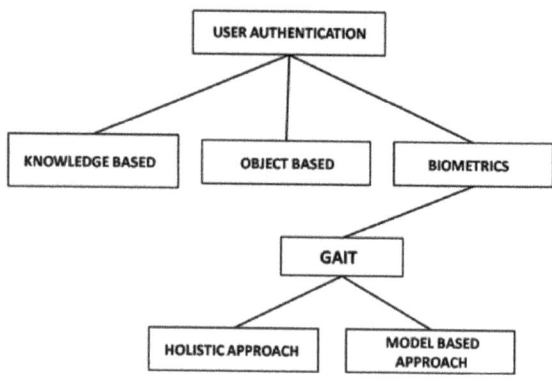

2 Literature Review

There are several features of gait which can be used in the recognition of an individual. Studies have suggested that 24 different parameters of human gait makes an individual different from others, makes them unique [6]. However, it is very difficult to measure accurately and detect some components of a human by a computer, e.g., angular displacements of foot, ankle, leg, thigh, and foot. So, for the extraction of accurate and precise measurements of different parts of the body and angles of the joint by vision systems is not an easy task because the motion of humans is not rigid, it possesses many transformations in its motion, since the human body has very flexible body structure. Moreover, other factors like clothing, errors in segmentation and different positions become a challenge for precise measurement and detection. Hence, the problem of recognizing a gait of an individual and then representing it is a challenging task.

2.1 Gait over Other Biometrics

When we compare gait with other biometrics, it has unique features. Other biometrics like facial recognition, fingerprints, palm, iris scan, voice recognition, and signature can be recognized only when in a physical contact of the subject or when the subject is in proximity of the biometric system. Gait has an advantage that it cannot be hidden, spoofed, fake, or stolen. As kinesiological parameters define the gait of a human that forms the base for identification, some limitations are there in capturing the gait which make it very difficult to recognize and record the parameters which affect the gait. Gait recognition relies on a sequence of a video. Even if we improve the accuracy of measuring gait parameters, then even the knowledge of the parameters are not sufficient to discriminate power to enable the expansion of gait recognition technology. However, gait changes with time which is affected by different factors like clothing, footwear, terrain, speed of walking, injury, etc. [7]. These factors are the limitations in accuracy of measuring the parameters of gait. The efficiency, uniqueness, and invariability of gait identification system makes it different from other biometrics.

2.2   Gait over Other Biometrics

Research driven in the field of gait recognition has proved that gait with the combination of other biometrics can be reliable. As we know that palm, retina, fingerprint, iris recognition methods belong to the obtrusive class of biometrics, i.e., it needs the awareness of the subject under observation. Biometrics like face, foot pressure [8] can be used in combination with gait in the multibiometric system. They can be used to limit the database of multibiometric systems. Whereas, face recognition can only be used for identification when there is a limited set of subjects. Otherwise, these three biometrics, i.e., face, foot pressure, and gait can be combined together to develop a multibiometric system. It has been observed that gait works more efficiently in a multibiometric framework when used in combination of facial features.

2.3   Related Work Done in Gait Analysis Techniques

In order to analyze and discuss the characteristics about what will be the key parameters which need to be gathered for gait analysis, basically it is crucial to understand the mechanism of walking in human beings. Basically, gait cycle comprises majorly two steps that can be categorized into events comprising various sequences of leg's activities. The events are classified in two phases, i.e., swing and stance phases. When the feet will be in contact along with the ground, this will comprise around 60% of the cycle time as compared to the time taken when the feet is not touching the ground. The cycle may vary due to the difference of right and left sides of the body [9].

Certain approaches use other characteristics to classify the human gait as a unique feature of a person. One of these approaches is the acoustic gait classification. Acoustic gaits use an inexpensive set of microphones with a computing device as an accurate and unintrusive gait analysis system. This is in contrast to the expensive and intrusive systems currently used in laboratory gait analysis such as the force plates, pressure mats, and wearable sensors, some of which may change the gait parameters that are being measured [10].

Recently, the improvement in machine learning field provides the novelty in vast variety of classification algorithms which prevails gait data that attracts the automation recognition and provide solutions for classification problem to improve the efficiency and performance of algorithms like SVM as Support vector machine employ to classify the gait pattern to assess the change in gait functions.

The manifold learning algorithm is used to provide the advanced power technique for the nonlinear dimensions reduction which now successfully applied in patterns like face acquisition and human motion detection. The general idea of the algorithm is to discover the intense part of the data with low dimensionality that preserves the geometric structure lying under manifolds in high dimensions. Moreover, consider the low dimension manifold as a significant gait feature embedded into high gait feature space that also provides the amount of information for the gait classification [11]. The results of the research show the nonlinear feature gathering using ISOMAP could capture relevant information for gait analysis while the individual was walking. Also, that could employ the train machine classifier to learn automatically by generalizing the human gait patterns with superior performance. Then, more relevant intrinsic features than principal component analysis (PCA) because finding a low dimension embedded of gait data preserves all pair points unlike Euclidean distance in principal component analysis are obtained. This work is a consistent partial part of the activity that deals with the spatial dimensional (3D) analysis of ordinary and pathological gait analysis established on the newly organized advanced systems for scientific approach applications, using below cost cellular accelerometers and the only signal processing algorithm. The systems automatically excerpt necessary gait actions such as heel strikes and the toe-offs which specialize the viewpoint or posture and the fluctuation levels of walking [12].

A person's gait has been identified by using the collaboration of various algorithms like SVM with K-NN and NN.

SVM

SVM or we would rather say this as support vector machine, it is a highly supervised ML model which is implemented to rectify the images between two classes or in other term we would say it as efficient algorithm to solve the group classification challenges. This model can be implemented by passing the labeled dataset of each category, hence with a motive that this model could give as accurate prediction as possible to identify the nature of the image that weather the image is forged or not. We have created the dataset as same way discuss above along with the csv file in which images with names they are labeled are stored inside image category, a class is defined weather the image is forged or not along with the representation that if the identified image is forged then we will mention that with 0 similarly if the image belongs to the real category then we are mentioning such types with 1.15 Now let's have a look toward how the working of the SVM model plays its role. This could easily be understandable with the help of the following example, consider the scenario you want differentiate which is orange and which one of them is apple. Now to reach the solution suppose you have two factors naming x, y, and we require a classifier which predicts the result out of the input pair. SVM consider the data nodes and then outputs that in in a spatial plane or we can say that a hyperplane which effectively separates the data, and this special hyperplane we termed this as the decision boundary and the parts which submerged on one side of that plane will be considered orange and the remaining one as apple.

K-Nearest Neighbors

K-nearest neighbors also known as K-NN classifier finds a dataset instantly based on the Euclidean distance. While using K-NN firstly the data given is being trained and then it is stored in the memory. Recognition or identification is done on a test data sample by the calculation of the Euclidean distance among the testing data and every other data which is present in the training set. K (here) indicates the total number of nearest neighbors that is being considered for the determination of the result. Then, the tested sample data is being checked so that it lies inside the sample majority. After that, it is considered to be—matched otherwise, it is considered as—not matched.

K-NN algorithm assumed the similarity features among the new dataset and the available cases and then put the new info into the category which is most familiar to the available classes. It stores all the available datasets and classifies a new key points based on the similarities. This means that when new dataset appeared, then it can be easily classify into a well suite categorical by using K-NN. It can be used both ways regression as well as classification but mainly it can be used for the classification problems. K-NN is the non-parametric method, which means it does not make any assumption on elemental datasets.

Neural Networks

In many other cases when a target is tested, it is not necessary that other data of training set should be near the target value. So, the technique of neural network is more famous technique that is used for the classification and identification or recognition in today's world. Neural network generally uses the theory of the nervous system of the human being. They generally use the neurons for the identification or recognition. They are very much relatable to the interconnected nodes. They are basically made up of three layers, i.e., the input layer, the hidden layer, and output layer. The nodes basically contain the output that is controlled by the weights.

Firstly, the video or live feed of a particular person is recorded and then it is divided into multiple frames. The frames are then pre-processed and after that unwanted part is removed from the background. Then, one feature is extracted using a model-based approach, i.e., Hanavan's model. Lastly, the person's gait which is captured is searched in the database of gait by using above-mentioned algorithms. All the steps mentioned above, i.e., capturing the video, then pre-processing, background subtraction, and finally, feature extraction is also performed on the videos of the database. For the implementation of this process, MATLAB software has been used [13]. Some equations are used for the removal of background noise or the unwanted parts (Fig. 3).

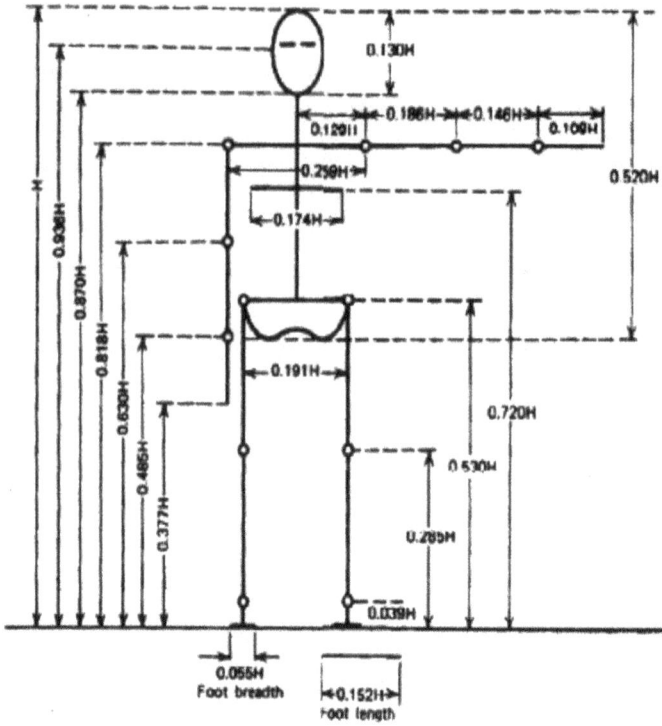

Fig. 3 Hanavan's model [13]

3 Proposed Model

The proposed model aims at successfully authenticating registered personnel on the basis of their gait. The initial model is targeted to work on a standard hardware/software system as mentioned below.

3.1 Project Requirements

Hardware

- *High-resolution camera*
- *Laptop with graphic card*
- *RAM 8 GB (Min.)*

Software

- *Windows 10 OS*

Fig. 4 Stances in gait cycle [14]

- *PyCharm Community Edition 2020.1*
- *Anaconda3*
- *VS Code*
- *Additional Packages—OpenCV, Tkinter, numpy, pandas, keras, etc.*

3.2 Standard Terminology

The approach is based on the prerequisite knowledge of the following terminology of the gait biometric.

Gait cycle: It can be defined as the interval among two mid-stances, i.e., left or right (consecutive).

Half cycle: The interval between two mid-stances (consecutive) is said to be half cycle.

Gait period: The time period in which the gait cycle is completed is called the gait period.

Gait frequency: The frequency of walking is known as the gait frequency (Fig. 4).

3.3 Approach

The initial approach of the model is based on capturing video data and processing it to extract features in order to train the ML model to recognize a registered user. The various steps involved in the process are mentioned below.

STEP 1: Live video of any person is recorded using a high-resolution camera. Then, the captured video is being converted into multiple frames.

STEP 2: Pre-processing involves converting the frames of input video feed from RGB to gray-scale components and then the images to binary form are also performed. After that the noise is removed from all the frames with the help of a median filter. After the pre-processing step, background subtraction is performed.

STEP 3: Hanavan's model is used for the feature extraction. For different classification techniques, all mentioned features are taken as input. All features are obtained for the database video and are also stored for the trained database.

STEP 4: Lastly, trained databases and above-mentioned features are classified and are matched using the machine learning classifiers such as SVM, NN, and K-NN. Recognition or identification basically consists of the classification and testing, and finally, the accuracy is calculated. The model will be using a selected random-seed in order to reproduce the randomized results.

STEP 5: The model is then tuned by fine-tuning of hyperparameters of the various algorithms incorporated. The fine-tuning is firstly done manually by tuning selected hyperparameters. After that the process is automated using the randomized search option for random modification of the hyperparameters in order to find the highest possible accuracy of the ML model.

Based on the above steps, the model aims at obtaining an accuracy of 95–99 ± 3%. The accuracy can, however, differ in cases of human interference/error (improper clothing, limp in gait due to injury, etc.) (Fig. 5).

4   Scope of the Proposed Model

Nowadays, security is the major concern as the existing biometric system such as fingerprint recognition, and face recognition can easily be manipulated which can compromise the security. So, the need of a new biometric system such as gait recognition is required. Several systems require a recognition system on which they can rely to recognize or identify an individual and confirm their identity. The main goal of such security biometric systems is to assure that only the authorized persons can access the offered services. The need of time is to develop such biometrics which can confirm the identity of an individual based on the fact that "who she is" instead of the fact that "what she has" like an ID card or "what she memorizes" like a password. Human beings have an ability to recognize a person from a distance simply by recognition of the way of walking of that person. This experience combined with the interests in biometrics security systems has led the research for the development of a gait recognition security system for biometric identification. Recent developments of gait biometrics and gait researches indicate that a lot of improvements, and still, there is a need to mature the gait technologies, and the practical applications of gait biometrics are expected to be used widely in the future. In the present scenario, there is a possibility that gait recognition biometric in combination with the other biometric systems can be deployed [15]. Although, future advancements in gait analysis, recognition, and gait biometrics along with other biometrics is a challenging and an open research field, and it is expected that in coming future a gait biometrics, and other gait technologies will be deployed widely not only in identification and surveillance, but also in other applications like in medical field, etc.

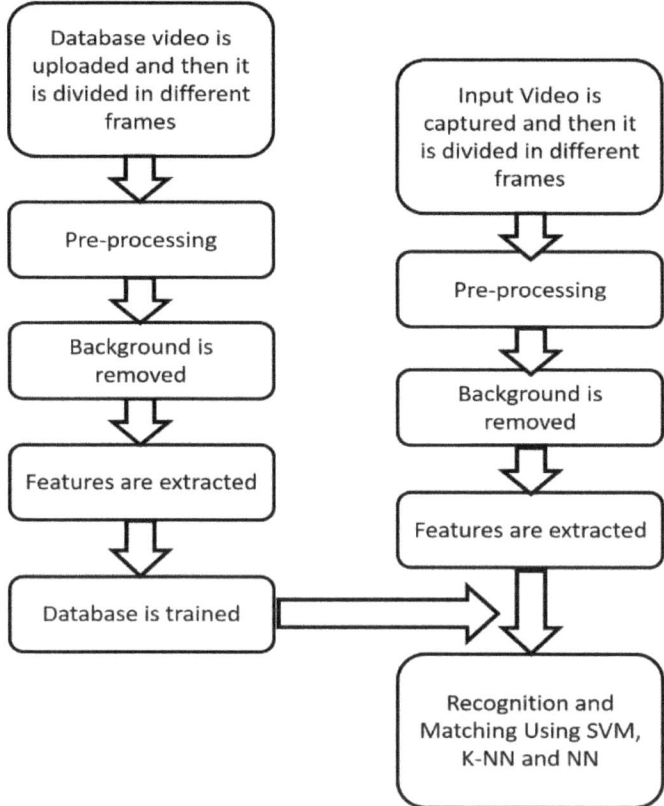

Fig. 5 Proposed model

5 Result

The gait recognition biometric features are so precisely selected that they are the most discriminant and distinctive features than others [16]. This research paper provided an overview of directions for basic researches in the field of gait analysis, recognition, and gait biometrics. It proposes a model for the recognition of human gait for biometric purpose and also provide an initial approach to implement the work of this paper. This paper also analyzes the gait to identify the person by using multiple MLs such as SVM, NN, and K-NN. Recent developments of gait biometrics and gait researches indicate that a lot of improvements and still there is a need to mature the gait technologies, and the practical applications of gait biometrics are expected to be used widely in the future. In the present scenario, there is a possibility that gait recognition biometric in combination with the other biometric systems can be deployed.

6 Conclusion

We hope that this research paper exposes the problems of gait analysis and recognition and challenges in gait biometrics problems that will encourage more involvement of researchers in the field or area of gait research in the future.

Although, future advancements in gait analysis, recognition, and gait biometrics along with other biometrics are a challenging and an open research field, and it is expected that in coming future a gait biometrics, and other gait technologies will be deployed widely not only in identification and surveillance, but also in other applications like in medical field, etc.

References

1. Hu K, Wang Z, Wang W, Ehgoetz Martens KA, Wang L, Tan T, Lewis SJS, Feng DD (2020) Graph sequence recurrent neural network for vision-based freezing of gait detection. IEEE Trans Image Process 29:1890–1901
2. Singh R, Chaudhary H, Singh AK (2019) A novel gait-inspired four-bar lower limb exoskeleton to guide the walking movement. J Mech Med Biol 19(04):1950020
3. Bajwa TK, Garg S, Saurabh K (2016) GAIT analysis for identification by using SVM with K-NN and NN techniques. In: Fourth international conference on parallel, distributed and grid computing (PDGC)
4. McGrath B (2003) The week in walks. The New Yorker, June 2, p 35
5. Tistarelli M, Bigun J, Grosso E (eds) (2005) Biometrics School 2003, LNCS 3161, 2005. Springer-Verlag Berlin Heidelberg, pp 19–42
6. Narendra KS, Thathachar MAL (1989) Learning automata: an introduction. Prentice-Hall Inc., Upper Saddle River, NJ, USA
7. Mason JE et al (2006) Machine learning techniques for gait biometric recognition. Springer International Publishing Switzerland, pp 9–16
8. Katiyar R, Pathak VK, Arya KV (2013) A study on existing gait biometrics approaches and challenges. IJCSI Int J Comput Sci Issues 10(1, No 1):135–142
9. Addlesee MD, Jones A, Livesey F, Samaria F (1997) ORL active floor. IEEE Pers Commun 4(5):35–41
10. Fang Q, Zhang Z, Tu Y, School of Electrical Engineering, RMIT University, Melbourne, 3000, Australia
11. Umair Bin Altaf M, Butko T, and Biing-Hwang (Fred) Juang, Fellow, IEEE 2015
12. Wu J (2012) Automated recognition of human gait pattern using manifold learning algorithm. In: 2012 8th international conference on natural computation (ICNC 2012)
13. Boutaayamou M, Schwartz C, Stamatakis J et al (2012) Validated extraction of gait events from 3D accelerometer recordings. In: 2012 International conference on 3D imaging (IC3D)
14. Tistarelli M, Bigun J, Grosso E (eds) (2005) Biometrics School 2003, LNCS 3161. Springer-Verlag Berlin Heidelberg, pp 19–42
15. Gupta et al (2011) Biometrics system based on human gait patterns. Int J Mach Learn Comput 1(4):377–379
16. Holzreiter SH, Kohle ME (1993) Assessment of gait patterns using neural net-works. J Biomech 26:645–651
17. Sudha R, Bhavani R (2008) Biometric authorization system using gait biometry, p 1

A New MCDM Approach to Solve a Laptop Selection Problem

Shankha Shubhra Goswami⬤**, Rajesh Kumar Moharana, and Dhiren Kumar Behera**⬤

Abstract The aim of this article is to recommend the best laptop model out of six possible alternatives using COPRAS and ARAS decision-making methodologies by implementing the concept of aggregated weightages. This study looks at applying various MCDM techniques to a particular decision-making problem and determining whether or not any of the approaches produce the same results. It has been discovered that most researchers use the AHP method to assess the parameters weights, but other methods such as SWARA, SMART, and others can also be used to determine their effectiveness in decision-making fields. In this case, the aggregated weights of SWARA-SMART are used to give equal importance to both methods, and newly developed methods such as COPRAS and ARAS are used as a ranking tool to extend their relevant areas. The relative choices and expectations for different laptop configurations are decided by communicating with some laptop users and reading customer feedback on various online shopping platforms. The criteria weights are first calculated using SWARA and SMART, and the combined aggregated weights from these two methods are then implemented in COPRAS and ARAS to recommend the final rating of the alternatives. This final alternative ranking from this study exactly matches with the previous researcher's findings, indicating that model 5 and model 4 are the most superior and inferior models among the list of six selected models.

Keywords AHP · ARAS · COPRAS · MCDM · SWARA · SMART · Laptop selection

1 Introduction

Nowadays, laptop has become an essential item in any fields and with these requirements, and their markets are also expanding day by day. It can be found from 2020 statistics that around 166 million laptops were shipped worldwide in 2019 and were

S. S. Goswami (✉) · R. K. Moharana · D. K. Behera
Indira Gandhi Institute of Technology, Sarang, Odisha 759146, India

© The Author(s), under exclusive license to Springer Nature Singapore Pte Ltd. 2022   41
D. Gupta et al. (eds.), *Proceedings of Data Analytics and Management*,
Lecture Notes on Data Engineering and Communications Technologies 90,
https://doi.org/10.1007/978-981-16-6289-8_5

expected to reach 171 million units in 2023 according to Statista Research Department [1]. Hence, there is a vital requirement to concentrate on this stuff, and the consumers should get some idea about the best laptop model that suits them the most while purchasing it among large quantities of available alternatives. Although it is not possible to consider all the available models in the whole world, but still the decision analysis can be carried out by considering few popular models among the laptop users.

For the last few years, several researchers have carried out MCDM analysis on computer selection problem by implementing various MCDM tools. For example, Ertugrul and Karakasoglu [2] addressed a laptop selection problem in a business based on the use of ELECTRE and FHP. Erpolat and Cinemre [3] researched two separate DEA models for comparing various laptop varieties and brands, where the 1st model has no weight constraints and the 2nd model's parameters are calculated using AHP. Srichetta and Thurachon [4] used FAHP to pick the best notebook computer product. Lakshmi et al. [5] used TOPSIS to choose the best laptop based on competing parameters, e.g., battery duration, warranty, capacity, and others. AHP approach was employed by Tampi et al. [6] to estimate a customer decision-making issue in choosing a laptop. Adali and Isik [7] used MULTIMOORA and MOOSRA to solve a laptop selection problem. Mitra and Goswami [8] used AHP-SAW to choose the best laptop model from six options, and Mitra et al. [9] used fuzzy-AHP methodology to interpret the same output results.

Not only this but also, many researchers introduced different MCDM tools in wide variety of areas apart from these applications which are as follows. Kaklauskas et al. [10] used COPRAS to pick low-e windows for public building retrofits. Zagorskas et al. [11] used COPRAS on the basis of GIS to conduct a sustainable development and urbanistic assessment of city compactness. Zavadskas et al. [12] used COPRAS-G to pick the most efficient dwelling house walls from four alternatives. Kersuliene et al. [13] chose a logical dispute resolution approach using the SWARA method. Zavadskas and Turskis [14] developed the ARAS system to assess the office room's microclimate. Zavadskas et al. [15] used the ARAS approach to choose the most suitable foundation installment alternative that is secure for building on aquiferous soil. Dadelo et al. [16] proposed a model for evaluating and rating elite security employee that is based on an expert assessment approach to evaluate requirements weights and ARAS to classify the best alternatives.

Keren et al. [17] used a combination of AHP and DEA methods to find a suitable project manager. Pitchipoo et al. [18] minimized the blind spot area by optimizing design parameters of rear-view mirrors in heavy vehicles using COPRAS. Karabasevic et al. [19] developed a staff selection MCDM prototype based on ARAS and SWARA methods. Adali and Isik [20] used COPRAS and ARAS methods to solve an air-conditioner selection problem. Parezanovic et al. [21] evaluated twenty-six sustainability mobility measures using fuzzy-COPRAS method. For the selection of the sales manager's assessment parameters, Karabasevic et al. [22] suggested a combined methodology focused on SWARA and the Delphi technique. Knezevic et al. [23] used a hybrid model that combined FAHP and TOPSIS to examine employee efficiency in a sample of D-Electrical power supply companies in Serbia.

Nasab and Anvari [24] used TOPSIS, COPRAS, and DEA to solve a material selection problem. Patel et al. [25] investigated the operation of the SMART MCDM system using a site selection case study for nuclear waste dumping among multiple options.

Zarbakhshnia et al. [26] suggested a MADM model focused on F-SWARA weight assessment and fuzzy-COPRAS rating and assortment of third-party inverse logistics providers. Bahrami et al. [27] used a novel MCDM hybrid approach called BWM-ARAS to integrate multisource geographical datasets in the Abhar region of NW Iran to define extremely Cu prospectively areas. Fu [28] chose the best catering provider in the airline industry by the combination of AHP and ARAS methods. Schitea et al. [29] created an IF set-based EDAS, COPRAS, and WASPAS for determining the best place in Romania to roll-out hydrogen agility. Balki et al. [30] used a SWARA and ARAS hybrid approach to optimize engine operating parameters in a small SI engine. Buyukozkan and Guler [31] used a combined HF linguistic ARAS-SAW system to resolve a smart watch assortment problem. Ghenai et al. [32] evaluated the sustainability metrics of a system for renewable energy using an expanded ARAS-SWARA hybrid approach. Roozbahani et al. [33] used COPRAS-G and F-COPRAS practices to test inter-basin water transmission projects in the central Iranian plateau. Yucenur et al. [34] used SWARA to obtain criteria weights and COPRAS to make their decision on the best city in Turkey for biogas production.

According to the aforementioned literatures, it is clear that ARAS and COPRAS methods are never used to execute a laptop selection problem, and the applications of SWARA and SMART are very limited and underrated. Hence, there is a vast opportunity to integrate these MCDM tools all together and utilize to execute a laptop selection problem. Furthermore, very few research projects have been completed addressing about the effects of implementing different subjective weighting methods to a single problem. Hence, this article fulfills the research gap of solving a laptop selection problem by COPRAS and ARAS using aggregated SWARA-SMART criteria weightages and comparing the outcomes with previous researcher's results.

2 Materials and Methods

The following is a synopsis of this segment. First, the parameters weights are calculated individually using the subjective weighting methods SWARA and SMART. Second, the criterion weights from both methods are combined together to compute the aggregated weights of each criterion. Finally, these weights are used in the COPRAS and ARAS methods to recommend the best model and rate the alternatives. Two individual alternative ranking is obtained from each technique, and the results are also compared to the earlier investigator's work. The complete structure of the analysis is depicted clearly by flowchart shown in Fig. 1.

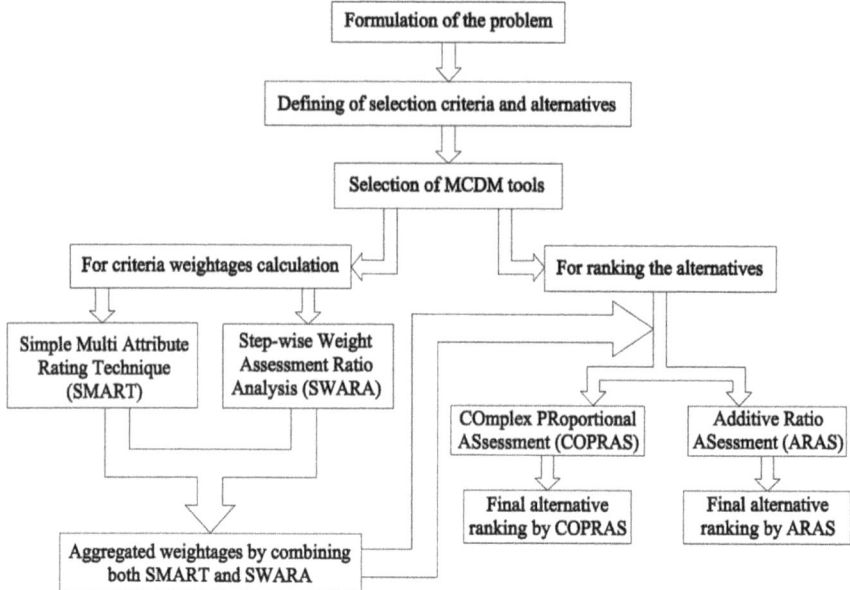

Fig. 1 Flowchart representation of the whole analysis

2.1 Step-Wise Weight Assessment Ratio Analysis (SWARA)

Kersuliene et al. [13] pioneered the SWARA process, which has found widespread use since then. It is a subjective weighting approach that relies on the DM's opinion to measure the requirements weights. The steps are explained briefly as follows [19, 35].

Step 1: Arrange the criterions in descending order according to their importance.
Step 2: According to Kersulienee et al. [13], the DM decides the comparative significance of the average value (s_j) of the jth criterion with respect to the $(j-1)$th criterion on a 1 scale, beginning with the second criterion.
Step 3: The co-efficient k_j is determined using Eq. 1.

$$k_j = \begin{cases} 1 & j = 1 \\ s_j + 1 & j > 1 \end{cases} \tag{1}$$

Step 4: The initial weights q_j is determined using Eq. 2.

$$q_j = \begin{cases} 1 & j = 1 \\ \frac{q_{j-1}}{k_j} & j > 1 \end{cases} \tag{2}$$

Step 5: Final weights w_j^{SWARA} is determined using Eq. 3.

Table 1 Criteria weightages by SWARA method

Criteria	s_j	k_j	q_j	w_j^{SWARA}
Processor (P)		1	1	0.31220
RAM (R)	0.33333	1.33333	0.75	0.23415
Screen size (SS)	0.5	1.5	0.5	0.15610
Storage capacity (SC)	0.33333	1.33333	0.375	0.11707
Brand (B)	0.5	1.5	0.25	0.07805
Operating system (OS)	0.33333	1.33333	0.1875	0.05854
Color (C)	0.33333	1.33333	0.14063	0.04390
Total			3.20313	1

(*Source* Author himself)

$$w_j^{SWARA} = \frac{q_j}{\sum_{j=1}^{n} q_j} \tag{3}$$

where $\{j = 1, 2..., n\}$. The comparative Importance (s_i) values revealed in Table 1 is based on the relative comparison matrix as proposed by Mitra and Goswami [8] and Mitra et al. [9]. The calculated criteria weights are shown in Table 1 along with the co-efficient (k_j) and initial weights (q_j).

2.2 Simple Multi Attribute Rating Technique (SMART)

It is also another arbitrary weighting approach and one of the simplest methods for assessing parameters weights [25]. This approach begins with the distribution of allocated weights (w_j^a) on a 100-point scale based on the DM's opinion, as shown in Table 2, and the final parameter weights (w_j^{SMART}) are determined using Eq. 4. Table 2 displays the final weights after they have been measured.

Table 2 Criteria weightages by SMART method

Criteria	w_j^a	w_j^{SMART}
Processor (P)	100	0.35088
RAM (R)	65	0.22807
Screen size (SS)	45	0.15789
Storage capacity (SC)	30	0.10526
Brand (B)	20	0.07018
Operating system (OS)	15	0.05263
Color (C)	10	0.03509
Total	285	1

(*Source* Author himself)

Table 3 Aggregated subjective criteria weightages

Criteria	$w_j{}^{\text{SWARA}}$	$w_j{}^{\text{SMART}}$	$w_j{}^{\text{SWARA}} * w_j{}^{\text{SMART}}$	$w_j{}^{\text{com}}$
Processor	0.31220	0.35088	0.10954	0.52160
RAM	0.23415	0.22807	0.05340	0.25428
Screen size	0.15610	0.15789	0.02465	0.11736
Storage capacity	0.11707	0.10526	0.01232	0.05868
Brand	0.07805	0.07018	0.00548	0.02608
Operating system	0.05854	0.05263	0.00308	0.01467
Color	0.04390	0.03509	0.00154	0.00733
Total			0.21001	1

(*Source* Author himself)

$$w_j^{\text{SMART}} = \frac{w_j^a}{\sum_{j=1}^n w_j^a} \quad \{j = 1, 2 \ldots, n\} \tag{4}$$

Now the criteria weights from both the methods are combined together to calculate the aggregated subjective criteria weightages (w_j^{com}) using Eq. 5 as shown in Table 3.

$$w_j^{\text{com}} = \frac{w_j^{\text{SWARA}} * w_j^{\text{SMART}}}{\sum_{j=1}^n \left(w_j^{\text{SWARA}} * w_j^{\text{SMART}} \right)} \quad \{j = 1, 2 \ldots, n\} \tag{5}$$

2.3   Complex Proportional Assessment (COPRAS)

Zavadskas, Kaklauskas, and Sarka firstly introduced the COPRAS method in 1994 [11, 12, 36]. According to Alinezhad and Khalili [35] 'This method is used to assess the maximizing and minimizing index values, and the effect of maximizing and minimizing indexes of attributes on the results assessment is considered separately' (pp. 87). The steps explained by Alinezhad and Khalili (pp. 87–89) [35] are as follows.

Step 1: Prepare a decision matrix based on Eq. 6. Table 4 depicts the decision matrix suggested by Mitra and Goswami [8] and Mitra et al. [9].

$$D(m_i \times n_j) = \begin{bmatrix} d_{11} & d_{12} & \ldots & d_{1n} \\ d_{21} & d_{22} & \ldots & d_{2n} \\ \ldots & \ldots & \ldots & \ldots \\ d_{m1} & d_{m2} & \ldots & d_{mn} \end{bmatrix} \tag{6}$$

Table 4 Decision matrix (COPRAS)

Types	Max	Max	Max	Max	Max	Max	Max
Models	P	SC	OS	R	SS	B	C
M 1	5	3	3	5	5	9	3
M 2	7	9	5	5	7	3	3
M 3	7	5	9	7	7	7	5
M 4	3	5	9	3	3	2	9
M 5	7	9	9	7	7	9	9
M 6	5	3	5	5	7	5	3
Sum	34	34	40	32	36	35	32

Mitra and Goswami (2019), Mitra et al. (2019)

where 'm' represents the no. of alternatives and 'n' represents the no. of parameters. The performance score of the 'ith' alternative and 'jth' criterion is denoted by 'd_{ij}.'

Step 2: Eq. 7 is used to normalize the decision matrix using linear normalization [10, 20]. Table 5 shows the normalized form of Table 4.

$$N_{ij} = \frac{d_{ij}}{\sum_{i=1}^{m} d_{ij}} \tag{7}$$

Step 3: Eq. 8 is used to measure the weighted values, and Table 6 displays the weighted form of Table 5.

$$Y_{ij} = w_j^{\text{com}} \times N_{ij} \tag{8}$$

Step 4: Using Eq. 9, measure the relative significances (Q_i) of each alternative.

Table 5 Normalized matrix (COPRAS)

Weights	0.52160	0.05868	0.01467	0.25428	0.11736	0.02608	0.00733
Models	P	SC	OS	R	SS	B	C
M 1	0.14706	0.08824	0.075	0.15625	0.13889	0.25714	0.09375
M 2	0.20588	0.26471	0.125	0.15625	0.19444	0.08571	0.09375
M 3	0.20588	0.14706	0.225	0.21875	0.19444	0.2	0.15625
M 4	0.08824	0.14706	0.225	0.09375	0.08333	0.05714	0.28125
M 5	0.20588	0.26471	0.225	0.21875	0.19444	0.25714	0.28125
M 6	0.14706	0.08824	0.125	0.15625	0.19444	0.14286	0.09375

(*Source* Author himself)

Table 6 Weighted normalized matrix along with relative significances and quantitative utility values

Models	P	SC	OS	R	SS	B	C	Q_i	U_i^c
M 1	0.07671	0.00518	0.00110	0.03973	0.01630	0.00671	0.00069	0.14641	68.597
M 2	0.10739	0.01553	0.00183	0.03973	0.02282	0.00224	0.00069	0.19023	89.128
M 3	0.10739	0.00863	0.00330	0.05562	0.02282	0.00522	0.00115	0.20412	95.638
M 4	0.04602	0.00863	0.00330	0.02384	0.00978	0.00149	0.00206	0.09513	44.569
M 5	0.10739	0.01553	0.00330	0.05562	0.02282	0.00671	0.00206	0.21343	100
M 6	0.07671	0.00518	0.00183	0.03973	0.02282	0.00373	0.00069	0.15068	70.599

(*Source* Author himself)

$$Q_i = S_{+i} + \frac{S_{-\min} \sum_{i=1}^{m} S_{-i}}{S_{-i} \sum_{i=1}^{m} \left(S_{-\min}/S_{-i}\right)} = S_{+i} + \frac{\sum_{i=1}^{m} S_{-i}}{S_{-i} \sum_{i=1}^{m} \left(1/S_{-i}\right)} \qquad (9)$$

where S_{+i} and S_{-i} are the addition quantities of the weighted normalized values for maximum and minimum parameters, as determined by Eqs. 10 and 11, respectively. The minimum value of S_{-i} is denoted by the symbol '$S_{-\min}$'.

$$S_{+i} = \sum_{j=1}^{n} Y_{+ij} \qquad (10)$$

$$S_{-i} = \sum_{j=1}^{n} Y_{-ij} \qquad (11)$$

The weighted values for the maximum and minimum parameters are Y_{+ij} and Y_{-ij}.

Step 5: Eq. 12 is used to measure the quantitative utility (U_i^c) of each alternative.

$$U_i^c = \left[\frac{Q_i}{Q_{\max}} \right] \times 100 \qquad (12)$$

where ($i = 1, 2..., m; j = 1, 2..., n$) and '$Q_{\max}$' is the maximum of Q_i. Table 6 shows the Q_i and the U_i^c values of each alternative.

2.4 Additive Ratio Assessment (ARAS)

The ARAS method was first introduced by Zavadskas and Turskis [14]. The performances of the alternatives are compared with an ideal alternative in this process, and

Table 7 Decision matrix (ARAS)

Types	Max	Max	Max	Max	Max	Max	Max
Models	P	SC	OS	R	SS	B	C
A_0	7	9	9	7	7	9	9
M 1	5	3	3	5	5	9	3
M 2	7	9	5	5	7	3	3
M 3	7	5	9	7	7	7	5
M 4	3	5	9	3	3	2	9
M 5	7	9	9	7	7	9	9
M 6	5	3	5	5	7	5	3
Sum	41	43	49	39	43	44	41

Mitra and Goswami (2019), Mitra et al. (2019)

the utility degree $(U_i{}^a)$ of each alternative is calculated to suggest the alternative rating [15, 16]. The ARAS steps are as follows [20–35].

Step 1: Decision matrix is formed according to Eq. 6. In addition, an ideal (optimal) alternative (A_0) is designed by considering the best values of each criteria as shown in Table 7.

Step 2: Linear normalization procedure is followed in ARAS also. The beneficial (maximum) criteria is normalized using Eq. 13.

$$R_{ij} = \frac{d_{ij}}{\sum_{i=1}^{m} d_{ij}} \tag{13}$$

The minimum criteria is normalized in two steps shown by Eqs. 14 and 15, respectively.

$$d_{ij}^* = \frac{1}{d_{ij}} \tag{14}$$

$$R_{ij}^* = \frac{d_{ij}^*}{\sum_{i=1}^{m} d_{ij}^*} \tag{15}$$

All of the parameters considered in this review are maximum in nature. As a consequence, Eq. 13 is used to normalize Table 7, and the normalized matrix is shown in Table 8.

Step 3: Estimate the weighted values using Eq. 16 and create the weighted normalized matrix as shown in Table 9.

Table 8 Normalized matrix (ARAS)

Weights	0.52160	0.05868	0.01467	0.25428	0.11736	0.02608	0.00733
Models	P	SC	OS	R	SS	B	C
A_0	0.17073	0.20930	0.18367	0.17949	0.16279	0.20455	0.21951
M 1	0.12195	0.06977	0.06122	0.12821	0.11628	0.20455	0.07317
M 2	0.17073	0.20930	0.10204	0.12821	0.16279	0.06818	0.07317
M 3	0.17073	0.11628	0.18367	0.17949	0.16279	0.15909	0.12195
M 4	0.07317	0.11628	0.18367	0.07692	0.06977	0.04545	0.21951
M 5	0.17073	0.20930	0.18367	0.17949	0.16279	0.20455	0.21951
M 6	0.12195	0.06977	0.10204	0.12821	0.16279	0.11364	0.07317

(*Source* Author himself)

Table 9 Weighted normalized matrix along with optimality function and degree of utility values

Models	P	SC	OS	R	SS	B	C	S_i	$U_i{}^a$
A0	0.08905	0.01228	0.00269	0.04564	0.01911	0.00533	0.00161	0.17572	–
M 1	0.06361	0.00409	0.00090	0.03260	0.01365	0.00533	0.00054	0.12072	68.700
M 2	0.08905	0.01228	0.00150	0.03260	0.01911	0.00178	0.00054	0.15685	89.263
M 3	0.08905	0.00682	0.00269	0.04564	0.01911	0.00415	0.00089	0.16836	95.812
M 4	0.03817	0.00682	0.00269	0.01956	0.00819	0.00119	0.00161	0.07823	44.518
M 5	0.08905	0.01228	0.00269f	0.04564	0.01911	0.00533	0.00161	0.17572	100
M 6	0.06361	0.00409	0.00150	0.03260	0.01911	0.00296	0.00054	0.12441	70.798

(*Source* Author himself)

$$Z_{ij} = w_j^{com} \times R_{ij}\left(\text{or } R_{ij}^*\right) \tag{16}$$

Step 4: Determine the optimality function (S_i) of each alternative using Eq. 17.

$$S_i = \sum_{j=1}^{n} Z_{ij} \tag{17}$$

Step 5: Determine the degree of utility ($U_i{}^a$) of each alternative using Eq. 18.

$$U_i^a = \left[\frac{S_i}{S_0}\right] \times 100 \tag{18}$$

where ($i = 1, 2..., m; j = 1, 2..., n$). '$S_0$' denotes the optimality value of the ideal alternative A_0. Table 9 displays S_i and $U_i{}^a$ values of the alternatives.

3 Result and Discussion

The U_i^c values from the COPRAS method and the U_i^a values from the ARAS method of each alternative are determined and displayed in Table 6 and Table 9. Now the rating of the laptop models is proposed by both methods according the descending order of these values which is portrayed in Table 10.

Table 11 compares the latest outcome rankings to the previous researcher's findings by SAW [8] and FAHP [9]. The ranking comparisons of the laptop models are also shown graphically in Fig. 2. The laptop models in Table 10 are listed in the order of choice.

$$Modal\ 5 > Model\ 3 > Model\ 2 > Model\ 6 > Model\ 1 > Model\ 4.$$

Table 10 Rating of the laptop models by COPRAS and ARAS

Models	COPRAS		ARAS	
	U_i^c	Rank	U_i^a	Rank
Model 1	68.597	5	68.700	5
Model 2	89.128	3	89.263	3
Model 3	95.638	2	95.812	2
Model 4	44.569	6	44.518	6
Model 5	100	1	100	1
Model 6	70.599	4	70.798	4

(*Source* Author himself)

Table 11 Ranking comparisons by SAW, FAHP, COPRAS, and ARAS

Models	SAW	FAHP	COPRAS	ARAS
Model 1	5	5	5	5
Model 2	3	3	3	3
Model 3	2	2	2	2
Model 4	6	6	6	6
Model 5	1	1	1	1
Model 6	4	4	4	4

(*Source* Author himself; Mitra and Goswami (2019), Mitra et al. (2019))

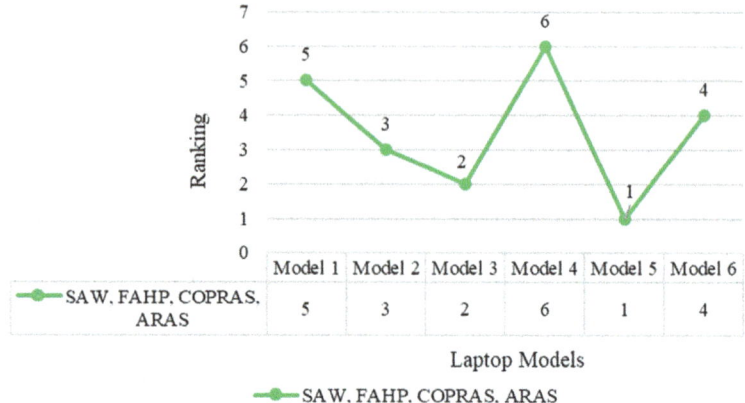

SAW, FAHP, COPRAS, ARAS

Fig. 2 Ranking comparisons of different MCDM tools

4 Conclusion

Table 10 clearly shows that model 5 is the best choice of the 6 available models, followed by models 3 and 2, and model 4 is the worst option and should be avoided. Table 11 also depicts that the rankings proposed by COPRAS and ARAS exactly matches with the previous results as shown in Fig. 2, and hence, it can also be concluded that there are not much differences among these MCDM tools, and any subjective weighting methods will provide the same output results.

Limitation There are some drawbacks, for example, when considering further selection parameters such as battery backup, weight, display resolution, and so on. In addition to these factors, the final ranking is subject to adjustment. Also, there may be some alterations in the final results, if objective weighting methods like entropy or CRITIC are used instead of subjective weighting methods, since objective methods is independent of the DM's opinion, and the weights are calculated using a decision matrix.

Future Scope Other MCDM systems can be used to solve the same problem, and more parameters can be added in the future. Furthermore, other weighting methods, such as Entropy and CRITIC, can also be used to calculate the criterion weights, and the results should be compared.

Acknowledgements We would like to express our gratitude to all of the professors, colleagues, and staff members at IGIT, Sarang for their invaluable assistance, encouragement, and cooperation. Our heartfelt thanks go to the authors of the reference papers from which we adapted some of the concepts and used them in this study. Finally, we want to thank the Almighty God for giving us the strength to finish this work on time and in good health.

References

1. Statista Research Department (2020) Forecast: global shipment of tablets, laptops, and desktop PCs 2010–2023. Statista (2020, March 2). Retrieved on 13th may, 2020 from https://www.sta tista.com/statistics/272595/global-shipments-forecast-for-tablets-laptops-and-desktop-pcs/
2. Ertugrul I, Karakasoglu N (2010) Computer selection for a business with ELECTRE and fuzzy-AHP methods. Dokuz Eylül Univ İktis İdari Bilimler Fak Derg 25(2):23–41. https://iibfdergi.deu.edu.tr/index.php/cilt1-sayi1/article/view/273/pdf_253
3. Erpolat S, Cinemre N (2011) Notebook selection in a hybrid approach: data envelopment analysis based on analytic hierarchy process. Istanb Univ J Sch Bus Adm 40(2):207–225. https://arastirmax.com/en/system/files/dergiler/2057/makaleler/40/2/arastirmax-notebook-sec iminde-hibrit-bir-yaklasim-analitik-hiyerarsi-yontemine-dayali-veri-zarflama-analizi.pdf
4. Srichetta P, Thurachon W (2012) Applying fuzzy analytic hierarchy process to evaluate and select product of notebook computers. Int J Model Optim 2(2):168–173. http://www.ijmo.org/papers/105-Q116.pdf
5. Lakshmi TM, Venkatesan VP, Martin A (2015) Identification of a better laptop with conflicting criteria using TOPSIS. Int J Inf Eng Electron Bus 6:28–36. http://www.mecs-press.org/ijieeb/ijieeb-v7-n6/ijieeb-v7-n6-5.pdf
6. Tampi YAN, Pangemanan SS, Tumewu FJ (2016) Consumer decision making in selecting laptop using analytical hierarchy process (AHP) method (Study: HP, Asus And Toshiba). J Ris Ekon Manaj Bisnis Akunt 4(1):315–322. https://media.neliti.com/media/publications/2903-en-consumer-decision-making-in-selecting-laptop-using-analytical-hierarchy-process.pdf
7. Adali EA, Isik AT (2017) The multi-objective decision making methods based on MULTI-MOORA and MOOSRA for the laptop selection problem. J Ind Eng Int 13:229–237. https://link.springer.com/content/pdf/10.1007/s40092-016-0175-5.pdf
8. Mitra S, Goswami SS (2019) Application of integrated MCDM technique (AHP-SAW) for the selection of best laptop computer model. Int J Res Eng Appl Manage 4(12):1–6. http://ijream.org/papers/IJREAMV04I1248003.pdf
9. Mitra S, Goswami SS, Parvej M (2019) Selection of the best laptop model by the application of Fuzzy-AHP methodology. I-manager's J Manage 14(1):33–43. https://doi.org/10.26634/jmgt.14.1.16044
10. Kaklauskas A, Zavadskas EK, Raslanas S, Ginevicius R, Komka A, Malinauskas P (2006) Selection of low e-windows in retrofit of public buildings by applying multiple criteria method COPRAS: a Lithuanian case. Energy Build 38(5):454–462. https://doi.org/10.1016/j.enbuild.2005.08.005
11. Zagorskas J, Burinskiene M, Zavadskas E, Turskis Z (2007) Urbanistic assessment of city compactness on the basis of GIS applying the COPRAS method. EKOLOGIJA 53:55–63. http://www.elibrary.lt/resursai/LMA/Ekologija/Eko72priedas/13.pdf
12. Zavadskas EK, Kaklauskas A, Turskis Z, Tamosaitiene J (2008) Selection of the effective dwelling house walls by applying attributes values determined at intervals. J Civ Eng Manage 14(2):85–93. https://www.tandfonline.com/doi/pdf/10.3846/1392-3730.2008.14.3?needAccess=true
13. Kersuliene V, Zavadskas EK, Turskis Z (2010) Selection of rational dispute resolution method by applying new stepwise weight assessment ratio analysis (SWARA). J Bus Econ Manage 11(2):243–258. https://www.tandfonline.com/doi/pdf/10.3846/jbem.2010.12?needAccess=true
14. Zavadskas EK, Turskis Z (2010) A new additive ratio assessment (ARAS) method in multicriteria decision-making. Technol Econ Dev Econ 16(2):159–172. https://journals.vgtu.lt/index.php/TEDE/article/view/5850/5093
15. Zavadskas EK, Turskis Z, Vilutiene T (2010) Multiple criteria analysis of foundation instalment alternatives by applying additive ratio assessment (ARAS) method. Arch Civ Mech Eng 10(3):123–141. https://doi.org/10.1016/S1644-9665(12)60141-1

16. Dadelo S, Turskis Z, Zavadskas EK, Dadeliene R (2012) Multiple criteria assessment of elite security personal on the basis of ARAS and expert methods. Econ Comput Econ Cybern Stud Res 46(4):65–87. http://www.ecocyb.ase.ro/20124pdf/Edmund%20Zavadskas%20(T).pdf
17. Keren B, Hadad Y, Laslo Z (2014) Combining AHP and DEA methods for selecting a project manager. Manage: J Sustain Bus Manage Solut Emerg Econ 71:17–28. http://management.fon. bg.ac.rs/index.php/mng/article/view/126/92
18. Pitchipoo P, Vincent DS, Rajini N, Rajakarunakaran S (2014) COPRAS decision model to optimize blind spot in heavy vehicles: a comparative perspective. Proc Eng 97:1049–1059. https://doi.org/10.1016/j.proeng.2014.12.383
19. Karabasevic D, Stanujkic D, Urosevic S (2015) The MCDM model for personnel selection based on SWARA and ARAS methods. Manage: J Sustain Bus Manage Solut Emerg Econ 77:43–52. http://management.fon.bg.ac.rs/index.php/mng/article/view/57/45
20. Adalı E, Işık A (2016) Air conditioner selection problem with COPRAS and ARAS methods. Manas J Soc Stud 5(2):124–138. https://www.semanticscholar.org/paper/AIR-CONDIT IONER-SELECTION-PROBLEM-WITH-COPRAS-AND-Adali-I%C5%9F%C4%B1k/b1f 99b4c8ff39137f5f1f044df9572e5f62331b3?p2df
21. Parezanović T, Bojković N, Petrović M, Tarle SP (2016) Evaluation of sustainable mobility measures using Fuzzy COPRAS method. Manage: J Sustain Bus Manage Solut Emerg Econ 21(78):53–62. http://management.fon.bg.ac.rs/index.php/mng/article/view/46/38
22. Karabasevic D, Stanujkic D, Urosevic S, Popovic G, Maksimovic M (2017) An approach to criteria weights determination by integrating the DELPHI and the adapted SWARA methods. Manage: J Sustain Bus Manage Solut Emerg Econ 22(3):15–25. http://management.fon.ac. rs/index.php/mng/article/view/163/121
23. Knežević S, Mandić K, Mitrović A, Dmitrović V, Delibašić B (2017) An FAHP-TOPSIS framework for analysis of the employee productivity in the Serbian electrical power companies. Manage: J Sustain Bus Manage Solut Emerg Econ 22(2):47–60. http://management.fon.bg.ac. rs/index.php/mng/article/view/86/112
24. Nasab SHM, Anvari AS (2017) A comprehensive MCDM-based approach using TOPSIS, COPRAS and DEA as an auxiliary tool for material selection problems. Mater Des 121:237–253. https://doi.org/10.1016/j.matdes.2017.02.041
25. Patel MR, Vashi MP, Bhatt BV (2017) SMART multi-criteria decision making technique for use in planning activities. Paper presented at the New Horizons in Civil Engineering (NHCE). https://www.researchgate.net/publication/315825133_SMART-Multi-criteria_decision-making_technique_for_use_in_planning_activities
26. Zarbakhshnia N, Soleimani H, Ghaderi H (2018) Sustainable third-party reverse logistics provider evaluation and selection using fuzzy SWARA and developed fuzzy COPRAS in the presence of risk criteria. Appl Soft Comput 65:307–319. https://doi.org/10.1016/j.asoc.2018. 01.023
27. Bahrami Y, Hassani H, Maghsoudi A (2019) BWM-ARAS: a new hybrid MCDM method for Cu prospectivity mapping in the Abhar area, NW Iran. Spat Stat 33. https://doi.org/10.1016/j. spasta.2019.100382
28. Fu YK (2019) An integrated approach to catering supplier selection using AHP-ARAS-MCGP methodology. J Air Transp Manage 75:164–169. https://doi.org/10.1016/j.jairtraman.2019. 01.011
29. Schitea D, Deveci M, Iordache M, Bilgili K, Akyurt IZ, Iordache I (2019) Hydrogen mobility roll-up site selection using intuitionistic fuzzy sets based WASPAS, COPRAS and EDAS. Int J Hydrog Energy 44(16):8585–8600. https://doi.org/10.1016/j.ijhydene.2019.02.011
30. Balki MK, Erdoğan S, Aydın S, Sayin C (2020) The optimization of engine operating parameters via SWARA and ARAS hybrid method in a small SI engine using alternative fuels. J Clean Prod 258. https://doi.org/10.1016/j.jclepro.2020.120685
31. Büyüközkan G, Güler M (2020) Smart watch evaluation with integrated hesitant fuzzy linguistic SAW-ARAS technique. Measurement 153. https://doi.org/10.1016/j.measurement. 2019.107353

32. Ghenai C, Albawab M, Bettayeb M (2020) Sustainability indicators for renewable energy systems using multi-criteria decision-making model and extended SWARA/ARAS hybrid method. Renew Energy 146:580–597. https://doi.org/10.1016/j.renene.2019.06.157

33. Roozbahani A, Ghased H, Shahedany MH (2020) Inter-basin water transfer planning with grey COPRAS and fuzzy COPRAS techniques: a case study in Iranian central plateau. Sci Total Environ 726. https://doi.org/10.1016/j.scitotenv.2020.138499

34. Yücenur GN, Çaylak S, Gönül G, Postalcıoğlu M (2020) An integrated solution with SWARA and COPRAS methods in renewable energy production: city selection for biogas facility. Renew Energy 145:2587–2597. https://doi.org/10.1016/j.renene.2019.08.011

35. Alinezhad A, Khalili J (2019) New methods and applications in multiple attribute decision making (MADM). In: Price CC, Zhu J, Hillier FS (eds) ISOR, vol 277. Springer, Cham. https://doi.org/10.1007/978-3-030-15009-9

36. Podvezko V (2011) The comparative analysis of MCDA methods SAW and COPRAS. Eng Econ 22(2):134–146. https://doi.org/10.5755/j01.ee.22.2.310

Fighting Media Hyper-partisanship with Modern Language Representation Models

Akshi Kumar, Utkarsh Tyagi, Tanish Grover, and Aheli Ghosh

Abstract The Internet has shaped how people gather knowledge, learn from their surroundings, form their individual opinions, and deal with socially relevant topics. In a time of polarization, when the news that we see is twisted according to one's view, extremely one-sided views aim to conquer the internet. In such a case, it is of utmost importance to devise an algorithm that can outperform and overcome such biases. We propose to build a convolutional neural network by utilizing sentence embeddings from language representation models like BERT, RoBERTa, DistilBERT, and XLNet, which would be able to classify whether an article displays a hyper-partisan narrative or not. We analyze the writing style of the author rather than depending on fact verification to prove an article's underlying bias. Our model gives an accuracy up to 88% with BERTweet-base. Such a model can actively prevent the spread of political propaganda through news outlets and can lead to the public consuming unbiased and accurate information.

Keywords Hyperpartisan · Sentence embedding · BERTweet · RoBERTa · XLNet · DistilBERT

1 Introduction

For many people, news articles are the first and major source to obtain unbiased and reliable information, which is necessary for forming a nuanced understanding of what is currently happening around us. News, TV, and social media articles have an important role in defining internal and public opinions.

Popular media frequently displays an inward bias in its news. Biased news articles can easily harm a whole nation as they may lead to the general public forming their views and opinions based on false or misleading information as presented in them. For example, during the coronavirus pandemic in India, WhatsApp was

A. Kumar · U. Tyagi · T. Grover (✉) · A. Ghosh
Delhi Technological University, New Delhi, India
e-mail: akshikumar@dce.ac.in

© The Author(s), under exclusive license to Springer Nature Singapore Pte Ltd. 2022   57
D. Gupta et al. (eds.), *Proceedings of Data Analytics and Management*,
Lecture Notes on Data Engineering and Communications Technologies 90,
https://doi.org/10.1007/978-981-16-6289-8_6

heavily monitored to curb the spread of fake news and violence. Therefore, it is now imperative to have an automatic tagging method for hyper-partisan news. There are several computational methods available to tackle fake news. However, the main focus of these models is on fact-checking, and they rarely examine the writing style of the article itself. Hyper-partisan articles contain factually correct news but framed in such a way that allows the media outlets to further push their political propaganda.

In this paper, we have utilized a small and well-labeled data set of articles to perform the classification task using some of the state-of-the-art (SOTA) language models for generating sentence embeddings. These embeddings facilitate the extraction of semantic information about the given text rather than just focusing on the word structures. Rather than relying solely on the primitive natural language processing techniques, these models consider contextual information for better semantic representation of our data set.

These language representation models, like BERT, RoBERTa, DistilBert, and XLNet help in learning language representations, which are further fine-tuned to cater to specific machine learning tasks. The models learn bidirectional relationships that help in calculating sentence embeddings, which can be further used to construct a binary classifier using a 1-D convolutional neural network for classifying news as neutral or hyperpartisan.

Roadmap: The rest of the paper comprises Sect. 2, which discusses the pre-existing work in this domain, Sect. 3 throws light on the data we have used for training, Sect. 4 details the entire methodology. This section is followed by Sect. 5, which highlights the results of our study, and Sect. 3 discusses the results. At last, Sect. 7 discusses the conclusion and future work.

2 Related Work

Several previous studies have been conducted for developing techniques to identify the presence of fake news. However, there are very few studies that are tackling the issue of detecting hyper-partisan arguments in written media. In this section, we discuss some of the techniques previously published by various authors.

Potthast et al. [6] focused on investigating whether the writing style similarities of news articles can distinguish hyper-partisan news from mainstream news. They devised a new approach based on unmasking to assess and visualize writing styles of text categories.

Alabdulkarim and Alhindi [1] used a Support Vector Machine (SVM) model on a feature vector to predict if the given text is hyper-partisan or not. They preprocessed the data set by removing punctuation, stop words, and non-alphabetical characters. Furthermore, they tokenized and lemmatized the text, and then finally used the TFiDF vectorizer to store the tokens as a vector. They used word vectors as features for the model, linguistic inquiry features, article structure features, and emotion features captured using NRC emotions lexicons.

The winners of the SemEval 2019 task for hyper-partisan detection were Jiang et al. [3]. They utilized word embeddings with the ELMo model and CNNs for predicting hyper-partisan news. The ensemble of models was used together to generate the final averaged predictions.

3 Training Data

For this paper, we used the data for PAN at SemEval 2019 Task 4: Hyper-partisan News Detection data set [4].

There are two parts of this data set: the first part contains a large number of articles labeled by the overall bias of the publisher as provided by BuzzFeed journalists and by MediaBiasFactCheck.com. The second part is a smaller set of articles, each of which is individually labeled through inputs from crowd-sourcing. For this paper, we have used the second part of the data set which contains more accurate, individually-labeled articles than the data set labeled according to the publisher. The "by article" set consists of 645 articles that are labeled as neutral or hyper-partisan. The data set has 37% articles which are marked as hyper-partisan and 63% articles which are marked as neutral. The articles in the data set are of varying lengths. Each sample in the data set is written in an HTML format containing the title and the article content. Figure 1 is an example of a hyper-partisan, hateful article. It is particularly targetting a specific political party and negatively influencing the readers.

4 Methodology

In this section, we go through the approach we take in this paper to build our model to detect hyper-partisanship. The complete process is shown in Fig. 2.

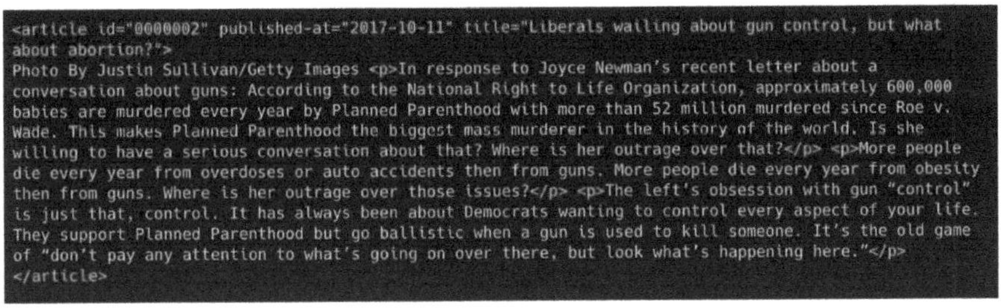
```
<article id="0000002" published-at="2017-10-11" title="Liberals wailing about gun control, but what
about abortion?">
Photo By Justin Sullivan/Getty Images <p>In response to Joyce Newman's recent letter about a
conversation about guns: According to the National Right to Life Organization, approximately 600,000
babies are murdered every year by Planned Parenthood with more than 52 million murdered since Roe v.
Wade. This makes Planned Parenthood the biggest mass murderer in the history of the world. Is she
willing to have a serious conversation about that? Where is her outrage over that?</p> <p>More people
die every year from overdoses or auto accidents then from guns. More people die every year from obesity
then from guns. Where is her outrage over those issues?</p> <p>The left's obsession with gun "control"
is just that, control. It has always been about Democrats wanting to control every aspect of your life.
They support Planned Parenthood but go ballistic when a gun is used to kill someone. It's the old game
of "don't pay any attention to what's going on over there, but look what's happening here."</p>
</article>
```

Fig. 1 Sample training article of hyper-partisan narrative

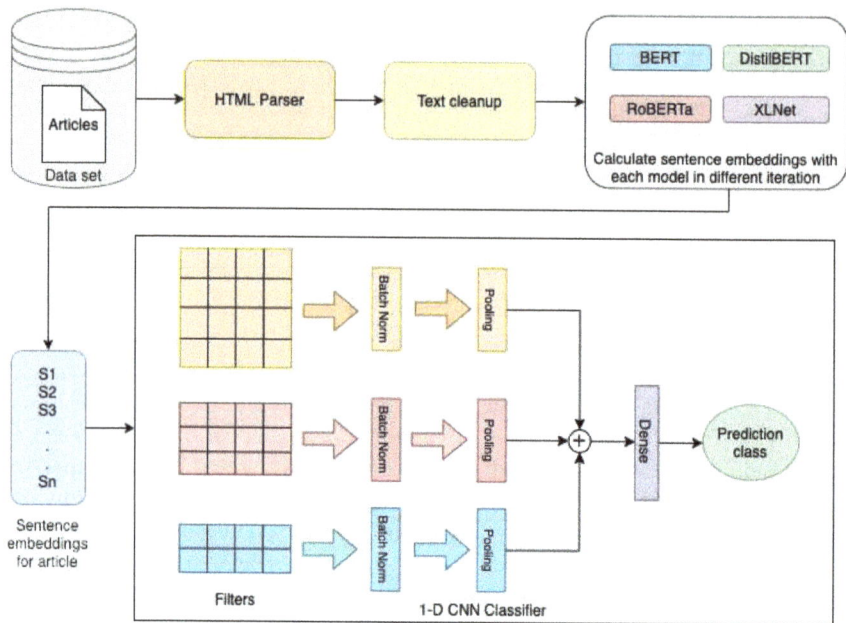

Fig. 2 Methodology followed

4.1 Data Preprocessing

The data is wrapped in an article tag with a title given as an attribute. The whole article is made up of paragraph tags along with anchor tags for hyperlinks. We use an HTML parser to extract the title and the article content. Each paragraph tag is considered a sentence break. We concatenate the title and article content into a combined text which we use as the input for our sentence embedding generation step (described in Sect. 4.2). The white spaces throughout the article have been replaced with single spaces. We also preserve the original case and the punctuation in the text as they can help convey the writer's intent.

4.2 Sentence Embeddings

Owing to the high variation in article length and the number of tokens in an article, we utilize sentence embeddings instead of using word embeddings to attain uniformity and avoid any unintentional dependency on sentence length. Word embeddings involve the representation of each word in our corpus as a vector while sentence embeddings represent the whole sentence as a vector. These sentence vectors together represent our whole article. We use the sentence-transformer[7] library to calculate

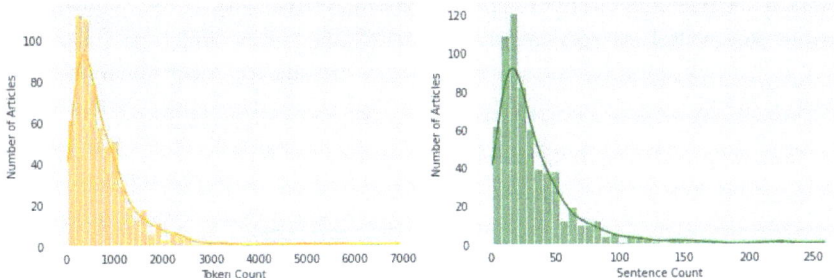

Token count distribution after pre-processing   Sentence count distribution after pre-processing

Fig. 3 Token and Sentence count distribution after pre-processing

these sentence embeddings for our article inputs using multiple pre-trained models—BERT, RoBERTa, DistilBERT and XLNet provided by the Transformers [9] Pytorch library.

Each of these models provide state-of-the-art performance in natural language processing tasks with certain caveats based on utility. These models have two variants differing in the number of stacked encoder layers- base and large; we have used the base variants in our study. The base versions for an input sentence, output an embedding vector containing 768 features. We go with the base versions as we have a small data set and the large variants often require a larger data set to perform well.

The models we are using for the sentence embedding generation task have been fined tuned on different data sets. We use models fine-tuned on the Stanford Sentiment Treebank v2(SST-2) data set [8] and the TweetEval dataset [2]. Both of these data sets are available for sentiment classification which relates to the bashful and neutral sentiment in hyperpartisan articles. For TweetEval tuned models, we are using the top-performing models on the sentiment task of this data set which were RoBERTa-base and bertweet-base [5] which is a modified version of BERT. These fine-tuned models will be able to accurately capture the sentiments of the writer in the embeddings and hence improve our classification performance of hyper-partisan detection.

Figure 3b shows the distribution of the count of a sentence per article. From this distribution, we are taking the first 200 sentence embeddings vectors representing our article as the input for our classifier. If an article has less than 200 sentences then the input is padded and in case it has more than 200 sentences then the rest of the sentence embedding vectors are discarded.

4.3 Binary Classification Model

We use the sentence embeddings generated from the language representation models as input to our classification model, which takes the sentence embedding as input to classify the document as hyper-partisan or neutral. We have six one-dimensional convolutional neural network layers of various kernel sizes, which have used the Leaky ReLU activation function. Their forward propagation mechanism has been explained in Eq. 1,

$$x_k^l = b_k^l + \sum_{i=1}^{N_{l-1}} conv1d(w_{ik}^{l-1}, m_{ik}^{l-1}) \tag{1}$$

x_k^l: input, b_k^l: the bias of the kth neuron at layer l, m_i^{l-1}: the output of the ith neuron at layer $l-1$, w_{ik}^{l-1}: is the kernel from the ith neuron at layer $l-1$ to the kth neuron at layer l.

Then, the outputs are normalized using BatchNormalization, which then go through MaxPooling, and are finally concatenated and flattened. The flattened output is passed into a dense layer activated by a sigmoid function(Eq. 2).

$$y = \text{sigmoid(flatten(concat(pooled\_output)))} \tag{2}$$

We use the binary cross-entropy function as our loss function and Adam as our optimizer. K-fold cross-validation is an appropriate data re-sampling technique for a small-sized data set to avoid over-fitting. This results in a more informed estimate of our accuracy. For this paper, we have used stratified K-fold cross-validation with K set to 10.

5   Results

We compare our results with the approach achieving the highest accuracy at SemEval 2019 for the hyperpartisan task. We will be comparing the accuracy achieved on the publicly available data set. The top submission by Jiang et al. [3] used ELMo based word embeddings with a deep learning classifier. They achieved an accuracy of 84.04% on the "by article" public data set. After testing out with various models, the best results we achieved with our approach are illustrated in Table. 1.

Table 1 Results achieved by our approach

Model used	Data set fine-tuned for	Accuracy(%)
DistilBERT-base	SST2	84.171
XLNet-base	SST2	84.185
RoBERTa-base	SST2	85.267
BERT-base	SST2	85.901
RoBERTa-base	TweetEval	87.440
BERTweet-base	TweetEval	**88.380**

Bold value indicates highest result

6 Discussion

We see that the BERTweet-base fine-tuned for TweetEval performs the best for our task with an accuracy of 88.380%, followed by RoBERTa-base which was also fine-tuned for the same data set. Other models also display good comparable accuracies.

Although the Bertweet and RoBERTa models displayed the best accuracy, these models, in reality, are huge and costly to use in production applications. Our motivation for writing this paper was fighting hyperpartisan articles spreading widely on social media websites. In such cases, a lighter model would be much more suitable. We note that the DistilBERT-base model, a much lighter distilled version of BERT also achieves a comparable accuracy of 84.171% and hence will be a good fit for use in applications.

7 Conclusion

A rapidly growing part of our population is accessing online media content these days, and the way people accumulate knowledge and engage with social topics is dictated by their media interaction. However, media coverage often reflects a bias in news articles, owing to its ownership, source of income, or political stance.

We propose the use of language representation models for deriving sentence embeddings, which capture the writing style of authors and extract their intent. We have tried models like BERT, RoBERTa, DistilBERT, and XLNet and compared the various tradeoffs of using them. Our BERT and RoBERTa models outperform the current best SemEval submission available in this arena (up to 88 % accuracy achieved).

Despite proving to be the most accurate, there are some caveats of using them like high memory usage and speed. Distilbert is an appropriate solution to that end. As the DistilBERT sentence embeddings also provided a very comparable accuracy while using only half of the parameters, and being lightweight, it will be best suited for production use. It should be the go-to choice for application developers trying to fight and contain the spread of hyper-partisan news.

Acknowledgements We are grateful to the Department of Computer Science and Engineering Delhi Technological University for providing us the labs and resources for performing our study. We are thankful to all other faculty members of our department for their guidance, and our parents for their encouragement.

References

1. Alabdulkarim A, Alhindi T (2019) Spider-Jerusalem at SemEval-2019 task 4: hyperpartisan news detection. In: Proceedings of the 13th International workshop on semantic evaluation. Association for Computational Linguistics, Minneapolis, Minnesota, USA, pp 985–989. https://doi.org/10.18653/v1/S19-2170. https://www.aclweb.org/anthology/S19-2170
2. Barbieri F, Camacho-Collados J, Espinosa-Anke L, Neves L (2020) TweetEval:Unified benchmark and comparative evaluation for tweet classification. In: Proceedings of findings of EMNLP
3. Jiang Y, Petrak J, Song X, Bontcheva K, Maynard D (2019) Team bertha von suttner at SemEval-2019 task 4: hyperpartisan news detection using ELMo sentence representation convolutional network. In: Proceedings of the 13th International workshop on semantic evaluation. Association for Computational Linguistics, Minneapolis, Minnesota, USA, pp 840–844. https://doi.org/10.18653/v1/S19-2146. https://www.aclweb.org/anthology/S19-2146
4. Kiesel J, Mestre M, Shukla R, Vincent E, Adineh P, Corney D, Stein B, Potthast M (2019) SemEval-2019 task 4: Hyperpartisan news detection. In: Proceedings of the 13th International workshop on semantic evaluation. Association for Computational Linguistics, Minneapolis, Minnesota, USA, pp 829–839. https://doi.org/10.18653/v1/S19-2145. https://www.aclweb.org/anthology/S19-2145
5. Nguyen DQ, Vu T, Nguyen AT (2020) BERTweet: a pre-trained language model for English Tweets. In: Proceedings of the 2020 Conference on empirical methods in natural language processing: system demonstrations, pp 9–14 (2020)
6. Potthast M, Kiesel J, Reinartz K, Bevendorff J, Stein B (2018) A stylometric inquiry into hyperpartisan and fake news. In: Proceedings of the 56th Annual meeting of the Association for Computational Linguistics (Volume 1: Long Papers). Association for Computational Linguistics, Melbourne, Australia, pp 231–240. https://doi.org/10.18653/v1/P18-1022. https://www.aclweb.org/anthology/P18-1022
7. Reimers N, Gurevych I (2019) Sentence-bert: sentence embeddings using siamese bert-networks. In: Proceedings of the 2019 Conference on empirical methods in natural language processing. Association for Computational Linguistics. https://arxiv.org/abs/1908.10084
8. Socher R, Perelygin A, Wu J, Chuang J, Manning CD, Ng A, Potts C (2013) Recursive deep models for semantic compositionality over a sentiment treebank. In: Proceedings of the 2013 Conference on empirical methods in natural language processing. Association for Computational Linguistics, Seattle, Washington, USA, pp 1631–1642. https://www.aclweb.org/anthology/D13-1170
9. Wolf T, Debut L, Sanh V, Chaumond J, Delangue C, Moi A, Cistac P, Rault T, Louf R, Funtowicz M, Brew J (2019) Huggingface's transformers: state-of-the-art natural language processing. CoRR **abs/1910.03771** (2019). http://arxiv.org/abs/1910.03771

Hyperparameter Tune for Neural Network to Improve Accuracy of Stock Market Prediction

Hiral R. Patel, Ajay M. Patel, Hiral A. Patel, and Satyen M. Parikh

Abstract As of late, hyperparameter tune (HPT) has become an inexorably significant issue in the field of AI for the advancement of more exact gauging models. In this examination, we investigate the capability of HPT in displaying stock returns utilizing a profound neural network. The capability of this methodology was assessed utilizing specialized pointers and basics analyzed dependent on the impact the regularization of dropouts and clump standardization for all info information. We found that the model utilizing specialized pointers and dropout regularization fundamentally outflanks three different models, demonstrating a positive consistency of 0.60% in-example and 2.11% out-of-test, consequently demonstrating the chance of beating the verifiable normal. We likewise exhibit the steadiness of the model as far as the adjustments in it include significance over the long run. The experiment will deploy on Python with its library support. The data crawler is ready to fetch latest live data and to store on local system. In previous papers we state that the NN gives remarkable performance so the HPT is apply and develop the model to forecast the price. The main aim of paper is to perform the hyper parameter tune for neural network in terms of hidden layer size, activation function, weight optimization, learning rate and iterations can be consider and analysed with different methods to maximize the accuracy of stock market prediction.

Keywords Hyperparameter tune · Hidden layers · Learning rate · Neural network · ROI · Stock market prediction

H. R. Patel (✉) · A. M. Patel · H. A. Patel · S. M. Parikh
Ganpat University, Kherva, India
e-mail: hrp02@ganpatuniversity.ac.in

A. M. Patel
e-mail: ajaykumar.patel@ganpatuniversity.ac.in

H. A. Patel
e-mail: hiral.patel@ganpatuniversity.ac.in

S. M. Parikh
e-mail: satyen.parikh@ganpatuniversity.ac.in

1 Introduction

Neural network has become a promising method to display the unpredictability of stock developments. It empowers us to catch non-direct developments, to relate enormous information, and to decrease commotion without a suspicion of a pre-indicated fundamental design. Simultaneously, it leaves us with a trouble in choosing various hyperparameters, which basically influences the presentation of the subsequent models. Most investigations managing a monetary time arrangement regularly pick pre-indicated hyperparameters and check the strength of the model dependent on little changes in the boundaries. This methodology expects specialists to invest a ton of energy into tuning various boundaries all the while, which frequently brings about a problematic model. Hyperparameter tune (HPT) can be utilized to moderate this issue via consequently looking for the most ideal hyperparameters in AI students and has been broadly used to recognize great arrangements all the more rapidly, for example, using a consecutive model-based calculation setup, tree-structure Parzen assessor (TPE), and Sprearmint [1]. HPT has additionally been exhibited to be an incredibly amazing methodology for programmed picture and discourse acknowledgment and offers preferences for managing with AI in an efficient way. To start with, it lessens the human exertion that is important in tuning the hyperparameters and opens up the chance of improving the exhibition of AI [2, 3]. Second, it improves the reproducibility and decency of logical examines in light of the fact that a robotized HPT is more reproducible than a hand-tuned approach utilizing experimentation searches to deliver an ideal conduct, consequently permitting us to think about various techniques all the more genuinely through a similar degree of tuning [4, 5].

In spite of such points of interest, monetary investigations have commonly not thought about this technique. HPT requires an enormous information scale to keep away from an overfitting happening in both the preparation and approval information. Stock-related information is gotten distinctly over a generally brief timeframe range, regularly from the year 1950 to the present. As demonstrated in Fig. 1, an irregular development of a stock return, for example, time-changing unpredictability and intermittent hops identified with crashes or unexpected upsurges, causes a period reliance of the model boundary set to explicit periods [6, 7].

Besides, cross-approval and rearranging, which are pivotal procedures for forestalling an overfitting, cannot be utilized on the grounds that stock-related information is time-requested, and a demonstrating measure requires saving the time requesting. Hence, the utilization of HPT has seldom been evaluated, and there is a helpless comprehension of its effectiveness in monetary information displaying. Therefore, experts need to focus harder on hyper parameter tuning and the coming about models to a great extent relying upon their experience.

The paper mainly focuses on the model implementation for stock market price prediction. In previous research, it is already stated that NN gives better performance by performing same techniques and other techniques on same tools as well

Fig. 1 Model
implementation steps

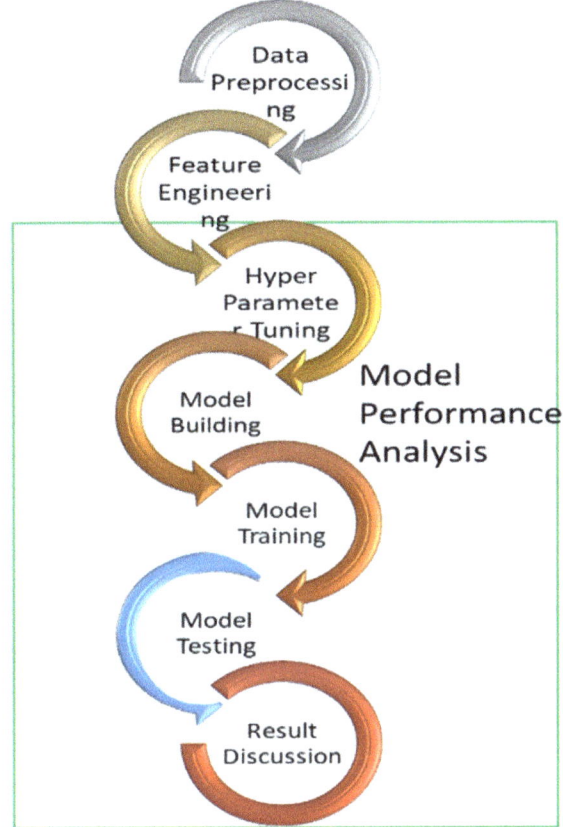

as other tools like Tanagra, orange and concludes that NN is better, so here the HPT is performed for NN implementation to achieve the best optimized accuracy [8, 9].

2 Data Processing

Technical indicators Technical analysis is a method for forecasting price movements using past prices and volume and includes a variety of forecasting techniques such as a chart analysis, cycle analysis, and computerized technical trading systems. Technical analysis has a long history of widespread use by participants in speculative markets [18–23], and there is a large body of academic evidence demonstrating the usefulness of a technical analysis, including theoretical support [24] and empirical evidence [25, 26], as well as their role in out-of-sample equity premium predictability [27, 10, 11]. The monthly market data for the S&P500 were obtained from Yahoo Finance and contain daily trading data, i.e., the opening prices, high prices, low prices, adjusted closing prices, and specialized pointers. Technical investigation is a strategy for anticipating value developments utilizing past costs and volume and incorporates an assortment of determining methods, for example, a diagram investigation, cycle examination, and electronic specialized exchanging frameworks.

Specialized investigation has a long history of inescapable use by members in speculative markets [18–23], and there is an enormous assortment of scholarly proof exhibiting the handiness of a specialized examination, including hypothetical help [24] and exact proof [25, 26], just as their job in out-of-test value premium consistency [27, 10, 11].

The month-to-month market information for the S&P500 was acquired from Yahoo Finance and contains day-by-day exchanging information, i.e., the initial costs, exorbitant costs, low costs, changed shutting costs, and end-of-day volumes. The information is from the time frame between January 1, 2015, and December 31, 2020. We utilized a full arrangement of 14 specialized pointers dependent on three sorts of famous specialized methodologies, moving normal hybrid guidelines, energy rules, and volume rules. Different indicators like MA, MOM, and VOL are implemented.

As a fundamental indicator, we use the financial indicators employed by [12] for the national stock market, which is available from NSE Web site which include dividend–price ratio, dividend yield, earning–price ratio, dividend payout ratio, book-to-market ratio, net equity expansion, long-term rate of return, etc [13–15].

3 Experimental Study

Profound feed-forward neural organizations were utilized in this examination. We applied TPE for robotized hyperparameter tuning with extra tests utilizing reproduced strengthening furthermore, an irregular hunt to additionally affirm our outcomes. The hyperparameters and their earlier disseminations are summed up. For hyperparameter determination, we prepared DNNs on an in-example preparing set and chose the model with the most reduced approval mistake. We restricted the quantity of capacity assessments for finding ideal hyper-boundaries to 50. Each assessment involved preparing the DNN models for 200 ages and choosing the model with the most minimal approval blunder. The following are the list of parameters for hyperparameter tuning.

- Hidden Layer Size: This parameter is used to define number of hidden layers. The default value is 100.
- Activation Function: This is dealt with different activation functions for hidden layer like identity, logistic, Tanh, and Relu.
- Weight Optimization: The weight optimization is used to train the neural.
- Learning Rate: Learning rate is used to consider the weight updates.
- Iterations: Number of iterations performed to evaluate the model.

Dataset: For performing stock market prediction case study, we have utilized the live data from BSE authorized Web source. The model has auto agent-based model to retrieve data from source and store in local formatted database storage. In this experiment, only few stocks from various sectors were experimented and applied for hyperparameter tune. The data is collected from Jan 2019 to Jan 2021. Here Cipla, HCL, Wipro, and Infosys-based stock experiments were discussed. MLPRegressor is applied and performs the hyperparameter tune to improve model accuracy, and after tuning the model, it is also compared with KNIME tool with same setup [16, 17].

4 Artificial Neural Network

An ANN is displayed as a portrayal of the people own organic neuron and how the neural organization capacities in a person. An ANN is an AI calculation strategy that is likewise called connectionism, equal conveyed preparing or neural calculation. One explanation why ANN is well known is because of its strength, adaptation to non-critical failure, capacity to learn and sum up, versatility, all-inclusive capacity estimation, and equal information handling. This empowers them to tackle complex non-straight and multi-input–yield (IO) relationship issues (Ortiz-Rodriguez et al. 2013; Gomes et al. 2011). Ghaffari et al. (2006) state that when contrasted with different strategies, ANN has been demonstrated to be predominant as displaying procedure for informational collections with non-direct connections for applications, for example, information fitting, and forecast.

Cerna et al. (2005) state that the most famous kinds of ANN are the Kohonen organization (self-organizing map) and the multilayer perceptron (MLP), which utilize solo and administered adapting separately. Ortiz-Rodriguez et al. (2013) state that multilayer perceptron prepared with back-propagation calculation is the 'most utilized ANN' in forecast, grouping, and demonstrating. Preparing can be completely administered learning as utilized in grouping or semi-managed learning as utilized in bunching. There is normally need to measure framework execution in run of the mill expectation applications. In other examination, Deng et al. (2011) estimated execution of their model on chosen stocks at the New York Stock Trade based on mean absolute percentage error (MAPE) and root mean square blunder (RMSE), while Neto et al. (2009) additionally estimated execution of their model for a Brazilian stock organization utilizing MAPE and mean square error (MSE).

A test in ANN configuration is the choice of the ideal number of units that are huge enough to fit the reason, however, not very huge that the ANN neglects to sum up the arrangement (overfitting). There are a few issues to consider in the plan of ANNs, for example, kind of organization (feed-forward, repetitive, and back-propagation), sort of preparing (administered, solo), extent of preparing and testing informational indexes (70:30 or 80:20), number of info and yield units (generally application subordinate), number and size of covered up layers ($2N + 1$, exploratory), number of reiterations during preparing (age), decision of enactment work (sigmoid, direct, exaggerated, and limit), and size of informational index (number of records). Other plan contemplations incorporate information volume, learning rate, and energy. In deciding fitting information volume, Devi et al. (2011) did tests over a long-term period, while Adhikari et al. (2013) utilized datasets of between 3 years and 4 years. On their part, Wong et al. (2012) thought about a long-term period, while Butler et al. (2011) utilized information that was up to 23 years. For preparing and testing information proportions, Ortiz-Rodrigues et al. (2013) utilized a 80:20 proportion, while Microsoft (2013) employs 70:30 on their SQL worker item.

Exploration has been done on different business sectors, where financial exchange value forecast has been endeavored. Deng et al. (2011) applied specialized examination for forecast of select stocks at the New York Stock Exchange. Aghababaeyan et al. (2011) attempted examination on forecast for the Tehran Stock Exchange, where they built up an instrument that accomplishing an exactness of 97%. Others, for example, Khan et al. (2011) led an investigation on the Bangladesh stock market where they built up a device with normal mistake of 3.7% and 1.5% in two reenactments, while Pan et al. (2005) built up an instrument for the Australian stock market with 80% exactness in foreseeing the value course. None of their devices appear to have been grown monetarily or focused on for the individual stockbrokers.

https://scikit-learn.org/stable/modules/generated/sklearn.neural_network.MLP Regressor.html#sklearn.neural_network.MLPRegressor

https://www.kdnuggets.com/2016/10/beginners-guide-neural-networks-python-scikit-learn.html

5 Methodology

Figure 1 shows the model implementation general steps. As per this experimental study is carried out.

In this paper, first data preprocessing will be incorporated. In this step, data gathering and cleaning will performed to make data clean and ready for experiment.

The next step is feature engineering where the input and target parameters identified.

In HPT, the NN parameter tuning was performed to make optimization in the context of accuracy.

In model building, the data splitting was done to divide test set in train set and test set in 70:30 ratio.

In model training, the model was developed with train set.

In model testing, the model given with test set and accuracy was calculated.

Lastly, result discussion is there.

So, this is the entire process how model works.

6 Result Discussion and Conclusion

The model is implemented using Python and Machine Learning Libraries. The final tune model is compared with KNIME tool for performance comparison. The dataset for performing experiment was collected from BSE, and here, only few stocks study was represented. The learning rate will be applied individually and perform the modeling. The constant learning rate working with default step size is 0.001. Invscaling gradually decreases the learning rate and count mean effective learning rate. Adaptive learning rate working with the function of increasing and decreasing form when model fails to model it decrease step and if success than increase the step size. So, adaptive gives better results (Table 1).

In this experiment, the following activation functions are applied for finding out the best suited function for stock data.

- Identity activation function means no any activation function is applied.
- Logistic activation function is the logistic sigmoid function.
- Tanh activation function is the hyperbolic tan function which utilized for activation.
- Relu, activation function is the rectified linear unit function.

The weight optimizers are also tuned for implementing the model which discussed below.

- LBFGS: It is based on quasi-Newton approach methods.
- SGD: It is stochastic gradient descent simplest approach.
- ADAM: Stochastic gradient-based optimizer which is utilized by optimizer pointers.

By tuning the neural network, the above results were generated and conclude that the logistic sigmoid activation function with LBFGS-based weight optimizer and adaptive learning rate give better results of forecasting with $r2$ score in between 0.8 and 0.9 for all selected stock.

After developing the model with hyperparameter tune, the same experiment with selected parameters is applied on selected stock using KNIME, and the model gets the $r2$ score as 0.98 which is better in Python but more accurate with tuned parameter.

Table 1 MLPRegressor hyperparameter tune

Stock data set	Techniques	Activation function	Weight optimizer	Learning rate	$R2$ error
Cipla	MLPRegressor	Identity	Lbfgs	'constant', 'invscaling', 'adaptive'	51.140
			Sgd	'constant', 'invscaling', 'adaptive'	Error overflow
			adam	'constant', 'invscaling', 'adaptive'	55.17
		Tanh	Lbfgs	'constant', 'invscaling', 'adaptive'	97.47
			Sgd	'constant', 'invscaling', 'adaptive'	112.38
			adam	'constant', 'invscaling', 'adaptive'	124.25
		Relu	Lbfgs	'constant', 'invscaling', 'adaptive'	8.22
			Sgd	'constant', 'invscaling', 'adaptive'	276.79
			Adam	'constant', 'invscaling', 'adaptive'	37.92
		Logistic	Lbfgs	'constant', 'invscaling', 'adaptive'	0.32
			Sgd	'constant', 'invscaling', 'adaptive'	132.20
			adam	'constant', 'invscaling', 'adaptive'	116.21
HCL	MLPRegressor	Identity	Lbfgs	'constant', 'invscaling', 'adaptive'	54.10
			Sgd	'constant', 'invscaling', 'adaptive'	Error overflow

(continued)

Table 1 (continued)

Stock data set	Techniques	Activation function	Weight optimizer	Learning rate	R2 error
			adam	'constant', 'invscaling', 'adaptive'	65.10
		Tanh	Lbfgs	'constant', 'invscaling', 'adaptive'	87.37
			Sgd	'constant', 'invscaling', 'adaptive'	122.33
			adam	'constant', 'invscaling', 'adaptive'	130.5
		Relu	Lbfgs	'constant', 'invscaling', 'adaptive'	9.2
			Sgd	'constant', 'invscaling', 'adaptive'	206.7
			adam	'constant', 'invscaling', 'adaptive'	40.2
		Logistic	Lbfgs	'constant', 'invscaling', 'adaptive'	0.42
			Sgd	'constant', 'invscaling', 'adaptive'	133.50
			adam	'constant', 'invscaling', 'adaptive'	118.34
Wipro	MLPRegressor	Identity	Lbfgs	'constant', 'invscaling', 'adaptive'	50.160
			Sgd	'constant', 'invscaling', 'adaptive'	Error overflow
			adam	'constant', 'invscaling', 'adaptive'	59.27
		Tanh	Lbfgs	'constant', 'invscaling', 'adaptive'	92.7

(continued)

Table 1 (continued)

Stock data set	Techniques	Activation function	Weight optimizer	Learning rate	$R2$ error
			Sgd	'constant', 'invscaling', 'adaptive'	110.30
			adam	'constant', 'invscaling', 'adaptive'	130.15
		Relu	Lbfgs	'constant', 'invscaling', 'adaptive'	7.92
			Sgd	'constant', 'invscaling', 'adaptive'	296.9
			adam	'constant', 'invscaling', 'adaptive'	34.99
		Logistic	Lbfgs	'constant', 'invscaling', 'adaptive'	0.36
			Sgd	'constant', 'invscaling', 'adaptive'	138.70
			adam	'constant', 'invscaling', 'adaptive'	111.11
Infosys	MLPRegressor	Identity	Lbfgs	'constant', 'invscaling', 'adaptive'	53.40
			Sgd	'constant', 'invscaling', 'adaptive'	Error overflow
			adam	'constant', 'invscaling', 'adaptive'	58.170
		Tanh	Lbfgs	'constant', 'invscaling', 'adaptive'	94.345
			Sgd	'constant', 'invscaling', 'adaptive'	110.18
			adam	'constant', 'invscaling', 'adaptive'	129.5

(continued)

Table 1 (continued)

Stock data set	Techniques	Activation function	Weight optimizer	Learning rate	R2 error
		Relu	Lbfgs	'constant', 'invscaling', 'adaptive'	9.99
			Sgd	'constant', 'invscaling', 'adaptive'	279.88
			adam	'constant', 'invscaling', 'adaptive'	38.88
		Logistic	Lbfgs	'constant', 'invscaling', 'adaptive'	0.33
			Sgd	'constant', 'invscaling', 'adaptive'	139.20
			adam	'constant', 'invscaling', 'adaptive'	115.21

References

1. Patel H, Parikh S (2012) Automated news based ULIP fund switching model. Presented and published with international conference GCEMP 2012, awarded as best technical paper
2. Patel H, Parikh S (2012) A comparative study on financial stock market prediction models. Int J Eng Sci (IJES) 1(2):188–191. Indexed in ANED (American National Engineering Database) Impact Factor 7.2, ISSN: 2319-1813 ISBN: 2319-1805
3. Patel H, Parikh S (2014) A proposed prediction model for forecasting the financial market value according different factors. Int J Comput Technol Appl (IJCTA) 5(1). Impact Factor—2.015 IC Value 5.17, ISSN 2229-6093
4. Patel H, Parikh S (2014) A technical and fundamental parameters analysis for financial market prediction using semantic analysis
5. Patel H, Parikh S (2015) Dynamic IS based asset allocation on crude trend analysis—exploring a hedging concept. Presented and publishing in GCEMP-15, GFJMR. ISSN 2229-4651
6. Patel H, Parikh S (2016) Comparative analytical study for news text classification techniques applied for stock market price extrapolation. In: SmartCom 2016 proceedings by Springer CCIS
7. Patel H, Parikh S (2016) Prediction model for stock market using news based different classification, regression and statistical techniques (PMSMN). ICTBIG 2016 IEEE publication, 18 and 19 Oct 2016 Indore
8. Tuli S, Tuli S, Tuli R, Gill SS (2020) Predicting the growth and trend of COVID-19 pandemic using machine learning and cloud computing. Internet Things 11:100222. https://doi.org/10.1016/j.iot.2020.100222
9. Shinde GR, Kalamkar AB, Mahalle PN, Dey N, Chaki J, Hassanien AE (2020) Forecasting models for coronavirus disease (COVID-19): a survey of the state-of-the-art. SN Comput Sci 1(4):197. https://doi.org/10.1007/s42979-020-00209-9
10. Patel H, Parikh S (2017) Experimental study on stock market to analyse the impact of the latest demonetization in India. CiiT international journal and publication

11. Patel H, Parikh S (2017) Prediction model based on NLP and NN for financial data outcome revelation. In: 3rd international young scientist congress (IYSC-2017), awarded as "Young Scientist Award". Res J Comput Inf Technol Sci E-ISSN: 2320-6527
12. Patel H, Parikh S (2016) Comparative analysis of different statistical and neural network based forecasting tools for prediction of stock data. Presented at ICTCS—2016, publication (ACM conference). http://dl.acm.org/citation.cfm?id=2905055.2905186
13. Liu W et al (2017) A survey of deep neural network architectures and their applications. In: Neurocomputing 234:11–26. ISSN: 0925-2312. https://doi.org/10.1016/j.neucom.2016.12.038. http://www.sciencedirect.com/science/article/pii/S0925231216315533 (visited on 04/12/2019)
14. Srivastava RK, Greff K, Schmidhuber J (2015) Training very deep networks. In: Cortes C et al (eds) Advances in neural information processing systems 28, Curran Associates, Inc. 2015, pp 2377–2385. http://papers.nips.cc/paper/5850-training-very-deep-networks.pdf (visited on 04/09/2019)
15. Hu J, Shen L, Sun G (2018) Squeeze-and-excitation networks. In: Proceedings of the IEEE conference on computer vision and pattern recognition, pp 7132–7141. http://openaccess.thecvf.com/content_cvpr_2018/html/Hu_Squeeze-andExcitation_Networks_CVPR_2018_paper.html (visited on 04/09/2019)
16. Schmidhuber J (2015) Deep learning in neural networks: an overview. In: Neural networks 61, pp 85–117. ISSN: 08936080. https://doi.org/10.1016/j.neunet.2014.09.003. http://arxiv.org/abs/1404.7828 (visited on 04/12/2019). Vaswani A et al (2017) Attention is all you need. In: Advances in neural information processing systems 30. Guyon I et al (eds) Curran Associates, Inc., pp 5998–6008. http://papers.nips.cc/paper/7181-attention-is-all-you-need.pdf (visited on 05/02/2019)

ANN-Based Handwritten Digit Recognition and Equation Solver

Amarjit Malhotra, Megha Gupta, Rishabh Yadav, Priyam Tokas, and Ankit Sharma

Abstract It is well known that handwritten digits recognition may be a complicated job that is incorporated in a variety of applications. It is been largely employed by machine learning enthusiasts and researchers for developing apps like bank check processing. The most important task for the identification of a mathematical expression is to segregate the characters and then classify those characters. Due to their random nature, handwritten digits are not recognized by hard computing, and if recognized, the complexity is way too much. In this paper, an artificial neural network has been enforced and trained using a digits data set, to acknowledge and identify written digits from zero to nine and mathematical operators to unravel a mathematical equation.

Keywords Digit recognition · ANN · Equation solver

1 Introduction

Convolutional neural network (CNN) is a deep learning ANN which is used for classifying images. The basic idea behind CNN is using pre-defined convolving filters to identify patterns in image edges, parts of objects and to build on this knowledge to identify various objects. CNN, essential in numerous tasks of computer vision, is designed to adaptively and automatically learn dimensional hierarchies of features through back propagation using convolution layers, pooling layers and fully connected layers. CNN performs better in terms of efficiency and computational cost when it comes to classifying digits and operators and thus solves an equation, as the convolutional layers benefit from the inherent properties of the images. CNN takes benefit of the local spatial coherence of the images [1–4]. The number of operations required to process an image by using convolution on patches of adjacent pixels is

A. Malhotra · R. Yadav · P. Tokas · A. Sharma
Department of Information Technology, Netaji Subhas University of Technology, Delhi, India

M. Gupta (✉)
Department of Computer Science, MSCW, University Of Delhi, Delhi, India

© The Author(s), under exclusive license to Springer Nature Singapore Pte Ltd. 2022
D. Gupta et al. (eds.), *Proceedings of Data Analytics and Management*,
Lecture Notes on Data Engineering and Communications Technologies 90,
https://doi.org/10.1007/978-981-16-6289-8_8

decreased dramatically since adjacent pixels together are meaningful which is known as local connectivity.

Convolutional neural network is better than a feed forward network since CNN has features parameter sharing and spatial property reduction. As a result of parameter sharing in CNN, the quantity of parameters is reduced so the computations are reduced as well. Learning from one part of the image is additionally helpful in another part of the image. Attributable to the spatial property reduction in CNN, the computational power required is reduced.

Multilayer perceptron network (MLP) is a class of feed forward ANN. Along with feed forward and back propagation, this work has implemented a multilayer perceptron neural network and evaluated its performance in classifying handwritten digits. The proposed work used regularization in the neural network so as to avoid the overfitting problem. The models that fit extremely well to the training data, but fits poorly to the testing data, have been avoided in this work.

Section 2 discusses some already published work similar to the concept. In Sect. 3, the proposed approach has been discussed. Section 4 presents the results.

2 Literature Survey

In work [5], the author recognized handwritten characters without feature extraction. It uses a multilayer feed forward neural network. The data set contains twenty-six alphabets. Fifty different data sets are considered for training the network. The trained neural network is then used for classification and recognition. It used a feed forward propagation neural network algorithm. The system yields good recognition rates which are at par with feature extraction-based techniques for handwritten character recognition. But handwriting recognition using basic algorithms used in soft computing has a low accuracy rate.

ANN is used to solve recognition problems like character recognition [6]. MATLAB's neural network toolbox is used to identify handwritten characters when they are projected on grids of different sizes. It was ascertained that accuracy of the character recognition is dependent on the resolution of projection of the characters. It is also observed that different writing styles cannot be identified using the same network, with the same exact precision.

Author in work [7] addresses the individual character recognition using neural networks, in the form of images. It provides the accuracy of each identification/recognition, as a part of the classification result. It is then used to customize the application according to the client's demands. The work [8] ascertains that large back propagation networks can be used in real image identification and recognition problems. It does not require a complex preprocessing stage that needs detailed engineering. A challenging problem in [9] was to recognize fully multi-oriented handwritten characters. There are several strategies to solve these types of classification tasks. Input features are obtained, and convolutional neural networks are used to extract higher-level features.

The authors [3] used the box method for extracting the features of the handwritten characters. Partitions are done on the character image into a pre-defined quantity of images known as boxes. Optical character recognition, an earlier used system, successfully identifies the handwritten characters having fixed size and fonts.

3 Proposed Work

The proposed technique is based on the following concepts:

Back propagation—It uses the delta rule to obtain the minimum value of the error function in weight space which is considered to be a solution to the learning problem.

Multilayer perceptron—It solves problem stochastically, giving the approximate solutions even for extremely complex problems.

Convolutional neural network—It is used in classification of image(s) and identification because of its high precision.

3.1 Data Set Description

The input to the neural network is the data set from MNIST which consists of images containing handwritten digits as shown in Fig. 1. The image size is 20×20, and they are converted to greyscale. This data set has 500 images out of which 350 are

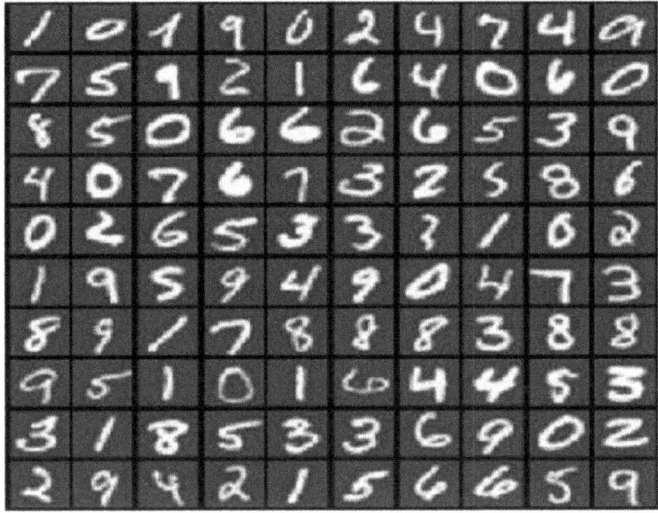

Fig. 1 Handwritten digits

considered and then used for training, 50 are considered for validation, and 100 are considered for testing.

These images cannot be sent into the neural network as such, so they are converted into a vector of size 400. This makes the processing easier and faster. After classifying every segmented image, this classification is converted to its corresponding digit or operator and concatenated to an expression string, and thus results have been calculated.

3.2 Methodology

Pooling helps in pulling down the computational cost/time and thereby allowing us to train the model to attain better efficiency. The image is downscaled using the pooling layers as shown in Fig. 2. It is feasible since features that are organized spatially like an image are retained throughout the network, and thus when they are downscaled, it helps to reduce the size of the image. It is not possible to downscale a vector on other inputs since there is no coherence between adjacent inputs.

The neural network consists of three layers that are input, hidden and output layer as shown in Fig. 3. The input layer consists of 400 neurons with 25 neurons in the hidden layer and ten output neurons (for digits 0–9). Each neuron in the first layer is connected to the neuron in the next layer, but there is no interconnection between neurons in the same layer. The given inputs are digit images having pixel values.

Gradient for the neural network is computed using the cost function. As soon as the gradient is computed using the advanced optimizer, then the neural network is trained to minimize the cost function. Firstly for the computation of gradient for the parameters of the neural network (un-regularized), back propagation algorithm is

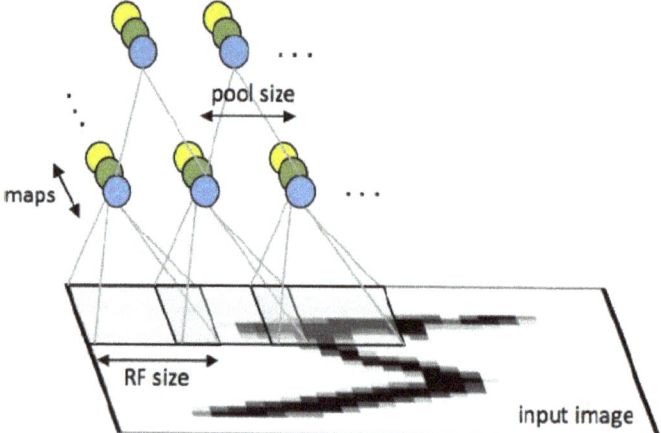

Fig. 2 First layer of a CNN with pooling

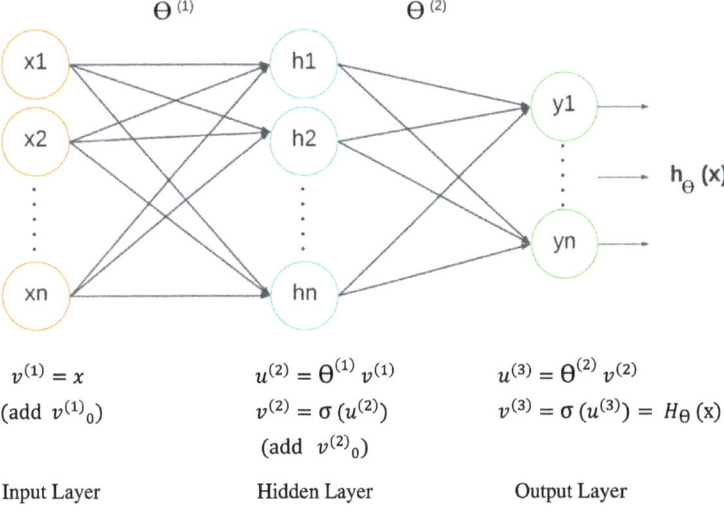

$$v^{(1)} = x$$
$$(\text{add } v^{(1)}{}_0)$$

$$u^{(2)} = \Theta^{(1)} v^{(1)}$$
$$v^{(2)} = \sigma(u^{(2)})$$
$$(\text{add } v^{(2)}{}_0)$$

$$u^{(3)} = \Theta^{(2)} v^{(2)}$$
$$v^{(3)} = \sigma(u^{(3)}) = H_\Theta(x)$$

Input Layer　　　　　Hidden Layer　　　　　Output Layer

Fig. 3 Model representation

implemented. After correctness of the computed gradient un-regularized case is verified, gradient for the regularized neural network is implemented. Back propagation technique as shown in Fig. 4 provides efficient network parameters.

The cost function for the network is given by

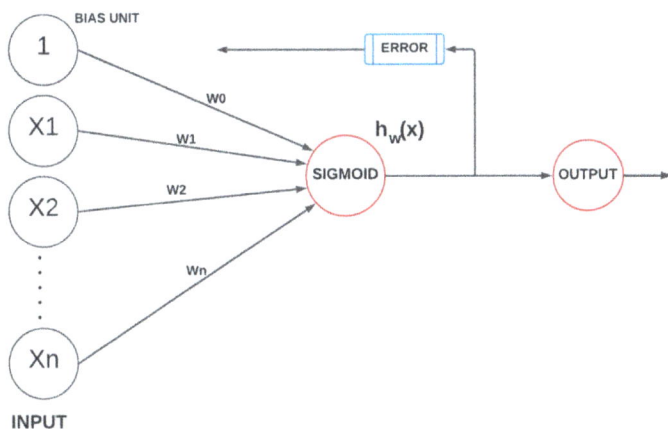

Fig. 4 Back propagation

$$C(\theta) = \frac{1}{n} \sum_{s=1}^{n} \cdot \sum_{t=1}^{T} \Big[-q_t^{(s)} \log\big(\big(H_\theta\big(p^{(s)}\big)\big)\big)_t$$

$$-\big(1 - q_t^{(s)}\big) log\big(1 - \big(H_\theta\big(p^{(s)}\big)\big)\big)_t\Big] \qquad (1)$$

T represents total number of possible labels.

$H_\theta\big(p^{(s)}\big)_t$ is the activation (output value) of the t-th output unit.

Variable q represents the labels for the neural network training.

Let us take an example, if $p^{(s)}$ is an image of the digit 5, then the corresponding $q^{(s)}$ which is used with the cost function should be a ten-dimensional vector with $q^5 = 1$, whereas the rest of the elements are set to 0. The feed forward computation that computes $H_\theta\big(p^{(s)}\big)$ is calculated for every example i, and at the end, it sums the cost over all examples.

4   Results and Discussion

This section discusses the results obtained with the trained model in terms of accuracy and execution time.

It can be inferred from Figs. 5, 6, 7, 8, 9 and 10 that, as the number of hidden units increases, the accuracy also increases but to a certain instant and after that the accuracy becomes constant. The maximum accuracy obtained is 94.03% when the number of hidden units was 50.

It can be inferred from Fig. 11 that maximum accuracy of 94.97% is achieved when the regularization factor is 0.2.

The graph in Fig. 12 depicts direct proportionality between execution time and number of hidden units. It can be seen as the number of hidden units increases, the

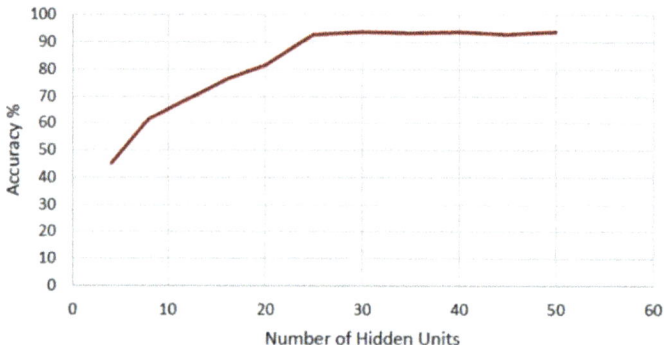

Fig. 5 Accuracy versus number of hidden units

Fig. 6 No. of hidden units are 10

Training set Accuracy:83.53%

Validation set Accuracy:82.49%

Test set Accuracy:84.15%

Fig. 7 No. of hidden units are 20

Training set Accuracy:91.622%

Validation set Accuracy:91.25999999999999%

Test set Accuracy:91.55%

Fig. 8 No. of hidden units are 30

Training set Accuracy:93.616%

Validation set Accuracy:93.42%

Test set Accuracy:93.32000000000001%

Fig. 9 No. of hidden units are 40

Training set Accuracy:94.132%

Validation set Accuracy:93.47999999999999%

Test set Accuracy:93.83%

Fig. 10 No. of hidden units are 50

Training set Accuracy:94.094%

Validation set Accuracy:93.28999999999999%

Test set Accuracy:94.03%

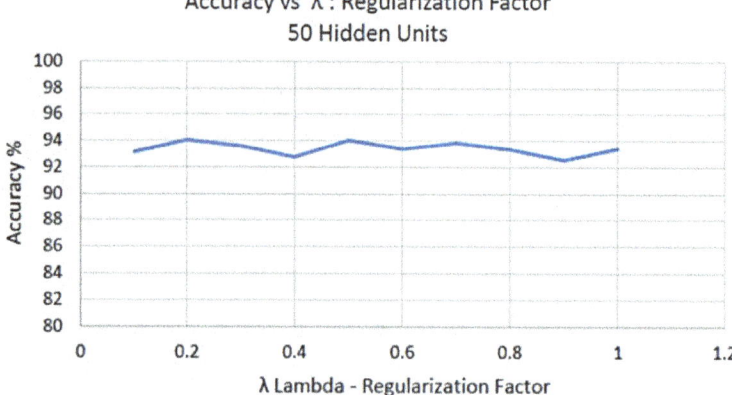

Fig. 11 Accuracy versus regularization factor

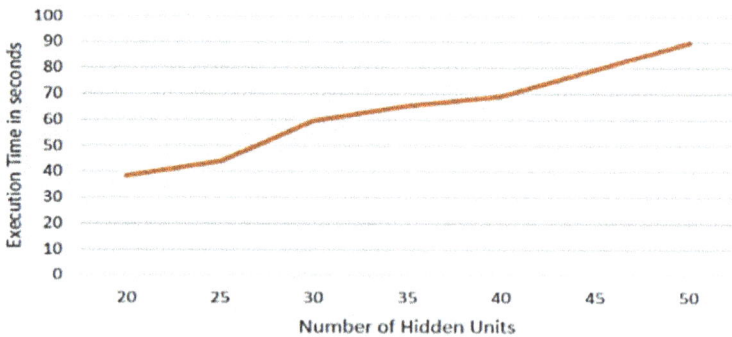

Fig. 12 Execution time versus number of hidden units

execution time also increases. That is why the number of hidden units was increased till 50 so that execution time is not increased.

Figures 13 and 14 discuss the performance measures for CNN. The model was trained with an accuracy of 99.34. And model is able to achieve testing accuracy of 98.90 to recognize the digits and operators.

It has been observed that increasing the number of hidden units from 16 to 50 led to increase in the accuracy from 76 to 94%. This significant increase in accuracy was

```
Epoch 1/2
1313/1313 [==============================] - 31s 24ms/step - loss: 0.0334 - accuracy: 0.9897 - val_loss: 0.0541
_accuracy: 0.9837
Epoch 2/2
1313/1313 [==============================] - 22s 17ms/step - loss: 0.0216 - accuracy: 0.9934 - val_loss: 0.0519
```

Fig. 13 Training accuracy

```
In [18]:    1  model.evaluate(X_test, y_test_cat)

            313/313 [==============================] - 2s 6ms/step - loss: 0.0359 - accuracy: 0.9890
```

Fig. 14 Testing accuracy

a driving reason for setting the number of hidden units to 50 and where the similar accuracy is achieved consistently.

Similarly, the direct proportionality of the no. of hidden units with the accuracy of the neural network was also observed. This stated that as the no. of hidden units increases, execution time also increases due to the increase in the number of computations to be performed. It has been observed that as the no. of hidden units increases from 30 to 50, the execution time increases from 61 to 85 s. The regularization factor λ can be used to avoid overfitting problem so that the algorithm produces similar results on testing data set. In this work, different values for λ are observed without regularization, and it has been found that the algorithm takes time to converge and learn light vectors. Therefore, different values of λ ranging between 0 and 1 are used and recorded for the corresponding accuracy and performance of the network. Moreover, it has been found very minor variation in accuracy against λ, i.e., an accuracy of 94.57% for $\lambda = 0.1$ and accuracy of 94.58% for $\lambda = 0.6$. Thus, regularization factor $\lambda = 0.2$ was chosen, which gave the highest accuracy of 94.97% when the number of hidden units was 50.

The proposed system delivered accuracy levels of above 90% and took less time in training due to efficient use of above-discussed methodology. Decent accuracy is being achieved with the available amount of data.

5   Conclusion and Future Work

The proposed technique is able to build an algorithm for handwritten digit recognition with the total efficiency to classify and recognize the digits and the operator and then finally solve the equation. The proposed method could be inferred as accurate while taking the minimum possible time for the computations. CNN performs better in terms of efficiency and computational cost when it comes to classifying digits and operators and thus solves an equation, as the convolutional layers benefit from the inherent properties of the images. The proposed model achieved a good training and testing accuracy. In future, the work will be compared with other works and extended to achieve wider results.

References

1. Shrivastava A, Jaggi I, Gupta S, Gupta D (2019) Handwritten digit recognition using machine learning: a review. In: 2019 2nd international conference on power energy, environment and intelligent control (PEEIC), Greater Noida, India, pp 322–326
2. Graves A, Liwicki M, Fernandez S, Bertolami R, Bunke H, Schmidhuber J (2019) A novel connectionist system for unconstrained handwriting recognition. IEEE Trans Pattern Anal Mach Intell
3. Ahmed S, Naz S, Swati S, Razzak MI (2019) Handwritten Urdu character recognition using one-dimensional BLSTM classifier. Neural Comput Appl 31:1143–1151
4. Balado J, Díaz-Vilariño L, Verbree E, Arias P (2020) Transfer learning for indoor object classification: from images to point clouds. ISPRS Annals of Photogrammetry, Remote Sensing and Spatial Information Sciences
5. Long M, Yan Z (2019) Detecting Iris liveness with batch normalized convolutional neural network. Comput Mater Contin 58:493–504
6. Chuangxia H, Liu B (2019) New studies on dynamic analysis of inertial neural networks involving non-reduced order method. Neurocomputing 325:283–287
7. Ahlawat S, Rishi R (2019) A genetic algorithm based feature selection for handwritten digit recognition. Recent Pat Comput Sci 12:304–316
8. Tabik S, Alvear-Sandoval RF, Ruiz MM, Sancho-Gómez JL, Figueiras-Vidal AR, Herrera F (2020) MNIST-NET10: a heterogeneous deep networks fusion based on the degree of certainty to reach 0.1% error rate. Ensembles overview proposal. Inf Fusion 62:73–80
9. Siddique F, Sakib S, Siddique MAB (2019) Recognition of handwritten digit using convolutional neural network in python with tensor-flow and comparison of performance for various hidden layers. In: 2019 5th international conference on advances in electrical engineering (ICAEE), Dhaka, Bangladesh, pp 541–546

Intrusion Detection Protocol Using Independent Outlier Ensembles

D. Divya⬤, M. Bhasi⬤, and M. B. Santosh Kumar

Abstract Numerous anomaly detection algorithms are used for detection of malicious networks'. Yet, a protocol to segregate suspicious and attacker networks is an essential enhancement in this domain. Therefore, this paper tries to generate a rule to differentiate suspicious, attacker networks and trusted networks. We used independent outlier ensemble algorithm for performing this segregation with random forest, AdaBoost, XGBoost and K nearest neighbours as the components of the independent ensemble design. When a user has to take caution about the network, the detection foot print is high. In such cases, logical OR operator helps to identify all the suspicious outlier points. However, for attacking scenarios, the detection foot print is low and AND operator detect only those points which are highly dangerous and action has to be taken by the user. We did evaluation of our algorithm with the help of CIDDS-001 dataset which an emulated dataset. Experiments show that our algorithm can successfully sort networks into suspicious, malicious and trusted ones. User can take caution for suspicious networks and can take action for malicious ones. Thus, this paper brings a new idea for classification of networks which is not addressed till now.

Keywords Intrusion · Outlier ensemble · Suspicious · Malicious

D. Divya (✉)
Research Scholar, Division of IT, School of Engineering, Cochin University of Science and Technology, Kochi, Kerala 682022, India
e-mail: divya.d@cusat.ac.in

M. Bhasi
Senior Professor, School of Management Studies, Cochin University of Science and Technology, Kochi, Kerala 682022, India

M. B. Santosh Kumar
Associate Professor, Division of IT, School of Engineering, Cochin University of Science and Technology, Kochi, Kerala 682022, India
e-mail: santo_mb@cusat.ac.in

© The Author(s), under exclusive license to Springer Nature Singapore Pte Ltd. 2022   87
D. Gupta et al. (eds.), *Proceedings of Data Analytics and Management*,
Lecture Notes on Data Engineering and Communications Technologies 90,
https://doi.org/10.1007/978-981-16-6289-8_9

1   Introduction

Network intrusion detection is the process of identifying abnormal activities in the network which cannot be identified by firewalls [1]. Although network intrusion systems have higher performance than traditional firewall systems, performance of these detectors depends upon the intrusion detection rate [2] and wrong detection rate. Nevertheless, emergence of new variety of threats coming in to the network may reduce the detection rate while increasing the false alarm rate [3]. Also, certain networks contain suspicious activities which is not a malicious ones. So user has to identify whether it is suspicious one where user has to take caution or malicious one where user has to take action to block that network.

Intrusion detection systems used in current systems can be network feature based or community-based methods [4]. In network feature-based techniques, the topology of the network is used to identify abnormal nodes; whereas in the latter one, it tries to detect distributed attacks [5]. Other than these systems, several dynamic defence model systems are also developed to detect the intrusions. Still there is a contradiction in deciding whether to block all the suspicious points or to allow these points [6].

Hence, objective of this paper is to discriminate the networks into normal, malicious and suspicious one by means of an independent outlier ensemble algorithm. This research work gives more clarity for the users either to proceed with the incoming network or completely block that network.

Rest of the paper is organised as follows. Section 2 details related works in this field, Sect. 3 illustrates data and methods, and Sect. 4 presents Experimental results.

2   Related Works

Various outlier detection problem techniques are evolved in literature to increase the outlier detection rate while maintaining the false alarm rate within limits. Outlier ensemble is one such method that achieves higher detection rate by taking outputs from various detectors. This section contains various independent outlier ensemble algorithms used in various domains and how they can be utilised in network intrusion problem.

2.1   Independent Outlier Ensembles

Ensemble analysis is widely used in data mining problems to decrease the dependency of models on a particular dataset [6]. Nevertheless, unavailability of ground truth makes it difficult to evaluate the performance of the ensemble systems and guarantee better precision of the systems. There can be sequential ensembles and independent ensembles for outlier detection [6]. In sequential ensemble, one or more outlier algorithms are applied sequentially, but in independent ensemble use different instantiations of the same algorithm or different instantiations of the data is used for the analysis. Here, the former one is known as model centred algorithm and the latter one can be called as data centred algorithm [6]. Combining outputs of different algorithms is also a major concern in parallel ensembles. Multiple methods like averaging, selecting maximum output out of all the results, weighted averaging are used for combining scores of independent algorithms [7]. All these independent outlier ensembles look at the same problem in different ways that helps to provide more robust results [8] and tries to maintain bias variance trade off of the system.

2.2 Network Intrusion Detection Using Outlier Detection Techniques

A huge number of recent works in this area utilises various clustering algorithms such as density-based clustering [9], incremental clustering [10], clustering algorithms with improved hybrid feature selection [11], and agent clustering [12] is used by researchers to detect anomalies in the system, and K nearest algorithms are used to categorise various attacks [12]. For delivering contextual information regarding the network work flow methods like service flow architectures has been proposed by researchers. These architectures incorporated linked visualisation dash boards to compare outputs [13].

3 Data and Methods

We used four independent supervised anomaly detection algorithms for identifying intrusions in the network data. The concept of independent outlier ensembles is used to get the final result. Experiments are conducted on an emulated dataset CIDDS-001.

3.1 Dataset

We used CIDDS-001 labelled flow-based dataset for evaluation of network intrusion detection systems. This dataset contains the attributes given in Table 1.

This dataset contains five class labels normal, attacker, victim, suspicious and unknown. We divided this dataset into two subsets. First subset label all the data points with label normal as '0' and rest of them labelled as '1'. This subset is used to evaluate detection rate of independent ensemble for suspicious networks. Thus, this subset acts as the suspicious database.

Table 1 Attributes within the CIDDS-001 dataset

A1	A2	A3	A4	A5	A6	A7	A8	A9	A10	A11
Src IP	Src port	Dest IP	Dest port	Proto	Date first seen	Duration	Bytes	Packets	Flags	Class

Second subset label all the data points with normal, suspicious and unknown points as '0' and attacker, victim points as '1'. This dataset is used for evaluating attacker networks. This makes the attacker database.

3.2 Methodology

We used random forest (RF), extreme gradient boosting (XGBoost), adaptive boosting (AdaBoost) and K nearest neighbour (KNN) algorithms as components of the independent ensemble. Random forest (RF) algorithm gives you better classification accuracies without over fitting [12, 14]. AdaBoost algorithm is the most commonly used algorithm for network intrusion detection due to its higher efficiency in dealing with complex behaviour of malicious activities [15]. XGBoost [16] also gained high accuracy for segregating malicious attackers and benign ones. K nearest neighbours have advantages of ease of implementation and better performance while dealing with binary classification problems [17]. Hence, we developed an independent ensemble using these four algorithms for detecting suspicious and attacker networks.

3.2.1 Flow Graph

Algorithm 3.2.2 illustrates algorithm for intrusion detection (Fig. 1).

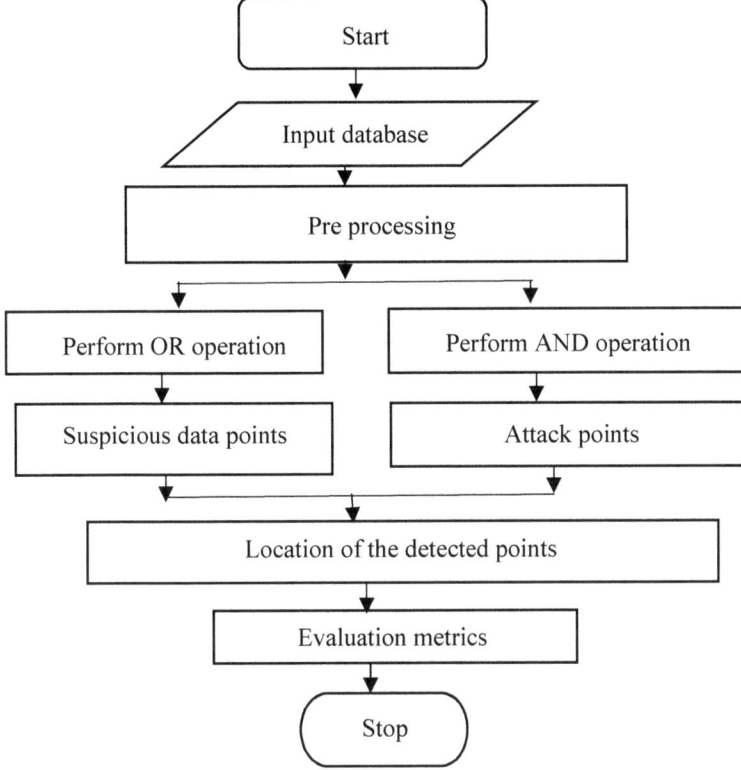

Fig. 1 Flow diagram for detection of suspicious and attacker networks

3.2.2 Algorithm

```
Begin
Input suspicious data subset SD, attacker subset AD
Call pre process
        Call label encoder (A1, A3, A5, A6)
                Return numerical A1, A3, A5, and A6
Call suspicious_point_detect ()
        Call RF, AdaBoost, XGBoost, Knn
        detected_suspiciousy = RFy | XGy | Aday | Knny
        loc_suspiciousy = np.where(detected_suspiciousy =1)
        Out_detection_rate= detected_sus /total_ sus
        False_alarm_rate=detectednormal_as_sus/total_normal
End
Call attack_point_detect ()
        Call RF,AdaBoost,XGBoost,Knn
        detected_attacky = RFy & XGy & Aday & Knny
        loc_attacky= np.where (detected_attacky =1)
        Out_detection_rate= detected_attack /total_ attack
        False_alarm_rate=detectednormal_as_attack/total_normal
End
```

In sub-routine suspicious_point_detect (), the four detection algorithms for intrusion detection are called independently. RFy, XGy, Aday and Knny contain outputs obtained from these algorithms. For identifying suspicious networks, these independent outputs are combined using 'OR' operator so that detected_suspiciousy contains anomalous points detected any of the detection algorithm.

In attack_point_detect (), instead of using 'OR' operator, we used 'AND' operator. This will help to identify those networks which are identified by all the individual algorithms as attacker networks. Thus by using OR and AND operator, we can clearly distinguish networks in which only caution is needed, and the ones where blocking of the malicious networks is required.

4 Experimental Results

Algorithm 3.2.2 returns points where a user need take caution or action. By using these values, outlier detection rate and false alarm rate of independent ensemble algorithm can be calculated where

$$\text{Outlier detection rate} = \frac{\text{Outliers identified}}{\text{Total outliers}} \tag{1}$$

$$\text{False alarm rate} = \frac{\text{normal points identified as outliers}}{\text{Total normal points}} \tag{2}$$

Here, outlier detection rate in suspicious database is the performance of the algorithm in detecting suspicious points. Same rate in attacker database is the performance of the algorithm in detecting attacker networks. Table 2 gives the performance comparison of the proposed algorithm with existing systems.

Table 2 Comparison with existing algorithms

Algorithm	Dataset used	Outlier detection rate	False alarm rate
Deep learning [3]	KDD99, UNSW-NB15	KDD99: 0.99996 UNSW-NB15: 0.89134	0.001
Incremental clustering [10]	KDD Cup 1999, UNSW-NB15	Cup 1999: 0.9953 UNSW: 0.8841	Cup 1999: 0.0047 UNSW: 0.1159
Mixed attribute outlier detection [11]	NSL KDD	0.9784	0.188
AdaBoost [15]	NSL KDD	0.73	0.14
Proposed model	**CIDDS-001**	**Suspicious database: 1.0 Attacker database0.998**	**0.00001**

Bold indicates the comparison of existing algorithms with proposed model

Our proposed algorithm outperforms existing algorithms in the same domain. The same concept can be applied on similar kind of problems [18].

5 Conclusion

Machine learning algorithms are widely used for the identification of network intrusions. Yet, an algorithm that can segregate identify suspicious and attacker networks is not available. Hence, we developed an independent ensemble algorithm combining four outlier detection techniques. In order to identify suspicious networks, we used a database with labelled suspicious data points and the second one with labelled attacker data points. In case of suspicious network identification, detection foot print is higher; hence, we used logical operator 'OR' for detecting all the suspicious points that is detected by any of the anomaly detection algorithm. While recognising attacker network detection foot print is low, thus user needs to block only those networks which are identified by all the four algorithms. 'AND' operator is used in this case. Here, caution has to be taken for suspicious points and network blocking is required for attacker networks. Comparison of the proposed algorithm with existing ones gives promising results. This paper presents a new idea for sorting the networks into normal, suspicious and trusted ones.

References

1. Selvakumar B, Muneeswaran K (2018) Firefly algorithm based feature selection for network intrusion detection. Comput Secur 81. https://doi.org/10.1016/j.cose.2018.11.005
2. Al-Qatf M, Lasheng Y, Al-Habib M, Al-Sabahi K (2018) Deep learning approach combining sparse auto-encoder with SVM for network intrusion detection. IEEE Access 6:52843–52856. https://doi.org/10.1109/ACCESS.2018.2869577
3. Khan FA, Gumaei A, Derhab A, Hussain A (2019) A novel two-stage deep learning model for efficient network intrusion detection. IEEE Access 7:30373–30385. https://doi.org/10.1109/ACCESS.2019.2899721
4. Bindu PV, Santhi Thilagam P, Ahuja D (2017) Discovering suspicious behavior in multi-layer social networks. Comput Human Behav 73:568–5820. https://doi.org/10.1016/j.chb.2017.04.001
5. Rehak M, Pechoucek M, Bartos K, Grill M, Celeda P (2007) Network intrusion detection by means of community of trusting agents. In: 2007 IEEE/WIC/ACM international conference on intelligent agent technology (IAT'07), Fremont, CA, USA, pp 498–504. https://doi.org/10.1109/IAT.2007.67
6. Yuan P, Wang B, Mao Z (2021) Using multiple classifier behavior to develop a dynamic outlier ensemble. Int J Mach Learn Cyber 12:501–513. https://doi.org/10.1007/s13042-020-01183-7
7. Zhao Y, Nasrullah Z, Hryniewicki MK, Li Z (2019) LSCP: 'locally selective combination in parallel outlier ensembles'. In: Proceedings of the 2019 SIAM international conference on data mining (SDM), pp 585–593
8. Aggarwal CC (2013) Outlier ensembles: position paper. ACM SIGKDD Explor 14(2):49–58

 9. Keyvanpour MR, Barani Shirzad M, Mehmandoost S (2021) CID: a novel clustering-based database intrusion detection algorithm. J Ambient Intell Human Comput 12:1601–1612. https://doi.org/10.1007/s12652-020-02231-4
10. Taheri S, Bagirov AM, Gondal I et al (2020) Cyberattack triage using incremental clustering for intrusion detection systems. Int J Inf Secur 19:597–607. https://doi.org/10.1007/s10207-019-00478-3
11. Beulah JR, Punithavathani DS (2020) An efficient mixed attribute outlier detection method for identifying network intrusions. Int J Inf Secur Privacy (IJISP) 14(3):115–133. https://doi.org/10.4018/IJISP.2020070107
12. Sandosh S, Govindasamy V, Akila G (2020) Enhanced intrusion detection system via agent clustering and classification based on outlier detection. Peer-to-Peer Netw Appl 13:1038–1045. https://doi.org/10.1007/s12083-019-00822-3
13. Laughlin B, Sankaranarayanan K, El-Khatib K (2020) A service architecture using machine learning to contextualize anomaly detection. J Database Manage (JDM) 31(1):64–84. https://doi.org/10.4018/JDM.2020010104
14. Santana FB, Neto W, Poppi R (2019) Random forest as one-class classifier and infrared spectroscopy for food adulteration detection. Food Chem 293. https://doi.org/10.1016/j.foodchem.2019.04.073
15. Mazini M, Shirazi B, Mahdavi I (2018) Anomaly network-based intrusion detection system using a reliable hybrid artificial bee colony and AdaBoost algorithms. J King Saud Univ. https://doi.org/10.1016/j.jksuci.2018.03.011
16. Husain A, Salem A, Jim C, Dimitoglou G (2019) Development of an efficient network intrusion detection model using extreme gradient boosting (XGBoost) on the UNSW-NB15 dataset. In: 2019 IEEE international symposium on signal processing and information technology (ISSPIT), Ajman, United Arab Emirates, pp 1–7. https://doi.org/10.1109/ISSPIT47144.2019.9001867
17. Li L, Zhang H, Peng H, Yang Y (2018) Nearest neighbors based density peaks approach to intrusion detection. Chaos, Solitons Fractals 110:33–40. https://doi.org/10.1016/j.chaos.2018.03.010
18. Sharma M, Pradhyumna SP, Goyal S, Singh K (2021) Machine learning and evolutionary algorithms for the diagnosis and detection of Alzheimer's disease. In: Khanna A, Gupta D, Pólkowski Z, Bhattacharyya S, Castillo O (eds) Data analytics and management. Lecture notes on data engineering and communications technologies, vol 54. Springer, Singapore. https://doi.org/10.1007/978-981-15-8335-3_20

Hybrid Context-Based Recommendation for Media

Ajay Kumar Rajpoot, Ankesh Krishna Prasad, Gaurav Tiwari,
Manish Rawat, and Mugdha Sharma

Abstract There are many recommendation engines present that work on providing users with recommendations based on a single media. This is our attempt to create a recommendation system that works on relationships between different media elements and provides users with the best new media item to choose from. "Hybrid context-based recommendation for media" works on pre-context filtering and a hybrid approach of content and collaborative-based filtering and given a media element as input and recommends the same or different media element that the user needs or demands. The given model has been prepared using the MovieLens 100 K Data set and Goodreads Book Data set. This model can be further enhanced through a systematic consideration of demographic attributes in the future.

Keywords Collaborative filtering · Content-based filtering · SVD · Hybrid approach · Media

1 Introduction

The use of recommendation systems has become very common these days. No matter what services we tend to pick, we find the applications of recommendation systems in them. Some of the important applications of recommendation systems that have become not a luxury but a necessity include Spotify and Amazon. The former is a music service which helps people find the right music for them exactly according to their tastes and preferences, whereas the latter is an online retailer that suggests people various commodities according to their previous shopping behaviours and choices. The online streaming service Netflix also uses such recommendation systems. The list might go on and on as there is no end to the application possibilities of the recommendation systems [1]. Thus, a recommendation engine filters the information on the basis of the user's specific preferences and choices, in order to provide the

A. K. Rajpoot · A. K. Prasad · G. Tiwari · M. Rawat · M. Sharma (✉)
Bhagwan Parshuram Institute of Technology, Delhi, India

© The Author(s), under exclusive license to Springer Nature Singapore Pte Ltd. 2022   95
D. Gupta et al. (eds.), *Proceedings of Data Analytics and Management*,
Lecture Notes on Data Engineering and Communications Technologies 90,
https://doi.org/10.1007/978-981-16-6289-8_10

user with the best next thing to choose from a vast list of items or commodities. They are primarily used in commercial applications [2].

Building an efficient recommender system has always been a challenge for developers due to the varying interests of people. With the evolution of recommender systems, it has now become possible to not only match the content of the items and their types to be considered but also the demographic patterns in which they are used. The diverse users can now be taken into account, and the pattern in which they use an item can be estimated on the basis of various independent aspects such as their attributes, choices and habits. This is, thus, the core of these recommender systems, which remains the same no matter where we use them. Building recommender engines has proved to be very crucial in the modern world where people are getting more confused about choosing the right items for themselves and, hence, need some guidance to make up their minds on what exactly they need [3]. Be it any kind of media, the more correct recommendations people tend to get, the more they tend to explore and get captivated by the media. Thus, such recommender systems have become a thing of complete necessity. Evidently, people have been benefiting from the usage of such recommendation engines as it makes their life much simpler by constantly providing them with the next item which best fits according to their requirements and needs. With the ever-growing demands of the increasing population, it can be safely assumed that the usage of the recommendation engines will be further enhanced and discovered in the regions where it has not been discovered yet. Hence, it is essential to try and take into account all these varying patterns and variables and put them together to build an efficient recommender engine.

2 Related Work

There are various techniques which can be implemented with different recommender systems depending upon the domain and area for which that recommender system is designed for [4]. When it comes to discussing the works related to "hybrid context-based recommendation for media", there have been several attempts to recommend a media item given the same media content as an input [5]. Several recommendations have been made where books have been given as input and books have been recommended in return; and movies have been recommended in similar ways [6, 7]. We include here a few points from some of the related papers that we have referred to. In their research, Dewi Soyusiawaty and Yahya Zakaria tried to solve the problem of making it easier for people to find relevant books in the given library, on the basis of their contents and not just the title [8]. Similarly, the paper by "SarunJuntui and PaweenKhoenkaw" presented the idea of applying the concepts of recommendation algorithms for book recommender shelves. The logs for the transaction between the customers and the shelves were used to serve as data and then were further used to predict which ones could be the next best sellers. The evaluation method chosen was based on the rankings of the predicted bestseller books. They showed that 0.43 times out of 1, the results were accurate, and the correct books were identified for the best sellers [9].

Meanwhile, "Adli Ihsan Hariadi and Dade Nurjanah" tried to tackle a basic problem by suggesting that recommendations could be made perfect only if there were similar persons or peers sharing the same interests as us by implementing their recommender system [10]. Interestingly, "Karol Waga, Andrei Tabarcea and PasiFranti" depicted in their paper the usage of a context-aware recommendation engine which they later used over the location-based data [11]. They tried to improve the already built recommendation engine by adding an extra layer of context analysis. There have been various studies which proved that using context-aware analysis, we can provide efficient recommendations to the users [12]. This opened up the scope of including it in the case of media as well. Many such papers have shown a significant improvement in the previously built recommendation engines which worked on content and collaborative filtering of the data, by using context analysis. This also provided users with the ability to add an extra layer of their preferences. Another such research paper that must be noted is the paper by "Mohammed F. Alhamid, MajdiRawashdeh and Abdulmotaleb El Saddik", which included "context-aware recommendations in an ambient intelligence environment" [13]. They put forward the idea of discussing context-aware analysis in mobiles and other smart devices. They discussed how the context-aware engine can help in making clear choices in today's world of growing enhancements in the field of multimedia and technologies. It proposed a framework that worked on an ambient intelligent environment and physiological parameters.

3   Methods Involved

3.1   Content-Based Recommendation

In content-based recommendation, media elements or items are recommended to users on the basis of the similarity of the items that the given users have liked earlier. The content-based analysis provides a basic version of comparison between the items and takes into account only the relevant items when providing recommendations to the users [14]. In this model, the similarity between the elements has been measured using cosine similarity. Such a recommendation mechanism may not suffice the necessity of the growing population and its requirements.

3.2   Collaborative Filtering

Collaborative filtering is a technique that comes to the rescue and provides a slight enhancement in its results as it takes into account other crucial aspects when making recommendations for some users [15]. When using this technique, the interests of multiple users are taken into account while making recommendations. Singular value decomposition (SVD) is used as one of the collaborative filtering techniques [16]. It provides a structure in the form of a matrix. We know that row and column attributes define the structure of a matrix. In SVD, rows represent different users participating in rating the media element, and the columns consist of unique media elements. SVD provides factorisation of this matrix formed and breaks it into three different matrices and is represented by

$$M = USV^{\mathrm{T}} \tag{1}$$

Here "M" is the utility matrix created between the users and items, "U" is an orthogonal left singular matrix, "S" is the diagonal matrix, and "V" is the diagonal right singular matrix. The orthogonal left singular matrix represents the relationship between users and the latent factors, the diagonal matrix represents the strength of each latent factor, and the diagonal right singular matrix depicts the relationship between items and latent factors. As the latent factors are extracted, the dimensions of the utility matrix are reduced, and the mapping of users and items is done on a latent space. The expected rating for any item by any user can be given as follows:

$$r = x_i^T \cdot y_i \tag{2}$$

where r is the form of factorisation, x represents the vector depicting users, y represents the vector representing items, and T depicts the transpose of the matrix. Some further modifications are made, some biases and regularisation terms are added, and the equations obtained lay the basis of the SVD-based recommendation system.

3.3 Pre-Contextual Filtering

Pre-contextual filtering refers to filtering the data much before any recommendations are made out of the given data [17]. This model uses the age recommendation attributes of the media being recommended, and if it has the apt content for the ages below 18, items are included in both solo and group experience recommendation lists. Otherwise the item is only a part of the solo experience list.

4 Proposed Hybrid Recommendation Engine

Hybrid recommender engine is the one in which the user can get benefits of all the three pre-contextual filtering, content-based and collaborative-based filtering [18, 19]. In this model, we have used a score distribution methodology in which all the elements being recommended by the content-based filter are given a score of 15 and all the elements being recommended by the collaborative filtering method are given a score of 30. We add the scores of all the elements which were recommended through both the methods, and the items which obtained the maximum score by adding both the scores are considered to be the better fits for the user.

4.1 Proposed Approach

The entire model works on the principle of recommendation of one media on the basis of the input fed on some different media choices, for example, Movie to Book, Movie to Manga, Manga to Anime, etc. The working of the model can be described in the following steps:

1. Take user details, first media item details as the input.
2. Pre-contextual filtering is applied over the user's choice whether they want to access the recommended media for a solo experience or a family experience, and the data is accordingly filtered out.
3. The user and media item details are made to pass through the content-based filtering model, and an individual score for all the items recommended is calculated.
4. Same input data as in step 3 is made to pass through the collaborative filtering model, and another individual score or all the items recommended are calculated.
5. The items with the better scores are recommended first to the user and are made to pass through a content-based filter of the second media.
6. A union of data for both the media is done, and a similarity matrix is generated for both the media.
7. This matrix helps in finding the most similar recommendations in comparison with the given inputs.

The above steps are depicted in the flow chart shown in Fig. 1.

4.2 Evaluation Metrics

Following two evaluation metrics are used to evaluate the performance of the proposed recommender system:

1. **RMSE**: Root mean square error helps in calculating the standard deviation of the errors caused due to prediction [20]. It is a measure of how far these prediction errors are spread.

$$\text{RMSE} = \sqrt{\frac{\sum_{i \in R_u}(p(i) - r(i))^2}{|R_u|}} \quad (3)$$

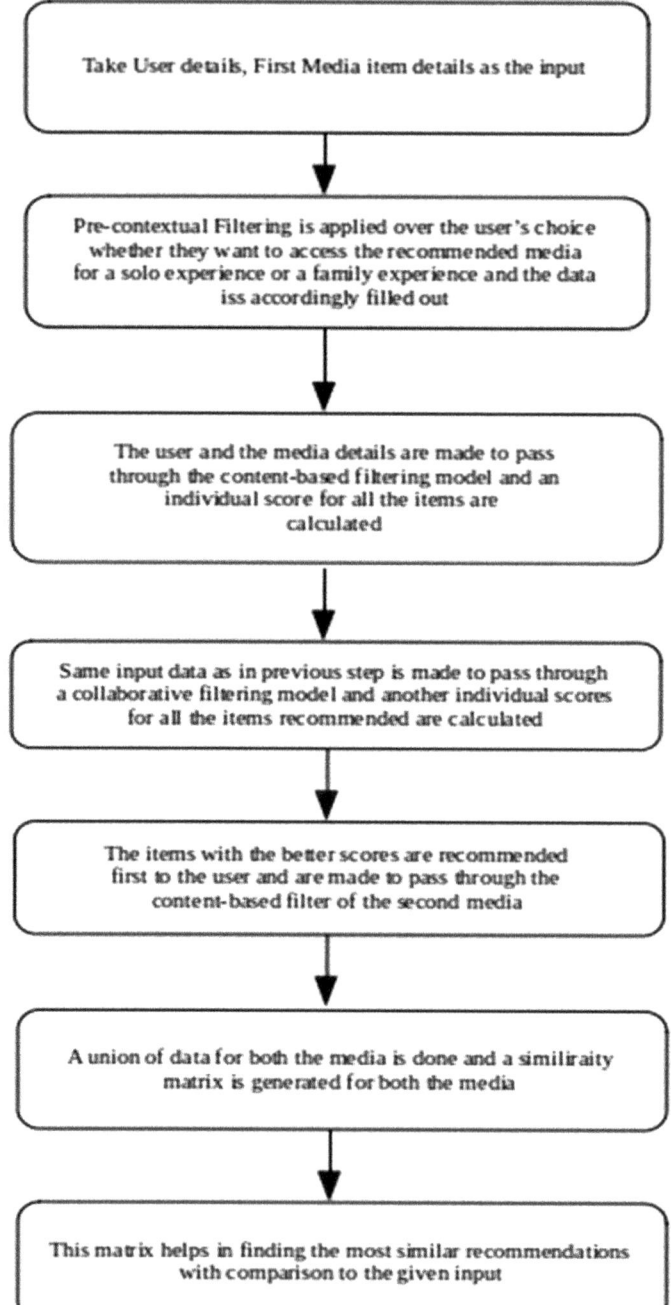

Fig. 1 Flow chart depicting proposed approach

Table 1 MAE and RMSE values obtained via SVD approach		MAE	RMSE
	SVD	0.7182	0.9089

Here, $p(i)$ depicts the predicted rating, $r(i)$ depicts the actual rating given by the user, and R_u represents the utility involved with these parameters.

2. **MAE**: Mean absolute error helps in calculating the difference between the predicted values and the observed values of the ratings [20].

$$\text{MAE} = \frac{\sum_{i \in R_u} p(i) - r(i)}{|R_u|} \qquad (4)$$

Again, $p(i)$ depicts the predicted rating, $r(i)$ depicts the actual rating given by the user, and R_u represents the utility involved with these parameters.

5 Experimental Results

The singular value decomposition algorithm that has been used in this model is the driver algorithm for the collaborative filtering that is being used to enhance the recommendations. The mean absolute error value obtained is 0.7182, the root mean square error value obtained by the model is 0.9089, and through the generation of top-n recommendations, the novelty obtained comes out to be 1.85 (Table 1).

Although the values of evaluation metrics are on a higher side, but our research deals with multiple media instead of a single media for generating recommendations, hence trying to pose as an enhancement over the already present recommendation models. We know that the lesser the value of errors, the better the algorithm is for the given model. Since the value of RMSE and MAE both come out to be low in their quantities, we can say that SVD proves to be a good choice when it comes to making recommendations on the basis of the media choices.

6 Conclusion and Future Scope

Our proposed approach has a modern approach for helping the users connect and switch between multiple media at the same time, and we take advantage of this relationship for enriching their experiences while choosing the next best entertainment content. Our approach obtains satisfactory results and tends to provide recommendations with less error. Even though our models do not perform better than the models already explored, dealing with multiple media and finding relationships between them makes it one of a kind.

For future scope, demographic attributes such as the location of users, age and chronology can be taken into account. These are some attributes that can play a significant role in determining recommendations for people. One does not have a significant proof that it improves the performance of a model which deals with different media elements getting involved. Hence, such explorations can prove to be a key to many opportunities in the field of multimedia recommendation. Exploration in the field of how people interact with multiple media could be done further, and data sets on the basis of those interactions could be updated. If such data could be collected through which one gets to know about people's interaction with different media simultaneously, then some more efficient aspects of building a recommender engine could be explored.

References

1. Shriver D (2018) Toward the development of richer properties for recommender systems. In: Proceedings of the 40th international conference on software engineering: companion proceedings (ICSE '18). Association for Computing Machinery, New York, NY, USA, pp 173–174.
2. Ricci F, Rokach L, Shapira B (2011) Recommender systems handbook. Springer, pp 1–35
3. Mohanty S, Chatterjee J, Jain S, Elngar A, Gupta P (2020) Recommender system with machine learning and artificial intelligence. Wiley-Scrivener
4. Sharma M, Ahuja L, Kumar V (2020) Study and classification of recommender Systems and their techniques: a Survey. In: Gupta D, Khanna A, Bhattacharyya S, Hassanien AE, Anand S, Jaiswal A (eds) International conference on innovative computing and communication. Advances in intelligent systems and computing, vol 1. Springer, Singapore
5. Paul D, Kundu D (2020) A survey of music recommendation systems with a proposed music recommendation system. In: Emerging technology in modelling and graphics. Springer, Singapore, pp 279–285
6. Darekar R, Dayma K, Parabh R, Kurhade S (2018) A hybrid model for book recommendation. In: Proceedings of the International Conference on Inventive Communication and Computational Technologies (ICICCT 2018), no. Icicct, pp 120–124
7. Sharma M, Ahuja L, Kumar V (2019) A hybrid context aware recommender system with combined pre and post-filter approach. Int J Inf Technol Project Manage 10(4):1–14
8. Soyusiawaty D, Zakaria Y (2018) Book data content similarity detector with cosine similarity. In: 12th international conference on telecommunication systems, services, and applications (TSSA)
9. Juntui S, Khoenkaw P (2018) Automatic non-personalized book recommender algorithm for bookstore shelf management. In: 2018 international conference on digital arts, media and technology (ICDAMT), pp 49–53
10. Hariadi I, Nurjanah D (2017)Hybrid attribute and personality based recommender system for book recommendation. In: 2017 international conference on data and software engineering (ICoDSE), pp 1–5
11. Waga K, Tabarcea A, Fränti P (2011) Context aware recommendation of location-based data. In: 15th international conference on system theory, control and computing, pp 1–6
12. Renjith S, Sreekumar A, Jathavedan M (2020) An extensive study on the evolution of context-aware personalized travel recommender systems. Inf Process Manage 57(1):102078
13. . Alhamid MF, Rawashdeh M, Saddik AE (2013)Towards context-aware recommendations of multimedia in an ambient intelligence environment. In: 2013 IEEE International Symposium on Multimedia, pp 409–414
14. Lops P, Jannach D, Musto C, Bogers T, Koolen M (2019) Trends in content-based recommendation—preface to the special issue on recommender systems based on rich item descriptions. User Model User-Adapt Interact 29(2):239–249

15. Zarzour H, Al-Sharif Z, Al-Ayyoub M, Jararweh Y (2018) A new collaborative filtering recommendation algorithm based on dimensionality reduction and clustering techniques. In: 2018 9th international conference on information and communication systems (ICICS), pp 102–106
16. Lahabar S, Narayanan P (2009) Singular value decomposition on GPU using CUDA. In: Parallel distributed processing, IPDPS 2009. IEEE international symposium, pp 1–10, May 2009
17. Ji L, Lin G, Tan H (2018) Neural collaborative filtering: hybrid recommendation algorithm with content information and implicit feedback. In: Yin H, Camacho D, Novais P, Tallón-Ballesteros A (eds) Intelligent data engineering and automated learning—IDEAL 2018. IDEAL 2018. Lecture Notes in Computer Science, vol 11314. Springer, Cham
18. Vall A, Dorfer M, Zadeh HE, Schedl M, Burjojee K, Widmer G (2019) Feature-Combination hybrid recommender systems for automated music playlist continuation. User Model User-Adap Inter 29:527–572
19. Hawashin B, Lafi M, Kanan T, Mansour A (2020) An efficient hybrid similarity measure based on user interests for recommender systems. Expert Syst 37(5):e12471
20. Herlocker JL, Konstan JA, Terveen LG, Riedl JT (2004) Evaluating collaborative filtering recommender systems. ACM Trans Inf Syst 22(1):5–53

Hyperspectral Imaging in Document Forgery

Sahima Srivastava, Vrinda Rastogi, Garima Jaiswal, and Arun Sharma

Abstract Document forgery is a crucial problem for this day and age due to the rampant utilisation of paper-based documents. The best type of technology must be utilised to the same. Hyperspectral imaging is a non-invasive and powerful tool used to capture images with several and contiguous bands. This imaging can create unique spectral signatures for each substance. The integration of the application of hyperspectral imaging with document forgery showcases tremendous results. Therefore, this work aims to present a literature review about the modern techniques employed to detect multiple types of forgeries in documents using hyperspectral imaging. This survey organised the various methodologies into various categories based on its central aim namely: writer detection, ink detection and miscellaneous. These methods were tabulated to display their similarities and contrasts. This paper aims to summarise recent advancements in the hyperspectral imaging field in regard to the document forgery application.

Keywords Hyperspectral imaging · Document forgery · Ink detection · Writer detection

1 Introduction

Document forgery is widespread and a significant problem today. Forgery is the process of fabricating, transforming or imitating writings, objects, or documents [1]. More often than not, forged documents are used to substantiate false claims [2]. For example, frauds committed using forgery have resulted in money being funeled out of the national treasury [3]. Signature-based forgery can be categorised into four broad brackets[4];

First two authors contributed equally to this work.

S. Srivastava · V. Rastogi (✉) · G. Jaiswal · A. Sharma
Indira Gandhi Delhi Technical University for Women, Delhi, Delhi, India

© The Author(s), under exclusive license to Springer Nature Singapore Pte Ltd. 2022   105
D. Gupta et al. (eds.), *Proceedings of Data Analytics and Management*,
Lecture Notes on Data Engineering and Communications Technologies 90,
https://doi.org/10.1007/978-981-16-6289-8_11

1. Simple Forgery: signatures belonging to a person who does not exist [5]
2. Forgery by Tracing: a genuine signature is traced to falsify [6]
3. Simulated or Copied Forgery: involves learning how to mimic the signature by trial and error and then use it to forge documents [7]
4. Forgery by trickery or by using built-up documents: a veritable signature is repositioned on a fake document via photocopying, scanning [8].

There are two primary ways to go about identifying forged documents; destructive and non-destructive. Destructive methods seek dissimilarity in the chemical composition of various inks. This kind of ink analysis is usually carried out in laboratories [9] and isn't always ethical [10]. Destructive techniques alter the current state of the document, either by removing a small portion of the document for analytical purposes or by disturbing the ink on the document [11]. The most common method of destructive ink testing is thin layer chromatography (TLC)[9].

A non-destructive analysis is preferred over destructive techniques as it doesn't change the document fundamentally [12]. The document is left intact, can be re-evaluated in the future, fast, and is usually low-cost [13]. The most prevalent technique for non-destructive analysis is spectrophotometry [14]. Broadly speaking, spectrophotometry lies under spectroscopy, which involves studying the interaction between matter and radiation[1] [15].

Hyperspectral Imaging(HSI) is a type of spectral imaging that analyses a vast part of the electromagnetic spectrum rather than just the visible region [16, 17]. This imaging attains both spectral and spatial information of a material as a function of the wavelength of light reflected [18, 19]. It consists of hundreds of thousands of bands of narrow and contiguous nature [20, 21]. HSI can be used to distinguish between different inks used in a document and thus contribute towards forgery detection [22]. This type of imaging has considerably changed over the years [23]. It has become an efficient tool for tackling ink ageing, forensic document analysis [24], and the ability to ameliorate readability of old documents [25]. Hence, hyperspectral imaging has dramatically improved document forgery techniques [26].

Regardless of the vast integration of technology globally, paper-based documents are integral to our daily lives [27]. It is a problem for a plethora of organisations ranging from banks, governmental sectors, hospitals, universities [28], legal documents [29], and so forth [30]. For instance, one could forge documents in order to procure a bank loan or receive some insurance money [31, 32]. Therefore, it is a diverse and vital problem [33].

This work aims to present the numerous ways in which hyperspectral imaging has been utilised to obtain useful characteristics of individual inks in a document. It outlines the various technologies used to analyse the disparity in inks leading to the detection of mismatch in ink, and consequently, detection of forgery.

This paper is organised is as follows: Sect. 2 explains the framework adopted in this survey. Section 3 elaborates on the methodologies employed for writer detection, whereas Sect. 4 explains processes for ink detection. Miscellaneous methods in

---

[1] https://socratic.org/questions/what-is-difference-between-spectrophotometry-and-spectroscopy.

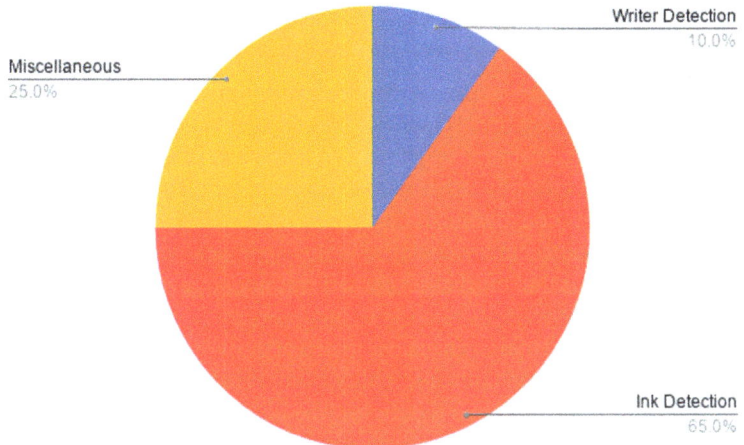

Fig. 1 Showcases the distribution of the types of methodologies

document forgery are showcased in Sect. 5. Section 6 draws a comparison between various works and Sect. 7 concludes the work.

2   Framework

This survey organises the work done in document forgery based on the specific goal. It focuses on research done in this domain from 2014–2020 using hyperspectral technology. Therefore, after analysing numerous papers, the techniques have been categorised as follows:

- Writer Detection: In this category, the methodology focuses on if the handwriting matches the correct writer.
- Ink Detection: In this type of process, research is done to make sure no new ink is applied to the document.
- Miscellaneous: Some works have adopted new goals to detect document forgery which can not be classified in one umbrella, hence this category was introduced.

Figure 1 displays the distribution of the classification done for processes detecting document forgery using hyperspectral imaging in the time range prescribed for this survey.

3   Writer Detection

In this section, various processes used writer detection to check for document forgery.

In [34], the UWA Writing Inks Hyperspectral Images (WIHSI) database was used to carry out their work. In order to garner hyperspectral data, spectral responses of the text pixels were taken by using neighbouring pixels. These response vectors were then fed to various convolutional neural networks (CNN) to detect the writer's identity. The highest test accuracy achieved in this approach was 71%. Yet, a severe limitation of this work is that the dataset chosen was not large enough to confirm the model's confidence.

Wang et al. [35] created their own dataset to detect the writer's identity. Their work focused on removing noise and dimensionality using Principal Component Analysis (PCA). Afterwards, pseudo colour synthesis was also applied. They analysed the recognition rate fluctuations based on various patterns of inks.

4   Ink Detection

In this section, various methodologies are explored to detect inks to check for document forgery. It is divided into two subcategories: supervised and unsupervised techniques. In supervised methods, the model is trained on specific inks whereas for unsupervised models can be used for various types of inks. Therefore, from an application point-of-view, it is more advisable to apply unsupervised techniques.

4.1   Unsupervised Techniques

Many research works have used the UWA Writing Inks Hyperspectral Images database which was created by [36]. Their major contribution is the creation of the first publicly available hyperspectral image database. This database consists of 33 visible band spectrums via 70 hyperspectral images. After creating the dataset, background segmentation was carried out to create a binary mask where each pixel is associated with the background. This was done using image thresholding. After-which, unsupervised clustering was implemented to establish membership between each class and all foreground pixels. This methodology found that using HSI, segmentation performed better in comparison with RGB. Another conclusion drawn was that most of the ink pixels were correctly identified in accordance with ground truth segmentation.

Khan et al. [37] implemented a varying technique to detect inks. They also utilised the WIHSI database but uses binarisation as their preprocessing technique. Another contrast is that their work used an unsupervised technique known as k-means clustering. A limitation of their work is that they fixed the number of clusters to two, as they only mixed two different types of inks. In real-world applications, this may not always be the case. In order to check their work, they visualised the clustering done. Their research also showcased how HSI imaging works better in comparison to RGB imaging.

Luo et al. [38] also implemented clustering, yet didn't specify the number of clusters. The cluster number estimation was done by using two different criteria namely: Schwartz's criterion and Point-to-Point distance criterion. After this estimation, inks were distinguished using three anomaly detection methods. These methods included Local Outlier Factor (LOF) anomaly detection, Connectivity-based Outlier Factor (COF) anomaly detection, and Influenced Outlierness (INFLO) anomaly detection. Their research showcased that the best results were found using the P2P criterion with INFLO anomaly detection algorithm.

Another unsupervised technique to detect ink mismatch was done by Devassy et al. [39]. In this method, t-Distributed Stochastic Neighbor Embedding (t-SNE) algorithm was applied to compare dimension reduction capability accompanied by Principal Component Analysis (PCA). They created a hyperspectral dataset using 60 different types of pens, varying from manufacturer and colour. After preprocessing and normalising the hyperspectral data, feature extraction was carried out using dimensional reduction. Afterwards, k-means clustering was applied. In order to check the accuracy of the model, both visualisation and quantitative methods were carried out. This technique yielded 0.94 clustering index for blue ink, 0.79 clustering index for black ink, and 0.97 clustering index for red ink.

In [40], a novel joint sparse band selection technique was proposed which detected crucial bands from hyperspectral images to detect inks. This work also used joint sparse PCA (JSPCA) which found the principal components but also separated the non-imperative bands in an unsupervised manner.

4.2 Supervised Techniques

Abbas et al. [41] utilised the WIHSI database for ink mismatch detection. This process utilised three main components: dimension reduction, endmember extraction, and inversion. In order to reduce complexity, dimension reduction was carried out whereas end member extraction was done to determine which chemical material was contributing to the spectra. Inversion estimated the endmember's abundance in the hyperspectral unmixing. This work performed especially well in regards to highly unbalanced ratios such as 1:8 or 8:1. Yet one of the limitations in this work was that was over-estimation in endmember estimation algorithms.

In [42], deep learning techniques have been implemented to detect ink. It was a novel approach as their work utilised both spatial and spectral data to analyse inks. WIHSI database has been used to both train and test their technique. For the purpose of training various types of CNN architectures, a target pixel was chosen. The spectral response was created using that target pixel and its neighbouring pixels in various types of strategies. These matrices were also later used to create mixed ink samples.

This method achieved an average of 99.6% accuracy for blue inks and 92% accuracy for black inks.

Devassy et al. [43] used a simpler version of CNN to detect ink mismatch. This method used a plethora of types of inks, specifically 29, in their self-created database. Their methodology implemented a one-dimensional convolutional neural network (1D CNN) which trains and tests faster, resulting in more applications such as using the model in mobile devices. This 1D CNN outperformed two standard classification methods used in hyperspectral imaging namely: Spectral Angle Mapper (SAM) and Spectral Information Divergence (SID). This technique achieved an accuracy of 98% using 1D CNN.

Morales et al. [44] proposed the use of hyperspectral analysis using Least Square Support Vector Machines (LS-SVM). They created their own dataset, removed the background, and then created hyperspectral curves. From the hyperspectral curve, smoothening techniques were applied to create an average filter of 21 pixels. Each ink curve was then represented using area and slope based features. Then the SVM classifier was applied on these feature pairs. Several types of experiments were performed such as without time lapse and with a one week time-lapse. This process showcased an accuracy of 80% when a similar type of ink is used, such as gel or marker.

In [45], five commonly used hyperspectral imaging classifiers were used namely: Spectral Information Divergence (SID), Spectral Angle Mapper (SAM), Binary Encoding (BE), Euclidean Distance (ED), and Spectral Correlation Mapper (SCM). These techniques were implemented for the first time in the field of ink classification. They also created their own dataset to carry forth this procedure. Via this experiment, it was concluded that SAM out-performed the other classifiers.

5   Miscellaneous Methods

In this section, diverse methodologies revolving around document forgery using hyperspectral imaging are explored.

Silva et al. [46] explored three types of common forgeries: obliterating text, adding text and approaching the crossing lines problem. For forgery by obliteration and forgery by adding text, this paper applied preprocessing techniques like; Standard Normal Variate (SNV), Multiplicative Signal Correction (MSC), Savitsky-Golay 1st and 2nd derivatives. This was followed by the application of PCA and Multivariate Curve Resolution-Alternating Least Squares (MCR-ALS). For intersecting lines, SNV, MSC, Savitsky-Golay 2nd derivative and auto-scaling techniques were applied on the data. Partial Least Squares-Discriminant Analysis (PLS-DA) was performed with the preprocessed data to select the main significant variables. It achieved 43%, 82%, 85% accuracy respectively for the three types of forgeries.

In [47], crossed ink lines were examined. These crossed ink imagery was created by ballpoint pens and gel pens. This work proposed to study the same using hyperspectral data which is near-infrared range (HSI-NIR). Their dataset comprised of 21 types of inks onto which various preprocessing techniques like PCA was applied to find the crucial parameters and remove excessive dimensions. After which Multivariate Curve Resolution-Alternating Least Squares (MCR-ALS) was applied. The technique's effectiveness was measured using two parameters namely: Lack of Fit (LOF) and percentage of explained variance.

6 Comparison Between Various Works

In this section, Table 1 showcases the similarities and contrasts between some techniques that implement hyperspectral imaging for document forgery.

7 Conclusion and Future Work

This work summarises the newest cutting edge technology employed to detect document forgery using hyperspectral imaging. After analysing various techniques, this work organised them into three categories based on their essential goal. It was observed that most of the work done has been focused on ink detection. Tabulation of a few critical methodologies was done to showcase various aspects of the work such as the dataset used, type of technique, results and limitations. This showcased that most processes generally adopted the UWA Hyperspectral Documents database. After carrying out the survey of the aforementioned papers, it was found that mostly clustering or deep learning techniques were most prevalent.

While carrying out this survey, several challenges were faced which should be addressed in the future. One such obstacle was the marginalised work done regarding document forgery in the hyperspectral domain. Due to the large potential of hyperspectral imagery and the importance of document forgery, more work can be done in this particular niche. Secondly, many methods concluded that their dataset was not large enough to give confident results. Hence, a larger public dataset should be created for document forgery using hyperspectral technology. Another solution for this shortcoming could be the utilisation of Generative Adversarial Networks (GAN) to synthesis fake hyperspectral data. Thirdly, it was found that black inks did not perform as well in comparison with blue inks. Therefore, more work should be done for black inks. In the future, more research can be done to implement more complicated CNN's such as 3D CNN or Recurrent Neural Networks (RNN) to detect document forgery.

Regardless of how technically advance our world becomes, paper-based documents are vital in all types of fields. It is imperative that any hints of forgery are detected to ensure nothing is amiss. Hyperspectral imaging is a powerful tool for the same. Therefore, hyperspectral imaging should be applied extensively in document forgery.

Table 1 The summarisation of recent techniques in document forgery using hyperspectral imaging

Paper	Detection	Dataset	Method	Results	Limitation
[34]	Writer	UWA hyperspectral documents	Deep Learning	71% was the highest test accuracy achieved	Requires a larger database
[35]	Writer	Self created	Statistical Mathematics	Depends on the pattern and fluctuates from 55 to 100%	The accuracy for black neutral pens need to improve
[36]	Ink	UWA Hyperspectral documents	Clustering	Depends on the various ratios of the hybrid images	Hardware limitation for signal to noise ratio
[37]	Ink	UWA hyperspectral documents	Clustering	Visualisation	Specifies only two clusters which may not always be the case in real-world applications
[38]	Ink	UWA Hyperspectral Documents	Clustering	Depends on the various ratios of the hybrid images	Assumption that maximum only two inks are present in a localised area
[39]	Ink	Self created	Clustering	Visualisation and quantitative methods	Needs to be compared to linear and non-linear techniques
[40]	Ink	Self created	Statistical mathematics	Depends on the various ratios of the hybrid images	Spatial neighbourhood should be taken into consideration and hyperspectral mismatching should be explored
[41]	Ink	UWA hyperspectral documents	Deep Learning	Average of 99.6% accuracy for blue inks and 92% accuracy for black inks	As it is supervised, it is not very robust or flexible
[42]	Ink	UWA hyperspectral documents	End-member extraction and inversion	Depends on the various ratios of the hybrid images	Overestimation in end member estimation algorithm

(continued)

Table 1 (continued)

Paper	Detection	Dataset	Method	Results	Limitation
[43]	Ink	Self created	Deep learning	98% accuracy has been achieved	More types of pens to be tested and a more complex neural network should be implemented
[44]	Ink	Self created	Machine learning	80%	Focuses on ink types, instead of ink identities
[45]	Ink	Self Created	Hyperspectral Imaging Classifiers	SAM Classifier performed the best	Improve classification for inks with near identical spectral signature
[46]	Obliterating text, adding text, and approaching the crossing lines problem	Self created	Statistical Mathematics	43%, 82%, 85% accuracy respectively for the three goals	Increase accuracy for obliterating text detection
[48]	Ink	UWA hyperspectral documents	Deep learning	98.2% accuracy, 88% accuracy for blue and black inks respectively	As it is supervised, it is not very robust or flexible
[49]	Ink	Self created	Statistical mathematics	Found crucial bands using hierarchical cluster analysis, PCA, and SAM	Implement UV or IR spectral range
[50]	Ink	UWA Hyperspectral documents	Graph orthogonal Non-negative matrix factorization	Depends on the various ratios of the hybrid images	Does not automatically estimate the number of inks nor finds the optimal hyper-parameters

References

1. Farid H (2009) Image forgery detection. IEEE Signal Process Mag 26(2):16–25
2. Cruz F, Sidere N, Coustaty M, D'Andecy VP, Ogier J-M (2017) Local binary patterns for document forgery detection. In: 2017 14th IAPR International Conference on Document Analysis and Recognition (ICDAR), vol 1. IEEE, pp 1223–1228
3. Bruna A, Farinella GM, Guarnera GC, Battiato S (2013) Forgery detection and value identification of euro banknotes. Sensors 13(2):2515–2529

4. Forged documents. http://nicfs.gov.in/wp-content/uploads/2017/01/Forged-Documents.pdf
5. Guo JK, Doermann D, Rosenfeld A (2001) Forgery detection by local correspondence. Int J Pattern Recogn. Artif Intell 15(04):579–641
6. Hilton O (1939) Detection of forgery. Am Inst Crim Law Criminol 30:568
7. Jessica Fridrich A, David Soukal B, Jan Lukáš A (2003) Detection of copy-move forgery in digital images. In: Proceedings of digital forensic research workshop. Citeseer
8. Ryu S-J, Lee H-Y, Cho I-W, Lee H-K (2008) Document forgery detection with svm classifier and image quality measures. In: Pacific-Rim conference on multimedia. Springer, pp 486–495
9. World of Forensic Science. Encyclopedia.com. 16 oct 2020, Dec 2020
10. Chandan Deep Kaur and Navdeep Kanwal (2019) An analysis of image forgery detection techniques. Statistics Optim Inform Comput 7(2):486–500
11. Craddock P (2009) Scientific investigation of copies, fakes and forgeries. Routledge
12. Springer E, Bergman P (1994) Applications of non-destructive testing (ndt) in vehicle forgery examinations. J Forens Sci 39(3):751–757
13. Polikreti K (2007) Detection of ancient marble forgery: techniques and limitations. Archaeometry 49(4):603–619
14. Causin V, Casamassima R, Marruncheddu G, Lenzoni G, Peluso G, Ripani Luigi (2012) The discrimination potential of diffuse-reflectance ultraviolet-visible-near infrared spectrophotometry for the forensic analysis of paper. Forens Sci Int 216(1–3):163–167
15. Braz A, López-López M, García-Ruiz C (2013) Raman spectroscopy for forensic analysis of inks in questioned documents. Forens Sci Int 232(1–3):206–212
16. Chang C-I (2003) Hyperspectral imaging: techniques for spectral detection and classification, vol 1. Springer, Berlin
17. El Masry G, Sun D-W (2010) Principles of hyperspectral imaging technology. In: Hyperspectral imaging for food quality analysis and control. Elsevier, pp 3–43
18. Manolakis D, Shaw G (2002) Detection algorithms for hyperspectral imaging applications. IEEE Signal Process Mag 19(1):29–43
19. Geladi Paul, Burger Jim, Lestander Torbjörn (2004) Hyperspectral imaging: calibration problems and solutions. Chemometrics Intell Lab Syst 72(2):209–217
20. Schultz RA, Nielsen T, Zavaleta JR, Ruch R, Wyatt R, Garner HR (2001)Hyperspectral imaging: a novel approach for microscopic analysis. Cytometry 43(4):239–247 (2001)
21. Barnes M, Pan Z, Zhang S (2015) Systems and methods for hyperspectral imaging, 25 Aug 2015. US Patent 9,117,133
22. Muhammad Jaleed Khan, Hamid Saeed Khan, Adeel Yousaf, Khurram Khurshid, and Asad Abbas. Modern trends in hyperspectral image analysis: a review. *IEEE Access*, 6:14118–14129, 2018
23. Sun D-W (2010) Hyperspectral imaging for food quality analysis and control. Elsevier, Amsterdam
24. Edelman GJ, Gaston E, Van Leeuwen TG, Cullen PJ, Aalders MCG (2012) Hyperspectral imaging for non-contact analysis of forensic traces. Forens Sci Int 223(1–3):28–39
25. Kim SJ, Deng F, Brown MS (2011) Visual enhancement of old documents with hyperspectral imaging. Pattern Recogn 44(7):1461–1469
26. Yaseen M, Ahmed RA, Mahrukh R (2020) Forgery detection in a questioned hyperspectral document image using k-means clustering. *arXiv preprint* arXiv:2006.16057
27. Elkasrawi S, Shafait F (2014) Printer identification using supervised learning for document forgery detection. In: 2014 11th IAPR International workshop on document analysis systems. IEEE, pp 146–150
28. Dlamini N, Mthethwa S, Barbour G (2018) Mitigating the challenge of hardcopy document forgery. In: 2018 International Conference on Advances in Big Data, Computing and Data Communication Systems (icABCD). IEEE, pp 1–6
29. Roy P, Bag S (2019) Detection of handwritten document forgery by analyzing writers' handwritings. In: International conference on pattern recognition and machine intelligence. Springer, pp 596–605

30. Cheddad A, Condell J, Curran K, Mc Kevitt P (2009) A secure and improved self-embedding algorithm to combat digital document forgery. Signal Process 89(12):2324–2332
31. Ahmed AGH, Shafait F (2014) Forgery detection based on intrinsic document contents. In: 2014 11th IAPR International workshop on document analysis systems. IEEE, pp 252–256
32. Bertrand R, Terrades OR, Gomez-Krämer P, Franco P, Ogier J-M (2015) A conditional random field model for font forgery detection. In: 2015 13th International Conference on Document Analysis and Recognition (ICDAR). IEEE, pp 576–580
33. Bertrand R, Terrades OR, Gomez-Krämer P, Franco P, Ogier J-M (2013) A system based on intrinsic features for fraudulent document detection. In: 2013 12th International conference on document analysis and recognition. IEEE, pp 106–110
34. Ul Islam A, Khan MJ, Khurshid K, Shafait F (2019) Hyperspectral image analysis for writer identification using deep learning. In: 2019 Digital Image Computing: Techniques and Applications (DICTA). IEEE, pp 1–7
35. Wang W, Zhang L, Wei D, Zhao Y, Wang J (2017) The principle and application of hyperspectral imaging technology in detection of handwriting. In: 2017 9th International Conference on Advanced Infocomm Technology (ICAIT). IEEE, pp. 345–349
36. Khan Z, Shafait F, Mian A (2013) Hyperspectral imaging for ink mismatch detection. In: 2013 12th International conference on document analysis and recognition. IEEE, pp. 877–881
37. Khan Z, Shafait F, Mian AS (2013) Towards automated hyperspectral document image analysis. In: AFHA, pp. 41–45
38. Luo Z, Shafait F, Mian AS (2015) Localized forgery detection in hyperspectral document images. In: 2015 13th International Conference on Document Analysis and Recognition (ICDAR). IEEE, pp 496–500
39. Devassy BM, George S (2020) Dimensionality reduction and visualisation of hyperspectral ink data using t-sne. Forens Sci Int, 110194
40. Khan Z, Shafait F, Mian A (2015) Automatic ink mismatch detection for forensic document analysis. Pattern Recogn 48(11):3615–3626
41. Abbas A, Khurshid K, Shafait F (2017) Towards automated ink mismatch detection in hyperspectral document images. In: 2017 14th IAPR International Conference on Document Analysis and Recognition (ICDAR), vol 1. IEEE, pp 1229–1236
42. Khan MJ, Khurshid K, Shafait F (2019) A spatio-spectral hybrid convolutional architecture for hyperspectral document authentication. In: 2019 International Conference on Document Analysis and Recognition (ICDAR). IEEE, pp 1097–1102
43. Devassy BM, George S (2019) Ink classification using convolutional neural network. NISK J 12 (2019)
44. Morales A, Ferrer MA, Diaz-Cabrera M, Carmona C, Thomas GL (2014) The use of hyperspectral analysis for ink identification in handwritten documents. In: 2014 International Carnahan Conference on Security Technology (ICCST). IEEE, pp 1–5
45. Devassy BM, George S, Hardeberg JY (2019) Comparison of ink classification capabilities of classic hyperspectral similarity features. In: 2019 International Conference on Document Analysis and Recognition Workshops (ICDARW), vol 8. IEEE, pp 25–30
46. Silva CS, Pimentel MF, Honorato RS, Pasquini C, Prats-Montalbán JM, Ferrer A (2014) Near infrared hyperspectral imaging for forensic analysis of document forgery. Analyst 139(20):5176–5184
47. Rodrigues e Brito L, Braz A, Honorato RS, Pimentel MF, Pasquini C (2019) Evaluating the potential of near infrared hyperspectral imaging associated with multivariate data analysis for examining crossing ink lines. Forens Sci Int 298:169–176
48. Khan MJ, Yousaf A, Abbas A, Khurshid K (2018) Deep learning for automated forgery detection in hyperspectral document images. J Electron Imag 27(5):053001
49. Chlebda DK, Majda A, Łojewski T, Łojewska J (2016) Hyperspectral imaging coupled with chemometric analysis for non-invasive differentiation of black pens. Appl Phys A 122(11):957
50. Rahiche A, Cheriet M (2020) Forgery detection in hyperspectral document images using graph orthogonal nonnegative matrix factorization. In: Proceedings of the IEEE/CVF conference on computer vision and pattern recognition workshops, pp 662–663

A Review: Trust Management Techniques Used for Cloud Computing

PoojaGoyal and Sukhvinder Singh Deora

Abstract Cloud computing can be defined as an on-demand self-service that measures according to usage metering and can be accessed via different modes. But with the expeditious advancement of distributed computing resources, it is profoundly indispensable to assure the quality of services provided by the service providers. For this objective, assurance is required between various entities in the cloud environment. Various trust evaluation models are developed by observing the historical interaction between entities either directly or indirectly. But so far, no trust evaluation system is developed that efficiently measures the optimal service provider for consumers. This paper explains the cloud computing landscape, presents issues in cloud computing and trust management system. This paper likewise incorporates a review of many existing trust management techniques and finds the various challenges that exist in the trust management system. This paper also defines various QoS parameters that will be evaluated during the evaluation of the trust degree of any service provider. The objective of this paper is to perform an extensive literature survey of various trust management methodologies used in cloud computing.

Keywords Trust · Cloud computing · Security · SLA (Service Level Agreement) · QoS (Quality of Service) · Reputation System

PoojaGoyal (✉) · S. S. Deora
DCSA, MDU, Rohtak, Haryana, India

© The Author(s), under exclusive license to Springer Nature Singapore Pte Ltd. 2022   117
D. Gupta et al. (eds.), *Proceedings of Data Analytics and Management*,
Lecture Notes on Data Engineering and Communications Technologies 90,
https://doi.org/10.1007/978-981-16-6289-8_12

1   Introduction

In today's scenario, Cloud Computing is the brain of internet computing. Distributed computing has brought colossal changes in the services provided by the internet. Cloud computing provides us a lot of benefits like resource pooling, on-demand services, broad network access, service measurement, and rapid elastically [1]. Cloud computing is mainly defined as a data storage, cluster, or supercomputer that delivers computing resources through the internet, on-demand from a remote location instead of dwelling on one's work area, mobile devices, or laptop. Cloud computing can be implemented through different deployment models such as public, private, community, and hybrid. The working principle of cloud computing is to make processing and storage infrastructure independent of time and location for the cloud user and provides trustworthy and efficient services. Cloud computing provides us with many services like infrastructure, platform, software, storage, expert, environment, data, and security on-demand at a lower cost.

But every good thing comes with some pitfalls. The imperfection of cloud computing stands with its environment whose nature is dynamic, complex, non-transparent, and open access. Consumers feel suspicious about what arises to their data once it goes into the cloud. They think about who can access their data and how it will be stored, transcript, shared, and used. Also, they perceive like they lose command of their data [2]. Due to this, consumers feel insecure at the time of using cloud services. Moreover, the selection of any cloud service depends on the QoS and function performed by service providers. But, the major issue is that we cannot measure QoS accurately as the nature of the cloud is dynamic, and also the requirement of consumers changes as per their needs. Also, the biggest source of measuring the performance of cloud service is feedback provided by its historical users. But, this feedback is affected by a malicious entity. To handle this entire obstacle, the concept of trust is used between a consumer and a provider. For building this faith, a Trust management system is used.

Trust is a mutual understanding between entities based on the context and base of any business relationship. When the base of security is trust management, then it creates a dynamic trustworthy relationship between the unknown object. Trust management is the most crucial task in a cloud environment. There are many trust evaluation models that exist but still, there is no universally accepted model. The largest source of information for the trust evaluation model is feedback provided by historical users [3]. But, many times users also do not provide genuine feedback and behave like malicious entities. Moreover, when the untrustworthy cloud service provider will be selected then it creates issues [4] like.

- The high job failure rate
- Less satisfaction of cloud user
- Loss of cost.

Due to these issues, it is mandatory to develop a framework that implements trust in successful ways in a cloud environment. In cloud computing, the main interest of researchers from all over the world is virtualization, architecture, and platform construction, but only a few of them have been focused on the study of trust management [5].

The main intent of this article is to

(1) The basic explanation of cloud computing. What are the cloud computing landscapes with pitfalls of cloud computing?
(2) How trust is necessary for distributed computing.
(3) How the trust concept helps requestors to overcome the hindrance of cloud computing.
(4) What are the basic characteristics of TMS, their requirements, and the technique of TMS?

The rest of this paper is categorized like so. The related concept of the Cloud computing landscape is introduced in Sect. 2. In Sect. 3, Trust management System is described. Section 4 represents the extensive literature survey of various trust management techniques. Finally, the paper ends with conclusions.

2 Cloud Computing Landscape

Cloud computing is a generic term that is applied to anything that involves obtaining services hosted on the internet [6]. The fundamental concept of cloud computing is paying as the per-use model used in the past in traditional utilities like water bills, electricity bills that are charged as per their consumption. This is the reason why organizations show their attention to this technology.

According to the American National Institute of Standards and Technology [7], "cloud computing is a model for qualifying agreeable on-request network access to a shared pool of customizable computing resources that can be quickly provisioned and delivered with nominal management effort on service provider association". The working principles of cloud computing are virtualization and SOA (service-oriented architecture) that supports the multi-tenancy concept where some services (infrastructure, software) are shared by a large set of consumers on various host platforms with heterogeneous implementation.

2.1 Service Delivery Model

We can also define cloud computing as the sum of three basic service models.

IaaS (Infrastructure as a Service): It deals with hardware resources like storage, computer, and network. For providing these physical resources, the virtualization

technique is used so that the cloud users can customize their infrastructure dynamically as per their requirements. **Example: Amazon, Google Compute Engine, EC2,
Microsoft Azure VM**.

PaaS (Platform as a Service): It delivers the platform as a service to its user. It
includes infrastructure resources as well as includes an operating system, database,
and program executable environment. It helps application developers to develop,
test, and run their applications. **Example: Google Application Engine, Amazon,
Elastic Beanstalk, Windows Azure compute**.

SaaS (Software as a Service): It provides running software as a service. Here
consumers use the provided application running on cloud infrastructure. Also,
consumers need only the web browser for accessing the application. Consumers are
free to manage infrastructure like networks, operating systems, storage. **Example:
Microsoft Office 365, Salesforce, Google Apps, Onlive, and AppExchange**
(Fig. 1).

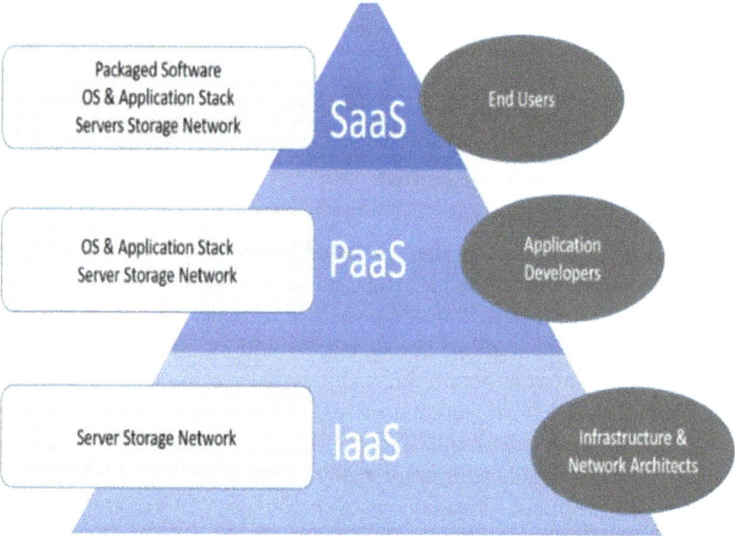

Fig. 1 Cloud service model

2.2 Cloud Building Deployment Model

Deployment model	Geographic covered area	Operated	Secure	Accessible	Cost	Example
Public cloud	Large	Third party	Less	Anyone	Free or at nominal charge	Goggle, IBM's (blue cloud computing platform)
Private cloud	Small like a one organization	Organization or user	More	User or employee of organization	More	Banking software or sales force
Community	Larger than private cloud	Third party	More, but less than private	More, but less private	More, but less than private	Banking software or sales force
Hybrid cloud	Combination of more than one cloud	Third party	Depends on its structure	Depends on its structure	Depends on its structure	Amazon web service

2.3 Cloud Entities

In the cloud environment, three major entities play a significant role. The service provider submits their service list and the consumer submits the task list as per their requirement to the trusted third party. Trusted third party evaluates the reputation of both a provider and a consumer and as per their choice, provides the best resource provider to consumers.

- **Cloud Service Provider**: Provides resources and services as a capability to consumers. They host and manage offered cloud services and infrastructures.
- **Cloud Service Consumers**: They make customized demands of resources and services for achieving their desired goal. Cloud Consumers are broadly divided into two parts:

1. **End Users** like government, business organizations, or educational institutes.
2. **Cloud-Based Service Providers**: They offer new services to consumers and these services completely host on the cloud. They develop their business model.

- **Cloud Brokers (Cloud Service Operator)** [8]: They set up an operational mechanism and define rules that include both technical process and business strategy. They managed the demand and supply of resources. They also provide add-on services over the services provided by service providers like Data as a service (DaaS), Network as a service (NaaS).

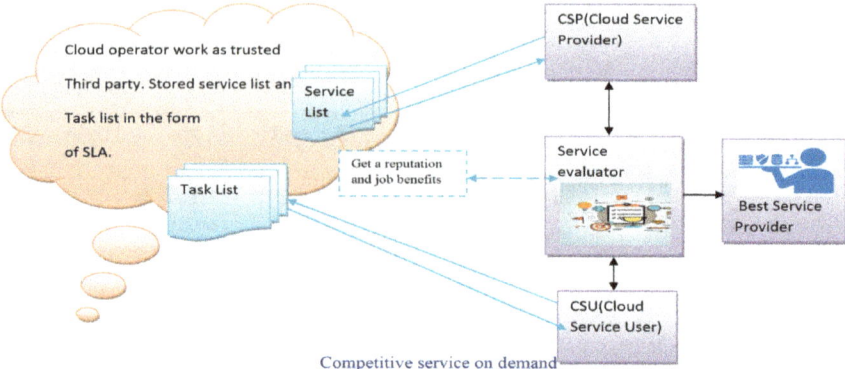

Fig. 2 Cloud entities

Among these three basic entities, the other entities are

- **Cloud Re-seller**: They provide services on the behalf of the Service provider.
- **Cloud Auditors**: They offered a certificate based on performance assessment, security %, and information system operation.
- **Cloud Carriers**: They provide connectivity (telecommunication, network) to other cloud entities to ensure better service provisioning (Fig. 2).

2.4 Issues with Cloud Computing

(1) **Privacy, Security**: With the technological advancement and popularity of cloud computing and service utility, the needs of privacy and trust management have also emerged as a new area of research. The primary issue related to cloud computing is security. It cannot handle security issues in a fast, flexible, and dynamic way [9], Because of this, users have less confidence in cloud computing. The insufficiency of consumer trust is a key obstruct to the acceptance and success of cloud services.

(2) **Control of Data**: Consumers feel apprehensive about what befalls their data once it goes into the cloud. They consider who can get to their information and how it will be put away, duplicated, shared, and utilized. Additionally, they have a feeling that they fail to keep a grip on their data [2]. Because of this, buyers feel uncertain at the hour of utilizing cloud administrations

(3) **Performance, Reliability, SLA**: In order to implement cloud computing successfully, cloud users must ensure that service providers have completed their predefined tasks according to the agreement (SLA) in a given duration. For selecting a cloud service, consumers are normally worried about predefined QoS (quality of service) parameters and capacities. But, in practice, it is

challenging to quantify the QoS values perfectly due to the dynamic nature of cloud and user specifications [10].

(4) **Identity Management**: The largest source of information for the trust evaluation model is the feedback provided by the historical users [11]. But, many times, users also do not provide genuine feedback and behave like a lunatic or a malicious user. Moreover, when the untrustworthy cloud service provider is selected, it creates issues [4] like loss of cost and consumer privacy.

Due to these present issues, it is mandatory to develop an appropriate framework that helps in the proper implementation of trust successfully in a cloud environment.

3 Trust Management System

M. Blaze introduced the concept of trust in the year 1996 (Matt Blaze et al. 1999). In layman language, faith is defined as a mutual understanding between two objects where one object plays a trustee, and another is the trustor. In such a system both the objects work together and share their resources with an aim of satisfying expectations of each other. Trust is imprecise just like human behavior and can't be described and evaluated quantitatively. Trust only depends upon understanding. "Trust is a state of mind that majorly defines three aspects expectation, risk, and belief [12]. Also, the essential feature of trust is that it is hard to achieve and easy to lose. Further, it is apposite to mention that the belief is dynamic because it has the power to increase and decrease the value as per trustee recent experience that they received from various sources. There are multiple methods for the estimation of the trust value. Although the methodologies would be different, the purpose of all mechanisms is the same and is known as Trust Model. The objective of TMS is to increase the confidence level between a consumer and service providers (Fig. 3).

Fig. 3 Trust aspects

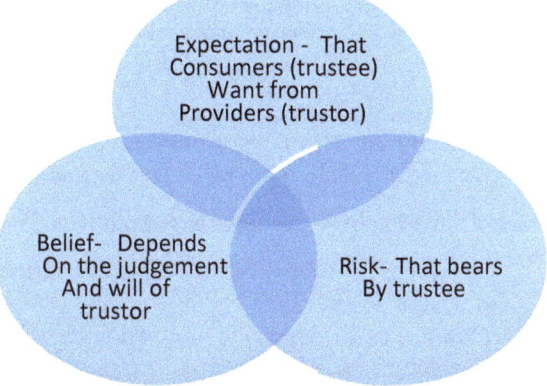

3.1 Nature of Trust [5, 13, 14]

(a) Based on perception
(b) Uncertain, fuzziness, and randomness
(c) Partial transitive
(d) Non commutative
(e) Non distributive
(f) Work in a risky environment and part of decision making
(g) One way relationship
(h) Non associative
(i) Non symmetric/Asymmetric
(j) Context depended
(k) Qualitative not Quantitative.

3.2 Principles of Trust for effective Trust Management System [15]

- Should be transitive.
- Should be measurable.
- Should be defined by time and situation.
- Should include formal and social tool for evaluation of trust.
- Should be able to handle uncertainty.

3.3 Trust Management Techniques

The method, protocols, mechanism or, technique used for evaluating the trust value is known as trust models. The fundamental for the trust model in the context of trust because dissimilar findings can have different security, privacy, and confidential stipulations [16]. To identify and analyze the current trust model based on their approach that is used to calculate trust degree, the trust model is systematically separated into four parts.

Agreement Based: This type of trust model is based on a contract or agreement. This trust model contains QoS parameter list and other security documents.

Steps of working in Agreement based model

(1) Cloud consumer submits their required Security and QoS parameter list to trust evaluation module.
(2) TMS makes an agreement negotiation request and forwards it to the service provider.

(3)   Service provider submits their capability list to contract parameter monitoring module that makes an SLA b/w service provider and service requestors and exchange their agreement.

(4)   After accessing the services monitoring module, evaluate the trustworthiness of SLA and update the trust score of service providers.

Broadly this type of model is divided into two parts

(a)   **SLA (Service Level Agreement Based)** This type of model is based on a mutual contract between a requestor and a resource allocator. This SLA (mutual contract) contains various QoS parameters that are specified by consumers.

(b)   **Service Policy-Based**: It contains service policies that are negotiated between a consumer and a service provider. So, broadly this service policy-based model is divided into two-layer

- If all the functioning of an association is executed under the power of the association then it is known as the internal layer and this layer uses identity and key management techniques to check the trustworthiness of the service provider.
- If all the operations of an association are not performed heavily influenced by the association at that point trust esteem is dependent on security and QoS boundaries and agreement level trust is considered.

Certificate-Based Model: For building trust, the concept of a certificate, trust ticket, endorsement key issued by a certified authority is used. Generally, this type of model is used when infrastructure, s/w, operating systems are issued as a service. Trust tickets are accommodated for improving the integrity and secrecy of data in the cloud environment so that consumers feel more confident in the cloud. Secret keys and certificates are used for transferring control over data from the cloud to consumers here data is transferred into encrypted form.

Feedback Based: This type of trust model collects responses and opinions from other consumers to measure the degree of trust of service providers. The Best example is in any digital shopping application where, after receiving or using the product, a customer gives a review of that product which is referred by another unknown customer at the time of purchasing. The provided feedback is either positive or negative depending on the perception of providers.

The perception of the feedback provider may be situation-specific or person-specific. A reputation has a direct or indirect impact on the trustworthiness of a particular entity.

Steps of Working of Feedback Based Model:

(a)   All service providers register with a trust model through the service registry module.

(b)   Feedback collector module collects feedback from other consumers who have access to the service of service providers in their historical records.

(c)   Trust Management System (TMS) evaluates the degree of trust of each service provider.
(d)   Consumer requests to TMS to send the score of the required service provider and the same is returned to the requesting consumer.

Domain Based Trust Model: Generally, this type of trust model is used with grid computing but several of the selective trust models are proposed below this category for cloud computing environments also. Here the cloud environment is divided into various autonomous domains and trust relationships are differentiated into inter and intra domain. Firstly trust score is checked by direct score if the direct score is not available, and then the recommendation score provided by other sources is considered. If trust value is evaluated from the direct and recommended trust from the same domain then it is known as intra domain and inter domain is considered comprehensive trust from other domains also.

3.4   Comparison of Trust Management Techniques [17]

Model	Deployment lease of applicability	Security (data confidentially)	Data availability	Performance (reliability) (detection of malicious)
Agreement based	Simple steps	Encryption	Replication	Credibility by weight
Certificate based	Complex management and distribution of keys	Encryption	No support	Ticket verification, certificates
Feedback based	Simple steps	No support	No support	No support
Domain based	Less computational complexity	No support	No support	Time decay function

3.5   QoS(Quality Of Service) For TMS

For determining the best service provider according to consumer need, QoS play a significant role. QoS represents a set of non-functional requirements used for judging the performance and trustworthiness of the service provider. Moreover, maximum service selection mechanism based on an only matching method for finding the best service provider, but in reality to determine the QoS of service is a very non-trivial and tedious task as QoS of any service depends on the environment of network, perception, and context of a consumer with time [18].

Typical QoS parameters are used for judging the performance and trustworthiness of the service provider.

Metrices	Description
Service availability	The ratio of No. of tasks accepted by cloud service and number of tasks submitted by consumer
Service reliability	The ratio of No. of tasks completed and number of tasks accepted by cloud service
Service response time	The delay time between service request and service completed
Service throughput	In a particular time unit, how much task can be processed by request allocator?
System elasticity	The ability of service provider to accept a change in its load during peak time
System scalability (adaptability)	The ability of resource allocator to perform well when there is a change in its load
Service integrity	The ratio of tasks completed in a secure manner and number of tasks completed
System accuracy	Amount of closeness to user awaited result with result generated by service provider
System interoperability	Degree of capability of service provider to interact with other services offered by same or different service provider

3.6 Challenges for TMS

(1) **Transferring Trust Between Contexts**: The consumer trust in the service providers in the cloud environment depends on the situation and scope of interaction, perception of a consumer. So, transferring feedback of historical consumers based on context and perception is a big challenge for TMS.

(2) **Trust Evaluation**: Multiple mechanisms exist for evaluating the trustworthiness of service providers. But selecting the best appropriate model is also a big challenge for TMS.

(3) **Attack Resistance**: When the confidence of the consumer grows on the TMS or Cloud services, then many malicious entities exist in the cloud environment that generates several attacks according to their requirements. Some different attacks that exist in the cloud environment are Sybil, whitewashing, playbook, proliferation.

(4) **Multi-faceted Trust Computation**: Multiple QoS parameters exist for evaluating the trustworthiness of service providers. However, some parameters are measured in qualitative terms and others in quantities. The summation of objective and subjective parameters is also a difficult task.

(5) **Customization and Aggregation**: Broadly, there are two mechanisms for storing and aggregating trust values that are centralized and distributed. Each

mechanism has its own positive and negative impact like the centralized approach is based on the trusted third party and the other hand distributed approach is based on trusted distributed rating so there is a challenge for the trust model which mechanism is used to evaluate the trust value. Also at which level customization is supported by the trust model is a big challenge.

4 Survey of Existing Trust Management Model

Title	References	Technique	+ve	Parameters
Trust management middleware for cloud service **preference** by prioritization	[19]	Objective trust evaluated using prioritized aggregation operation. Subjective trust assessment performed using covariance analysis	Weight is assigned to QoS in a dynamic manner based on priority class and satisfaction level	Availability, reliability, security, privacy, and customer support
A reliability based trust management mechanism for cloud service	[20]	Filter mechanism is proposed based on historical feedback with frequency of usage with time instance	Proposed mechanism is very simple and based on average method	Reliability that based on consistency, familiarity
Performance prediction model for cloud service selection from smart data	[21]	Proposed an automatic performance prediction model dependent on naive bayes classifier to anticipate the performance metric concerning various choices for configuration of node resources	Used a naive bayes classifier to train the dataset so that unlabelled information or node is predicted through this malicious entity is controlled	Memory usage, response time, CPU Utilization

(continued)

(continued)

Title	References	Technique	+ve	Parameters
SMI cloud: a framework for computing and ranking cloud service	[22]	Proposed a CSMIC (cloud service measurement index consortium) model for judging a best resource provider dependent on fundamental and superfluous QoS according to consumer need. The proposed model utilized AHP strategy for appointing QoS	Consumer specify their own essential and non-essential requirement according to their need with each QoS parameter have different dimension units(Boolean, Numeric, Unordered, Range)	Response time, sustainability, cost, elastically, suitability, accuracy, stability, availability, usability, transparency, interoperability, reliability, adaptability
Towards a trust evaluation middleware for cloud service selection	[18]	Proposed a model known as TRUSS that is a middleware approach for assessing the trustworthiness of resource provider based on both subjective as well as an objective approach	Reputation of consumer also evaluated for judging either consumer is trustworthy or not. QoS based on user preference	Objective trust dependent on QoS checking and SLA of administration and performance of service provider is monitored at time interval
Two way ranking based service mapping in cloud environment	[23]	Proposed a model known as TRCSM based on perspective of both service provider and consumer using AHP	For service provider ranking model considered SMI and QoS of consumers For consumer ranking, QoS of service providers and behavior attributes are considered	QoS of service providers are accountability, agility, confirmation, financial, performance, security and protection, ease of use QoS of consumers are turnover, duration, and transaction

(continued)

(continued)

Title	References	Technique	+ve	Parameters
A cloud service trust evaluation model based on combining weights and gray correlation analysis	[24]	Proposed a cloud service trust evaluation model (CSTEM). Objective trust is evaluated by using rough set theory and subjective trust based on AHP (analytical hierarchical process)	Here penalty mechanism was introduced to control the malicious entities	Cost, performance, reputation
Multilevel trust agreement in cloud environment	[25]	Proposed MLTA (multilevel trust agreement) framework that worked on hierarchical level and used the ABC (artificial bee colony) algorithm for evaluating trust	The proposed model performs functions-feedback assessment, risk observing, data accessibility, reward/punishment selection, and time factor investigation	Availability, time factor, risk monitoring, reward, and punishment
A service satisfaction-based trust evaluation model for cloud manufacturing	[5]	Proposed a trust evaluation method based on indirect and direct trust	Used punishment and reward function to control malicious entity and comprehensive trust value is updated dynamically	Direct trust: service response rate, service cost deviation, service reliability, delivery timelines, and service success rate Indirect trust: level of service, service cooperation rate, service energy efficiency rate, and recent activities
Trust management in cloud computing	[26]	They proposed RTMA model that includes 4 modules: (1) feedback collector (2) feedback segregator (3) compliance generator report (4) trust evaluation	Very simple method was introduced for evaluating trust value based on only feedback	Considered feedback as parameters. Take feedback from cloud user and service provider

(continued)

(continued)

Title	References	Technique	+ve	Parameters
A new method for trust and reputation evaluation in the cloud environment using the recommendations of opinion leader's entities and removing the effect of troll entities	[13]	Proposed a mechanism that evaluates trust in the cloud environment, identifies and removes the troll entities, and gives more preference to the opinion leader	If we increase the threshold of reputation than fewer percent of opinion leaders are selected it can identify more troll entities	availability, reliability, data integrity, identity capability, and behavior

5 Conclusion

In this paper, we have presented an outline of the cloud environment with its feature, deployment model, service model, basis entities of the cloud environment, and the concerns related to security and privacy issues of consumers. The paper covered the present issue of cloud computing and why trust plays a crucial role in distributed computing. We discussed trust management techniques with their comparison, trust management system, types of trust, nature, and factors that affect the trust value. This paper presented different trust assessment models with their QoS parameters for distributed computing. The overview introduced in this article shows the fact that energizing advancement has been made toward a careful survey about the execution practices of different Cloud services, there is as yet a wide range of open issues in every classification that need further examination; consequently offering fascinating points for future exploration.

References

1. Noor TH, Sheng QZ, Zeadally S, Yu J (2013) Trust management of services in cloud environments 46(1):1–30
2. Pearson S, Privacy, security and trust in cloud computing.
3. S. based T. M. for C. E. Machhi (2016) Feedback based trust management for cloud environment
4. Varalakshmi P, Judgi T, Balaji D (2018) Trust management model based on malicious filtered feedback in cloud. Commun Comput Inf Sci 804:178–187. https://doi.org/10.1007/978-981-10-8603-8_15
5. Yang X, Wang S, Yang B, Ma C, Kang L (2019) A service satisfaction-based trust evaluation model for cloud manufacturing. Int J Comput Integr Manuf 32(6):533–545. https://doi.org/10.1080/0951192X.2019.1575982
6. Manvi SS, Krishna Shyam G (2014) Resource management for Infrastructure as a Service (IaaS) in cloud computing: a survey. J Netw Comput Appl 41(1):424–440. https://doi.org/10.1016/j.jnca.2013.10.004

7. Kristiani E, Yang CT, Wang YT, Huang CY (2019) Implementation of an edge computing architecture using openstack and kubernetes. Lect Notes Electr Eng 514:675–685. https://doi.org/10.1007/978-981-13-1056-0_66

8. Alzubi JA et al (2020) Hashed Needham Schroeder industrial IoT based cost optimized deep secured data transmission in cloud Meas. J Int Meas Confed 150:107077.https://doi.org/10.1016/j.measurement.2019.107077

9. Thirukkumaran R, Muthu Kannan P (2019) Survey: security and trust management in internet of things. In: Proceedings of 2018 IEEE Global Conference on Wireless Computing and Networking, GCWCN 2018, pp 131–134. https://doi.org/10.1109/GCWCN.2018.8668640

10. Habib SM, Hauke S, Ries S, Mühlhäuser M (2012) Trust as a facilitator in cloud computing: a survey. J Cloud Comput 1(1):1–18. https://doi.org/10.1186/2192-113X-1-19

11. Kotha HD, Mnssvkr Gupta V (2018) IoT application, a survey. Int J Eng Technol 7:891–896. https://doi.org/10.14419/ijet.v7i2.7.11089

12. Huang J, Fox MS (2006) An ontology of trust—formal semantics and transitivity. In: ACM international conference on proceeding series, pp 259–270. https://doi.org/10.1145/1151454.1151499

13. Chiregi M, Navimipour NJ (2016) A new method for trust and reputation evaluation in the cloud environments using the recommendations of opinion leader's entities and removing the effect of troll entities. Comput Human Behav 60:280–292. https://doi.org/10.1016/j.chb.2016.02.029

14. Nagarajan R, Thirunavukarasu R, Shanmugam S (2018) A fuzzy-based intelligent cloud broker with Mapreduce framework to evaluate the trust level of cloud services using customer feedback. Int J Fuzzy Syst 20(1):339–347. https://doi.org/10.1007/s40815-017-0347-5

15. Damera VK, Nagesh A, Nagaratna M (2020) Trust evaluation models for cloud computing. Int J Sci Technol Res 9(2):1964–1971

16. Duan Q (2017) Cloud service performance evaluation: status, challenges, and opportunities—a survey from the system modeling perspective. Digit Commun Networks 3(2):101–111. https://doi.org/10.1016/j.dcan.2016.12.002

17. Jaswal S, Malhotra M (2019) A detailed analysis of trust models in cloud environment. In: ACM international conference on proceeding series, pp 1–5. https://doi.org/10.1145/3368691.3368740

18. Tang M, Dai X, Liu J, Chen J (2017) Towards a trust evaluation middleware for cloud service selection. Futur Gener Comput Syst 74:302–312. https://doi.org/10.1016/j.future.2016.01.009

19. Smithamol MB, Rajeswari S (2019) TMM: trust management middleware for cloud service selection by prioritization. J Netw Syst Manag 27(1):66–92. https://doi.org/10.1007/s10922-018-9457-0

20. Fan W, Perros H (2013) A reliability-based trust management mechanism for cloud services. https://doi.org/10.1109/TrustCom.2013.194

21. Al-faifi AM, Song B, Mehedi M (2018) Performance prediction model for cloud service selection from smart data Performance prediction model for cloud service selection from smart data. Futur Gener Comput Syst. https://doi.org/10.1016/j.future.2018.03.015

22. Garg SK, Versteeg S, Buyya R (2011) SMICloud : a framework for comparing and ranking cloud services. https://doi.org/10.1109/UCC.2011.36

23. Yadav N, Goraya MS (2017) Two-way ranking based service mapping in cloud environment. Futur Gener Comput Syst. https://doi.org/10.1016/j.future.2017.11.027

24. Wang Y, Wen J, Wang X, Tao B, Zhou W (2019) A cloud service trust evaluation model based on combining weights and gray correlation analysis. Secur Commun Networks 2019. https://doi.org/10.1155/2019/2437062

25. Kaushik S, Gandhi C (2019) Multi-level trust agreement in cloud environment. In: 2019 12th international conference on contemporary computing IC3 2019, pp 1–5. https://doi.org/10.1109/IC3.2019.8844933

26. Vwhp et al PVV (2017) 7Uxvw 0Dqdjhphqw Lq &Orxg &Rpsxwlqj, vol 6, pp 295–298

Optimized Usability Features of Academic Websites Using Chicken Swarm and Cat Swarm Optimization Algorithm

Vritesh Gera, Charu Mangla, Gagan Deep Bhatia, Deepak Gupta, Kalpana Sagar, and Tariq Hussain Sheikh

Abstract **Aim**: The aim of this research is to recognize the intricate usability patterns and characteristics from the top 50 websites of academic institutes listed on the National Institutional Ranking Framework using Chicken Swarm and Cat Swarm Optimization algorithm. **Method**: This study proposes a novel approach based on the feature selection using evolutionary algorithms. It comprises of three main steps: collecting the qualitative usability data for the dataset, applying Chicken Swarm Optimization (ChSO) and Cat Swarm Optimization (CSO) for feature selection to identify the troublesome usability patterns, and then applying machine learning classifiers like SVM, Naive Bayes, Decision Tree, and AdaBoost after removing these characteristics from the dataset to compare the accuracy of the usability of the academic websites. **Result**: The complex features identified using Chicken Swarm Optimization belonged to several different categories including design process and evaluation, links, graphics, images, and multimedia, optimizing the user experience, page layout, search, navigation and heading, titles and labels while using Cat Swarm Optimization the features identified majorly belonged to three categories search, security and social media. The results on the two subsets of data obtained by removing the features received from ChSO and CSO are then compared by applying the same machine learning classifiers on the data on which feature selection has not been performed to validate the performance of feature selection algorithms on our dataset. **Conclusion**: All the academic institutions have their websites providing various functionalities for students, teachers, and even potential candidates. As a result, in order to resolve the issues, it is important to design usable websites in the current academic websites.

V. Gera (✉) · C. Mangla · G. D. Bhatia · D. Gupta
Maharaja Agrasen Institute of Technology, Delhi, India

D. Gupta
e-mail: deepakgupta@mait.ac.in

K. Sagar
KIET Group of Institutions, Ghaziabad, India

T. H. Sheikh
Government Degree College, Poonch, JK UT, India

D. Gupta et al. (eds.), *Proceedings of Data Analytics and Management*,
Lecture Notes on Data Engineering and Communications Technologies 90,
https://doi.org/10.1007/978-981-16-6289-8_13

Identifying and eliminating the characteristics that reduce the usability of websites is helpful in software development.

Keywords Usability · Evolutionary algorithms · Feature selection · Classification · Decision Tree · Support vector machine · Naive Bayes · AdaBoost

1 Introduction

Usability is considered as the level of simplicity with which items in a software or web application can be used to successfully and efficiently achieve the desired results. The difficulty level associated with using a user interface is measured by ease of use. Even though the ease of use can only be assessed using aberrant measures and is, therefore, a nonfunctional requirement, it is inextricably linked to an item's functionality. Usability assessment incorporates investigations of the clearness of websites and programs. These investigations are led by usability experts. At the point when an item is regarded to have good usability, this implies it is anything but difficult to learn, and effective, and fulfilling to utilize. The word "user-friendly" gave rise to the principle of usability. Usability is an important feature of User Experience design that ensures that a product's or website's user interface does not cause consumer or end-user any discomfort or hardship. Usability is defined by IEEE Std.610.12 [1] as "the ease with which a user can learn to operate, prepare inputs for, and interpret outputs of a system or component." ISO 9241-11 Usability is described as "the extent to which a product can be used by specified users to achieve specific goals with effectiveness, satisfaction, and efficiency in a specified context of use." "Context of use" signifies the depiction of genuine circumstances in which an intelligent framework is being assessed. Context analysis is somewhat necessary to carry out research on the topic of usability. With so many definitions of the term "usability" and every expert making their own interpretation, it becomes difficult to evaluate an application's degree of ease of access. One more challenge in defining the usability of any application is the detection of such usability features which take into consideration the whole context in which the usability of the application is being estimated. Here, the context of use overall alludes to the Simultaneous investigation of ease of the use cases across various frameworks inside a similar domain. To conquer this test, ease of use testing and heuristic assessment are performed. This research experiment is done in an attempt to tackle such problems. We use 94 ISO 9241-151 guidelines in 16 different areas in this experiment and get 500 users to respond to these guidelines in the form of questions for the top-50 educational institutions of India. Then, we utilize feature selection algorithms to pick out the perplexing characteristics which need to be eliminated for determining the usability of the website. The two evolutionary feature selection algorithms were used, Chicken Swarm Optimization (ChSO) and Cat Swarm Optimization (CSO).

Evolutionary algorithms are based on Darwin's theory of evolution. An initial population is selected, as well as a fitness function, which is then determined for each data point in the population. These algorithms aim to maximize the fitness of the population. Some of the better candidates are chosen which become the parents and crossover and mutation functions are applied to them which lead to a new generation of offsprings known as their children or descendants which will compete with the previous generation for a place in the population-based on their fitness levels. Crossover is the combining of two or more population members and producing one or more fresh candidates. A mutation is carried out to protect genetic variability in the population, and the mutation likelihood is applied to candidates.

ChSO is a Swarm Intelligence Optimization method used for feature selection which mimics the hierarchy among the groups of chickens. In a swarm of chickens, there are three varieties of chickens: roosters, hens, and chicks. All the features are considered to be chickens and a fitness function is evaluated for each feature. The roosters have the maximum fitness followed by the hens and the least fitness is possessed by the chicks. This population of chickens is broken down into groups with each group consisting of one head rooster along with some hens and among those hens are some mother hens that have some children chicks. The chicks are assigned randomly to a mother hen in the group. The positions of the hens, roosters, and chicks are updated according to the type of chicken they are, that is, rooster, hen, or chick until the end of iterations is reached.

CSO algorithm mimics the actions of the cats. Cats can be considered to be in one of the modes at any instance of time that is seeking mode or the tracing mode. In seeking mode, cats are resting and they are observing their environment in search of prey and seeking which area to visit next. In tracing mode, the cats have a velocity and are chasing their prey. The positions of the cats are updated according to the mode that they are in currently, that is, seeking mode or racing mode.

We have employed four ML algorithms, namely SVM, Naive Bayes, Decision Tree, and AdaBoost to substantiate the performance of feature selection algorithms on the dataset.

A support vector machine (SVM) is a discriminative classifier with an isolating hyperplane as its formal definition. As a result, given labeled training data (administered learning), the algorithm produces an ideal hyperplane that divides the data points into different classes. This hyperplane is a two-dimensional line that divides a plane into two parts, with each class on either side.

Naive Bayes is an AI classifier that uses probability at its core by making use of Baye's theorem. The features have powerful independent assumptions.

Decision Trees are a form of supervised machine learning in which data is separated on a regular basis based on a parameter. The tree consists of two main sub-parts, in particular nodes and leaves. The leaves represent the options or the final outcomes. The nodes are the place where the information is split.

Adaboost classifier consolidates feeble classifier algorithms to frame a reliable classifier. A solitary classifier may characterize the items ineffectively. Yet, in the event that we consolidate different classifiers with a selection of training set at each cycle and allotting appropriate measure of weight inconclusive voting, we can have a great precision score for the overall classifier. In short, Adaboost iteratively retrains the algorithm by selecting a training set depending on prior training accuracy and the weight of each prepared classifier at each iteration is determined by the accuracy achieved.

The following is how the rest of the paper is structured: The paper is divided into six sections: Sect. 2 presents related work, Sect. 3 contains methods, Sect. 4 depicts implementation, Sect. 5 depicts findings and discussion, and ultimately, the paper comes to a close with Sect. 6.

2 Related Work

This section contains information about related work in the field of software usability and evolutionary algorithms. Sagar in [2] proposes an approach based on conventional usability testing and heuristic evaluation using the data-mining knowledge discovery process. Steps involved are: Formulating Qualitative Usability Feature Selection with Ranking (QUFSR), Choosing participants, Formulation of heuristic questions, Data collection for QUFSR process, Implementing Data-mining knowledge discovery process. The Weka workbench is used to implement three data-mining techniques, including association rules (Predictive Apriori algorithm), Decision Trees (ID3 algorithm), and attribute selection (Filtered Attribute Evaluator with Ranker algorithm), among others. They want to be able to distinguish between fully functional and problematic usability features. A list of comprehensive association rules was generated and problematic usability attributes were identified. Gupta in [3] proposes a Modified Crow Search Algorithm (MCSA) is a method for extracting usability features from a hierarchical model with the best or optimal solution when looking for traits that are beneficial. MCSA is an expansion of the crow search algorithm (CSA). The Modified Crow Search Algorithm's results are compared to those obtained by the Binary Bat Algorithm, the original Crow Search Algorithm, and the Modified Whale Optimization Algorithm. Results show that the suggested MCSA outperforms the regular BBA and original CSA because the algorithm presented produces 17 feature selections instead of 18 in the BBA, 23 in the CSA, and 19 in the MWOA. Gupta proposes a Modified BBA for usability characteristics selection in [4], which aims to find the best solution for finding useful usability features from a collection of usability features. MBBAT is an extension of the BBA, which takes inspiration from the actions of bats and is as accurate as possible. When the proposed MBBAT algorithm's selected characteristics and precision is compared to the original BBA, because it creates fewer selected features and has low accuracy, the suggested metaheuristic algorithm beats the original BBA. BBA chooses 18 attributes, whereas MBBAT chooses only 12 attributes. Jain et al. [5] suggest a Modified Whale Optimization Algorithm for which aims to reduce usability features and performs better than the Whale Optimization Algorithm in terms of both precision and competence at the same cost. Using MWOA, the authors select 19 features out of 23 in the original dataset. The accuracy for these 19 features is observed to be 80% over 10 iterations. Comparison of MWOA with other algorithms shows that

MWOA outperforms algorithms like Whale Optimization Algorithm, BBA, MBBA up to 5 iterations but WOA's accuracy is highest for 6 iterations.

Kabir et al. [6] analyze 10 software quality models which include McCall, Boehm, Shackel, FURPS, Nielsen, SUMI, ISO 9242-11, ISO 9126, QUIM, and SEM. After analyzing the aforementioned models, the authors propose a new quality model which includes 12 quality factors. Gumussoy in [7] proposes usability guidelines particular to banking software applications. The approach can be divided into five parts: data on usability from three banking software is used as input, rating problems in terms of severity by three experts, rating each usability criteria for how well it explains the usability problem, cross-tabulation and categorization of heuristics according to the severity level of usability problems. In the analysis, the author found out 13 heuristics and how well they explain the usability problems. Gupta et al. [8] propose a fuzzy hierarchical usability model which can predict accurate usability values. This model can be easily integrated into popular engineering practices. The validation of the model is done on six SDLC models which are then ranked according to their predicted values. The ranking order of the SDLC models is found out to be Spiral, Iterative, RAD, Evolutionary, Waterfall and Build & Fix where the Spiral model is at rank 1 and Build & Fix is at rank 6.

Meng proposes in [9] an algorithm to mimic the hierarchical order in the chicken swarm and the behaviors of the chicken swarm, including roosters, hens, and chicks, so that CSO can efficiently extract the chickens swarm intelligence to optimize problems. The chicken swarm is divided into several groups, each of which has one rooster and a large number of hens and chicks. Different chickens follow different laws of motion. Under the basic hierarchical order, there are competitions between different chickens. Taie in [10] suggests a method to detect brain tumors from MRI scans. The framework consists of 4 steps, namely segmentation, feature extraction, feature reduction, and classification. The authors use Chicken Swarm Optimization to enhance the parameters of SVM. The findings reveal that without using swarm optimization, 71.63% accuracy with sevenfold and 84.55% accuracy with tenfold was achieved but with using Chicken Swarm Optimization, 94.39% accuracy was obtained, and by further increasing the training data, 99.9% accuracy was attained. Tian et al. [11] present a deadlock-free migration ChSO algorithm to solve the virtual machine consolidation (VMC) problem. The results for real and synthetic datasets reveal that Chicken Swarm Optimization has a higher convergence rate than other deadlock-free migration algorithms. Ahmed et al. [12] propose a Chicken Swarm Optimization-based Adaptive Neuro-Fuzzy Interface System (CSO-ANFIS) to evaluate the performance of posted data on Facebook. The experiment was carried out on 790 cosmetic articles, with Lifetime Post Consumers serving as the criterion for accuracy. The results revealed that CSO-ANFIS outperformed ANFIS, with the least root mean squared error of 0.07531.

Chu in [13] presents a swarm intelligence algorithm named Cat Swarm Optimization. It has been observed that alertness in cats is very high. They are constantly observing their environment even when they look lazy while lying around. There are two sub-models of cat behavior: seeking mode and tracking mode. The former is used to mimic the circumstance of a cat reclining, gazing around, and determining where to move to next. The four major components in seeking mode have been identified: seeking memory pool (SMP), seeking a range of the specified dimension (SRD), counts of dimension to change (CDC), and self-position taking into account (SPC). The second mode of a cat is tracing. During this, the cat follows and pounces on its prey. Experimental results demonstrate that Cat Swarm Optimization was able to outperform Particle Swarm Optimization. Lin et al. [14] state two methods to modify the existing Cat Swarm Optimization. The results of the Improved Cat Swarm Optimization (ICSO) and Cat Swarm Optimization (CSO) for feature selection are evaluated utilizing SVM. The problem chosen for evaluation by the authors is text classification utilizing term frequency-inverse document frequency. The results demonstrate that utilizing either of the improvement methods over CSO yields better accuracy than CSO. Zeng et al. [15] propose an improved Cat Swarm Optimization to extract and classify objects in an image. The author compares the accuracy of ICSO to CSO, Ant Colony Optimization, and Particle Swarm Optimization. The results conclude that ICSO has better performance than its counterparts. Kumar et al. [16] propose an evolutionary filter approach using Cat Swarm Functional Link Artificial Neural Network (CS-FLANN) to remove Gaussian noise from CT images. The authors compare the results of CS-FLANN to Least Mean Squared FLANN (LMS-FLANN) and present that the peak signal to noise ratio obtained using CS-FLANN was higher than LMS-FLANN showing its superiority. Although the time taken (training + testing) by CS-FLANN was higher than LMS-FLANN.

3   Methodology

Usability is an important quality factor of a software product. ISO standard 9241 describes various characteristics or guidelines that evaluate a software's usability. These guidelines are presented as questions for the users to answer. The responses to these guidelines were recorded. If the end-user can see the specific function on the website, the response can be "Yes," "No" if the end-user does not find the characteristic or "Sometimes" if the user finds the characteristic but it is not functional on the website. Table 1 shows that 65% of users belonged to the 18–24 age group, 22% to

Table 1 End-user distribution

Details of end-users	Percentage
Age distribution	
18–24	65
25–29	22
30–35	13
Gender	
Male	68
Female	32

25–29 and the remaining 13% were between 30 and 35 years. In addition, 68% of the participants were male and 32% were female.

To reduce the dimensionality of the data, which consists of 94 questions, feature selection algorithms were used. The aim of feature selection is to remove 15% of the guidelines that are troublesome for the website's usability. Chicken Swarm Optimization and Cat Swarm Optimization have been applied for feature selection. Figures 1 and 2 depict the use of these algorithms.

3.1 Preprocessing

The preprocessing step consists of converting the categorical values in the dataset to numerical values. Label Encoder provided by sci-kit learn, a free machine library for Python language, is used to convert the string values "No," "Sometimes," and "Yes" to numerical values 0, 1, and 2, respectively. The resultant data is used in further steps.

3.2 Chicken Swarm Optimization

Chicken Swarm Optimization (ChSO) [9] imitates the behavior and hierarchical ordering among the chicken swarm, which includes hens, roosters, and chicks. The swarm is split into smaller groups, each containing one rooster, a few hens, some of which are mother hens, and a few chicks. The rooster is the head of the group and is responsible for finding food for the group and thus has the highest fitness value. The chicks are dependent on their mothers for food and have the least fitness value in the group. The remaining chickens in the group are assigned to be hens and a random relationship is created among chicks and mother hens.

$$x_{i,j}^{t+1} = x_{i,j}^t * \left(1 + \text{Rand}\, n\left(0, \sigma^2\right)\right) \tag{1}$$

$$\sigma^2 = \begin{cases} 1, \text{if } f_i \leq f_k \\ \exp\left(\frac{f_k - f_i}{|f_i| + \varepsilon}\right), \text{otherwise}, k \in [1, N], k \neq i \end{cases} \tag{2}$$

$$x_{i,j}^{t+1} = x_{i,j}^t + S1 * \text{Rand} * \left(x_{r1,j}^t - x_{i,j}^t\right) + S2 * \text{Rand} * (x_{r2,j}^t - x_{i,j}^t) \tag{3}$$

$$S1 = \exp((f_i - f_{r1})/(\text{abs}(f_i) + \varepsilon)) \tag{4}$$

$$S2 = \exp((f_{r2} - f_i)) \tag{5}$$

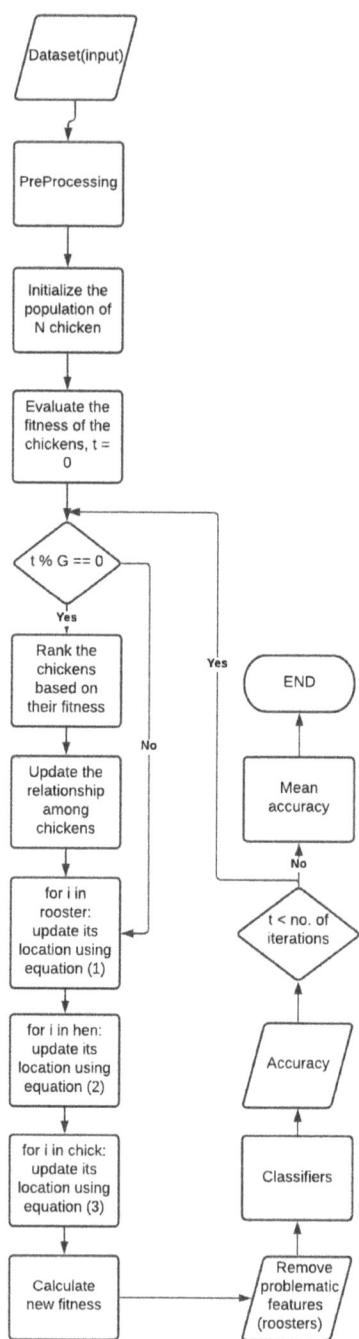

Fig. 1 Chicken swarm feature selection flowchart

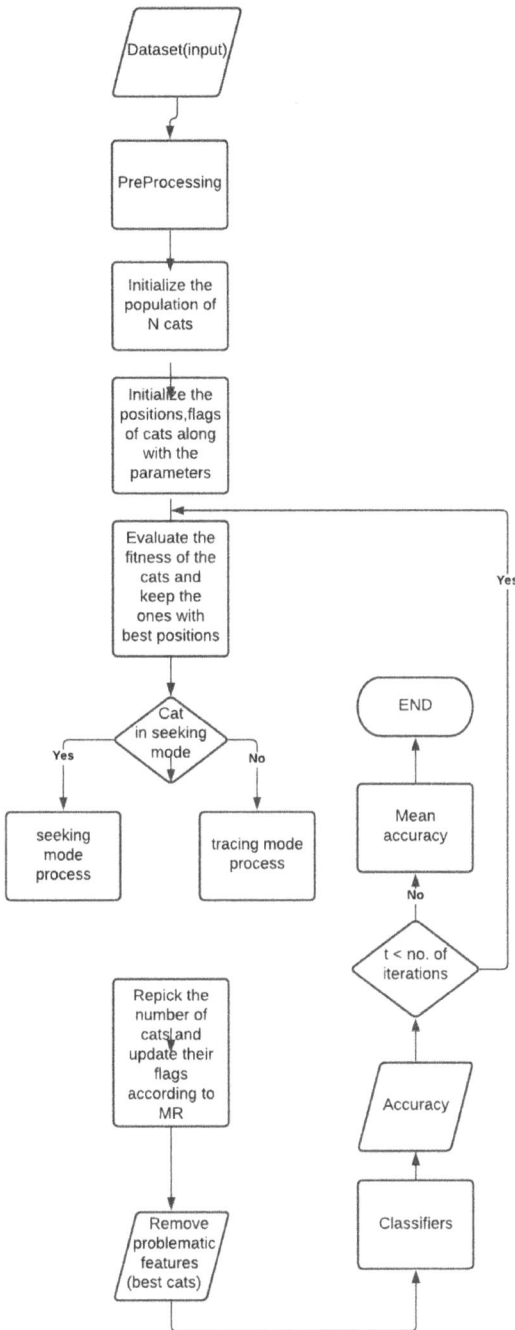

Fig. 2 Cat swarm feature selection flowchart

$$x_{i,j}^{t+1} = x_{i,j}^t + FL * (x_{m,j}^t - x_{i,j}^t) \tag{6}$$

The roosters with better fitness have a priority for food and can look for it in a more extensive scope. This is depicted in Eqs. (1) and (2) where $\mathrm{Rand}\, n(0, \sigma^2)$ is a Gaussian distribution with mean 0 and standard deviation σ^2. ε, is used to avoid zero division error. K, a rooster's index, is chosen randomly from the rooster group, f is the fitness value corresponding to x.

The hens hunt for food behind the roosters in their flock. The prevailing hens would have a benefit in going after food than compliant ones. This is depicted through Eqs. (3)–(5). Rand is a uniform random number over [0, 1]. $r1 \in [1, \ldots, N]$ is the index of the rooster, which is the ith hens' group-mate, while $r2 \in [1, \ldots, N]$ is an index of the chicken (rooster or hen), which is randomly chosen from the swarm. $r1 \neq r2$.

The chicks roam about their mother and are depicted in Eq. (6).

Figure 1 illustrates the application of the Chicken Swarm Optimization feature selection algorithm.

3.3 Cat Swarm Optimization

Cat Swarm Optimization (CSO) [13] is a program that replicates cat behavior. The cats can be considered to exist in either of the two modes, seeking mode or tracing mode. In the seeking mode, the cats are stationary. They are observing their environment in search of their next position. There are four factors defined in seeking mode: The extent of each cat's seeking memory range, which reflects the positions the cat is looking for is set by the seeking memory pool (SMP), seeking a range of selected dimension (SRD) to indicate a mutation ratio of the selected dimensions, counts of dimensions to change (CDC) to determine how many dimensions to vary and self-position considering (SPC) which is a boolean variable that decides whether the point, where the cat is currently standing is to be considered as a candidate to move to. The position in the seeking mode is updated according to Eq. (7) where FS is the fitness value. In tracing mode, the cat is moving and is tracking some targets. A mixture ratio is known as the ratio of the number of cats in tracing mode to the number of cats in seeking mode to combine the two modes of the cats. Equation (8) describes the update of velocity for every dimension $v_{k,d}$, $x_{\mathrm{best},d}$ is the position of the cat who has the best fitness value, $x_{k,d}$ is the position of the cat k. c_1 is a constant and r_1 is a random value in the range [0, 1]. Equation (9) describes how the positions of the cats are updated in tracing mode. The mixture ratio is kept small to ensure that more cats are in seeking mode than tracing mode, simulating real-life cat behavior in which they spend the majority of their time resting.

$$P_i = \frac{|\mathrm{FS}_i - \mathrm{FS}_b|}{\mathrm{FS}_{\max} - \mathrm{FS}_{\min}}, \quad \text{where } 0 < i < j \tag{7}$$

Table 2 Parameter description

Algorithm	Parameters
Chicken swarm optimization	$G = 5$, $FL = 0.5$
Cat swarm optimization	$MR = 10$, $SMP = 2$, $SPC = $ False, $CDC = 1$, $SRD = 0.1$
SVM	kernel $=$ rbf
Decision tree	criterion $=$ "gini," splitter $=$ best
Naive Bayes	priors $=$ none
AdaBoost	n_estimators $= 50$, learning_rate $= 1$

$$v_{k,d} = v_{k,d} + r_1 * c_1 * \left(x_{\text{best},d} - x_{k,d}\right), \quad \text{where } d = 1, 2, \ldots, M \tag{8}$$

$$x_{k,d} = x_{k,d} + v_{k,d} \tag{9}$$

4 Implementation

The tests were carried out on a system with an Intel Core i5 6th gen 2.4 GHz, 12 GB of RAM, and Windows 10 installed. Python 3.6.3::: Anaconda, Inc. was used to implement the algorithms.

4.1 Input Parameters

The details of the parameters given to the various algorithms used are displayed in Table 2.

4.2 Dataset Description

The dataset is collected for the top 50 academic websites listed on National Institutional Ranking Framework. ISO 9241-151 and other heuristic guidelines have been used to evaluate the usability of the end-users. The responses to these guidelines have been recorded using Google forms where the users have to invest minimal time. The response to each question can be either of the three, "Yes," "No" or "Sometimes." If the user finds the characteristic asked by the question, the user marks a "Yes" and if the user is unable to find the characteristic, the user responds with "No." "Sometimes" is used in a situation where the user recognizes the feature in the website but it is not functional. In Fig. 3, the column headings represent a usability attribute

	Keeping the cont	Making the date	Enabling commu	Accepting online	Multi-language:	Scope: Is Multi-li	News Comprehe	Contact Data: Dc	Analysing the tar	Appropriateness	Completeness of	Text Des
	A	B	C	D	E	F	G	H	I	J	K	L
1												
2	Yes	No	No	No	No	No	Yes	Yes	Yes	Yes	No	Yes
3	Yes	Yes	Yes	Sometimes	Yes	Yes	Yes	Yes	Yes	Yes	Yes	Yes
4	Sometimes	Yes	Yes	Sometimes	Yes	Sometimes	Yes	Sometimes	Yes	Yes	Sometimes	Yes
5	Yes	No	Yes	Yes	No	No	Yes	Yes	Yes	Yes	Yes	Yes
6	Yes	Sometimes	Yes	No	No	No	Yes	Yes	Yes	Yes	Yes	Yes
7	Yes	Yes	Yes	No	Yes	Yes	Yes	Yes	No	No	Yes	Yes
8	Yes	Yes	Yes	Yes	Yes	Sometimes	Yes	Yes	Sometimes	Yes	Yes	Yes
9	Yes	Yes	Yes	No	Yes	No	Yes	Yes	No	Yes	Yes	Yes
10	Yes	Yes	Sometimes	Sometimes		Yes	Yes	Yes	Sometimes	Sometimes	Sometimes	Yes
11	Yes	Yes	Yes	No	No	No	Yes	Yes	Yes	Sometimes	Yes	Sometin
12	No	No	Yes	No	No	No	No	No	Yes	Yes	No	Yes
13	Yes	No	No	Yes	No	No	Yes	No	Yes	Yes	No	Yes
14	Yes	Yes	No	No	Sometimes	Yes	No	Yes	Yes	No	Yes	Yes
15	Yes	Yes	Yes	Yes	No	No	Yes	No	Sometimes	Yes	Yes	Yes
16	Yes	Sometimes	Yes	Yes	Yes	Yes	Yes	Yes	Yes	Sometimes	Yes	Yes

Fig. 3 A partial view of the usability attributes

belonging to the "search" category and the rows depict the website evaluation done by a user. The dataset in [2] consisted of the responses of 100 users. Additional 400 users were reached out in this study, and their responses were collected for the same guidelines prescribed by ISO 9241-151 with the help of Google forms and stored in CSV format. The participants gave honest feedback on the websites because they believe it will help to eliminate inconsistencies in website design and growth.

5 Results and Discussion

The dataset comprises 500 tuples that belonged to either one of the two categories, a website that is usable or not. We were able to extract 14 attributes from the 94 attributes using feature selection algorithms, ChSO and CSO. The description of the extracted features using ChSO and CSO is presented in Table 3. Subsets of data are created by removing these identified non-significant attributes and these subsets are injected into 4 ML classifiers: SVM, Decision Tree, Gaussian Naive Bayes, and AdaBoost. The metrics used to evaluate the results of the classifier are accuracy, precision, and recall. Equations 10, 11, 12, and 13 describe the formula for accuracy, precision, and recall, respectively, where TN is the number of True Negatives, FN is the count of False Negatives, TP is the number of True Positives, and FP is the count of False Positives.

$$Accuracy = \frac{TP + TN}{TP + FP + TN + FN} \tag{10}$$

$$Precision = \frac{TP}{TP + FP} \tag{11}$$

$$Recall = \frac{TP}{TP + FN} \tag{12}$$

$$F - score = \frac{2 \times Precision \times Recall}{Precision + Recall} \tag{13}$$

Table 3 Problematic features identified by ChSO and CSO

Features identified by ChSO	Features recognized by CSO
Compatible for a physically disabled person: is the site open to people who are physically disabled?	Is there some kind of warning message about malicious software on the university's website?
Redundant links	Giving suggestions for unsuccessful searches
Quantity of text per information unit/page (e.g., quantity of text must be sufficient enough and self-descriptive)	Image and video-based search: do the website support images and videos-based search features?
The naming of URLs (it should be small and easy to remember)	Search time: is it easy for the user to find search results?
Selecting appropriate media objects	Advanced search
Writing style (e.g., it should be simple and easy to read)	Blog: is there a blog on the website to engage users and improve online visibility?
Use of "white space"	Error-tolerant search
Desired search results: is the user getting the results he or she wants?	Facebook page: is there a Facebook page for the website? Is your university socially engaged on social media?
Showing users where they are	Twitter account: is there a Twitter account on the university's website?
Identifying all pages of a site	Provide a simple search facility
Distinguishing navigation links from action links (e.g., action links generally appear in yellow or red color that shows the latest news)	SSL secure: is the university's website protected by an SSL certificate, allowing for a secure transaction or encrypted link between users and the server?
General page information	Full-text search
Error pages	Media to help: is there any interactive storytelling media on the website to assist users?
Placing title information consistently	Desired search results: Is the user getting the results he or she wants?

The comparison of the metrics of various classifiers is displayed in Figs. 4, 5, 6 and 7. We observe that the highest accuracy with feature selection obtained by ChSO was by using AdaBoost classifier which is 98.16% with CSO not close behind with 98.13% accuracy. The highest accuracy reached without feature selection is 99.20% with the use of the SVM while comparing precision, we witness that without using feature selection we get 100% precise results in two cases by using Decision Tree and AdaBoost classifier, on the other hand, maximum precision obtained with ChSO is 98.07 and 97.95% with CSO. This indicates that without using feature selection, there were no false positives and with feature selection, some examples were incorrectly classified as positives. Peak recall achieved with ChSO is 98.57% and CSO is 98.61%, however, 100% recall was obtained without the use of feature selection with SVM.

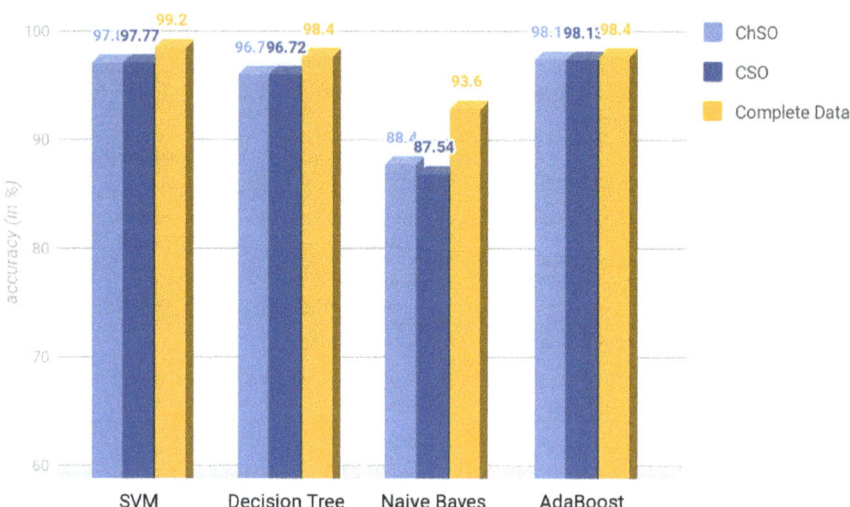

Fig. 4 Accuracy comparison

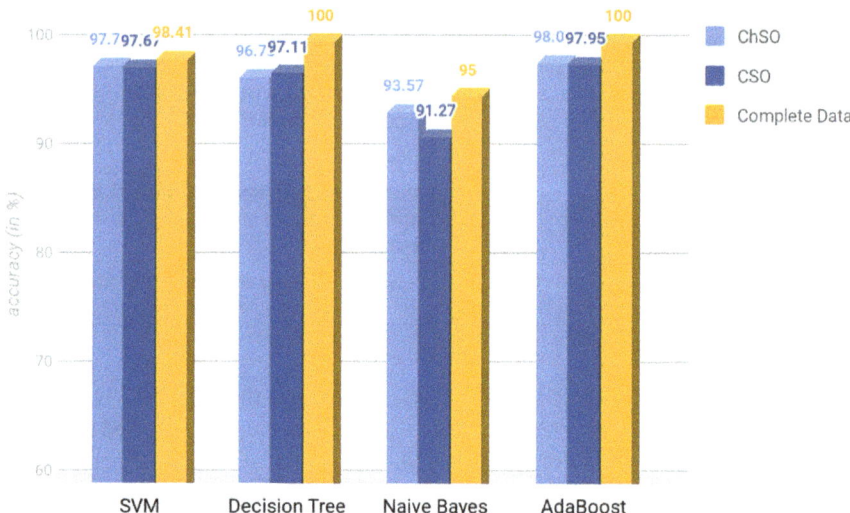

Fig. 5 Precision comparison

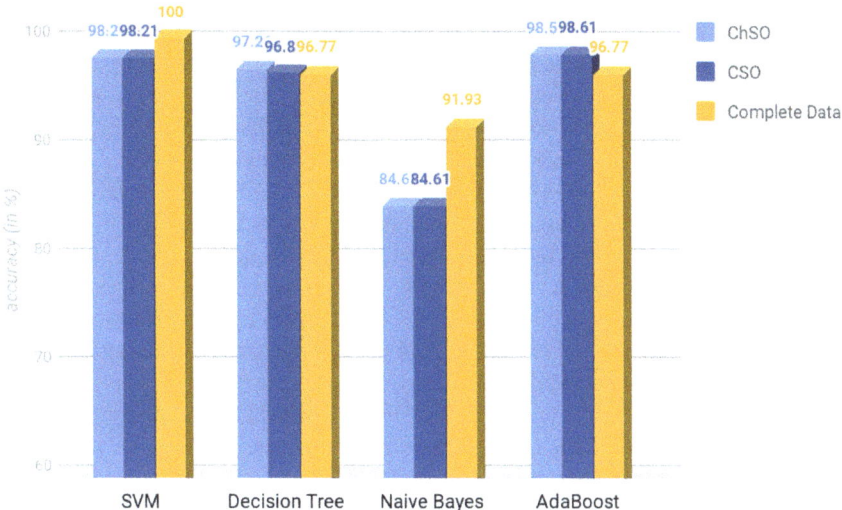

Fig. 6 Recall comparison

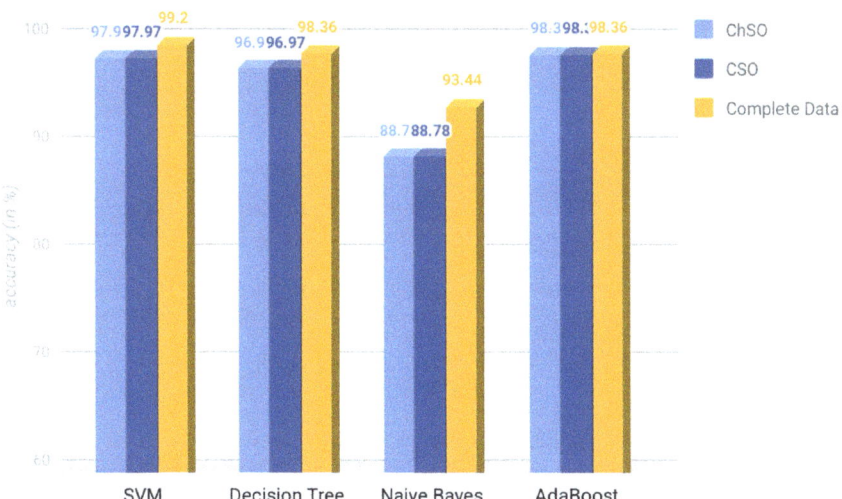

Fig. 7 *F*-score comparison

This shows that some tuples were incorrectly classified as negative while they were actually positives. All the results are summed up in Figs. 4–7.

6   Conclusion and Future Work

Usability of software is regarded as the ease of use without any particularized training. It is inextricably related to the software's efficacy and efficiency. In this paper, we were able to devise an approach to extract the features that affect the usability of academic websites using Chicken Swarm Optimization and Cat Swarm Optimization algorithm. The particulars of these characteristics are displayed in Table 3. Our system, when performing classification without these difficult features, yielded the highest accuracy of 98.33% using CSO and 98.23% using ChSO with AdaBoost Classifier. However, higher accuracy was obtained on the complete dataset without the use of feature selection which attained 98.4% accuracy by using AdaBoost Classifier. The advantages of feature selection are that it reduces the overfitting by eliminating the redundant and noisy data and it also improves the accuracy over traditional approaches as it is evident from this study. Thus, we conclude that the use of evolutionary algorithms to evaluate qualitative usability in a sense of use as a whole can be effectively implemented in real-world problems.

Our model is sturdy which produces better results without the use of feature selection. In the future, an even greater dataset can be collected considering the attributes that have not been examined in this research. Deep learning algorithms can be engaged to achieve more accurate results by creating more comprehensive and exhaustive results. This study can help software developers/engineers to focus on the features that the users consider to boost the usability of an application.

References

1. Kajko-Mattsson M (1990) Preventive maintenance! Do we know what it is? In: Institute of electrical and electronics engineers (1990) IEEE standard glossary of software engineering terminology. IEEE Std., pp 610.12
2. Sagar K, Saha A (2017) Qualitative usability feature selection with ranking: a novel approach for ranking the identified usability problematic attributes for academic websites using data-mining techniques. Human-Centric Comput Inf Sci 7–29
3. Gupta D, Rodrigues J, Sundaram S, Khanna A, Korotaev V, Albuquerque V (2018) Usability feature extraction using modified crow search algorithm: a novel approach. Comput Aided Med Diagn
4. Gupta D, Ahlawat A (2017) Usability feature selection via MBBAT: a novel approach. J Comput Sci 23:195–203
5. Jain R, Gupta D, Khanna A (2018) Usability feature optimization using MWOA. In: International conference on innovative computing and communications, pp 453–462
6. Kabir MA, Rehman M, Majumdar S (2016) An analytical and comparative study of software usability quality factors. In: 7th IEEE international conference on software engineering and service science, pp 800–803
7. Gumussoy C (2016) Usability guideline for banking software design. Comput Hum Behav 62:277–285
8. Gupta D, Ahlawat A, Sagar K (2017) Usability prediction & ranking of SDLC models. In: Open engineering, pp 161–168
9. Meng X, Liu Y, Gao X, Zhang H (2014) A new bio-inspired algorithm: chicken swarm optimization. In: International conference on swarm intelligence, pp 86–94

10. Taie S, Ghonaim W (2017) CSO-based algorithm with support vector machine for brain tumor's disease diagnosis. In: IEEE international conference on pervasive computing and communications workshops (PerCom Workshops)
11. Tian F, Zhang R, Lewandowski J, Chao KM, Li L, Dong B (2017) Deadlock-free migration for virtual machine consolidation using Chicken Swarm Optimization algorithm. J Intell Fuzzy Syst 1389–1400
12. Ahmed K, Hassanien A-E, Ezzat E, Bhattacharyya S (2018) Swarming behaviors of chicken for predicting posts on Facebook branding pages. In: The international conference on advanced machine learning technologies and applications, pp 52–61
13. Chu S-C, Tsai P, Pan J-S (2006) Cat swarm optimization. Trends Artif Intell
14. Lin K-C, Zhang K-Y, Huang Y-H, Hung J, Yen N (2016) Feature selection based on an improved cat swarm optimization algorithm for big data classification. J Supercomput 3210–3221
15. Zeng Z, Yang F, Wen Z, Liu L, Guan L (2015) Objects extraction and classification based on an improved cat swarm optimization algorithm. J Inf Comput Sci 5053–5061
16. Kumar M, Mishra S, Sahu S (2016) Cat swarm optimization based functional link artificial neural network filter for gaussian noise removal from computed tomography images. Appl Comput Intell Comput

Traffic Signal Control Methods: Current Status, Challenges, and Emerging Trends

Ishu Tomar, S. Indu, and Neeta Pandey

Abstract Traffic management at intersections is a challenging problem within the transport system. Various traffic signal control strategies have been used to manage the traffic at intersections in real time but are not able to deal with the congestion at road intersections. They are intended to acknowledge the continuous flow of vehicles on heavy traffic routes. However, the process of networking traffic lights of adjacent junctions is a complex matter because of many boundaries. Variable flows entering the junctions are not controlled by traditional systems. Moreover, there is no common intervention in the current traffic signal framework between the adjacent traffic light systems, the difference in the movement of cars over time, injuries, the passage of emergency vehicles, and pedestrian crossings. This paper provides a systematic literature review of the existing methods and algorithms for traffic signal control at intersections. Furthermore, this paper discusses the open challenges with the purpose of synchronizing Traffic Light for Intelligent Vehicles.

Keywords Traffic signal control (TSC) · Intersections · Traffic light synchronization

1 Introduction

In order to ensure traffic protection and smooth traffic flow, management of traffic at intersections is a difficult issue within the transport system. Intersections are the meeting places for the movement of pedestrians, cyclists, and cars. Intersections are the points where a vehicle user changes its direction. Intersections are the obstacle of

I. Tomar (✉) · S. Indu · N. Pandey
Department of Electronics and Communication Engineering, Delhi Technological University, Delhi 110042, India

S. Indu
e-mail: s.indu@dce.ac.in

N. Pandey
e-mail: neetapandey@dce.ac.in

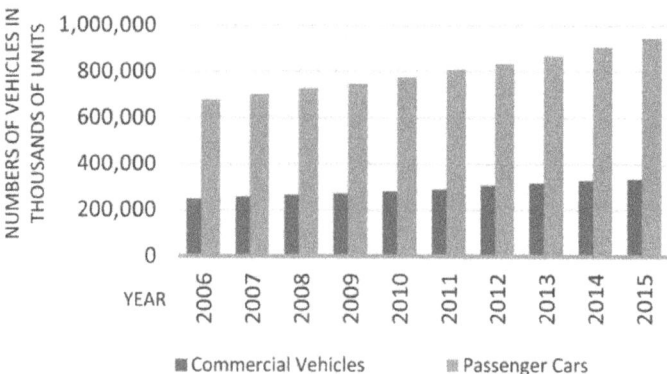

Fig. 1 Number of vehicles in use globally from 2006 to 2015 [5]

routes in metropolitan cities, significantly leading to loss of travel time due to traffic interruption, control, and management [1, 2]. The intersection delay doesn't only affect the travel efficiency of road users, it also affects the signal control logic [3]. Traffic signal delays are estimated to contribute to 10% of traffic delays all over the world. In the United States, it was observed that over 295 million traffic hour latency was found to have occurred on main roads last year [4]. The number of vehicles in use globally between 2006 and 2015 is shown in Fig. 1 [5]. For the real time control and management of traffic, it is therefore important to have accurate estimation of time-dependent delays at intersections on urban roads.

One of the applications targeted by ITS is traffic control management at intersections, a difficult and complex area of research. Intersections require a balance between safety and efficient traffic control. Maximum number of vehicles and users should pass through an intersection to reduce the congestion without compromising the safety of the parties involved [6]. Traffic at sensitive intersections is generally controlled nowadays by traffic light signaling [7]. The timings of traffic lights are set in accordance with the average road traffic every day. The complex design of the traffic does not integrate this. Congestion and delays are mainly due to the allocation of the same green light schedules regardless of the density of traffic of that route. The imbalanced traffic flow situation in cities such as New Delhi is reasonably prevalent since most of the working people live in nearby places. When people drive to their jobs and residency, the imbalance of traffic is usually high, i.e., in the morning and evening. Therefore, a traffic signal control strategy is needed in metropolitan cities, which updates the timing of traffic signals based on real time data. Various traffic signal control strategies have been used to manage the traffic at intersections in real time. This paper provides a comprehensive survey of the developed traffic signal control strategies. It highlights the research gaps and provides possible directions for future work for researchers interested in the field. It also summarizes the implementation status of Traffic Light Synchronization.

This paper's framework is organized as follows. Initially, we briefly introduce the development background of Traffic Light control technology and the technical

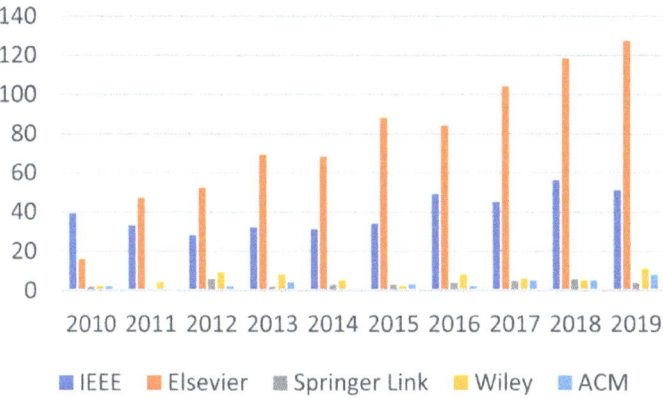

Fig. 2 Number of research works published between 2010 and 2019 [8]

benefits of the current traffic light systems. Then, we discuss the different traffic signal control methods in detail. Then, we discuss the future opportunities and challenges for the research and development of Synchronization of traffic lights for Intelligent Vehicles. Finally, we conclude the whole paper.

2 Overview of Traffic Light Control Technology

The design of intelligent traffic control systems is an active research topic. To solve this stressful issue, many researchers are developing new methods and advanced systems. Regional strategies for coordinated traffic control can synchronize traffic signals to enhance system performance at coordinated intersections. Many researchers have suggested regional organized traffic management techniques, which primarily include algorithms based on fuzzy logic, artificial intelligence, swarm intelligence, evolutionary algorithms, neural network, image processing, linear programming, and data fusion approach.

Different algorithms have been studied in the literature to solve the Traffic Signal Control (TSC) problem. The number of research works published between 2010 and 2019 is shown in Fig. 2 [8]. The rising trend reflects the growing interest in the subject and emphasizes the importance of current research.

3 Existing Algorithms for Traffic Signal Control

In this section, we review different traffic signal control methodologies and algorithms.

3.1 Traditional Approach

Traffic collection and intelligent traffic light systems is an area in which many researchers are working all around the world. It has long been recognized that traffic-responsive management strategies can significantly minimize intersection delays. Different adaptive traffic systems, such as SCOOT [9], SCATS [10] OPAC [11], RHODES [12], PRODYN [13], and MOTION [14] have been developed and introduced in some cities in the past few decades.

These systems optimize, based on current traffic demand, the traffic signals offset values in the network and produce 'green-wave' for major car movement. Both actual traffic arrivals and the forecast of potential arrivals are used by these systems and a signal phase schedule is chosen to optimize the objective functions. At the same time, these systems have major limitations with effective functionality.

Generally, these systems have a very high implementation cost. The system's initial cost is $20,000–$30,000 per intersection with a per mile/year cost of $28,800 [15] for the installation. These devices are normally costly and difficult to install and maintain because of the use of video cameras and loop detectors for tracking vehicles. In the meantime, they are centralized control systems, indicating the need for a large communication infrastructure that can accommodate high-data-rate centralized control. All these variables add to a very expensive strategy that clarifies the question that why high-cost systems have been built so far by very few cities in the world.

For fixed signal control at isolated junctions, Webster procedure was used in [16] for real time operation based on measured flows. For undersaturated traffic conditions, the evolved real time demand based strategy is a feasible signal control strategy. Inside the TUC, it can actually be used to remove the requirement for a pre-specified static signal schedule. Compared to the original TUC process, it can lead to more successful outcomes. But the problem is that this procedure is very much complicated.

3.2 Sensor Based Approach

Authors of [17] proposed a method to design a mobile app. that allows users to collect GPS data and use it to assign equal green light timings, plan fair lane departures and recommend lanes for users approaching an intersection accordingly.

A data fusion approach has been proposed in [18] to collect traffic information in a vital urban road network. The speed, location, and direction were gathered from GPS devices in the vehicles on the road. This approach only used the data to enhance the flow of traffic on current road network, not to manipulate the timing of traffic lights at road intersections.

In [19], detectors are used to keep the count of arrival and departure of vehicles and this count was used to prolong or end the current signal phase. On-board GPS unit [20], Big Data technology [21] are some other techniques which are used to collect information about vehicles so that traffic congestion can be reduced in urban areas.

In cars, embedded technology is also used to capture GPS data and push it via GSM/GPRS [22] to the Traffic Monitoring System. While sensors and traffic servers also capture and use local traffic data to monitor local traffic in the area, they are also used. Although the work is effective, implementation on a wide scale is costly.

For the signal control problem, an adaptive framework based on a linear programming model was implemented in [23] that minimizes the waiting time of vehicles at intersections. The model was focused on traffic information generated at each intersection by real time sensors deployed.

Authors of [24, 25] proposed Wireless Sensor Network (WSN) based traffic light control systems. A time-fragment based control system is proposed in [24]. It divides the time into pieces, and the green ratio (percentage of green time all the time) is calculated in each lump.

3.3   Vision Based Approach

In [26], a video-based adaptive traffic signaling system is proposed to grant equal green light times using video camera data and then to allocate sufficient time using a fair weight and optimal weight calculator. This method is simple but very expensive because of the installation of video camera at every intersection.

In [27], a dynamic Bayesian networks approach has been used for predicting and monitoring the traffic using a video camera. This method is very powerful and efficient but it is computationally light.

For the identification and monitoring of moving vehicles in real time, feature-based methods [28] are also used. They track characteristics that are less susceptible to partial occlusion. While the strategy worked for both daytime and evening, as it had low precision, and required great space for storage, it was not useful for real time signal control. Vehicles are identified using binarization, rule-based logic, cameras, and classified road conditions. The calculation involved in vehicle detection is very complex and tedious, so the system is inappropriate for operation in real time [28]. In addition, neural networks are also used for traffic prediction in real time [29].

Due to their computational complexity and greater execution time, algorithms used for image segmentation and vehicle tracking get to be inefficient for the online/real time operation. Some methods detect and monitor vehicles during daylight hours, but nighttime and varying weather conditions do not work well. In [30], the author takes a 15-min daytime volume of traffic on expressway and attempts to predict the next 15-min traffic volume using neural networks. This strategy has high complexity and memory requirements, making it difficult to scale this work to a full-fledged network. Hence, for real time systems, this algorithm would not be viable.

3.4 Evolutionary Algorithm and Swarm Optimization Based Approach

In [31], a genetic algorithm (GA) based approach in integration with urban traffic microscopic simulation model (UTMSM) was used to improve the traffic flow at intersections.

Authors of [32, 33] used GA for optimizing Traffic lights and finding the best routes for vehicles in real time.

Authors of [34] improved GA with the help of a simple neighborhood algorithm was used as a local search technique to reduce the average traffic delay as much as possible.

In [35], GA was used to propose a multipurpose framework to solve the signal control dilemma. This model reduced the overall delay, number of stops, and vehicle emissions by optimizing the green times of each phase.

Authors of [36] proposed a multipurpose GA for traffic optimization in a 9-junction network. But this algorithm was not based on real time traffic data.

Authors of [37–39] also used various GA based models for solving the TSC problem in real time.

A swarm intelligence approach is used in [40–43] for traffic light scheduling. These strategies use particle swarm optimization (PSO) algorithm to control traffic signals so that the vehicles can reach their destination at earliest as possible.

Ant Colony Optimization [44, 45] and Harmony search [46, 47] are some other swarm intelligence algorithms which have been used for TSC.

3.5 Connected Vehicle Based Approach

An algorithm based on Connected Vehicle technology is proposed in [48]. It is used at isolated intersections for the optimization of the order of discharge of vehicle's platoons based on their proximity.

Virtual Traffic Light is another approach to adaptive decision making [16]. It is a technology for V2V traffic control that has tremendous potential as it can raise traffic flows by more than 30% and stops the use of costly traffic lights. Unfortunately, VTL may require full penetration of DSRC (Dedicated Short Range Communication) technology into vehicles that may not occur instantly. Furthermore, V2V communications in VTL could encounter non-LoS circumstances that could make it very difficult to make timely decision making [49].

In the meantime, the Virginia Transportation Department (VDOT) has implemented a connected and autonomous vehicle program (CV) [50, 51]. This plan focuses on DSRC technology-based various V2V and V2I applications. CV aims to reduce the costly infrastructures of roadside guide signs and traffic signals based on DSRC technologies, somewhat similar to VTL. Connected Corridors are also set up as part of the initiative, enabling the real world production and application of connected vehicle technology. Although this program is helpful in removal of intersection traffic signal infrastructures, there is a requirement for full penetration of DSRC.

In [52], a DSRC based intersection traffic control scheme was proposed. This methodology gives priority to roads that involve DSRC-equipped vehicles. Using this priority mechanism, the average waiting time and average travel time can be reduced on each TL during rush hours. This method can work well even if low percentage of vehicles are equipped with DSRC technology. This approach gives a cost-effective economical solution for urban control because it only needs DSRC RSUs as compared to other control systems like cameras and loop detectors.

Autonomous intersection management (AIM) with multi-agent systems has been presented in [53] where drivers and intersections are treated as independent agents. In this mechanism, intersections use a brand new reservation-based approach designed on an in-depth communication protocol.

3.6 Fuzzy Logic Based Approach

Authors of [54] proposed an approach that uses autonomous vehicles to reduce congestion at the junctions. Using V2V communication, the speed and location of other neighboring vehicles can also be determined.

For solving the coordination and complex instability issue in heavy traffic, authors of [55, 56] (based on fuzzy and Q-learning) proposed a type-2 FLC system for organized traffic.

In [57], traffic signal synchronization system based on real time, with multi-agent fuzzy logic and Q-learning method, has been used to smoothen the traffic flow all over the city. Some other fuzzy logic based approaches have been used in [58, 59] for TSC.

Authors of [60, 61] designed a real time Fuzzy logic based control system to minimize the waiting time of vehicles at junctions.

3.7 Learning Based Approach

Many researchers have suggested that reinforcement learning (RL) could be used to control traffic signals and ameliorating traffic congestion [62, 63]. RL can learn directly from the feedback, unlike conventional transportation methods that rely heavily on predefined models. In RL, every junction is modeled as an agent which optimizes its input based travel time from the environment after setting the traffic signals (i.e., the action) [64].

Authors of [65] proposed a deep Reinforcement Learning based delay tolerant ITSC algorithm for the intelligent TSC. This algorithm mitigates the performance loss and resolves the basic security concerns in cloud-based ITSC systems caused by network latency. This algorithm is suitable for real world deployment and is suitable for infrastructure-based as well as infrastructure-free ITSC systems based on cloud computing as well as other technologies that require a remote computation resource.

Authors of [66] proposed a new FRAP model based on the universal concept of phase competitive modeling, through which invariance can be achieved in situations like traffic flow flipping and rotation. Priority is given to the road with high volume of traffic. This method finds better solutions and achieves better performance than other learning methods in complex traffic conditions and multi-junction structures.

Yau et al. [67] reviews various RL models and algorithms applied to traffic signal control in the aspects of the representations of the RL model (i.e., state, action, and reward), performance measures, and complexity to establish a foundation for further investigation in this research field.

In [68], a new MARL (Multi-agent reinforcement learning), called Co-DQL, i.e., cooperative double Q-learning technique is presented to solve the traffic signal control problem. Co-DQL uses a double Q-learning approach which is based on the upper confidence bound and double estimators policy, which avoids the overestimation issue that plagues classical independent Q-learning. TSC is subjected to Co-DQL, which is then evaluated on a variety of traffic flow scenarios using TSC simulators. In addition, it also introduces a new reward allocation mechanism for the stability and robustness of agents.

In [69, 70], the author has investigated the new advancements in reinforcement learning (RL) approaches that can be used for resolving the problems faced in controlling traffic lights.

Different RL algorithms have been investigated in [71]. This method discusses the use of algorithms like PPO, A2C, and ACKTR so that the problem in partial detection of vehicles can be resolved.

A deep Q-learning with partial detection of vehicles is proposed [72] for the traffic signal control. This learning approach partially detects vehicles with the help of DSRC technology. The advantage of this technique is that it can also detect the vehicles using other detection methods like BLE 5.0, LTE/5G, and RFID.

4 Networking of Traffic Lights: Where Are We Now?

Traffic light control frameworks help to synchronize the time of traffic lights in a region of a city. These frameworks limit the waiting time experienced by vehicles going through a junction network. Regardless of all the research studies done in this field in the last fifty years, there is as yet a need for a more successful administration of traffic signal control framework.

In [73], a linear programming based adaptive system was used to control traffic signals to minimize the waiting time of vehicles at the junctions. This model was based on the real time traffic information collected from the sensors at every junction.

Authors of [23] used a LGLR, i.e., Long green and long red strategy to synchronize the traffic signals within a region so that the traffic flow efficiency can be maximized at the high-density intersections.

In [74], a real time traffic signal synchronization system with multi-agent fuzzy logic and Q-learning method was used to smoothen the traffic flow all over the city.

Authors of [75] proposed an algorithm to increase the efficiency of traffic synchronization by dynamically adjusting the values of the timer. This system obtains density of vehicles at every junction using traffic cameras and then performs operations with images using OpenCV.

A self-organized traffic control algorithm based on sensors is proposed in [76]. It increases the traffic flow efficiency and minimize waiting time of vehicles at the junctions. This system provides more information in comparison with the conventional traffic control system. It first gives priority to the approaches with high density of vehicles at a junction. But it does not connect all the intersections, i.e., it does not give any account about the networking of junctions.

For transmission of information on infrastructure like road changes, speed limits, traffic light information to smart vehicles, V2I communication have been used by Liang et al. [77] for the networking of intersections and other signs at road.

A large portion of these works is confined to one intersection or convergence where the impact of the neighboring junctions has not been investigated. In this manner, the problem turns out to be more complex and broadly focused. In order to achieve maximum modeling, control, and monitoring of numerous synchronized junctions, more efforts should be made.

5   Discussion and Open Issues

The initial intentions of researching traffic signal control technologies are to safeguard traffic safety, enhance travel quality, and reduce pollutant emissions. However, in the last 20 years, because of the rise of numerous issues in the marketing process, even developing countries have used only the most basic traffic signal control technologies.

Due to the increase in vehicular traffic of metropolitan urban areas, traffic jam has become a fundamental issue. The issue is surprisingly more dreadful at road junctions because of the pre-timed traffic lights which cause delay, increment in vehicle activity cost, and air pollution. The state of one light at a crossing point, which affects the stream of motion at neighboring convergences, is also responsible for delay. In addition, the conventional traffic scheme does not think about the circumstance of mischance, road works, and vehicle breakdown compounding blockage of travel. In addition, the smooth flow across crossing points of Emergency vehicles with higher needs like fire brigades, ambulances, police, and VIP people, is a key problem. Therefore, to resolve the issue of traffic congestion and to reduce the waiting times, to reduce travel time, to increase vehicle safety and efficiency, to increase environmental and economic benefits, the normal traffic system needs to be strengthened.

In most of the methods used in traffic monitoring at intersections, very expensive cameras and sensors are used which are hard to install and maintain. The methods used to control traffic signals at road intersections are computationally costly.

Real time Signal control strategies need to be improved at road intersections so that the junctions can be made jam free. The traditional techniques do not synchronize the traffic signals of an area which is a great drawback.

Henceforth to overcome these issues, traffic light synchronization is the procedure where vehicles beginning toward one side of road and going at explicit speed can go to opposite ends ceaselessly by getting maximum number of green lights at any junction. Synchronization of traffic signal is needed such that the green light cycles for a sequence of intersections may be matched, allowing the most number of cars to drive through and eliminating waits and delays.

6   Conclusion

Synchronization of Traffic Lights for Intelligent Vehicles is a promising innovative technology with extraordinary potential in the field of intelligent transportation systems. In this survey, we have presented a systematic literature review of the existing methods and algorithms for traffic signal control at intersections. In addition, we have discussed many emerging applications of networking of Traffic Lights, the open challenges, and future opportunities for the research and development of traffic signal control at junctions in detail. Furthermore, this paper provides motivation to researchers with the purpose to develop novel algorithms for synchronizing Traffic Lights for Intelligent Vehicles.

References

1. Nielsen OA, Frederiksen RD, Simonsen N (1998) Using expert system rules to establish data for intersections and turns in road networks. Int Trans Oper Res 569–581
2. Homburger WS, Hall JW, William RR, Edward CS, Michelle D, Loretta H, John JL, Matthew R, Vernon HW (2007) Fundamentals of traffic engineering; institute of transportation studies. University of California, Berkeley, CA, USA
3. Heidemann D (1994) Queue length and delay distributions at traffic signals. Transp Res Part B Methodol 377–389
4. Denney RW, Curtis E, Olson P (2012) The national traffic signal report card. ITE J
5. Passenger Cars in Use Worldwide 2005–2015|Statistic. Accessed on 5 Aug, 2020. [Online]. Available: https://www.statista.com/statistics/274365/passenger-cars-in-use-worldwide/
6. Chen L, Englund C (2016) Cooperative intersection management: a survey. IEEE Trans Intell Transp Syst 17(2):570–586
7. Zaghal R, Thabatah K, Salah S (2017) Towards a smart intersection using traffic load balancing algorithm. In: Computing conference, London, UK
8. Shaikh PW, El-Abd M, Khanafer M, Gao K, A review on swarm intelligence and evolutionary algorithms for solving the traffic signal control problem. IEEE Trans Intell Transp Syst. https://doi.org/10.1109/TITS.2020.3014296.

9. Bing B, Carter A (1995) SCOOT: the world's foremost adaptive traffic control system. In: Traffic technology international'95; UK and International Press, Surrey, UK
10. Sims AG, Dobinson KW (1980) The Sydney coordinated adaptive traffic (SCAT) system philosophy and benefits. IEEE Trans Veh Technol 29:130–137
11. Gartner NH (1983) OPAC: a demand responsive strategy for traffic signal control. Transp Res Record 906:75–81
12. Mirchandani P, Head L (2001) A real-time traffic signal control system: architecture, algorithms, and analysis. Transp Res Part C Emerg Technol 9:415–432
13. Henry JJ, Farges JL, Tuffal J (1984) The Prodyn real time traffic algorithm. Control Transp Syst 16:305–310
14. Brilon W, Wietholt T (2013) Experiences with adaptive signal control in Germany. Transp Res Rec J Transp Res Board 2356:9–16
15. Intelligent traffic system cost. https://www.itskrs.its.dot.gov/costs/about, online. Accessed on 23 Nov 2017
16. Kouvelas A, Aboudolas K, Papageorgiou M (2011) A hybrid strategy for real-time traffic signal control of urban road networks. IEEE Trans Intell Transp Syst 12(3)
17. Zheng W, Lee D-H, Shi Q (2006) Short-term freeway traffic flow prediction: Bayesian combined neural network approach. J Transp Eng 132(2):114–121
18. Sharma KK, Indu S (2019) GPS based adaptive traffic light timings and lane scheduling. In: IEEE intelligent transportation systems conference (ITSC) Auckland, NZ, Oct 2019
19. Indu S, Nair V, Jain S, Chaudhury S (2013) Video based adaptive road traffic signaling. In: Seventh international conference on distributed smart cameras (ICDSC)
20. Lin Y (2014) The design and simulation of intelligent traffic signal control system based on fuzzy logic. Found Intell Syst 965âˆ’A ¸S973
21. Tanwar R, Majumdar R, Sidhu GS, Srivastava A (2016) Removing traffic congestion at traffic lights using GPS technology. In: 6th international conference—cloud system and big data engineering (Confluence)
22. Liu Y (2018) Big data technology and its analysis of application in urban intelligent transportation system. In: International conference on intelligent transportation, big data and smart city (ICITBS)
23. Coll P, Factorovich P, Loiseau I, Gómez R (2013) A linear programming approach for adaptive synchronization of traffic signals. Int Trans Oper Res 20(5):667–679
24. Liping X, Dangying L (2014) A novel embedded vehicle terminal for intelligent transportation. In: Fifth international conference on intelligent systems design and engineering applications
25. Li Q, Liu B (2010) Research on traffic light control system based on wireless sensor network. Sci Technol Assoc Forum 6:71–73
26. Tiedong W, Jingjing H (2014) Applying floating car data in traffic monitoring. In: 2014 IEEE international conference on control science and systems engineering
27. Dresner K, Stone P (2008) A multiagent approach to autonomous intersection management. J Artif Intell Res 31:591–653
28. Chaudhary S, Indu S, Chaudhury S (2018) Video-based road traffic monitoring and prediction using dynamic Bayesian networks. IET Intel Transport Syst 12(3):169–176
29. Hadi RA, Sulong G, George LE (2014) Vehicle detection and tracking techniques: a concise review. Signal Image Process Int J (SIPIJ) 5(1)
30. van Lint JWC, Hoogendoorn SP, van Zuylen HJ (2005) Accurate freeway travel time prediction with state-space neural networks under missing data. Transp Res C Emerg Technol, 347–369
31. Zhang B, Shang L, Chen D (2010) Traffic intersection signal-planning multi-object optimization based on genetic algorithm. In: Proceedings of 2nd international workshop intelligent system and applications, May 2010, pp 1–4
32. Dezani H, Marranghello N, Damiani F (2014) Genetic algorithm-based traffic lights timing optimization and routes definition using Petri net model of urban traffic flow. In: Proceedings of 19th World Congress of the International Federation of Automatic Control, pp 11326–11331
33. Dezani H, Bassi RDS, Marranghello N, Gomes L, Damiani F, Nunes da Silva I (2014) Optimizing urban traffic flow using genetic algorithm with Petri net analysis as fitness function. Neurocomputing 124:162–167

34. Sabar NR, Kieu LM, Chung E, Tsubota T, Maciel de Almeida PE (2017) A memetic algorithm for real world multi-intersection traffic signal optimisation problems. Eng Appl Artif Intell 63:45–53

35. Kou W, Chen X, Yu L, Gong H (2018) Multiobjective optimization model of intersection signal timing considering emissions based on field data: a case study of beijing. J Air Waste Manage Assoc 68(8):836–848

36. Li X, Sun J-Q (2018) Signal multiobjective optimization for urban traffic network. IEEE Trans Intell Transp Syst 19(11):3529–3537

37. Chin YK, Yong KC, Bolong N, Yang SS, Teo KTK (2011) Multiple intersections traffic signal timing optimization with genetic algorithm. In: Proceedings of IEEE international conference on control system, computing and engineering, Nov 2011, pp 454–459

38. Li X, Sun J-Q (2019) Multi-objective optimal predictive control of signals in urban traffic network. J Intell Transp Syst 23(4):370–388

39. Ma C, Liu P (2019) Intersection signal timing optimization considering the travel safety of the elderly. Adv Mech Eng 11(12):1–8

40. Dong C, Huang S, Liu X (2010) Urban area traffic signal timing optimization based on sa-PSO. In: Proceedings of international conference on artificial intelligent and computer intelligent, Oct 2010, pp 80–84

41. García-Nieto J, Alba E, Carolina Olivera A (2012) Swarm intelligence for traffic light scheduling: application to real urban areas. Eng Appl Artif Intell 25(2), 274–283

42. Yong Z, Peng W (2015) Optimal control for region of the city traffic signal based on selective particle swarm optimization algorithm. In: Proceedings of international conference on intelligent transportation, Big Data Smart City, Dec 2015, pp 544–550

43. Geda Pasek Suta Wijaya I, Uchimura K, Koutaki G (2015) Traffic light signal parameters optimization using particle swarm optimization, case study of OOE Toroku road network optimization. In: Proceedings of international seminar intelligent technology and its application, pp 11–16

44. Baskan O, Haldenbilen S (2011) Ant colony optimization approach for optimizing traffic signal timings. In: Proceedings of ant colony optimization methods application, pp 205–220

45. Renfrew D, Yu X-H (2012) Traffic signal optimization using ant colony algorithm. In: Proceedings of international joint conference on neural network (IJCNN), June 2012, pp 1–7

46. Dell'Orco M, Baskan O, Marinelli M (2013) A harmony search algorithm approach for optimizing traffic signal timings. Traffic Transp 25(4):349–358

47. Gao K, Zhang Y, Sadollah A, Su R (2016) Optimizing urban traffic light scheduling problem using harmony search with ensemble of local search. Appl Soft Comput 48:359–372

48. Yousef KM, Al-karaki JN, Shatnawi AM (2010) Intelligent traffic light flow control system using wireless sensors networks. J Inf Sci Eng 26(3):753–768

49. Zhang R, Schmutz F, Gerard K, Pomini A, Basseto L, Hassen SB, Ishikawa A, Ozgunes I, Tonguz OK (2018) Virtual traffic lights: system design and implementation. In: Vehicular technology conference (VTC-Spring), 2018 IEEE 87th IEEE

50. Tonguz OK (2018) Red light, green light—no light. IEEE Spectr Mag 55(10):24–29

51. Vdot Launches Smarter Roads. Accessed on 23 Nov 2017. [Online]. Available: http://www.virginiadot.org/newsroom/statewide/2017/vdot_launches_smarterroads118667.asp

52. Virginia Connected Corridors. Accessed on 23 Nov 2017. [Online]. Available: https://www.vtti.vt.edu/facilities/vcc.html

53. Tonguz OK, Zhang R (2020) Harnessing vehicular broadcast communications: DSRC-actuated traffic control. IEEE Trans Intell Transp Syst 21(2)

54. Milanes V, Perez J, Onieya E, Gonzalez C (2010) Controller for urban intersections based on wireless communications and fuzzy logic. IEEE Trans Intell Transp Syst 11(1):243–248

55. Bi Y, Lu X, Srinivasan D, Sun Z (2018) Optimal type-2 fuzzy system for arterial traffic signal control. IEEE Trans Intell Transp Syst, 1–19

56. Moghaddam MJ, Hosseini M, Safabakhsh R (2015) Traffic light control based on fuzzy Q-learning. In: 2015 the international symposium on artificial intelligence and signal processing (AISP) Mashhad, Iran, pp 124–128

57. Iyer V, Jadhav R, Mavchi U, Abraham J (2016) Intelligent traffic signal synchronization using fuzzy logic and Q-learning. In: 2016 international conference on computing, analytics and security trends (CAST), Pune, India, 2016, pp 156–161
58. Askerzada IN, Mahmood M (2010) Control the extension time of traffic light in single junction by using fuzzy logic. Int J Electr Comput Sci 10(2)
59. Protri S, Wuttidittachotti P, Thajchayapong S (2015) Traffic signal control using fuzzy logic. In: International conference on electrical engineering/electronics, computer, telecommunication and information technology (ECTI-CON), Hua Hin, Thailand, pp 1–6, Aug. 2015
60. Firdous M, Din Iqbal FU, Ghafoor N, Qureshi NK, Naseer N (2019) Traffic light control system for four-way intersection and T-crossing using fuzzy logic. In: 2019 IEEE international conference on artificial intelligence and computer applications (ICAICA), Dalian, China, pp 178–182
61. Hawi R, Okeyo G, Kimwele M (2017) Smart traffic light control using fuzzy logic and wireless sensor network. In: 2017 computing conference, London, UK, pp 450–460
62. Wei H, Zheng G, Gayah V, Li Z (2019) A survey on traffic signal control methods. arXiv preprint arXiv:1904.08117
63. Li L, Lv Y, Wang F (2016) Traffic signal timing via deep reinforcement learning. IEEE/CAA J Autom Sin 3(3)
64. Mousavi SS, Schukat M, Howley E (2017) Traffic light control using deep policy gradient and value-function-based reinforcement learning. IET Intell Transp Syst 11(7):417–423
65. Zhang R, Zhou X, Tonguz OK (2020) Using AI for mitigating the impact of network delay in cloud-based intelligent traffic signal control. Proc J. arXiv preprint arXiv:2002.08303
66. Zheng G, Xiong Y, Zang X, Fang J, Fie H, Zheng H, Li Y, Xu K, Li Z (2019) Learning phase competition for traffic signal control. In: CIKM '19: proceedings of the 28th ACM international conference on information and knowledge management, pp 1963–1972, Nov 2019
67. Yau K-LA, Qadir J, Khoo HL, Ling MH, Komisarczuk P (2017) A survey on reinforcement learning models and algorithms for traffic signal control. ACM Comput Surv 50(3):1–38
68. Wang X, Ke L, Qiao Z, Chai X (2020) Large-scale traffic signal control using a novel multiagent reinforcement learning. IEEE Trans Cybern, 1–14
69. Wei H, Zheng G, Gayah V, Li Z (2021) Recent advances in reinforcement learning for traffic signal control: a survey of models and evaluation. ACM SIGKDD Explor Newslett 22(2)
70. Joo H, Ahmed SH, Lim Y (2020) Traffic signal control for smart cities using reinforcement learning. Comput Commun 154:324–330
71. Zhang R, Leteurtre R, Striner B, Alanazi A, Alghafis A, Tonguz OK (2019) Partially detected intelligent traffic signal control: environmental adaptation. In: 2019 18th IEEE international conference on machine learning and applications (ICMLA), Boca Raton, FL, USA, 2019, pp 1956–1960
72. Zhang R, Ishikawa A, Wang W, Striner B, Tonguz OK (2021) Using reinforcement learning with partial vehicle detection for intelligent traffic signal control. IEEE Trans Intell Transp Syst 22(1):404–415
73. Liang XJ, Guler SI, Gayah VV (2020) An equitable traffic signal control scheme at isolated signalized intersections using connected vehicle technology. Transp Res Part C Emerg Technol 110(2020):81–97
74. Hu X, Lu J, Wang W, Zhirui Y (2015) Traffic signal synchronization in the saturated high-density grid road network. In: Computational intelligence and neuroscience, pp 1–11
75. Iyer V, Jadhav R, Mavchi U, Abraham J (2016) Intelligent traffic signal synchronization using fuzzy logic and Q-learning. In: 2016 international conference on computing, analytics and security trends (CAST), Pune, 2016, pp 156–161
76. Amogh AS, Pujari S, Gowda SN, Nyamati VM (2016) Traffic timer synchronization based on congestion. In: 2016 international conference on computation system and information technology for sustainable solutions (CSITSS), Bengaluru, India, 2016, pp 255–260
77. Khan A, Ullah F, Kaleem F, Rahman S, Cho Y (2017) Sensor-based self-organized traffic control at intersections. In: International conference on information and communication technology convergence (ICTC), Oct 2017

Study and Development of Self Sanitizing Smart Elevator

Satyan Gupta, Saniya Tyagi, and Kaushal Kishor

Abstract The year 2020 will always be remembered as the year of a global pandemic. With many places gradually emerging out of lock-downs, the thought of concerted existence amid COVID-19 is being revisited. One such problem is the use of elevators in any building, use of elevators in the times of COVID can be very risky in the transmission of virus. Based on this hypothesis, this paper presents a methodology for maintaining sanitization and security, by incorporating facial recognition in elevators. These elevators are designed to work using voice commands making them completely contactless, also this system includes a thermal sensor to measure the body temperature of every user and notify everyone around them. Using all these different sensors embedded in machine, the device can do much more than an ordinary elevator and can act as a modern smart device that is capable of solving all the problems single handedly. The key aim of this research paper is to demonstrate how various modules function together to accomplish a common goal: to render the experience of moving between floors smart and sanitary during COVID.

Keywords Automatic sanitation · Facial recognition · Voice command · Thermal scanning · Smart elevator · Innovation · Contactless · COVID-19 and security

1 Introduction

Elevators (also known as "lifts" in some places) are used in buildings to meet the inhabitants' vertical transportation needs. When it comes to a building's performance as an operational, working, or utility center, the total vertical transportability may be a critical factor. Elevators must be simple to use, open at all times, offer excellent support, and be dependable.

Any possible transmission while using elevators is a major problem and bringing people back to a safer and healthier working environment and keeping them confident and productive are part of a healthier building strategy. Adapting integrated security

S. Gupta (✉) · S. Tyagi · K. Kishor
ABES Institute of Technology, Ghaziabad, U.P, India

to supply monitoring, oversight, and control of critical health, safety, and security factors can assist you sustain compliance, manage risk. Maintaining sanitization in elevators is a task that isn't handled very well by the person in charge, while reviewing the research papers published over the last few years, we cumulated that we can provide a new and better idea that includes facial recognition being used in the field of elevators.

In order to make the travel even safer and completely contactless, feature of speech analysis is used to take input destination floor, thus avoiding any chance of possible contact. In any industry, having secured facilities is an issue with alarming concern. Using facial recognition systems in elevators offers secure environment and the ability to monitor the whereabouts of everyone using them. Face recognition is a crucial application of Image processing. Monitoring and recording the users can play a vital role in the security sector. The purpose of developing smart elevators is to computerize the traditional way of traveling among floors. The idea of smart elevator is to monitor people accessing the building by maintaining facial identity.

This proposed system not only discusses the building of a completely contactless elevator and increasing its security by including facial recognition, but also how is it sanitized automatically reducing the transmission of virus and creating healthier surroundings.

In the presented paper an elaborate explanation on the working of the device has been explained, use of different sensors to obtain a singular goal of automatic sanitization and smarter travel between floors is done. Thorough study of different research papers related to our own goal has been done and mentioned below. The paper presents methodology explaining how all the different modules of these devices work together and deliver the expected results mentioning all the advantages and changes that have been made in regard to the older methods being used.

2   Literature Review

In Kortli et al. [1] appeared in 2020, which deals with the various approach in facial recognition systems alongside the comparison between the technologies benefits and disadvantages in terms of strength, accuracy, and complexity. This paper's key feature is a description of the most widely used databases for facial recognition, as well as their possible directions.

This research is still at the emerging state. The similar concept was found, In Chudgar et al. [2] proposed facial recognition system which is able to categorize or validates an individual by a picture via video sources. Technologies like the Internet of things, IoT are used for problem solving.

Few years earlier, in Mallikarjuna Reddy [3] authored an overall paper especially focusing on facial recognition systems and its existence in real life. This study deals with the comparative study of many methods in this system.

In papers above [1–3], basics of our proposed system are mentioned. For next module of the proposed system, a compact study on infrared sensors on core body temperature monitoring is published also in the year 2020 in Chen et al. [4]. A similar investigation into the operation of infrared thermometers (IRT) and their importance during the COVID-19 era. For head temperature measurement became the most widely used practice to screen people for the illness. The efficiency of tympanic infrared thermometers and industrial infrared thermometers is contrasted in this report.

Incorporating temperature measurement techniques in elevators is a challenging issue of the whole system. In Krišto and Ivašić-Kos [5], alongside providing security to users, we found a combined study on better recognition of surveillance system growth. During this paper, a study of its strength and weaknesses is given. Also includes its application in real world.

Last year, a related idea was introduced. Hussain Mallhi and Khan collaborated on the construction of walk-through sanitization gates to sanitize citizens walking by and avoid potential transmission in [6]. However, clinical evidence of the efficacy of such walk-through gates in containing COVID-19 is lacking. Furthermore, these walk-through gates have the ability to cause public health problems. Because of the risk for eye and skin inflammation, bronchospasm during inhalation, and stomach symptoms including nausea and vomiting, spraying citizens with disinfectant chemicals are highly prohibited by several health officials across the world. The dangers associated with using such walk-through gates, according to this report, outweigh any possible benefits.

In Tavagad et al. [7] present a survey of smart security tracking systems based on Raspberry Pi. This study demonstrated the critical need for these systems in terms of stability. High-end functioning cameras are needed in commercial spaces, schools, and other indoor and outdoor environments.

A less expensive experiment by using a low-cost single Raspberry Pi for image processing. Similar concept is used in our proposed model for security and maintaining records.

Speech recognition (SR) is the conversion of spoken words into text, as defined by Trivedi in [8]. It's also known as "automatic speech recognition" (ASR), "machine speech recognition," or simply "speech to text" (STT). A secret Markova model (HMM) is a mathematical model in which the mechanism under consideration is a Markov method with hidden states. A simple dynamic Bayesian network may be used to describe an HMM. Dynamic time warping is a well-known technique for determining the best fit between two defined sequences under such constraints; the sequences are twisted in a non-linear manner to balance one another. The non-linear knowledge-driven self-adaptive method is known as ANN. It finds, establishes, and learns co-related trends between the input dataset and the target values. When training ANNs, the end effects of the most recent independent input file are normally predicted.

The voice recognition technology is the subject of Das et al. [9]. The speech recognition program helps the consumer to monitor machine functions and then dictates text by voice. This system consists of two parts: a microphone for recording our speech and a signal interpretation machine.

Finally, the signal is mapped into text, and the voice recognition device converts speech to text. Mel Frequency Cepstral Coefficients (MFCC) and vector quantization methods are used to recognize speech (voice).

Ahmad et al. used captured photographs as the basis for biometric research on facial recognition in [10]. Face-based authentication has a range of benefits over other biometrics. A human face is a complex entity with a lot of variation in any aspect, which makes it a difficult problem in computer vision. The performance of the system and its speed of operation are the key concerns in this case.

In this article, numerous face detection and recognition approaches are evaluated, with the goal of offering solutions that are more accurate and effective. It also addresses proposals for improving the response time of video surveillance as a first move.

In Cao et al. [11] did a collective study introducing a large-scale dataset. This dataset contains an average of hundreds of images for each subject. Images that are downloaded have large variations in attributes such as age, pose, profession, ethnicity, and illumination. The goals while collecting the dataset were to have large number of identities and the number of images corresponding each identity, pose, age and ethnicity. Also, we need to reduce the label noise for ensuring high accuracy for the images to each identity.

Now we train ResNet-50 Convolution Neural networks using the new dataset with and without Squeeze-and-Excitation blocks leading to improved recognition performance. Finally, we use the models that were trained using these datasets. All the datasets and models are available for public use.

Wang et al. investigated the causes and effects of mark noise in current datasets in [12].

We profile and evaluate mark noise properties using cleaned subsets of common face databases, as well as the original datasets and cleaned subsets. The analysis of the interaction between various forms of noise. We're looking for ways to clean up our records.

In Shi and Jain [13] emphasis are given on probabilistic face embeddings. The probabilistic face embedding methods have wide application in face recognition as through these methods we could describe facial features of a person and the person could be identified.

However when a blurred face is there as image then recognition of facial features learned by embedding methods is ambiguous.

In latent space, probabilistic face embedding may be thought of as a Gaussian distribution. As a result, the variation represents the volatility of the face value, while the mean of the distribution estimates the most likely characteristic of the face.

Zhang et al. collaborated on a dataset and large-scale multi-model face anti-spoofing analysis in [14]. Face anti-spoofing aids in deciding if a face captured by a face recognition device is genuine or not. PAD [face presentation attack detection] is used to guarantee the protection of the face recognition device. Because of the help of face anti-spoofing data sets, the face PAD algorithm has received a lot of attention. Both current databases, for example, RGB, have a small number of subjects and just one model.

As a result, a wide scale multi-modal face anti-spoofing dataset called CASJA —SURF with 1000 subjects and 21,000 video dips with three modalities was introduced for the advancement of this topic of face anti-spoofing (RGB, Depth, IR). For each modality, a multi-model fusion approach was used to integrate (merge) the three modalities and conduct modal based function re-weighting to pick the more informative and minimize the less efficient. This implemented dataset was subjected to extensive testing in order to confirm its significance.

In Bala et al. [15] presented a paper in which they used the Mel Frequency Cepstral Coefficient (MFCC) and Dynamic Time Warping (DTW) function matching to construct a voice command recognition method. Our speech is a signal that contains an endless amount of data. For high-speed and precise automated voice recognition, digital processing of speech signals is needed. The details must be processed and evaluated in order to derive valuable knowledge from the speech signal.

For the transfer of digital signals, The latest problems for researching voice signals are function extraction and feature matching.

Marchini et al. contrasted the precision of an infrared thermometer to a normal optical thermometer while calculating kidney temperature during arterial clamping with and without renal cooling in [16].

Since clamping the main artery with the IRT and a digital thermometer, the renal temperature was measured at 0, 2, 5, and 10 min. A total of 20 pigs were used in the experiment. Ten pigs with frost, ten pigs without ice.

Guangli Long proposes a non-contact infrared thermometer for fast body temperature calculation in [17]. The primary function of an infrared body temperature measurement sensor is to transform infrared from the human body into a voltage signal, which is then amplified using an operating amplification. The MCU processes details, and the LCD show and voice record body temperature and time.

Keerthi Nayani et al. worked on enhancing image resolution and de-noising using auto-encoder in [18]. This leads to improvement in quality of image and better classification of results.

In Agrawal and Gupta [19] published a research paper for the analysis of COVID-19 data using machine learning techniques to better understand the care and measures required for this pandemic.

Rani and Solanki [20] mention the loss of some sequence of data while transmitting through wireless sensor network which is a prevalent issue. This paper examines various techniques available in deep learning.

3   Feature Extraction

3.1   Hardware Requirements for Different Processes

- Raspberry Pi
- Infrared Thermal Sensor

- Proximity Sensor
- Motion Sensor
- Camera Lens
- Microphone and Speaker
- ESP Chip.

3.2 Methodology

Facial Recognition System. Face recognition software uses pattern recognition methods to better identify people in photos and images. The HOG is a feature descriptor that takes a photo and transforms it into feature vectors for use in computer vision and image processing to detect objects. This approach counts the number of times a gradient orientation appears in different sections of an image. The first move is to double-check the color and gamma values that have been normalized. The cell histograms are created in the second phase. Based on the values contained in the gradient computation, each pixel inside the cell casts a weighted vote for an orientation-based histogram channel. The gradient strengths should be domestically normalized to account for variations in lighting and contrast, which necessitates grouping the cells into wider, spatially linked blocks. These blocks also overlap, implying that each cell relates to the final descriptor in some way.

Encode_Faces.py

1. Parse the following arguments by constructing an argument parser

    - dataset
    - encodings
    - detection-method.

2. In our dataset, get the paths to the input images.
3. Create a catalog of all recognized encodings and names.
4. Iterate across the picture directions.
4.1. Extract the person name from the image path
4.2. Load the input picture and transform it to dlib ordering from RGB (OpenCV ordering) (RGB)
4.3. Identify the (x, y)-coordinates of the bounding boxes in the input picture that refer to each face.
4.4. Facial embedding for the face are computed
4.5. Loop over all the encodings
4.5.1. Add each encoding + name to the set of known usernames and encodings
5. Dump all the facial encodings + usernames to disk.

Pi_Face_Recognition.py

1. Parse the following arguments by constructing an argument parser
   - cascade
   - encodings

2. Load embeddings and the known faces fort with OpenCV's Haarcascade for face detection
3. Start the video stream which allows the camera sensor to get ready
4. Initialize the FPS counter
5. Iterate over all frames from the video file stream
5.1. To speed up loading, remove the frame from the video stream and scale it to 500px.
5.2. The input frame is converted from (1) BGR to grayscale for face detection and (2) BGR to RGB for face recognition.
5.3. Faces are detected in the greyscale frame
5.4. We needed an extra bit for reordering because OpenCV returns bounding box coordinates in (x, y, w, h) order, but we needed them in (top, right, bottom, left) order.
5.5. For each face bounding box, compute the facial embeddings.
5.6. Loop over the facial embeddings
5.6.1. Map each face in the input picture to our established encodings as best you can.
5.6.2. Check to see if we have found a match
5.6.2.1. Find the indexes of all paired faces, then build a dictionary to count how many times each face has been matched.
5.6.2.2. Maintain a count for each recognized face by looping through the matched indexes.
5.6.2.3. The most famous recognizable face is determined by the amount of votes received.
5.6.3. Update the list of names
5.7. Loop over the faces you meet.
5.7.1. On the picture, draw the projected face name.
5.8. Display the picture on our computer screen.
5.9. Break from the loop if the q key was pushed.
5.10. Update the FPS counter
6. Stop the timer to see how many frames per second you're having.

Voice Command System. Speech analysis is a method of analyzing the speech signal in order to obtain relevant data of the signal in a very additional and compact form than the speech signal itself. This system allows us to replace the traditional way of entering floor numbers with the feature of speech analysis, thus making the elevator a germ free area. For the purpose of speech to text conversion, we are using Google Speech API; it converts the speech recorded by the mic into text. The converted text is then compared by the commands already stored in the system, if the commands are matched with recorded text then the command gets executed otherwise the system will ask the user to give a correct command.

Voice_Command.py

1. Record audio
2. Speech recognition using Google Speech Recognition

    2.1   Compare the value stored in 'r.recognize_google (audio)' with different test cases using dictionary mapping

    2.1.1   If any test case matches with the 'r.recognize_google (audio)' then enter that floor number in the elevator queue

    2.1.2   Else ask the user to enter a valid command.

Body Temperature Measurement. Speech analysis is a method of analyzing the speech signal in order to obtain relevant data of the signal in a very additional and compact form than the speech signal itself. This system allows us to replace the traditional way of entering floor numbers with the feature of speech analysis, thus making the elevator a germ free area. For the purpose of speech to text conversion, we are using Google Speech API; it converts the speech recorded by the mic into text. The converted text is then compared by the commands already stored in the system, if the commands are matched with recorded text then the command gets executed otherwise the system will ask the user to give a correct command.

3.3   Our Contribution on the Subject

The idea of managing these technologies simultaneously and the area of application, i.e., ELEVATORS turns out to be the most efficient and possible way to eliminate any chance of transmission amid COVID-19. Facial recognition alone is a very new and innovative idea in the field of engineering. With thermal scanning attached to it, could not only be a prominent step towards building smart elevators but also enhances security which is an alarming concern. In order to reduce the transmission the most important task is to reduce any contact, so we have used the technique of voice command allowing the user to input floor number just using speech. The impact of our approach that addresses the limitations is to extend the work above to its maximum capacity under different circumstances.

MODULE 1: Facial recognition

In [1–3] the papers provide a thorough discussion of possible directions in terms of facial recognition techniques. This approach provides a vital answer to our issue statement in our article.

We read more about comparative analysis of different methods of Face Recognition Systems from [10, 11]. Papers [12, 18] helps in understanding the importance of noise cancelation and its importance in the improvement of the image quality.

MODULE 2: Temperature measurement and Sanitization

The crisis of pandemic demands temperature measurement to screen people for the illness. From paper [13, 14] apart from the various techniques to perform this process, we include this alongside our face recognition system to notify the area if anyone is having more than suggested temperature. Paper [6] described the use of walk-through sanitization and use of IRT sensors to measure temperature is well informed in [16, 17].

With the help of different papers, we include all of them collectively in our system, so sanitizes the area before any transportation in between the floors takes place, ensuring safe and sanitized travel.

MODULE 3: Voice Command

The process of converting speech into text for voice command system in [8] gives a glimpse of idea for replacing traditional method of button system in elevators. With this technique and technology in [9], helps the user to communicate through voice with the help of a microphone.

Collectively this module could eliminate the chances of transmitting the virus.

MODULE 4: Security

The paper [7] examines a smart security tracking device centered on the Raspberry Pi. In today's world, video monitoring is crucial in terms of protection. Exclusive cameras are needed in commercial facilities, classrooms, universities, clinics, and other demanding indoor and outdoor environments. This paper discusses how to use the Raspberry Pi, a low-cost single-board machine.

This initiative has a greater effect since it uses a less costly monitoring device. Keeping track of different users improves stability, which is just as important as the other features.

4 Proposed System Overview

The platform on which this device is going to work on is Raspberry Pi. All the features of this system and the flow of data from one process to another have been shown in the following flowchart.

The first step in the system would be to call the elevator using the input from the proximity sensors installed on every floor. After elevator is called and stops on the floor to open it automatically sanitizes and makes sure to eliminate and possibilities of virus.

As the user enters inside the lift the main functioning of machine begins, it runs three different processes simultaneously which include—voice command, facial recognition and measuring the body temperature of the users.

When any user the elevator and enters the floor number on which they intend to leave using the voice command, the system also runs a facial recognition program to check the identity of the user alongside measuring their body temperature and stores in the system and different details such as floor number they entered and exited, time of usage, facial identity, and their body temperature.

If the body temperature of any user is higher than 99 °F then all the users are alerted using a beeping noise (Fig. 1).

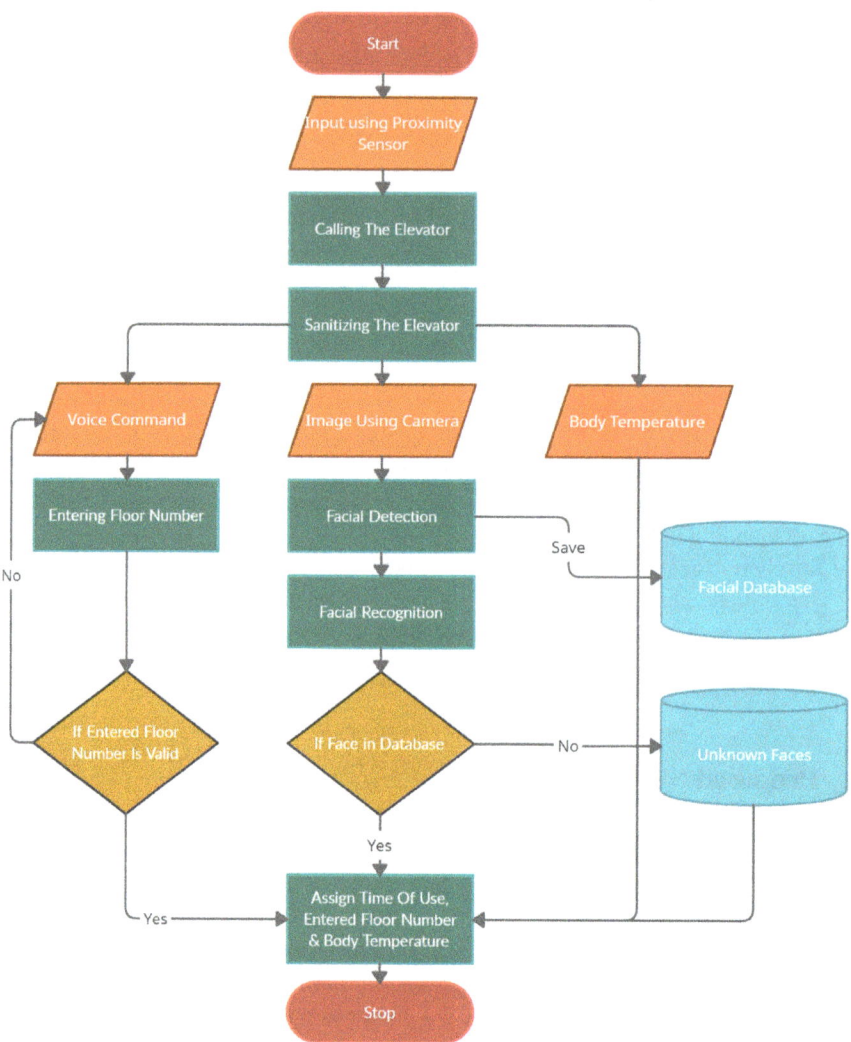

Fig. 1 Dataflow diagram of the proposed system

4.1 Calling the Elevator

The first step is to call the elevator. A proximity detector is ready to sight the presence of near objects with no physical contact. These sensors are installed on every floor. The user just needs to wave their hand in proximity of 1 inch and the system uses that input to call the elevator.

4.2 Sanitizing the Elevator

Nozzles implanted inside the elevator automatically start sanitizing before any use. To activate the process of sanitization a motion sensor is used. The sensitivity is adjusted accordingly so no wastage of sanitizer happens.

4.3 Entering the Floor Number

This process requires a machine that runs on voice command and is also attached from a microphone in order to take input from the user, making it completely contactless and hassle free.

4.4 Facial Recognition and Body Temperature Measurement

Now for the most important part of this product, we run a facial recognition algorithm on every user and record their body temperature parallelly using an infrared thermal sensor. If in any case the user is not identified by the device due to insufficient parameters then only the temperature is recorded with their picture and is stored. d.

4.5 Notifying the User

If the system identifies any user with more than normal body temperature, it makes an alerting sound, and data is stored accordingly.

5   Results

This section includes depiction of the expected output of the system that is being extracted from the continuous video stream and the thermal sensor installed. As you can see in Fig. 2 identity of the user is recognized using the facial recognition algorithm and also the body temperature gets recorded.

All the attributes related to any user are going to be stored in the following manner so the security of the parameter in which the elevator is placed could be improved and a proper record of everyone could be maintained (Fig. 3).

Fig. 2 Facial recognition system along with body temperature measurement

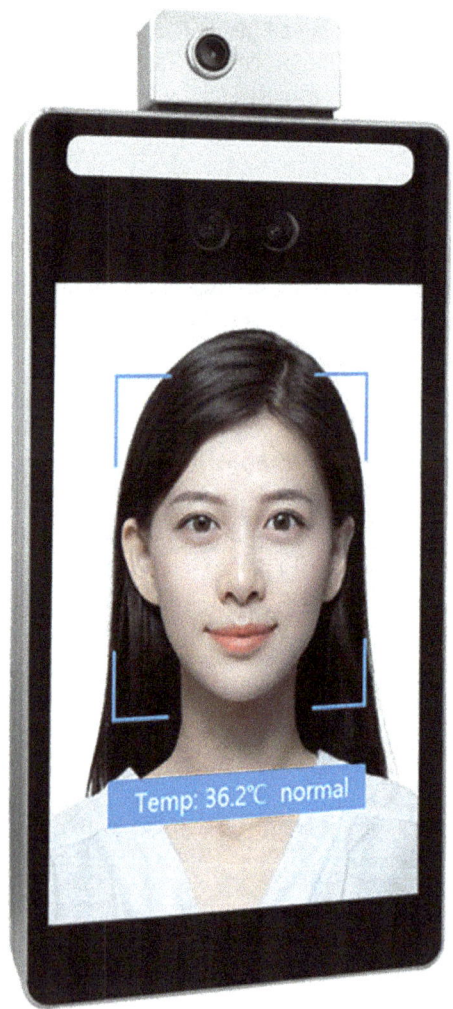

Name	Temperature	Time	Entry	Exit
Saniya Tyagi	98.5 °F	4:20 pm	Floor 0	Floor 4
Rahul Sharma	98.6 °F	5:43 pm	Floor 1	Floor 5
Satyan Gupta	99.2 °F	6:09 pm	Floor 6	Floor 0
Sanchi Mahajan	98.6 °F	7:16 pm	Floor 5	Floor 3

Fig. 3 Database managed to keep entry of every username

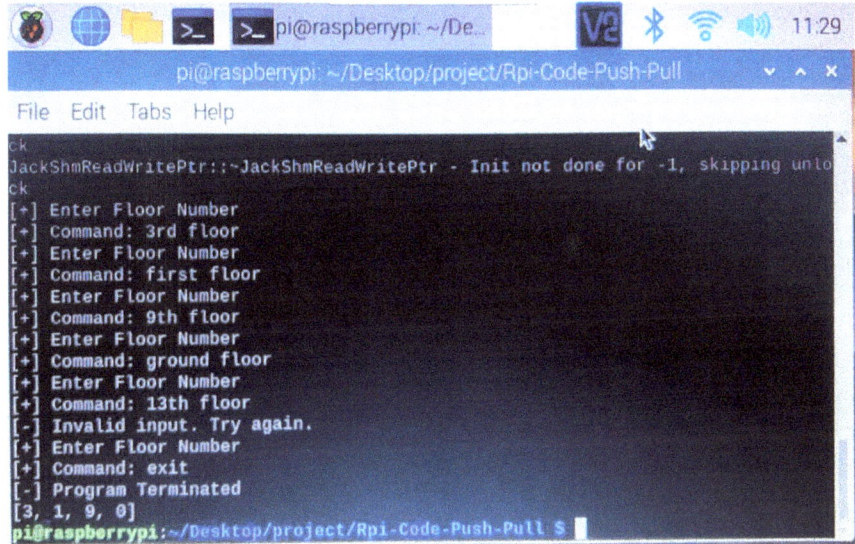

Fig. 4 Entering floor numbers using voice command

In the end, Fig. 4 display input is being taken by the voice command system which works on google speech recognition API. Whenever user enters an invalid command the system gives output as "invalid command". For example in this case we are constructing the functioning of elevator for a building of ten floors, therefore when a user tries to input floor number 13, the system shows it as an invalid command.

6 Conclusion

This system is a leap forward taken in the modernization of vertical transportation among floors. Through this paper, we try to propose our idea of smart elevation by means of facial recognition and we also added proper safety measures as our primary feature for a smooth and hassle free travel even in the times of COVID.

Facial recognition can be used for a large number of applications, from security to advertisements. This system when indulged in elevators can transform into a smart one while recording temperature at the same time. This system is able to recognize an elevator passenger's face in a line of vision of a camera in close proximity to the elevator. A lot of people use elevators in public places, therefore using this system offers a sanitized environment and the ability to monitor the whereabouts of everyone using them.

The various advantages of the implemented method or system are mentioned below:

- Can be used as an effective machine in the time of pandemic.
- Improved level of Security and Sanitization.
- This method is practically applicable.

References

1. Kortli Y, Jridi M, Falou AA, Atri M (2020) Face recognition systems. AI-ED Department, Yncrea Ouest, 20 rue du Cuirassé de Bretagne, 29200 Brest, France. MDPI, 7 Jan 2020
2. Chudgar S, Fernandes J, Poduval A (2019) Smart elevator using facial recognition. Vishwakarma Institute of Technology, Pune, Maharashtra. Int J Adv Res 5(6)
3. Mallikarjuna Reddy A, Ravi Kishore M, Sreenivasulu P, Jyothi V (2018) A survey paper for face recognition system, Assistant Professor, Dept. of CSE, Anurag Group of Institutions, Hyderabad, T.S., India. Int J Eng Res Comput Sci Eng (IJERCSE) 5(4)
4. Chen H-Y, Chen A, Chen C (2020) Investigation of the impact of infrared sensors on core body temperature monitoring by comparing measurement sites. Department of Materials Science and Engineering, University of California, San Diego, CA 92093, USA. Published 19 May 2020
5. Krišto M, Ivašić-Kos M (2018) An overview of thermal face recognition methods. Department of Informatics, University of Rijeka, Rijeka, Croatia Published, 1 May 2018
6. Hussain Mallhi T, Khan YH (2020) Walkthrough sanitization gates for COVID -19" University Sains Malaysia
7. Tavagad S, Bhosale S, Singh AP, Kumar D (2016) Survey paper on smart surveillance system. Computer Department, Savitribai Phule Pune University, Maharashtra, India, vol 03(02)
8. Trivedi PA, Introduction to various algorithms of speech recognition: hidden Markov model, dynamic time warping and artificial neural networks. V.V.P. Engineering College Rajkot, Gujarat, India, vol 2(4)
9. Das P, Acharjee K, Das P, Prasad V (2015) Voice recognition system: speech-to-text, vol 1(2)
10. Ahmad F, Najam A, Ahmed Z, Image-based face recognition and detection 'State of he Art'" IJCSI paper
11. Cao Q, Shn L, Xie W, Parkhi OM, Zisserman A (2018) VGGFace2: a dataset for recognizing faces across pose and age, vol 2. IEEE
12. Wang F, Chen L, Li C, Huang S, Yanjie C, Qian C, Loy CC (2018) The devil of face recognition is the noise. arXiv:1807.11649v1 [cs.CV]
13. Shi Y, Jain AK (2019) Probabilistic face embeddings. arXiv:1904.09658v4 [cs.CV]
14. Zhang S, Wang X, Liu A, Zhao C, Wan J, Escalera S, Shi H, Wang Z, Lee SZ (2019) A dataset and benchmark for large-scale multi-modal face anti-spoofing. arXiv:1812.00408v3 [cs.CV]
15. Bala A, Kumar A, Birla N (2010) Voice command recognition system based On MFCC and DTW. Int J Eng Sci Technol 2(12):7335–7342
16. Marchini GS, Duarte RJ, Mitre AI, Tiseo BC, Cassão VD, Torricelli FCM, Arap MA, Srougi M (2013) Infrared Thermometer: an accurate tool for temperature measurement during renal surgery. Int Braz J 39(4) Rio de Janeiro July/Aug. 2013

17. Long G (2016) Design of non-contact infrared thermometer. Int J Smart Sens Intell Syst 9(2)
18. Keerthi Nayani AS, Sekhar C, Srinivasa Rao M, Venkata Rao K (2021) Enhancing image resolution and denoising using autoencoder. In: Khanna A, Gupta D, Pólkowski Z, Bhattacharyya S, Castillo O (eds) Data analytics and management. Lecture notes on data engineering and communications technologies, vol 54. Springer, Singapore. https://doi.org/10.1007/978-981-15-8335-3_50
19. Agrawal R, Gupta N (2021) Analysis of COVID-19 data using machine learning techniques. In: Khanna A, Gupta D, Pólkowski Z, Bhattacharyya S, Castillo O (eds) Data analytics and management. Lecture notes on data engineering and communications technologies, vol 54. Springer, Singapore. https://doi.org/10.1007/978-981-15-8335-3_45
20. Rani S, Solanki A (2021) Data imputation in wireless sensor network using deep learning techniques. In: Khanna A, Gupta D, Pólkowski Z, Bhattacharyya S, Castillo O (eds) Data analytics and management. Lecture notes on data engineering and communications technologies, vol 54. Springer, Singapore. https://doi.org/10.1007/978-981-15-8335-3_44

Predict COVID-19 with Chest X-ray

Ankit Sharma, Nikhil Jha, and Kaushal Kishor

Abstract Covid-19 is an increasingly growing infective virus which really infects humans that interacted with it. Whilst these clinicians have mostly been infected with such a respiratory tract disease whenever they come in contact with both the disease, it was revealed in a clinical trial of COVID-19 treated persons that they had been mostly diagnosed with a respiratory tract infection when they made contact with the disease. A chest x-ray (also recognised as radiography) is a somewhat complicated imaging technique for detecting concerns in the respiratory system. Artificial intelligence is the most widely used accomplished machine learning algorithm for examining a substantial array of chest x-ray images, but it has the capacity to have a significant impact on Covid-19 testing. In this study, we have used PA interpret of x-rays tests both for covid-19 patients and safe individuals. We tested CNN templates and deep learning strategies. To review ResNeXt models and examine their performance in order to ascertain the presentation, 6432 chest x-ray scan specimens were taken from of the Kaggle database. To make no health claims, these researches primarily focus on potential alternative treatments for cluster covid-19 infected patients.

Keywords Deep learning · Radiography images · COVID-19 classification · Convolution neural network

1 Introduction

Covid-19 may be a serious illness that takes the lives of a huge amount of citizens every day. This outbreak has affected not only one nation, but also the whole planet as a consequence of this virus disease. The whole planet is already riddled with Covid-19 disease [1], and the biggest explanation for this is that no single country's scientists have been able to produce a lasting vaccine. Safe people and Covid-19

A. Sharma · N. Jha (✉) · K. Kishor
ABES Institute of Technology, Ghaziabad, UP, India

K. Kishor
e-mail: kaushal.kishor@abesit.in

D. Gupta et al. (eds.), *Proceedings of Data Analytics and Management*,
Lecture Notes on Data Engineering and Communications Technologies 90,
https://doi.org/10.1007/978-981-16-6289-8_16

impacted people's X-ray photographs were made open for analysis on the internet in March 2020 in a number of Collections include Github and Kaggle. Covid-19 outbreak illness that has turned into an outbreak which presents a worldwide danger to humanity [2]. The novel corona virus disease first manifested itself as strep throat, and people began to have breathing difficulties. This virus uses a sequence method [3] to spread from one human to another as they come into contact with covid-19 infected people. Hospital employees, clinicians, physicians and health laboratories also perform a critical part in the epidemic's identification. Many methods have been used to reduce the symptoms of the corona virus. The results of covid-19 on the anatomy can also be studied and predicted using medical imaging. Fair individuals and Covid-19 clinicians are often studied together in this study using a chest X-ray picture [4, 5]. We typically compile X-ray images of stable and Covid-19 affected patients from multiple sources and send three separate versions to arrive at a Covid-19 description (InceptionV3, Xception and ResNeXt) [6]. CNN, a machine learning application, is used to evaluate the data that has been obtained. This study looks at how CNN models should be used to identify chest X-ray images of patients that have the corona virus. We tried to make a contrast to previous task templates by including the sector and appearance for potential task templates, which could be tested further to demonstrate continuity with practical scenarios. In most system vision and medical image analysis tasks, deep learning-based models (particularly convolution neural networks (CNN)) have been shown to outperform conventional AI approaches in recent years [7–9].

Inception, xception and resnet are often used for image detection. Inception: It's just a convolutionary network design that aids in image recognition and object identification, and it began as a Googlenet module. It is the third instalment of Google's Innovation Convolutional Neural Network, which was first unveiled mostly during ImageNet Evaluate Information.

Francois Chollet suggested the Xception Model. Xception is an Inception Architecture extension that substitutes the regular convolution layers with depth wise separable convolutions.

ResNet, shortened for Residual Networks, is a traditional neural network that serves as the foundation for many computer vision applications. In 2015, this concept captured the ImageNet contest. The fundamental advancement with ResNet was that it allowed us to effectively train exceptionally Deep 150-layer genetic algorithms. Trained very deep learning models were problematic prior to ResNet due to the issue of disappearing gradients.

This article, which was further divided into parts, discussed the perspectives of various scholars on the effect of the covid-19 illness on countries and people. In the data set and model formulation, different matrices and algorithms were used. The following move is to examine the findings in terms of model planning and simulation for the ambiguity matrices that would be used.

2 Literature Review

We found and read a few literature reviews posts, which are mentioned below. Ses articles have common implementations and proposals, but they also have their own collection of limitations. The following articles were sent to us in order for us to draw certain assumptions and describe their shortcomings in comparison with our approach:

Afshar et al. [10] the authors proposed a structure model allowing Capsule Networks to use X-ray images to diagnose Covid-19 (i.e. COVID-CAAPS) illness. This study utilises a combination of layers and capsules to overcome class imbalance. They found that whilst smaller types of trainable parameters are used, the projected model has 95.7% accuracy, 90% sensitivity and 95.80% accuracy as a consequence of the experimental study.

In Alqudah et al. [11], the author developed a hybrid approach to allow computation using convolutional neural networks algorithms (i.e. Convolution Neural Network (CNN)). Criminal inquiries expressly follow the proposed scheme. Any Covid-19 instances will use chest X-ray scans.

Hassanien et al. [12] the authors suggested utilising a deep learning-based algorithm to identify Covid-19-infected patients using X-ray images (with a vector interface classifier). Hospital doctors can use this procedure to examine cases of covid-19 infection early on. Using different matrices criteria, they achieve a precision of 97.48% for the proposed model for respiratory organ classification.

The writers Ilyas et al. [13] spoke about the different strategies for detecting covid-19 disease and sweet-faced obstacles. They also mentioned that an automated system for analysing the Covid-19 virus should be developed in the near future in order to prevent the disease from spreading by interaction. They have looked at some chest X-rays to see whether they could detect respiratory disorder, although they come to the conclusion that it's impossible to determine if Covid-19 affects respiratory disease or whether other signs are to blame.

Ozkaya et al. [14] the authors proposed a model that uses chest X-ray images to mechanically diagnose the Covid-19. On two entirely separate classification models, the intended one is used to express right medicine (i.e. binary and multi-class). They define the duration of the target detection process utilising the DarkNet model.

3 Proposed System Overview

The provided model depicts how the proposed framework will operate, as well as all of the measures we'll take to put it in place, and how we'll use it to forecast data based on its results.

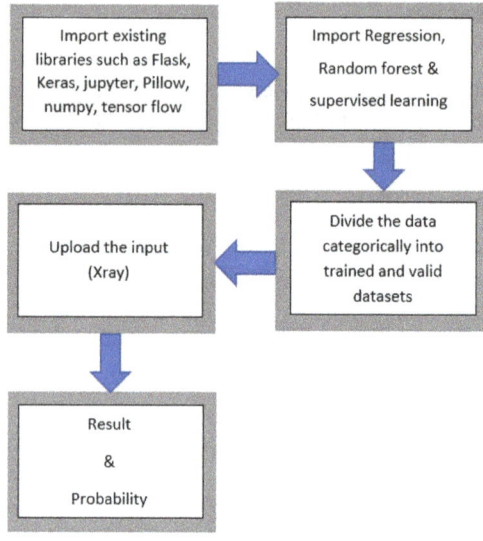

4 Feature Extraction

4.1 Methodology

4.1.1 Dataset

We obtained the data from the Kaggle archive [15], which includes regular and impacted Covid-19 chest X-ray scans. The cumulative number of chest X-ray photographs in the dataset is 6432. A normal, covid and pneumonia preparation (i.e. 5467) and validation (i.e. 965) package is included in this data collection. Norm is 1345, covid is 490 and pneumonia is 3632 in the training kit. 238 Regular case tests, 86 covid specimens and 641 pneumonia measurements were collected during the validation process. The scans were scaled down to 128,128 to support our model's quick coaching (Table 1).

Table 1 Demonstrates the distribution of data for preparation and research

Patient type	Train	Test
Healthy person	1345	238
Covid-19 infected person	490	86
Pneumonia infected person	3632	641

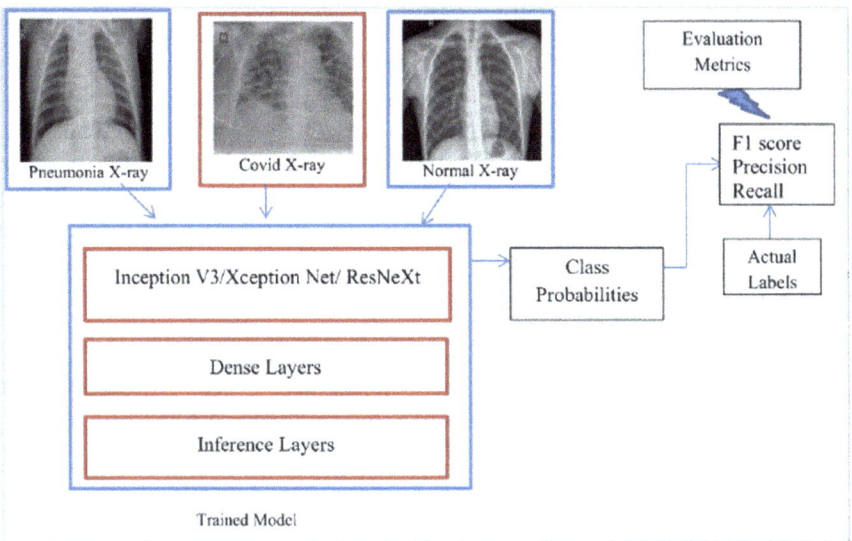

Fig. 1 The architecture of CNN model

4.2 Model Formulation

The dataset's aim is to avoid overfitting [15]. Amongst the improvements were rotation, zooming and picture sharing. To support the model [16] generalise and prevent overfitting, the data was shuffled. The configured datasets were used to generate the results. In a more thorough study, three different models were used, and their output evaluated to assess accuracy. Inside the listed models, we recommend using LeakyReLU induction instead of previously used reactivation, rendering it a special technique. This approach shortens the learning loop, avoiding neuron dying (i.e. Relu membranes inactivate due to nil declines). The suggested model for collecting chest x-ray images as seen in Fig. 1.

4.3 Inception Net V3

Original conception internet V3 is a grouping CNN-based system. Uses starting components with a concatenated layer of 1 * 13 * 3 and 5 * 5 convolutions and has 48 layers. As a result, we will reduce the number of standards and increase the coaching rate [17]. It is also recognised as the GoogLeNet style.

4.4   XCeption Net

It is a reworking of the first internet. The origination modules have been substituted with depth wise divisible convolutions in this model. Its parameter size is comparable to that of the origination internet [18], but it performs marginally better than the origination internet.

4.5   ResNeXt

RESNeXt is a deep residual network associate degree extension architecture. The consistency remaining blocks are substituted in this model with one that uses the break—remodel—merge technique used in the Inception models [19].

Algorithm

Stage 1: Ready the file, e.g. image $= X$.
   (For this goal, we used Keras data generator: to reshape the picture (X) (128, 128, 3) 10° random rotation spectrum True Zoom Range $= 0.4$ Horizontal Flip.
   Stage 2: Add the image to the model's first input.
   Stage 3: Get the performance of the model's last convolution sheet.
   Stage 4: Reduce n size to $n - 1$ to flatten size.
   Stage 5: For Inception Net, use a dense layer unit of 256 and for ResNeXt, use a dense layer unit of 128.

$$Z = W(A + bZ) = W(A + b) = W(A + b)$$
$$= W(A + b) = W(A + b) = W(A + b)$$

Stage 6: Apply $\dfrac{A = \text{Leaky ReLU}(Z) A = \text{Leaky ReLU}(Z) A = \text{Leaky ReLU}(Z)}{A = \text{Leaky ReLU}(Z) A = \text{Leaky ReLU}(Z)}$.

Stage 7: For abstract thinking, adds a Thick Layer.

$$Z = W(A + bZ) = W(A + b) = W(A + b) = W(A + b)$$
$$= W(A + b) = W(A + b)$$

Stage 8: Use Softmax to classify your results.

$$\text{Softmax}(Z_i) = \left(e^{zi}\right) / \sum_{j=1}^{k} e^{zj}$$

5 Experimental Result

Matrices used for result evaluation

As seen in the below equations, the proposed model was tested using a range of parameters, including accuracy, recall, $F1$ score [19], quality, sensitivity [20] and specificity.

$$Precision = \frac{True\ Positive\ True}{True\ Positive\ False + Positive\ Precision}$$

$$Recall(orSensitivity) = \frac{True\ Positive}{True\ Positive + False\ Negative}$$

$$F1\ Score = 2 * \frac{Accuracy * Recall}{Accuracy + Recall}$$

$$Accuracy = \frac{True\ Positive + True\ Negative}{True\ Positive + False\ Negative + True\ Negative + False\ Positive\ Accuracy}$$

$$Specificity = \frac{True\ Negative}{True\ Negative + False\ Positive}$$

Xception net

It is a tweaked version of the first internet. The origin modules were substituted with depth wise dissociable convolutions in this model. It has a parameter scale that is equivalent to the original internet, but it works marginally better than the original internet (Table 2).

The uncertainty matrix is seen in Fig. 2 from the Xception model's training and testing results.

Table 2 $f1$-score for the XCeption net model's training and training dataset

Patient type	Train		Test
Healthy person	1345		238
Covid-19 infected person	490		86
Pneumonia infected person	3632		641
Lable	**Precision**	**Recall**	**F1-score**
Normal	0.98	0.93	0.95
Covid-19	0.99	0.92	0.95
Pneumonia	0.97	0.99	0.98
Accuracy			0.97

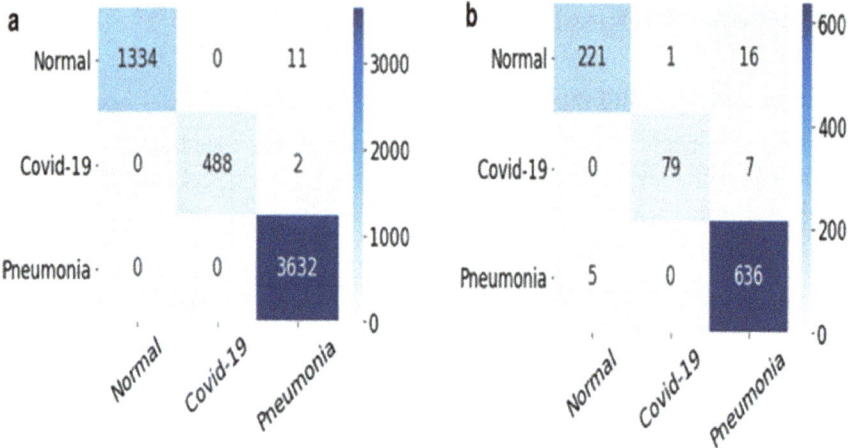

Fig. 2 **a** The Xception model's train data confusion chart. **b** The Xception model's test data confusion matrix

Inception net V3

It is a CNN classification network that's cutting-edge. It uses starting modules and has 48 layers, with a concatenated layer comprising 1 * 13 * 3 and 5 * 5 convulsions. As a result, the amount of criteria will be reduced, and the training speed will be improved. GoogLeNet style is another name for it (Table 3).

The uncertainty matrix is seen in Fig. 3 from the Inception V3 model's training and testing results.

ResNeXt

It's a deep network expansion for nursing associates. Residual high-quality blocks are substituted by those using a split-rework-merge technique in this model. Models of Inception allow using this technique (Table 4).

The uncertainty matrix is seen in Fig. 4 from the ResNeXt Training and checking outcomes for Model.

Typical chest X-ray scans were connected to Covid-19 tainted research participants in the study. Original conception Net V3, Hopfield Net and ResNeXt accuracy matrices are studied. After that the results were contrasted to evaluate the simplest

Table 3 $f1$-score for the inception V3 model on the training and testing dataset

Lable	Precision	Recall	$F1$-score
Normal	0.98	0.93	0.95
Covid-19	0.99	0.92	0.95
Pneumonia	0.97	0.99	0.98
Accuracy			0.97

Fig. 3 **a** Confusion matrix of Inception V3 train info. **b** Confusion matrix of Inception V3 research data

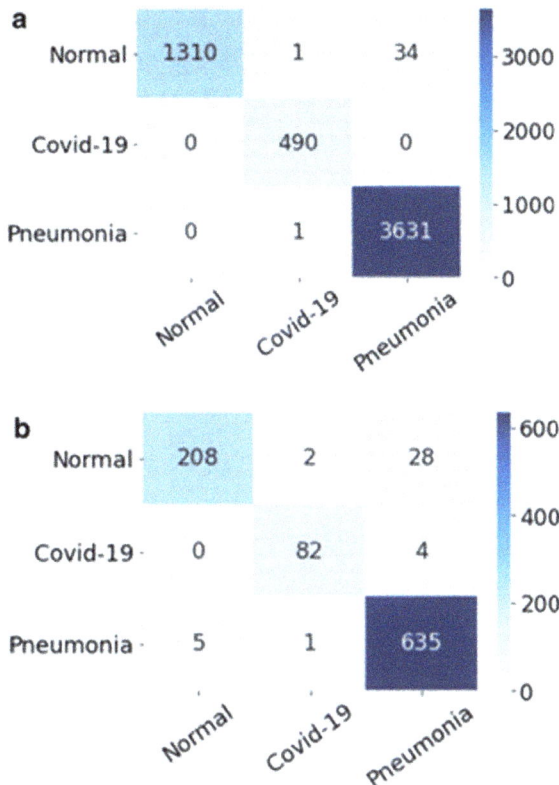

Table 4 f1-scores for the ResNeXt Model's data source focussing on implementing

Lable	Precision	Recall	F1-score
Normal	0.99	0.98	0.98
Covid-19	1.00	0.90	0.95
Pneumonia	0.98	1.00	0.99
Accuracy			0.98
Normal	0.91	0.89	0.90
Covid-19	0.97	0.78	0.86
Pneumonia	0.97	0.97	0.95
Accuracy			0.93

model. We advocate validating the execution for future dataset improvements because the model's predictions are extremely reliable. The model is only trained on 1560 samples due to a lack of evidence.

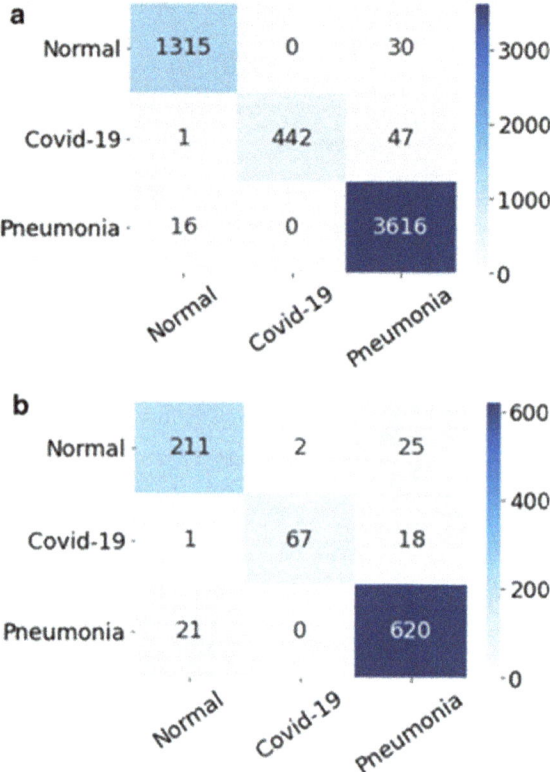

Fig. 4 a The ResNeXt model's train data confusion matrix. **b** Confusion matrix of ResNcXt model test results

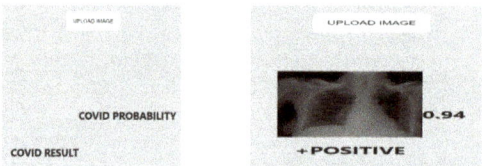

6   Advantages

The below are some of the benefits of the applied process or system:

1. Detecting COVID-19 with a chest x-ray is less expensive than using an RTPCR kit.
2. The outcome of the RTPCR Kit requires 24 h, although it takes less time if the procedure is begun with a chest X-ray.

7 Conclusion

A covid-19 pandemic is sending shockwaves. With the growing number of cases, bulk case testing can become essential. In order to explain the Covid-19 impacted patients' mistreatment of their chest X-ray test, we have a propensity to play with multiple CNN models in this work. Furthermore, we have an inclination to conclude that the XCeption web edition provides the greatest performance and is the most appropriate for use of the three. We were able to classify covid-19 scans effectively, which explains the potential for utilising such a method to simplify study practises in the immediate future. The high precision obtained may also be reason for worry, since it is the consequence of overfitting. This will be confirmed by applying it to new proof that will be released shortly. The broad dataset of chest X-rays is often suggested for potential testing of our proposed model. It is always a good idea to seek medical guidance on any practical applications of this programme. We build a faultless diagnostic system, but our focus is on discovering the most cost-effective ways to cure this disorder. More research into these approaches can be pursued in order to demonstrate their viability in the real world.

References

1. Ooi GC, Khong PL, Müller NL, Yiu WC, Zhou LJ, Ho JC, Lam B, Nicolaou S, Tsang KW (2004) Severe acute respiratory syndrome: temporal lung changes at thin-section CT in 30 patients. Radiology
2. Wang Y, Hu M, Li Q, Zhang XP, Zhai G, Yao N (2020) Abnormal respiratory patterns classifier may contribute to large-scale screening of people infected with COVID-19 in an accurate and unobtrusive manner
3. Koo HJ, Lim S, Choe J, Choi SH, Sung H, Do KH (2018) Radiographic and CT features of viral pneumonia. Radiographics
4. Narin A, Kaya C, Pamuk Z (2020) Automatic detection of coronavirus disease (covid-19) using x-ray images and deep convolutional neural networks
5. A Jain K Kishor 2020 Financial supervision and management system using Ml algorithm Solid State Technol 63 6 18974 18982
6. Apostolopoulos ID, Mpesiana TA (2020) Covid-19: automatic detection from x-ray images utilising transfer learning with convolutional neural networks
7. Yan L, Zhang HT, Xiao Y, Wang M, Sun C, Liang J, Li S, Zhang M, Guo Y, Xiao Y, Tang X (2020) Prediction of criticality in patients with severe Covid-19 infection using three clinical features: a machine learning-based prognostic model with clinical data in Wuhan
8. Kanne JP (2020) Chest CT findings in 2019 novel coronavirus (2019-nCoV) infections from Wuhan, China: key points for the radiologist
9. Tyagi D, Sharma D, Singh R, Kishor K (2020) Real time 'driver drowsiness' & monitoring & detection techniques. Int J Innov Technol Explor Eng 9(8):280–284. ISSN 2278-3075
10. Afshar P, Heidarian S, Naderkhani F, Oikonomou A, Plataniotis KN, Mohammadi A (2020) Covid-caps: a capsule network-based framework for identification of covid-19 cases from x-ray images
11. Alqudah AM, Qazan S, Alqudah A (2020) Automated Systems for Detection of COVID-19 using chest X-ray images and lightweight convolutional neural networks

12. Hassanien AE, Mahdy LN, Ezzat KA, Elmousalami HH, Ella HA (2020) Automatic X-ray COVID-19 lung image classification system based on multi-level Thresholding and support vector machine

13. Ilyas M, Rehman H, Nait-ali A (2020) Detection of Covid-19 from chest X-ray images using artificial intelligence

14. Ozkaya U, Ozturk S, Barstugan M (2020) Coronavirus (COVID-19) classification using deep features fusion and ranking technique

15. Prashant Patel, Chest X-ray (Covid-19 & Pneumonia), Accessed at: https://www.kaggle.com/prashant268/chest-xray-covid19-pneumonia

16. Wang S, Kang B, Ma J, Zeng X, Xiao M, Guo J, Cai M, Yang J, Li Y, Meng X, Xu B (2020) A deep learning algorithm using CT images to screen for Corona virus disease (COVID-19).

17. Szegedy C, Vanhoucke V, Ioffe S, Shlens J, Wojna Z (2016). Rethinking the inception architecture for computer vision

18. Chollet F (2017) Xception: deep learning with depthwise separable convolutions

19. Sokolova M, Japkowicz N, Szpakowicz S (2006) Beyond accuracy, F-score and ROC: a family of discriminant measures for performance evaluation.

20. Parikh R, Mathai A, Parikh S, Sekhar GC, Thomas R (2008)

A Systematic Approach to mHealth in COVID-19: Patient Generated Health Data on Opportunities and Barriers for Transforming Healthcare

Vibha Taneja, Smriti Mishra, and Ankur Saxena

Abstract Wearable gadgets, portable mHealth applications, and geo-location advancements have the capacity to track, screen, and report information related to COVID-19 pandemic. These developments make an "associated wellbeing," where people gather information outside of the medical care experience and acknowledge caretakers for it. Assortment of this PGHD or Patient Generated Health Data can possibly sway conveyance of medical care through distant observing, and by permitting patients and medical care groups to give focused on and productive consideration that lines up with the wellbeing status of individual patients. We examine the idea of a participatory computerized contact notice way to deal with help following of contacts who are presented to affirmed instances of Covid infection (COVID-19); It includes two types of approaches; one is based on apps and other on data collection. The proposed tool fills in as a supplemental agreement following way to deal with check the scarcity of medical services staff. To comprehend the worth and boundaries related to clinical reconciliation of PGHD, this study took data of stakeholders, looking at their viewpoints and encounters of PGHD use. Moreover, this research looked to exhibit the utilization of a cell phone and tablet application that upholds PHR (patient health records) based wellbeing perception by coordinating checking capacities explicit to Coronavirus. Interpreting progresses in innovation and data following into effective clinical usage requires seeing how stakeholders conceptualize and utilize PGHD, the potential worth that PGHD can add to mind, and the difficulties that may restrict PGHD's guarantee. Results represent the worth and difficulties related to health-framework execution of PGHD.

Keywords Covid-19 · PGHD · WHO · mHealth · Healthcare

V. Taneja · S. Mishra · A. Saxena (✉)
AUUP, Amity Institute of Biotechnology, Amity University, Sector-125, Noida, Uttar Pradesh 201313, India
e-mail: asaxena1@amity.edu

© The Author(s), under exclusive license to Springer Nature Singapore Pte Ltd. 2022   193
D. Gupta et al. (eds.), *Proceedings of Data Analytics and Management*,
Lecture Notes on Data Engineering and Communications Technologies 90,
https://doi.org/10.1007/978-981-16-6289-8_17

1 Introduction

In healthcare, it has been observed that patients' pathophysiological events and data trends are relevant to their disease as well as their health. In the era of "big data", we can use these PGHD data to predict model status of disease and notify health promoting interventions. Certainly, many latest consumer precise data fields yield information that is either biomedically pertinent or could update clinical care and research. PGHD is "data created, recorded, and gathered from patients which, after evaluation by medical experts, sent to them" often through the use of technology like wearable devices or Smartphones [1]. It includes health records, biometrics, treatment history, lifestyle choices, etc. to address a health concern.

With time advancement in PGHD has risen, we can now see a converging consumer-driven individual expertise and health-related uses. Many bigger firms like Apple and Samsung have recently declared major digital health initiatives, apple has novel feature integration as "HealthKit" and it has proclaimed its partnership with Mayo Clinic and EPIC electronic health record [2].

PGHD also can facilitate improved reach to statistics concerning an individual's fitness, and enhanced employment and contact with healthcare teams as well as providers [3]. The format for tracking PGHD data is explained in Fig. 1.

From a research perspective, PGHD generated by the wearable devices including prominent signs like changing mood, physical activity, stress levels, weight gain or loss, nutrition diet, and sleep patterns, blood pressure, medicines, exposure to alcohol or tobacco, and environmental exposures. Within research context, patient-related history, diaries, health reports and status, risk assessments also contribute valuable information for the data collection. Data that is not associated with healthcare could be selected to create health associated insights, for example, geolocations, financial and social data [4]. Key features indicating patient generated healthcare data include capturing and recording the data obtained from patients. PGHD is longitudinal, with a

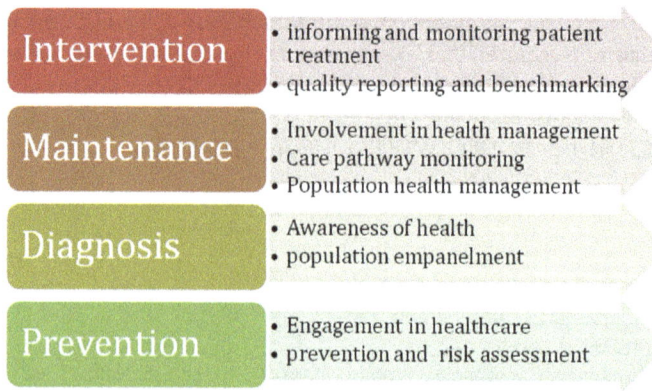

Fig. 1 PGHD formats and types tracked by healthcare providers and consumers

Fig. 2 PGHD and its role in monitoring healthcare data for COVID-19 patients

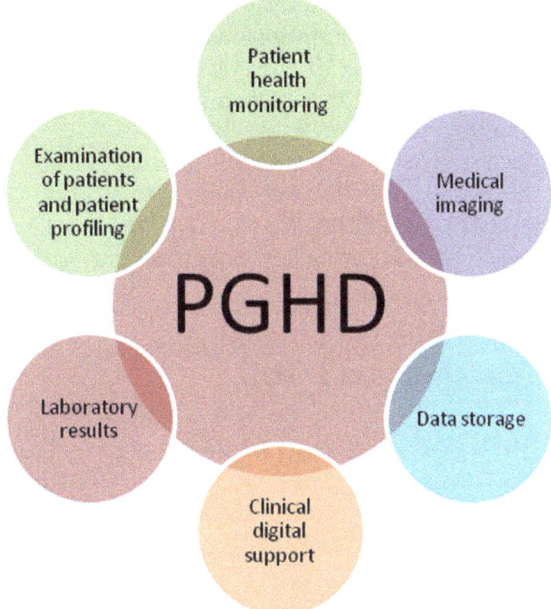

capacity for repetitive events over time. Data can be collected frequently, this enables continuous data streams for a long time, depending upon metric of interest.

While PGHD offers a chance to give a more powerful perspective on a person's wellbeing and health, medical services frameworks have been delayed to officially incorporate PGHD into clinical work processes and care change as explained in Fig. 2 [5].

Separately, it is becoming increasingly obvious that clinical trial design and conduct in biomedical research are in desperate need of improvement. In the current scenario, clinical trials are expensive, inefficient, and time-consuming. A lot of studies have been done on this topic (Institute of Medicine 2010), these limitations have concrete consequences, including increase in pressure for large number of clinical trials groups, and internal systems among drug and device producers to lower research and growth budgets. Perhaps most crucially, the basic scientific foundations in many of the most crucial areas of human suffering and disease, such as cancer, are not well understood. Disease states are being better understood at a faster rate than ever before [6]. Clinical trials to address management development and execution factors to think about are also increasing. There are several chances of mismatch in science and practical data which becomes easily outdated so there is a loss of opportunity for improvement in patient outcomes.

Against the proposed challenges mentioned above, several opportunities may be provided by PGHD to meet the shortcomings in the health care trials. Information regarding design consideration (for future reference) and hypothesis can be generated by PGHD. Merely by increasing the contribution of each patient on clinical trial

PGHD can maximize information that is extracted from every trial. Not only this but results in the reduction in sample sizes held for future reference [7]. Potentially PGHD improves the number of observations from trials and data points of each patient, resulting in advanced scientific insights related to negative and positive impacts in healthcare system.

In the rest of this survey, we will examine potential contemplations identified with the integration of PGHD into future clinical advances, GENERATION OF PGHD, CAPACITY OF PGHD. We will show how PGHD can add to discoveries that are created by data collected in various trials and medical surveillance of patients.

2 Generation of PGHD

PGHD is generated through the use of mobile applications (Smart phones), wearable devices (Wrist bands, Smart watches), registered medical devices, and surveys and questionnaires. Each of these avenues for the cohort of PGHD is utilized in a patient's everyday life, outside of the traditional venue of a medical practice or other healthcare centers [8] (Fig. 3).

The movement of PGHD from the patient/consumer (creation and collecting stages), through intermediaries (communication, division, and explanation stages), and back to the patient in the form of prevention, health promotion, and interaction (operation and influence stages). is depicted in the above framework.

3 Capacity of PGHD

There are three principal purposes for which practices should utilize PGHD: Expanded precision of patient wellbeing data more prominent information on a patient's general wellbeing condition a look at a patient's wellbeing status between office visits, with the objective of being more involved in treatment designs and dodging the requirement for incidental office visits. Practices that are prepared to execute the utilization of PGHD in their training would be astute to begin a little scope, likewise with any new undertaking. To start with, it's ideal to devise a methodology for how you will gather and use PGHD, and teach your office staff what PGHD is, the reason it's significant for your office to utilize it, and how it's intended to improve tolerant results [9]. Next, consider what your training may as of now be doing that can take into account the assortment of PGHD, search for where you need more deliberate PGHD assortment, and settle on how you'll make that progress. At that point, likewise, with any program in your training, you should execute clear techniques and approaches for the program. Next, acquire your patients on the arrangement, telling them that you will use PGHD for their advantage, how they can give this data, and what they can expect similarly as how your training will utilize the information they give. At last, actualize the starting system and screen for where changes might be

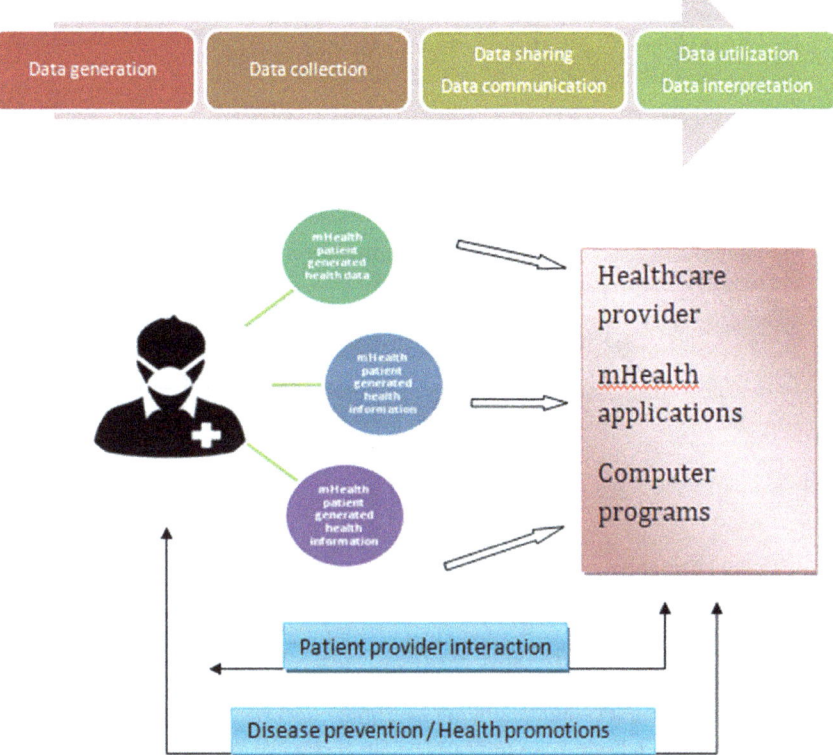

Fig. 3 Framework for PGHD flow and context for anticipation and health elevation

required, and work from that point. Figure 4 shows the steps to use PGHD in clinical research [10, 11].

We will likewise offer a plan for methodological exploration that we accept is basic to illuminating future, addressing a portion of the current boundaries and restrictions and challenges to the utilization of PGHD.

4 Methodology

I. **App based approach**—Many covid-19 applications are used to track and trace the health records of patient. PGHD is integrated with contact tracing that is the basis of app based tracking of coronavirus patients. Scandit has built up a portable application to catch quiet ID and clinical example information, rapidly and securely. This is used to contain PGHD and it can be used further for COVID-19 patient monitoring [12]. This free application is accessible for

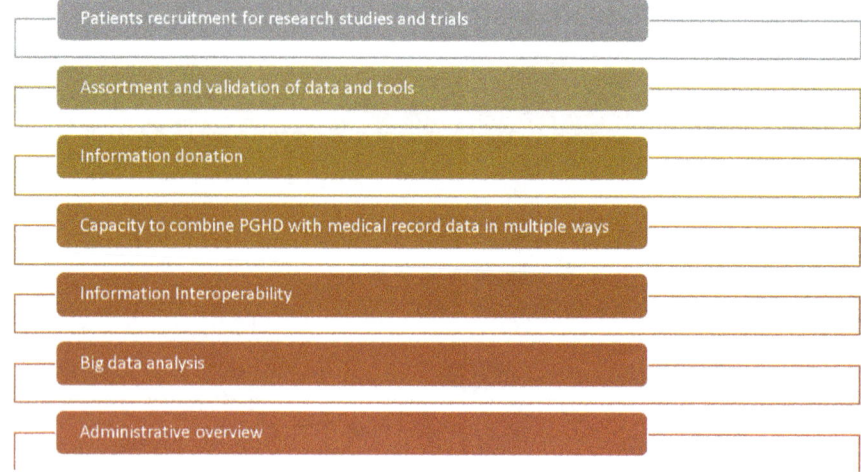

Fig. 4 Steps to use PGHD in clinical research

download presently, can be conveyed quickly and its advantages incorporate the accompanying:

- No integration or setups is required.
- It helps with the field testing measure anywhere, including tents and drive-through, where speed, accuracy, and health are essential concerns.
- Medical services laborers can filter persistent ID reports and scanner tags on tests with any shrewd gadget, without physical contact.
- No manual information section is required during test organization, caught information can be essentially exported.
- The application is HIPPA compliant– no information is kept inside the application on the gadget or imparted to Scandit or 3rd party (Fig. 5).

How it accomplishes the application work?

1. Download the application for nothing on any iOS or Android gadgets.
2. Select the Patient Data Capture mode in the application.
3. Output quiet ID archives.
4. Output the comparing standardized identification on the example assortment tubes.
5. Print or fare the gathered and parsed information immediately.

Table 1 specifies the mHealth tools that can be used in tracking and tracing covid-19 so that we can come up with effective solution to cope up with it

II. **Data collection approach**—Data was collected from ongoing work directed by the AMIA Affiliation investigating latent arrangement systems and related methodologies for PGHD. Technology in joint effort with ANA consultations was the focal point of this work as they are moderately best in class in public

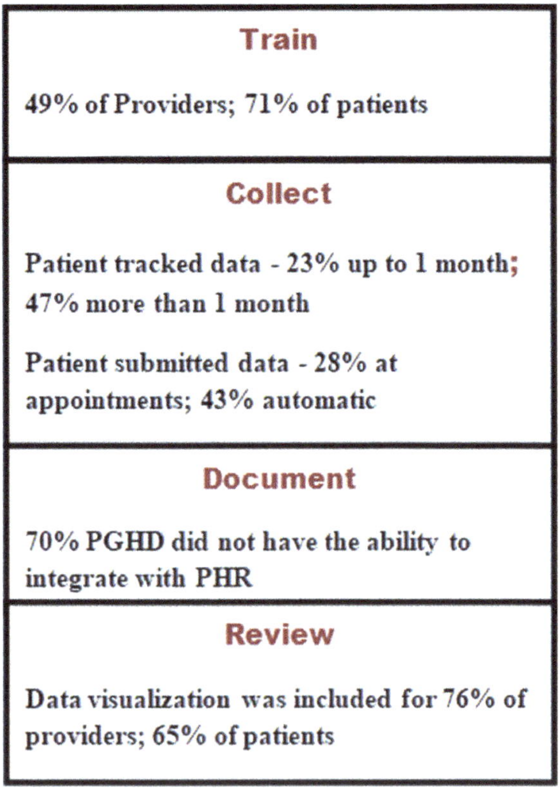

Fig. 5 Workflow for PGHD
in healthcare

Train

49% of Providers; 71% of patients

Collect

Patient tracked data - 23% up to 1 month;
47% more than 1 month

Patient submitted data - 28% at
appointments; 43% automatic

Document

70% PGHD did not have the ability to
integrate with PHR

Review

Data visualization was included for 76% of
providers; 65% of patients

Table 1 List of mHealth tools used in the tracking of COVID-19

Criteria for COVID-19	App	Wearable's (Wrist bands) and sensors	Thermal camera	GPS
Temperature			✓	
Proximity and contact tracing	✓	✓		✓
Quarantine compliance (geo-fence)				✓
Surveys	✓			
Alerts	✓	✓	✓	✓
Flow modeling (mobility report)	✓			✓
Symptom checker	✓	✓	✓	
Privacy	✓	✓		✓

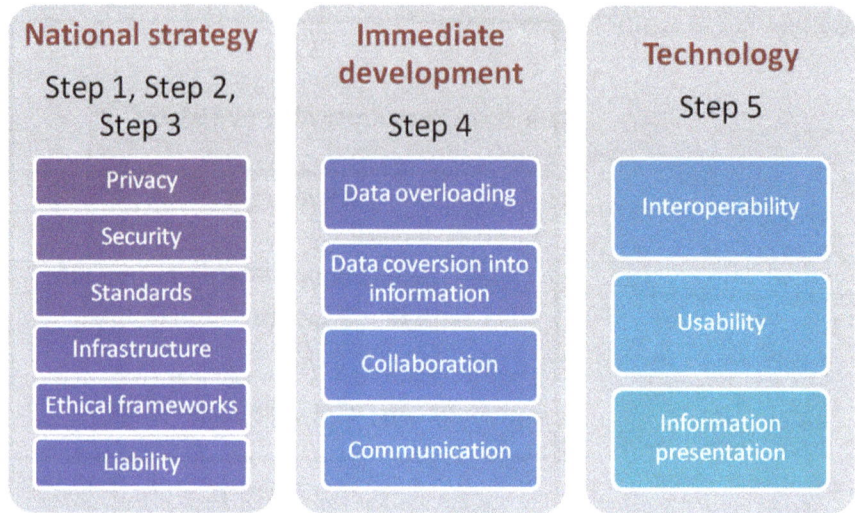

Fig. 6 Existing challenges in association with key strategic priority

strategy consultations comparable to PGHD. We have outlined 5 key need zones, rising up out of these structures, placed them in the UK setting to outline how public system could expand on prevailing difficulties, activities, advances, ability, and frameworks [13]. Difficulties and suggestions are summed up in Fig. 6.

The steps involved in data collection for PGHD to work are as follows:

Priority 1: Promoting information integration across data sources with innovations by creating regulatory environments

Information sources without smothering advancement in PGHD might be seen as an expansion to existing wellbeing data foundations that are assembled, made, and kept up by stakeholders, including patients themselves. Creating information norms that empower organized information sections furthermore, data sharing across different applications is vital to accomplishing a level of data reconciliation across data frameworks. There is presently an additionally NHS Digital Apps Library with NHS approved patient confronting functionality [14]. Henceforth, sharing the PGHD is frequently restricted to messaging data to applicable partners, be they specialists, family members, or caretakers. Therefore, information related to health is held in storehouses and in this way, presently an expanding worldwide drive to build APIs which needs endorsement by Medicines and Healthcare Items Administrative Agency—an exceptionally extensive process. Figure 7 shows the working of PGHD and how data is collected. It uses real-time, structured data and has targeted action.

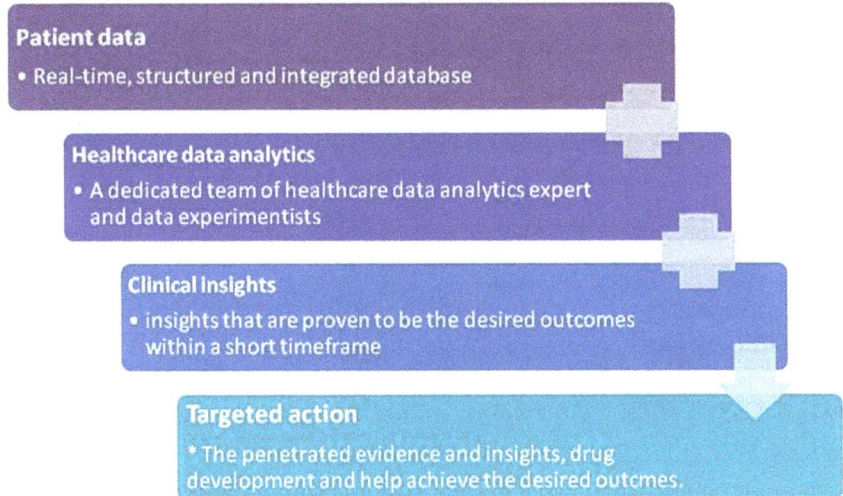

Fig. 7 Data collection approach and working of PGHD

Priority 2: Developing ethical frameworks as well as data governance that allow sharing data across settings

Corresponding to PGHD, there is no recent advanced planned system to share information across settings and agencies based on infrastructural contemplations talked about above. These frameworks should be adequately adaptable. For example, existing rules treat all medical care information similarly in spite of the fact that not all information is similarly sensitive, e.g.,—levels of patient activity [15]. Frameworks additionally should guarantee stability between persistent information being satisfactorily secured, while as yet advancing information flexibility.

Furthermore, there is an earnest worldwide need to create robust and ethical frameworks, firmly adjusted for lawful systems, for rising clinical advances and related PGHD. Setting up a public moral working gathering will be significant, working cooperatively with framework engineers to envision developing moral difficulties and relieving potential dangers early.

As shown in Fig. 8 the information sharing across apps will lead to establishing synchronization between patients and stakeholders.

Priority 3: Standardize person-centered methods and related outcomes

Even with questions encompassing the legitimacy and exactness of different PGHD gadgets and rising idea of endeavors to fuse EHRs with PGHD, there is a significant need to reinforce the observational proof base encompassing different functions and utilities. In request to total proof also, line up with various functionalities. Here, beginning endeavors could base on evidence-based approach (for example applications) and settings, where PGHD is tried on a pilot scale (for example asthma, diabetes) [16].

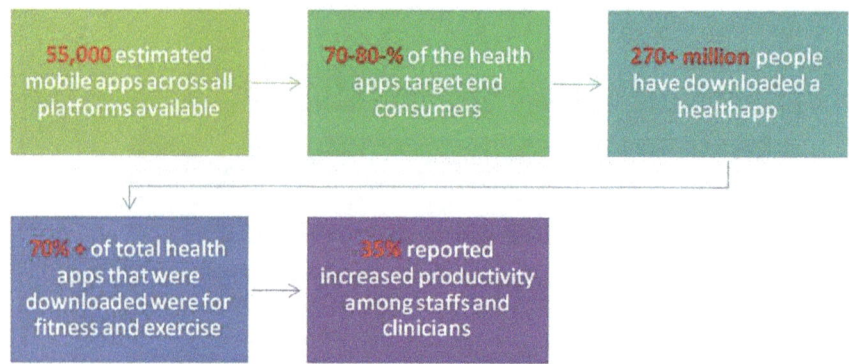

Fig. 8 Information sharing across apps

This work should necessarily guarantee suitable inclusion encompassing patients, so as to characterize core datasets that are pertinent to explicit health results, and clinician-educated, all together to guarantee that data produced by patients is utilized by different clinicians (refer Priority 4). Focusing on hypothetic ways to deal with assessment will be critical in this regard, so as to learn from experience and sum up among settings and across conditions. Such work ought to likewise draw and expand on continuous worldwide evaluation efforts.

Priority 4: Incentives for various stakeholders for creating, using, and reusing PGHD

Generation and applications of PGHD should be induced for boosting more data collection from patients. It is important to educate patients on the significance of PGHD for themselves.

Furthermore, developing tools that are simple for clinicians, patients, scientists to deliver and use PGHD. These data are further used for another patient. They need to need to make quite confronting applications that are intended to incorporate with clinical frameworks, while as yet making a budgetary benefit [17]. Here, it additionally should be incentive plans for developers and clinicians also. Therefore, it must be noted that challenges prevail along with increased security risks, unexpected releases, risk with data protection.

Priority 5: Collaboration across stakeholder groups to create tools that work for all

Patients and suppliers confronting advancements should be usable as else they will be dismissed or utilized in a manner that does not yield adequate top-notch information utilized in different manners. However, existing EHRs have been found to need ease of use/inerrability and frequently experience issues incorporating adequately with medical services proficient work practices. The improvement of innovation needs to be educated by patients focused plan standards which need close cooperation in

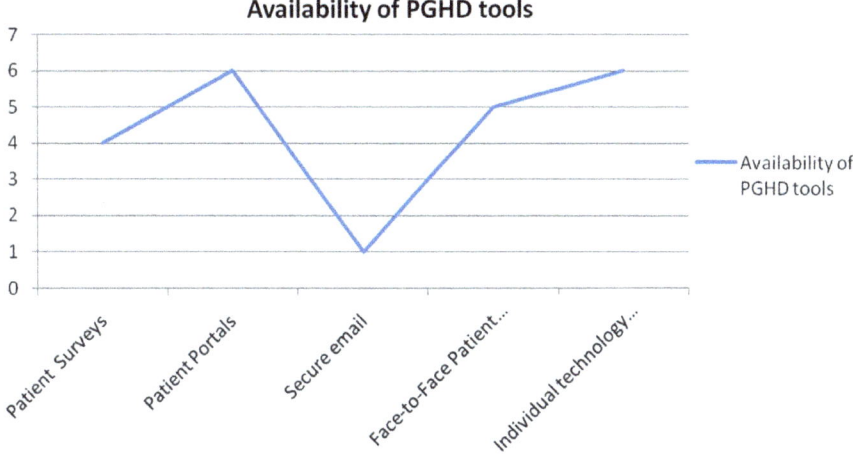

Fig. 9 Survey result for PGHD tool from healthcare providers. Question asked was—what PGHD tools are you most interested in learning more about?

partners (engineers, clinicians as well as patients) to guarantee that data has been collected, introduced in a configuration that lines up with patient's needs [18].

Data introduction of PGHD in clinical frameworks is prone to represent the greatest test, as there is a lot of spontaneous information in electronic frameworks, which may occupy clinicians [19]. In this regard, setting up measures for when data is shown in clinical frameworks can be amazingly supportive and this has global importance. In this way, machine learning may facilitate self-management helping patients to make decisions as to when to consult a clinician.

Various availability and usage of PGHD tools by healthcare providers are explained in Fig. 9 as taken from a survey result. Most available PGHD tool was patient portals and individual technology as it may improve patient assignation by permitting patients to entree their EMRs and enabling protect patient provider communication.

5   Challenges to PGHD

Obviously, interruption doesn't happen without difficulties. While quite produced information is being utilized in numerous territories of medical services today, it is still in the generally beginning phases. A typical concern is that tolerant produced information is a short-lived pattern and that the utilization of advanced innovation will in the end tumble off. Notwithstanding, in an ongoing interview, Fitbit CEO James Park tended to this worry clarifying that "we noticed that out of 18 million new enlisted gadget clients included 2015, 72% were as yet dynamic clients at year end". In all actuality as innovation keeps on progressing, gets less expensive, quicker,

Table 2 Barriers for PGHD integration

Barriers	Provider	Consumer
Feasibility	✓	✓
Problem of tracking	✓	✓
Absence of accountability		✓
Absence of data integration	✓	✓
Availability issues	✓	✓
Burden of tracking	✓	✓
Absence of resources	✓	✓
Absence of evidence	✓	

and substantially more open to a more extensive scope of patients and suppliers, it normally gets woven into the regular texture of our lives [20]. Table 2 shows barriers for PGHD integration into clinical healthcare so that they may be overcome in near future.

Another dread is that tolerant created information will supplant conventional assessments and clinical skills, eventually reducing the function of the medical services supplier. Actually, the capacity to gather ongoing information supplements the ability specialists have, moving where they center their endeavors, not supplanting their clinical information [21]. Admittance to information and the capacity to tweak the patient experience is a differentiator, not a substitute, for the medical services supplier.

Various other challenges such as not having proper technology to collect data, communicating the purpose of PGHD to patients, not having proper infrastructure to analyze and integrate the data into clinical workflows are given in Fig. 10 [22].

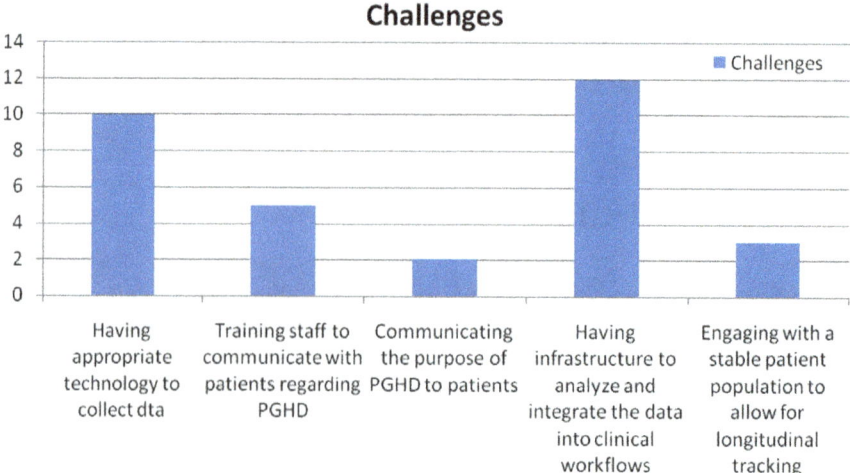

Fig. 10 What are your biggest challenges relate to PGHD?

They are significant in identifying the main causes for weakening of PGHD to some extent in healthcare.

6 Result

Interpretingdevelopments in technology and information following into positive clinical usage involves a robust comprehension of how partners hypothesize and make utilization of PGHD, the potential worth of PGHD in the care system, and the difficulties that might restrict PGHD's guarantee. The results revealed here give amusing context also, comprehension of the encounters and requirements of the medical care purchasers, medical services suppliers, and medical care directors who interface with PGHD. These PGHD approaches can be used further for the treatment of Coronavirus patients as well as for other various diseases. Patients counted a scope of advantages for utilizing PGHD for COVID-19, demonstrating wide arrangement that PGHD is a significant expansion to clinical consideration [23]. There is no leakage of data via application-based approach or data collection approach. With it, there is also an advantage of saving time as this infection is spreading at a higher rate in many developing countries. This features the chance to adjust PGHD reconciliation with care paths that can profit from the kind of granular, continuous accessibility of information that PGHD empowers. Our outcomes assemble on these discoveries, further showing the possible ways for PGHD to advise and uphold objective accomplishment and patient provider correspondence across medical issues [24].

Briefing out the biggest opportunity's application in the healthcare segment (Fig. 11) is stated by a graphical representation: biggest opportunity as analyzed by Eighty-one percent of expertise is improving care coordination. However, not with major difference in stats, about seventy nine percent of participants believe that data

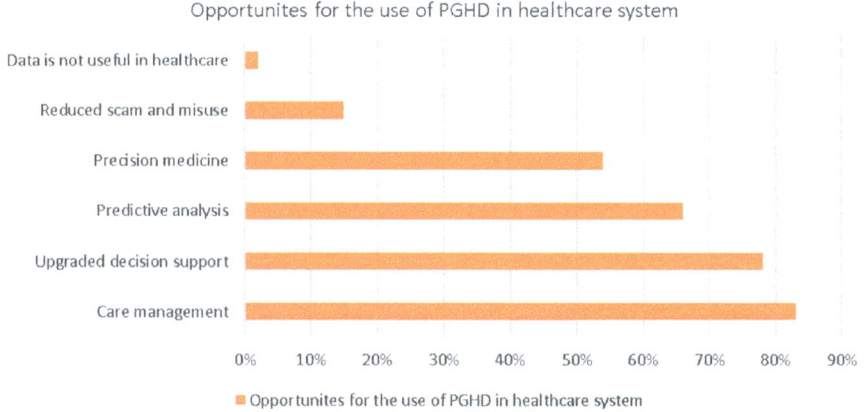

Fig. 11 Biggest opportunities for data application in the healthcare segment

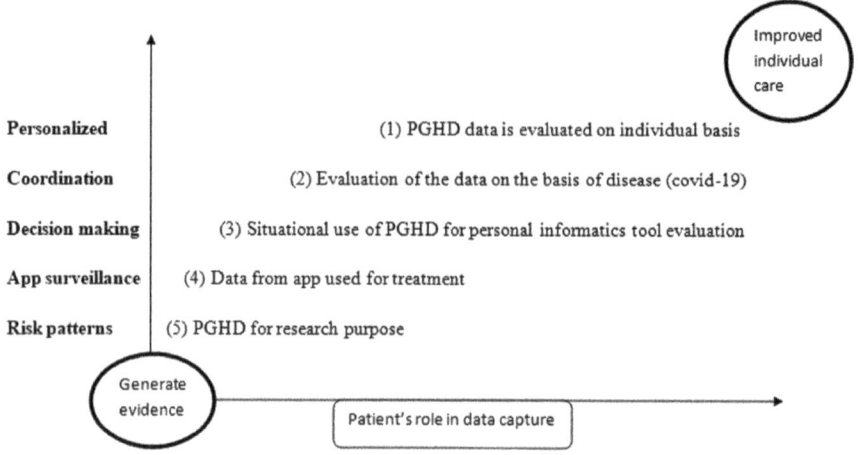

Fig. 12 Graphical representation about PGHD in improving healthcare sector

analytics will improve the decision support. Sixty eight percent are looking forward to more accurate predictive analytics [25]. Moving on, about forty five percent of participants think that precision medicine is also a necessary aim, although there is an evident growth in the interest in leveraging genomics and personalized patient data for decision making. Further low statistics report that only fourteen percent see the use of data analytics or patient generated data in reducing fraud and abuse.

It is recommended that health observing of PHRs is helpful for conveying out productive wellbeing perception of developing irresistible infections for people outside of a conventional emergency clinic setting. PHRs, which are intended to promote health through daily observation, were utilized as a counter-cluster tactic against large-scale infectious diseases in this study. In the event of highly infectious disorders, infection is frequently discovered in people who have been in close contact with a known COVID-19 infected patient. The graphical representation below in Fig. 12 determines the role of a patient in PGHD and its role in improving healthcare sector [26].

7   Discussion

We undertook this study to understand the opportunities and barriers to PGHD that can be implemented in healthcare systems. In previous studies, we have come across various barriers to PGHD being used in the healthcare sector which is studied in this paper thoroughly [27]. We discuss some of the concerns that must be addressed when PGHD is included in the proper medical treatment process. Our work contributes to study what motivates patients to collect data about themselves, as well as how this information may be used in clinical care.

Our outcomes fabricate on these discoveries, moreover showing the potent for PGHD for informing then upholding objective accomplishment and communication of patient provider with health conditions. We attempted this investigation to comprehend what is expected to support health system frameworks execution of PGHD. Straightforwardly drawing in individuals with experience utilizing PGHD gives significant knowledge on the chances and difficulties introduced by fusing PGHD to advanced experimental care. Our consultations additionally originate that while there is force to incorporate PGHD into experimental consideration, extra show of its advantages might be required.

Preferably, the manner in which data is introduced is adaptable and delicate to singular settings/jobs so extraordinary clinical clients just see patient-produced data they need for the main job. In this regard, setting up measures for when data is shown in clinical frameworks can be incredibly accommodating and this has worldwide significance [28]. For model, PGHD might be most fittingly specifically introduced to clinicians when the patient encounters intensifications or in various degrees of deliberation that can be chosen as per singular requirements/inclinations. In this regard, AI may encourage self-administration (for example by assisting patients with settling on choices regarding when to counsel a clinician) and it might likewise assist with fitting bits of knowledge from information to parts of clinical clients.

More examination is expected to build up the viability of utilizing PGHD information in clinical consideration, to decide the best procedures for executing this information into clinical consideration measures, and to think about the moral ramifications of these various systems. Future work will aim to identify the most effective ways for estimating the effect of PGHD based on the perspectives of several stakeholders, in order to enable resourceful evidence translation and inform decision making on the scope and spread of PGHD use throughout healthcare systems [29].

8 Conclusion

The outcomes of this PGHD indexing review give a system to see in what way PGHD is presently utilized inside medical services frameworks. The potential advantage that advanced heath can add to medical services lies in its capacity to improve our comprehension of disease courses, push us past our existing infection standards, and backing better connectedness among patients and care providers. The current scenario addresses a fascinating and advanced chance for the utilization of these new types of PGHD into clinical preliminaries. Through late declarations by significant customer innovation organizations, and expanding interest in PGHD all through academic organizations and governmental units, here is an extraordinary assembly of resources and interest around here. Notwithstanding, we are alive in the early days of this new time—however, we have distinguished a few potential chances in this audit, much work has to be done in order to combine, acquire, identify, authenticate and perfect pertinent PGHD streams with the goal that this information can be helpful in the clinical research consideration.

What does the drawn-out fate of PGHD resemble? In the ebb and flow of immense data era, we are progressively perceiving that new system biological approaches and computational methodologies will be important to mine and figure out the genomic, proteomic, and phenotypical information that we are currently generating in clinical consideration and examination. PGHD addresses a significant new type of information to be added to this better approach for taking a gander at wellbeing and infection. From a more extensive viewpoint, another worldview is arising: every individual can produce an analyzable "personal data cloud of billions of data points" that will at last assist with inventorying the changes between, and predictors of, wellbeing and illness. In spite of the fact that this information can be accumulated to drive populace experiences and insights, the sum and intricacy of information permits every individual to fill in as their own authority after some time, making a progression of "*n* of 1" studies.

The development of PGHD offers a sure shot chance to enhance the use of "big data" in the setting of clinical malignancy research. Later on, utilizing the force of different constant, customized information streams will permit us, as an exploration and clinical local area, to get maximal knowledge from every participant of each trial. Such a methodology, we accept, will improve preliminary effectiveness, create organic and clinical experiences into disease conduct and therapy reaction, and, eventually, regard the significant responsibility that each patient in a healthcare system.

References

1. Wood WA, Bennett AV, Basch E (2014) Emerging uses of patient generated health data in clinical research. https://doi.org/10.1016/j.molonc.2014.08.006. 1574–7891/ª 2014 Federation of European Biochemical Societies. Published by Elsevier B.V
2. Johnston NW, Lambert K, Hussack P et al (2013) Detection of COPD exacerbations and compliance with patient-reported daily symptom diaries using a smart phone-based information system. Chest 144(2):507e514
3. World Health Organization (2020) WHO Director-General's opening remarks at the media briefing on COVID-19 11 March 2020. https://www.who.int/dg/speeches/detail/who-director-general-sopening-remarks-at-the-media-briefing-on-covid-19—11-march-2020. Accessed on 8 Mar 2020
4. Peeples MM, Iyer AK, Cohen JL (2013) Integration of a mobile-integrated therapy with electronic health records: lessons learned. J Diabetes Sci Technol 7(3):602–611 [FREE Full text] [Medline: 23759392]
5. Hoffmann CL, Marie JD, Uncover beliefs about PGHD. webMed
6. Cohen DJ, Keller SR, Hayes GR, Dorr DA, Ash JS, Sittig DF (2016) Integrating patient-generated health data into clinical care settings or clinical decision-making: lessons learned from project HealthDesign. http://humanfactors.jmir.org/2016/2/e26/
7. Act on the prevention of infectious diseases and medical care for patients with infectious diseases. Act No. 114. Ministry of Health, Labor and Welfare. URL:http://www.japaneselawtranslation.go.jp/law/detail/?vm=04&re=01&id=2830. Accessed on 20 June 2020
8. National Institute of Infectious Diseases (2020) URL: https://www.niid.go.jp/niid/en/2019-ncov-e/2484-idsc/9472-2019-ncov-02-en.html. Accessed on 20 June 2020

9. Mobi Health News. NHS healthcare organization Salford Royal taps Validic to integrate patient-generated health data into EHR. Available from: http://www.mobihealthnews.com/content/nhshealthcare- organization-salford-royal-taps-validic-integratepatient- generated-health. Accessed on 12 Apr 2018

10. NHS Digital (2018) NHS Digital Launches New API Lab. Available from: https://digital.nhs. uk/article/7927/NHS-Digital-launchesnew-API-Lab. Accessed on 12 Apr 2018

11. SMART: Tech Stack for Health Apps. Available from http://docs.smarthealthit.org/. Accessed on 12 Apr 2018

12. GOV.UK (2018) Medical Devices: Software Applications (apps). Available from:https://www. gov.uk/government/publications/medical-devices-software-applications-apps. Accessed on 12 Apr 2018

13. Cheng KG, Hayes GR, Hirano SH, Nagel MS, Baker D (2015) Challenges of integrating patient-centered data into clinical workflow for care of high-risk infants. Pers UbiquitComput 19(1):45–57. https://doi.org/10.1007/s00779-014-0807-y

14. Grudin J (1994) Groupware and social dynamics: eight challenges for developers. In: Commun ACM. United States: association for computing machinery, Jan 1994, pp 92–105. https://doi. org/10.1145/175222.175230

15. Using Patient-Generated Health Data to Transform Healthcare: Milestone B000061603 Evaluation Report. July 2017. Available online: https://www.pcori.org/research-results/2017/using-patient-generated-health-data-transform-healthcare

16. Dedoose Version 7.0.23, web application for managing, analyzing, and presenting qualitative and mixed method research data 2016. Los Angeles, CA: SocioCultural Research Consultants, LLC. Available online: www. dedoose.com

17. Scott K, Lewis CC (2015) Using measurement-based care to enhance any treatment. Cogn Behav Pract 22:49–59

18. Chung CF, Dew K, Cole A et al (2016) Boundary negotiating artifacts in personal informatics: patient-provider collaboration with patient-generated data. CSCW Conf Comput Support Coop Work 2016:770–786

19. Petersen C, DeMuro P (2015) Legal and regulatory considerations associated with use of patient-generated health data from social media and mobile health (mHealth) devices. Appl Clin Inform 6:16–26

20. Saxena A, Sharma S, Analysis of Hadoop and MapReduce tectonics through hive big data. Int J Control Theory Appl 9/14:3811–3911

21. Saxena A, Kaushik N, Kaushik N, Dwivedi A (2016) Implementation of cloud computing and big data with Java based web application. In: 2016 3rd international conference on computing for sustainable global development (INDIACom), New Delhi, pp 1289–1293

22. Saxena A, Chaurasia A, Kaushik N, Dwivedi A, Kaushik N (2018) Handling big data using map-reduce over hybrid cloud. In: International conference on innovative computing and communications. Springer, pp 135–144

23. Saxena M, Saxena A (2018) Advancements in systems medicine using big data analytics. In: 4th international conference on computers and management (ICCM) 2018 ELSEVIER-SSRN, pp 1–7. ISSN: 1556-5068

24. Saxena M, Saxena A (2020) Evolution of mHealth eco-system: a step towards personalized medicine. In: International conference on innovative computing and communications. Advances in intelligent systems and computing, vol 1087. Springer, Singapore, pp 351–370

25. Nagpal D, Sood S, Mohagaonkar S, Sharma H, Saxena A (2019) Analyzing viral genomic data using Hadoop framework in big data. In: 2019 6th international conference on computing for sustainable global development (INDIACom), New Delhi, India, pp 680–685

26. Saluja MK, Agarwal I, Rani U, Saxena A (2020) Analysis of diabetes and heart disease in big data using MapReduce framework. In: Gupta D, Khanna A, Bhattacharyya S, Hassanien AE, Anand S, Jaiswal A (eds) International conference on innovative computing and communications. Advances in intelligent systems and computing, vol 1165. Springer, Singapore. https://doi.org/10.1007/978-981-15-5113-0_3

27. Saxena M, Deo A, Saxena A (2020) mHealth for mental health. In: Gupta D, Khanna A, Bhattacharyya S, Hassanien AE, Anand S, Jaiswal A (eds) International conference on innovative computing and communications. Advances in intelligent systems and computing, vol 1165. Springer, Singapore. https://doi.org/10.1007/978-981-15-5113-0_84
28. Mohanty S, Sharma R, Saxena M, Saxena A (2021) Heuristic approach towards COVID-19: big data analytics and classification with natural language processing. In: Khanna A, Gupta D, Pólkowski Z, Bhattacharyya S, Castillo O (eds) Data analytics and management. Lecture notes on data engineering and communications technologies, vol 54. Springer, Singapore. https://doi.org/10.1007/978-981-15-8335-3_59
29. Srivastava S, Prithivi PPR, Srija1 K, Vaishnavi P, Savitha HSS, Grover A, Saxena M, Chandra S, Saxena A (2021)Analysis and visualization of the pandemics using artificial intelligence. In: IOP conference series: materials science and engineering, 1st international conference on computational research and data analytics (ICCRDA 2020), vol 1022, 24th Oct 2020, Rajpura, India

Handwritten Offline Devanagari Compound Character Recognition Using CNN

Juhee Sachdeva and Sonu Mittal

Abstract Character recognition is the most challenging research topic due to its diverse applicable environment. Numerous research on Devanagari basic characters has been conducted, but due to difficulties associated, research on handwritten compound characters has received very little attention. The dilemma becomes much more complicated as a result of the different authors writing styles and moods. The traditional machine earning approach of character recognition focuses more on feature extraction, whereas the deep learning approach is a subset of machine learning that uses deep neural networks for learning. For current research work, we have created our own dataset for handwritten Devanagari compound characters. Our dataset has 5000 instances of 50 classes of compound characters collected from various writers of different age groups. This paper presents a convolutional neural network model for the recognition of Devanagari compound characters. We have implemented the ResNet model of CNN and used ReLu as an activation function as it effectively trains deep neural networks. We have implemented three-layer CNN, four-layer CNN, and five-layer CNN on our dataset, and its results are compared. We have achieved the highest accuracy of 100% on our dataset.

Keywords Handwritten character recognition · Devanagari compound characters · CNN · ResNet · ReLu

1 Introduction

Artificial intelligence (AI) is a specialized field that seeks to make systems that imitate human intelligence. The main goal of AI is to empower a computer so that it can see, understand, and read documents. Optical character recognition (OCR) system is used for this purpose. OCR system helps computers to utilize human-like

J. Sachdeva (✉) · S. Mittal
Jaipur National University, Jaipur, India

S. Mittal
e-mail: dr.sonumittal@jnujaipur.ac.in

capabilities like reading documents. Enough research has been conducted on printed offline OCR compared to handwritten offline OCR, due to the diverse nature of handwritings. As the character set in the Devanagari script is vast, also the presence of Shirorekha, modifiers, and compound characters made the handwritten Devanagari character recognition a complicated problem. Traditionally, OCR system for character recognition has the following steps like data collection, preprocessing, feature extraction, and image classification. Traditional OCR system utilizes various machine learning algorithms and extracts features from image to train the classifier and at last, it makes predictions about the class of the image [1], whereas the deep learning method uses layers of neural networks stacked one upon another [2]. The feature extraction process is done automatically between the layers, and in each layer process, some set of weights to perform unsupervised feature learning. For the current article, we have used convolutional neural network (CNN). This technology has outperformed in areas like computer vision and image processing [3]. For the proposed work, we aim to develop a systematic approach for the recognition of offline handwritten Devanagari compound characters using a convolutional neural network. We have implemented CNN residual network (ResNet) architecture. For multilayer CNN, the ReLu activation function is applied between the layers to increase model performance [4].

2   Literature Review

Acharya et al. [5] have used deep CNN for Devanagari script recognition and show an accuracy rate of 98.47%. They have created a dataset of 92,000 images having 46 classes of Devanagari characters and split data into 85% for training and 15% for the testing phase. The author has reported that deep CNN has performed better than traditional neural networks. Pius et al. [6] proposed a system for comparison of models ResNet and LeNet for Malayalam character recognition. The author has performed data augmentation on the dataset using data scaling and transformation, they concluded that LeNet has performed better. Rabby et al. [7] created an EkushNet dataset of 368,776 images of Bangla alphabets and digits, they reported an accuracy of 97.73% using CNN. Rahman et al. [8] presented a CNN-based Bangla character Recognition, they have used a dataset with 20,000 images and reported an accuracy of 85.96%. Ahlawat et al. [9] proposed a recognition system for handwritten digits using CNN. They have used MNIST handwritten digits dataset and reported an accuracy of 99.89%. Chakraborty et al. [10] developed a dataset with 300,000 images of Bangla compound characters and get an accuracy rate of 89.20%. Sonawane et al. [1] have conducted experiments using AlexNet on a dataset of 16,870 instances of 22 classes of Devanagari consonants. They concluded that using CNN with transfer learning on handwritten Devanagari characters gives an accuracy of 95.46% on test data and 94.49% accuracy on the training dataset.

Fig. 1 Devanagari vowels

Vowels	अ	आ	इ	ई	उ	ऊ	ऋ
	ॠ	ए	ऐ	ओ	औ	अं	अः

	क	ख	ग	घ	ङ	च	छ	ज	झ	ञ	ट
Consonants	ठ	ड	ढ	ण	त	थ	द	ध	न	प	फ
	ब	भ	म	य	र	ल	व	श	ष	स	ह

Fig. 2 Devanagari consonants

क+क=क्क	घ+न=घ्न	त+य=त्य	ज+व=ज्व
श+म=श्म	च+छ=च्छ	व+य=व्य	ल+ल=ल्ल

Fig. 3 Devanagari compound characters

3 Devanagari Script

Devanagari script has 14 "Swaras" or vowels shown in Fig. 1 and 33 "vyanjanas" or consonants as shown in Fig. 2. These vowels can be written as an individual basic character in Devanagari script or they can be written in the form of "matras" or diacritical marks that are attached to consonants. Devanagari script also includes complex structure known as compound characters, and these characters are formed when two basic characters are joined together. These characters are combined in such a way that, the first character is split into its half form conjoint with the full form of the second character. Few Devanagari compound characters are shown in Fig. 3.

4 Proposed Methodology

The proposed model uses system architecture as shown in Fig. 4.We have developed a dataset of handwritten Devanagari compound characters. All handwriting samples are collected from people of various age groups. Figure 6 shows 50 classes of compound characters that are selected from the Devanagari script. These characters are the most frequently used compound characters in the script. All sheets for data collection are preformatted having boxes. Users have to write characters inside

Fig. 4 Proposed model

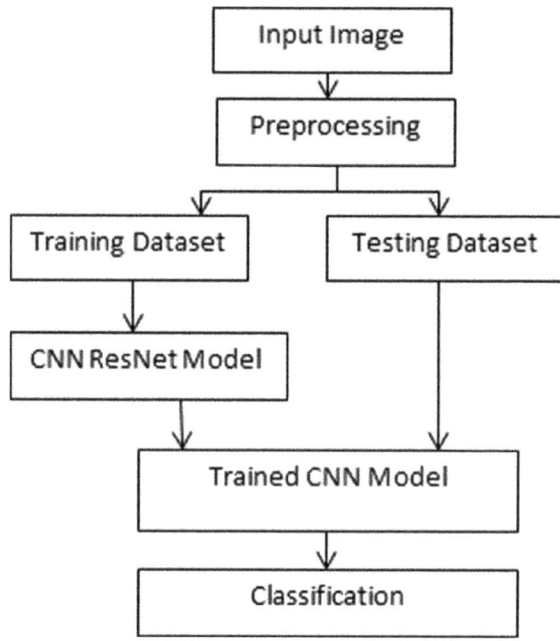

the box without touching its edges. Sample of handwritten Devanagari compound characters as shown in Fig. 5. A dataset of 5000 samples is created for the proposed system. The preprocessing phase of compound character recognition refers to the phase where all handwriting samples collected are resized, and the quality of the image is enhanced. Handwritten sheets are scanned using a good quality scanner at a high resolution of 500dpi and saved in Jpeg format. Each image is then cropped manually in 28 * 28 pixel size and saved in the respective image folder. During image acquisition, many noises may occur in the image sample so noise elimination is performed for the removal of such noise in images.

The dataset is then divided into two parts such as training dataset and testing dataset. Out of whole data instances present in the database, 70% of the random dataset is used for training purposes and the rest 30% of the dataset is used for testing the model. At the training phase, the CNN model is created.

A. **CNN model**—Convolutional neural network is a deep learning algorithm. It is widely used for natural language processing and pattern recognition field. CNN is a fully connected network [9] having multiple layers shown in Fig. 7. Each layer calculates weight and generates an activation function. The output of one layer serves as input to another layer. The first layer extracts few basic features, the second layer of the network extracts more complex features of the input image and the process continues till it classifies the input image and generates its output. The convolutional network consists of convolutional network and pooling (or subsampling layer) [3]. The convolutional layer extracts imported

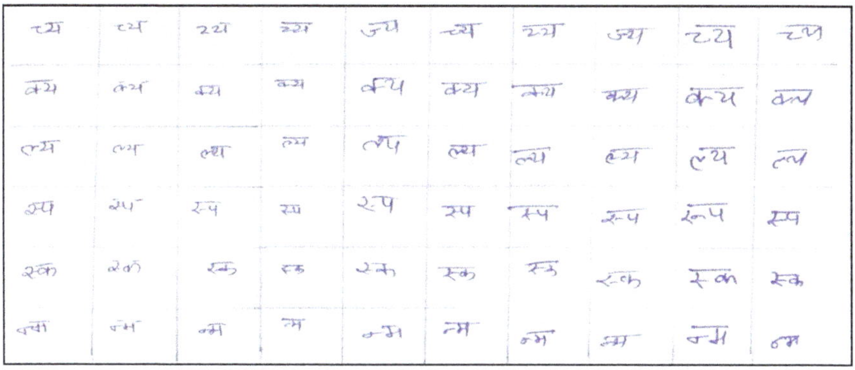

Fig. 5 Handwritten Devanagari compound characters

Fig. 6 50 class of
Devanagari compound
characters

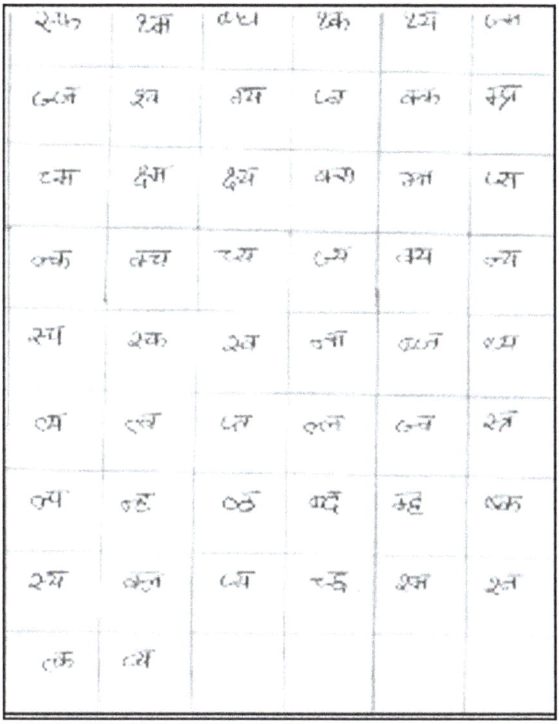

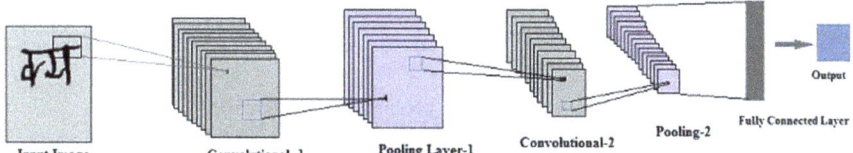

Fig. 7 Convolutional neural network

Fig. 8 ResNet model

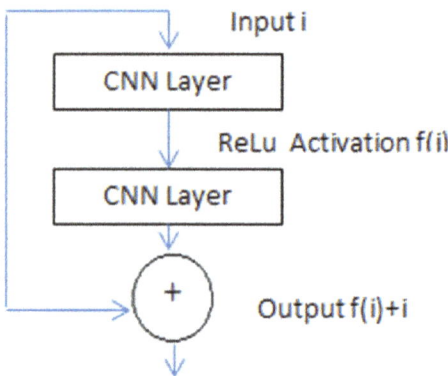

features from the input image, and each convolutional layer is followed by a subsampling layer. The pooling layer selects only relevant features and passes them to the next layer. The last layer after the convolutional and pooling layer combination is a fully connected layer, it combines all significant characteristics and classifies the image as output [11].

B. **ResNet**—Residual network is proposed in the year 2015. ResNet is an extremely deep CNN with around 150 layers. It applies a feedforward network. To overcome the vanishing gradient problem in neural networks, ResNet is the most effective model [12]. The residual network uses skip connections, where the training of few layers of the network is skipped and it connects directly to the output as shown in Fig. 8. The input i in Fig. 8 skip training from CNN layer 1 and CNN layer 2 and directly connects to the output function, this technique is termed as skip connection [13].

5 Experimental Results

We have carried out experiments using multiple layered convolutional networks. Details of all layers and parameters are given in Table 1. For three-layer CNN architecture, we have used convolutional layer followed by pooling layer and output layer. Four-layer CNN has convolutional layer and pooling layer followed by convolutional and output layer. For five-layer CNN, we have two convolutional layers, pooling,

Table 1 Convolutional neural network

No. of CNN layers used	Layers description
Three-layers CNN	Convolutional layer, pooling layer, output layer
Four-layers CNN	Convolutional layer, pooling layer, convolutional layer, output layer
Five-layers CNN	Convolutional layer, pooling layer, dense layer, convolutional layer, output layer

dense, and output layer. For network training, the dataset is divided into two random sets of a training set (70%) and a testing set (30%). Table 2 shows the accuracy rate of different layer CNN architecture. Different parameters like accuracy, precision, recall, and F-measure are calculated for all CNN models, and the result is analyzed. Four-layer CNN architecture gives the highest accuracy of 100%. Figure 9 shows a screenshot of class wise accuracy parameters of three-layer CNN. Screenshot of the confusion matrix of four-layer CNN is shown in Fig. 10.

6 Results and Discussion

The proposed system is to create a model for a handwritten Devanagari compound character recognition system using a convolutional neural network. Research for

Table 2 Accuracy rate for different layer CNN architecture

Statistic	Three-layer CNN	Four-layer CNN	Five-layer CNN
Accuracy (%)	95.56	100	99.98
Precision	95.8	100	100
Recall	95.6	100	100
F-measure	95.6	100	100

```
Classifier output

      TP Rate  FP Rate  Precision  Recall  F-Measure  MCC    ROC Area  PRC Area  Class
      1.000    0.000    1.000      1.000   1.000      1.000  1.000     1.000     ba_da
      1.000    0.000    1.000      1.000   1.000      1.000  1.000     1.000     ba_ja
      1.000    0.000    1.000      1.000   1.000      1.000  1.000     1.000     ch_chh
      1.000    0.000    1.000      1.000   1.000      1.000  1.000     1.000     ch_ya
      1.000    0.000    1.000      1.000   1.000      1.000  1.000     1.000     dh_ya
      1.000    0.000    1.000      1.000   1.000      1.000  1.000     1.000     Ja_va
      1.000    0.000    1.000      1.000   1.000      1.000  1.000     1.000     ja_ya
      1.000    0.000    1.000      1.000   1.000      1.000  1.000     1.000     ka_la
      1.000    0.000    1.000      1.000   1.000      1.000  1.000     1.000     ka_ya
      0.990    0.000    1.000      0.990   0.995      0.995  1.000     1.000     la_la
      1.000    0.000    0.990      1.000   0.995      0.995  1.000     1.000     la_pa
      1.000    0.000    1.000      1.000   1.000      1.000  1.000     1.000     la_ya
      1.000    0.000    1.000      1.000   1.000      1.000  1.000     1.000     ma_ha
      1.000    0.000    1.000      1.000   1.000      1.000  1.000     1.000     na_ha
      1.000    0.000    1.000      1.000   1.000      1.000  1.000     1.000     na_ma
      1.000    0.000    1.000      1.000   1.000      1.000  1.000     1.000     pa_ta
      1.000    0.000    1.000      1.000   1.000      1.000  1.000     1.000     pa_ya
      1.000    0.000    1.000      1.000   1.000      1.000  1.000     1.000     sa_ka
      1.000    0.000    1.000      1.000   1.000      1.000  1.000     1.000     sa_pa
      1.000    0.000    1.000      1.000   1.000      1.000  1.000     1.000     sa_tra
      1.000    0.000    1.000      1.000   1.000      1.000  1.000     1.000     sa_va
      1.000    0.000    1.000      1.000   1.000      1.000  1.000     1.000     sa_ya
      1.000    0.000    1.000      1.000   1.000      1.000  1.000     1.000     sha_ka
      1.000    0.000    1.000      1.000   1.000      1.000  1.000     1.000     sha_ma
      1.000    0.000    1.000      1.000   1.000      1.000  1.000     1.000     sha_na
      1.000    0.000    1.000      1.000   1.000      1.000  1.000     1.000     ta_ka
```

Fig. 9 Class wise accuracy of three-layer CNN

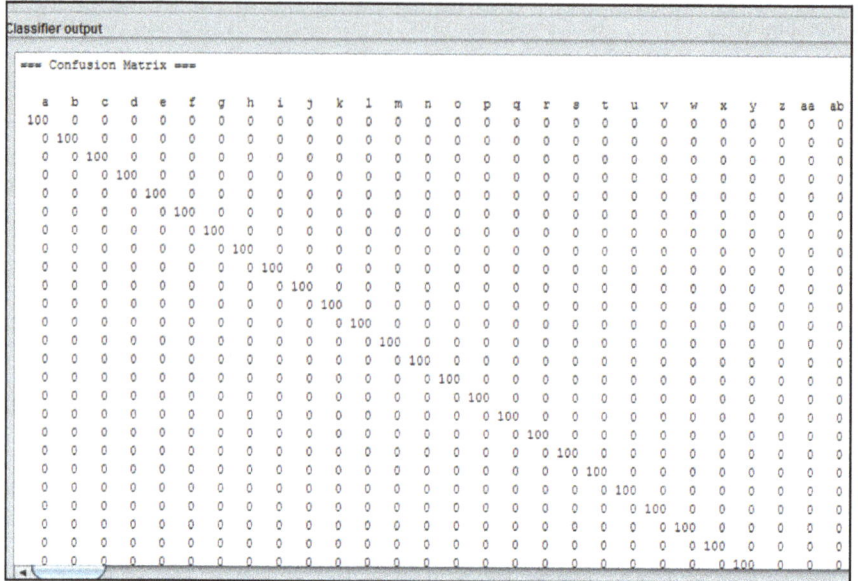

Fig. 10 Confusion matrix of four-layer CNN

Devanagari compound characters is still in its initial stage and no standard dataset is available for research. So we have developed our dataset of 5000 handwriting samples of 50 different classes of Devanagari compound characters. All images are preprocessed to enhance the quality of an image. After preprocessing phase, the dataset is randomly divided into a training set of 70% of dataset images and the remaining 30% for the testing set. Image dataset is then passed to CNN Model. For the current article, we have implemented multiple layers of CNN and compared its results. We have implemented three-layer CNN, four-layer CNN, and five-layer CNN. Using suitable layers for convolutional neural network make classification of image data more accurate. Convolutional layer with pooling layer, dense layer, and fully connected layer is implemented as a hidden layer to generate more accurate and significant feature map of images. Using ResNet for three-layer CNN gives 95.56% accuracy. Four-layer CNN gives the highest accuracy of 100%, whereas five-layer CNN gives an accuracy of 99.98% accuracy shown in Fig. 11. Table 3 indicates a comparison of accuracy achieved of present work with all other prior research done. It is clear from Table 3 that we have achieved the highest accuracy rate.

7 Conclusion

The present article proposes a handwritten Devanagari compound character recognition system using the CNN ResNet model. Experiments are carried out on a dataset

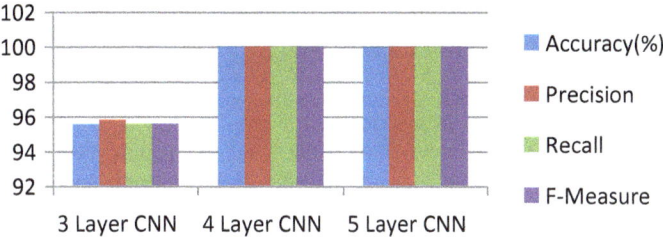

Fig. 11 Accuracy rate of different layer CNN

Table 3 Accuracy comparisons of current approach with other approaches in the previous research

Author	Script	Model used	Result (%)
Ahlawat et al. [9]	Digit	CNN	99.76
Pius et al. [6]	Malayalam	LeNet, ResNet	LeNet performs better
Rabby et al. [7]	Bangla	CNN	97.73
Shelke et al. [1]	Devanagari	AlexNet	72.46
Proposed model	Devanagari	ResNet	100

of 5000 samples. Dataset is randomly divided into 70% of the training set and 30% of the test set. Experiments are carried out on three-layer CNN, four-layer CNN, and five-layer CNN architecture, and results are analyzed. We have achieved an accuracy of 100% for four-layer CNN and 99.98% accuracy for five-layer CNN. In future, we can increase dataset size and also include other Indian language scripts for experiments.

References

1. Sonawane PK, Shelke S (2018) Handwritten Devanagari character classification using deep learning. In: 2018 International conference on information, communication, engineering and technology (ICICET), Pune, India, pp 1–4. https://doi.org/10.1109/ICICET.2018.8533703
2. Narang SR, Kumar M, Jindal MK (2021) DeepNetDevanagari: a deep learning model for Devanagari ancient character recognition. Multimed Tools Appl. https://doi.org/10.1007/s11042-021-10775-6
3. Zainudin Z, Shamsuddin SM, Hasan S (2019) Deep learning for image processing in WEKA environment. Int J Adv Soft Compu Appl 11(1)
4. Deore SP, Pravin A (2020) Devanagari handwritten character recognition using fine-tuned deep convolutional neural network on trivial dataset. Sādhanā 45:243. https://doi.org/10.1007/s12046-020-01484-1
5. Acharya S, Pant AK, Gyawali PK (2015) Deep learning based large scale handwritten Devanagari character recognition. In: 2015 9th international conference on software, knowledge, information management and applications (SKIMA), pp 1–6. https://doi.org/10.1109/SKIMA.2015.7400041

6. Pius NK, Johny A (2020) Malayalam handwritten character recognition system using convolutional neural network. Int J Appl Eng Res 15(9):918–920
7. Rabby AKMSA, Haque S, Abujar S, Hossain SA (2018) EkushNet: using convolutional neural network for Bangla handwritten recognition. Proc Comput Sci 143:603–610. ISSN 1877-0509. https://doi.org/10.1016/j.procs.2018.10.437
8. Rahman MM, Akhand MAH, Islam S, Shill PC, Rahman MH (2015) Bangla handwritten character recognition using convolutional neural network. Int J Image Graphics Signal Process 7(8):42
9. Ahlawat S, Choudhary A, Nayyar A, Singh S, Yoon B (2020) Improved handwritten digit recognition using convolutional neural networks (CNN). Sensors 20(12):3344
10. Chakraborty S, Paul S (2021) Bengali handwritten character transformation: basic to compound and compound to basic using convolutional neural network. In: 2021 2nd international conference on robotics, electrical and signal processing techniques (ICREST), pp 142–146. https://doi.org/10.1109/ICREST51555.2021.93
11. Jangid M, Srivastava S (2018) Handwritten Devanagari character recognition using layer-wise training of deep convolutional neural networks and adaptive gradient methods. J Imaging 4(2):41. https://doi.org/10.3390/jimaging4020041
12. Mhapsekar M, Mhapsekar P, Mhatre A, Sawant V (2020) Implementation of residual network (ResNet) for devanagari handwritten character recognition. In: Advanced computing technologies and applications. Springer, Singapore, pp 137–148. https://doi.org/10.1007/978-981-15-3242-9_14
13. He K, Zhang X, Ren S, Sun J (2016) Deep residual learning for image recognition. In Proceedings of the IEEE conference on computer vision and pattern recognition, pp 770–778

GA–JAYA: A Novel Hybridization Technique to Solving Job Scheduling Problems

Biswaranjan Acharya and Sucheta Panda

Abstract An extensive scope of application is utilized in different optimization algorithms to find nominal and best value, in respect of used function. Depending upon the application, the computational time and behavior of the algorithm can consider as serious issues. The traditional job scheduling problem has been a research topic. It has high computational complexity and well known as NP-hard in nature. The hybrid flow shop problem (HFSP) belongs basic JSSP family. HFSP need more than one machine to process a job which creates more complexity. In the last few years, numerous metaheuristic optimization algorithms have been proposed just to get the best solution. A novel hybridization technique GA–JAYA combination of genetic algorithm (GA) and JAYA algorithm has been proposed to solve the job scheduling problem. Because the JAYA algorithm does not require more parameter. This GA–JAYA is a solution over the traditional JAYA algorithm. Hence, it never gets caught at the local optima problem. The proposed JAYA algorithm performance is compared with the various metaheuristic algorithms, and it shows better experimental outcomes.

Keywords JAYA optimization algorithm · GA · Metaheuristics · Job scheduling · Local search

1 Introduction

For the successful working of an association, scheduling is one of the main components. Scheduling will help in expanding the operational productivity with which creation is completed. The same condition is considered in manufacturing industries. The effective scheduling technique improves the overall performance by optimum

B. Acharya (✉) · S. Panda
Department of Computer Application, Veer Surendra Sai University of Technoogy, Burla, Odisha, India

S. Panda
e-mail: suchetapanda_mca@vssut.ac.in

© The Author(s), under exclusive license to Springer Nature Singapore Pte Ltd. 2022   221
D. Gupta et al. (eds.), *Proceedings of Data Analytics and Management*,
Lecture Notes on Data Engineering and Communications Technologies 90,
https://doi.org/10.1007/978-981-16-6289-8_19

utilizing all the resources in the system. The scheduling technique can meet the deadline in real-time task execution for various works like aircraft scheduling, medical diagnosis scheduling, missile testing, and also in production sectors. The manufacturing industry uses scheduling technologies at a various level such as providing raw material and different stages of process for getting the final product [1]. For the scheduling process, many optimizing algorithms are also developed. Every single optimizing algorithm has a unique aim. In the time of pandemic also needs scheduling for vaccination on a priority basis using cloud [2]. Now, secure and distributed computing is essential because of risk factor. There are many tasks like healthcare or the defense sector that must need secure computing. To this end, we may prefer one of the advanced and trustworthy technique as blockchain [3]. Aircraft also uses scheduling mechanism for up and departure of the airplanes, and in that place or any public place to prevent lives and wealth of the nation, the surveillance system will be advantageous [4].

Every optimization algorithms have a target to identify an ideal or best value in response to given criteria inside a specific area. Although, every function is upgraded with highly complex and many different design factors. Undoubtedly, numerous functions have local minima problem [5]. Overcoming it and finding the best solution are difficult due to latency time between tasks. Generally, these techniques mainly depend upon straight polynomial math. Deterministic methodologies can give general apparatuses to tackling optimization problems to get the best or an estimated nearest best solution [6].

Almost at the same time, the JAYA algorithm was introduced as an optimizing algorithm. It utilizes one stage at each cycle. This JAYA algorithm is a gradient-free algorithm that repeatedly updates the population or we can say input to provide the optimum solution. An updated and improved version of the JAYA algorithm has been proposed as resent. These changed algorithms incorporate the selective JAYA, the JAYA itself, and the semi-oppositional-based GA–JAYA algorithm. Consequently, a hybrid GA–JAYA algorithm with a new and improved local optima finding method is proposed in this paper and gave a better result in comparison to compared algorithms.

The remaining paper follows this structure. In Sect. 2, the recent work is done using the JAYA optimization algorithm and its advantages. Section 3 describes the identification of the problem. In Sect. 4, JAYA algorithm discussed. Section 5 defines the proposed work, and Sect. 6 defines the experimental result and performance of the JAYA algorithm. Conclusions are discussed in Sect. 7.

2   Related Work

In the previous years, there has been wide research done in the space of hybrid flow shop problem or flexible job shop scheduling problem (FJSSP). The proposed GA-JAYA utilizing few methodologies to produce starting populace, determination of chromosomes and a crossover method. Because of the shortfall of definite techniques to tackle huge occurrences of the hybrid flow shop problem, research has progressively started to utilize metaheuristics to take care of the issue [7]. After that genetic algorithm along with other local search method implemented to increase the capacity to discover a superior optimized solution. A better solution found from hybridized particle swarm optimization (PSO) with the random search algorithm in task scheduling problem of hybrid flow shop problem (HFSP) [8].

The HFSP considering machine failures also utilized a GA–JAYA which uses two phases of search optima to take care of the minimization of makespan. It stabilizes as streamlining standards.

3 Identify the Problem

A hybrid flow shop problem also is recognized as flexible job shop scheduling (FJSSP). Multiple jobs to be handled on m_j machines. Each work comprises o_t various tasks to be handled in arrangement on the accessible machines.

In this case, the hybrid flow shop problem was considered. For a mathematical model of the hybrid flow shop problem, we consider some condition as referred to in literature [9]. The detailed assumption is formulated as follows:

- Assumption.
- Total active machines m_j at every stage along with stage number s_n are determined on a prior basis.
- Every machine is accessible during the entire processing time of each job, and these are determined and well known in advance.
- Processing time consists of transportation times with setup times for each job.
- No interruption is allowed when the process starts in specific pre-emption's not allowed in processing time.
- Several machines can be used to perform each operation of the jobs with diverse processing epochs
- The machine can take one job at a time for processing.
- One job allows accessing the machine only once no circulation is allowed.

4 JAYA Algorithm

JAYA algorithm is an advanced metaheuristic algorithm, and it is a modified version of teaching learning-based optimization (TLBO) proposed by Rao in 2016 [10]. It deals with the rule of persistent improvement. It follows a single condition to estimate and get new and best solutions at each iteration. Also, the JAYA algorithm has the benefit of not having any algorithm explicit factor that can affect the searching methods. So this algorithm is simple to understand and execute in, respectively, other metaheuristics algorithms. In every iteration, the best and worst solution is getting dependent on the target work esteem. Some steps are discussed below those are followed by the JAYA algorithm (Fig. 1).

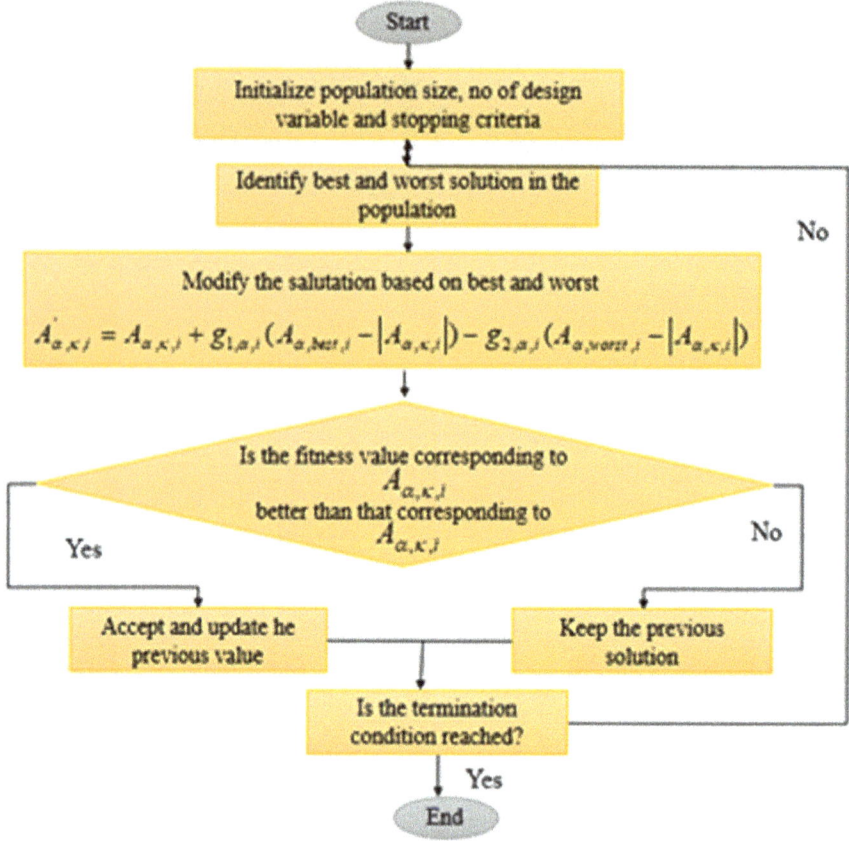

Fig. 1 defines the flowchart of the JAYA algorithmdefines the flowchart of the JAYA algorithm

Those solutions are left in the populace are adjusted by the accompanying following condition.

5 Proposed Method

In this section, we discussed the improved JAYA algorithm. Even though the convergence of the JAYA algorithm is acceptable, but it is seen that when this is applied on the HJSP, the JAYA algorithm get stuck at the local optimum due to the absence of variety in the populace. For this reason, result gets no improvement. To overcome this problem, variety in population mutation is incorporated from the genetic algorithm. As mutation suggests, get variety from the same population using crossover the position. This advanced step is added after basic local search space. We can expect after incorporation of traditional JAYA which will work better. The next section describes

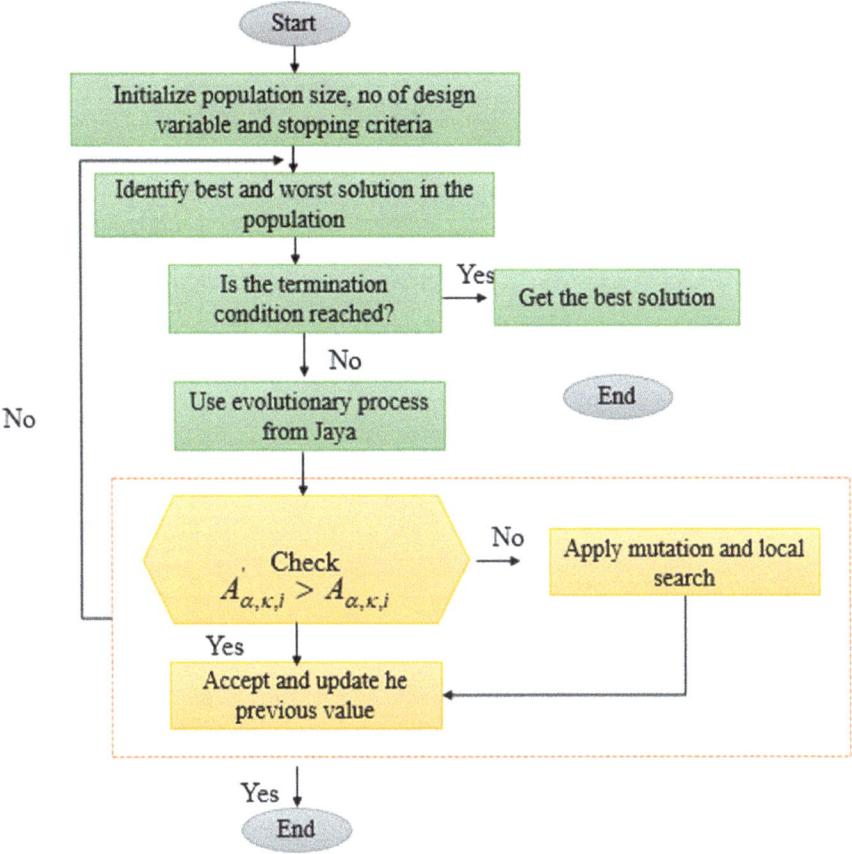

Fig. 2 Workflow diagram of improved JAYA algorithm

details about stating populace, proposed JAYA, the local search system, and mutation. Figure 2 defines the workflow diagram of the hybrid JAYA algorithm.

6   Result and Discussion

This section deliberates the computational implementation used to assess the presence of the proposed algorithm (GA–JAYA). The proposed algorithm (GA–JAYA) was executed in MATLAB 2019a on an Intel Core i5 3.4-GHz PC with 8 GB of memory on Windows 10. For each algorithm, 20 free runs taken to make reasonable correlations with others. So, 20 free runs are taken for each metaheuristic algorithm to check the efficiency and productivity of the proposed algorithm, surprisingly proposed algorithm perform better result among other. Table 1 processing time of every machine.

Table 1 Processing time of every machine

Jobs	Task	Machine	
		$m1$	$m2$
Job1	Task1	10	15
	Task2	16	9
Job2	Task1	–	11
	Task2	15	–
Job3	Task1	14	–
	Task2	–	9

The chosen implementation measure using relative percentage increase (RPI), which is determined as follows:

$$RIP = \frac{fitness_{min} - fitness_{best}}{fitness_{best}} \times 100 \qquad (2)$$

$fitness_{best}$ define the best value, and $fitness_{min}$ define nominal fitness value got using an algorithm. We consider a different number of randomly produced instance; those are $j = \{15, 25, 35, 45, 55, 80\}$ several jobs with a $m = \{5, 6, 7, 8\}$ number of the machine.

Figures 3 and 4 show the convergence curve of instance 1 and 10, respectively. Only two graphs are displayed among 20.

Table 2 shows the convergence curve examinations for the various kinds of cases. It is seen from this experiment and performance that the proposed GA–JAYA technique displays the better result with others. Figure 5 shows speedup capacity using JAYA and GA–JAYA algorithm. Figure 6 shows efficiency using JAYA and GA–JAYA algorithm. Here, in the graphical model, only five instances are considered.

7 Conclusion

The JAYA algorithm has been demonstrated to be a powerful optimization algorithm. This hybrid GA–JAYA is proposed to tackle the hybrid job shop problem with makespan minimization. It has the benefit of being straightforward and simple to execute yet provide not full converge. JAYA algorithm has less diversity in population, so it is stuck in the local minima problem. To get rid of this situation, this hybridized algorithm proposed with a productive local search procedure and mutation mechanism acknowledgment which adjusts searching and misuse of the working space in a successful way. A state-of-the-art comparison is done between other competitive metaheuristic algorithms with the proposed algorithm. The efficiency of the proposed algorithm was illustrated. In the future, proposed algorithm can apply in

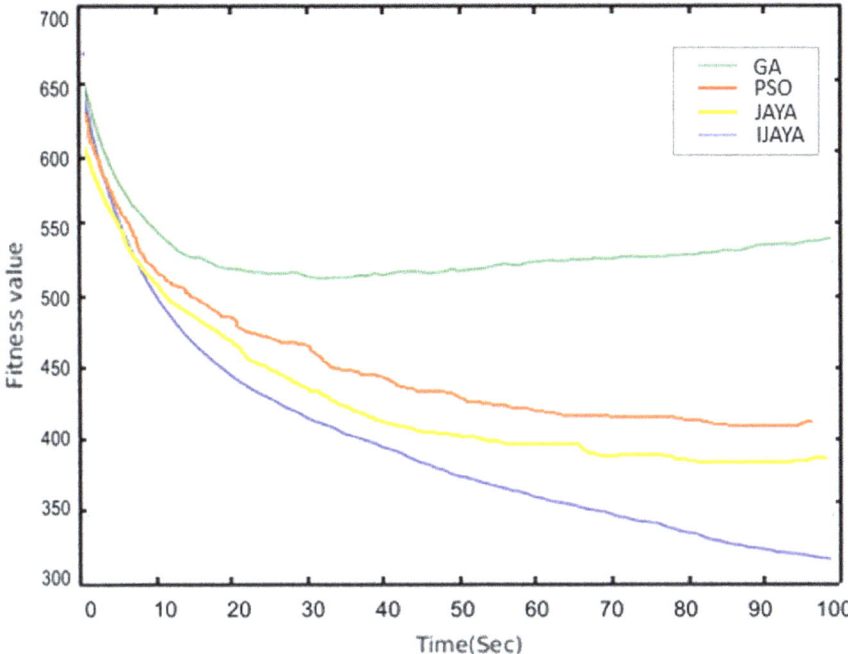

Fig. 3 Convergence curve for instance 1

a real-world application. The proposed technique can be utilized to address other scheduling related case in a different aspect.

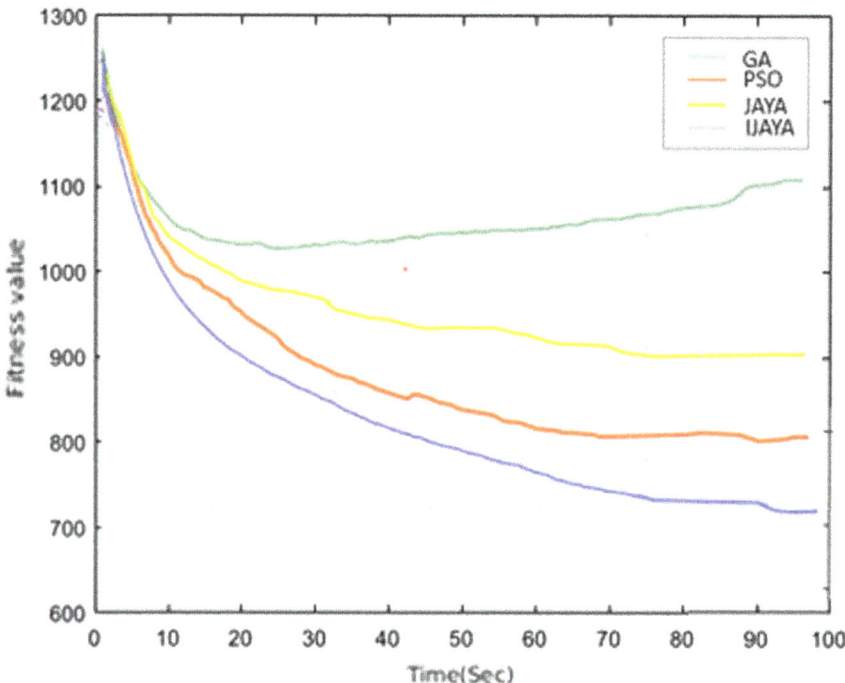

Fig. 4 Convergence curve for instance 10

Table 2 Comparison result between the proposed model and others metaheuristic model

Instances	Best solution	Fitness			
		GA	PSO	JAYA	GA–JAYA
Instance 1	323.8	345.2	346.6	335.80	323.8
Instance 2	291.9	322.1	341.7	447.0	291.9
Instance 3	353.2	547.9	448.9	504.8	353.2
Instance 10	715.9	1167.7	753.0	803.4	715.9
Instance 11	885.0	1265.8	927.4	1001.9	885.0
Instance 20	1666.1	3023.3	1742.2	1872.0	1666.1
Mean	882.655	1370.61	967.026	1019.24	884.745

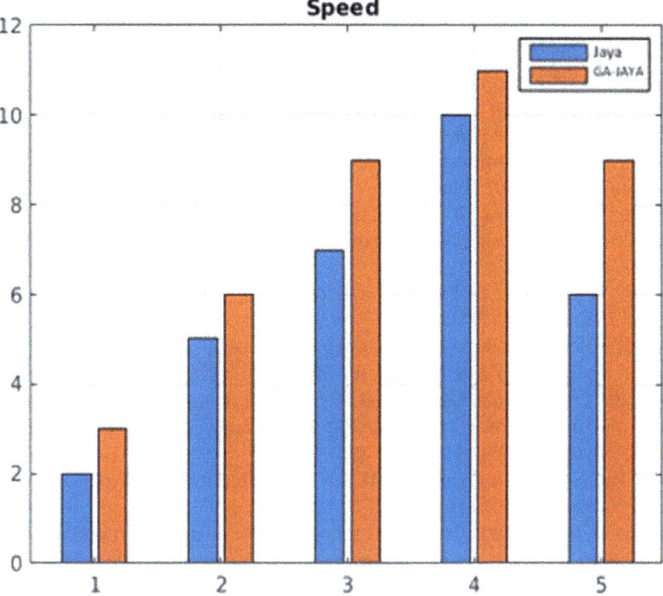

Fig. 5 Speedup capacity using JAYA and GA–JAYA algorithm

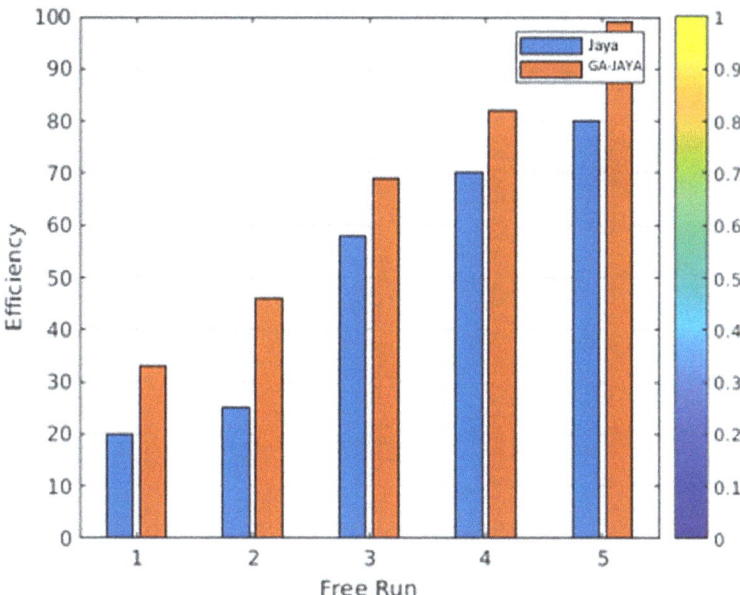

Fig. 6 Efficiency using JAYA and GA–JAYA algorithm

References

1. Rao R (2016) Jaya: a simple and new optimization algorithm for solving constrained and unconstrained optimization problems. Int J Ind Eng Comput 7(1):19–34
2. Rani P, Singh PP, Balyan A, Shokeen J, Jain V, Sangwan D (2021) A secure epidemic routing using blockchain in opportunistic Internet of Things. In: Data analytics and management. Springer, Singapore, pp 101–110
3. Jamader AR, Das P, Acharya BR (2019) BcIoT: blockchain-based DDos prevention architecture for IoT. In: 2019 international conference on intelligent computing and control systems (ICCS). IEEE, pp 377–382
4. Acharya BR, Gantayat PK (2015) Recognition of human unusual activity in surveillance videos. Int J Res Sci Innov (IJRSI) 2(5):18–23
5. Caldeira RH, Gnanavelbabu A (2021) A Pareto based discrete Jaya algorithm for multi-objective flexible job-shop scheduling problem. Expert Syst Appl 170:114567
6. Kato ERR, de AguiarAranha GD, Tsunaki RH (2018) A new approach to solve the flexible job shop problem based on a hybrid particle swarm optimization and Random-Restart Hill Climbing. Comput Indus Eng 125:178–189
7. Nayak SK, Padhy SK, Panigrahi SP (2012) A novel algorithm for dynamic task scheduling. Futur Gener Comput Syst 28(5):709–717
8. Huang X, Li C, Chen H, An D (2019) Task scheduling in cloud computing using particle swarm optimization with time-varying inertia weight strategies. Cluster Comput 1–11
9. Abdel-Basset M, Mohamed R, Abouhawwash M, Chakrabortty RK, Ryan MJ (2021) A simple and effective approach for tackling the permutation flow shop scheduling problem. Mathematics 9(3):270
10. Rao R (2016) Review of applications of TLBO algorithm and a tutorial for beginners to solve the unconstrained and constrained optimization problems. Decis Sci Lett 5(1):1–30

Image Segmentation Techniques: A Survey

Riya Yadav and Manish Pandey

Abstract Image segmentation plays a crucial role in digital image processing. It is the process of dividing an image into various segments called image objects which in turn reduces the complexity of an image to analyze them efficiently. The purpose is to represent an image free from noise and distortion. Moreover, it is used to specify the position of objects and their boundaries. Numerous segmentation techniques have been proposed in the past with varied approaches and little diversity. This paper presents a comprehensive analysis of different segmentation techniques like edge-based segmentation, region-based segmentation, clustering, thresholding, soft computing-based segmentation.

Keywords Segmentation · Edge detection · Region-based · Clustering · Thresholding · Soft computing techniques · Neural network · Genetic algorithm

1 Introduction

Digital image processing is the process of applying computer algorithms to a digital image. It consists of different steps like image acquisition, segmentation, feature extraction, classification. In this, image segmentation refers to a technique of splitting the image into different objects or dividing the images into regions in the form of pixels. The first course of action in segmentation is preprocessing which is done to carry out any computational challenge like classification, recognition, analysis, retrieval on an image. As the next course of action on this preprocessed image, we impose various segmentation techniques. This output is then fed as an input to a classifier or any other model. The overall process is depicted in Fig. 1.

The procedure of segmentation makes it easy to distinguish between various regions of an image by assigning a value to each pixel. This differentiation among various segments of the image is carried out based on three important properties, i.e.,

R. Yadav (✉) · M. Pandey
Department of Computer Science and Engineering, Maulana Azad National Institute of Technology, Bhopal, India

© The Author(s), under exclusive license to Springer Nature Singapore Pte Ltd. 2022   231
D. Gupta et al. (eds.), *Proceedings of Data Analytics and Management*,
Lecture Notes on Data Engineering and Communications Technologies 90,
https://doi.org/10.1007/978-981-16-6289-8_20

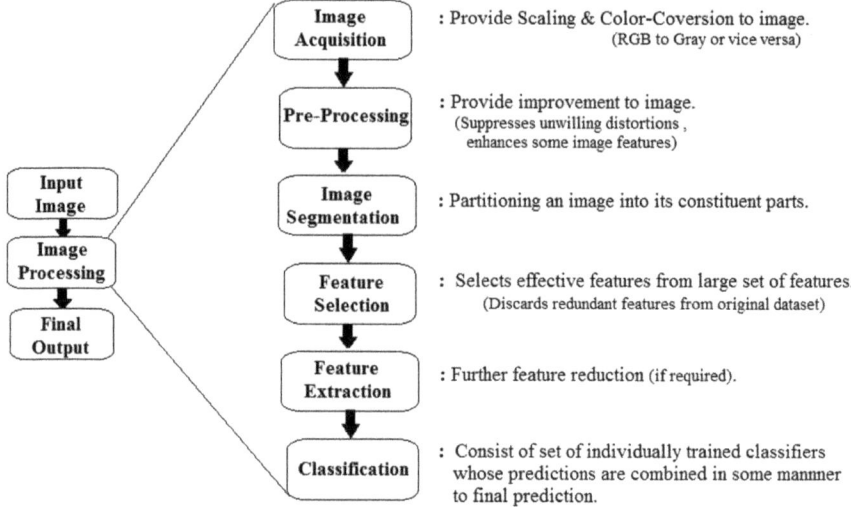

Fig. 1 Steps involved in image processing

color, intensity, and texture of that image. Hence, any image segmentation technique is selected after having a closer look at the problem domain.

The purpose of this paper is to describe various existing techniques for image segmentation. Various methods have been proposed as the input images vary in terms of texture, scaling, color, and rotation. The different papers are reviewed and discussed along with their merits and demerits.

2 Related Work

We have discussed various existing image segmentation techniques along with their applications in this section. Different techniques have been discussed and evaluated along with their advantages and limitations. We have many segmentation techniques, some of the latest and most widely used techniques are discussed below, and a summary of the same is present in Table 1.

2.1 Edge-Based Image Segmentation Technique

This technique is focused on the rapid change of intensity value of an image. The edges are detected and connected together to form the object boundaries to segment the required regions. The feature of an image is observed for a significant change in gray level. Several operators are used in the edge detection method. These are

Table 1 List of discussed methods

S. No.	Technique	Dataset	Result
1	Edge-based phase symmetry [1]	TRUS images	Accuracy at least 87%
2	Watershed transform [2]	Digital panoramic radiograph of 678 children	Accuracy = 94%
3	Improved watershed [3]	Sixty-six sound and 84 infected orange images	Success rate of sound oranges = 96%
4	IRG [4]	Four lung CT images	Accuracy = 98%
5	K-mean clustering [5]	Ecoil dataset	Accuracy = 98.25%
6	HMS [6]	Eleven images	P-value = 110.91
7	Integrated region growing and thresholding [7]	Alpert dataset	Average recall = 0.91 Average precision = 0.65
8	FCMs [8]	MRI images	Less noise than FCM
9	Genetic algorithm [9]	Eight images	PSNR = 5.08 (Tsallis), 6.09 (Renvi)
10	GA for shape extraction [10]	CE-shape-1, INRIA horses	Accuracy = 84.68%
11	Mean-c thresholding [11]	DRIVE dataset	Accuracy = 95.5%
12	MLP [12]	Forty Abdomen,95 Colorectal images	Average accuracy = 0.88 ±0.11
13	Deep ConvNets [13]	CT images of 82 patients	DSC = 83.6 ± 6.3% (training)
14	Deep multi-channel neural network [14]	One hundred and sixty-five colorectal cancer images	F1-score = 0.893 (part A)

Robert's operator, Sobel operator, Prewitt operator, Kirsch, Robinson, LoG operator, Marr-Hildreth operator, and Canny operator.

Zaim [1] proposed an edge-based technique that is based on feature phase symmetry for the segmentation of the prostate which requires no human intervention. Different energy models have been proposed in the past, but they needed human intervention.

Muthukrishnan [15] discussed the performance of various edge detection operators in their paper. Among all the edge detectors they used, it was found that the Canny edge detector produced a superior edge map. Moreover, the edge map produced by Marr-Hildreth, LoG, and Canny edge operators were almost similar.

2.2 Region-Based Image Segmentation Technique

Region-based image segmentation is a technique in which similar images are segmented into various regions so that the regions can be determined directly.

Adjacent pixels of an image are grouped to form regions with the help of image characteristics. In this section, some of the region-based segmentations are reviewed.

Mohammad [2] used watershed transform technique to segment the first molar tooth on a panoramic radiograph. This technique uses the notion of topological interpretation. Xi [3] presented an improved version of watershed segmentation to remove the limitation of the traditional watershed method as their results were not satisfactory because of over-segmentation caused by irregular gray disturbance and noise in the image.

Jamshid Soltani [4] proposed an improved region growing (IRG) technique to increase the accuracy in a shorter period and achieved 98% accuracy in comparison to basic algorithm. Region growing technique segments the image into numerous regions on the basis of growing of seeds which are controlled by the connectivity between pixels and by using prior knowledge of the problem.

2.3 Clustering-Based Image Segmentation Technique

Clustering is the process in which datasets are replaced by clusters. These clusters are group of data points or pixels which belong together. Such pixels belong together as they have the same color or texture. The principle behind clustering is that it groups input datasets with similar characteristics into the same cluster, keeping the dissimilar data points into different clusters. Clustering-based image segmentation is classified as k-means clustering, mountain clustering, and subtractive. Some of the widely used clustering-based segmentation techniques are discussed in this section.

Purohit [5] proposed a new and better approach for k-mean clustering algorithm. Their methodology removed the limitations of the existing k-mean algorithm by deciding the initial centroids systematically. The data points are selected in such a way that the distance between them is the shortest. The algorithm improved the efficiency of the k-mean clustering technique.

Dhanachandra [16] used subtractive clustering algorithm which is an extension of mountain clustering method. The density of surrounding data points is used for generating clustering. Images were segmented into k number of clusters by using k-mean clustering algorithm and then a median filter was used to remove noise from the images.

Mousavirad [6] proposed a human mental search (HMS) algorithm which is a population-based algorithm and used k-mean clustering for bid clustering. The experimental result demonstrated that the proposed algorithm is robust and converges faster as compared to conventional clustering methods.

2.4 Threshold-Based Image Segmentation Technique

It is based on the consideration that the pixels of the objects of interest (foreground) have a property value significantly different from those of the pixels of the object (background). Hence, thresholding is very useful in distinguishing between foreground and background. Threshold values are mostly selected in an ad hoc manner when a single property is considered. Threshold segmentation techniques are divided into three categories, i.e., global threshold technique, variable threshold technique, and multiple threshold technique.

Thresholding [17] can be expressed as:

$$T = T[x, y, p(x, y), f(x, y)] \tag{1}$$

where T denotes the threshold value. Coordinates of the threshold value point are represented by x, y. $p(x, y), f(x, y)$ represent points of the gray-level image pixels. Threshold image $g(x, y)$ can be defined as:

$$g(x, y) = \begin{cases} 1 & \text{if } f(x, y) > T \\ 0 & \text{if } f(x, y) \leq T \end{cases} \tag{2}$$

Hore [7] combined the region growing segmentation with threshold-based segmentation to locate the segmented region of interest of an image. The proposed work suggested a homogeneity based on intensity of pixel. Methods like the iterative method, Otsu's method, local thresholding have been used to procure the best threshold value. J Dash [11] segmented digital retinal images of a blood vessel using mean-c thresholding. The main feature of this technique is that it can be used when the images are poorly illuminated. Here, the threshold is calculated for each pixel in the image based on some local statistics.

2.5 Soft Computing-Based Image Segmentation Techniques

Soft computing techniques use the thinking ability of the human brain to deal with complex real-time problems. Thus, they are extremely powerful. Because of their adaptive nature, high performance, steady solutions to complex problems, these are preferred by the researchers [18, 19]. These were introduced by Zadeh [20] in the 1990s and categorized techniques into three subgroups, i.e., fuzzy approach, genetic algorithm approach, and neural network approach.

2.5.1 Fuzzy Approach

Various new segmentation techniques have been integrated with the fuzzy logic proposed by Zadeh. Some of the widely used fuzzy approaches are discussed below.

Dunn proposed the fuzzy c-means algorithm (FCM) in 1973, and Bezdek improved it in 1981. FCM is an unsupervised clustering technique based on fuzzy sets and needs a priori knowledge of the number of clusters. FCM algorithm is a nonlinear iterative optimization approach that assigns pixels to each category by using fuzzy memberships or objective function and minimizes the cost function as follows.

$$J_{\text{FCM}} = \sum_{i=1}^{c} \sum_{k=1}^{N} u_{ik}^{p} = \|x_k - v_i\|^2 \tag{3}$$

In the above equation, c represents the number of clusters, u_{ik} denotes the membership property of pixel x_k in the i-th cluster, v_i is the i-th center of cluster,. represents the standard Euclidean distance, and m is a constant. The parameter m controls the amount of fuzziness generated by the new partition. Conventional FCM does not has the spatial function within its objective function which in turn makes FCM sensitive against image artifacts. Because of this, fuzzy c-mean with spatial information (FCMS) [8] was proposed.

2.5.2 Genetic Algorithm (GA) Approach

The genetic algorithm was invented by John Holland and is based on natural selection. These are efficient and dynamic in nature and are regarded as optimization methods. GA uses a function called fitness function for searching. Some of the real-world segmentation approaches are discussed below.

Abdel-Khalek et al. [9] presented a method based on genetic algorithm for 2D segmentation by using the representation of Tsallis and Renyi entropies. Jaffar [21] also presented a GA incorporated with ANN and SOM networks for color images.

Wei and Tang [10] also presented a novel technique based on GA that can identify the most common contour fragments by using FCSs. The shape representation illustrated the important structural logic of a contour. The proposed technique is better than the old representation based on chain codes.

2.5.3 Neural Network Approach

The neural network is a multi-layer network of variable weighted neurons connected by links. ANN learns from the input datasets and does not use any kind of rule sets. These methods have high computational complexity but can learn directly from the input data automatically. It consists of three layers which are input, hidden, and

Fig. 2 Neural network approach

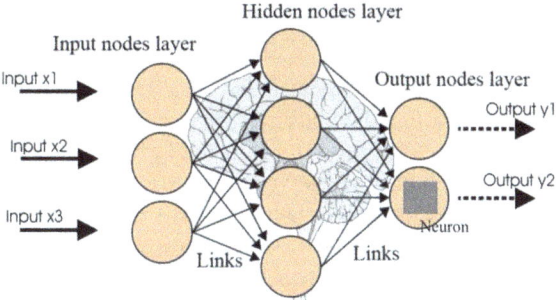

output layers as shown in Fig. 2. Some of the neural network approaches for CT images are discussed.

Vorontsov [12] proposed a deformable model which interprets the CT image using multi-layer perceptron (MLP). MLP is a feedforward neural network that uses a nonlinear activation function for the input vector. Roth et al. [13] used multi-level deep convolutional networks (ConvNets) technique for the segmentation of pancreas in abdominal computed tomography.

Xu et al. in [14] used deep learning approach for the segmentation of glands in colon histology images. They used a deep multi-channel convolutional neural network for this purpose.

3 Application and Limitation of Methods

Thresholding is the oldest and widely used segmentation technique. Though it is simple and fast, it does not consider the spatial information of an image. Thus, it is sensitive to noise and intensity inhomogeneity. Variants of edge detection methods are very less resistant to noise and do not work well if the edges of the image are not defined properly. On the contrary, region-based techniques are immune to noise, but it performs poor in terms of memory and time. However, it works best when the homogeneity of the region is simple. *K*-means clustering has proven to have a good advantage of easy implementation and is computationally faster than hierarchical clustering. Furthermore, the output depends on the selection of initial centroids. Besides this, it can handle a large number of variables. The subtractive clustering algorithm is preferred when the dimension of data is high so that the computation complexity can be reduced. The neural network approach is very efficient as it takes the advantage of its parallel nature. The drawback is that the amount of operations of such a model is comparatively huge. In addition to this, the training time of a model is quite high. SOM, PCNN, MLP, BPN, etc., are collaborating with several different methods to improve the efficiency of the techniques. Hybridization of image segmentation techniques is preferred as they have increased accuracy and efficiency.

4   Conclusion and Future Scope

In this paper, we have discussed numerous image segmentation techniques which help in extracting the objects of our interest so that processing can be achieved. We have discussed the pros and cons of each technique and covered some of the recent work on such techniques. Segmentation allows translation, rotation, and scaling of an image. Visualizing a particular part of an image has become easy. The output of segmentation is affected by various parameters like continuity, homogeneity of images, texture, color, spatial structure character of an image. Depending upon the problem domain, different segmentation techniques are employed to attain the desired accuracy.

In the future, we plan to integrate different deep learning models with a reinforcement learning algorithm for image segmentation. We will try to exploit the deep reinforcement learning algorithm so that an efficient, accurate, scalable, fast, high performance in the high dimensional state space image segmentation technique can be presented which can process 3D image datasets and can run in real world.

References

1. Zaim A (2008) An edge-based approach for segmentation of prostate ultrasound images using phase symmetry. In: Proceedings of 3rd international symposium on communications, control and signal processing, pp 10–13
2. Mohammad N, Yusof MYPM, Ahmad R, Muad AM (2020) Region-based segmentation and classification of Mandibular First Molar Tooth based on Demirjian's method. In: International conference on telecommunication, electronic and computer engineering, vol 1502
3. Tiana X, Fana S, Huanga W, Wanga Z, Li J (2020) Detection of early decay on citrus using hyper spectral transmittance imaging technology coupled with principal component analysis and improved watershed segmentation algorithms. Postharvest Biol Technol 161:111071
4. Soltani-Nabipour J, Khorshidi A, Noorian B (2020) Lung tumor segmentation using improved region growing algorithm. Nucl Eng Technol. https://doi.org/10.1016/j.net.2020.03.011
5. Purohit P, Joshi R (2013) A new efficient approach towards k-means clustering algorithm. Int J Comput Appl (0975–8887) 65(11)
6. Mousavirad SJ, Ebrahimpour-Komleh H, Schaefer G (2019) Effective image clustering based on human mental search. Appl Soft Comput J 209–220
7. Hore S et al (2016) An integrated interactive technique for image segmentation using stack based seeded region growing and thresholding. Int J Electr Comput Eng. 6(6):2773
8. Chuang K-S, Tzeng H-L, Chen S, Wu J, Chen T-J (2006) Fuzzy c-means clustering with spatial information for image segmentation. Comput Med Imaging Graph 30:9–15
9. Abdel-Khalek S, Ben Ishak A, Omer OA, Obada ASF (2017) A two-dimensional image segmentation method based on genetic algorithm and entropy. Optik 131:414–422
10. Wei H, Tang X-s (2015) A genetic-algorithm-based explicit description of object contour and its ability to facilitate recognition. IEEE Trans Cybernet 45(11):2558–2571
11. Dash J, Bhoi N (2017) A thresholding based technique to extract retinal blood vessels from fundus images. Future Comput Inf J 2:103–109
12. Vorontsov E, Tang A, Roy D, Pal CJ, Kadoury S (2017) Metastatic liver tumor segmentation with a neural network-guided 3D deformable model. Med Biol Eng Comput 55:127–139

13. Roth HR et al (2015) DeepOrgan: multi-level deep convolutional networks for automated pancreas segmentation. In: MICCAI 2015: medical image computing and computer-assisted intervention, pp 556–564
14. Xu Y et al (2017) Gland instance segmentation using deep multichannel neural networks IEEE Trans Biomed Eng 99
15. Muthukrishnan R, Radha M (2011) Edge detection techniques for image segmentation. Int J Comput Sci Inf Technol 3(6):259
16. Dhanachandra N, Manglem K, Chanu YJ (2015) Image segmentation using K-means clustering algorithm and subtractive clustering algorithm. In: Eleventh international multi-conference on information processing-2015 (IMCIP-2015)
17. Al-amri SS, Kalyankar NV, Khamitkar SD (2010) Image segmentation by using thershold techniques. J Comput 2(2)
18. Senthilkumaran N, Rajesh R (2009) Image segmentation—a survey of soft computing approaches. In: International conference on advances in recent technologies in communication and computing
19. Chouhan SS, Kaul A, Singh UP (2018) Soft computing approaches for image segmentation: a survey. Multimedia Tools Appl
20. Zadeh LZ (1993) Fuzzy logic, neural networks and soft computing. Microprocessing Microprogramming 38(1–5):13. https://doi.org/10.1016/0165-6074(93)90117-4
21. Khan A, Jaffar MA (2015) Genetic algorithm and self organizing map based fuzzy hybrid intelligent method for color image segmentation. Appl Soft Comput 32:300–310

Low-Cost IoT Framework for Indian Agriculture Sector: A Compressive Review to Meet Future Expectation

Ashish Verma and Rajesh Bodade

Abstract This paper reviews the complete framework of IoT-based agriculture systems with a specific emphasis agricultural sector of India. Agriculture is a major contributor to the economies of developing countries, and it assists with satisfying the fundamental necessities of food, pay, and work for the population. Minimal effort with low cost is a significant factor in making any IoT network valuable and satisfactory to farmers. LoRa is a fewer power consuming, long-range remote systems administration innovation, reasonable for small-rate large area of applications in the Internet of Things. An IoT-based agriculture framework provides better resource management, crop management, improved production quality and quantity, cost-effective farming, crop monitoring, and field tracking. The IoT equipment and networking methods related to wireless sensors work in agricultural applications are thoroughly investigated. Advance image processing methods and the Internet of Things together produce a new way of smart agriculture systems. IoT-based framework system includes monitoring of crop status, field irrigation, insect and pest detection on field, weather monitoring, actuator intervention, expert suggestions and warning system for farmers, and automations. This study aims to determine the better method for a low-cost IoT-based agriculture system.

Keywords IoT · LoRaWAN · Image processing · Insect and pest · Irrigation · Machine learning · Deep learning

A. Verma (✉)
Institute of Engineering and Technology College Indore (IET-INDORE), Military College of Telecommunication Engineering MHOW-Indore (MCTE-MHOW), Mhow, India

R. Bodade
Faculty of Communication Engineering, Military College of Telecommunication Engineering MHOW-Indore (MCTE-MHOW), Mhow, India

© The Author(s), under exclusive license to Springer Nature Singapore Pte Ltd. 2022   241
D. Gupta et al. (eds.), *Proceedings of Data Analytics and Management*,
Lecture Notes on Data Engineering and Communications Technologies 90,
https://doi.org/10.1007/978-981-16-6289-8_21

1 Introduction

The agriculture sector is the prime factor of the Indian economy. The agriculture sector in India accounts for 18% of the country's GDP and employs more than half of the population. According to recent data, the population of whole world is anticipated 9.8 billion people by 2050, requiring a 25% rise in food production from current levels [1, 2]. In India, however, farming practices are still predominantly traditional [3]. The major challenges of the agricultural sector are extreme weather conditions, climate change, environmental impact, crop diseases, etc. [4]. The IoT is a technology with far-reaching consequences, including advanced networking of end-users, networks, and utilities [5].

The IoT framework for agricultural monitoring should be low cost and low power to allow for a long network existence that farmers can afford. Data is collected from nodes and uploaded to the server by the sync sensor [6]. Because the majority of IoT end equipment's sensor nodes are battery-power operated, power consumption must be kept to a minimum and power life profiles must be carefully built to extend battery life. End systems contain vast areas such that communication devices have to travel from a few meters to several kilometers. Only low-performance wide-range network (LPWAN) technology is able to do this [7]. In addition, IoT enables well-organized resource planning, which ensures optimum use of IoT to improve efficiency [8]. Figure 1 depicts a well-structured diagram of agricultural patterns that enable simple

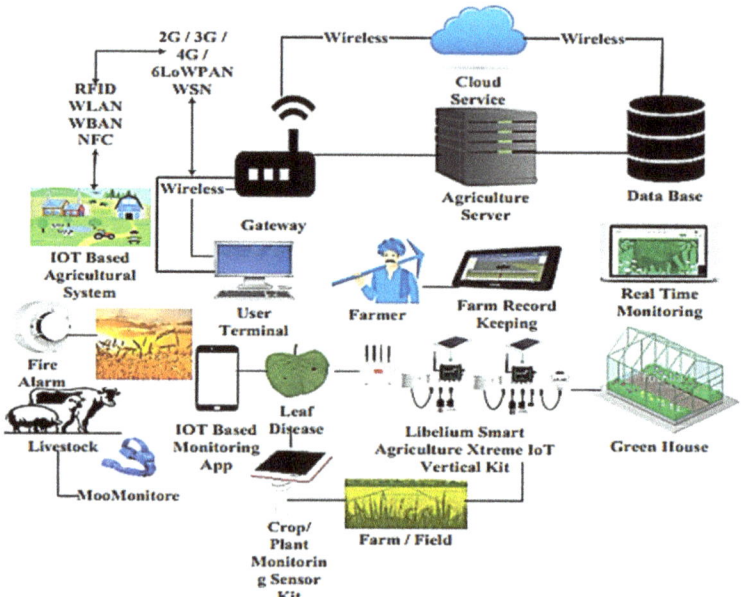

Fig. 1 Agriculture trends framework of connectivity across farmers, individual greenhouse, livestock, and field monitoring [5]

and cost-effective interactions by ensuring secure and uninterrupted communication through farmers, field monitoring, individual greenhouse, and livestock [5, 9]. This system uses Libelium vertical kit and MooMonitore which can replace to LoRaWAN and RFID, respectively in our system. LoRa is a "long-range" communication technology ensuring broad-area radio coverage, but due to its low data rate, it is unsuitable for transmitting multimedia data [10, 11]. Internet-based field monitoring system uses a number of sensors to track a particular area. Short-range radio systems such as ZigBee and Bluetooth are not ideal for wide-range transmission coverage. Mobile networks 2G–4G and other wireless connectivity systems can offer greater coverage, and they use too much energy and raise operational costs [12, 13]. Several LPWAN methods innovations are now on the market: LoRaWAN, SigFox, or NB-IoT [7]. This system will be used to prevent excessive use of various resources such as water and pesticides during land irrigation. Plant pathologies can be found in a variety of ways. Some diseases have no signs or only occur when it is too late to do something about them. In such cases, advanced analysis is needed, which is typically accomplished using powerful microscopes.

Remote sensing techniques that detect multi- and hyperspectral image capture are a popular approach in this situation. To accomplish their objectives, this approach often uses digital image processing tools [14–16]. Low-cost IoT-based framework system includes monitoring of crop status, field irrigation, insect and pest detection on the field, and this system also includes weather monitoring, actuator intervention, expert suggestions and warning system for farmers to create a complete framework for agriculture system especially for Indian scenario.

2 Literature Review

2.1 Automatic Recognition of Soybean Leaf Diseases Using UAV Images and Deep Convolutional Neural Networks

Tetila, Everton Castelão et al., this paper suggested a computer vision framework for detecting soybean diseases from UAV images. Leaves in captured images are planted into segments, and this approach uses the basic linear-iterative-clustering (SLIC) superpixels method. SLIC is simpler and requires less memory than superpixels-based approaches. SLIC superpixels are equated with k-means, with the k-parameter that refers to the number and scale control of superpixels in the image [17, 18]. Figure 2 The suggested scheme involves image dataset, image acquisition, SLIC segmentation, and classification of foliar diseases.

When segments are classified of images of plants, the computer vision system includes a postprocessing. This method determines the degree of disease infestation in each planting area, enabling efficient pathogen control over the target field. To detect plant leaves from images, SLIC superpixels procedure is used by the researchers to compare four deep learning models whereas ResNet-50, Inception-v3, Exception,

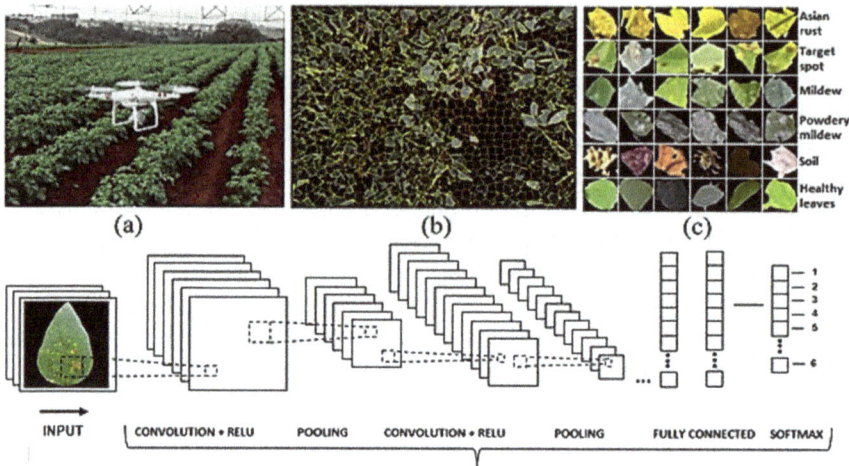

Fig. 2 Computer vision system **a** UAV **b** SLIC segment **c** Image dataset **d** Classification

and VGG-19. Compared four deep neural network models for fine-tuning (FT) that were trained with different parameters like training time, learning error, and accuracy. This technique will be tested for higher resolution and multi-hyperspectral cameras sensors in the future [17, 19].

2.2 An IoT-Based Smart Solution for Leaf Disease Detection

Apeksha-Thorat et al., this work included features such as leaf disease detection, soil moisture sensing, humidity sensing, and temperature sensing . Raspberry Pi-3 was used as a platform for mounting various sensors. A camera and RPI can be used to detect leaf disease. Farmers receive immediate farm status, such as a leaf disease, as well as other crop-affecting environmental factors including humidity, temperature, and moisture sensing through RPI with Wi-Fi gateway [2, 20]. Image analysis methods may be used to assess the condition of the plant. The steps are shown below (Fig. 3).

On the server side, data like humidity (DHT11 sensor), temperature, and soil moisture value are validated and matched with ideal values. A message is sent to the farmer via his cell phone or Web site if there is a discrepancy on the predefined threshold value and the real value. Image processing methods may be used to quantify the plant's state. Image processing is used to detect plant disease using red, blue, green (RGB) images, k-mean clustering, hue, saturation, value, brightness (HSV or HSB) images, and green pixel masking. The results of the leaf disease detection algorithm were compared to the results of the manual method. The results of image recognition in this paper are dependent on sunlight, and there is no power backup functionality.

Fig. 3 Steps to detect status of leaf [2]

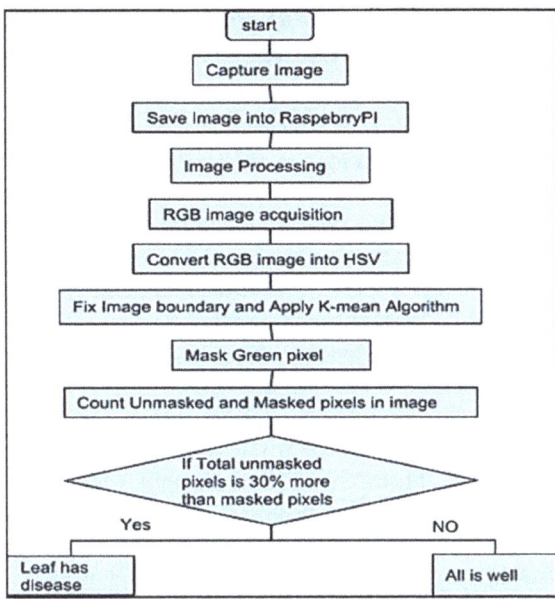

It can be scaled up to a large quantity in the future, with additional sensors such as soil quality monitoring, a buzzer for enemies, fertilization, and for a specific crop growth process [2].

2.3 A Low-Power Wide-Area Network Information Monitoring System by Combining NB-IoT and LoRa

X. Zhang et al., this study made a suggestion that LPWAN has low speed, less power consumption, wide coverage, and less-expensive wireless networking technology. The combination of NB-IoT and LoRa raises coverage area and lowers the operating expenditure of a WAN information monitoring system. Information processing, transmission, low voltage, solar power supply, node and sub-node features are all achieved using a hardware architecture system [21]. LoRa works on physical layer that uses spread spectrum modulation with sub-GHz ISM band frequency. Implementing the least recently used (LRU) communication algorithm and improving the LRU to process data will effectively reduce frame transfer rate of error while saving storage space [10]. Transmitting range of LoRa system can be propositional to transmit power and inverse propositional to the airborne rate. The sub-node serves as the hardware platform for the LoRa multi-hop communication [12, 22]. To increase the system's efficiency and functionality, future work can either improve the algorithm architecture in the programmed. For future aims, software and hardware designs are used in fields such as data collection from grassland and lake.

2.4   Internet of Things (IoT)-Based Smart Agriculture: Toward Making the Fields Talk

Muhammad-ayaz et al., this study shows significance of various wireless technologies. LPWAN techniques, in order to make the agriculture smartly, handle more efficient way to fulfilled future desire. The work provides an in-depth analysis of production driving factors on smart agriculture, primary agricultural technology generators, and technology adoption barriers. Many traditional agricultural problems will avoid by IoT including pest control, land suitability, irrigation and yield optimization. The work cover application hierarchy, issues of increased urban population, soil— sampling, fertilization, soil mapping, irrigation, pest control, yield tracking, crop disease control, forecasting, and harvesting, vertical farming (VF), hydroponics, and phenol-typing are being used. FAN is farm area network that represents framework in real time. Node-to-node wireless connectivity networking can be provided by BLE and ZigBee. It provides LPWAN communication between wireless sensors and the cloud with low power consumption. Farmers have an advantage of cloud computing because it allows them to access repositories which are based on knowledge that provide a wealth information, equipment choices available on the market, experiences about farming activities with all the requisite data. In the absence of a stable communication infrastructure, UAVs have an option [23].

2.5   A Low-Power IoT Network for Smart Agriculture

S. Heble et al., this paper proposes and advances a contact device that is built on embedded technology. In this approach, it uses special-use sensors for carbon dioxide and monitoring total solar radiation (TSR). They allowed the node to communicate in wireless protocols ZigBee, Bluetooth, and Wi-Fi. These techniques are allowing to reach monitoring, networking opportunities, lower power consumption. Amount of light, humidity, and atmospheric temperature are among the environmental parameters assessed. The proposed agricultural monitoring network consists of distributed nodes on specific area equipped with an IEEE 802.15.4 wireless domain, cameras, and solar-powered control circuit. The nodes sensors have an 80-m LOS range, to make reliable transmission range that covers the whole field and can easily approach sink node. The sink and sensor node's the IITH mote (Fig. 4), IEEE-802.15.4 compatible, built at the IITH(IITHWireles) Networks (WiNet) Lab.

Created a gateway for deliver data on server and enable availability over the Internet. An Intel Edison and 4G router make gateway, which connects to an IITH mote that serves as a drain. Future work suggested developing a LoRa-based portal to address the problem of power lines while also covering wide area for agriculture field. In the future, they expect to use remote control from drones for precision agriculture [24].

 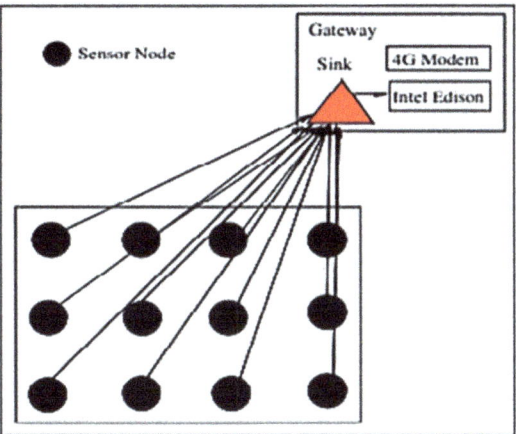

A. IITH Mote   B. Agricultural Monitoring Network Architecture

Fig. 4 IIIH mote and agricultural monitoring network architecture [24]

2.6 Internet of Things in Agriculture, Recent Advances, and Future Challenges

A. Tzounis et al., this paper starts with a review of recent advances in IoT with near-field communications (NFC) like technique radio-frequency identification (RFID), WSN, addressing of equipment's and cloud-based applications. It also gives a broad overview of IoT hardware, IoT wireless protocols, farming applications, and challenges. The three layers that make up the I data storage and manipulation system are the vision layer, network layer, and device layer. At the vision point, designers come across technologies such NFC, WSN. WSNs are commonly used in programmes that include the control and temperature regulation of storage. Logistics facility provided by RFID readers with help of Electronic Product Code (EPC), triggering, interpret, tags numbers. At the network layer, communication protocols are built on top of IEEE-802.15.4 wireless standards. It accepts networking through gateways to cover all end nodes. ZigBee, BLE, Sigfox, LoRa, and low-power Wi-Fi are examples of such protocols. Data points include sensors on all sides of structures, weather stations, and documentary evidence from archive on data libraries [25]. Heterogeneity is a big issue in the Internet of Things. Distributed computation and detailed facility control can help create the best environments for both animals and plants. Autonomous propose systems will be capable to control production in accordance with market conditions, optimizing profit, and minimizing costs [26].

2.7 An Automated Approach for Classification of Plant Diseases Toward Development of Futuristic Decision Support System in Indian Perspective

Y. Dandawate and R. Kokare et al., this paper focuses on a technique for detecting diseases problem in soybean plants by the use of image processing. In India, diseases and pests are a major cause of lower agricultural output, and protecting crops of fields from diseases and insect pests is the most difficult task for increasing production in farming [27]. Automated plant disease identification and recognition are a promising area, as it may be useful for monitoring vast fields, and recognize plant diseases basis on symptoms identified on the leaves of crop and plants [28] (Fig. 5).

Farmers can photograph plant leaves with any smartphone camera with a resolution of more than 2 megapixels. Context clutter, lighting shifts, shade, scale, direction, and broad categories of resolution are all problems addressed in the proposed work. During preprocessing, resize the file and render it smaller for RBG and HSV transfer. Image data acquisition for leaves of soybean, separation of soybean leaves and classified them from complex data history, statistical analysis, and classified disease are the four-key level in the proposed technique. For identifying plant organisms, the SIFT, scale-invariant feature transform is the methods used in the proposed work to extract key points from the input image for matching feature. If and when possible, DSS will provide adequate assistance to farmers over the phone. The SVM classifier shows its ability to automatically and correctly distinguish pictures. Using this approach, the leaves can be labeled with an average accuracy of 93.79%. Under the new scheme, farmers will be able to obtain agricultural specialist advice with minimal effort [14, 28]. One of the expectations is to incorporate stem of plants, fruit, and root level-based image processing techniques into future work. The suggested research can be applicable to a variety of agricultural crops [27].

Fig. 5 Block diagram of disease detection system [27]

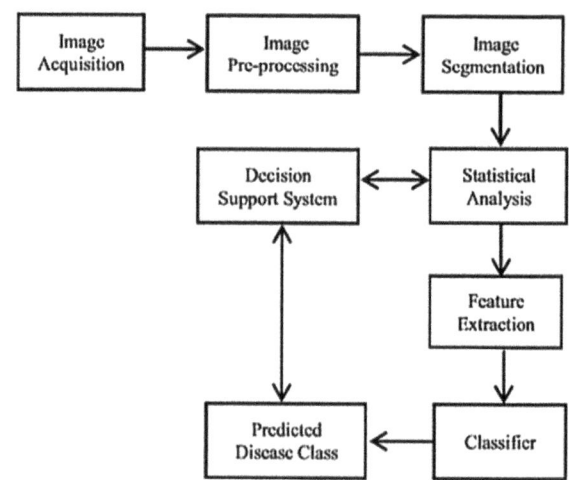

2.8 Thorough Study of LoRaWAN Performance Under Different Parameter Settings

D. Magrin et al., this paper describes how many network parameters like energy efficiency, transparency, and capacity affected by different network performance indices have been configured. This helps the network activity in principle to be adapted to the application scenario's particular requirements. Both gateway (GW) and the end devices that handle activities and capabilities with the networks are available. Depending on how packets of DL are handled, the three types standard of end devices are Class A, i.e., all, Class B, i.e., beacon, and Class C, i.e., continuous. Class A is used to receive packets of DL directly, then after receiving packets of UL, Class B devices will set the ping interval for DL packets. Class C devices can accept packets only before they send. Slightly altering the ACK protocol and prioritizing receipt over transmission at the GW will greatly enhance system performance in terms of energy consumption, system capacity, packet distribution ratio, and fairness, particularly in the presence of congestion [29].

2.9 Analysis of Latency and MAC-Layer Performance for Class A LoRaWAN

R. B. Sorensen et al., under regulatory obligation cycling restrictions, considering exponential inter-arrival times, this paper suggests computational models for analyzing the efficiency of the LoRaWAN uplink such as throughput, latency, and collision rate. LoRa spreading factors are 7–12 assisted by LoRaWAN. If LoRaWAN is loaded, but no MAC instructions are optional, overhead is 13 byte applied. In a star topology, a gateway supports several devices when relaying messages to a central server. The network processor in LoRaWAN uses an adaptive data rate to select the data rate and accessibility of channels to each node. Propose statistical models for evaluating LoRaWAN parameters throughput, latency, rate of collision, inter-arrival times, fixed payload size, duty cycling [7]. The UL model mentioned in this paper can be used in conjunction with DL models for LoRaWAN devices in Classes A, B, and C, as well as more complex models [30].

2.10 On the Use of LoRaWAN for Indoor Industrial IoT Applications

M. Luvisotto et al., LoRaWAN is used to evaluate the efficiency of industrial Internet of Things (IIoT) schemes, like as indoor industrial surveillance applications, in this article. A LoRaWAN network is made up of three types following are gateways, end devices, and network servers. Functional classes are divided into three categories of the LoRaWAN requirements: A, B, and C, with the first being required for all LoRaWAN end devices. To enter the channel at random, Class A EDs use an ALOHA-like device. Class B EDs open for predetermined intervals, while class C EDs hold them open all the time, sacrificing energy performance for low latency. The LoRaWAN requirements depict authentication and encryption schemes at different stages to ensure communication secrecy and security. Both licensed and unlicensed bands are available for LPWAN technologies [13]. IEEE 802.15.4 nodes consume more resources on average than LoRaWAN nodes. In this regard, it is worth remembering that the IEEE 802.15.4 protocol was introduced in its original form, with no particular power-saving technique implemented [8, 31]. Since EDs in LoRaWAN networks do not have direct Internet connections, they must be reached remotely by GWs. Though there have been plans to implement IPv6 over LoRaWAN, their feasibility and effect on network efficiency must be extensively researched [32, 33].

2.11 Identification of Plant Diseases Using Convolutional Neural Networks

Sachin B. Jadhav et al., using pre-trained AlexNet and GoogLeNet convolutional neural networks, this paper proposes an efficient soybean CNN-based disease detection method. The suggested pre-trained GoogLeNet CNN architecture fed the preprocessed images in the first phase of the study. The suggested models were retrained to classify artifacts from the set of disease data into four classes. In this work, 80 test datasets of soybean leaf images were tested by using AlexNet and GoogLeNet CNNs. Training datasets AlexNet's included 199 images of bacterial blight disease, 150 images of brown spot disease, 200 for "FLS" disease, 100 healthy, i.e., non-disease images. Same as, GoogLeNet training reports 150 bacterial blight disease, 150 FLS disease, 100 brown spot, and 150 healthy images. Black blight Class 1 disease, brown spot Class 2 disease, frogeye leaf spot Class 3 disease, and stable Class 4 disease [28]. Convolution layers use a series of convolutional filters to process the input data, different features unlocked from the images. A convolutional layer's output can be expressed as Eq. (1).

$$M_j^p = f\left(\sum_{i \in M_j} M_j^{p-1} * K_{ij}^p + N_j^p\right) \tag{1}$$

where p shows the pth sheet, convolutional kernel k_{ij}, Nj for bias, and Mj input maps. The kernel's bias and weight are two architecture parameters that are normally learned unsupervised. Reducing the network must learn number of parameters by pooling layers. According to E, the likelihood p for each area j in stochastic pooling should be computed first.

$$P_i = \frac{\alpha_i}{\sum k \epsilon S_{j\alpha_k}} \qquad (2)$$

Sj stands for the pooling area *j*, *F* for the function map, and *I* for the index of each vector within region *j*. For potential map *F* for each, the pooling procedure employs stochastic (St), which is denoted by

$$\alpha_{xy}^{p,k} = \mathrm{St}(m_i n, x, y) \varepsilon P\left(\alpha_{m,n}^{P-1,F} w(x, y)\right) \qquad (3)$$

where a *p*, *k x*, *y*; neuron activation at coordinate (*x*, *y*), feature map *F* in *p*th layer, and weighing function is *w* (*x*, *y*). According to the findings, in the classification of soybean diseases, the theoretical deep CNN model outperformed the machine learning model. The model's performance rate could be improved in the future by changing the weight minibatch size, bias learning rate [34, 35].

2.12 Disease Detection in Crops Using Remote Sensing Images

L. Shanmugam et al., this paper describes a system for detecting diseases automatically using remote sensing videos. The most crucial part of our research is the use of remote sensing imagery to detect the disease it appears on the leaves top layer. The first step in this process is to train data images for both stable and diseased images of leaves. After the collection of stable and contaminated photographs of samples has been obtained, the threshold is calculated. The RGB values are collected after that, and the image threshold is compared. During the training period, the MATLAB technique is used to isolate the layers of RBG from the image of the stable sample. Layer differentiation is critical in disease detection if the derived threshold values are shown to be greater or smaller than the threshold values of healthy samples. Convert the input sample to grayscale if the isolated sample varies from the normal values and is not attributable to age, which is following histogram method for analysis to diagnose a specific disease [28, 35].

The focus of future studies will be on satellite high-resolution images and sophisticated color extraction features that cover a wide area, and it will be expanded to provide satellite images in the future [36].

The Algorithm

- Training of system by healthy and infected crop.
- Each division has its own set of reference images.
- From the RGB image, separate the layers.
- From the picture, extract the threshold values.
- Take periodic crop images and compare them to the training image.
- If there is a significant difference in the threshold, proceed to the fourth step.
- If not, proceed to the disease identification procedure.

Disease Detection Algorithm

- Compare the two samples if the threshold is not due to aging.
- Convert a grayscale copy of the image.
- Use CANNY's edge detection method.
- Get the values from the histogram.
- Use the reference picture to identify the specific disease.
- Farmer receives an alert message.

2.13 Low-Cost Green Power Predictive Farming Using IOT and Cloud Computing

A. H. Adam, R. Tamilkodi, and K. V. Madhavi, this work is based on low-cost power solution made up of the IoT and cloud computing infrastructure. Predictive farming is a method of improving assets and growth by collecting essential data and deploying sensors on WSN. The proposed model is divided into two parts: The first part includes all IoT hardware equipment, sensors that produce real-time data, and the second part includes all IoT software devices. IoT applications like cloud computing services are included in the second segment. It has capability of storing real data on a scalable and reliable predictive farming system cloud, accessing real value on time generated data to analyze and function on predefined impairments to achieve the desired outcome. The nodeMCU is a multi-local node with an integrated Wi-Fi module that allows it to communicate with WSN via Wi-Fi. Multiple sensors mounted in agricultural fields will provide real-time data to the nodeMCUs. The data acquisition from each sensor is sent to the master node, which includes DHT22, GPS, soil moisture, light sensors, and a number of other sensors. Raspberry Pi serves as the master node. Cayenne collects data from RPI, processes it, and sends the production or outcome to the farmer's Web site as well as a text message to their computer [32, 37].

2.14 Wheat Disease Detection Using Image Processing

V. P. Gaikwad and V. Musande et al., this paper discusses numerous methods for detecting wheat disease, which is an important aspect of development. According to the first histogram form, good wheat has the highest peak rate relative to unhealthy wheat, implying that natural color occurs more often than other colors. The k-function aids in data clustering and returns an index based on the aid in clustering. For large amounts of data, k-means clustering is appropriate. In the second approach, employ two techniques: neural networks and SVM, i.e., support vector machines. The objective of paper is to create software that can detect and identify plant disease automatically. It comprises four steps: image data acquisition, preprocessing, fragmentation, and extraction of feature, which takes color, form, and size into account. This case used a neural network-based classifier for classification. The help vector machine outperforms the neural network in terms of accuracy. The proposed method effectively identifies and classifies wheat fungal disease. This study's future work will concentrate on improving the proposed algorithm in order to minimize classification error [38].

2.15 Smart Farming: Pomegranate Disease Detection Using Image Processing

M. Bhange and H. Hingoliwala et al., this work uploads fruit images to the framework in this report, and the authors suggest a web-based Web site to assist farmers in detecting fruit disease. The approach to "bacterial blight" disease of the "pomegranate fruit" is proposed as a web image processing approach. The image is clustered with the k-means algorithm after the image has been resized and features such as color, morphology, and CCV have been removed. SVM is then used to decide whether the image is damaged or not. The best results were obtained from the three characteristics derived. This condition affects plant stem, leaves, and berries, but fruit is the most distractive. Disease appears to be dark brown covered by dark yellow. This turns dark brown on the skin, fruit disease with water-swept lesions [38, 39]. The experimental findings indicate differing degrees of disease detection accuracy depending on the state of the input picture and the stage of the disease [15, 39]. The device's average precision is estimated to be 82%. In the future, the technology can be enhanced by incorporating additional features such as training the system to detect diseases in other fruits and increasing dataset size to boost overall system efficiency and more reliably detect diseases [39].

3   Discussion and Inference on Literature Survey

Smart agriculture and farming are now possible thanks to the Internet of Things (IoT). IoT networks reduce human physical working requirements by remotely tracking crop health and the field climate. A WSN serves as mainstay of the IoT, collecting data for monitoring and control applications. Sensor/actuator integration, digital transmission, low power consumption, scalability, and protection enable us to use in a variety of IoT applications. Smart farming includes irrigation, field surveillance, fertilizer control, RFID tags on vegetable or fruit plants for analytical monitoring, soil monitoring, disease detection, and water quality monitoring. Deep learning and machine learning algorithms with analytical algorithms are used to detect and identify diseased plant leaves and classify them using image processing techniques. On the LoRa gateway, all sensing and monitoring nodes are connected. Agriculture data from sensor network serves two types, first is real-time monitoring and controlling, and second is analysis by expert government agency with suggestion via cloud.

4 Inference on Literature Survey

S. No	Title of paper	Technology/methods	Future work/research gap
A	Automatic recognition of soybean leaf diseases using UAV images and deep convolutional neural networks	Use UAVs, and deep learning models were used to detect soybean leaf diseases • SLIC superpixels algorithm • k-means algorithm	Learning error and training time two problems; for better results, higher resolution and multi-hyperspectral camera will be used
B	An IoT-based smart solution for leaf disease detection	Detect leaf disease • Various sensors • Raspberry Pi • RGB, k-mean	It will be created for a large-scale structure. There is no power backup, and the work is dependent on sunlight
C	A low-power wide-area network information monitoring system by combining NB-IoT and LoRa	Centered on NB-IoT and LoRa, an LPWAN information monitoring method has been developed • LRU algorithm • Improved LRU	Future suggestions for grassland and lake tracking to increase accuracy and functionality by improving algorithm architecture
D	Internet of Things (IoT)-based smart agriculture: toward making the fields talk	Review discusses on all aspect • Soil sampling, mapping • Vertical farming • LPWAN technique • Unmanned aerial vehicles	Review the available new things that have been suggested for future smart farming to improve production and lower costs
E	A low-power IoT network for smart agriculture	IoT-based agriculture • Wi-Fi, ZigBee, and Bluetooth • IITH mote • LoRa-based gateway	Drone-based remote control was proposed as part of a LoRa network for precision agriculture
F	Internet of Things in agriculture, recent advances, and future challenges	Recent trends review • NFC, RFID, EPC, WSN • Micro-precision model • ZigBee, BLE, Sigfox, LoRa, LP Wi-Fi	Recent technological trends and cloud IoT solutions, climate optimization based on cloud analytics services is a viable option
G	An automated approach for classification of plant diseases toward development of futuristic decision support system in Indian perspective	Detection of plant disease using image processing • RGB image, SVM • k-mean clustering • SIFT • DSS	Future work includes stem, fruit, image of root founded processing and analysis with wide range of crops

(continued)

(continued)

S. No	Title of paper	Technology/methods	Future work/research gap
H	A thorough study of LoRaWAN performance under different parameter settings	LoRaWAN performance parameters • UL-PDR, CPSR • DL transmissions • Full-duplex GW	Best performance parameter variation on multi-gateways with depth analysis of networks
I	Analysis of latency and MAC-layer performance for Class A LoRaWAN	MAC-layer performance and analysis • Different nodes • ADR • UL model	More sophisticated collision models to better understand LoRaWAN's bidirectional efficiency
J	On the use of LoRaWAN for indoor industrial IoT applications	Industrial IoT networks • LoRaWAN network, ED • Classes A, B, and C • ALOHA–like scheme	Find better sophisticated collision models for bidirectional performance in LoRaWAN
K	Identification of plant diseases using convolutional neural network	Disease detection using deep learning model • CNN • Train CNN AlexNet and GoogLeNet	Future work, the model's success rate will be improved by manipulating minibatch size, weight, bias learning rate
L	Disease detection in crops using remote sensing images	Automated diseases detection • RGB method • CANNY's edge detection • Histogram analysis • Training of images	Future work will be on very high-resolution satellite images and advanced color extraction features that cover a wide area
M	Low-cost green power predictive farming using IOT and cloud computing	Predictive farming cloud • Wireless sensor networks • Cloud computing platform • Raspberry Pi DHT22, GPS, soil moisture, light sensors	The master node is a RPI. Cayenne gathers data from the RPI, processes it, and then sends the result to the farmer's Web site and their phone
N	Wheat disease detection using image processing	Disease detection • Histogram-based method • Comparing trained dataset • k-means cluster	This study's future work will concentrate on improving the proposed algorithm to reduce classification error
O	Smart farming: pomegranate disease detection using image processing	Propose a web-based tool • CCV • SVM • Training the system	This study's future work will concentrate on improving the proposed algorithm to minimize error

5   Conclusion

This paper explores the part of IoT in agricultural sector. Farmers spend a lot of time waiting for disease diagnosis on the crop because these diseases are difficult to identify until they occur over a wide area. After investigating literature, the smart farming is combination of heterogeneous technologies, and integration of information from various sources is ensuring increase of production with less use of resources. Early recognition of leaf diseases is important for prevention of crop. To achieve their goals, these methods frequently use digital image processing tools such as k-means clustering, SVM, RBG, CNN, and ANN with learning techniques. Implementation of the IoT network to address the challenges of smart agriculture main node linked sub-node end nodes centralized communication and managing, cloud connectivity, processing, and analysis, as well as everything can be done in a simple and convenient manner anywhere at any time. This paper took into account all of these factors, which emphasized the role of various methodologies, especially the Internet of Things with LoRa to conquer the problem of cost and power requirement, to make agriculture that is more intelligent and efficient effective in order to keep up with future demands. This research work focuses on privately one of the major objectives of our research work to develop the low cost and robust framework for Indian agriculture sector including the sensors and hardware interface and application implementation.

References

1. Newsletter of United Nation's Department of Economic and Social Affairs. World population projected to reach 9.8 billion in 2050, and 11.2 billion in 2100. 21 June 2017, New York. https://www.un.org/development/desa/en/news/population/world-population-prospects-2017.htm
2. Thorat A, Kumari S, Valakunde ND (2017) An IoT based smart solution for leaf disease detection. In: 2017 international conference on big data, IoT and data science (BID)
3. Raj S, Sehrawet S, Patwari N, Sathiya KC (2019) IoT based model of automated agricultural system in India. In: 2019 3rd international conference on trends in electronics and informatics (ICOEI)
4. Bhupal Naik DS, Ramakrishna Sajja V, Jhansi Lakshmi P, Venkatesulu D (2021) Smart farming using IoT. In: Bhattacharyya D, Thirupathi Rao N (eds) Machine intelligence and soft computing. Advances in intelligent systems and computing, vol 1280. Springer, Singapore. https://doi.org/10.1007/978-981-15-9516-5_34
5. Farooq MS, Riaz S, Abid A, Abid K, Naeem MA (2019) A survey on the role of IoT in agriculture for the implementation of smart farming. IEEE Access 7:156237–156271
6. Heble S, Kumar A, Prasad KVVD, Samirana S, Rajalakshmi P, Desai UB (2018) A low power IoT network for smart agriculture. In: 2018 IEEE 4th world forum on internet of things (WF-IoT)
7. Haxhibeqiri J, Poorter ED, Moerman I, Hoebeke J (2018) A survey of LoRaWAN for IoT: from technology to application. Sensors 18(11):3995
8. Rajalakshmi P, Devi Mahalakshmi S (2016) IOT based crop field monitoring and irrigation automation. In: 10th international conference on intelligent systems and control (ISCO), 7–8 Jan 2016 published in IEEE Xplore Nov 2016

9. Ahmed L, Nabi F (2021) Agriculture 5.0—the future. In: Agriculture 5.0: Artificial Intelligence, IoT, and Machine Learning, CRC Press, 2021, pp. 187–203, https://doi.org/10.1201/978100 3125433-9.
10. Ji M, Yoon J, Choo J, Jang M, Smith A (2019) LoRa-based visual monitoring scheme for agriculture IoT. In: 2019 IEEE sensors applications symposium (SAS)
11. LoRa world coverage available in: www.lora-alliance.org/
12. Zhang X, Zhang M, Meng F, Qiao Y, Xu S, Hour S (2019) A low-power wide-area network information monitoring system by combining NB-IoT and LoRa. IEEE Internet Things J 6(1):590–598
13. Mekki K, Bajic E, Chaxel F, Meyer F (2019) A comparative study of LPWAN technologies for large-scale IoT deployment. ICT Express 5(1):1–7
14. Jain P, Sarkar R (2018) IoT based smart field monitoring system with disease identification. Int J Pure Appl Math 118(22):703–707. ISSN: 1314-3395
15. Barbedo JGA (2013) Digital image processing techniques for detecting, quantifying and classifying plant diseases. SpringerPlus 2(1)
16. Halder S, Kumar Singh S (2021) Knowledge-based expert system for diagnosis of agricultural crops. In: Bhattacharjee D, Kole DK, Dey N, Basu S, Plewczynski D (eds) Proceedings of international conference on frontiers in computing and systems. Advances in intelligent systems and computing, vol 1255. Springer, Singapore. https://doi.org/10.1007/978-981-15-7834-2_33
17. Tetila EC et al (2020) Automatic recognition of soybean leaf diseases using UAV images and deep convolutional neural networks. In: IEEE geoscience and remote sensing letters, no 5, institute of electrical and electronics engineers (IEEE), pp 903–07. https://doi.org/10.1109/ lgrs.2019.2932385
18. Tetila EC, Machado BB, Belete NA, Guimaraes DA, Pistori H (2017) Identification of soybean foliar diseases using unmanned aerial vehicle images. IEEE Geosci Remote Sens Lett 14:2190–2194
19. Lottes P, Khanna R, Pfeifer J, Siegwart R, Stachniss C (2017) "UAV" based crop and weed classification for smart farming. In: 2017 IEEE international conference on robotics and automation (ICRA), pp 3024–3031
20. Prathibha SR, Hongal A, Jyothi MP (2017) IOT based monitoring system in smart agriculture. In: 2017 international conference on recent advances in electronics and communication technology
21. Mekki K et al (2019) A comparative study of LPWAN technologies for large-scale IoT deployment. In: ICT Express, no 1, Elsevier BV, pp 1–7. https://doi.org/10.1016/j.icte.2017. 12.005
22. Georgiou O, Raza U (2017) Low power wide area network analysis: can LoRa scale. IEEE Wirel Commun Lett 6(2):162–165
23. Ayaz M, Ammad-Uddin M, Sharif Z, Mansour A, Aggoune E-HM (2019) Internet-of-things (IoT)-based smart agriculture: toward making the fields talk. IEEE Access 7:129551–129583
24. Tetila EC, Machado BB, Menezes GK, Oliveira ADS, Alvarez M, Amorim WP, Belete NADS, Silva GGD, Pistori H (2020) Automatic recognition of soybean leaf diseases using UAV images and deep convolutional neural networks. IEEE Geosci Remote Sens Lett 17(5):903–907
25. Agrawal H, Dhall R, Iyer KSS, Chetlapalli V (2019) An improved energy efficient system for IoT enabled precision agriculture. J Ambient Intell Humanized Comput 1–12
26. Tzounis A, Katsoulas N, Bartzanas T, Kittas C (2017) Internet of things in agriculture, recent advances and future challenges. Biosys Eng 164:31–48
27. Dandawate Y, Kokare R (2015) An automated approach for classification of plant diseases towards development of futuristic decision support system in Indian perspective. In: 2015 international conference on advances in computing, communications and informatics (ICACCI)
28. Yang L et al (2018) Identification of rice diseases using deep convolutional neural networks. Neurocomputing 267:378–384
29. Magrin D, Capuzzo M, Zanella A (2020) A thorough study of LoRaWAN performance under different parameter settings. IEEE Internet Things J 7(1):116–127

30. Sorensen RB, Kim DM, Nielsen JJ, Popovski P (2017) Analysis of latency and MAC-layer performance for class A LoRaWAN. IEEE Wirel Commun Lett 6(5):566–569
31. "The Constrained Application Protocol (CoAP)." [Online]. Available: https://tools.ietf.org/pdf/rfc7252.pdf. [Accessed: 01-June-2019]
32. Pianini D, Salvaneschi G (2018) IoT architectural framework: connection and integration framework for IoT systems. In: First workshop on architectures, languages and paradigms for IoT EPTCS 264, pp 1–17. https://doi.org/10.4204/EPTCS.264.1
33. Luvisotto M, Tramarin F, Vangelista L, Vitturi S (2018) On the use of LoRaWAN for indoor industrial IoT applications. Wirel Commun Mob Comput 2018:1–11
34. Jadhav SB et al (2020) Identification of plant diseases using convolutional neural networks. Int J Inf Technol (Springer Science and Business Media LLC). https://doi.org/10.1007/s41870-020-00437-5
35. Practical Deep Learning Examples with MATLAB (2018) Math Works, Inc., pp 1–33
36. Shanmugam L, Adline ALA, Aishwarya N, Krithika G (2017) Disease detection in crops using remote sensing images. In: 2017 IEEE technological innovations in ICT for agriculture and rural development (TIAR)
37. Adam AH, Tamilkodi R, Madhavi KV (2019) Low-cost green power predictive farming using IOT and cloud computing. In: 2019 international conference on vision towards emerging trends in communication and networking (ViTECoN)
38. Gaikwad VP, Musande V (2017) Wheat disease detection using image processing. In: 2017 1st international conference on intelligent systems and information management (ICISIM)
39. Bhange M, Hingoliwala H (2015) Smart farming: pomegranate disease detection using image processing. Procedia Comput Sci 58:280–288

Machine Learning-Based Breeding Values Prediction System (ML-BVPS)

S. V. Vasantha and B. Kiranmai

Abstract Understanding the connection between a genotype and its phenotype is a key challenge in predicting the breeding values. However, genotype-to-phenotype prediction presents significant challenges for machine learning algorithms, limiting their use in this context. The data's high dimensionality makes generalization difficult and limits the scalability of most learning algorithms. The accurate prediction of phenotypes is needful in improving crop breeding. We analyzed GWAS and implemented a strategy (ML-BVPS) for the prediction of rice plant height based on its genotype. We implemented Machine learning algorithms for the classification of rice subpopulation and phenotype prediction. We achieved 94% accuracy in classifying the rice population. We achieved an accuracy range of 0.82–1.0 in the prediction of phenotype value based on its Lead SNP markers. We also recommend genotype and its corresponding GWAS information for each subpopulation category to obtain a better breeding value.

Keywords Genotype · Phenotype · GWAS · Machine learning · Breeding value

1 Introduction

Rice (*Oryza sativa* L.) is a major food crop that provides food for more than half of the world's population [1]. Isozyme analysis [2] was the first widely used molecular classification of rice groups, which found six varietal groups. Following DNA analysis, five subpopulations of rice were discovered within the two Oryza subspecies. The aus subpopulation is a group of people who live in Australia [3]. The most popular rice subpopulation, indica and japonica, are differentiated by genetic information.

Intermediate between wild family and cultivars, it provides important information for crop improvement by creating a genetic reservoir that can adapt to environmental changes and increase crop sustainability. Garris et al. [4] proposed that *O. sativa* can be divided into five different groups based on model-based structure analysis:

S. V. Vasantha · B. Kiranmai (✉)
Department of CSE, KMIT, Narayanguda, India

© The Author(s), under exclusive license to Springer Nature Singapore Pte Ltd. 2022   259
D. Gupta et al. (eds.), *Proceedings of Data Analytics and Management*,
Lecture Notes on Data Engineering and Communications Technologies 90,
https://doi.org/10.1007/978-981-16-6289-8_22

indica, aus, aromatic, temperate japonica, and tropical japonica rice. As a result, there are still differences in the genetic structures or taxa of cultivated rice species. Each subpopulation has desirable cultivar characteristics, breaking down genetic barriers between them will allow rice breeders to freely introduce desirable genes from different subpopulations into an elite line. This will help to reduce breeding costs and ensure the long-term viability of rice production, which is threatened by ongoing climate change [5, 6].

The reduction of genotyping cost and availability of high-throughput genotyping services has made feasible to deploy various genomic selection approaches in regular plant breeding programs [7]. Oryza Sativa genetic markers are used as a genotyping tool for genetic analysis and marker assisted breeding. The restriction fragment is the first type of molecular marker used in genetic analysis [8].

Genomic Selection (GS) is a plant breeding technique that uses genome-wide marker data to predict the genetic value of untested lines. The technique has been extensively tested in both simulated and real-world plant breeding programs.

The data used in the GS model, such as the size of the training population, relationships between individuals, marker density, and use of pedigree information, has been shown to affect the accuracy of GS [9]. In hybrid breeding, genomic selection (GS) is more effective than traditional phenotype-based methods [10].

Genomic prediction (GP) using single nucleotide polymorphism (SNP) markers has emerged as a useful tool in a variety of human healthcare settings [11]. Singh et al. [12] looked at allelic sequence variation in three Sub1 genes in a panel of 179 rice genotypes and came up with some interesting results.

Kim [6] used phenotyping variables and the logistic regression model (LRM) to classify indica/japonica. They tailored LRM to classify indica and japonica based on seven phenotypic factors. Jeon et al. [13] used GP modeling to create a prediction model with the best marker set. The interaction effects of the SNP markers are considered indirectly through modeling, and statistical, machine, and/or deep learning techniques are used and put to the test on actual datasets with four different phenotypes. The prediction results were better than those obtained by using the entire set of markers or the GWAS-top markers, which are commonly used in prediction studies. Despite its flaws, the GMStool is expected to help researchers predict quantitative phenotypes in a variety of studies. Li et al. [14] noticed that the RF and particularly GBM algorithms are efficient in identification of SNPs subset having direct correspondence to the candidate genes that are responsible for growth trait.

Wang et al. [10] assess the predictive ability of GS for hybrid performance in rice using the NC II scheme, in which 115 inbred rice lines were crossed with 5 male sterile lines, and to show how predicting characteristics can be used to predict potential crosses between the 115 inbred lines and other genotyped varieties. They have worked on GBLUP for predicting genetic values and phenotypes using whole genome markers. Yan et al. [11] devised subpopulation classification using Phylogenetic Tree and predicted Genomic Estimation Breeding values for four phenotypes.

2   Problem Statement

Significant work carried out for predicting Breeding Values of *O. Sativa* populations such as, Yan et al. [11] solution for predicting *O. Sativa's* Genomic Breeding values and its accuracy highly varied from 0.23 to 0.90. Prediction accuracy of solutions presented by Joen [13] and [10] are 0.52 and 0.88, respectively. Thus, there is a need for improving the prediction accuracy of Breeding Values (BVs) of Oryza species. Hence, the ML based Breeding Values Prediction System (ML-BVPS) for O. Sativa is designed.

3   Proposed Model

It mainly focuses on improving prediction accuracy of Breeding Values (BVs) in *O. Sativa*. It involves a strategy that initially finds out the type of subpopulation of each given sample. And then based on the identified type, it predicts the BVs. It also recommends genotypes for better breeding values for each subpopulation type.

The proposed model comprises of three main modules.

(i)   Classification of Oryza samples based on subpopulation types
(ii) BVs Prediction
(iii) Genotype Recommendation system for Gene Editing.

Data Gathering and Preprocessing

Rice Cultivars data related to phenotype, GWAS, and its genotype information is collected from the Data-Source: [15] http://ricevarmap.ncpgr.cn/. Phenotype values and GWAS information are extracted for the plant height. Next its corresponding imputed genotype information is extracted. Datasets are prepared by normalizing and scaling the extracted data for further classification and BVs prediction tasks.

Classification of Oryza samples based on subpopulation types

This module deals with classification of given *O. Sativa* samples based on its subpopulation type. It uses imputed SNP markers which are spread across twelve chromosomes. Usually, SNP markers, which improve classification accuracy, are selected for identifying the type of subpopulation, but the proposed model is designed to improve BVs prediction accuracy, thus LEAD SNP markers specific to phenotype are selected.

Various popular ML algorithms are applied to classify the given cultivars such as logistic regression, Naive Bayes, decision trees, random forests, and many more. Among these algorithms, random forest classifier exhibited most promising results. Data is partitioned into training sets and testing sets out of which, 80% is used for training and 20% is used for testing.

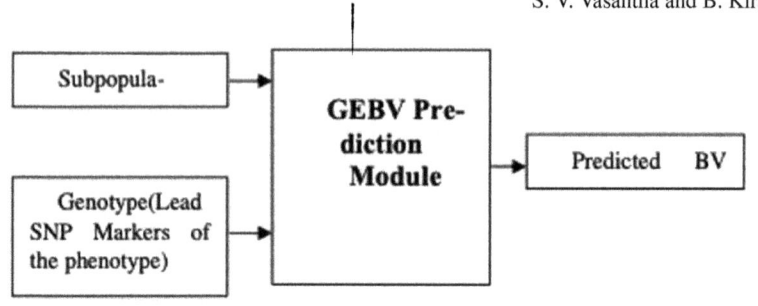

Fig. 1 BV prediction module

BVs Prediction

This module is designed to predict BVs of the given samples with improved prediction accuracy, which is depicted in the Fig. 1. It takes genotype, that is, the Lead SNP markers set specific to the phenotype to be predicted along with subpopulation type predicted in first module as inputs. These inputs are processed by ML algorithms, to predict a phenotypic value range. Here the ML model is trained based on the genotype along with its formulated phenotypic value range of the samples for each subpopulation category so as to improve BV prediction accuracy.

K-means Algorithm is used to formulate phenotypic value range of the samples. Various well known ML techniques are applied to identify appropriate phenotype range based on the given sample's genotype. 80% of dataset is considered for training and 20% of it is considered for testing. Random forest classifier expressed most promising prediction accuracy among other ML techniques.

Genotype Recommendation system for Gene Editing

This module provides suggested genotype for given phenotype with respect to each subpopulation category, which is shown in the Fig. 2. Here one of the significant phenotypic traits of *O. Sativa* known as plant height is considered and for which the genotype is recommended for better BV.

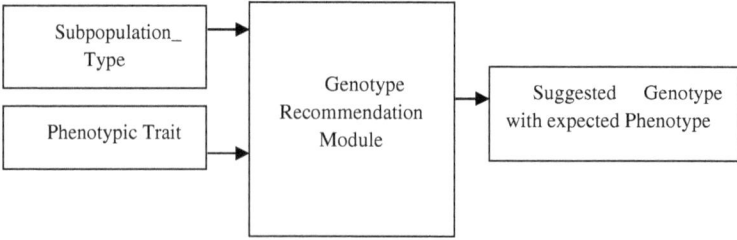

Fig. 2 Genotype recommendation module

Suggested genotype is framed based on the subpopulation's best samples with respect to given phenotype. Thereby, it facilitates Gene editing by providing its corresponding GWAS information.

4   Results Discussion and Comparative Analysis

This section gives details about normalization and scaling of data, results of subpopulation classification and BV prediction and also their comparative analysis with other ML techniques and some contemporary models.

(i)   **Data Normalization and Scaling**

The genotype data is normalized as follows:

A—1, C—2, G—3, T—4, N—5, H—6

Plant height value ranges are formulated with respect to each subpopulation as follows:

Aromatic	129, 130–158, 160–189
Aus	111–159, 161–190
Tropical Japonica	84–141, 143–183
Temporate Japonica	74–98, 99–129, 131–207
Indica	69–132, 133–204

(ii)   **Subpopulation Classification Results**

The subpopulation classification accuracy of given *O. Sativa* samples when RF Classifier is applied along with the parameters tuning information is expressed in below Table 1.

Its Classification Report is as follows:

Table 1   Subpopulation classifier accuracy

Test size	Random_state	Classifier_Accuracy	Classifier prediction accuracy of test data
0.1	800	0.94	0.84
0.15	800	0.93	0.81
0.2	800	0.91	0.78

	precision	recall	f1-score
Aus	0.80	1.00	0.89
Indica I	0.83	0.83	0.83
Indica II	0.82	0.90	0.86
Indica III	0.00	0.00	0.00
Indica Intermediate	0.71	0.56	0.63
Intermediate	1.00	0.50	0.67
Japonica Intermediate	1.00	0.67	0.80
Temperate Japonica	0.92	1.00	0.96
Tropical Japonica	1.00	1.00	1.00
VI/Aromatic	1.00	1.00	1.00
accuracy			0.85
macro avg	0.81	0.75	0.76
weighted avg	0.87	0.85	0.85

(iii) Comparative Analysis of Subpopulation Classifier Accuracy

Various popular ML techniques such as random forest, neural network MLP, ridge classifier, ridge classifierCV, Gaussian Naïve Bayes, and decision tree are applied for classifying samples based on the type of subpopulation. Classifier Accuracy of these techniques are expressed in the Figure 3. Random forest classifier is exhibiting the better accuracy compared to other techniques.

(iv) Phenotype Prediction Results

Phenotype value of plant height trait is predicted for the whole population, but its accuracy is comparatively less than the accuracy of each subpopulation classified by its type and mean accuracy of all the subpopulations is 0.92. Phenotype prediction accuracy of each subpopulation and whole population along with parameter tuning information is expressed in the Table 2

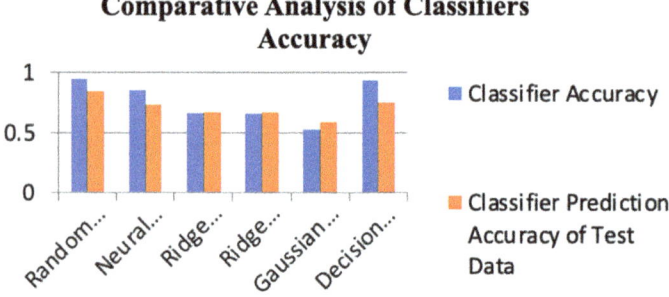

Fig. 3 Comparative analysis of classifiers accuracy

Table 2 Phenotype prediction accuracy of subpopulations

Subpopulation type	Test size	Random_State	Classifier accuracy	Classifier prediction accuracy of test data
Indica	0.25	350	0.99	1.0
Aromatic	0.25	50	1.0	1.0
Temporate Japonica	0.2	350	0.91	0.82
Aus	0.25	500	0.93	0.91
Tropical Japonica	0.2	1800	0.95	0.88
Whole population	0.25	350	0.85	0.57

Table 3 Comparative analysis of various ML techniques

Classifier/regressor name	Classifier/regressor accuracy	Prediction accuracy of test data
Random forest classifier	0.99	1
Ridge classifier	0.78	0
Ridge classifierCV	0.73	0
Logistic regressor	0.65	0.63
GLM	0.45	0.75

(xxii)   Comparative Analysis of Phenotype Prediction Accuracy

Plant height phenotype is predicted by applying various ML techniques such as random forest classifier, ridge classifier, ridge classifierCV, logistic regressor, and GLM. Among all these techniques, random forest classifier expressed most promising prediction accuracy which is shown in Table 3.

Comparative Analysis of Phenotype Prediction Accuracy with contemporary solutions is expressed in Table 4. Our ML-BVP method, mean prediction accuracy is outperforming when compared to other solutions. Prediction accuracy of ML-BVP ranges from 0.82 to 1.0 which shows drastic enhancement when compared to [11] solution.

Table 4 Comparative analysis of contemporary techniques

Method	Prediction accuracy
ML-BVP (Proposed)	0.92
Joen et al.	0.52
Wang et al.	0.88
Yan et al.	0.2–0.9

5 Conclusion

The proposed ML-BVP system is used to classify *O. Sativa* samples based on their population type, the same is used as a key input for predicting BVs. It is observed that the prediction accuracy is improved over whole population samples and other solutions. RF exhibited most promising results when compared other ML models.

References

1. Khush GS (2005) What it will take to feed 5.0 billion rice consumers in 2030. Plant Mol Biol 59(1):1–6
2. Glaszmann JC (1987) Isozymes and classification of Asian rice varieties. Theor Appl Genet 74(1):21–30. https://doi.org/10.1007/BF00290078 PMID: 24241451
3. Ahmed ZU, Panaullah GM, Gauch H, McCouch SR, Tyagi W, Kabir MS, Duxbury JM (2011) Genotype and environment effects on rice (*Oryza sativa* L.) grain arsenic concentration in Bangladesh. Plant Soil 338(1):367–382
4. Garris AJ, Tai TH, Coburn J, Kresovich S, McCouch S (2005) Genetic structure and diversity in *Oryza sativa* L. Genetics 169(3):1631–1638
5. Kim B (2018) Classifying Asian rice cultivars (*Oryza sativa* L.) into Indica and Japonica using logistic regression model with publicly available phenotypic data. bioRxiv. 1 Jan 2018 (470351)
6. Kim B (2019) Classifying *Oryza sativa* accessions into Indica and Japonica using logistic regression model with phenotypic data. PeerJ 7:e7259
7. Vinayan MT, Seetharam K, Babu R, Zaidi PH, Blummel M, Nair SK (2021) Genome wide association study and genomic prediction for stover quality traits in tropical maize (*Zea mays* L.). Sci Rep 11(1):686. Published 12 Jan 2021. https://doi.org/10.1038/s41598-020-80118-2
8. Tuhina-Khatun M, Hanafi MM, Rafii Yusop M, Wong MY, Salleh FM, Ferdous J (2015) Genetic variation, heritability, and diversity analysis of upland rice (*Oryza sativa* L.) genotypes based on quantitative traits. BioMed Res Int
9. Robertsen CD, Hjortshøj RL, Janss LL (2019) Genomic selection in cereal breeding. Agronomy 9(2):95
10. Wang X, Li L, Yang Z, Zheng X, Yu S, Xu C, Hu Z (2017) Predicting rice hybrid performance using univariate and multivariate GBLUP models based on North Carolina mating design II. Heredity 118(3):302–310
11. Yan J, Zou D, Li C, Zhang Z, Song S, Wang X (2020) SR4R: an integrative SNP resource for genomic breeding and population research in rice. Genomics Proteomics Bioinform 18(2):173–185
12. Singh A, Singh Y, Mahato AK, Jayaswal PK, Singh S, Singh R, Yadav N et al (2020) Allelic sequence variation in the Sub1A, Sub1B and Sub1C genes among diverse rice cultivars and its association with submergence tolerance. Sci Rep 10(1):1–18
13. Jeong S, Kim JY, Kim N (2020) GMStool: GWAS-based marker selection tool for genomic prediction from genomic data. Sci Rep 10(1):1–12
14. Li B, Zhang N, Wang Y-G, George AW, Reverter A, Li Y (2018) Genomic prediction of breeding values using a subset of SNPs identified by three machine learning methods. Front Genet 9:237. https://doi.org/10.3389/fgene.2018.00237
15. http://ricevarmap.ncpgr.cn/
16. http://variation.ic4r.org/
17. http://www.ricediversity.org/
18. http://ricepedia.org/rice/rice-as-a-plant/rice-species

Machine Learning Based Supraventricular Tachycardia Detection Model of ECG Signal

Pampa Howladar, Khokan Mondal, and Manodipan Sahoo

Abstract The study of electrocardiogram (ECG) signals is an important method for diagnosing cardiac illness, as the ECG procedure is non-invasive and simple to use. This work presents an ECG model for the prediction that consists of a few steps such noise filtering, unique set of ECG features and machine learning based classifier model aiming to detect supraventricular tachycardia arrhythmia. We de-trend and denoise the signal before features extractions to remove noise for better detect functionality. The requisite features were extracted and parameters correlated with this features are calculated. Using these parameters, we developed a machine learning based model with a classification system which can effectively detect supraventricular tachycardia arrhythmias. Our findings show that the most powerful models of machine learning for supraventricular tachycardia arrhythmia are the logistic regression and decision tree models. By using this approach, this work solves the issue of reducing the critical signal misclassification of the supraventricular tachycardia with great efficiency. Experimental data shows satisfactory improvements and show a strong algorithm robustness we suggested.

Keywords Electrocardiography (ECG) · ECG signals · Filtering · Data classification · Feature extraction · Supraventricular tachycardia arrhythmia

1 Introduction

The supraventricular tachycardia (SVT) is an abnormally rapid heart-rhythm that arises from incorrect electrical activity in the upper part of the heart. Speeded-up rhythms can panic the patient whether they are chronic or constant, which can induce severe illness. The supraventricular tachycardia (SVT) is a type of arrhythmia that

P. Howladar · K. Mondal
Indian Institute of Engineering Science and Technology, Shibpur, Howrah, India

M. Sahoo (✉)
IIT (ISM), Dhanbad, India
e-mail: manodipan@iitism.ac.in

© The Author(s), under exclusive license to Springer Nature Singapore Pte Ltd. 2022   267
D. Gupta et al. (eds.), *Proceedings of Data Analytics and Management*,
Lecture Notes on Data Engineering and Communications Technologies 90,
https://doi.org/10.1007/978-981-16-6289-8_23

occurs above the Bundle of His level, which includes normal atrial, irregular atrial, and regular atrioventricular tachycardias. In this arrhythmia, ECG displays narrow, complex tachycardia in the absence of an aberrant conduction (e.g., bundle of a branch block) [1, 2].

Symptoms and signs can appear all of a sudden and can heal without medication. Stress, exercise and emotion may contribute to natural or physiological changes in heart rate but can precipitate SVT more seldom. Episodes will last for a couple of minutes to a couple of days, often lingering until treated. The fast heart rate limits the possibility for the pump to fill the cardiac supply between beats and the subsequent blood pressure. The symptoms are typical of 150–270 beats or more per minute. The symptoms observed are shortness of breath, rapid breathing, chest pain, dizziness, pounding heart, loss of consciousness (in only the worst serious cases) [3, 4].

In the last few decades, there has been an unprecedented pace of surgical, clinical and technological developments. Since then, substantial attempts have been made to take advantage of the technical advances and computer applications in the medical field. Since the electrocardiogram (ECG) is used to study the most important organ in the human body, cardiologists are especially interested in the heart's cardiac effects, ECG analysis with the greatest precision [5]. In the field of ECG analysis, several studies have been performed to automatically detect signals at almost perfect speeds. Efforts to model the skills of cardiologists and specialists using computers have been focused over the last thirty years. This issue has been addressed by many researchers who have developed various ECG recognition, and QRS detection algorithms which have become widely recognized in the literatures [6–10]. Among the approaches analyzed and evaluated, the machine learning approach has taken special interest because of its features, such as nonlinearity, learning capacity and a universal approach allowing it to overcome complicated signals such as QRS detection and SVT diagnostics. This paper explores our approach to developing and applying a better approach to machine learning for supraventricular tachycardia (SVT) detection.

1.1 Electrocardiogram Signals

The electrocardiogram (ECG) tracks the heart's induced electrical activity on the body surface. The ECG was first noted by Waller in 1899. In 1903, electrical waves with 398 electrophysiological components were named by Einthoven, are still in use. His P-to-U letters were attributed to waves, avoiding inconsistencies with other physiological waves he researched. Figure 1 displays the standard ECG signal. The ECG signals normally range from $\pm$ 2 mV amplitude and bandwidth 0.05–150 Hz. ECG wave morphology is based on the amount of tissue involved in a time unit and the relative heart activation rate and direction [9].

Consequently, ECG cannot detect the physiological potential of the pacemaker, i.e., the SA-nodal, which is created by a comparatively small myocardial mass. The first ECG wave in the cardiac cycle is the P-Wave representing atrial depolarization.

Fig. 1 A diagram of a
normal ECG signal

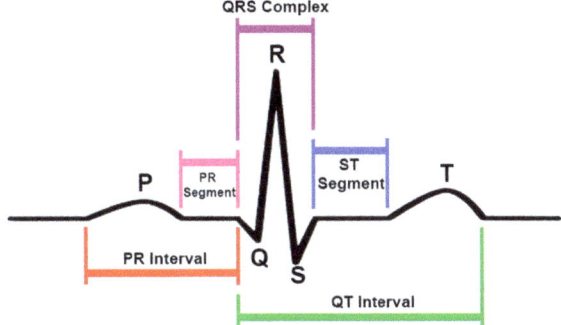

The cardiac impulse is directed by a sequence of specialized heart structures from the atria to the ventricles (the AV node and the His-Purkinje system). The tiny isoelectric segment following the P wave is the PQ interval, which is related to the AV node propagation delay of 0.2 s. The ECG surface is easily and extensively modified when the ventricle's large muscle mass is excited after the QRS complex there is a second isoelectric portion, the ST interval. After ventricular cell activation, the ST Interval shows the depolarizing time, usually from 0.25 to 0.35 s. If the ST segment is ended, the ventricular cells return to their resting electrical and mechanical position and complete the period of repolarization as a low-frequency T wave signal. Some people have a light peak called the U wave at the end right after the T wave [11]. Its origin has never been entirely identified, but is called a repolarization potential.

1.2 Supraventricular Tachycardia Arrhythmias

Heart defects that cause cardiac beats irregular are due to arrhythmias. Rhythmic heart disorders (heart arrhythmias) occur if electric pulses that are not properly coordinated resulting in making the heartbeats too fast, too slowly or irregularly. The location from which tachycardia originates in the heart may be classified into two categories. One is supraventricular tachycardia (SVT) that begins in upper part of the heart, usually called the atria and another is ventricular tachycardia (VT) which begins in the lower chamber of the heart which is called the ventricles. The diagnosis of bradycardia is made if less than 60 beats a minute are detected and when the heart beats over 100 beats per minute at cardiac level, tachycardia is detected [3, 4].

Several kinds of arrhythmias are present. One of the major cardiovascular arrhythmias is supraventricular tachycardia (SVT). This type of arrhythmia begins in the atria (upper chambers of the heart). "Supra" represents above, "ventricular" means to the lower chambers of the heart, or ventricles. In this type of arrhythmia, the ECG will display a narrow-complex tachycardia in the absence of aberrant conduction (e.g., bundle branch blocks).

Symptoms: Symptoms usually associated with Supraventricular Tachycardia includes dyspnea, chest discomfort or pressure, light headedness or dizziness, fatigue, palpitations (including possible pulsations in the neck), chest pain (more severe than discomfort) and sudden death (may occur with Wolff-Parkinson-White syndrome).

Treatment: Medication, certain maneuvers, catheter-based procedures (ablation) and an electrical shock to the heart (cardioversion) can help slow the heart.

1.3 Related Works

During these years, numerous cardiac arrhythmia classification algorithms were developed. These include set of rules prepared by cardiologists [12, 13], SVM [14–16], optimal path forest [17], hidden Markov model (HMM) [18–20], artificial neural networks [21, 22], auto-regressive modeling [23], etc. Although these strategies have shown benefits in the diagnosis of supraventricular arrhythmia, they have certain restrictions. Some methods are too difficult to implement or quantify, some of which do not differentiate between normal and abnormal conditions and all maintain late detection time, which is generally insufficient for intervention.

1.4 Our Contributions

In this paper, we propose cardiac arrhythmia prediction model for supraventricular arrhythmia detection. Major contributions of this paper are as follows:

(a)   During preprocessing stage, we de-trend and denoise the signal to remove the noise in order to detect features properly.

(b)   After that required features have been extracted using our proposed technique.

Once these features have been extracted, it's related parameters like, *RR* interval, QRS duration and HBR (Heart beat rate), RMSSD (Root mean square of the successive differences) and SDSD (The standard deviation of the differences between successive *RR* intervals) are calculated.

(c)   Using these parameters, effective machine learning based model has been developed for prediction and accurate detection of supraventricular tachycardia arrhythmias.

To the best of our knowledge, it is the first to evaluate different classification approaches based on machine learning and one strategy has been chosen based on their high efficiency in order to diagnose the supraventricular arrhythmias of ECG-signal.

The remainder of this paper is also arranged accordingly. In Sect. 2, proposed method for efficient ECG signal modeling has been presented. Result and discussion are presented in Sect. 3. Finally, in Sect. 4, concluding remarks are discussed.

2 Method

The following steps compose our proposed method (Fig. 2).

(a) *Data Preprocessing*

A range of noises from extremely high and low frequencies [9] can also be reported in electrocardiograms which cause baseline drifts and signal noise in the ECG and which are very difficult to diagnose clinically. For a successful ECG diagnosis, noise from signal must be removed.

Figure 3 shows the steps of ECG signal preprocessing before feature extraction. A procedure to remove the drift of the baseline signal [8, 11] is known as a de-trending and to remove the signal noise is known as denoising. These two methods come under the field of preprocessing of the ECG signal. An ECG signal must be preprocessed in order to minimize noise artifacts. In Fig. 4a, baseline drift of record 826 signal is shown and after removing baseline drift final signal is shown in Fig. 4b.

Frequencies of 0–0.5 Hz are needed to remove to lower the baseline drift. It is also important to take into account low-frequency noise as it can impact the peak detection process. Noise may not affect the R-peak detection as concerned peak amplitude is high. However, since the wave detection of P, Q, S and T is of low amplitude, noise can affect these portions. In order to eliminate baseline drift and noise from the signal, butterworth filter with two poles comprising both low pass

Fig. 2 A block diagram of proposed method's stage

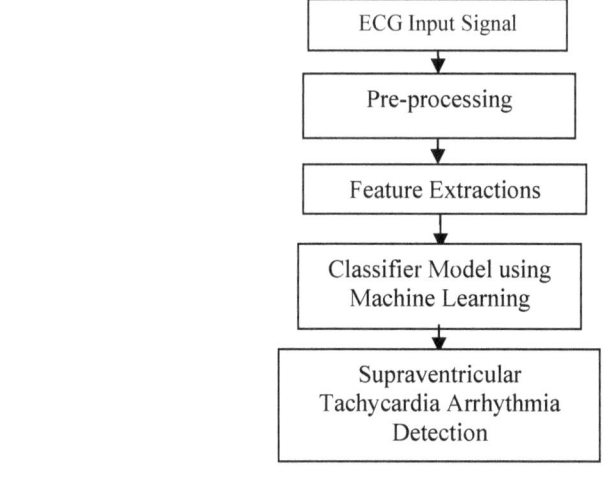

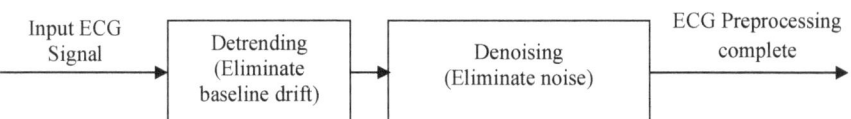

Fig. 3 ECG signal preprocessing before feature extraction

Fig. 4 Record 826 signal **a** with baseline drift and noise **b** after removal of baseline drift and noise

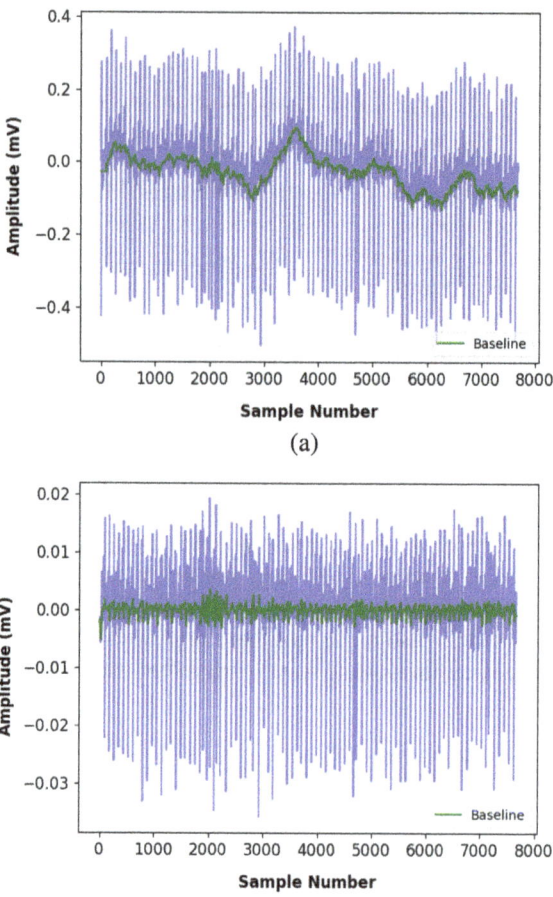

(a)

(b)

and high-pass unidirectional filter having cut-off frequencies of 0.5 and 49.5 Hz has been applied in our work, shown in Fig. 5. Once signal is preprocessed, it can be used for further processing.

(b)   *Feature extraction*

The signal beats received from ECG are classified into the critical and non-critical groups. Necessary features (Q, R and S) are extracted from the received signal. These features are used to measure the necessary parameters afterward. These values are compared to the ECG standard signal values in order to define the criticality of the signal. For supraventricular tachycardia detection using machine learning, following parameters considered are:

- QRS duration [Normal range: Upto 0.10 s]
- RR interval [Normal range: 0.6–1 s]
- Heart beat rate (HBR) [Normal range: 60–100 BPM]

Fig. 5 Record 821 **a** signal with noise **b** signal after noise removal

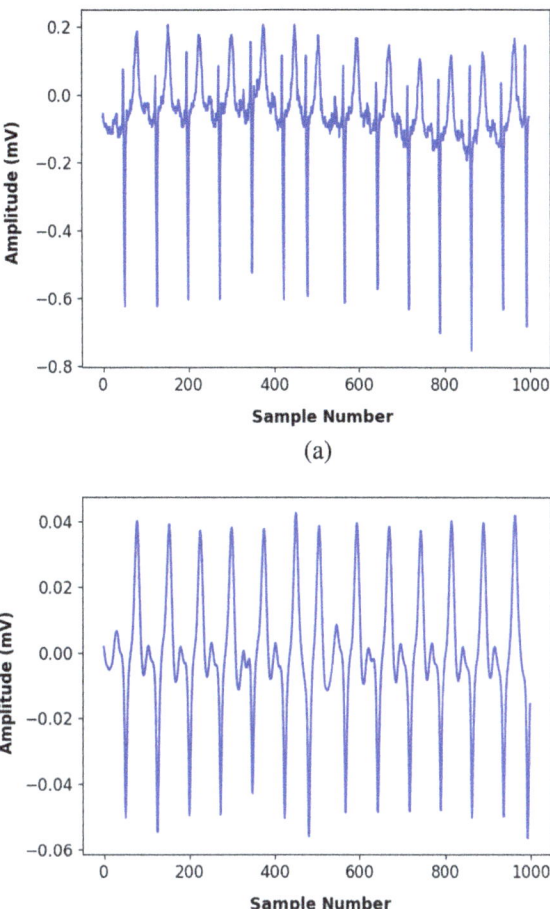

(a)

(b)

- Root mean square of the successive differences (RMSSD)
- The standard deviation of the differences between successive *RR* intervals (SDSD).

We proposed the peak detection algorithm in order to detect QRS complex as described in following steps of our proposed algorithm.

Step 1 First, we calculate total number of recorded samples (RS_{ECG}) and total time (Rt_{ECG}) with respect of RS_{ECG} of a given signal.

Step 2 We already knew that standard range of *RR* interval is 0.6–1 s. Let's say we assumed t_{RR} where $0.6 < t_{RR} < 1$. Therefore, total Rt_{ECG}/t_{RR} number *R*-peak value should be present in one non-critical signal.

Step 3 Now store Rt_{ECG}/t_{RR} number maximum values of this signal in an array and calculate the mean value. This mean value is considered as threshold value Th_S of this signal.

Step 4 Now each point (sample) of the given recorded signal is compared with threshold value Th_S. If any point of the signal is greater than this Th_S then this value is considered as R-peak value. In this way, R-peak value is detected.

Once peak value is detected, QRS value is measured using wavelet transform [24]. Using the feature R peak, we can easily calculate heart rate beat (HBR), shown in Fig. 6. Figure 6 also shows heart beat rate calculation and an effect of filtering of it.

$$RMSSD = \sqrt{\frac{\sum_{i=1}^{N-1} (RR_i - RR_{i+1})^2}{N-1}} \tag{1}$$

Fig. 6 Heart beat rate (HBR) calculation **a** before de-trending and denoising **b** after de-trending and denoising

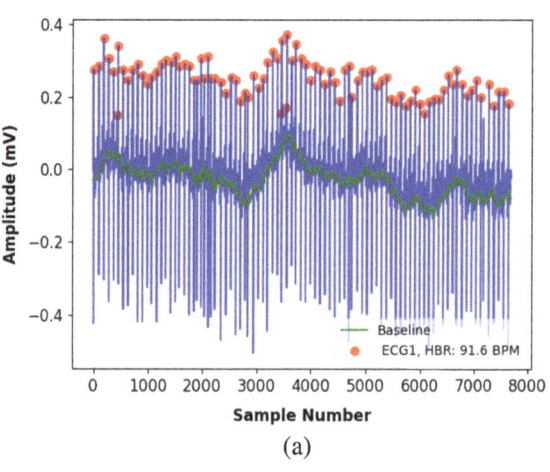

(a)

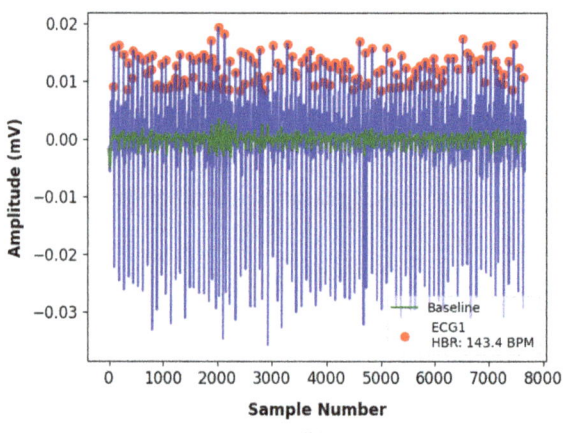

(b)

The RMSSD is the root mean square of the successive R–R interval difference. Equation 1 is the formula for calculating RMSSD where N is the total number of samples. The corresponding standard deviation of successive RR intervals (SDSD) is a short-term variability only.

$$RR_i - RR_{i+1} = D_i \tag{2}$$

This variation is then used in order to find out the equation of SDSD.

$$SDSD = \sqrt{\frac{\sum_{i=1}^{i=n-1} (D_i - D_{\text{mean}})^2 \sum}{n-1}}$$

where:

i interval index.
n total number of intervals.
$n-1$ number of interval differences.

$$D_{mean} = \frac{1}{n-1} \sum_{i=1}^{i=n-1} D_i$$

(c)   *Classifier model*

We focus here on the proper classification of the ECG signal, which leads to the development of a machine learning system based on supraventricular tachycardia arrhythmia identification. Identifying the ECG sample beat correctly and reducing errors is our prime objective. It is therefore extremely important that critical signals are not misclassified as non-critical and can lead to serious device problems. Precision, sensitivity, or recall and the F1 score are also used for the estimation of the performance of machine learning systems, such as SVM, KNN, decision tree and logistic regression. The weight matrix of neural networks is also analyzed in order to distinguish important characteristics that can affect the model.

3   Result and Discussion

3.1 *Material*

In PhysioNet, there are a variety of ECG databases. For this work paper, ECG data are collected from Cudb, MIT-BIH nsrdb, and MIT-BIH svdb of PhysioNet database [25] and analyzed in order to assess any variations in ECG signals. Each record of cudb,

MIT-BIH nsrdb and MIT-BIH svdb, includes 127,232 samples, 11,730,944 samples and 230,400 samples, respectively. These sets represent different subject categories and recording conditions including sampling rates (128 and 250 Hz) and interferences. The ECG1 data of every record is used without exclusion. The frequency of supraventricular arrhythmias can be diagnosed with the extraction of information about Heart Beat Duration, R–R intervals, QRS amplitudes.

3.2 Simulation Results

Python 3.7 is used in all simulations for ECG Signal Filters in Python 3.7 as well as for training and testing. Different signal characteristics observed in these studies are seen in Table 1 by a small number of random chosen records. Abnormalities observed in some of these selected signals among all signals of Table 1, are shown in Table 2.

Sensitivity, precision, F1 score and accuracy are measured as assessment of our work.

These definitions are given below.

(1) The precision (PR): $PR = \frac{TP}{(TP+FP)}$

(2) The sensitivity (SE): $SE = \frac{TP}{(TP+FN)}$

(3) The specificity (SP) or F1 score: $SP = \frac{TN}{(TN+FN)}$.

Table 1 Results of various extracted features with a few number of random selected signals before filtering

Features	Record 802	Record 824	Record 826	Record 856
RR interval (s)	0.96	0.76	0.66	0.62
QRS width (s)	Normal	Normal	Normal	Normal
HBR (BPM)	63	79	92	95
Supraventricular Tachycardia	No	No	No	No

Table 2 Results of various extracted features with a few number of random selected signals of Table 1 after filtering

Features	Record 802	Record 824	Record 826	Record 856
RR interval (s)	0.49	0.38	0.42	0.62
QRS width (s)	Narrow	Narrow	Narrow	Normal
HBR (BPM)	121	157	143	95
Supraventricular Tachycardia	Yes	Yes	Yes	No

Table 3 Performance analysis

Algorithm	Precision	Sensitivity	F1 score	Accuracy (%)
KNN	0.86	0.83	0.79	83
SVM	0.89	0.88	0.85	88
Logistic regression	0.96	0.96	0.96	96
Decision tree	0.96	0.96	0.96	96
MLP (Neural network)	0.91	0.90	0.87	89

Table 4 Comparative study

Author	Precision	Recall	F1 score (%)	Accuracy (%)
Zihlmann et al. [26]	–	–	79	82
Goodfellow et al. [1]	84%	85%	85	88
Jalali et al. [27]	86%	86%	85	89
Proposed method	96%	96%	96	96

Where: FP = False Positives; TP = True Positives; FN = False Negatives; TN = True Negatives; and N = FP + FN + TP + TN.

Precision is the percentage of ECG signal that is already identified as a vital one. The proportion of the vital ECG signal that is labeled as critical is called sensitivity or recall. The F1 score or specificity is the harmonic mean of precision and sensitivity which is more suitable than accuracy in cases of uneven class distribution. In the case of F1, it is found from Table 3 that both logistic regression and decision tree have the highest F1 scoring, and then comes into decreasing order MLP (neural network), SVM and KNN. Table 3 also states that decision tree and logistic regression surpassed KNN, SVM and MLP (neural network), while considering accuracy results. Decision tree and logistic regression perform equally better. Hence, logistic regression and decision tree are preferable to detect critical ECG signal bits.

In Table 4, our experimental results are compared with [1, 26, 27]. Experimental outcomes show satisfying improvements and great algorithm robustness that we have proposed.

4 Conclusion

In this work, we suggested a machine learning approach to diagnose supraventricular tachycardia arrhythmias of ECG signal. The preprocessing stage, feature extraction and classifier model follow our proposed method. We de-trended and denoised the signal in the preprocessing phase to remove the noise for accurate detection of features. Thereafter, R-peak and QRS detection have been conducted with our proposed method. Once this R-peak is observed, the associated parameters, such as the RR interval, QRS durations, HBR, RMSSD, SDSD are determined. We have developed an effective machine learning model based on these parameters which can detect supraventricular tachycardia with high performance. This paper addresses the problem of reducing the misclassification of the supraventricular tachycardia based critical signal very effectively. Experimental findings show that our suggested method has developed well and seems extremely stable.

As far as we know, it is the first report where different machine learning classification methods are analyzed and then a high efficiency approach is selected to detect supraventricular tachycardia arrhythmias of the ECG signal. Our findings show that the most effective machine learning models for diagnosis of supraventricular tachycardia arhythmias are logistic regression and decision tree models. Since perfection is very much expected in medical domain, with more advanced algorithms, we can learn to achieve better results with more data in future.

References

1. Haddad SAP, Houben RPM, Serdijn WA (1989) The evolution of pacemakers. In: IEEE engineering in medicine and biology magazine. In: Young M (ed), (2006) The technical writer's handbook. Mill Valley. University Science
2. Qiu P, Liu KJR (2008) A robust method for QRS detection based on modified p-spectrum. In: IEEE International conference on acoustics, speech and signal processing, ICASSP 2008
3. Chouhan VS, Mehta SS (2008) Threshold-based detection of P and T-wave in ECG using new feature signal. Int J Comput Sci Netw Secur (IJCSNS) 8(2)
4. Exarchos TP, Tsipouras MG, Nanou D, Bazios C, Antoniou Y, Fotiadis DI (2005) A platform for wide scale integration and visual representation of medical intelligence in cardiology: the decision support framework. Comput Cardiol 167–170 (IEEE)
5. Tsipouras MG, Fotiadis DI, Sideris D (2002) Arrhythmia classification using the RR-interval duration signal. Comput Cardiol 485–488 (IEEE)
6. Song MH, Lee J, Cho SP, Lee KJ, Yoo SK (2005) Support vector machine based arrhythmia classification using reduced features. Int J Control Autom Syst 3(4):571–579
7. Nasiri JA, Naghibzadeh M, Yazdi HS, Naghibzadeh B (2009) ECG arrhythmia classification with support vector machines and genetic algorithm. In: Third UKSim European symposium on computer modeling and simulation, 2009. EMS'09, pp 187–192. IEEE
8. Ye C, Coimbra MT, and Kumar BVKV (2010) Arrhythmia detection and classification using morphological and dynamic features of ECG signals. In: 2010 Annual international conference of the IEEE engineering in medicine and biology society (EMBC), pp 1918–1921. IEEE
9. Luz EJS, Nunes TM, De Albuquerque VHC, Papa JP, Menotti D (2013) ECG arrhythmia classification based on optimum-path forest. Expert Syst Appl 40(9):3561–3573
10. Andreão RV, Dorizzi B, Boudy J (2006) ECG signal analysis through hidden Markov models. IEEE Trans Biomed Eng 53(8):1541–1549
11. Li Z, Derksen H, Gryak J, Hooshmand M, Wood A, Ghanbari H, Gunaratne P, Najarian K (2018) Supraventricular tachycardia detection via machine learning algorithms. IEEE Int Conf Bioinf Biomed (BIBM) 2419–2422 (IEEE)
12. Coast DA, Stern RM, Cano GG, Briller SA (1990) An approach to cardiac arrhythmia analysis using hidden Markov models. IEEE Trans Biomed Eng 37(9):826–836
13. Ceylan R, Özbay Y (2007) Comparison of FCM, PCA and WT techniques for classification ECG arrhythmias using artificial neural network. Expert Syst Appl 33(2):286–295

14. Ozlem Ozcan N, Gurgen F (2010) Fuzzy support vector machines for ECG arrhythmia detection. In: 2010 20th international conference on pattern recognition (ICPR). IEEE, pp 2973–2976

15. Ge D, Srinivasan N, Krishnan SM (2002) Cardiac arrhythmia classification using autoregressive modeling. Biomed Eng Online 1(1):5

16. Lin C-C, Chang H-Y, Huang Y-H, Yeh C-Y (2019) A novel wavelet-based algorithm for detection of QRS complex. Appl Sci 9(10). https://doi.org/10.3390/app9102142

17. http://www.physionet.org/physiobank

18. Zihlmann M, Perekrestenko D, Tschannen M (2017) Convolutional recurrent neural networks for electrocardiogram classification. Computing 44:1

19. Goodfellow SD, Goodwin A, Greer R, Laussen PC, Mazwi M, Eytan D (2018) Towards understanding ECG rhythm classification using convolutional neural networks and attention mappings. In: Machine learning for healthcare conference, pp 83–101

20. Jalali A, Lee M (2020) Atrial fibrillation prediction with residual network using sensitivity and orthogonality constraints. J Biomed Health Inf 24(2)

21. Rojo-Alvarez JL, Arenal-Maiz A, Artes-Rodriguez A (2002) Discriminating between supraventricular and ventricular tachycardias from EGM onset analysis. Eng Med Biol Mag 21(1):16–26 (IEEE)

22. Sohinki D, Obel OA (2014) Current trends in supraventricular tachycardia management. Ochsner J 14(4):586–595

23. https://www.medicalnewstoday.com

24. https://www.webmd.com

25. Charafeddine F, Itani M, Shublaq N (2006) "Application of artificial Neura and Fuzzy neural networks to QRS detection and PV diagnosis," final project, University of Beirut

26. Amann A, Tratnig R, Unterkofler K (2007) Detecting ventricular fibrillation by time-delay methods. IEEE Trans Biomed Eng 54(1):174–177

27. Mehta SS, Lingayat NS (2007) Comparative study of QRS detection in single lead and 12-lead ECG based on entropy and combined entropy criteria using support vector machine. J Theor Appl Inf Technol, 2007 JATIT

Quantum-Inspired Support Vector Machines for Human Activity Recognition in Industry 4.0

Preeti Agarwal and Mansaf Alam

Abstract In Industry 4.0, robots and humans share roles and responsibilities. Many industries, such as manufacturing and logistics, still depend heavily on manual labor. However, by integrating sensor technology and data processing, different processes can be appropriately tracked and streamlined. Comprehensive knowledge of the occurrence, duration, and properties of related human activities is needed to draw inferences on enhancing employee performance. The increased use of numerous wearable sensors to monitor human activities has resulted in an unexpected data explosion. The pace at which this data is being generated is expected to outpace conventional technology's ability to handle it by 2030. At present, quantum computing is the most promising market solution. This paper proposes the use of a quantum support vector machines for human activity recognition. As compared to various state-of-the-art machine learning algorithms, such as linear discriminant analysis (LDA), logistic regression, k-nearest neighbor (kNN), and conventional support vector machines (SVMs) with three different kernels, the proposed approach takes the lead by achieving an accuracy of 98% with execution time of 19 s.

Keywords Human activity recognition (HAR) · Machine learning · Quantum support vector machines (QSVMs) · Quantum computing · Industry 4.0

1 Introduction

Responsibilities and roles are pooled between human workers and robots in the Industry 4.0 vision [1, 2]. Nonetheless, with a steady number of workers, manual tasks are likely to remain prevalent in manufacturing and logistics [3]. Robots are unlikely to fully replenish manual labor in manufacturing industries [4, 5]. This is

P. Agarwal (✉) · M. Alam
Department of Computer Science, Jamia Millia Islamia, New Delhi, India
e-mail: rs.preeti.agw@jmi.ac.in

M. Alam
e-mail: malam2@jmi.ac.in

© The Author(s), under exclusive license to Springer Nature Singapore Pte Ltd. 2022   281
D. Gupta et al. (eds.), *Proceedings of Data Analytics and Management*,
Lecture Notes on Data Engineering and Communications Technologies 90,
https://doi.org/10.1007/978-981-16-6289-8_24

because computers still have a hard time emulating human cognitive and motor skills [6].

Ambient-assisted living, lifestyle monitoring, smart homes, health care, and industrial environment have all benefited from HAR methods [7–11]. In order to recognize human behaviors in industrial settings, they must first be identified, for example, locomotion, retrieval, and service use [12, 13]. There is no predetermined set of activities that can be clearly separated in industrial environments since their meaning varies depending on the use case. For example, in logistics delivery, human activity recognition may be used to verify delivery parameters such as timings and tasks performed by a delivery person to ensure a smooth and effective logistics process. HAR can also recognize productive gestures [14] and locomotion movements [15–17].

The advent of sensors and their integration with various cloud storage has further enhanced the field [18, 19]. It has opened the door for new applications of autonomous HAR [20]. The pace at which data from wearable sensors is growing soon will outpace the capacity of traditional technologies [21]. The most promising solution at present available is quantum computing.

This paper presents the use of quantum-inspired support vector machines for HAR. The following are the major contributions of this paper:

i.   Firstly, the quantum support vector machine (QSVM) architecture is developed to be deployed on quantum computers using quantum libraries.
ii. Secondly, IBM's Qiskit library [22] is used to train and test the proposed model in a quantum environment using configured parameters.
iii. Thirdly, an experiment is performed on sensor data from 30 participants engaged in six different forms of physical activity from the UCI-HAR repository [23].
iv. Fourthly, the proposed model's performance is compared to six current state-of-the-art models in terms of accuracy, precision, recall, and F-score, as well as execution time.

The paper is arranged as follows: Section 2 provides a review of the literature on HAR practices based on industrial applications. Section 3 gives the description of the quantum-inspired support vector machines. Section 4 presents experimentation and results. Section 5 concludes the paper with discussions and future potential.

2 Literature Review

Numerous studies on the application of HAR for industrial purposes have been conducted, including those in the manufacturing sector [24, 25], data warehousing [26], construction industry [27], and maintenance industry [7, 28]. These studies mainly employ IMU sensors connected to various parts of the body for activity recognition. Smartphone sensors are also becoming more important for activity recognition. Locomotion activities are used explicitly for behavior identification.

Koskimaki et al. [58] and Tao et al. [86] emphasized the significance of HAR while conducting assembly operations. Koskimaki et al. [58] employ a single IMU device mounted on the wrist and employ kNN to recognize hand movements during car assembly. Tao et al. [86] employ one IMU sensor and one surface electromyography (sEMG) sensor on the forearm, as well as CNN to identify hand movements while performing different operations.

Hammererla et al. [26] use wearable inertial sensors for HAR and use convolutional and recurrent neural networks to identify locomotion movements such as walking, standing, sitting, jumping, and jogging. Smart logistics takes advantage of the recognition of these activities. Deep neural networks are used by [29] to identify everyday activities. HAR was used on both stationary and low-mobility work processes. Tao et al. [25] use deep CNN architectures on acquired IMU sensor data to examine HAR for order picking.

Zhao and Obonyo [90] showed the use of HAR in the construction industry for injury prevention. Ordonez and Roggen [7] use wearable sensors and CNN and LSTM recurrent networks for HAR in the maintenance industry. Zeng et al. [28] use a novel CNN with weight sharing to monitor worker activity.

3 Description of Quantum-Inspired Support Vector Machine (QSVM)

QSVM is a quantum-inspired support vector machine that leverages the disadvantages of the classical support vector machine. HAR is a multi-class classification problem. The supervised learning algorithm's goal is to train it to predict which activity group the data from the dataset belongs to [30].

When the number of classes in the SVM increases, so does the complexity of the classical SVM. Identifying a large number of planes to classify the data into different hyperplanes becomes more difficult. As a consequence, the problem is solved using quantum machine learning. Quantum machine learning uses the properties of quantum physics like entanglement and superposition to enable parallel computing [31].

The multi-class QSVM is an extension of binary QSVM that uses a one-versus-one approach for separating data into multiple classes. If we need to classify a activities, then we need to classify activity data into one of the a classes. The one-versus-one constructs $\frac{a(a-1)}{2}$ binary classifiers for all possible distinct pairs, and the class with the maximum votes is the final prediction result.

In multi-class classification problems, the aim is to find the largest margin between each class using the basic strategy of SVM. To find the largest margin, we assume that the hyperplane is defined by Eq. 1.

$$w.\vec{x} - b = \pm 1 \tag{1}$$

where w and $\vec{x}$ are vectors, and b is numeric.

The distance between two hyperplanes is expressed as $\left\|\frac{2}{w}\right\|$. Minimizing w and adding constraints lead to maximum margin. Therefore, this can be represented as an optimization problem, as shown in Eq. 2.

$$\min_{w,b} \frac{1}{2}\|w^2\| \tag{2}$$

subject to the constraint

$$y^i\left(wx^i - b\right) \geq 1$$

for all training samples $i = \{1, 2, \ldots, N\}$ and $y^i = \{-1, 1\}$.

The objective function is formulated using Lagrange's multiplier α^i as shown in Eq. 3.

$$F = \min_{w,b} \max_{\alpha^i \geq 0} \left(\frac{1}{2}w^2 - \sum_{i=1}^{N}\left[\alpha^i\left(wx^i - b\right) - 1\right] \right) \tag{3}$$

To solve this maximization objective function, set the derivates to zero. $\frac{\partial F}{\partial w^i} = w^i - \alpha^i y^i x^i = 0$

$$\frac{\partial F}{\partial b} = \sum_{i=1}^{N} \alpha^i y^i = 0$$

as a consequence, $\vec{w}$ can be represented as Eq. 4

$$\vec{w} = \sum_{i=1}^{N} \alpha^i y^i x^i \tag{4}$$

and the objective function can be expressed as Eq. 5

$$\min_{\alpha^i} \left\{ \frac{1}{2}\sum_{i,j} \alpha^i \alpha^j y^i y^j x^i x^j - \sum_{i=1}^{N} \alpha^i \right\} \tag{5}$$

The optimization can be generalized to any arbitrary kernel function $K\left(x^i, x^j\right)$. The equation can be written as shown in Eq. 6.

$$\min_{\alpha^i} \left\{ \frac{1}{2}\sum_{i,j} \alpha^i \alpha^j y^i y^j K\left(x^i, x^j\right) - \sum_{i=1}^{N} \alpha^i \right\} \tag{6}$$

The quantum minimization searches the objective function space to find the optimal set of α^i, which solves the parameters w and b and find global minima.

The final predicted output t_{predict} will be the class with the highest probability for all the binary classes according to Eq. 7.

$$t_{\text{predict}} = t_{\text{argmin}} c_{N_c} \tag{7}$$

where $N_c = \left\{ 1, 2, \ldots, \frac{a(a-1)}{2} \right\}$ and c is cost function.

Significant steps followed for QSVM are:

i. Initialize kernel function and its parameters.
ii. Conversion of classical data to quantum data. The classical data is represented as bits, which are converted to qubits that can be represented in Hilbert space.
iii. Scan the data, and find the optimal set of learning parameters. Different optimizer functions can be selected to minimize the error rate.
iv. Apply measurement to read the output.

4 Experiment and Results

The experiment was conducted using data from the UCI machine learning repository dataset [23], which includes data for six physical activities. Each activity is carried out by a group of 30 volunteers, who are all wearing a Samsung Galaxy smartphone around their waist. The smartphone's accelerometer data and gyroscope data are collected at a rate of 50 Hz. The data is preprocessed with noise filters before being segmented into 2.56 s sliding windows with 50% overlap, yielding 128 readings per window. Figure 1 depicts the activity graph. The collected dataset is randomly divided into 70% for training and 30% for testing.

For activity classification, quantum-inspired support vector machines are used. The experiment is performed using the Qiskit library developed by IBM, which uses a quantum simulator as a backend [22]. All analysis is performed in Python using Jupyter notebook. The ZZFeaturemap is used to convert classical data to quantum data. The feature dimensions were set to 12 with two shots. Simultaneous perturbation stochastic approximation (SPSA) is used as the optimization function. For this multiclass classification problem, Qiskit's all pairs multi-class extension is used.

The four most commonly used metrics are used to compare the performance of the classifiers [32]. The results show that QSVM outperformed various traditional methods such as LDA, logistic regression, kNN, and SVM with three kernels. The QSVM achieved an accuracy of 98% in 17 s. The performance parameters are shown in Table 1, and performance graphs are shown in Fig. 2.

The confusion matrix for different classifiers for six class of activities is shown in Fig. 3.

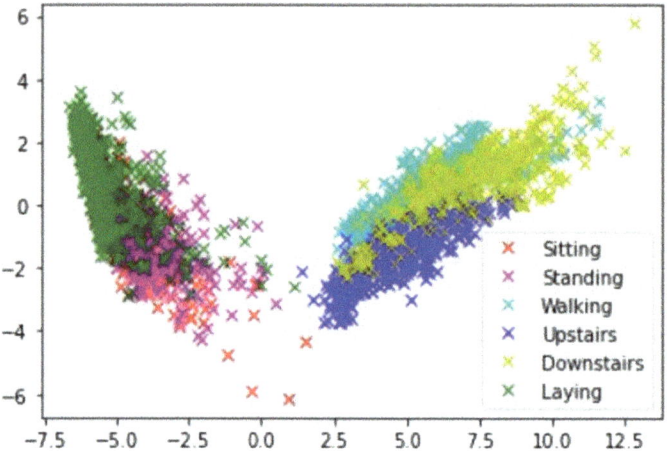

Fig. 1 Activity plot

Table 1 Comparison of performance metrics

	Accuracy	Precision	Recall	F-score	Time (s)
LDA	96.23	96.42	96.18	96.30	22
Logistic regression	96.2	96.44	96.13	96.28	25
kNN	90.74	91.21	90.28	90.74	23
SVM-rbf	94.03	94.1	93.74	93.92	32
SVM-linear	96.4	96.59	96.35	96.47	30
SVM-polynomial	90.74	90.36	90.21	90.28	33
QSVM (Proposed)	**98**	**97.97**	**98**	**97.98**	**19**

The performance parameters of our proposed QSVM is highlighted in bold in the table for comparison with other state of the art classifiers

5 Conclusions

A quantum-inspired support vector machine is deployed for HAR. The model is deployed in quantum environment using the Qiskit library and quantum simulator at the backend to obtain the advantages of quantum computing. Compared to several already existing machine learning algorithms and classic support vector machines with three different kernels, the built model produces better performance.

This research may be expanded in the future to classify other complex activities. Many industrial applications could benefit from using HAR, including complex packaging, heavy industry for loading and unloading, and in-house activity monitoring. Multiple multimodal sensors can be deployed. The performance of the QSVM can be further improved by experimenting with different QSVM kernel functions.

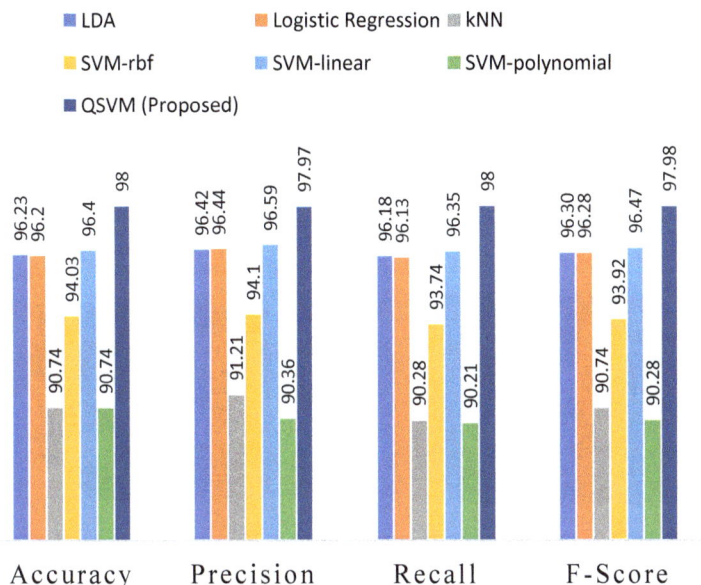

Fig. 2 Performance graph

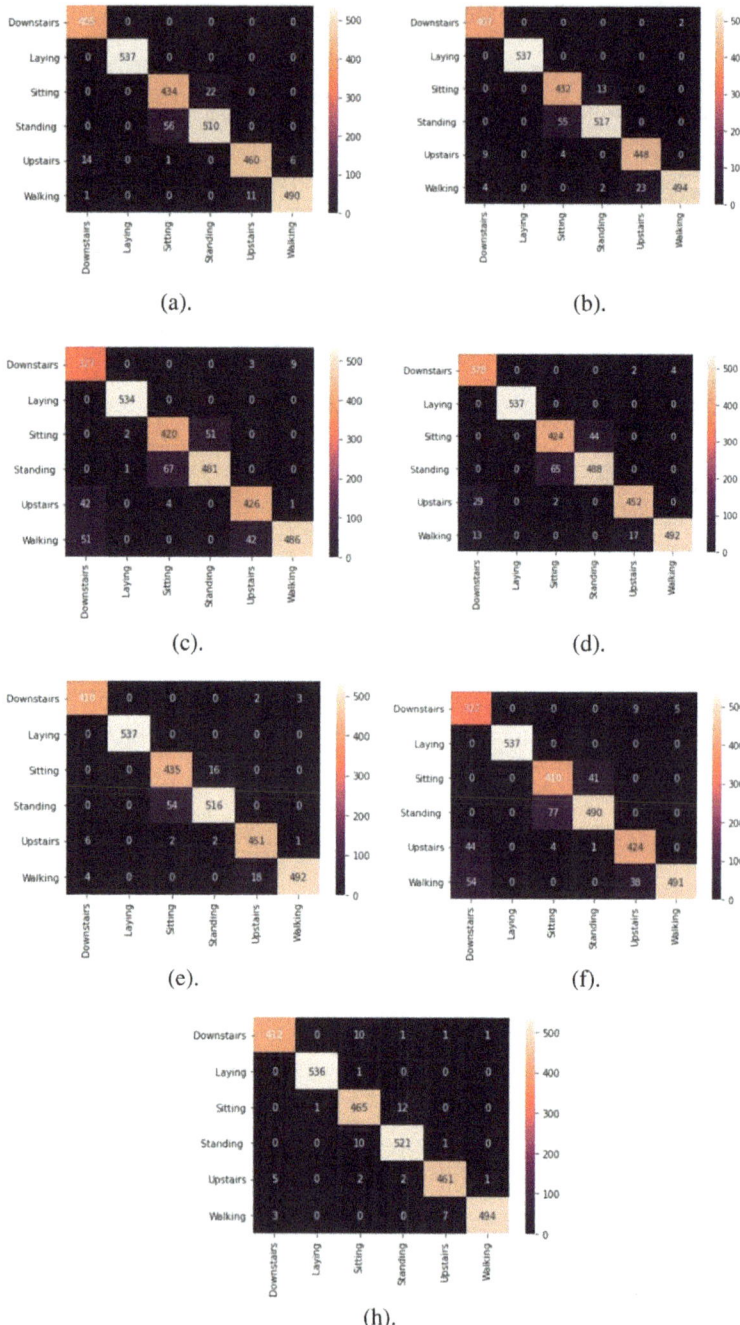

Fig. 3 Confusion matrix **a** LDA, **b** logistic regression, **c** kNN, **d** SVM with rbf kernel, **e** SVM with linear kernel, **f** SVM with poly kernel, and **h** QSVM

References

1. Dregger J, Niehaus J, Ittermann P, Hirsch-Kreinsen H, ten Hompel M (2018) Challenges for the future of industrial labor in manufacturing and logistics using the example of order picking systems. Procedia cirp 67:140–143
2. Hofmann E, Rüsch M (2017) Industry 4.0 and the current status as well as future prospects on logistics. Comput Ind 89:23–34
3. Schlögl D, Zsifkovits H (2016) Manuelle Kommissioniersysteme und die Rolle des Menschen. BHM Berg-und Hüttenmännische Monatshefte 161(5):225–228
4. Liang C, Chee KJ, Zou Y, Zhu H, Causo A, Vidas S et al (2015) Automated robot picking system for e-commerce fulfillment warehouse application. In: The 14th IFToMM World Congress
5. Oleari F, Magnani M, Ronzoni D, Sabattini L (2014) Industrial AGVs: toward a pervasive diffusion in modern factory warehouses. In: 2014 IEEE 10th international conference on intelligent computer communication and processing (ICCP). IEEE, pp 233–238
6. Grosse EH, Glock CH, Neumann WP (2015) Human factors in order picking system design: a content analysis. IFAC-PapersOnLine 48(3):320–325
7. Ordóñez FJ, Roggen D (2016) Deep convolutional and LSTM recurrent neural networks for multimodal wearable activity recognition. Sensors 16(1):115
8. Haescher M, Matthies DJ, Srinivasan K, Bieber G (2018) Mobile assisted living: smartwatch-based fall risk assessment for elderly people. In: Proceedings of the 5th international workshop on sensor-based activity recognition and interaction, pp 1–10
9. Agarwal P, Alam M (2020) A lightweight deep learning model for human activity recognition on edge devices. Procedia Comput Sci 167:2364–2373
10. Hölzemann A, Van Laerhoven K (2018) Using wrist-worn activity recognition for basketball game analysis. In: Proceedings of the 5th international workshop on sensor-based activity recognition and interaction, pp 1–6
11. Reining C, Niemann F, Moya Rueda F, Fink GA, ten Hompel M (2019) Human activity recognition for production and logistics—a systematic literature review. Information 10(8):245
12. Feldhorst S, Aniol S, ten Hompel M (2016) Human activity recognition in der Kommissionierung–Charakterisierung des Kommissionierprozesses als Ausgangsbasis für die Methodenentwicklung. Logistics J Proc (10)
13. Alam MAU, Roy N (2017) Unseen activity recognitions: A hierarchical active transfer learning approach. In: 2017 IEEE 37th international conference on distributed computing systems (ICDCS). IEEE, pp 436–446
14. Zappi P, Lombriser C, Stiefmeier T, Farella E, Roggen D, Benini L, Tröster G (2008) Activity recognition from on-body sensors: accuracy-power trade-off by dynamic sensor selection. In: European conference on wireless sensor networks. Springer, Berlin, Heidelberg, pp 17–33
15. Feldhorst S, Masoudenijad M, ten Hompel M, Fink GA (2016) Motion classification for analyzing the order picking process using mobile sensors. In: Proceeding of international conference on pattern recognition applications and method, pp 706–713
16. Reyes-Ortiz JL, Oneto L, Samà A, Parra X, Anguita D (2016) Transition-aware human activity recognition using smartphones. Neurocomputing 171:754–767
17. Ronao CA, Cho SB (2017) Recognizing human activities from smartphone sensors using hierarchical continuous hidden Markov models. Int J Distrib Sens Netw 13(1):1550147716683687
18. Agarwal P, Alam M (2020) Investigating IoT middleware platforms for smart application development. In: Smart cities—opportunities and challenges. Springer, Singapore, pp 231–244
19. Naqvi K, Hazela B, Mishra S, Asthana P (2021) Employing real-time object detection for visually impaired people. In: Data analytics and management. Springer, Singapore, pp 285–299s
20. Agarwal P, Alam M (2020) Open service platforms for IoT. In: Internet of things (IoT). Springer, Cham, pp 43–59
21. Bhatia M, Sood SK (2020) Quantum computing-inspired network optimization for IoT applications. IEEE Internet Things J 7(6):5590–5598

22. Aleksandrowicz G, Alexander T, Barkoutsos P, Bello L, Ben-Haim Y, Bucher D et al (2019) Qiskit: an open-source framework for quantum computing. Accessed on 16 Mar
23. Anguita D, Ghio A, Oneto L, Parra X, Reyes-Ortiz JL (2013) A public domain dataset for human activity recognition using smartphones. Esann 3:3
24. Koskimäki H, Huikari V, Siirtola P, Röning J (2013) Behavior modeling in industrial assembly lines using a wrist-worn inertial measurement unit. J Ambient Intell Humaniz Comput 4(2):187–194
25. Tao W, Lai ZH, Leu MC, Yin Z (2018) Worker activity recognition in smart manufacturing using IMU and sEMG signals with convolutional neural networks. Procedia Manuf 26:1159–1166
26. Hammerla NY, Halloran S, Plötz T (2016) Deep, convolutional, and recurrent models for human activity recognition using wearables. arXiv preprint arXiv:1604.08880
27. Zhao J, Obonyo E (2018) Towards a data-driven approach to injury prevention in construction. In: Workshop of the European group for intelligent computing in engineering. Springer, Cham, pp 385–411
28. Zeng M, Nguyen LT, Yu B, Mengshoel OJ, Zhu J, Wu P, Zhang J (2014) Convolutional neural networks for human activity recognition using mobile sensors. In: 6th International conference on mobile computing, applications and services, pp 197–205. IEEE
29. Moya Rueda F, Grzeszick R, Fink GA, Feldhorst S, Ten Hompel M (2018) Convolutional neural networks for human activity recognition using body-worn sensors. In: Informatics, vol 5, no 2. Multidisciplinary Digital Publishing Institute, p 26
30. Kopczyk D (2018) Quantum machine learning for data scientists. arXiv preprint arXiv:1804.10068
31. Dema B, Arai J, Horikawa K. Support vector machine for multi-class classification using quantum annealers
32. Sokolova M, Lapalme G (2009) A systematic analysis of performance measures for classification tasks. Inf Process Manage 45(4):427–437

A Master Data Management Solution for Building Frameworks: A Constructive Way to Pilot the Implementation

Dilbag Singh and Dupinder Kaur

Abstract In today's age of information technology, the frameworks like enterprise resource planning (ERP) and customer relationship management (CRM) collect the data from various sources. The data can be duplicate, fragmented and obsolete. Master data management (MDM) is a true standalone application which guarantees that uniformity, semantic consistency, accuracy and accountability of the organization's data assets will be shared by having "single view of customer data." It generates a central repository called "master file" with independent application resources. This information can be used to motivate the organization for promoting business, decision making and reporting. The present study has been conducted to propose a constructive approach for the implementation of MDM. The paper is structured into three sections. First, the fundamental technical components of MDM are mapped into six key parameters K1–K6 to highlight their importance in different application domains. Second, an analysis has been made to know the latest trends and future extents of MDM using Gartner's hype cycle. Final, a MDM solution is suggested by constructing a framework, roadmap design and data flow diagram to pilot the implementation.

Keywords Master data management(MDM) · Golden record · Hype cycle · Framework

1 Introduction

The intention behind initiatives of deploying various applications and system (ERP, CRM) is to consolidate organization's data into a centralized information asset. Despite the fact that this data could provide analysis and reports related to various assets of business, it introduced different challenges. The data may not be synchonized at original sources or may generate inconsistence, inaccurate and patchy data [1]. The demand for accurate, consistent and timely information is always at

D. Singh · D. Kaur (✉)
Department of Computer Science and Engineering, Chaudhary Devi Lal University, Sirsa, Haryana, India

© The Author(s), under exclusive license to Springer Nature Singapore Pte Ltd. 2022   291
D. Gupta et al. (eds.), *Proceedings of Data Analytics and Management*,
Lecture Notes on Data Engineering and Communications Technologies 90,
https://doi.org/10.1007/978-981-16-6289-8_25

higher priority as wrong data can destroy business, particularly since the vast majority of the business choices depend on data. **Master data management** is a technique that provides integrated and single view of essential business entities to offer better help in measuring business processes [1]. It incorporates procedures, guidelines, arrangements, policies and standards that recognize and maintain the organization's essential data. The main idea behind MDM is "management" to provide following abilities:

- Collecting information in subsystems and sharing the information as data service.
- Generating a unified version of truth for customers, products and suppliers.
- Conveying master data between frameworks utilizing mechanized distributing and membership models.
- Establishing different data management policies and validation rules for various departments and businesses.
- Better view point on planning and spend analysis of future [2].

Master data or golden record is consistent and uniform set of identifiers that represent the business objects that are shared across value-based applications. MDM creates master record or *golden record* that includes necessary information upon which organization relies and made decisions [3]. Master data is always non-transactional in nature. The value of master data does not change frequently as compared to transactional data [4].

In the light of the examination and investigation, following are the major contributions of this paper.

- Understanding the fundamental ideas of master data management, its components and Gartner's hype
- Comprehending the implementation of designed key parameters in various application domains.
- Observing the latest technological trends and future scope of MDM, by analyzing recent hype cycles.
- Providing a solution for the implementation of MDM by proposing a framework, roadmap and DFD.

2   MDM Catalog

The purpose of MDM is to provide continuous data quality improvement, synchronizing multiple data items, managing and monitoring the results, etc. To get insights into MDM, the study brought MDM catalog with constituents: MDM components and hype cycle. The following section presents detailed description of these constituents.

Table 1 MDM components and service layer (Source MDM by Loshin [5])

Component name	Services provided by each component
Architecture	In order to relate the structure and control, this component represents model such as MDM master data model and system architecture
Governance	The function of this component is to build set of policies, principals and methods to ensure high data quality. The other areas of concentration are consolidation of metadata management systems, data stewardship and setting up standards for data
Operations management	Some specific management services such as identity management for unified view, hierarchy management, relationships and any migration plans are provided throughout the complete life cycle of data
Identification and consolidation	Setting match and merge rules, identity resolution, linking the hierarchies and survivorship rules to create the best version of data
Integration	Service of this layer is to provide integrated master data back into tactical and analytical use for rapid and effective implementation. Also, a forecast on future is made
Business process management	Various business process requirements, their integration, rule-based operation system and MDM solutions as per the problem are specified

2.1 MDM Components

For unique representation of variant versions of instances of core data objects, there is need to hire MDM fundamental component or service layer model. In general, these layers provide a platform for board spectrum in MDM creation and associations [5]. The following table describes the fundamental blocks of MDM.

Table 1 depicts that majorly six components are crucial for implementing MDM program.

2.2 Hype Cycle

The hype cycle is a graphical depiction developed by the firm "Gartner" to understand and track technology maturity and future potential [6].

Figure 1 illustrates a 5-stage hype cycle pattern practically followed by almost every new technology. At each phase, certain decisions can help to adopt the technology when it is the right time for particular use case and business prerequisites [6]. The description of these five stages is mentioned below.

Table 2 gives a profound understanding into each phase of Gartner's hype cycle.

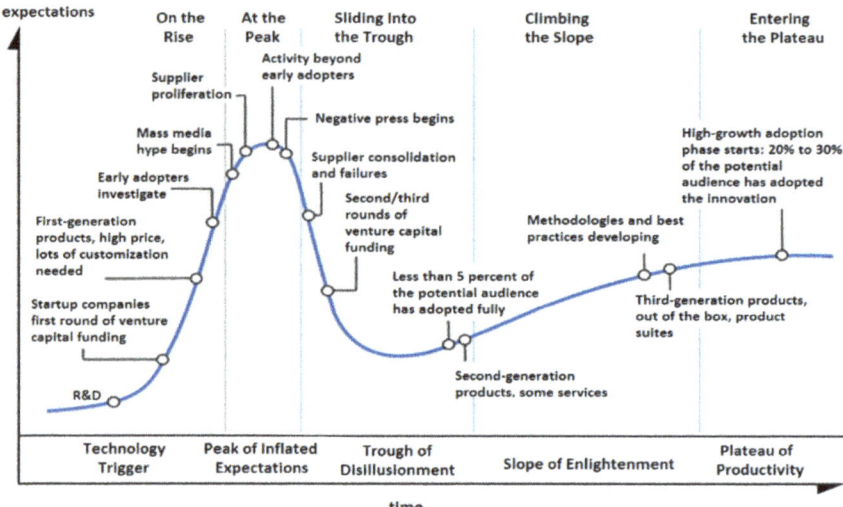

Fig. 1 Structure of a hype cycle [*Source* Gartner Hype Cycle]

Table 2 Description of hype cycle stages

Name of stage	Description
Technology trigger	Presents new innovations—things like AI, chatbots, block-chain which catch the innovation and create a rapid rise in expectations
Peak of inflated expectations	Early exposure delivers various examples of success stories—regularly joined by scores of disappointments. A few organizations make a move and most do not
Trough of disillusionment	Reality before long sets in, however, as individuals understand that the guarantees of that publicity are not working out as expected
Slope of enlightenment	More cases of how the innovation can profit the endeavor begin to solidify with second- and third-generation products launched by innovation suppliers
Plateau of productivity	At long last, the desires for the innovation are ingested into regular day-to-day existence, with well-defined practices

3 Research Methodology

Research methodology provides a path to formulate the problem and objectives. It provides a deep insight into research method, design and approach to be used throughout the study. Taking consideration, the study offers background study for finding ways to improve business performance with MDM. Hence, an exploratory research has been conducted to have a better understanding of existing problem and its future studies. This situational motivation arises from everyday needs of MDM system to support various processes, applications and business. The present research comprises of study of existing work, analyzing technical trends and a strategic solution for implementing an MDM.

4 Key Parameters for MDM Implementation

Loshin (2009) provided the guidance to implement a roadmap for MDM through fundamental building components of MDM. Thus, in this study, these are mapped into six key parameters K1–K6 for analysis. Each component has a wide spectrum of terms, but this study is limited to specific scope as shown in table below. The main idea is to identify the existence of these key parameters in various application domains to guide the implementation and enhance the business productivity.

Table 3 signifies the mapping of basic MDM components and key parameters used for analysis.

To execute the plan, a review study has been performed on existing work.

Table 4 demonstrates the review and analysis of key parameters on existing study, in different application areas. The existence and non-existence of key parameters are represented as Y for yes and N for no. It is discovered that maximum five key parameters are implemented in health care and banking area, while there is still need of improvement in proper implementing these key parameters in other domains.

Further, it can be said that K6 parameter is most implemented, followed by K5, K1, K2, K3 and K4. For the successful deploy of an MDM project, inclusion of all key parameters is necessary. Hence, an emphasis on inclusion of all the key parameters has been made in this study.

Table 3 Key parameters considered for mapping of MDM components

MDM components		Index	Scope of key parameters covered in this study
Architecture	Mapped into →	K1	Existence of MDM model or architecture
Governance		K2	Inclusion of policies and rules for data quality and standards
Management		K3	Presence of data cycle, hierarchy or any migration plan
Integration		K4	Method for merging, record linkage or resolution
Identification		K5	Any back propagation method or prediction of future
Business process management		K6	Coverage on business rules and MDM solution

Table 4 Review study with key parameters

Author and year of study	Application area covered	Description	K1	K2	K3	K4	K5	K6
			Analyzed key parameters					
Gualo et al. [1]	Software quality certification	A solution for evaluating "functional suitability" of MDM applications by considering functional requirements from part 100–140 of ISO 8000 is provided	Y	N	N	N	Y	Y
Zhao et al. [2]	Data networking	A model for evaluating the effectiveness and rationality of master data network is proposed. Case study showed less un-certainty, high response, timely update of data in cloud manufacturing environment	Y	Y	Y	N	Y	Y
Mrigen et al. [3]	Pharmaceutical industries	An approach to create singularity and hierarchies in pharmaceutical company is presented. Support of MDM vendors, challenges and issues in pharmaceuticals are given	N	Y	N	N	N	Y
Aditya Rahman et al. [4]	Health care service	Set of policies needed to implement MDM is described. Case study revealed 90% of activities related to MDM have been implemented	Y	Y	Y	N	Y	Y
Prokhorov et al. [7]	Banking sector	Conceptual model for creating centralized MDM with single regulatory and reference information system is proposed	Y	N	N	N	N	Y

(continued)

Table 4 (continued)

Author and year of study	Application area covered	Description	Analyzed key parameters					
			K1	K2	K3	K4	K5	K6
Pratama et al. [8]	Education and culture	Measurement of MDM implementation with MD3M in an organization of ministry of education and culture showed that 50% of capabilities in MD3M have been implemented	Y	Y	Y	N	Y	Y
Murti et al. [9]	XYZ institute	Method to implement personnel information system (attendance, licensing information system and employee performance goals) is given	Y	N	Y	N	Y	Y
Zuniga et al. [10]	Microfinance sector	Implementation of MD3M by framing a series of formal requirements and a comprehensive analysis with consolidated evaluation criteria is performed	Y	Y	Y	N	Y	Y
Qodarsih et al. [11]	Supreme court	The extent of maturity of MDM employees of Supreme Court Republic Indonesia showed that overall maturity level is 1	Y	Y	Y	N	Y	Y
Ng et al. [12]	City planning	Critical review on MDM solution for open-source MDM system, smart city concept model is presented that improves accuracy, integrity and transparence of infrastructure asset data	Y	Y	Y	N	Y	Y

(continued)

Table 4 (continued)

Author and year of study	Application area covered	Description	Analyzed key parameters					
			K1	K2	K3	K4	K5	K6
Arthofer et al. [13]	Health care service	Role of data quality, maturity levels and management cycle for creating master record in hospital is covered	N	Y	Y	N	N	Y
Vilminko et al. [14]	Public sector	Issues and challenges in establishing and developing MDM functions are mentioned. Among 15 identified challenges, eight issues are proved to be more generic toward implementing MDM	N	N	N	N	Y	Y
Ray et al. [15]	Financial banks	A framework for enhancing business intelligence through sounds MDM approach is presented	Y	Y	Y	Y	N	Y
Myung [16]	Enterprise PLM system	Classification of master data for product life cycle management (PLM) system deployment along with its implementation is highlighted	Y	N	N	N	N	Y

5   Study of Recent Hype Cycles

The hype cycle is a marked apparatus which is frequently utilized in reference for promoting and revealing new technology. Organizations can utilize this cycle to direct technology choices as per their degree of solace with risk [17]. Thus, to have a consistent and productive view point on latest trends, the scope and future of MDM, following Gartner's MDM hype cycle for recent years, are determined.

Table 5 shows the movement of MDM terms on hype cycle's five stages from year 2012 to 2020. This analysis is performed to generate the latest and future trend in MDM.

Figure 2 represents year movements from 2015 to 2020 on X-axis and five stages of Gartner's hype cycle on Y-axis . By analyzing complete graph for recent years, it can be predicted that in the future, scope of cloud MDM and multi-domain MDM will be at high priority.

6   Proposed Strategic Solution for Implementing MDM

MDM is significant for all kinds of organization as it empowers to make a total start to finish arrangement that drives advancement and accomplishes better business results. An effective MDM implementation empowers better utilization of critical information present in organization. Thus, it requires an approach to drive real and practical MDM. In order to accomplish this target, the proposed strategy is divided into three sections: MDM framework, a roadmap and data flow diagram providing details of step-by-step execution of complete process. These are described in the following sections.

6.1   Framework for MDM

A framework supports and empowers more compelling utilization of the MDM technology. It focuses on determining how master data fits into overall business strategy by surveying the current abilities. Afterward, it utilizes this information to construct effective and successful system, governance and process for implementing MDM. To fulfills this demand, a framework for implementing MDM is proposed in this study as shown below.

Figure 3 depicts a proposed framework for implementing MDM. The description of each layer is mentioned below.

1. ***Plan, Build and Configure***: Once the data is stacked into framework, beginning appraisal plans of organization are resolved. This gathered data is stored into system database from which a data model is fabricated that utilizes the essential business and information rules to generate a single copy of the master data.

Table 5 Study on recent Gartner's hype cycle

Name of graph and year	Stage 1 Technology trigger	Stage 2 Peak of inflated expectations	Stage 3 Trough of disillusionment	Stage 4 Slope of enlightenment	Stage 5 Plateau of productivity
Hype cycle for data and analytics and governance and MDM (2020) [18]	• Financial data risk assessment • Data security governance • Adaptive data and analytics governance • AI governance	• Augmented data quality • Cloud MDM	• 360° view • Metadata management solutions • Multi-domain MDM solutions • MDM	• MDM of customer data • MDM of product data	
Hype cycle for data and analytics governance and MDM (2019) [19]	• Financial data risk assessment • Data security governance • Adaptive data and analytics governance • AI governance	• Multi-vector MDM solutions • 360° view • Cloud MDM hub service	• Metadata management solutions • Multi-domain MDM solutions • MDM	• MDM of customer data • MDM of product data	
Hype cycle for information governance and MDM (2018) [20]	• AI governance • Adaptive data and analytics governance • Machine learning-Enabled data quality	• Multi-vector MDM solutions • 360° view • Cloud MDM hub service	• Metadata management solutions • Multi-domain MDM solutions • MDM	• MDM of customer data • MDM of product data	
Hype cycle for information governance and MDM (2017) [21]	• Cloud MDM hub service • Analytics governance • 360° view	• Multi-vector MDM solutions • Enterprise metadata management	• Multi-domain MDM solutions • MDM	• MDM of customer data • MDM of product data	• Data quality tools • Information exchanges
Hype cycle for information governance and MDM (2016) [22]	• Cloud MDM hub service • Analytics governance	• Multi-vector MDM solutions • Reference data management	• Multi-domain MDM solutions • MDM	• MDM of customer data • MDM of product data	• Data quality tools • Information exchanges

(continued)

Table 5 (continued)

Name of graph and year	Stage 1 Technology trigger	Stage 2 Peak of inflated expectations	Stage 3 Trough of disillusionment	Stage 4 Slope of enlightenment	Stage 5 Plateau of productivity
Hype cycle for enterprise information management (2015) [23]	• Analytics governance • Multi-vector MDM solutions	• Reference data management • Cloud business rule service	• Multi-domain MDM solutions • MDM	• MDM of customer data • MDM of product data	
Hype cycle for enterprise information management (2014) [24]	• Reference data management • Cloud business rule service	• Enterprise metadata management	• Multi-domain MDM solutions • MDM	• MDM of customer data • MDM of product data	
Hype cycle for enterprise information management (2013) [25]	• Reference data management • Multi-vector MDM solutions	• Enterprise metadata management • Multi-domain MDM solutions	• MDM • MDM of product data • MDM of customer data		
Hype cycle for MDM (2012) [26]	• Cloud-based data identification and enrichment services • MDM solutions in the cloud	• Enterprise metadata management • Multi-domain MDM solutions	• MDM of product data solutions • MDM of customer data solutions	• Data quality tools	

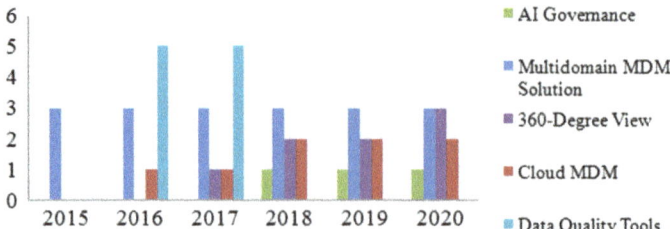

Fig. 2 Latest trends in MDM

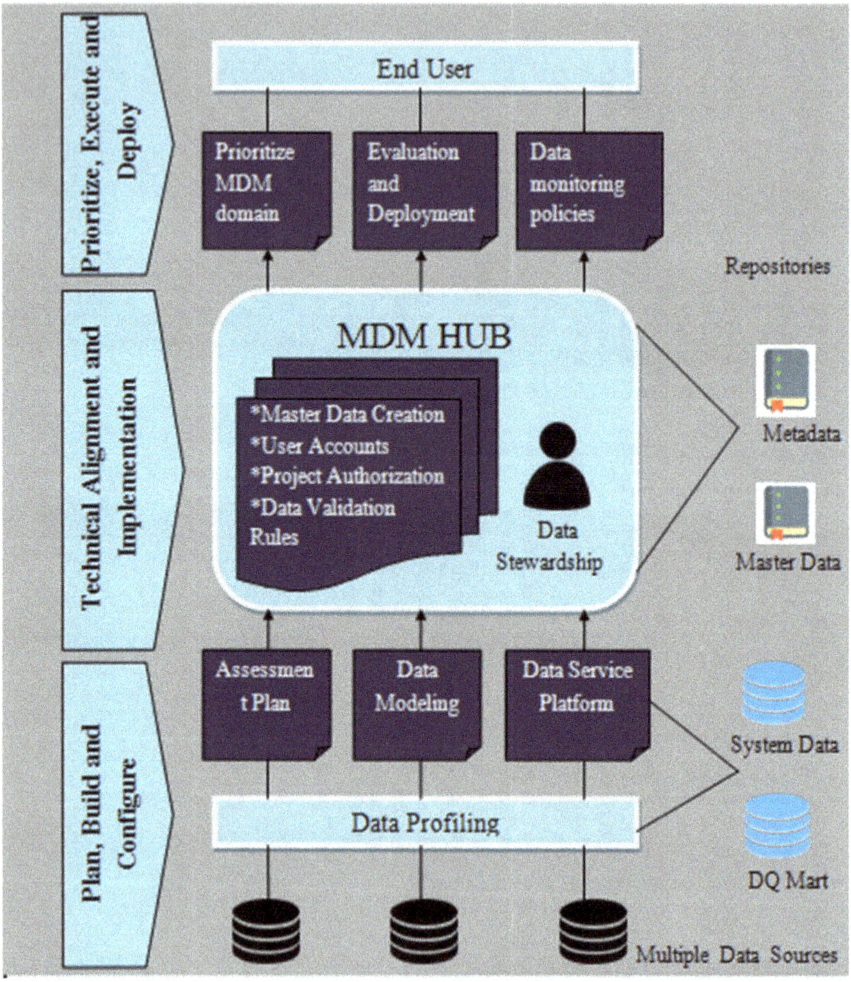

Fig. 3 Proposed framework for implementing MDM

Then, data service platform cleanse the data by applying data standardization rules to enrich the business data. Along with this, a database named as data quality (DQ) mart is used to verify and govern the data quality.

2. **Technical Alignment and Implementation**: Even cleansed and standardized records are loaded into system, it still contains duplicate records. The process of creating master record is actually implemented here. Application server will store information such as user accounts, project authorization, data validating rules and business model routines. A data stewardship provides the management for high quality data in a consistent manner. On the basis of this information, a repository is created which contains master data, admin and project metadata.

3. **Prioritize, Execute and Deploy**: Once golden copy of record is created, a third party can consume data directly from MDM system. Prioritization of MDM domain is necessary to understand organization's efficiency, revenue growth and cost reduction in various domains such as service provider, customer and products. After evaluating the results, final product is deployed to target system. Data monitoring and maintenance polices are consistently applied to get the best outcomes.

6.2 Roadmap for Implementing MDM

A roadmap is a strategic plan that characterizes an objective and includes the major steps to be followed to achieve desire milestone. Thus, for implementing MDM, this roadmap serves as a communication tool for guiding the devolvement. In this study, a 3-phase roadmap is designed for MDM as shown below.

Table 6 explains 3-phase roadmap for MDM.

Phase 1 is plan, build and configure in which objectives are initialized, plan is build, and all about data is discovered. Phase 2 is technical alignment and implementation where master record is generated and polices rules and security mechanisms are defined. Phase 3 is prioritization, execution and deployment where the final results are deployed to end system. Monitoring and maintenance polices are regularly applied with any changes made in requirement.

6.3 Data Flow Diagram

The step-by-step execution of roadmap is explained below with following data flow diagram (DFD). It covers all the terms and condition required in implementing MDM for an organization. It portrays the mandatory steps for executing MDM. Initially, objectives and need of creating master data in the organization are characterized. Data is collected from various sources.

Figure 4 expresses the DFD for the proposed framework of MDM. With dataset requirements, Data Model 1 is created, consisting of source data model that is stored

Table 6 Roadmap for MDM

3-phase roadmap for implementing MDM

Phase 1: Plan, Build and Configure
- Identify organization's objective for creating master data
- Produce an initial assessment plan to determine explicit and implicit business rules in the domain of application
- Collect data from various data sources
- Build a standardized data model for integrating and managing key information objects (**K1**)
- Evaluate its quality with predefined standards and polices (**K2**)
- Construct, Integrate and format the data with quality improvement techniques if it is not proper

Phase 2: Technical Alignment and Implementation
- Provide an input to produce a unified view of data
- Apply match and merge technique for an existing record to produce a golden record (**K4**)
- Form relationship hierarchies between various master records created above (**K3**)
- Implement MDM architecture and raise the awareness of issues to be addressed during next steps
- Establish standards and polices as per organization's need and objectives
- Apply data validation rules to maintain adequate level of data quality
- Provide optimize mechanism for workflow, information sharing, communication and coordination among the organization
- Use of security mechanism to protect the metadata
- Go to Phase 1 to determine new input from source to synchronize the records

Phase 3: Prioritize, Execute and Deploy
- Develop and improve the change control for master data, metadata and models
- Provide an MDM solution by assessing and evaluating the results (**K6**)
- Deploy the project to target system
- Suggest the future scope of relevant work (**K5**)
- Apply data monitoring and maintenance polices regularly for successful results

in a database named as "system data." Set of rules as per the objectives and initial plans are defined in the system data database to validate the model. Validated model is given for the process of record fetching. The quality of consolidated data is verified with quality verification process. On the off chance that the quality is adequate that implies data is consistent, uniform and up to date, at exactly, that point data is submitted to database "DQ mart"; otherwise, quality improvements techniques like data cleaning, enrichment or scrubbing are applied under cleaning and standardization process to make data clean and consistent. Once the data is enriched, quality verification is processed again. The cycle is rehashed until the satisfactory quality is not accomplished. After getting adequate quality, the record is submitted to record fetching section. The condition for the record is generated to check for new or existing one. If selected record is new, direct validation rules, policies, authentication rules are applied on it; otherwise, match and merge algorithm is applied. Data Model 2 consists of model for MDM as per user requirements. Database "metadata" contains information like business terms, data owners, database system name, table rows and columns, etc. "Master data" database contains unique records that are given to evaluation process which executes the results. The results are regularly monitored and

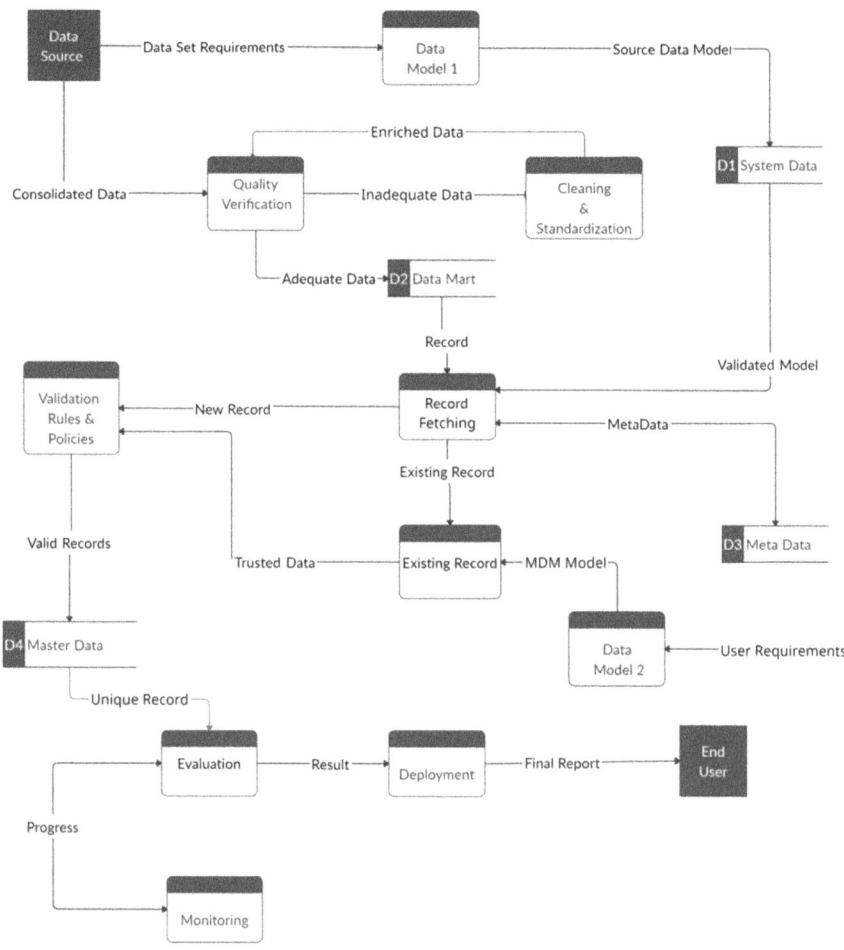

Fig. 4 DFD for implementing MDM

maintained under "maintenance" section. The final report is deployed to end user. This key method of implementing MDM may assist a business with demonstrating benefit and effective outcomes.

7 Conclusion

To supply the best products, fulfilling expanded customer needs and generating superior results, an adaptable methodology is required to make a total and extensive a 360° view customer record. That is where MDM exists. To deliver consistent, unique and accurate records, a constructive way to pilot the MDM implementation is of utmost priority. To have a better view point on this, a review study on existing work is conducted by narrowing the extent of fundamental building blocks of MDM: architecture, governance, management, integration, identification and business process management. These blocks are mapped into six key parameters K1–K6. The study determines the inclusion of these key parameters into various application domains. As per analysis, it is discovered that maximum five key parameters are implemented in health care and banking area, while there is still need of improvement in proper implementing these key parameters in other domains. Hence, an emphasis is made on inclusion of all key parameters for MDM implementation in this study.

To know the latest technical trends in MDM, an examination on recent Gartner's hype cycle for MDM reveals that among the five stages of hype cycle, AI governance and cloud MDM are at Stage 2, while multi-domain MDM and 360° view are at Stage 3 in year 2020. Thus, by analyzing complete graph for recent years, it can be predicted that in the future, scope of cloud MDM and multi-domain MDM will be at high priority.

Further, to get a deep insight into MDM, a solution for building framework, roadmap and DFD designs is proposed. This technique can help to pilot the implementation process by generating, handling, validating and monitoring master data. Thus, an organization can use this approach to convey operational data to each point in the organization and, eventually, can make better and quick choices that lower the expenses and increase the productivity.

References

1. Gualo F et al (2020) Towards a software quality certification of master data-based applications. Software Qual J 28(3):1019–1042
2. Zhao C, Ren L et al (2020) Master data management for manufacturing big data: a method of evaluation for data network. World Wide Web, Springer 23:1407–1421
3. Mrigen S et al (2020) Relevance of master data management in pharmaceutical industries. Int J Res Appl Sci Eng Technol 8(6):190–197
4. Aditya Rahman A et al (2019) Master data management maturity assessment: a case study of A Pasar Rebo Public Hospital. In: 2019 international conference on advanced computer science and information systems (ICACSIS), IEEE, Bali, Indonesia, pp 497–504
5. Loshin D (2009) Master data management. Morgan Kaufmann, Burlington, MA
6. https://en.wikipedia.org/wiki/Hype_cycle, Last accessed on 16 Sept 2020
7. Prokhorov I et al (2018) Development of a master data consolidation system model (on the example of the banking sector). In: Post proceedings of the 9th annual international conference on biologically inspired cognitive architectures, BICA, vol 142. Elsevier, Prague, Czech Republic, pp 412–417
8. Pratama FG et al (2018) Master data management maturity assessment: a case study of organization in ministry of education and culture. In: International conference on computer, control, informatics and its applications (IC3INA), IEEE, Tangerang, Indonesia, pp 1–6
9. Z. Murti et al (2018) Master data management planning: (Case study of personnel information system at XYZ Institute). In: International conference on information management and technology (ICIMTech). IEEE, Jakarta, pp 160–165

10. Zuniga DV et al (2018) Master data management maturity model for the microfinance sector in Peru. In: Proceedings of 2nd international conference on information system and data mining, USA, pp 49–53

11. Qodarsih N et al (2018) Master data management maturity assessment- a case study in the supreme court of the republic of Indonesia. In: 6th international conference on cyber and IT service management (CITSM), pp 1–7

12. Ng ST et al (2017) A master data management solution to unlock the value of big infrastructure data for smart, sustainable and resilient city planning. Procedia Eng 196:939–947 (Elsevier)

13. Arthofer K et al (2017) Data quality and master data management—a hospital case. Study Health Technol Inform 236:259–266

14. Vilminko R et al (2017) Master data management and its organizational implementation: an ethnographical study within the public sector. J Enterp Inf Manag 30(3):454–475

15. Kekwaletswe RM et al (2016) A framework for improving business intelligence through master data management. J South African Bus Res 1–12

16. Myung S et al (2016) Master data management in PLM for the enterprise scope. In: Advances in information and communication technology, vol 467. Springer, Cham, pp 771–779

17. Singh D, Kaur D (2020) Data profiling model for assessing the quality traits of master data management. Int J Recent Technol Eng 8(6):446–450

18. Gartner Research Home page. https://www.gartner.com/en/documents/3987607/hype-cycle-for-data-and-analytics-governance-and-master. Last accessed 14 Sept 2020

19. Gartner Research Home page. https://www.gartner.com/en/documents/3947320/hype-cycle-for-data-and-analytics-governance-and-master. Last accessed 14 Sept 2020

20. Gartner Research Home page. https://www.gartner.com/en/documents/3883694/hype-cycle-for-information-governance-and-master-data-ma. Last accessed 16 Sept 2020

21. Gartner Research Home page, https://www.gartner.com/en/documents/3769863/hype-cycle-for-information-governance-and-master-data-ma, Last accessed 2020/09/16.

22. Gartner Research Home page. https://www.gartner.com/en/documents/3372217/hype-cycle-for-information-governance-and-master-data-ma. Last accessed 18 Sept 2020

23. Gartner Research Home page. https://www.gartner.com/en/documents/3096424/hype-cycle-for-enterprise-information-management-2015. Last accessed 18 Sept 2020

24. Gartner Research Home page. https://www.gartner.com/en/documents/2816517. Last accessed 19 Sept 2020

25. Gartner Research Home page. https://www.gartner.com/en/documents/2571515. Last accessed 19 Sept 2020

26. Gartner Research Home page. https://www.gartner.com/en/documents/2100815/hype-cycle-for-master-data-management-2012. Last accessed 19 Sept 2020

A Heuristic Approach to Extract Knowledge from the Text Considering Explicit and Implicit Features Both

Sartaj Ahmad, Ashutosh Gupta, and Neeraj Kumar Gupta

Abstract The important thing to extract knowledge from text data (comments, chat, blogs, news articles, etc.) is how to convert unstructured data into structured data sometime called metadata and further how to get implicit features from it. Implicit features are very important for semantic understanding of any sentence. Generally, features are opinionated in terms of adjectives. Researchers find such features easily using natural language processing and the concept of association between adjectives and high frequency nouns, sometime synonyms of nouns and adjectives also play very important role in the acquisition of such features. In this paper, adjectives and their synonyms are considered to have highly relevant features and opinions in terms of pointwise mutual index. This paper presents the introduction, proposed method, framework, result, and finally conclusion.

Keywords Knowledge extraction · Explicit feature · Implicit features · Pointwise mutual index · Natural language processing

1 Introduction

Extraction of implicit features is very important for better understanding of users' feedbacks and becomes difficult if we consider Chinese like language for the same task. It is due to the complexity of this language [1, 2]. The authors present a hybrid association rule minding approach for this task. They consider different dataset for their work and present comparison of their result with the existing results. However,

S. Ahmad (✉) · N. K. Gupta
KIET Group of Institutions, Delhi NCR, Ghaziabad, India
e-mail: sartaj.ahmad@kiet.edu

N. K. Gupta
e-mail: neeraj.gupta@kiet.edu

Affiliated to AKTU, Lucknow, Uttar Pradesh, India

A. Gupta
School of Science in UPRTU, Allahabad, India

© The Author(s), under exclusive license to Springer Nature Singapore Pte Ltd. 2022
D. Gupta et al. (eds.), *Proceedings of Data Analytics and Management*,
Lecture Notes on Data Engineering and Communications Technologies 90,
https://doi.org/10.1007/978-981-16-6289-8_26

finding best parameters and reasonable rules is still problem for future work. Zhang and Zhu [3] describe implicit feature mining and propose a novel co-occurrence association-dependent approach. They describe how to find best rule using confidence and support on the given dataset. Such approach of mining provides fine-grained results. However, they did not give any concrete approach to deal with such type of feedbacks. Schouten and Frasincar [4] show mining of implicit features in their work. For this purpose, they propose a method that also depends on association rules. They improve recall using customers' reviews. In future, a classifier can be used for preprocessing of the sentences whether these are having implicit sentences or not. Consider these sentences for further processing once these are classified. Khairudin et al. [5] present a temporal attribute's effect in association rule mining for Web log data. They use temporal relational rule mining approach to achieve it. They discuss detailed mechanism for modeling to find temporal relationships from the Web log data that help in the discovery of interesting rules. It further helps in extracting the hidden information from the Web user log. They describe about the importance of developing a tool as future work to find temporal aspects from Web log dataset during preprocessing. Goel [6] proposes a technique for cleaning of data on the server log. Log files store information about users. These files are used to know the surfing behavior of customers and used to recommend product or service to the end user. They also explain data preprocessing, pattern discovery, and pattern analysis. The proposed technique helps in decreasing the size of log files and the number of records that in turn enlarges the quality of data. However, they do not describe about the problems as path completion and session identification. Authors are combining the data mining methods for the analysis of textual data to see if there is an evidence of future financial achievement. Therefore, Zia et al. [7] exploit the same concept and they consider each review in both directions. They increase size of training and test data by considering reverse side of each sentence. They also propose an algorithm to consider both side of each sentence. This way they improve performance of classification model through precision and recall. In future, such algorithm can be generalized to cover more sentiment analysis task and to cover more polarity shift in sentences. Panchendrarajan et al. [8] focus on finding implicit features from the reviews about restaurant. They propose novel approach with high F-measure. Same approach can be considered for another domain. In [9–12], authors discuss about finding of implicit aspects and their orientation. They propose technique and apply the same technique on text written in the Turkish language. They explain that some improvement can be proposed for the aspect sentiment mappings by considering implicit features. Therefore, proposed research work is related to following tasks but different in some aspects and approach. In case of document classification, textual content available in the different sources, like news, articles, editorial, stories, novel, books etc., can be used. These documents are classified in terms of + ve and − ve opinions as mentioned by Li and Zhong in [13]. Here, documents are replaced with reviews. Similarly, in case of sentiment classification, there is consideration of sentences and further featurewise sentiments mapping. It includes explicit and implicit features both. There are different algorithms and approaches to find explicit features and implicit features. However, in this case approach is little bit different.

In this work, data is collected and converted from unstructured form to structured form, then the data mining technique is applied to get highly precise information. Features are two types; one is frequent which can be found as explicit features and second type is expressed indirectly (implicit features).

2 Proposed Method

Finding of explicit feature: According to Hu and Liu [14] algorithm as shown below takes a tag file which is one of the files produced by NLP parser as input, pick noun (as per threshold value) corresponding adjective from each sentence and send this information to a file.

For example, in the sentence like "Battery is long lasting" **battery** can be selected as frequent noun if it satisfies the targeted threshold and **lasting** as adjective. For this task, they use association mining to get all frequent items sets. However, they left infrequent items sets. The infrequent items sets can play an important role to increase precision and recall of extracted information and can affect the process of finding implicit features and finally summary. For example, sentences like

for each review
    for each sentence(S)
        if (S has frequent feature) then
        Pick feature and adjective
        Store feature and adjective in a file

"Overall, it is nice product."
"It is a good product."
"It is a good item."
After applying association mining, such sentences will have rules like.
nice → product.
good → product.
good → item.
In their perspectives, these rules are not considered strong rules because these have less support. However, by the observation it is found that these rules are telling about a product, which is automatically a feature. Such rules increase count of sentiment for the product (feature) and can affect implicit feature. Next step is showing how to deal with implicit feature.

Finding of implicit feature: Few features may be opinionated without mentioning their name. Therefore, in such case feature is found based on association. Before finding association, synonyms of each adjective are found, then using PMI method to find association between adjective and frequent nouns. Noun having high value of PMI will be considered in this case. For example, "It is long lasting."

Table 1 Adjective versus noun

	Adjective			
Noun	Lasting	Durable	Stable	Unchangeable
Battery	10	5	2	2
Product	5	3	1	0

This sentence is opinionated, but noun is not mentioned. However, it can be found in terms of adjective as follows (Table 1).

$$
\begin{aligned}
\textbf{PMI (Battery, Adjective)} &= \text{PMI (Battery, lasting)} + \text{PMI (Battery, durable)} \\
&\quad + \text{PMI (Battery, stable)} + \text{PMI (Battery, unchangeable)} \\
&= 10/20 + 5/9 + 2/3 + 2/5 \\
&= 2.13
\end{aligned}
$$

$$
\begin{aligned}
\textbf{PMI (Product, Adjective)} &= \text{PMI (Product, lasting)} + \text{PMI (Product, durable)} \\
&\quad + \text{PMI (Product, stable)} + \text{PMI (Product, unchangeable)} \\
&= 5/20 + 3/9 + 1/3 + 0/5 \\
&= 0.92
\end{aligned}
$$

$$
\text{As PMI (Battery, Adjective)} > \text{PMI (Product, Adjective)}
$$

Therefore, lasting adjective will present a feature named Battery not Product. This process increases the number of records in the dataset. Furthermore, this dataset can be used to find the orientation of feature using the following algorithms [14]. This adjective orientation list is used at the time of finding the orientation of each sentence. This list works as one of the inputs for the algorithm number 2 as presented by Ahmad et al. in [15]. In some cases, PMI approach fails. For example, some common adjective can be used for more than one feature. It means any combination having more PMI value than other combinations do not guarantee a correct association.

For example, in case of sentence like (Fig. 1).

"It is fast."

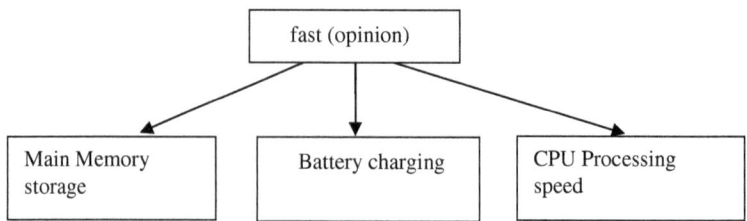

Fig. 1 Relation of "fast" word with the different features

It is difficult to judge fast is related to which feature.

Therefore, the method discussed can be used to extract implicit features from such sentences, but it is difficult to ensure that it is the same feature as the user is talking about. Next paragraph presents a framework that shows how data is converted from unstructured to structured form and how to get knowledge from structured metadata.

3 Framework for Knowledge Extraction

As shown in the figure, this process consists of different subtask as mentioned below (Fig. 2).

- Collect data from online Web sites. For this purpose, Web's scrapping tool can be used like Google's scrapper, but it requires manual interference. Therefore, Web crawler is better solution to do it automatically. In this specific data is collected and copied in the file.
- It requires removal of stopping words, stemming, lemmatization and converting abbreviation with its full form.
- Keep all sentences in a common format means in the upper or lower form.
- All preprocessed data as file is sent to the parser. This parser gives three files as output.

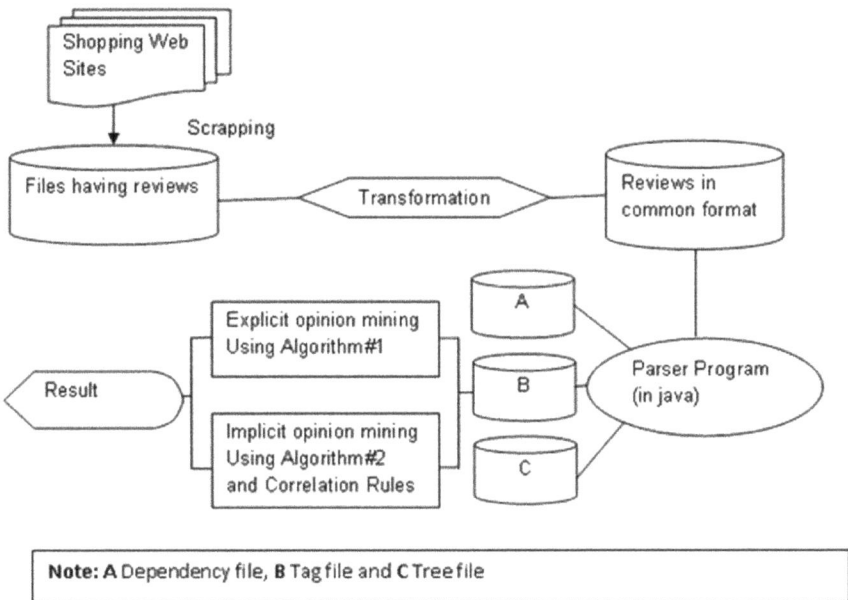

Fig. 2 Framework required for knowledge extraction

- Out of these three files, one file (B) is passed to the algorithm #1 [15]. This algorithm gives output in terms of positive and negative opinions with reference to the corpus related to the negatives and positives words provided.
- Same file (B) is also passed to the algorithm #2 [15]. This algorithm returns positive and negative opinions and returns metadata (structured in nature) in terms of noun, verb, adverb, and adjectives. Few records for implicit features are also generated using approach that is mentioned above. This metadata is used to find an association between words and correlation in terms of support and confidence.

Like A → B [Support, Confidence, Correlation].

4 Experiment and Result

Apply algorithm #2 on the tagged file as mentioned in [15] to get metadata that is structured in nature as shown in Table 2. This data is expanded further using WordNet (lexical database). Further synonyms of nouns and adjectives are considered to obtain more combinations that help in finding associations. Furthermore, adjectives not having any feature are also considered to find relevant feature in terms of PMI.

Figure 3 describes the association of adjectives with features that are mainly expressed in the reviews. Such datasets help in finding the more features from implicit sentences. Frequent nouns are considered in finding the association with adjectives (not having features). This process increases size of training set. Now orientation corresponding to each noun (feature) depending on the adjective can be set manual or automatic as expressed in algorithm#2. As a result, we get data as shown in Table 3. Same algorithms as mentioned earlier are applied on the different datasets to get different metadata sets. Before classification, it is essential that there must be large number of records in the training set. Therefore, these metadata are processed further

Table 2 Noun versus adjective

Nouns	Adjective
Battery	Best
Camera	Decent
Camera	Great
Camera	Good
Camera	Decent
Display	Great
Display	Colorful
Interface	Brilliant
Interface	Unintuitive
Keypad	Best
Keypad	Excellent

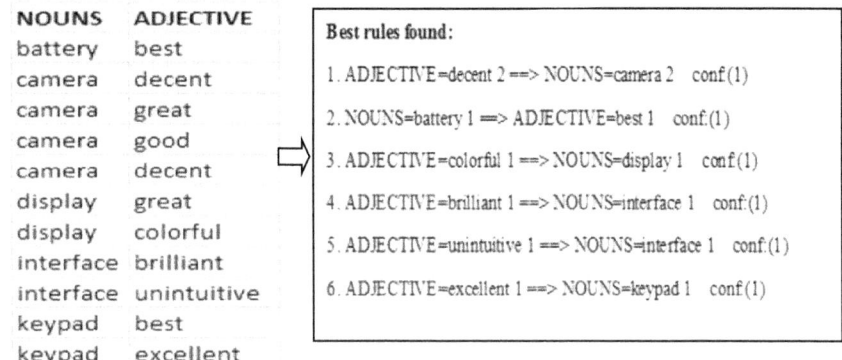

Fig. 3 Association between noun and adjective

Table 3 Result of classification

Product	Precision	Recall	F-measure	Class
Mobile	0.902	1	0.997	P
	1	0.625	0.769	N
Weighted Avg	0.90	0.93	0.897	
Camera	0.92	0.96	0.94	P
	0.98	0.7	0.82	N
Weighted Avg	0.95	0.83	0.88	
Router	0.94	0.98	0.96	P
	1	0.68	0.81	N
Weighted Avg	0.97	0.83	0.89	
Diaper	0.94	0.96	0.95	P
	0.98	0.80	0.88	N
Weighted Avg	0.96	0.88	0.92	

to enhance number of records using infrequent nouns (features) and considering both sides of each sentence as mentioned in [7]. Selection of an algorithm for the classification purpose is also an important task. Therefore, Naïve Bayes algorithm is found suitable that provides following improved result as far as precision, recall, and F-measure are concerned.

Therefore, infrequent features play very important role in case of precision. It can be seen in the following Fig. 4 considering positive sentiments.

Results show high precision in case of all the datasets. Recall may be less due to some impurity in the dataset that can be in terms of sarcasm or spam. These are major issues where researchers are also working.

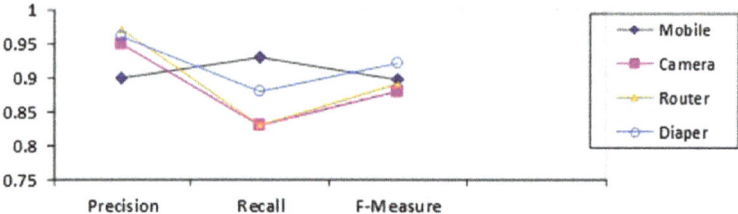

Fig. 4 Performance of positive sentiments

5 Conclusion

In this paper gap of the research is presented knowledge extraction from the Web considering reviews as dataset. After a lot digging, it is found that infrequently used features and implicit features both are very important for obtaining accurate information. Infrequently used features support implicit features, which can further improve the results. Framework and some basic algorithms are also presented that will help and guide in such type of work. In future, some generalized heuristic algorithms can be developed to obtain more precise information from domain-independent datasets.

References

1. Wang W, Xu H, Wan W (2013) Implicit feature identification via hybrid association rule mining. Expert Syst Appl 40(9):3518–3531
2. Tubishat M, Idris N, Abushariah MA (2018) Implicit aspect extraction in sentiment analysis: review, taxonomy, opportunities, and open challenges. Inf Process Manage 54(4):545–563
3. Zhang Y, Zhu W (2013) Extracting implicit features in online customer reviews for opinion mining. In: Proceedings of the 22nd international conference on World Wide Web, pp 103–104
4. Kim S (2014) Implicit feature detection for sentiment analysis. In: Proceedings of the 23rd international conference on World Wide Web, pp 367–368
5. Mohd Khairudin N, Mustapha A, Ahmad MH (2014) Effect of temporal relationships in associative rule mining for web log data. Sci World J
6. Goel R (2014) Enhanced web mining technique to clean web log file. Int J Comput Appl 96(16)
7. Xia R, Xu F, Zong C, Li Q, Qi Y, Li T (2015) Dual sentiment analysis: considering two sides of one review. IEEE Trans Knowl Data Eng 27(8):2120–2133
8. Panchendrarajan R, Pemasiri A (2016) Implicit aspect detection in restaurant reviews using cooccurrence of words. In: Proceedings of the 7th Workshop on computational approaches to subjectivity, sentiment and social media analysis, pp 128–136
9. Lazhar F (2019) Implicit feature identification for opinion mining. Int J Bus Inf Syst 30(1):13–30
10. Kama B, Ozturk M, Karagoz P, Toroslu IH, Kalender M (2017) Analyzing implicit aspects and aspect dependent sentiment polarity for aspect-based sentiment analysis on informal Turkish texts. In: Proceedings of the 9th international conference on management of digital EcoSystems, pp 134–141
11. Eldin SS, Mohammed A (2020) An enhanced opinion retrieval approach via implicit feature identification. J Intell Inf Syst 1–26

12. Dang N, Khanna A, Allugunti VR (2021) TS-GAN with policy gradient for text summarization. In: Khanna A, Gupta D, Pólkowski Z, Bhattacharyya S, Castillo O (eds) Data analytics and management. Lecture notes on data engineering and communications technologies, vol 54. Springer, Singapore
13. Li Y, Zhong N (2004) Web mining model and its applications for information gathering. Knowl-Based Syst 17(5–6):207–217
14. Hu M, Liu B (2004) Mining and summarizing customer reviews. In: Proceedings of the tenth ACM SIGKDD international conference on knowledge discovery and data mining, pp 168–177
15. Ahmad S, Gupta A, Gupta NK (2020) Design, development, and comparison of heuristic driven algorithms based on the crossed domain products' reviews for user's summarization. Recent Adv Comput Sci Commun 13(5):884–892

Hybrid System Based on Genetic Algorithm and Neuro-Fuzzy Approach for Neurodegenerative Disease Forecasting

Haneet Kour⊙, Jatinder Manhas, and Vinod Sharma

Abstract Artificial intelligence (AI) tools have been discovered to be quiet significant in the diagnosis and prediction of different medical disorders. Medical diagnostic systems are introduced using a variety of soft computing techniques to assist medical practitioners in accurately anticipating disorders. However, these soft computing technologies suffer from some limitations when dealing with ambiguous and complex data. To address these challenges, hybrid AI techniques have been developed, which combine two or more classical methodologies to create expert systems that produce better outcomes. In this study, a hybrid system called GANFIS has been created by merging Genetic Algorithm approach and Adaptive Neuro-Fuzzy Inference System to better diagnose neurodegenerative disorders (NDs). In order to validate the presented model, the experimental study has been conducted using benchmark datasets of two neurodegenerative disorders, i.e. Dementia and Parkinson's. The evaluation metrics such as accuracy, precision, recall, f-score and kappa coefficient have all been applied to evaluate the proposed hybrid technique. As per experimental results, GANFIS outperformed the traditional neuro-fuzzy technique in the diagnosis of dementia and Parkinson's disease.

Keywords ANFIS · Dementia · Genetic algorithm · Medical diagnosis · Parkinson

1 Introduction

Neurodegenerative Diseases (ND) is an umbrella term for a set of ailments in which brain neuronal cells degrade over time, altering the structure and operation of the nervous system. It results in Dementia, Amyotrophic lateral sclerosis, Parkinson's, Lewy body and other neuronal ailments. The first stage involves the degeneration of neurons, which causes problems with coordination and remembering names. Later on, many neurons continue to degenerate, resulting in a loss of cognitive abilities, judgement abilities, orienting skills, language issues, and so on [1]. ND affects

H. Kour (✉) · J. Manhas · V. Sharma
Department of Computer Science and IT, University of Jammu, Bhaderwah Campus, Jammu, India

© The Author(s), under exclusive license to Springer Nature Singapore Pte Ltd. 2022   319
D. Gupta et al. (eds.), *Proceedings of Data Analytics and Management*,
Lecture Notes on Data Engineering and Communications Technologies 90,
https://doi.org/10.1007/978-981-16-6289-8_27

millions of people worldwide. As per statistics reported in [2], approximate 5.4 million people are affected with Alzheimer's disorder in United States and 5 million suffer from Parkinson's disease. Neurodegenerative illnesses are anticipated to affect more than 12 million Americans by 2030. Approximately, 3 crore people in India go through a range of neurological diseases [3]. As per statistics published in [4]; approximate 122,019 deaths have been recorded due to Alzheimer's in 2018; and reported deaths from Alzheimer's has been increased by 146.2% during the period 2000–2018.

Thus, it is critical to diagnose ND patients as soon as possible in order to improve their life quality. Neuroimaging techniques [5] and psychological testing [6] can be used to diagnose these disorders. Despite significant advancements in the medical field, these disorders continue to affect millions of people around the world. Hence, more study is needed in this area to reduce the global prevalence rate of ND. In last few years, machine learning (ML) approaches have been implemented to ND diagnosis for assisting neuropsychiatrists in their diagnosis. These strategies have been proved to be effective in assisting doctors in making correct diagnosis.

Traditional machine learning methods, on the other hand, endure few problems when working with imprecise and enormous amounts of data. To address these challenges, hybrid methodologies were created by combining two or more standard machine learning algorithms using cascaded/ensemble techniques. These hybrid strategies have demonstrated to be significant particularly in the case of large, multi-dimensional, and imprecise medical data [7]. This influenced the current authors to propose a hybrid framework formulated on *attribute selection* and *prediction approach* to better anticipate the ND diseases.

The current research study implements a hybrid model called *GANFIS* by coupling *Genetic Algorithm* with *Fuzzy-Neural Network* for ND prediction. The proposed system sought to get better the diagnostic efficiency of the regular neuro-fuzzy (NFS) approach for ND disease diagnosis.

2 Related Work

In past, machine learning techniques were used for the prediction and classification of diverse health issues [8]. Over the years, various researchers applied soft computing approaches for predicting neurodegenerative diseases [9]. Traditional neuro-fuzzy approaches have been introduced for the diagnosis of Autism [10], Schizophrenia [11] and Ischemic Stroke [12] on *clinical data*. In order to diagnose ND from *imaging data* (i.e. CT Scans, MRI images, etc.) or *biomedical signals* (i.e. EEG signals, Voice signals, etc.); NFS based hybrid models have been developed by researchers for the diagnosis of Epilepsy from EEG signals [13], Dementia from MRI scans [14] and Alzheimer's from MRI scans [15].

Alvarez et al. [16] developed a framework based on principal components analysis (PCA) and different classification approaches including naïve bayes, decision tree, Kstar, Sequential minimal optimization (SMO), J48 and others; for the diagnosis of primary progressive aphasia (PPA). Weka tool was used to carry out experimental investigation on FDG-PET scans of 91 patients with PPA, and 28 healthy controls. PCA-SMO was found to have achieved the best performance amongst all the classifiers taken in the study.

Parkinson's diagnostic tool was introduced by Oliveira et al. [17]. The system was implemented by applying partial directed coherence with random forest approach to the dataset of EEG signals. The presented model achieved 99% classification accuracy and kappa coefficient of 0.98.

Fan et al. [18] introduced an approach for the diagnosis of Alzheimer from MRI images using support vector machine (SVM). The model was trained on a total of 1203 MRI scans; and then validated on 281 instances; and it had 75.26% prediction accuracy on test data.

An intelligent medical system was employed by Jain et al. [19] for detecting Parkinson's from voice characteristics. The presented system was built using ensemble classifier along with oversampling techniques. The experimental findings predicted maximum accuracy of 91.5%.

Lamba et al. [20] implemented hybrid approach by integrating random forest with synthetic minority oversampling technique (SMOTE) to discover Parkinson's from voice signals. The experimental results predicted accuracy of 95.58%.

3 Proposed Hybrid Model

The goal of this research study is to introduce GANFIS, a hybrid framework for neurodegenerative disease prediction. GANFIS framework has been implemented by hybridization of Genetic Algorithm with Neuro-fuzzy System. GANFIS cascades *feature selection* with *classification* task. It works by selecting the optimal features in the data using Genetic algorithm. The selected features are then fed to ANFIS to classify the data. Frequently, all of the attributes in large medical datasets are not significant for disease detection. The less important attributes result in more computational complexity to the model, which causes *model overfitting* and reduces its effectiveness. As a result, attribute selection has a significant impact on the learning strategy for achieving *low space and time complexity*. It identifies the most contributing attributes towards knowledge discovery within the dataset and removing the irrelevant attributes that don't contribute towards information mining or reduces classification accuracy. The dataset with selected features will then be used to train the model, thereby enhancing the classification accuracy [21].

3.1 Genetic Algorithm

Genetic algorithm (GA) is one of the most advance technique used for feature selection. It's a method of feature optimization formulated on the concept of natural evolution. In nature, organisms' genes persist to progress to better adjust to the environment over successive generations. It includes the Darwin's theory of survival of the fittest idea into search algorithm to find out the optimum solution amongst all the feasible solutions. For a particular problem, it retains the population of potential solutions termed as *individuals. Selection, crossover* and *mutation* are the genetic operators performed to regulate these individuals. This approach improves its solution over the course of several generations. Every individual gets assessed using *fitness function* that plays a significant role in optimizing the solution [22]. Fitness function can be of any type based on the problem domain. The general steps involved in genetic algorithm are presented in Algorithm 1.

Algorithm 1 Genetic Algorithm

Step 1   Set up the node population.
Step 2   Assess node population using *Fitness Function.*
Step 3   Perform *selection* of parent nodes for reproduction.
Step 4   Compute *crossover* and *mutation.*
Step 5   Estimate population survival.
Step 6   Steps 3 through 5 to be repeated until the stopping criteria are met.

In the current study, GA involves identification of optimal subset of features in the data where each *individual* represents *the subset of features* to be selected in order to enhance the efficiency of the learning model. The *genes* refer to all the attributes present in the data set. The feature set is depicted by a binary vector, with genes having binary values indicating whether a feature in the learning model is included (1) or excluded (0).

3.2 Adaptive Neuro-Fuzzy Inference System (ANFIS)

Neuro-Fuzzy technique is a smart hybrid approach that concatenates the human interpretation capability of fuzzy logic with learning expertise of neural network to adjust the premise and consequent factors of the fuzzy inference system. This research work employed ANFIS application of neuro-fuzzy approach [23]. ANFIS is presented as a five-layered design as shown in Fig. 1 and is mathematically stated as follows:

Assume A_1 and A_2 to be two inputs and D be the output, each having two fuzzy if–then rules as given below:

Rule 1: If A_1 is R_1 and A_2 is S_1, then $D_1 = l_1x + m_1y + n_1$.
Rule 2: If A_1 is R_2 and A_2 is S_2, then $D_2 = l_2x + m_2y + n_2$.

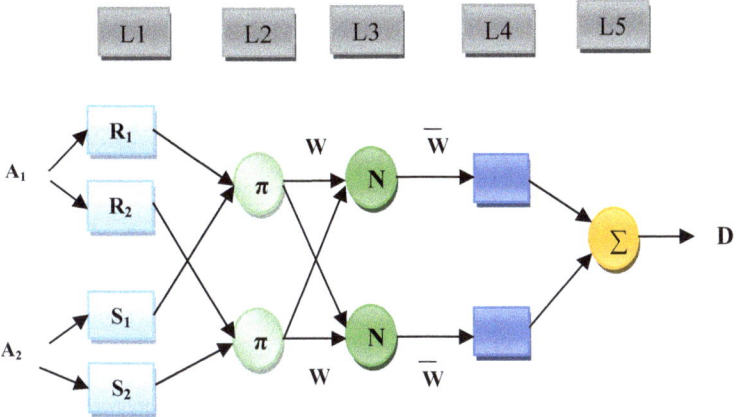

Fig. 1 ANFIS architecture

Here, R_k, S_k refer to fuzzy sets of the antecedent factors, and l_k, m_k, n_k represent the consequent factors to be optimized during the learning task.

$L1$: It represents the first step having kth node as an adaptive node and performs fuzzification operation on linguistic variables:

$$d_k^1 = F_{R_k}(A_1) \quad k = 1, 2$$

$$d_k^1 = F_{S_{k-2}}(A_2) \quad k = 3, 4$$

Here, $F()$ denotes *fuzzification function* such as Gaussian or any other; that is selected based on the problem at hand.

$L2$: This step computes the firing strength of each fuzzy-rule unit through multiplication of the input signals, and it generates the output given as:

$$d_k^2 = W_k = F_{R_k}(A_1).F_{S_k}(A_2) \quad k = 1, 2$$

$L3$: This step performs the normalization on the firing strengths of each unit as:

$$d_k^3 = \overline{W_k} = \frac{W_k}{W_1 + W_2} \quad k = 1, 2$$

$L4$: This step evaluates the node function using equation given below:

$$d_k^4 = \overline{W_k}.D_k = \overline{W_k}.(l_k x + m_k y + n_k) \quad k = 1, 2$$

$L5$: The final output is calculated as:

$$D = d_k^5 = \sum \overline{W_k}.D_k = \sum \overline{W_k}.(l_k x + m_k y + n_k) \quad k = 1, 2$$

4 Methodology

The current research study implements a hybrid model called *GANFIS* to fore-cast neurodegenerative disorders. This two-step hybrid framework cascades Genetic Algorithm with the Neuro-Fuzzy method. The first step applies *Genetic Algorithm* to identify optimal features in ND Diseases; and the second one entails using ANFIS to categorize the selected optimal features. The experiments were run on Matlab platform using two benchmark datasets i.e. *Dementia* and *Parkinson's disease* (PD). The methodology for the undertaken research work is depicted in Fig. 2.

The datasets for Dementia and Parkinson's were collected from *Kaggle* [24] and *UCI Machine Learning Repository* [25], respectively. The former contains 336 records, whereas the latter has 195 instances. Table 1 presents the description of both datasets taken in this research study. Both datasets were split into *train set* and *test set* as per 75:25 ratios. Data pre-processing has been applied on the collected data to encode categorical and nominal attributes into integer values. The evaluation

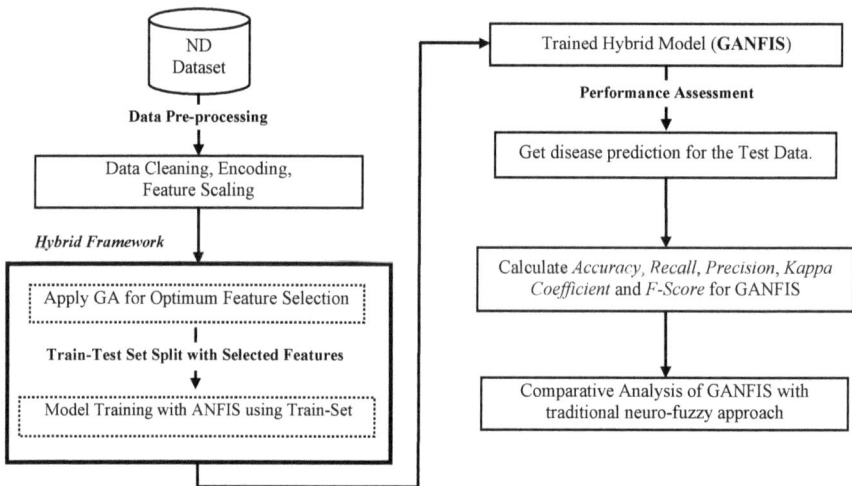

Fig. 2 Methodology for the current study

Table 1 Datasets details and optimal features selected by GA

S. No	Dataset	Number of input attributes	Number of instances	Output class	Number of features selected by GA
1	Dementia	9	336	Demented-146 Non-demented-190	3
2	Parkinson	22	195	PD-147 Control-48	11

metrics such as *accuracy, precision, recall, f-score*, and *kappa coefficient* have all been applied to evaluate the validity of proposed hybrid technique.

The proposed hybrid model, i.e. GANFIS works in two phases. In the first phase, GA is applied to select the most important features from the dataset. To implement GA, the value of *population size, maximum number of generations, crossover* and *mutation rate* has been set to 10, 100, 0.8 and 0.01, respectively. The *fitness function* has been implemented by *k-nearest neighbour classifier* ($k = 5$) to optimize the error rate. In case of Dementia dataset, out of 9, three features have been selected by GA and eleven features have been selected out of 22 in PD dataset. These selected features have been fed to ANFIS as inputs in the next phase.

In the second phase, ANFIS is applied to classify the data with selected features in order to diagnose ND. *Subtractive Clustering* was used to employ ANFIS in order to generate *Sugeno FIS* and then *hybrid learning algorithm* has been applied to optimize the fuzzy-rule base. This model was trained for *100 epochs. Gauss* and *linear* functions were used for the input and output membership functions, respectively.

5 Results and Discussion

The current research work introduced a hybrid framework by cascading Genetic Algorithm with neuro-fuzzy system for Neurodegenerative Disorders Prediction. After performing the experiments, it has been found out that the presented hybrid system GANFIS can diagnose ND effectively. The *GANFIS models* generated for Dementia and PD are illustrated in Figs. 3 and 4, respectively, whereas Figs. 5 and 6 point to rule base generated for both NDs taken in this study.

Table 2 depicts the experimental results predicted by GANFIS and the standard NFS technique to identify Dementia as well as Parkinson's disorder. The presented hybrid framework has been validated using various evaluation metrics: *accuracy, recall, precision*, f-*score* and *kappa coefficient*. Doctors seek a diagnostic system that can predict reliable findings when it comes to medical diagnostics. This diagnostic approach is also expected to avoid disease misclassification by preventing both false positives and false negatives.

Figure 7 presents the comparative analysis of GANFIS model with traditional NFS approach. It improved *accuracy* of traditional NFS from 96.43 to 98.81% in case of Dementia, and 75 to 91.67% in PD. Since dementia dataset already has only 9 attributes thus feature selection enhanced its performance a little bit. But PD dataset is higher dimensional dataset with 22 attributes, thus GA brought a significant improvement in its performance. Apart from this, *Recall* and *F-score* values for traditional NFS have been increased from 94.60% to 100%, and 0.9581 to 0.9867, respectively, in case of Dementia disorder. GANFIS further enhanced *F-score* value from 0.8235 to 0.9487 for Parkinson's disease. *Kappa coefficient* values have been enhanced by GANFIS model from 0.927 to 0.976 for Dementia, and 0.406 to 0.727 for PD. GANFIS also improved the space complexity since it optimized the size of rule base in both ND disorders. Also, precision has been increased from 97.22% to

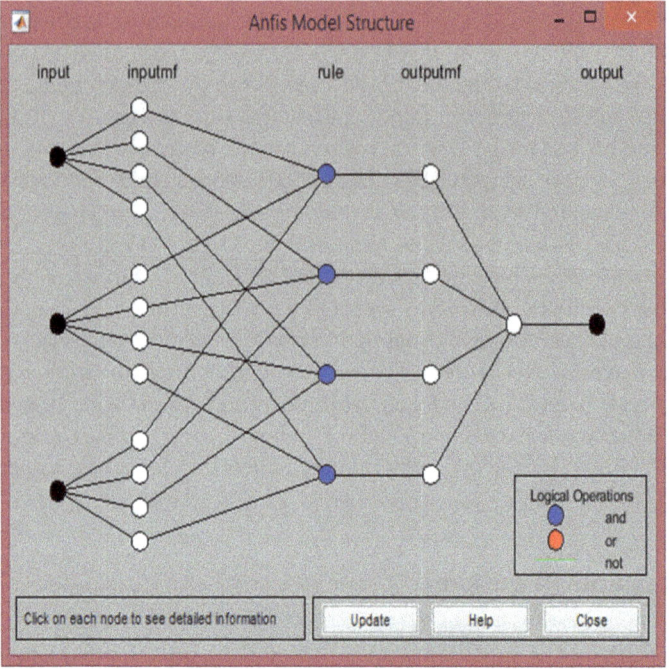

Fig. 3 GANFIS model generated for Dementia

97.37% in case of Dementia, and it has been found to be increased from 90.3% to 92.5% in case of PD. Class imbalance has also been found to affect the performance of learning model. It can be analyzed from experimental results that GANFIS optimized the performance of traditional neuro-fuzzy system in terms of all the *performance metrics* taken in this study for the diagnosis of both ND disorders. The experimental findings indicated that GA enhanced the performance of NFS model by selecting the optimum features in the dataset, thus reducing its computational complexity.

6 Conclusion and Future Scope

The proposed model GANFIS can be efficiently employed for the diagnosis of neurodegenerative illnesses, according to the results of the experimental study. Currently, this presented model is limited to accept input in csv format. In the future, this model can be upgraded to accept inputs in *imaging data* or *biomedical signals* in order to make a diagnosis of health conditions. Traditional NFS can be explored with other soft computing techniques to optimize its performance.

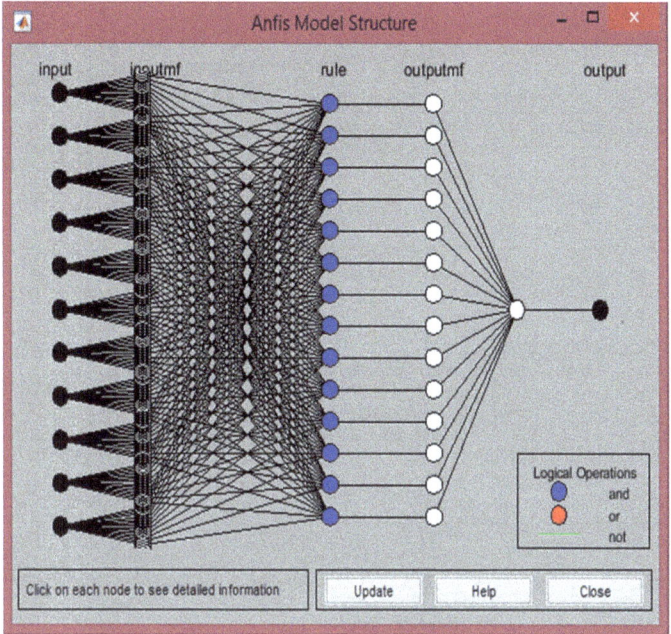

Fig. 4 GANFIS model generated for PD

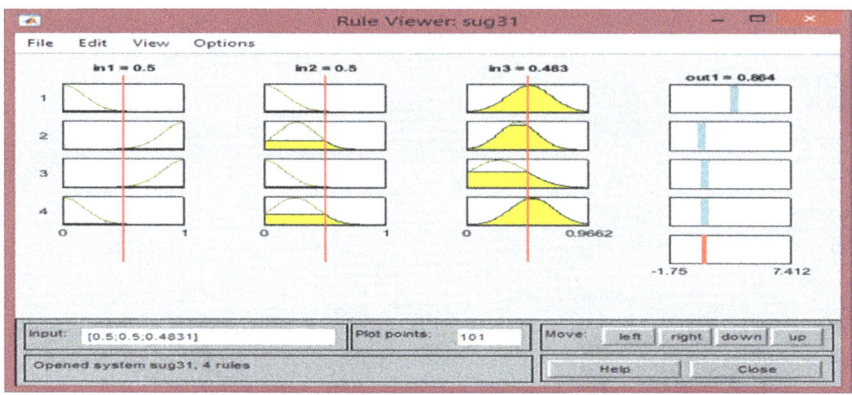

Fig. 5 Rule base generated by GANFIS for Dementia

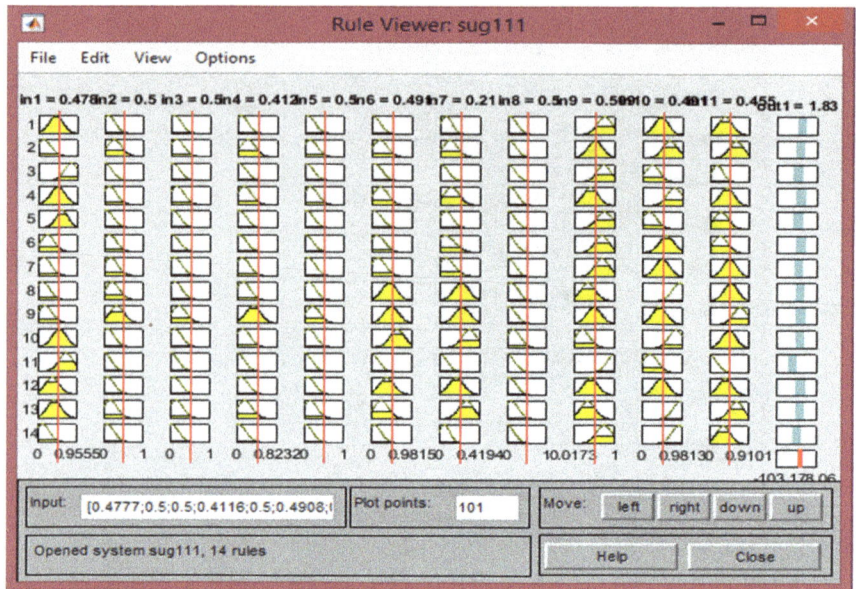

Fig. 6 Rule base generated by GANFIS for PD

Table 2 Experimental results for GANFIS and traditional NFS

Parameter	Dementia		Parkinson's	
	GANFIS	Traditional NFS	GANFIS	Traditional NFS
Accuracy (%)	98.81	96.43	91.67	75
Recall (%)	100	94.60	97.37	75.67
Precision (%)	97.37	97.22	92.5	90.3
F-score	0.9867	0.9581	0.9487	0.8235
Kappa coefficient	0.976	0.927	0.727	0.406
Fuzzy rules	04	30	14	71

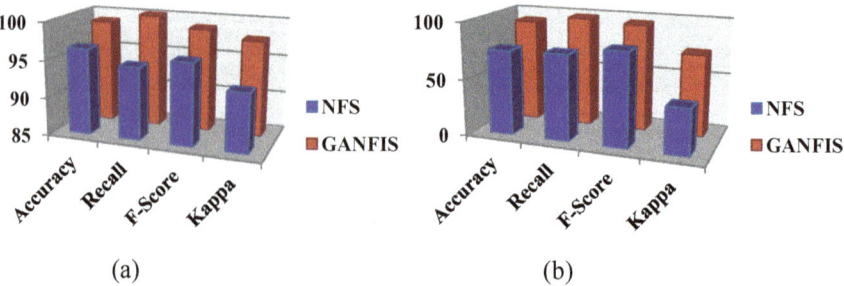

Fig. 7 Comparative analysis of GANFIS with traditional NFS: **a** Dementia and **b** Parkinson's disease

References

1. Neurodegenerative Diseases. http://www.neurodegenerationresearch.eu/about/what/, 2 Feb 2021
2. Neurodegenerative Diseases. https://www.niehs.nih.gov/research/supported/health/neurodegenerative/index.cfm, 3 Feb 2021
3. https://www.deccanchronicle.com/lifestyle/health-and-wellbeing/120418/over-3-million-people-in-india-suffer-from-various-forms-of-neurologi.html, 6 Feb 2021
4. 2020 Alzheimer's Disease Facts and Figures. https://alz-journals.onlinelibrary.wiley.com/doi/full/10.1002/alz.12068, 7 Feb 2021
5. Neuroimaging in Dementia. https://www.ncbi.nlm.nih.gov/pmc/articles/PMC5823524/, 10 Feb 2021
6. Neuropsychological Testing. https://dailycaring.com/diagnosing-alzheimers-or-dementia-neuropsychological-testing/, 4 Feb 2021
7. Ardabili S, Mosavi SA, Annamaria R (2020) Advances in machine learning modeling reviewing hybrid and ensemble methods. In: Proceedings of the international conference on global research and education, lecture notes in networks and systems, 101:215–227. https://doi.org/10.1007/978-3-030-36841-8_21
8. Waring J, Lindvall C, Umeton R (2020) Automated machine learning: review of the state-of-the-art and opportunities for healthcare. Artif Intell Med 104. https://doi.org/10.1016/j.artmed.2020.101822
9. Myszczynska MA, Ojamies PN, Lacoste AMB et al (2020) Applications of machine learning to diagnosis and treatment of neurodegenerative diseases. Nat Rev Neurol 16:440–456. https://doi.org/10.1038/s41582-020-0377-8
10. Arthi K, Tamilarasi A (2008) Prediction of autistic disorder using neuro fuzzy system by applying ANN technique. Int J Dev Neurosci 26:699–704. https://doi.org/10.1016/j.ijdevneu.2008.07.013
11. Lan TH, Loh EW, Wu MS, Hu TM, Chou P, Lan TY, Chiu HJ (2008) Performance of a neuro-fuzzy model in predicting weight changes of chronic schizophrenic patients exposed to antipsychotics. Mol Psychiatry Nat 13:1129–1137. https://doi.org/10.1038/sj.mp.4002128
12. Cpalka K, Rebrova O, Rutkowski L (2009) A new method for complexity reduction of neuro-fuzzy systems with application to differential stroke diagnosis. In: Proceedings of the international conference on artificial neural networks, lecture notes in computer science. Springer, 5769, pp 435–444. https://doi.org/10.1007/978-3-642-04277-5_44
13. Ubeyli ED (2009) Automatic detection of electroencephalographic changes using adaptive neuro-fuzzy inference system employing Lyapunov exponents. Expert Syst Appl 36:9031–9038. https://doi.org/10.1016/j.eswa.2008.12.019
14. Shanthi KJ, Sasikumar MN, Kesavadas C (2010) Neuro-Fuzzy approach toward segmentation of brain MRI based on intensity and spatial distribution. J Med Imaging Radiat Sci 41(2):66–71. https://doi.org/10.1016/j.jmir.2010.03.002
15. Sampath R, Saradha A (2015) Alzheimer's disease classification using hybrid neuro fuzzy Runge-Kutta (HNFRK) classifier. Res J Appl Sci Eng Technol 10(1):29–34. https://doi.org/10.19026/rjaset.10.2550
16. Álvarez JD, Matias-Guiu JA, Cabrera-Martín ML, Jose LRM, Ayala JL (2019) An application of machine learning with feature selection to improve diagnosis and classification of neurodegenerative disorders. BMC Bioinform 20(491). https://doi.org/10.1186/s12859-019-3027-7
17. De Oliveira APS, De Santana MA, Andrade MKS, Gomes JC, Rodrigues MCA, Wellington PS (2020) Early diagnosis of Parkinson's disease using EEG, machine learning and partial directed coherence. Res Biomed Eng 36:311–331. https://doi.org/10.1007/s42600-020-00072-w
18. Fan Z, Xu F, Qi X, Li C, Yao L (2020) Classification of Alzheimer's disease based on brain MRI and machine learning. Neural Comput Appl 32:1927–1936. https://doi.org/10.1007/s00521-019-04495-0

19. Jain D, Mishra AK, Das SK (2021) Machine learning based automatic prediction of Parkinson's disease using speech features. In: Bansal P, Tushir M, Balas V, Srivastava R (eds) Proceedings of international conference on artificial intelligence and applications, Advances in intelligent systems and computing, 1164. https://doi.org/10.1007/978-981-15-4992-2_33
20. Lamba R, Gulati T, Alharbi HF, Jain A (2021) A hybrid system for Parkinson's disease diagnosis using machine learning techniques. Int J Speech Technol. https://doi.org/10.1007/s10772-021-09837-9
21. Cai J, Luo J, Wang S, Yang S (2018) Feature selection in machine learning: a new perspective. Neurocomputing 300:70–79. https://doi.org/10.1016/j.neucom.2017.11.077
22. Koehler GJ (1997) New directions in genetic algorithm theory. Ann Oper Res 75:49–68. https://doi.org/10.1023/A:1018928017332
23. Fuller R (2000) Introduction to neuro-fuzzy systems. Adv Intell Soft Comput. ISBN 978-3-7908-1256-5. https://doi.org/10.1007/978-3-7908-1852-9
24. Dementia Dataset. https://www.kaggle.com, 5 Feb 2021
25. Parkinsons Dataset. https://archive.ics.uci.edu/ml/datasets/parkinsons, 7 Feb 2021

Envisaging Industrial Perspective Demand Response Using Machine Learning

Nabeela Hasan and Mansaf Alam

Abstract The utilization of Internet of Things (IoT) sensors in industry which generates the data that is used in a variety of analytics for the acquisition of valuable information is characterised as Industrial Internet of Things (IIoT). In predictive analytics, the primary aspect is typically the kind of data provided by the sensors. Various kinds of data are collected while performing predictive modelling, for, e.g., area, season, energy, cost, etc. In this paper, a case study is presented, in which a dataset is used to examine the functioning of equipment and to analyse the demand and response of that equipment. The main aim of this paper is to employ various machine learning algorithms in order to devise of predictive models in industrial IoT environment using the dataset mentioned above.

Keywords Industrial Internet · Demand response analytics · Wireless sensor network

1 Introduction

Current developments and major contributions in the areas of artificial intelligence, automation and robotic technology have had a strong effect on organizational processes that forms the foundation to perform predictive analytics in the manufacturing, healthcare, residential, monetary, business sector, social media networks and various others mainly to achieve smart surroundings [1–4]. In addition, Industrial Revolution led to the initiation of Smart Production, i4.0, Industrial Internet, Smart manufacturing that has led to the tremendous advancement of industrialization globally. The main components that make the whole Industrial IoT paradigm are Internet

N. Hasan (✉) · M. Alam
Jamia Millia Islamia, New Delhi 110025, India
e-mail: nabeela1910394@st.jmi.ac.in

M. Alam
e-mail: malam2@jmi.ac.in

© The Author(s), under exclusive license to Springer Nature Singapore Pte Ltd. 2022 331
D. Gupta et al. (eds.), *Proceedings of Data Analytics and Management*,
Lecture Notes on Data Engineering and Communications Technologies 90,
https://doi.org/10.1007/978-981-16-6289-8_28

of Things, Cyber Physical Systems (CPSs), Wireless Sensor Networks (WSNs) and cloud Computing [5–7].

The main requirement of industrial Internet paradigm is to be flexible, intelligent, dynamic and also have capability to function in real-time environment [8–10]. These technological innovations have revived the significance of academics, businesses and governments to apply predictive analytics in the world of industries. This area of data processing established tools and techniques for the creation of models capable of analysing events in the future, faults or actions [11, 12]. Machine learning (ML), computational approaches or data analyses are the prediction models used to discover activity patterns found in a dataset, and state whether openings and risks are established in those patterns [13, 14].

In order to obtain valuable insights into decision-making, such as faults prediction, criteria, performance, market volume, and a vast number of data, which remain a task, must be processed [15]. This paper is broken down as follows:

- First section provides a brief description of Industry 4.0, machine learning (ML) deployment for Industry 4.0 and performance evaluation.
- The case study is then clarified.
- Finally, the findings show the accuracy of the logistic regression and the Random Forest and show the conclusion and potential experiments.

2 Related Work

In order to improve the efficiency of industrial machines, demand and response is one of the solutions by reacting easily quickly to supply or system results. The study in paper [3] includes presenting a multi-period DR challenge with multiple services as a Stackelberg billing minimization game. The research in [4] entails automated task scheduling-based optimization process and price determination to reduce the peak to average ratio (PAR). In [16], simulation strategies have been suggested for day-to-day planning of industrial and commercial buildings. The real-time preparation of a microgrid scenario [17] takes into account the uncertainties generated by renewable energy sources. One of the earlier works [5] is the DR models as a Stackelberg Game with utilities as leaders and users. An optimization model and an overview of load change for home consumers are given in [6]. Distributed Active DR (DADR) load shifting structures with or without minimum contact was analysed in [7] for PAR reduction. A fascinating research is conducted in the game theoretical approach to residential market DR [8], where consumer preferences are modelled on the expense of inconvenience, along with the load motivation framework. The standard model provides for well-known DR cases. In this way, DR is seen as an instrument for holding the total load of the grid below a utility threshold, perhaps to prevent overcapacity or reduce the effects or combinations of high energy prices. The standard DR model and its variants in literature [3] have established a considerable amount of importance and analysis, e.g., [3–8, 16, 17]. Research performed by Piette et al. (2015) has been evaluating cost data from some 50 ADR systems that have been deployed during the last decade in major commercial and industrial facilities. Study in [3, 4] investigated the cost enabling and the reliability of limited customer loads and end uses that could provide fast DR services. Paper [4] is one of the first research pieces to comprehensively report the cost of enabling DR-technology capable of providing different bulk power system services. This study, however, was unique to California.

3 Industry 4.0

The idea of Industrial IoT had emerged in the year 2011 in Germany when Bosch used to lead the business sector and the government. A research group had been formed to discover a collective system to administer the applications of new technological advancements. The first study was written in 2012 and was unveiled to the public at the Hannover Fair in 2013. This led to the emergence of the concept which is now identified as the fourth industrial revolution, with various nations referring to it in various terminology as per the policies established in those nations. It is applied both at the macro-level to small and medium-sized businesses within the industrial ecosystem [5, 18].

IoT, IIoT and Industry 4.0 definitions are loosely interconnected but are not synonymous. In this section, we include a rough classification of these words. As far as IoT is concerned, there are also other meanings that want to catch one of their basic characteristics. It is also considered to be a type of web for machines which underlines the purpose of data exchange. However, the application fields are so diverse that some criteria (particularly communication aspects) can vary widely based on the goals and end users, the business model underlying and the technical solutions implemented. What is generally known as IoT may be better known as consumer IoT than commercial IoT [19, 20].

In determinism, latency and performance, IoT focuses more on designing new communications standards, which can flexibly and user-friendly connect new devices with the Internet ecosystem [21]. The current design of IIoT, however, emphasizes that once isolated plants, work islands and also machinery can be integrated and interconnected, thereby providing better production and new services [19]. For this reason, IIoT can be considered more an evolution than a revolution compared to IoT.

With regard to criticality and connectivity, IoT is more adaptable, enables mobile and ad hoc network structures and has lower reliability and timing needs (except for medical applications). On the other hand, IIoT typically uses network solutions based on fixed infrastructure and designed to fit the needs of communications and coexistence. In IIoT, communications are machine-to-machine connections which must meet stringent timeliness and reliability requirements.

3.1 Predictive Maintenance in Machine Learning

The idea of predictive analytics is not a new one. However, as a growth axis for the implementation of the Industry 4.0 system, predictive analytics is the topic of interest. The goal is to obtain frameworks that reduce the instability of evaluation. Ballesteros, in [22] addresses the basic requirements that must be fulfilled to examine that the corporation has a predictive analysis system:

- When controlling and evaluating the functioning of any machine, it is supposed to be carried out in a non-intrusive manner under standard test circumstances.
- The parameter to be calculated to carry out the estimations have to comply with the following requirements: replicability, interpretation, diagnosis, and parametrization.
- The outcomes and values of the measurements which be expressed in correlated indexes or in physical units.

The main goal of predictive maintenance is precise maintenance planning and the avoidance of unexpected failures. Knowing when a particular machine needs to be maintained simplifies the planning of maintenance resources such as replacement staff or parts. Moreover, system availability can be increased by transforming "unplanned stops" into "planned stops" ever shorter. Additional benefits include potentially longer plant life, increased plant safety, fewer accidents with adverse environmental impacts and optimized replacement parts handling.

There is a movement to expand the research, which makes it possible to apply predictive models in industrial applications that introduce a predictive power supply management system, which other authors propose to implement in the wind energy generator or detect anomalies in triaxial machinery [23, 24]. These studies share similar elements: they concentrate on machine learning methods (ML) and aim to advance ML algorithms that increase prediction performance [11, 25].

3.2 Predictive Analysis Using Machine Learning

Significant contributions have been made in the area of different approaches of artificial intelligence, such as Case-Based Reasoning (CBR) [8, 19, 20, 26, 27] and various machine learning methods, in order to analyse the data, boost performance and make the dataset more comprehensive. It is therefore important to build models that make prediction and analysis simpler for decision-making [16, 28–30].

Artificial intelligence approaches have transformed a wide range of fields in recent decades, such as industrial, social networks [16, 31], technology, healthcare [32], neuroscience and number of household activities, and a variety of research works have been conducted on this topic. This is an indicator of its significance [31, 33, 34]. Based on features extracted, predictive analytics models can be created using ML algorithms and statistical analysis [19, 20].

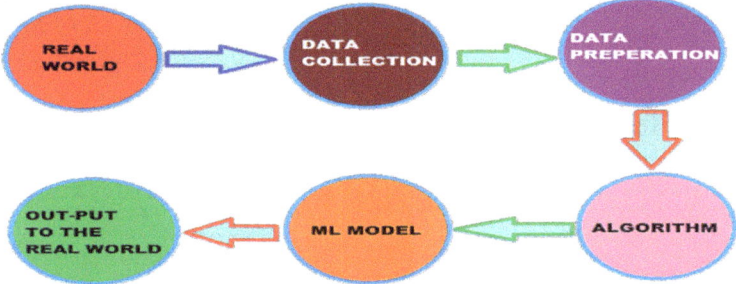

Fig. 1 ML stages

In the existing real-world situation, a dataset must be collected before ML techniques can be implemented on that data [26–28]. Successively, they will undergo various stages, such as data pre-processing, training of processed data and the implementation of the learning model and eventually the assessment phase. As discussed in [32], a sequence of ML stages is presented in Fig. 1 in a series in which they are executed. Data pre-processing is performed in order to transform raw data into meaningful data. At this point, the data is noisy, unstructured, inconsistent and incomplete, and transformed to be used as inputs data for the models selected for training. Successively, the training dataset will be used to train the designed model, as well as the predictions derived from the new test data will be acquired.

In order to validate the model, the results of the statistical tests and the error estimation data are evaluated, these estimations are used to modify the constraints of the algorithms used and to decide whether the use of other algorithms is required [34].

4 Industry 4.0 Environment Case Study

The area, season, energy, cost, pair no and distance are the parameters of the equipment that created an ideal circumstance in the context of industry. Since these technological industrial equipment plays a vital role in implementing Industry 4.0 in real-world, it is necessary to setup a scheduled maintenance regulator that assist in identifying the failures of these equipment regularly.

The goal of predictive analytics in Industrial perspective is to preserve the quality of equipment using various methods and tools to detect irregular behaviour such as temperature, vibration or balance. Corresponding with the significance of predictive analysis, a case study is presented.

Table 1 Dataset

Column	Description
Area	Area required by equipment to be installed
Season	Amount of work done by equipment in particular season
Energy	Energy consumed by an equipment per unit of time
Cost	Cost incurred on equipment installation
Pair no	Combination of equipment working together
Distance	Distance of equipment from main server
DEMAND_RESPONSE	Energy demand/response of equipment

4.1 Data Description

A dataset structured by columns was used in this case study. It comprises of data about several of an equipment that is used in an industry. It was used to investigate various features of an equipment and rate of demand of that equipment based on the features discovered from the dataset. This dataset includes a total of 16,382 (seventeen thousand three hundred and eighty-two) sensor equipment features and the independent variable has forecast the DEMAND RESPONSE. The structure of the dataset and the variables is presented in Table 1.

4.2 Data Pre-processing

Many researchers have used techniques to perform pattern recognition and pre-processing [19, 20]. This system established a range for the demand response of the consumers for the industrial equipment. The data variable that has been used is described below:

- *DEMAND_RESPONSE*: It has maximum value of 5 and lowest of 0. It is in decimals and also our target variable. It is a self-made scale to measure the demand of a particular equipment based on other data provided. Std deviation of 1.706 represents not very high fluctuations in values.
- *Area*: This might be area which the equipment needs to be placed or maybe the area of field or factory in which the equipment was installed. It has minimum value of 1954.2300 and maximum of 50,895.5.
- *Energy*: It is the energy consumed by an equipment when operated for some fix amount of time. Since values are negative too, this represents head dissipated.

- *Season*: Has very large values that need to be scaled before feeding to model. Very high standard deviation shows high variance. Graph will shed light on the skewness of the data.
- *Cost*: It is a cost for installing the equipment, since minimum Cost is 0 it depicts cost is correlated to area.
- *Pair no*: Since maximum value is not equal to number of rows, we can say that pair number can be repetitive. Looks like takes whole number as value only. One entry denotes a pair of two of more entries.
- *Distance*: It is a distance between the equipment from the main server or distance of given pair from any other pair.

Figure 2 shows the distributions plot of the attributes that we have used in the data to determine the skewness and the probabilities of specific data values within the distribution. The skewed dataset on the right has a long tail extending to the right. Another way to talk with the distorted data collection to the right is to suggest it is skewed favourably. The mean and median of this case are both bigger than the mode. The data are usually biased to the right, and the mean is higher than the median. When dealing with the left distorted results, the situation is reversed. The mean and the median in this case are also lower than mode. In general, with data tilted to the left most of the time, the average is less than the median. In order to solve this issue of skewness, we can simply use the log transformation and Min–Max scaling to convert the skewed graph into normal distribution or Gaussian distribution. Min–max scales the data in the range 0 and 1. The formula used is:

$$X_{\text{norm}} = \frac{X - X_{\text{min}}}{X_{\text{max}} - X_{\text{min}}} \tag{1}$$

It helps get the data closer and simplify it so that there are no enormous values that will alter the value of the cost function significantly at once. We won't apply Min–Max transformation here since most of the values does not lie in a small range.

When the values of an attribute are over a very long interval, log transformation is used. Log transformation makes data easier to compare and view and also helps prevent overflows. Hence, we will apply log transformation for normalization of distribution plots.

$$\log(Y) = \log(A) + \alpha \log(L) + \beta \log(K) \tag{2}$$

In Fig. 2a, the distribution plot of cost variable is skewed towards right which usually acts as an outlier for a statistical model and can also affect the model's performance adversely.

To solve this issue, we performed log transformation to fit a very skewed distribution into normal distribution. We can easily see pattern in the data after log transformation in Fig. 2b. Similarly, we performed log transformation for area, distance and season variable as they are highly skewed towards right as we can see in Fig. 2c, e and g. After log transformation, these skewness have been converted into normal

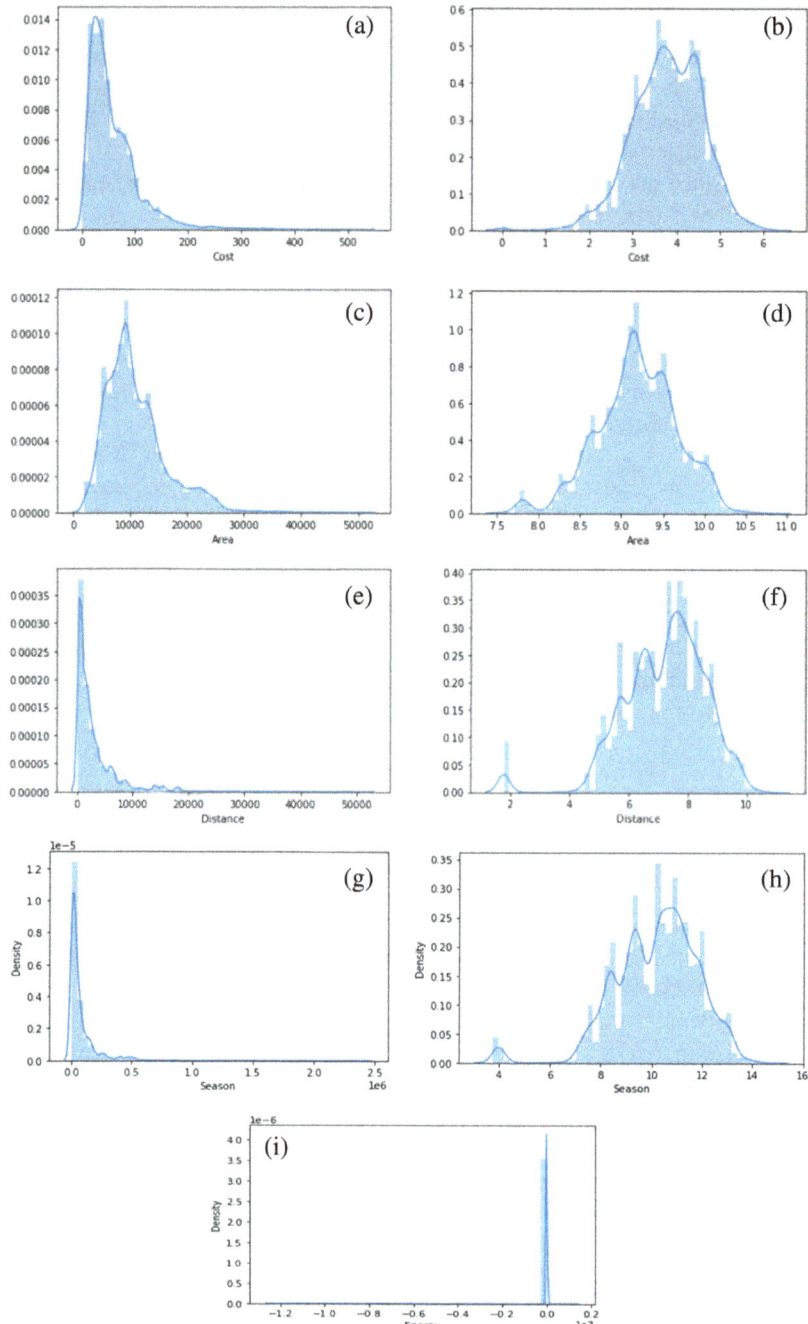

Fig. 2 **a** Cost distribution plot. **b** Cost distribution plot. **c** Area distribution plot. **d** Log area distribution plot. **e** Distance distribution plot. **f** Log Distance distribution plot. **g** Season distribution plot. **h** Log Distance distribution plot. **i** Energy distribution plot

distribution that are shown in Fig. 2d, f and h. It is not possible to take log transformation for the variables which contains negative values. Since energy distribution plot graph is skewed towards left and also contains negative values, it is not possible to solve it using log transformation. And at last, we capped the features with high numbers and took log of them.

5 Results

Once the data has been pre-processed, an improved dataset was used to break the data into a data train and data evaluation, and a predictive model was used for the first to implement machine learning algorithms. This model was tested with the data test. Two supervised learning algorithms have been used for the training of the data which are: Random Forest (RF) and Logistic regression to determine the accuracy of each estimation.

The X train and X test values showed the corresponding 12,286 and 4096 values. After applying the regression model and achieving its consistency, the Random Forest classification algorithm was implemented into the dataset. Several authors [29, 30] state that a random forestry classifier is an effective measurement method [35, 36]. This classification is also known as a non-parametric mathematical method that allows the regression and classification of two or more categories to be calculated. Recent analysis by Scornet et al. [29], quoted by [10], reveals that RF is consistent and has extremely low efficiency parameters. Authors of the Random Forest mentioned in [26] are as follows: "… a random forest is a classifier consisting if a collection of tree-structured classifiers $\{h\ (x,\ \Theta k),\ k\ =\ 1,\}$ where the $\{\Theta k\}$ are independent identically distributed random vectors and each tree casts a unit vote for the most popular class at input x…".

We have applied various regression models and conclude that Random Forest regression show an accuracy of 0.781 for a total of 12,286 records of training dataset and 4096 for the test dataset model. The comparative analysis of various approaches with their accuracy has been shown in Table 2. Using estimators $= 336$ in Random

Table 2 Comparative analysis of accuracy

Model	Accuracy	Mean absolute error
Linear regression	0.604	1.4195
Decision tree regression	0.661	0.4736
Random forest regression	0.781	0.4788
Extra trees regression	0.574	0.5331
Adaboost regression	0.152	1.3497
Xgb regression	0.504	0.9564
Bagging regression	0.7572	0.4970

Forest we obtained accuracy of ~80% and Mean Absolute Error = 0.4700 which is least till now.

6 Conclusion and Future Work

The model proposed is still in its early stages of growth. This allows the deployment of many other learning machines and the application of broader datasets in the order setting of the sensors network. The datasets used in this case study show that the precision of the logistic regression model is near the Random Forest when forecasting consumer demand for industrial machinery. Modelling and integration of vast volumes of machinery-produced and sensor acquired industrial data is a significant challenge that future research needs to address. Validation of other training, estimation and classification methods for machine learning. These tests are the basis for the creation of algorithms which generate predictive models for industry 4.0 organizations.

References

1. Antonopoulos I, Robu V, Couraud B, Kirli D, Norbu S, Kiprakis A, Flynn D, Elizondo-Gonzalez S, Wattam S (2020) Artificial intelligence and machine learning approaches to energy demand-side response: a systematic review. Renew Sustain Energy Rev 130:109899. ISSN 1364-0321 https://doi.org/10.1016/j.rser.2020.109899
2. Chamoso P, Rivas A, Martín-Limorti JJ, Rodríguez S (2018) A hash based image matching algorithm for social networks. In: Advances in intelligent systems and computing, vol 619, pp 183–190
3. Vázquez-Canteli JR, Nagy Z (2019) Reinforcement learning for demand response: a review of algorithms and modeling techniques. Appl Energy 235:1072–1089. ISSN 0306-2619. https://doi.org/10.1016/j.apenergy.2018.11.002
4. Krč R, Kratochvílová M, Podroužek J, Apeltauer T, Stupka V, Pitner T (2021) Machine learning-based node characterization for smart grid demand response flexibility assessment. Sustainability 13(5):2954. https://doi.org/10.3390/su13052954
5. Sittón I, Rodríguez S (2017) Pattern extraction for the design of predictive models in industry 4.0. In: International conference on practical applications of agents and multi-agent systems. Springer, Cham, pp 258–261
6. Ballesteros F (2017) La Estrategia Predictiva en el mantenimiento industrial. In: Grupo Álava, España, Predictécnico, vol 23, pp 36–45
7. Canizo M, Onieva E, Conde A, Charramendieta S, Trujillo S (2017) Real-time predictive maintenance for wind turbines using Big Data frameworks. In: IEEE international conference on Prognostics and health management (ICPHM), pp 70–77
8. García-Valls M (2016) Prototyping low-cost and flexible vehicle diagnostic systems. In: ADCAIJ: Advances in distributed computing and artificial intelligence journal, Salamanca, vol 5, no 4
9. Monino JL, Sedkaoui S (2016) The algorithm of the snail: an example to grasp the window of opportunity to boost big data. In: ADCAIJ: Advances in distributed computing and artificial intelligence journal, Salamanca, vol 5, no 3

10. Baruque B, Corchado E, Mata A, Corchado JM (2010) A forecasting solution to the oil spill problem based on a hybrid intelligent system. Inf Sci 180(10):2029–2043
11. Corchado JA, Aiken J, Corchado ES, Lefevre N, Smyth T (2004) Quantifying the Ocean's CO_2 budget with a CoHeL-IBR system. In: Advances in case-based reasoning, vol 3155, pp 533–546
12. Fernández-Riverola F, Corchado JM (2003) CBR based system for forecasting red tides. In: Knowledge-based systems, vol 16, (5–6 SPEC), pp 321–328
13. Corchado JM, Borrajo ML, Pellicer MA, Yáñez JC (2004) Neuro-symbolic system for business internal control. In: Industrial conference on data mining, pp 1–10
14. Fyfe C, Corchado JM (2002) A comparison of Kernel methods for instantiating case-based reasoning systems. Adv Eng Inf 16(3):165–178
15. Fyfe C, Corchado JM (2001) Automating the construction of CBR systems using kernel methods. Int J Intell Syst 16(4):571–586
16. Agarwal P, Alam M (2020) Investigating IoT middleware platforms for smart application development. In: Ahmed S, Abbas S, Zia H (eds) Smart cities—opportunities and challenges. Lecture notes in civil engineering, vol 58. Springer, Singapore. https://doi.org/10.1007/978-981-15-2545-2_21
17. Khan S, Ali SA, Hasan N, Shakil KA, Alam M (2019) Big data scientific workflows in the cloud: challenges and future prospects. In: Das H, Barik R, Dubey H, Roy D (eds) Cloud computing for geospatial big data analytics. Studies in big data, vol 49. Springer, Cham. https://doi.org/10.1007/978-3-030-03359-0_1
18. Carneiro D, Araujo D, Pimenta A, Novais P (2016) Real time analytics for characterizing the computer user's state. In: ADCAIJ: Advances in distributed computing and artificial intelligence journal, vol 5, no 4, pp 01–18
19. Li T, Sun S, Corchado JM, Siyau MF (2014) Random finite set-based Bayesian filters using magnitude-adaptive target birth intensity. In: FUSION 2014—17th international conference on information fusion
20. Méndez JR, Fernández-Riverola F, Díaz F, Iglesias EL, Corchado JM (2006) A comparative performance study of feature selection methods for the anti-spam filtering domain. In: Lecture notes in computer science (including subseries lecture notes in artificial intelligence and lecture notes in bioinformatics), 4065 LNAI, pp 106–120
21. Bandyopadhyay D, Sen J (2011) Internet of things: applications and challenges in technology and standardization. Wireless Pers Commun 58(1):49–69
22. González-Peña D, Díaz F, Hernández JM, Corchado JM, Fernández-Riverola F (2009) geneCBR: a translational tool for multiple-microarray analysis and integrative information retrieval for aiding diagnosis in cancer research. In: BMC bioinformatics, vol 10, pp 187
23. Alvarado-Pérez JC, Peluffo-Ordóñez DH, Theron R (2015) Bridging the gap between human knowledge and machine learning. In: ADCAIJ: Advances in distributed computing and artificial intelligence journal, vol 4, no 1, pp 54–64
24. Goyal S, Goyal GK (2013) Machine learning ANN models for predicting sensory quality of roasted coffee flavoured sterilized drink. In: ADCAIJ: Advances in distributed computing and artificial intelligence journal, vol 2, no 3, pp 09–13
25. Li T, Sun S, Bolić M, Corchado JM (2016) Algorithm design for parallel implementation of the SMC-PHD filter. Sig Process 119:115–127
26. Méndez JR, Fernandez-Riverola F, IglesiasEL, Díaz F, Corchado JM (2006) Tracking concept drift at feature selection stage in spam hunting: an anti-spam instance-based reasoning system. In: Lecture notes in computer science (including subseries lecture notes in artificial intelligence and lecture notes in bioinformatics), 4106 LNAI, pp 504–518
27. Corchado JM, Fyfe C (1999) Unsupervised neural method for temperature forecasting. Artif Intell Eng 13(4):351–357
28. Corchado JM, Fyfe C, Lees B (1998) Unsupervised learning for financial forecasting. In: Proceedings of the IEEE/IAFE/INFORMS, conference on computational intelligence for financial engineering (CIFEr) (Cat. No.98TH8367), pp 259–263

29. Román JA, Rodríguez S, de la Prieta F (2016) Improving the distribution of services in MAS. In: Communications in computer and information science, vol 616, pp 37–46
30. Hortonworks (2017) Analyse HVAC machine and sensor data. https://es.hortonworks.com/ha-doop-tutorial/how-to-analyze-machine-and-sensor-data/#section-2
31. Genuer R, Poggi JM, Tuleau-Malot C, Villa-Vialaneix N (2015) Random forests for big data. Big Data Res 9:28–46
32. Janitza S, Tutz G, Boulesteix A (2016) Random forest for ordinal responses: prediction and variable selection. Comput Stat Data Anal 96:57–73
33. Scornet E, Biau G, Vert JP (2015) Consistency of random forests. Ann Stat 43(4):1716–1741
34. Breiman L (2001) Random forests. Machine Learn 45(1)5–32
35. Akerberg J, Gidlund M, Bjorkman M (2011) Future research challenges in wireless sensor and actuator networks targeting industrial automation. In: Proceedings of the 9th IEEE International conference on industrial informatics, pp 410–415
36. Hasan N, Chamoli A, Alam M (2020) Privacy challenges and their solutions in IoT. In: Alam M, Shakil K, Khan S (eds) Internet of things (IoT). Springer, Cham. https://doi.org/10.1007/978-3-030-37468-6_11

K-Prototype Algorithm for Clustering Large Data Sets with Categorical Values to Established Product Segmentation

Ritu Punhani, V. P. S. Arora, and A. Sai Sabitha

Abstract Analyst classically looks for general physiognomies like communal desires and safeties, alike lifestyles, or even alike demographic outlines in dividing or segmenting marketplaces. The global purpose of segmentation is to recognize high produce segments—that are the most lucrative or have growth potential—to select for special consideration. Business-to-business organizations might slice the marketplace into dissimilar kinds of trades or nations. Business-to-consumer organizations might slice the marketplace into different areas such as firmographics, geographic, demographic, behavioral, psychographic, or any other meaningful fragment. This paper researcher identified 36 parameters by studying various researches and selecting 11 parameters like a product, total revenue, profits, sales channel, order priority, country, region, unit sold, unit cost, unit price (MRP), and total cost. By using these parameters, fetch the data from an online e-commerce website—Autofurnish.com, develops the clusters for product segmentation by using K-prototype to generate results like which product is popular among customers and generates more revenue in a particular region.

Keywords Product segmentation · Categorical clustering · K-prototype · Mixed type data · Customers

1 Introduction

E-commerce is the platform for electronically buying or selling of products online over the Internet. E-commerce trades shape how people shop for products. Typical e-commerce communications include purchasing household items, clothes, jewelry,

R. Punhani (✉) · A. Sai Sabitha
Department of Information Technology, ASET, Amity University, Uttar Pradesh, Sector 125, Noida, India

V. P. S. Arora
Sharda University, Greater Noida, Gautam Buddha Nagar, Uttar Pradesh, India

© The Author(s), under exclusive license to Springer Nature Singapore Pte Ltd. 2022   343
D. Gupta et al. (eds.), *Proceedings of Data Analytics and Management*,
Lecture Notes on Data Engineering and Communications Technologies 90,
https://doi.org/10.1007/978-981-16-6289-8_29

books, cosmetics, accessories, etc., and, to a less extent, customized, or personalized customer purchases by suggesting the products online [1]. The areas of e-
commerce are the exchange of products from a business-to-consumer (B2C), the
sale of products from a business-to-business (B2B), mobile payments and check-
outs on e-commerce sites, seamless shipping, virtual and augmented reality, chat-
bots, personalized marketing, omnichannel experiences, social commerce and media,
impeccable customer service, service personalization, and target markets.

Our primary focus is on the target product market, the customers' location, and the
total sales amount generated from the specified areas. The target product market decision is choosing which product to aim its marketing program positioning strategy in a
product-market. This decision is one of organization management's most demanding
challenges. Would an organization effort to aid all keen and able to manufacture or
keep a complete inventory of the product's specific cluster in their warehouse? In
advance, a thought of a product-market is vital to make the target product market
choice.

The phases in choosing a target product market strategy are: Choose how to form
clusters in the product-market, label the organization's product in an apiece cluster,
assess target product market substitutes, and elect a target product-market strategy.

The probabilities for picking, organization's target cluster of customers and
the product in a product-market, assortments from endeavoring to appeal to most
customers in need called a merchandisable approach to go after new-found clusters
inside the marketplace. Organization management must identify possible product
clusters to regulate which marketing positioning approach will attain the most
extraordinary profitable contributions. Following are the four criteria: (1) Identify
the truthfulness of customer in the product-market, (2) It must be achievable to
classify two or more distinct product clusters, (3) In terms of profits produced and
cost incurred, segmentation need is essential, (4) The segments must exhibit good
constancy over the period so that the organizations' efforts through segmentation
will have sufficient time to reach expected performance heights.

Figure 1 shows that segmenting the product market has described the entire
process, including Segmentation → Targeting → Positioning [2], as an extensive
framework for streamlining the process. Segmentation encompasses recognizing the
product to be segmented for the expansion of customer profiles. Targeting includes an
assessment of each segment's allure and the assortment of the segment to be targeted.
Positioning encompasses the identification of the optimum place and growth of the
market.

The Diana function in R suite group [3] implements divisive clustering. The
process agnes in package cluster [4] and the function hclust implemented in each
of the package's stats, fastcluster [5], and flashClust are commonly used functions
for agglomerative clustering. A compromise between agglomerative and divisive
clustering developed by Chipman and Tibshirani and implemented in R package
hybridHclust of Chair et al. [6]. Partitioning methods generally involve a single split
of the data set into mutually exclusive and exhaustive clusters.

The k-means methods or alternatives such as the k-medoids algorithms, for
example, partitioning around medoids method of Kaufman and Rousseeuw [7].

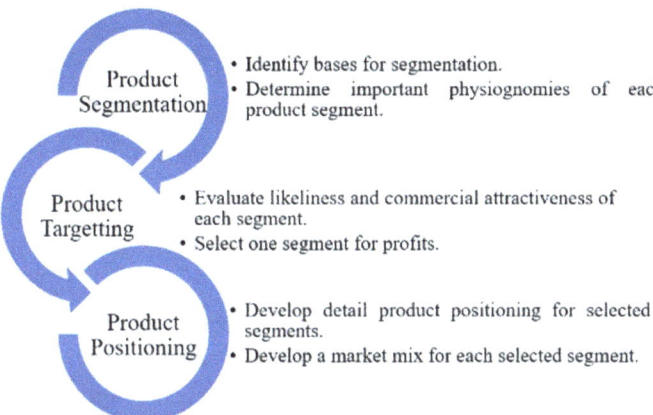

Fig. 1 Segmentation—targeting—positioning approach

The k-means function implements the Hartigan–Wong, Lloyd, and MacQueen k-means algorithms in the stats package [8]. Kernel k-means use kernels to plan the data into a nonlinear feature space before applying k-means and implementing the kkmeans function in the kernlab suite [9]. The purpose of kcca in package flexclust implements generalizations of k-means using random centroid statistics and distance measures [10].

Trimmed k-means, which shrinks a specified proportion of extreme values in each cluster, are implemented in packages trimcluster and tclust [11]. The pam function in the package cluster [4] implements k-medoids. The Clara function in the same package implements a computationally efficient version of k-medoids. The partitioning steps are executed on subsampled data and then applied to the entire data set. This efficient algorithmic structure makes Clara well-suited for large data sets stored in RAM [4].

The finite mixture model is another partitioning method in which each cluster is assumed to follow some parametric distribution [12].

Segment a set of items in databases to form segments [13] is a vigorous data analytical process. This is helpful in numerous methods, such as classification means unsupervised data, segmentation, and aggregation or dissection [13].

Clustering [7] is a method through which customer and product segmentation can be done [8].

'Market segmentation is to divide a marketplace into smaller segments of customers with dissimilar wants, physiognomies, or behaviors who might require distinct products or marketing mixes.'.

Statistical clustering methods [7], segments their customer and product dovetailing to some similarity or dissimilarity factors, while conceptual clustering methods, cluster their customer and product dovetailing to the outcome of the idea carry [7].

Azmat and Lakhani (2015) researched the impact of brand positioning approaches on the customer's insight [14].

2 Methodology

The research paper aims to analyze the practical data set from AutoFurnish using the K-Prototype [15] Clustering Algorithm for mixed datatypes [16]. Figure 2 gives the methodological analysis espoused for implementing the research work in small stages plan.

I.   Identification of parameters of E-Commerce Site—AutoFurnish

II.   The data set 'ProductAutoFurnish.csv' (Refer Fig. 3) from AutoFurnish.com taken.

III. The preprocessing and cleaning of the data set were conducted to remove the rows with 'not set' values or missing values.

IV. Applying the K-Prototype Clustering Algorithm [17] [Refer 4(a)], an appropriate K-value selected defines the number of clusters for the data set analysis.

    a.   K-prototype algorithm [18] was conceptualized by Huang [19] which is a method used to cluster the mixed type data sets. This algorithm is used

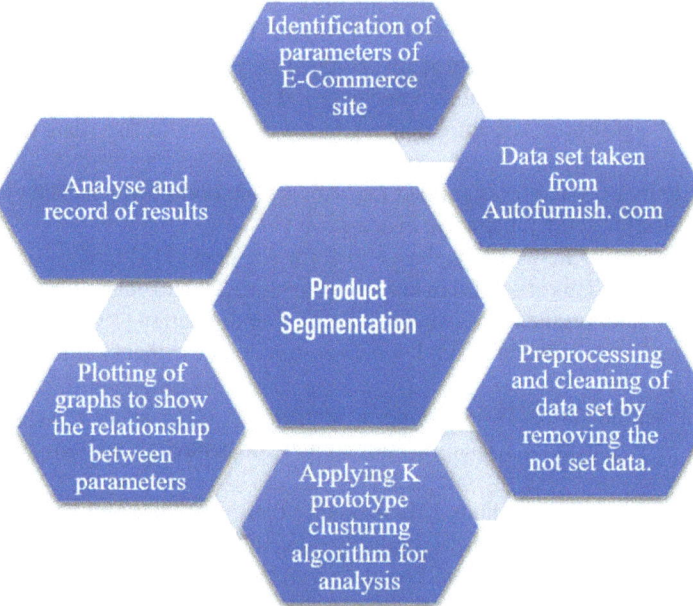

Fig. 2 Methodological analysis

Region	Country	Item Type	Sales Channel	Order Priority	Order Date	Order ID	Ship Date	Units Sold	Unit Price (MRP)	Unit Cost	Total Revenue (Rs)	Total Cost (Rs)	Total Profit (Rs)
0	Sub-Saharan Africa	Chad	3D Car Auto Seat Back Multi Pocket Storage Bag...	Online	L	1/27/2011	292494523	2/12/2011	4484	651.210	524.960	2920025.640	2353920.640	566105.000
1	Europe	Latvia	3D Car Foot Mats (Black) Complete Set For Hyun...	Online	C	12/28/2015	361825549	1/23/2016	1075	47.450	31.790	51008.750	34174.250	16834.500
2	Middle East and North Africa	Pakistan	3M Complete Car Care Kit	Offline	C	1/13/2011	141515767	2/1/2011	6515	154.060	90.930	1003700.900	592408.950	411291.950
3	Sub-Saharan Africa	Democratic Republic of the Congo	Autofurnish 3D Car Auto Seat Back Multi Pocket...	Online	C	9/11/2012	500364005	10/6/2012	7683	668.270	502.540	5134318.410	3861014.820	1273303.590
4	Europe	Czech Republic	3D Car Foot Mats (Black) Complete Set For Hyun...	Online	C	10/27/2015	127481591	12/5/2015	3491	47.450	31.790	165647.950	110978.890	54669.060

Fig. 3 Data set 'ProductAutoFurnish.csv'

for both string and numeric data types. It is a partition-based clustering algorithm which combines k-means and k-modes to the domains includes mixed data [20] and categorical data [21] domains. Its procedure describes as

i.   Pace 1: Select k number of initial prototypes [22] from a data set 'ProductAutoFurnish.csv' and assign one for each cluster.

ii. Pace 2: Assign every item in 'ProductAutoFurnish.csv' to a cluster that has the nearest prototype. After each allocation, update the prototype of the cluster.

iii. Pace 3: The items have been assigned to a cluster, test again the items' likeness aimed at current sample. If any of the entities found, i.e., its nearest sample belongs to the other cluster than its existing cluster, then reassign it to the same cluster and update prototypes for both clusters.

iv. Pace 4: Reiteration Pacc 3 until no item changes its cluster after the testing sequence of 'ProductAutoFurnish.csv'Y.

V.   Plotting the graphs to show the relationship between the identified parameters to find analysis.

VI. After analysis, send recommendations to the organization for better revenue generation [23].

3 Experimental Setup

3.1 Data Set

The information related to the selected data set named 'ProductAutoFurnish.csv' is given in the table below:

Name: 'ProductAutoFurnish.csv'

Year: 2019

Site: AutoFurnish.com

Total Number of Rows: 10,000

Total Columns based on parameters: 11

Parameters identified [24]: product, total revenue, profits, sales channel, order priority, country, region, unit sold, unit cost, unit price (MRP), and total cost.

According to the K-Prototype clustering method [25], in the data set few columns were omitted from the analysis. The unique values for parameters considered are—Region column (7), Country column (185), Item Type column (12), Sales Channel column (2), Order Priority column (4) [1].

3.2 Tools Used

Tools used for analysis are listed below, along with its definition [3]:

- Seaborn is a library of python data visualization, which uses the concept of matplotlib and offers an interface of high-level view for drawing informative and interactive statistical graphics.
- Matplotlib is used to plot graphs and has libraries for plotting the python software programming concepts and its mathematical extension NumPy.
- NumPy is a library of python programming language which focuses on operating matrices and multi-dimensional arrays with the help of high-level functions present in the library.
- Pandas is used for data analysis and manipulation of data. It is a library developed in python language. It is an open-source library which offers different data operations for manipulating numeric data tables.
- Scikit-learn is a free library included in python language. Its topographies several classification methods, regression methods, and clustering methods.

	Units Sold	Unit Price (MRP)	Unit Cost	Total Revenue (Rs)	Total Cost (Rs)	Total Profit (Rs)
count	10000.000	10000.000	10000.000	10000.000	10000.000	10000.000
mean	5002.856	268.143	188.807	1333355.131	938265.784	395089.347
std	2873.246	217.944	176.446	1465026.174	1145914.069	377554.961
min	2.000	9.330	6.920	167.940	124.560	43.380
25%	2530.750	109.280	56.670	288551.078	164785.530	98329.140
50%	4962.000	205.700	117.110	800051.210	481605.840	289099.020
75%	7472.000	437.200	364.690	1819143.390	1183821.520	566422.708
max	10000.000	668.270	524.960	6680026.920	5241725.600	1738178.390

Fig. 4 Statistical calculative values

- Jupyter is an interactive tool. It is an open source and free tool available on the Internet. In this researcher used to associate the code, output, instructions, and different resources of multimedia in a solitary for form further use.

4 Analysis of Data Set

4.1 Simple Statistical Analysis

According to the statistics, group the raw data into categories [26] and then visualizing it for drawing conclusions and facts. They include thoughtful and handling of results. Figure 4 describes the statistical calculative values (count, mean, standard deviation, min, max for each column.

4.2 Distribution of Sales in Each Region

Figures 5 and 6 describe Asia as a significant contributing region because of the maximum revenue generated (see Total Revenue) though the units sold in Asia are comparatively more minor. Europe with the highest number of purchases does not contribute to the total profit [7].

4.3 Developing Clusters

The **Elbow Method** is logically decisive method to find the optimal number of clusters [27] in each data set. This method calculates the values of k by iterative process. Then plot the variation of k values in a graphical form to find out the number

	Region	Total	Units Sold	Total Revenue	Total Cost	Total Profit
5	North America	215	5373.358	1559778.805	1097008.967	462769.838
1	Australia and Oceania	797	4986.769	1317192.267	910578.451	406613.816
2	Central America and the Caribbean	1019	5081.063	1369509.041	973672.093	395836.948
4	Middle East and North Africa	1264	5116.219	1357304.982	953884.180	403420.803
0	Asia	1469	5015.488	1365082.080	965215.983	399866.097
6	Sub-Saharan Africa	2603	4967.808	1287190.079	903155.468	384034.611
3	Europe	2633	4920.385	1322207.396	932158.169	390049.226

Fig. 5 Aggregation on columns concerning region

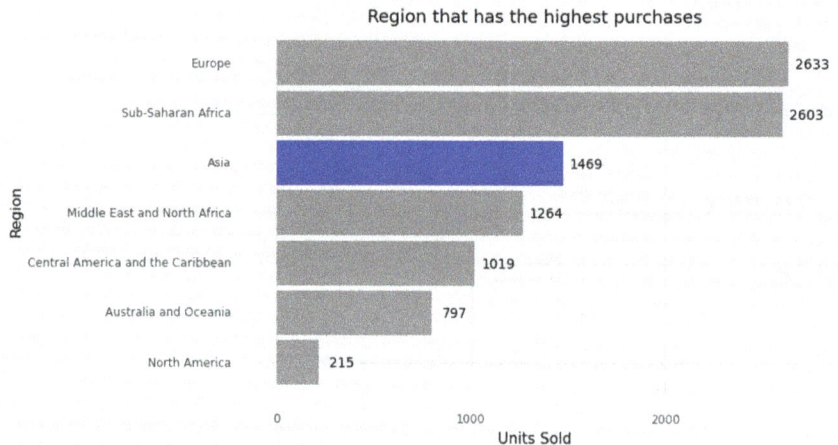

Fig. 6 Distribution of sales in each region

of clusters which are selected by the elbow curve (see Fig. 7). The **Elbow Method** is the most popular approach to determine this optimal value of k [24].

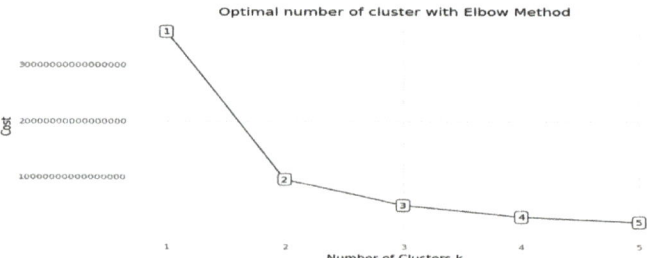

Fig. 7 Clusters (k) with elbow method

With the help of the Elbow Method, $K = 3$ is optimal for our defined data set. Davies–Bouldin index and silhouette method also verify the value of K [28]. Figure 8 shows three clusters—Cluster 0 (First Segment in Red), Cluster 1 (Second Segment in Blue), and Cluster 2 (Third Segment in Green) [23].

From the results obtained and visualized in Figs. 8 and 9, Table 1 is used to summarize the results.

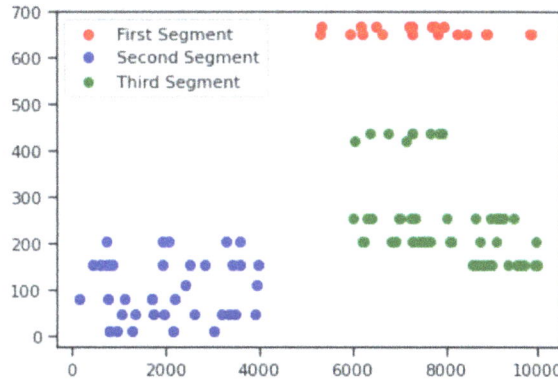

Fig. 8 Clusters 0 (Red), Cluster 1 (Blue), Cluster 2 (Green)

Segment	Total	Region	Item Type	Sales Channel	Order Priority	Units Sold	Unit Price	Total Revenue	Total Cost	Total Profit
0	First	1190	Europe	Autofurnish 3D Car Auto Seat Back Multi Pocket...	Offline	L	7904.366	593.527	4622760.548	3559121.123	1063639.424
1	Second	6381	Sub-Saharan Africa	Autofurnish Microfiber Car Cleaning & Polishin...	Online	C	4046.669	163.259	467709.452	281142.955	186566.497
2	Third	2429	Europe	Autofurnish Microfiber Car Cleaning Glove hand...	Online	H	6093.275	384.265	1995888.123	1380539.527	615348.596

Fig. 9 Results

Table 1 Summary of results

Cluster 0	1190 rows (cluster size)
Overall conclusion: 7904 Units of Product—Autofurnish 3D Seat Back Multi-Pocket Storage Bag Organizer Holder Hanger Accessory is sold and has generated the highest revenue of Rs. 4,622,760.548 in Region Europe	
Cluster 1	6381 rows (cluster size)
Overall conclusion: 4046 Units of Product—Autofurnish Set of 4 Microfiber Car Cleaning and Polishing Towel Cloth is sold and has generated the highest revenue of Rs. 467,709.452 in Region Sub-Saharan Africa	
Cluster 2	2429 rows (cluster size)
Overall conclusion: 6093 Units of Product—Autofurnish Microfiber Car Cleaning Glove hand washcloth is sold and has generated the highest revenue of Rs. 1,995,888.123 in Europe	

5 Conclusion

Analysis in this paper recommends [29, 30] that Product—(1) Autofurnish 3D Seat Back Multi-Pocket Storage Bag Organizer Holder Hanger Accessory, (2) Autofurnish Set of 4 Microfiber Car Cleaning and Polishing Towel Cloth, (3) Autofurnish Microfiber Car Cleaning Glove hand washcloth is the most sold car accessories as per our data set. The scope of analysis is not limited to the show results. The organization can perceive the work in other ways to maximize its profits and production [31, 32].

References

1. Su X, Khoshgoftaar TM (2009) A survey of collaborative filtering techniques. Adv Artif Intell
2. Moutinho L (2000) Segmentation, targeting, positioning and strategic marketing. Strateg Manag Tourism 121–166
3. Foss AH, Markatou M (2018) Kamila: clustering mixed-type data in R and Hadoop. J Stat Softw 83(1):1–44
4. Maechler M, Rousseeuw P, Struyf A, Hubert M, Hornik K (2021) Cluster: cluster analysis basics and extensions. R package version 2.1.2 — for new features, see the 'Changelog' file (in the package source), https://CRAN.R-project.org/package=cluster
5. Müllner D (2013) Fastcluster: fast hierarchical, agglomerative clustering routines for R and python. J Stat Soft 53(9):1–18. https://doi.org/10.18637/jss.v053.i09
6. Chair S, Charrad M, Ghazzali N (2016) A new R package for multi-SOM clustering
7. Kaufman L, Rousseeuw PJ (1990) Finding groups in data: an introduction to cluster analysis. 8th March 1990. ISBN: 9780471878766
8. Morissette L, Chartier S (2013) The k-means clustering technique: general considerations and implementation in mathematica. Tutorials in Quant Methods Psychol 9:15–24. https://doi.org/10.20982/tqmp.09.1.p015
9. Karatzoglou A, Smola A, Hornik K, Zeileis A (2004) Kernlab - an S4 package for Kernel methods in R. J Stat Soft 11(9):1–20
10. Leisch F (2006) A toolbox for K-centroids cluster analysis. Comput Stat Data Anal 51:526–544. https://doi.org/10.1016/j.csda.2005.10.006
11. Fritz H, García-Escudero LA, Mayo-Iscar A (2012) tclust: an R package for a trimming approach to cluster analysis. J Stat Soft 47(12):1–26. https://doi.org/10.18637/jss.v047.i12
12. Mclachlan G, Basford K (1988) Mixture models: inference and applications to clustering. https://doi.org/10.2307/2348072
13. Klösgen W, Zytkow JM (eds) (2002) Handbook of data mining and knowledge discovery. Oxford University Press, Inc., USA
14. Azmat M, Lakhani A (2015) Impact of brand positioning strategies on consumer standpoint (A consumer's perception). J Mark Consum Res 14:109–116
15. Kim B (2017) A fast k-prototypes algorithm using partial distance computation. Symmetry 9(4):58
16. Ji J, Bai T, Zhou C, Ma C, Wang Z (2013) An improved k-prototypes clustering algorithm for mixed numeric and categorical data. Neurocomputing 120:590–596
17. Ahire SR, Landge L. K Prototype clustering with efficient summarisation for topic evolutionary tweet stream clustering
18. Pham DT, Suarez-Alvarez MM, Prostov YI (2011) Random search with k-prototypes algorithm for clustering mixed datasets. Proc Royal Soc A: Math Phys Eng Sci 467(2132):2387–2403

19. Huang Z (1998) Extensions to the k-means algorithm for clustering large data sets with categorical values. Data Min Knowl Disc 2(3):283–304
20. Weißhuhn P (2019) Regional assessment of the vulnerability of biotopes to landscape change. Glob Ecol Conserv 20:e00771
21. Szepannek G (2018) clustMixType: user-friendly clustering of mixed-type data in R. R J 10(2):200
22. Byoungwook KIM (2017) A fast K-prototypes algorithm using partial distance computation
23. Zheng Z, Gong M, Ma J, Jiao L, Wu Q (2010, July) Unsupervised evolutionary clustering algorithm for mixed type data. In: IEEE congress on evolutionary computation. IEEE, pp 1–8
24. McParland D, Gormley IC (2016) Model based clustering for mixed data: clustMD. Adv Data Anal Classif 10(2):155–169
25. Gubu L, Rosadi D (2020) Robust Mean-Variance portfolio selection using cluster analysis: a comparison between Kamila and weighted K-mean clustering. Asian Econ Finan Rev 10(10):1169–1186
26. Hummel M, Edelmann D, Kopp-Schneider A (2017) Clustering of samples and variables with mixed-type data. PloS One 12(11):e0188274
27. Aschenbruck R, Szepannek G (2020) Cluster validation for mixed-type data. Arch Data Sci Ser A 6(1):02
28. Caruso G, Gattone SA, Balzanella A, Di Battista T (2019) Cluster analysis: an application to a real mixed-type data set. In: Models and theories in social systems. Springer, Cham, pp 525–533
29. Lakshmanaprabu SK, Shankar K, Gupta D, Khanna A, Rodrigues JJ, Pinheiro PR, de Albuquerque VHC (2018) Ranking analysis for online customer reviews of products using opinion mining with clustering. Complexity
30. Sivasankar E, Vijaya J (2019) Hybrid PPFCM-ANN model: an efficient system for customer churn prediction through probabilistic possibilistic fuzzy clustering and artificial neural network. Neural Comput Appl 31(11):7181–7200
31. Gupta R, Pathak C (2014) A machine learning framework for predicting purchase by online customers based on dynamic pricing. Procedia Computer Sci 36:599–605
32. Chen C, Wang L (2008) Integrating rough set clustering and grey model to analyze dynamic customer requirements. Proc Inst Mech Eng Part B: J Eng Manuf 222(2):319–332
33. Szepannek G, Szepannek MG (2017) Package 'clustMixType'

Decentralized Library Management System Using Blockchain Technology

Sonakshi, Sujal Garg, Taanvi Jain, Priyanka Rani, and Neha Batra

Abstract Blockchain technology provides a secure and decentralized nature. The existing library management system are completely centralized, i.e., based on a central database solely. This means that in case the database crashes or maybe attacked by some intruder, the data stored in it will be at risk. To conquer, this we are proposing a blockchain-based decentralized and secure library management system. Not only this cover the book owning organization but also this can be implemented to propose a system where any person owning a book can give his book to someone who is in need. The addition of new book owner and new book borrower will be managed by verification at different nodes, hence providing it a secure and authenticated way. We have implemented it in python with the help of its advanced libraries (https://manavrachna.edu.in).

Keywords Blockchain · Hash · Nonce · Timestamp · Proof of work

1 Introduction

Blockchain technology has a wide range of application in current world scenario. Blockchain is not limited to Bitcoin but is now used by many leading tech companies in various field of application. Many fields are advancing in term of security just by the use of blockchain technology. Blockchain technology provides a distributed ledger for the data which is stored in the form of blocks that are added to the blockchain only after the approval at various nodes. In blockchain, each block has an information and a hash value that is generated with the help of the information stored in it. If some intruder tries exploiting its content, the hash value associated with it changes and

Sonakshi (✉) · S. Garg · T. Jain · P. Rani · N. Batra
Manav Rachna International Institute of Research and Studies, Faridabad, Haryana, India

P. Rani
e-mail: Priyankagrover.fet@mriu.edu.in

N. Batra
e-mail: nehabatra.fet@mriu.edu.in

the blockchain encounters that change, and hence, it reduces the possibility of data breach.

Today, when information is massive, library is a home to uncountable information resources. Managing which is not an easy task. Considering the fact, that the library management systems that are currently running are all centralized and driven by a central database. In case of database crash and some attacker hacks the database, the information stored in it is at risk. Hence, our decentralized system for library management will help provide security to the data by the use of blockchain technology in python. Also in case of data breach in database, neither the fault recognition is easy in the vast database nor the replacement of corrupted information. But this problem also has a solution in the name of blockchain, as it has a chain of all the blocks connected by a chain hash values. On data breach, the hash value also changes and hence the chain breaks which makes the data breach recognizable.

The paper has been organized as follows: Section 2 presents information about the blockchain technology. Related works are mentioned in Sect. 3. Section 4 shows the proposed work, Sect. 5 contains implementation of proposed methodology, and Section 6 concludes the paper.

2 Blockchain

Blockchain is the technology behind the very vast and well-known Bitcoin. Blockchain is a system that is decentralized and consist of a chain of blocks. Each block consists of data, hash value and previous hash.

Data: Information of each transaction stored in the blockchain.

Hash: Unique hash value assigned to each block that is generated from the data itself.

Previous hash: Hash of the previous block stored to keep the chain going and to traverse the blocks of the blockchain (Fig. 1).

Because hash calculation is a thing of few seconds the hackers can easily corrupt the blockchain and replace the hash of the whole blockchain to conquer this risk we have something called proof of work. Proof of work algorithm is very useful

Fig. 1 Blockchain mechanism

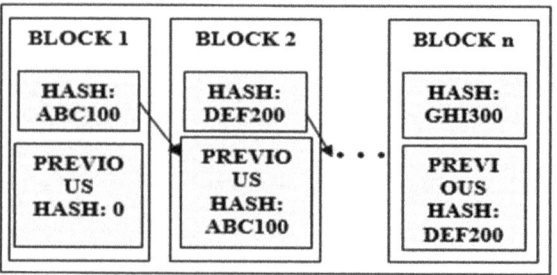

to enhance the blockchain's security [1]. It ensures that out of all possible hashes, and the blocks of a blockchain should have a hash that starts with a fixed number of zeros. The number of zeros is also known as difficulty of the blockchain. This increases the time taken by each block to get the hash value and hence makes it difficult for the intruder to crack the blockchain out. The number of times the cycle of hash calculation runs before getting the desired hash is known as nonce.

2.1 Applications

Banking and finance. Blockchain provides a way to securely and efficiently create log of activities, transactions, etc. It makes banking services more secure in international payment and money transfers. It removed the intermediaries from the money transfer system by automating entire process by blockchain.

Trade finance. Earlier trade finance has a major drawback of slow processes. Using blockchain deals and processes are simplified across border. It makes transaction easy, fast and secure.

Healthcare. Blockchain in healthcare field can be used to store data like medical history, name, age, blood group and vital signs. Details specific to certain customers can be stored and accessed without privacy issues.

Voting. Blockchain technology can be used to make voting process more easily accessible and secure. Hackers would not be able to crack the chain and corrupt the voting data as making changes in blockchain needs a lot of resources and is not everyone's cup of tea. Also, by increasing the difficulty of proof of work, we can increase the time for calculation of hash, and hence, activity of hacker would be easily encountered.

Cyber security. Cyber security is the biggest advantage of blockchain as it removes the risk of single point of failure. It also provides end to end encryption.

3 Related Work

MunibaMemon, Syed ShahbazHussain, Umair Ahmed Bajwa and AsadIkhlas in their work "Blockchain beyond Bitcoin: Blockchain Technology Challenges and Real-World Applications" have made the concept of blockchain much easier for us to understand with its various applications [2]. There are many work going on in the field of cyber security using blockchain. Some of them are Solar Energy Distribution Using Blockchain and IoT Integration [3], Decentralized E-Voting Portal Using Blockchain [4], A Blockchain-based Secure IoT Control Scheme [5], Customer Data Sharing Platform: A Blockchain-Based Shopping Cart [6], A Study of Blockchain Technology in Farmer's Portal [7], Smart FIR: Securing e-FIR Data through Blockchain within Smart Cities [8], Electronic Voting Recording System Based on Blockchain Technology [9], Blockchain-based Management of Video Surveillance Systems [10], A Blockchain-based File-sharing System for Academic Paper Review [11], Organ Donation Decentralized Application Using Blockchain Technology [12], A Decentralized Cryptographic Blockchain Approach for Health Information System [13], A Blockchain-Based Framework for Secure Log Storage [14]. These are the existing systems referred for the proposed system. Ali Akbar Movassagh used an integrated algorithm to determine the neural network input coefficients. [15]. M. Vasim Babu proposed an Integration of Distributed Autonomous Fashion with Fuzzy If–then Rules (IDAF-FIT) algorithm [16].

Other than this, blockchain and IOT are the most attractive duo these days. Some of the examples are "Dependable IoT using blockchain-based technology" [17] and Toward Secure and Decentralized Sharing of IoT Data [18]. Jafar A. Alzubi proposes a blockchain-assisted highly secure system for medical IoT devices using LamportMerkle Digital Signature [19].

4 Proposed System

In this research, we are focusing upon providing a decentralized library management system that is more secure and has lesser probability of data breach. This security is provided using blockchain technology and is implemented in python. In our proposed system, all the details of borrower and the book borrowed will be stored in the blockchain. Although the details of all the books and user will be stored in the database. Also, we are proposing a platform for students to share their books among each other. Someone who has a book that is of no use to him/her can give that book to the library for someone to borrow and the owner of the book can also borrow at the same time. This way students can take benefit of library to make their idle book useful. And the giving and borrowing process will work safe and sound with the help of blockchain technology.

4.1 Framework

The basic framework of our proposed system is laid on blockchain technology. Blockchain is very secure, but hash calculation is a deal of not more than 2–3 s. Hence, any attacker can easily corrupt the data and change the hash values of the blocks ahead. This can be coped up by using proof of work mechanism. Proof of work mechanism ensures that only those hash values are used that starts with some required number of 0's. The number of 0's are decided by the difficulty of block chain.

In the proposed system, user's information, book information and Nonce has been used to create a Hash Function, and on that basis decision has been made that whether

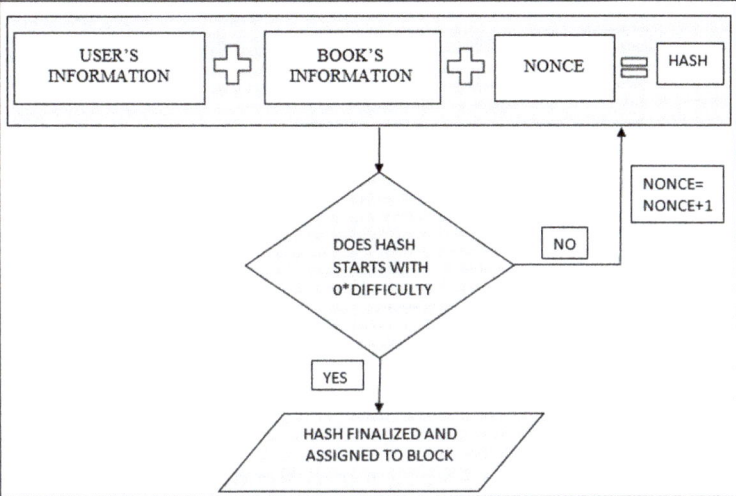

Fig. 2 Hash calculation mechanism using proof of work

hash starts with 0*difficulty or not, as the hash values starting with 0 or not, if the answer comes out to be no then Nonce is further incremented and it repeats the same process again, else it will be assigned a block. The working of proof of work is shown in Fig. 2

4.2 Implementation and Result

Case 1: If the user is owner of a book (Fig. 3).

In just a few moment, the book gets added to our database and is shown to the borrower when someone asks for the available books.

Case 2: If the user is a borrower (Fig. 4).

In just a few second of interaction, the user gets added to a new block of the blockchain. This is the beauty of blockchain that is extremely secure, and its decentralized behavior makes it very reliable. The blockchain is running on localhost: 5000 by default but we can change its port as per our convenience. In this way, the centralized behavior of database can be coped up using blockchain.

Fig. 3 User interface for owner

```
Do You Want To give a book???(y/n)y
Enter your namexyz
Enter name of your bookThe Secret
Record added successfully
```

```
Do You Want To give a book???(y/n)n
enter your nameabc
The available books are:
book_id,book_name,owner
1,The Secret,xyz
Enter the book's ID in which you are interested
1
The Secret Selected
enter your phone number10000000
enter email IDxyz@gmail.com
information added to blockchain
 * Serving Flask app "block" (lazy loading)
 * Environment: production
   WARNING: This is a development server. Do not use it in a production deployme
nt.
   Use a production WSGI server instead.
 * Debug mode: on
 * Running on http://127.0.0.1:5000/ (Press CTRL+C to quit)
```

Fig. 4 User interface of a borrower

5 Comparison Between Existing and Proposed System

The database model of library management system has an organization involved in it, which means that there is a copyright of a single governing authority on all the books. Management of old edition of books is also a challenge in this case. Other than this, these days SQL injection attacks are very common and can easily be performed. If our database is threatened by any such attack the problem arises with the data protection as the whole system is completely dependent on the database. Our crucial information can be erased or corrupted which is the main point of concern.

Considering our proposed system, which holds the data in the blockchain at the localhost server. Our proposed decentralized library management system serves as a solution to management of old edition of books, since the books are no longer owned by a single organization. Also, the issue of data protection from an attack or data breach is also prevented in the proposed system by the use of blockchain technology. The proposed system proves to be better than the existing system in terms of security and dependability (Table 1).

Table 1 Comparative analysis of existing and proposed system

Parameters	Existing system	Proposed system
Nature	Centralized	Decentralized
Security	More prone to attacks like SQL injection	Secure as compared to existing system
User verification	None	Distributed
Book owner	The organization	General people

6 Conclusion

Existing library management system are database driven. Databases are prone to cyber-attacks which puts our data at risk. The crucial data which has the record of complete library activities and books needs to be stored in a secure way. Protection from data breach is our main concern. Our objective is served with the help of blockchain technology that is a somehow the best way of dealing with crucial information and hence storing it in a secure way. We have proposed a system which takes information of user and stores it in the blockchain, any intrusion in it can lead to change in hash value which can be easily detected. This system provides data security and also removes centralization from library management systems that are currently running. Some web application or website can also be created in order to make this system more user friendly.

References

1. Gemeliarana IGAK, Sari RF. Evaluation of Proof of Work (POW) blockchains security network on selfish mining. In: 2018 international seminar on research of information technology and intellegent systems (ISRITI). https://doi.org/10.1109/ISRITI.2018.8864381
2. Memon M, Hussain SS, Bajwa UA, Ikhlas A (2018) Blockchain beyond Bitcoin: blockchain technology challenges and real-world applications. In: 2018 international conference on computing, electronics and communications engineering (iCCECE). https://doi.org/10.1109/iCCECOME.2018.8658518
3. Jain R, Dogra A (2019) Solar energy distribution using blockchain and IoT integration. IECC'19, 7–9 July 2019, Okinawa, Japan. https://doi.org/10.1145/3343147.3343163
4. Kritipatidar, Swapniljain (2019) Decentralized e-voting portal using blockchain. In: 10th ICCCNT 2019, 6–8 July 2019, IIT—Kanpur, Kanpur, India. https://doi.org/10.1109/ICCCNT45670.2019.8944820
5. Choi SS, Burm JW, Heo YJ (2018) A blockchain-based secure IoT control scheme. In: 2018 international conference on advances in computing and communication engineering (ICACCE-2018), Paris, France, 22–23 June 2018. https://doi.org/10.1109/ICACCE.2018.8441717
6. Shrestha AK, Joshi J, Vassileva J. Customer data sharing platform: a blockchain based shopping cart. In: 2020 IEEE international conference on blockchain and cryptocurrency (ICBC). https://doi.org/10.1109/ICBC48266.2020.9169421
7. Talreja R, Chouksey R, Verma S. A study of blockchain technology in farmer's portal. In: Proceedings of the second international conference on inventive research in computing applications (ICIRCA-2020). https://doi.org/10.1109/ICIRCA48905.2020.9182969
8. Khan ND, Chrysostomou C, Nazir B. Smart FIR: securing e-FIR data through blockchain within smart cities. In: 2020 IEEE 91st vehicular technology conference (VTC2020-Spring). https://doi.org/10.1109/VTC2020-Spring48590.2020.9129428
9. Agbesi S, Asante G. Electronic voting recording system based on blockchain technology. In: 2019 12th CMI conference on cybersecurity and privacy (CMI). https://doi.org/10.1109/CMI48017.2019.8962142
10. Jeong Y, Hwang DY, Kim KH. Blockchain-based management of video surveillance systems. In: 2019 international conference on information networking (ICOIN). https://doi.org/10.1109/ICOIN.2019.8718126
11. Zhou I, Makhdoom I, Abolhasan M, Lipman J, Shariati N. A blockchain-based file-sharing system for academic paper review. In: 2019 13th international conference on signal processing and communication systems (ICSPCS). https://doi.org/10.1109/ICSPCS47537.2019.9008695

12. Dajim LA, Al-Zuraib AA, Al-Farras SA, Mathew RM, Al-Shahrani BS. Organ donation decentralized application using blockchain technology. In: 2019 2nd international conference on computer applications and information security (ICCAIS). https://doi.org/10.1109/CAIS.2019. 8769459
13. Sosu RNS, Quist-Aphetsi K, Nana L. A decentralized cryptographic blockchain approach for health information system. In: 2019 international conference on computing, computational modelling and applications. https://doi.org/10.1109/ICCMA.2019.00027
14. Huang W. A blockchain-based framework for secure log storage. In: 2019 IEEE 2nd international conference on computer and communication engineering technology-CCET. https://doi. org/10.1109/CCET48361.2019.8989093
15. Movassagh AA, Alzubi JA, Gheisari M, Rahimi M, Mohan S, Abbasi AA, Nabipour N. Artificial neural networks training algorithm integrating invasive weed optimization with differential evolutionary model. J Ambient Intell. Humanized Comput. https://doi.org/10.1007/s12 652-020-02623-6
16. Babu MV, Alzubi JA, Sekaran R, Patan R, Ramachandran M, Gupta D. An improved IDAF-FIT clustering based ASLPP-RR routing with secure data aggregation in wireless sensor network. Mobile Netw Appl Oct 2020. https://doi.org/10.1007/s11036-020-01664-7
17. Zorzo AF, Nunes HC, Lunardi RC, Michelin RA, Kanhere SS. Dependable IoT using blockchain-based technology. In: 2018 eighth latin-american symposium on dependable computing (LADC). https://doi.org/10.1109/LADC.2018.00010
18. Thu Truong HT, Almeida M, Karame G, Soriente C. Towards secure and decentralized sharing of IoT data. In: 2019 IEEE international conference on blockchain, https://doi.org/10.1109/ Blockchain.2019.00031
19. Alzubi JA (2021) Blockchain-based LamportMerkle digital signature: authentication tool in IoT healthcare. Computer Commun 170:200–208

Latent Dirichlet Allocation (LDA) Based on Automated Bug Severity Prediction Model

Ritu Bibyan, Sameer Anand, and Ajay Jaiswal

Abstract The software systems prevail various types of bugs reported by users. On the basis of severity levels of bugs, it is decided which bug needs to be fixed first. If the bug is crucial, then it is fixed immediately, and if the bug is minor, then its fixing procedure can be postponed. This research seeks to present an automated bug severity prediction model by using topic modelling and classification techniques. The summary of historical bug reports of the Eclipse project is pre-processed to extract features. Later feature extraction is done by using term frequency and latent Dirichlet allocation (LDA): a topic modelling technique. The topics extracted from LDA are then used to train the classifiers. Our approach also endorses the severity level for the newly disclosed bug. After training the model, the results actively exhibit admissible accuracy scores.

Keywords Bug severity prediction · Machine learning · Feature extraction · Topic modelling · Latent dirichlet allocation (LDA)

1 Introduction

The intricacy of software is increasing eventually day by day, making it certain that software faults will occur. In an open-source project named Eclipse, near about 300 bug reports are disclosed or reported daily [1]. These reports are efficiently handled by using a bug tracking system (BTS) [2]. The software systems unavoidably have defects which consist of error leading to failure. So it becomes necessary to fix these bugs, and it is one of the arduous works in the software maintenance phase. Users

R. Bibyan (✉)
Department of Operational Research, University of Delhi, Delhi, India

S. Anand · A. Jaiswal
S.S. College of Business Studies, University of Delhi, Delhi, India
e-mail: sameeranand@sscbsdu.ac.in

A. Jaiswal
e-mail: ajayjaiswal@sscbsdu.ac.in

© The Author(s), under exclusive license to Springer Nature Singapore Pte Ltd. 2022
D. Gupta et al. (eds.), *Proceedings of Data Analytics and Management*,
Lecture Notes on Data Engineering and Communications Technologies 90,
https://doi.org/10.1007/978-981-16-6289-8_31

or developers generally draft these bugs. The BTS (e.g. Bugzilla) thoroughly checks errors in lines of code to help software project business to fix the bugs and achieve their goal [3]. As soon as the bug is reported in the bug repository, bug triager allots the bug to a developer who fixes the bug using BTS. A bug report gives the brief about bug-like how it occurred and under what circumstances. It also has information related to the regeneration of bug.

In Bugzilla, the bug report typically has various labels or attributes such as priority, summary, severity, Bug ID, date, description and status. One of the most substantial attributes is severity as it decides how instantly a report should be fixed. There are several levels of severity: Blocker, Critical, Major, Normal, Minor, Trivial and Enhancement. Many kinds of researches have been proposed to automate the prediction of bug severity [4–9]. The approaches considered traditional classification algorithms like naïve Bayes, support vector machine, decision tree. The severity is manually assigned by the users while reporting a bug that requires the knowledge of this field. This manual work leads to false assessment of the severity because

1. A considerable amount of reports are recorded everyday which amplifies the traigers workload.
2. Lack of experience of users in the domain. Lack of knowledge regarding bug report due to which traiger might make a mistake while assigning developer for fixing the bug.

In this paper, we have built an automatic severity prediction model by recommending topic modelling. The topics of all bug reports are extracted using the summary to make a base of topics. We seek to incorporate latent Dirichlet allocation (LDA) to find out topics as features for prediction. We affirm that this new technique can boost the performance of the bug triage and severity prediction. The research questions proposed for our experimental setup are:

RQ1: To what level the proposed model is effective while predicting the severity?
RQ2: When comparing the proposed models and traditional models, what is the improvement in terms of performance is observed?

The paper is presented as follows: Sect. 2 discusses the literature survey and the research gap. Section 3 gives a brief knowledge about the methodology in discussing how the data is pre-processed and how LDA is applied. Later different classifiers used are explained which also included description related to evaluation matrices. In Sect. 4, we have shown result in performance analysis. The paper is concluded in Sect. 5 with future work in Sect. 6.

2 Literature Survey

In the past few years, many researchers have given their inputs in the field of bug severity prediction. The concept of bug triaging using text mining was introduced by Murphy and Cubranic [10] using naïve Bayes (NB) classifier. His work was extended by Anvik, Hiew [11] using supervised learning algorithms on the dataset of Eclipse and Firefox. Menzies and Marcus [12] worked on NASA projects for severity prediction of bug reports using textual summary. Another model was bought up by Ahsan, Ferzund [13] in which weighted term methods term frequency–inverse document frequency (TF-IDF) was used. The concept of semantic indexing was introduced along with support vector machine (SVM) in the field of automated bug triaging. Lamkanfi, Demeyer [8] attempted to predict bug reports as severe or non-severe by using the summary of newly reported bugs. They used three different open-source project Eclipse, GNOM, and Mozilla and classified reports using naïve Bayes. This work was extended by comparing three classification algorithms, namely multinomial naïve Bayes, SVM and k-nearest neighbour on same dataset [14].

In 2010, Rastkar, Murphy [15] summarized the software artefacts into smaller summaries of bug reports using conversation-based classifiers. These conversation-based classifiers were trained on email threads, meetings and bug report corpus. They considered different open-source projects to validate generated summaries and observed that classifiers show more precision. In 2014, they extended their work by using the generated summaries to classify duplicate bug reports [16]. Chaturvedi and Singh [5] proposed an automated classification model on the basis of severity attribute. The projects considered were taken from PROMISE repository for validation purpose using NB, k-NN, SVM, RIPPER and J48. Later, an approach was given by Yang, Chen [17] which considered four indicators based on the quality for severity prediction model. The model was trained on the dataset of Eclipse and the results showed improvement in accuracy. Zhou, Tong [18] proposed a multi-stage approach by linking data and text mining together. In the initial stage, textual pre-processing steps are applied on summary of bug reports and classify them into three levels. In the second stage, feature extraction is done to train the machine learning classifiers. This experiment was performed on four projects Firefox, Mozilla, OpenFOAM and JBoss and obtained increased f-measures value.

Pandey, Sanyal [19] used NB, SVM, random forest (RF), k-NN classification algorithms to classify the bug reports as severe and non-severe. An accuracy from 75 to 83% was obtained using this approach. In 2016, Zhang, Chen [20] applied REP_{topic} and k-NN algorithm on historical bug reports and extracted features to classify them into five different severity levels. This approach was performed on GCC, Eclipse and Mozilla datasets to measure the improvement in severity prediction. Goseva-Popstojanova and Tyo [21] used supervised and unsupervised classifiers on NASA dataset and found out that supervised approach showed better performance than unsupervised. Hamdy and El-Laithy [22] used hierarchical topic modeller to automate the severity assignments of bug reports. In 2020, Bibyan, Anand [23] proposed an approach by using TF- IDF on summary of bug reports and later classified new reports by neural network as classifier.

3   Methodology

We propose a model to predict bug severity by using text mining, topic modelling and machine learning. The overview of the model is shown in Fig. 1. In this study, we extract bug reports of different components of Eclipse project from Bugzilla. Words are then extracted from bug reports summary and their frequencies using CountVectorizor. These words are then used to create topics using LDA. The topics contain tokens or terms related to topic. As soon as a new bug is reported, we compare tokens between constructed topics and a new bug report. Then, the distribution score is computed in each topic. At last, we find out the similar topic related to the new bug report. And using the similar topic we predict the severity of bug reports.

3.1   Data Description

The experiment is performed on an open-source project named Eclipse. The bug reports which are reported in Bugzilla are collected for different components of Eclipse projects, as shown in Fig. 2. The components which are considered in our experiments are: UI, Debug, Core, text and IDE with 42,719, 9056, 54,420, 6225 and 1169 reports, respectively. The reports are downloaded from Bugzilla in CSV format with status as resolved, verified, closed and resolution as fixed and worksfrome.

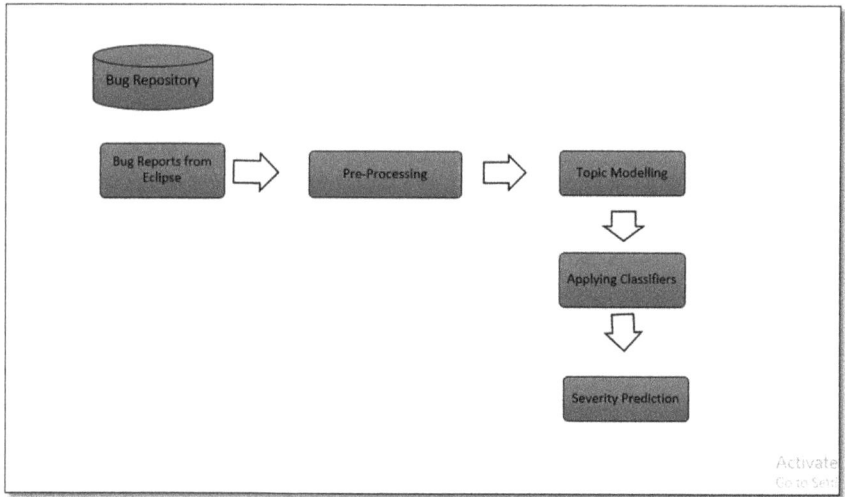

Fig. 1 Bug severity prediction model using topic modelling flow diagram

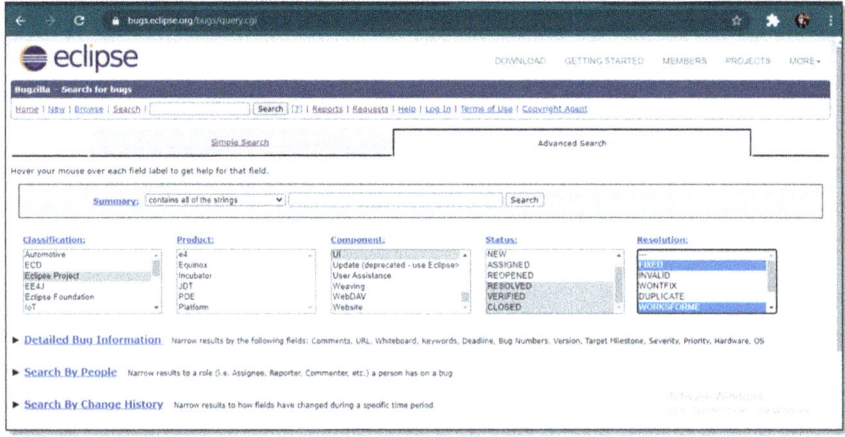

Fig. 2 Bug report extraction for eclipse project

3.2 Pre-Processing of Data

The reports contain bugs with different severity levels: Blocker, Critical, Major, Normal, Minor, Trivial and Enhancement. The bugs with severity level "normal" are not considered in our study because these type of bugs creates confusion for classifiers. The normal bugs indicate that the reporter is not sure about the severity level. Also, bugs with severity level enhancement are not considered as they point towards the incorporation of new feature or improvement in current features. The enhancement reports can be studied separately for enhancement prediction models. After collecting files for each component, we modify the reports manually. The bugs with severity level Blocker, Critical and Major are combined and named as severe reports. The bugs with severity level Minor and Trivial are combined and named as non-severe reports.

The machine learning algorithms require structured text for the processing purpose. Therefore, the next step is to convert unstructured text into structured text by pre-processing which is an essential tool of text mining. The structured text is more manageable for representation purpose and is suitable for analysis [24]. The text mining has three major activities [25]:

- Tokenization: this action breaks the sentence into a sequence of terms or tokens. A token does not contain any space or punctuation, and it is a string of characters.
- Stop Word Removal: this action removes the words which are commonly used and do not have and useful information related to the context. Such words are prepositions, articles, verbs, noun, pronouns, conjunctions and adverbs.
- Stemming: this action converts the derived words into their stem word; i.e. it brings the word back to base form, for example, studying and studied s converted to study.

Once the data is pre-processed, it is ready for application of models: vector extraction models, vector weighting models and document similarity methods. In this study, we have used a document similarity method technique: topic Model. It is a well-known statistical model which recognizes topics from bunch of documents. Each topic contains terms occurring in the documents. And, each document might be associated with one or more than one topics. There are various techniques under topic modelling like—latent Dirichlet allocation (LDA), latent semantic analysis (LSA), correlated topic model (CTM) and probabilistic latent semantic allocation (PLSA).

3.3 Topic Modelling

Topic modelling is used for the unsupervised classification of documents. The task of the topic modelling is to discover hidden topics in the data containing documents. It also classifies the documents using these hidden topics.

LDA is one of the most famous techniques of topic modelling based on generative probabilistic model. A document contains numerous words, and these random mixture of words are used to create latent topics. The basic idea behind LDA is to find topics from document entirely based on the words probabilities. This technique was first brought up in 2003 by Blei, Ng and Jordan [26]. The three parts on which LDA works are:

1. Words belonging to the document.
2. Words belonging to topic.
3. The probability of words that it belongs to a topic which needs to be calculated.

When we establish LDA, word distribution is used to create topics and their frequency of co-occurrence. There are four parameters for LDA:

α: it is the rate at which a document can be covered in numerous topics.
β: it is the rate at which various words can be covered in classified topics.
γ: it defines the iterations that how many times topic model building is repeated.
k: it is number of topics assigned randomly according to our desire.

As soon as the new bug is reported, we look for the matching topic(s). Then, we find out the frequencies of topic terms, which are present in the new bug report. The new bug report belongs to topic(s), if the frequency of topic terms is highest. Later, historical bug reports belonging to same topic(s) and the new bug report are considered for the severity prediction.

3.4 Classification Techniques

The machine learning algorithms are applicable on datasets having various instances (reports) y_i, where i-$\{1......,\}$. Each instance consists of independent variables and one dependent variable. The dependent variable is output attribute named as label ore class, whereas the independent variables are input attributes generally named as features. There are various classification techniques such as

k-nearest neighbour (k-NN):- it acknowledges the historical instances (neighbours) and uses the similarity measure to k-nearest neighbour for classifying new instance. Generally, Euclidian distance is used to calculate the distance between neighbours. The new instance is classifies based on the labels of its neighbours. If the new report belongs to more than one label, it will be assigned to the class having majority vote.

Naïve Bayes (NB): it uses Bayesian theorem for predicting to which class a report belongs. The probabilities of a report belonging to each classes C_k given that the report "y" is $P(C_k|y)$. It assumes that, the value of a specific feature is self-reliant of the value of any other feature [27].

Multinomial Naïve Bayes (MNB): it not only considers the presence and absence of words to decide output label, but it also considers the frequency of the words in the report which is expressed as a multinomial distribution of words [6, 28]. It acknowledges this report as an ordered sequence of word frequencies, with each word frequency as an independent trial [29].

Decision Tree (DT): it is based on hierarchical model with a number of internal and leaf nodes in a tree. The internal nodes symbolize as features and leaf node as class label. It has tree-like structure organized in the form of test questions and conditions. It guides the decision-making process to determine to which class the instances belong to.

Artificial Neural Network (ANN): it is based on the biological neural network of a human being [30]. It consists of three layers: input, hidden and output. The input layer consists of neurons which sends information to hidden layer. The hidden layer converts the information for output layer which is processable by the output layer. These layers have neurons or nodes connected to each other consecutively. The neural network nodes are linked with weights, which defines the strength of the connection between them [31].

Support Vector Machine (SVM): this technique tries to separate training instances into given labels in an optimal way based on the maximum margin hyperplane [32]. The output is a hyperplane that maximizes the distance between the feature vectors of reports of class labels. The features can be changed using the kernel function depending on the type of dataset.

Random Forest (RF): it follows two principles: (i) in a single model, it creates multiples of decision tress and their combination, (ii) the end decision is dependent on the ruling of the majority of forming trees [33].

Table 1 Confusion matrix

		Predicted	
		Negative	Positive
Actual	Negative	True negative (TN)	False positive (FP)
	Positive	False negative (FN)	True positive (TP)

P: bug report is severe
N: bug report is not severe
TP: bug report is severe and predicted as severe
TN: bug report is non-severe and predicted as non-severe
FP: bug report is non-severe and predicted as severe
FN: bug report is severe and predicted as non-severe

Table 2 Performance measures

Performance measures	Formula	Description
Accuracy	$\frac{TP+TN}{TP+FP+TN+FN}$	Percentage of correctly classified instances
Precision	$\frac{TP}{TP+FP}$	Positively classified severity level within the dataset
Recall	$\frac{TP}{TP+FN}$	Classifiers completeness
F–measure	$\frac{2 \times recall \times precision}{recall + precision}$	Harmonic mean of the precision and recall

3.5 Performance Measure

The most commonly used performance measures are accuracy, precision, recall and F-measure. These measures are evaluated using confusion matrix given as given in Table 1.

Using the confusion matrix, the performance measures can be obtained by using formulae provided in Table 2.

4 Performance Analysis

This section of the paper explains the experimental setup using different components of the Eclipse project. We discuss the experimental setup and design with the results obtained after applying our model.

The bug reports are collected from Bugzilla repository for an open-source project Eclipse. The Eclipse project has various components for different products. In this experiment, we have considered five components, namely UI, Core, debug, text and IDE. After extracting the documents containing reports, we apply some pre-processing steps. Once the data is pre-processed, we use countvectorizor in Python to convert data into vector of terms count for each component and extract the top 150 words or features.

Once the 150 features are obtained, the LDA as topic model is applied with $k = 10$ (number of topics) on each component. The topics extracted for each component are shown in Table 3. The term frequency and LDA topics information are used as feature set to train the classifiers. The model is validated using naïve Bayes (NB), k-nearest neighbour (k-NN), random forest (RF), decision tree (DT), artificial neural network (ANN) and support vector machine (SVM). We have evaluated the performance measures using 10-cross-validation.

The performance measure values for all the components using seven different classifiers are shown in Table 4.

5 Conclusion

This paper purposes the bug severity prediction model by training data from Eclipse project. We conducted experiments with five components of Eclipse projects and seven machine learning techniques. The topic modelling technique LDA is used to extract topics from the data. The term frequency and topic(s) information is then used as feature for training the classifiers. We have applied naïve Bayes (NB), k-nearest neighbour (k-NN), random forest (RF), decision tree (DT), artificial neural network (ANN) and support vector machine (SVM) to predict the severity of new bug reports.

In order to answer the RQ1, we evaluated the performance of the model for each classifier. The results show that the k-NN performance is better than other classifiers for components text and debug in terms of accuracy and F1-score. And, MNB has outperformed from other classifiers for components UI, IDE and core in terms of accuracy and F1-score. The evaluation results show that our model using LDA effectively accomplishes severity prediction.

In order to answer RQ2, we compared the proposed model with the existing models. The existing models majorly used term frequency–inverse document frequency (TF-IDF) to extract features. The proposed model used TF and LDA to extract features and has shown better results.

6 Future Scope

In future, the work can be extended for a multi-class classification problem and the problem of imbalanced data can also be considered. Additionally, we plan to implement and execute this model on the commercial projects.

Table 3 Topics extracted through LDA

Components	Topic 0	Topic 1	Topic 2	Topic 3	Topic 4	Topic 5	Topic 6	Topic 7	Topic 8	Topic 9
UI	View	View	Eclipse	Eclipse	Dialog	Page	File	File	Junit	Workbench
	Editor	Markers	Project	Workbench	Type	Preference	Method	Workbench	Error	Junit
	Eclipse	Problems	Dialog	View	Open	Preferences	Refactoring	Import	Test	Package
	Project	Hierarchy	Workbench	Editor	Error	Code	Class	Export	Dialog	Explorer
	Dialog	Menu	Java	Launch	Method	Junit	Quick	Error	Project	Test
	Java	Junit	File	Internal	Refactoring	Key bindings	Type	Dialog	Plugin	Window
	File	Tasks	Build	Runtime	Preference	Target	Code	Menu	Build	Method
	Menu	Package	Import	Application	Workbench	Workbench	Junit	Window	Wizard	Quick
	Open	View mgmt	Wizard	Contribution	Dialogs	Build	Extract	Plugin	Path	Plug
	Error	Progress	Explorer	Junit	Eclipse	Wizard	Error	Wizard	Tests	Work
Text	Assist	Editor	Editor	Preferences	Code	File	Text	Typing	Templates	Misc
	Content	Java	Java	Code	Line	Preferences	Preferences	Line	Selection	Selection
	Editor	Preference	Preferences	Page	Text	Misc	Content	Selection	Template	Typing
	Code	Text	Text	Templates	Selection	Open	Page	Java	Block	Code
	Java	File	File	Preference	File	Implementation	Replace	Block	Implementation	Editor
	Preference	Page	Page	Template	Mining	Templates	Preference	Work	Replace	Block
	Text	Misc	Misc	Dialog	Typing	Changed	Selection	File	Java	Replace
	Javadoc	Implementation	Implementation	Pref	Block	Page	Eclipse	Replace	Dialog	Assist
	File	Selection	Selection	Mining	Replace	Dialog	Dialog	Content	Eclipse	Mode
	Line	Line	Line	Line	Eclipse	External	Typing	Mode	Work	Dialog
IDE	Eclipse	Project	View	Import	File	Resource	Error	Java	Workspace	Markers
	Error	View	Markers	Export	Open	Linked	Java	Update	Page	Workspace
	Java	Markers	Problems	Resource	Java	Open	Export	Resource	Preference	Open
	Crashes	Explorer	Content	Open	Resource	Dialog	Import	Linked	Error	Dialog

(continued)

Table 3 (continued)

Components	Topic 0	Topic 1	Topic 2	Topic 3	Topic 4	Topic 5	Topic 6	Topic 7	Topic 8	Topic 9
	Project	Problems	Display	Workspace	Menu	Error	Internal	Using	Linked	Problem
	Start	Import	Help	File	Dialog	Page	Core	Crashes	Mnemonic	Import
	Unable	Resource	Selection	Menu	Mnemon	Preference	File	Markers	Folder	Unable
	Workspace	Folder	Missing	Linked	Folder	Linked resources	Message	Launch	File	Wizard
	Open	Dialog	Menu	Projects	Work	Resources	Menu	Preference	Using	Quick
	Crash	Export	Views	Mnemonic	Unable	Unable	Opening	Opening	Linked resources	Projects
Debug	Debug	Debug	Launch	Launch	Console	Java	Dialog	Eclipse	Eclipse	Page
	View	Eclipse	Dialog	Configuration	Eclipse	Page	Variable	Error	Variables	Installed
	Variables	Launch	Breakpoint	Variables	Output	Variables	Variables	Breakpoint	Breakpoint	Jres
	Breakpoint	Java	Java	View	Java	Installed	Error	Launching	Error	Breakpoints
	Breakpoint	Page	Configuration	Launching	Error	Ures	Edit	Java	Exception	Error
	Eclipse	Configuration	Launching	Config	Stack	Preference	Select	Class	Variable	Preference
	Java	Launching	Breakpoints	Console	Missing	Variable	Help	Internal	Page	Pref
	Launch	Preference	Error	Configuration	Source	Project	Preferences	Source	Internal	Preferences
	Menu	Perspective	Exception	Variable	Program	Source	Button	Dialog	Running	Message
	Error	Source	Project	Save	Page	Logical	Configuration	Running	File	Scrapbook
Core	Eclipse	View	Eclipse	View	Project	Type	Java	Code	Method	Error
	Java	Editor	Editor	Eclipse	Java	Dialog	Type	Method	Class	Java
	Compiler	Menu	Java	Workbench	Package	Open	Hierarchy	Formatter	Project	Dialog

(continued)

Table 3 (continued)

Components	Topic 0	Topic 1	Topic 2	Topic 3	Topic 4	Topic 5	Topic 6	Topic 7	Topic 8	Topic 9
	Error	Package	Workbench	Junit	Explorer	Eclipse	Open	Class	File	Method
	Editor	Open	Project	Core	Build	Method	View	Assist	Package	File
	View	Project	Core	Markers	Dialog	Hierarchy	Search	Refactoring	Explorer	Refactoring
	Type	Explorer	File	Menu	Import	Refactoring	Compiler	Javadoc	Refactoring	Class
	Project	Dialog	Plugin	Problems	File	Class	Perspective	Quick	Editor	Search
	File	Problems	Build	Package	Path	Workbench	Quick	Import	Extract	Message
	Class	File	Launch	Launch	Wizard	Quick	Model	Extract	Quick	Perspective

Table 4 Performance measure for all components

Components	Measures	Classifiers						
		k-NN	ANN	NB	DT	SVM	RF	MNB
UI	Accuracy	0.84	0.877	0.882	0.887	0.882	0.882	**0.89**
	F1-Score	0.806	0.777	0.753	0.766	0.753	0.753	**0.843**
	Recall	0.84	0.89	0.882	0.887	0.882	0.882	0.877
	Precision	0.796	0.842	0.665	0.933	0.665	0.665	0.912
text	Accuracy	**0.759**	0.668	0.687	0.700	0.681	0.75	0.723
	F1-Score	**0.778**	0.733	0.796	0.647	0.781	0.718	0.75
	Recall	0.759	0.668	0.687	0.700	0.681	0.75	0.723
	Precision	0.758	0.673	0.715	0.726	0.698	0.751	0.713
IDE	Accuracy	0.848	0.892	0.892	0.892	0.892	0.892	**0.921**
	F1-Score	0.765	0.765	0.765	0.765	0.765	0.765	**0.875**
	Recall	0.848	0.892	0.892	0.892	0.892	0.892	0.921
	Precision	0.734	0.678	0.678	0.678	0.678	0.678	0.9
Debug	Accuracy	**0.832**	0.802	0.802	0.806	0.802	0.802	0.783
	F1-Score	**0.81**	0.652	0.652	0.661	0.652	0.652	0.667
	Recall	0.832	0.802	0.802	0.806	0.802	0.802	0.783
	Precision	0.818	0.562	0.562	0.962	0.562	0.562	0.682
Core	Accuracy	0.853	0.888	0.889	0.873	0.889	0.889	**0.894**
	F1-Score	0.83	0.801	0.761	0.784	0.761	0.763	**0.848**
	Recall	0.853	0.888	0.889	0.873	0.889	0.889	0.88
	Precision	0.821	0.834	0.674	0.842	0.674	0.986	0.873

References

1. Yang G, Zhang T, Lee B (2014) Towards semi-automatic bug triage and severity prediction based on topic model and multi-feature of bug reports. In: 2014 IEEE 38th annual computer software and applications conference. IEEE
2. Zimmermann T, et al (2009) Improving bug tracking systems. In: 2009 31st international conference on software engineering-companion volume. IEEE
3. Jalbert N, Weimer W (2008) Automated duplicate detection for bug tracking systems. In: 2008 IEEE international conference on dependable systems and networks with FTCS and DCC (DSN). IEEE
4. Sharma M, et al (2014) Multiattribute based machine learning models for severity prediction in cross project context. In: International conference on computational science and its applications. Springer
5. Chaturvedi K, Singh V (2012) Determining bug severity using machine learning techniques. In: 2012 CSI sixth international conference on software engineering (CONSEG). IEEE
6. Lamkanfi A, et al (2011) Comparing mining algorithms for predicting the severity of a reported bug. In: 2011 15th European conference on software maintenance and reengineering. IEEE
7. Menzies T, Marcus A (2008) Automated severity assessment of software defect reports. In: IEEE international conference on software maintenance, 2008. ICSM 2008. IEEE

8. Lamkanfi A, et al (2010) Predicting the severity of a reported bug. In: 2010 7th IEEE working conference on mining software repositories (MSR). IEEE
9. Tian Y, Lo D, Sun C (2012) Information retrieval based nearest neighbor classification for fine-grained bug severity prediction. In: 2012 19th working conference on reverse engineering. IEEE
10. Murphy G, Cubranic D (2004) Automatic bug triage using text categorization. In: Proceedings of the sixteenth international conference on software engineering and knowledge engineering. Citeseer
11. Anvik J, Hiew L, Murphy GC (2006) Who should fix this bug? In: Proceedings of the 28th international conference on Software engineering. ACM
12. Menzies T, Marcus A (2008) Automated severity assessment of software defect reports. In: 2008 IEEE international conference on software maintenance. IEEE
13. Ahsan SN, Ferzund J, Wotawa F (2009) Automatic software bug triage system (bts) based on latent semantic indexing and support vector machine. In: 2009 Fourth international conference on software engineering advances. IEEE
14. Lamkanfi A, et al (2011) Comparing mining algorithms for predicting the severity of a reported bug. In: 2011 15th European conference on software maintenance and reengineering (CSMR). IEEE
15. Rastkar S, Murphy GC, Murray G (2010) Summarizing software artifacts: a case study of bug reports. In: 2010 ACM/IEEE 32nd international conference on software engineering. IEEE
16. Rastkar S, Murphy GC, Murray G (2014) Automatic summarization of bug reports. IEEE Trans Software Eng 40(4):366–380
17. Yang C-Z, et al (2014) Improving severity prediction on software bug reports using quality indicators. In: 2014 IEEE 5th international conference on software engineering and service science. IEEE
18. Zhou Y et al (2016) Combining text mining and data mining for bug report classification. Journal of Software: Evolution and Process 28(3):150–176
19. Pandey N et al (2017) Automated classification of software issue reports using machine learning techniques: an empirical study. Innovations Syst Softw Eng 13(4):279–297
20. Zhang T et al (2016) Towards more accurate severity prediction and fixer recommendation of software bugs. J Syst Softw 117:166–184
21. Goseva-Popstojanova K, Tyo J (2018) Identification of security related bug reports via text mining using supervised and unsupervised classification. In: 2018 IEEE international conference on software quality, reliability and security (QRS). IEEE
22. Hamdy A, El-Laithy A (2020) Semantic categorization of software bug repositories for severity assignment automation. In Integrating research and practice in software engineering. Springer. p 15–30
23. Bibyan R, Anand S, Jaiswal A (2020) Assessing the severity of software bug using neural network. In: Strategic system assurance and business analytics. Springer. p 491–502
24. Feldman R, Sanger J (2007) The text mining handbook: advanced approaches in analyzing unstructured data. Cambridge university press
25. Williams G (2011) Random forests. In: Data mining with rattle and R. Springer. p 245–268
26. Blei DM, Ng AY, Jordan MI (2003) Latent dirichlet allocation. J Mach Learn Res 3:993–1022
27. Murphy KP (2012) Machine learning: a probabilistic perspective. MIT press
28. Kibriya AM, et al (2004) Multinomial naive bayes for text categorization revisited. In Australasian joint conference on artificial intelligence. Springer
29. Kim S-B et al (2006) Some effective techniques for naive bayes text classification. IEEE Trans Knowl Data Eng 18(11):1457–1466
30. Haykin S (1998) Learning in neural networks. A comprehensive foundation. Prentice Hall
31. Zhou J, Zhang H, Lo D (2012) Where should the bugs be fixed? More accurate information retrieval-based bug localization based on bug reports. In: 2012 34th International conference on software engineering (ICSE). IEEE

32. Vapnik V, Golowich SE, Smola AJ (1997) Support vector method for function approximation, regression estimation and signal processing. In Advances in neural information processing systems
33. Breiman L (1996) Bagging predictors. Mach Learn 24(2):123–140

A Novel Approach of Transfer Learning for Satellite Image Classification

Rohit Bharti, Dipen Saini, and Rahul Malik

Abstract Many different approaches have been tried for the classification of the remote sensing satellite images, and great results are produced. In this paper, we have applied a novel technique of transfer learning on remote sensing satellite images. The pre-trained models were used with different augmentation parameters and optimization of loss by using different loss functions. The results which we have achieved are promising. The datasets which we used are publicly available, namely SIRI-WHU, WHU-RS-19, AID, Brazilian coffee scene, and NWPU-RESISC45. We tried three different pre-trained models as VGG16, ResNet50, and Xception for the implementation. By using these pre-trained models, we were able to achieve accuracy up to 98% with a kappa index score of 0.95.

Keywords VGG16 · ResNet50 · Xception · Deep learning · Transfer learning

1 Introduction

Advancement in deep learning approaches using various new algorithms and techniques for remote sensing image scene understanding has made the deep learning field more efficient. Satellite image scene understanding is becoming more and more popular among the new emerging researchers or the ones who are already experienced in this particular field. As satellite images are being used for the scene understanding in remote areas where it is not possible to go physically and the conditions could be checked by studying the images, natural calamities prevention and their prediction are also possible now. Firstly, there was only a simple convolutional neural network [1] which created a benchmark for analyzing different remote sensing images. CNN becomes a great technique like some milestone technique that could extract a great deal of information from the images and feature extraction from the images became

R. Bharti (✉) · D. Saini · R. Malik
SCSE, Lovely Professional University, Phagwara, Punjab, India

R. Malik
e-mail: rahul.23360@lpu.co.in

© The Author(s), under exclusive license to Springer Nature Singapore Pte Ltd. 2022   379
D. Gupta et al. (eds.), *Proceedings of Data Analytics and Management*,
Lecture Notes on Data Engineering and Communications Technologies 90,
https://doi.org/10.1007/978-981-16-6289-8_32

more efficient and less complex. There are different methods through which different classification could be performed on the satellite images; for example, there is scene-level classification which is the most basic, then retrieval of information from the images is the next method and at last object detection which is scene guided is the third one. In the starting phases, the algorithms which were proposed depended only on pixel-level [2] and object-level [3] but now scene-level classification [4] is preferred. Also, different methods were proposed by different authors like Cheng et al. [5] have tried to provide a summary of the different methods used for scene-level classification. Ref. [6] have made a review on retrieving the scene information, and Bank et al. [7] also studied and reviewed different techniques for object detection using remote sensing satellites.

Many algorithms could be used for scene-level classification using remote sensing satellite images. There are mainly two types of categories of these algorithms namely supervised algorithms and unsupervised algorithms. Some supervised algorithms are artificial neural networks [8] used for regression and classification, convolutional neural networks [1] used for computer vision, recurrent neural networks [9] used for time series analysis, and some unsupervised algorithms are self-organizing maps [10] used for feature detection, deep Boltzmann machines [11] used for recommendation systems, and autoencoders [7] used for recommendation systems. Some more algorithms like generative adversarial networks (GANs) [12], cycle GANs [13], and many more could also be used for remote sensing images.

One different approach is known as transfer learning in which different pre-trained neural networks are used which need not be trained from scratch as they are already trained on a benchmark dataset ImageNet [14]. There are different pre-trained neural networks like VGG16 [15], VGG19 [16], ResNet50 [17], Xception [18], GoogleNet [19], and many more are increased and could be customized according to the need of the dataset. These networks play a vital role as we could use the pre-trained weights and then apply those to our dataset. One could use their classifier on top and train it. Also, fine-tuning could be performed in which we retrain a limited number of weights of the pre-trained network and get better results. In some cases, transfer learning performs better than other networks which have to be trained from the scratch.

In this paper, transfer learning techniques are used for the scene-level classification of the remote sensing satellite images. The rest of the paper is divided into four sections, Sect. 2 comprises of literature review of the papers on which most of this research is based for the implementation purpose, Sect. 3 comprises the methodology and various techniques which are used for the implementation purpose, Sect. 4 comprises of the results, and Sect. 5 comprises of the conclusion and future scope of this implementation.

2 Literature Review

Several papers were considered and followed thoroughly, and a literature review of some papers is shown in Table 1.

The architecture which we propose consists of the pre-trained neural network and some data augmentation with preprocessing which is important for any image dataset. The flowchart that how the architecture will work is shown in Fig. 1.

3 Methodology

In this paper, we have implemented transfer learning on remote sensing satellite images and used three datasets, namely SIRI-WHU [28], WHU-RS-19 [45], AID [46], Brazilian coffee scene [47], and NWPU-RESISC45 [24]. All the images are preprocessed according to the pre-trained neural network in which we are using them and then different augmentation techniques are applied. If needed fine-tuning is also performed in which we retrain some of the weights and the classification accuracy increases. The pre-trained neural network architecture and the networks which we applied are VGG16, ResNet50, and Xception Network. The architecture of this pre-trained network is explained below.

3.1 Datasets

The datasets used are publicly available. The dataset's name and their important specifications are shown in Table 2.

3.2 The Architecture of the Pre-trained Model

A pre-trained neural network is trained on a benchmark dataset ImageNet. It consists of more than 14 million images with 1000 classes and tech giants like Google, Facebook, and many other big companies participate in the Global competition 'ImageNet' and create complex networks which are trained for weeks on GPUs and TPUs. Then we could harness that massive power of these networks by utilizing the already trained weights in our less complex networks and access its full power. We select some pre-trained models as our base model, then we delete some last few last layers as they are specifically designed for the ImageNet dataset. We use the complete base model until its head and change the head with our custom head for the custom task. In fine-tuning which is better suited for the bigger datasets, we still define a custom head, but we do not freeze the whole base model, we only freeze the

Table 1 Findings of the literature review

Sr. No.	Author	Year	Findings
1	[20]	2020	Different types of new methods of machine learning and deep learning for getting the benchmark results are discussed
2	[21]	2020	Using different augmentation techniques for better accuracy on satellite datasets
3	[22]	2020	Using the TensorFlow features and different augmentation methods for improving the result on the hyperspectral EuroSat dataset
4	[23]	2019	A review on different types of remote sensing as scene guided information retrieval, image understanding, and how dataset are available for performing these tasks are discussed
5	[24]	2017	Different available datasets which are public datasets available are discussed and a review is provided
6	[25]	2016	A new classifier which is an update of a bag of words is introduced as local–global feature bag-of-visual words is used for the classification of the high spatial resolution images and the results are promising
7	[26]	2016	A new approach in multi-bag-of-visual words, by adding multiple steps for the feature extraction and separating them at different levels is considered which outperforms the native bag-of-visual words
8	[27]	2016	They proposed a new method in which the feature extraction method is used before the fisher kernel coding method, and the results achieved for the images of high spatial resolution are great
9	[28]	2016	For overcoming the gap which exists in the semantic analysis of the satellite images of high spatial resolution, a new algorithm Dirichlet-derived multiple topic models were proposed and feature extraction was performed by eliminating various factors like probabilistic gap, topography understanding problem
10	[29]	2016	In the paper, the authors combined different aspects of the images which are of high spatial resolution and proposed a hierarchical structure–color binary partition tree (CBPT) so that with the fusion more great features could be extracted
11	[30]	2016	The features of the images are learned by combining one weak learner and one supervised strong learner algorithm and with this combination stacked sparse autoencoder was used for the outcome of different features
12	[31]	2016	A mixture of a bag-of-visual words with the combination of fisher-coding, a new layer was created for a better understanding of the land images called hierarchical coding vectors. They tested on two datasets 21-Class Land use and the other one as RSSCN7 with great accuracies reaching up to 91%
13	[32]	2016	For classifying the features of the images, a spatial pooling method was used on datasets UIUC-Sports, 21-Land use, Scene 15, and great results were achieved
14	[33]	2016	A comparison of seven publicly available datasets was performed for computing the retrieve time, changing the object detection, and all the results are compared

(continued)

Table 1 (continued)

Sr. No.	Author	Year	Findings
15	[34]	2015	They stated seven different methods for the better remote sensing of the satellite images and their applications
16	[35]	2015	For selecting more useful features in the image deep belief network was proposed, and the results achieved on the 21-Land use dataset are promising
17	[36]	2015	For the less expert in the field of remote sensing images, for extracting the features and making a cluster of the important feature a new method called the active clustering method was introduced which can effectively annotate the images
18	[37]	2015	An improved method for showing the multi-correlation method was introduced, and it performed better than previous methods like a bag-of-visual words
19	[38]	2015	For extracting the land used images feature, more efficiently a new method compressive sensing fusion was introduced and the accuracy attained is much better than the previous results
20	[39]	2013	A new method helped compensate the cost and computational time for the frequency domain processing of the images by including GPUs and CPUs for the multiprocessing of the data
21	[40]	2013	A method for creating the datasets from the remote sensing earth observation satellites for the development of the urban areas more practically is discussed
22	[41]	2013	For efficiently detecting and analyzing the landslide in remote areas, two methods were used, namely bag-of-visual words combined with probabilistic latent semantic analysis, and for classifying, the results k-nearest neighbor is used
23	[42]	2011	Extraction of the information from hyperspectral images in gigabytes and studying several different aspects of the satellite images from NASA
24	[43]	2011	For the prevention of landslides, a new and efficient method with 77.77% accuracy is proposed. Proper outline segmentation and high-resolution images of the sensors namely resources at-1 linear imaging and self-scanning sensor IV were used
25	[44]	2011	A new technique of object-oriented image analysis is used for a better understanding of the already affected by landslides like Italy, France, and very high-resolution images of these affected areas from datasets Quickbird, aerial photographs were used, and stable results were achieved with accuracy reaching up to 87%

starting half of it and the remaining layers are unfrozen and ready to be trained on a custom dataset. The simple architecture of the pre-trained model is shown Fig. 2.

Fig. 1 Architecture of the
proposed model

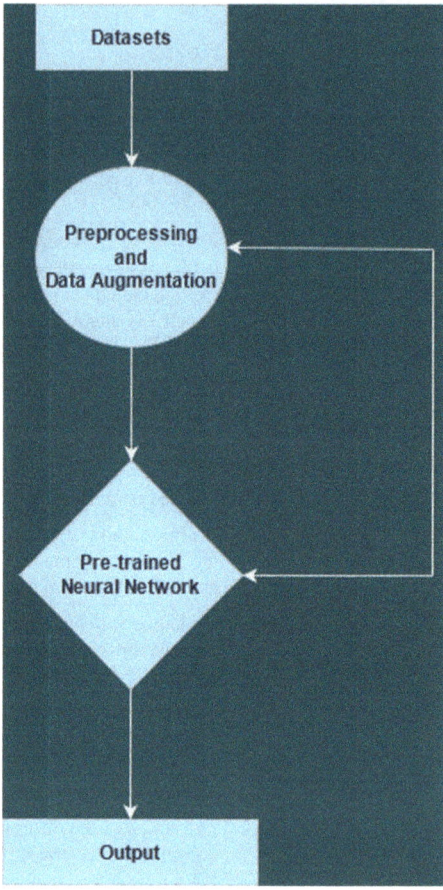

Table 2 Dataset table

Name of the dataset	Image size	Total classes	Year
WHU-RS-19	600×600	19	2010
SIRI-WHU	200×200	12	2016
AID	600×600	30	2017
Brazilian coffee scene	64×64	2	2005
NWPU-RESISC45	256×256	45	2017

3.3   The Architecture of the VGG16 Network

In the field of computer vision, the VGG network has achieved a benchmark place
while classifying several datasets. Its different versions came after its first successful
benchmark result on the ImageNet dataset in 2014. The basic architecture of VGG
is based on CNN architecture and every model which is pre-trained accepts image

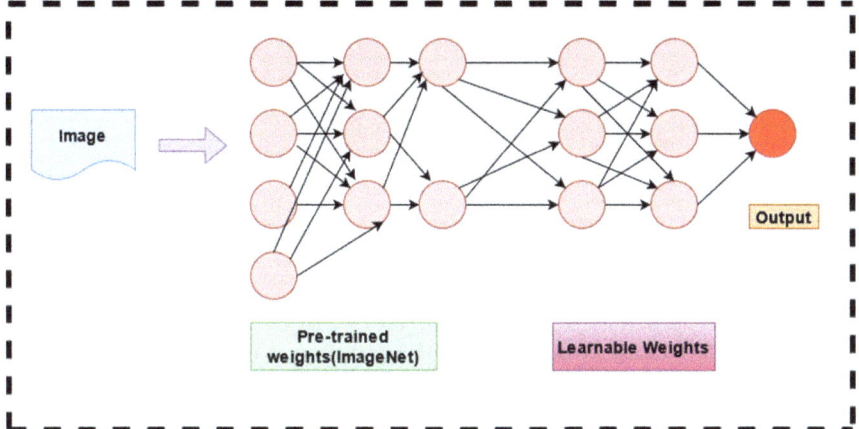

Fig. 2 Basic architecture of a pre-trained model

size which is quite fixed and VGG16 is 224 × 224. There is VGG16, VGG18, and VGG19. The basic architecture is the same consisting of several convolution layers, pooling layers, and finally fully connected dense layers which are responsible for the output. In VGG16, the idea is the same in which the head or the dense layers are not considered and we use our custom head and then a classifier for the classification. The basic architecture of VGG16 is shown in the Fig. 3.

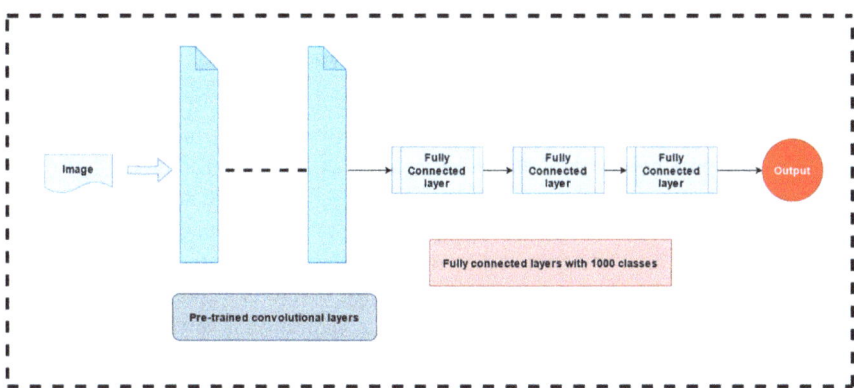

Fig. 3 Basic architecture of a VGG16 model

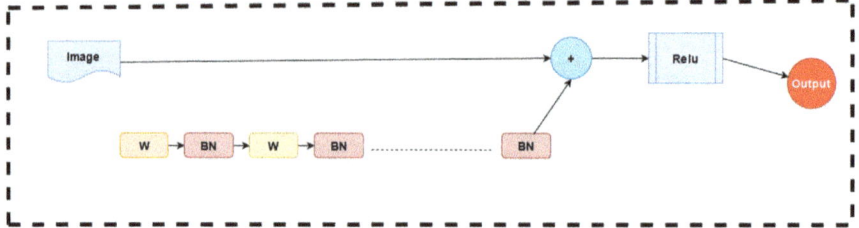

Fig. 4 Basic architecture of a ResNet50 model

3.4 Architecture of ResNet50

There are many architectures of the ResNet model which are inspired by the basic architecture consisting of several residual blocks with batch normalization layers in between them. There are 16 ResNet blocks and 1 convolutional layer with 1 dense layer aggregating to a total of 50 layers. The intuition for this model is that sometimes when there are many hidden layers and other layers, then it becomes difficult for getting the exact output that we require. So, to solve this problem one layer of input is passed to the output neuron with a softmax activation function. By using this method, the network now knows what to predict and the network performs better and we could have as many layers in between as much we want. The basic architecture of ResNet50 is shown in Fig. 4.

3.5 The Architecture of Xception Network

The Xception network is a type of extreme InceptionV3 [48] model. These both have the same parameters, the only difference is that the Xception model has more optimized parameters and the results obtained are also better in several cases. The Xception network works on depthwise separable convolution. It covers spatial characters of the image first, and then at the output, it again uses depthwise convolution. In both operations, ReLU is used for nonlinearity. There is a total of 36 convolutions structured into 14 more modules, and each convolution is surrounded by linear residual modules except the first and the last. The basic architecture of the Xception network is shown in Fig. 5.

4 Results

In the implementation of datasets, the VGG16 model performed better than ResNet50 and Xception models. There is a small difference in the performance but VGG16

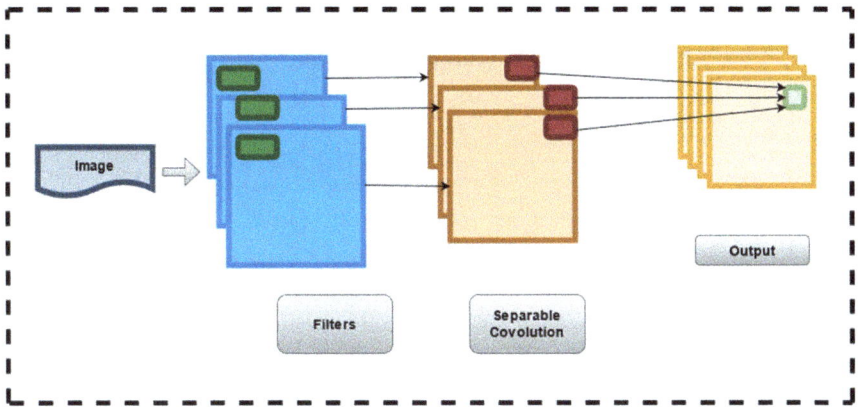

Fig. 5 Basic architecture of Xception network

shows promising results. The performance of the models is summarized in the tabular form and is shown in Tables 3, 4, 5, and 6.

The WHU-RS-19 different parameters, accuracy, and kappa index are shown in Table 2.

The SIRI-WHU dataset different parameters, accuracy, and kappa index are shown Table 3.

Table 3 Models performance on WHU-RS-19

Model	Kappa index	Accuracy	Loss	Best optimizer
VGG16	0.95	0.98	0.02	Adamax
ResNet50	0.89	0.81	0.04	Adamax
Xception	0.88	0.79	0.05	Adam

Table 4 Models performance on SIRI-WHU

Model	Kappa index	Accuracy	Loss	Best optimizer
VGG16	0.88	0.79	0.01	Adamax
ResNet50	0.85	0.76	0.01	Adamax
Xception	0.81	0.74	0.03	Adamax

Table 5 Models performance on Brazilian coffee scene

Model	Kappa index	Accuracy	Loss	Best optimizer
VGG16	0.94	0.85	0.02	Adam
ResNet50	0.92	0.79	0.04	Adamax
Xception	0.91	0.81	0.04	Adam

Table 6 Models performance on NWPU-RESISC45

Model	Kappa index	Accuracy	Loss	Best optimizer
VGG16	0.96	0.99	0.04	Adamax
ResNet50	0.85	0.79	0.05	Adam
Xception	0.88	0.81	0.04	Adamax

Also, the graphs showing accuracy and loss are shown in Figs. 6, 7, 8, and 9.

On the WHU-RS-19 dataset, the maximum accuracy reached is 0.98 with the kappa index 0.95 whose graphical representation is shown in Fig. 6.

On the SIRI-WHU dataset, the maximum accuracy reached is 0.79 with the kappa index 0.88 whose graphical representation is shown in Fig. 7.

On the Brazilian Coffee scene dataset, the maximum accuracy reached is 0.85 with the kappa index of 0.94 whose graphical representation is shown in Fig. 8.

On the NWPU-RESISC45 dataset, the maximum accuracy reached is 0.99 with the kappa index of 0.96 whose graphical representation is shown in Fig. 9.

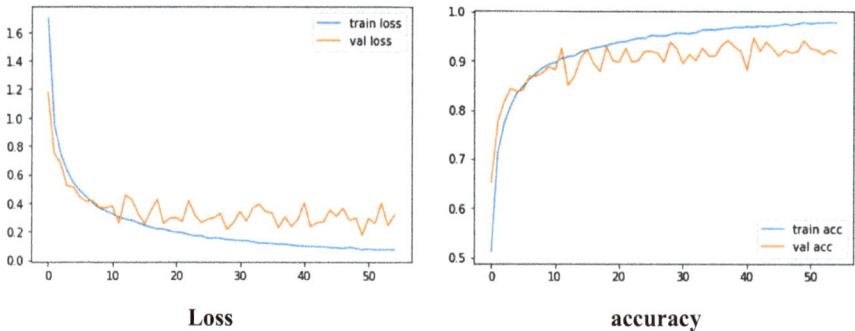

Fig. 6 Loss and accuracy graph of VGG16

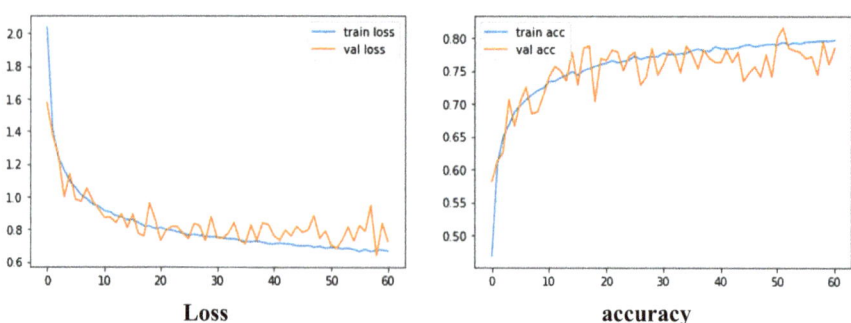

Fig. 7 Loss and accuracy graph of VGG16

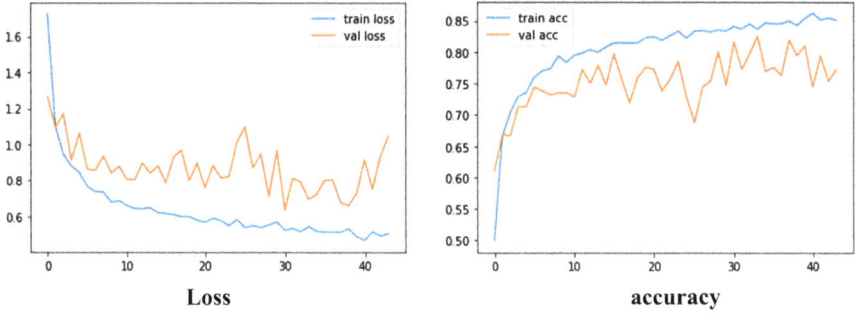

Fig. 8 Loss and accuracy graph of VGG16

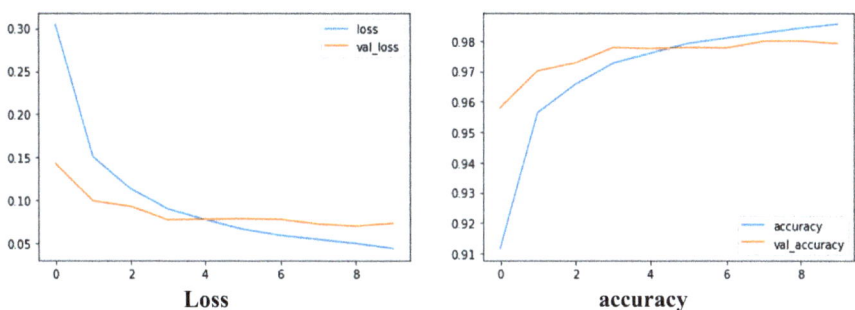

Fig. 9 Loss and accuracy of VGG16

There are not many research papers in which transfer learning is used for the remote sensing satellite images on the datasets which we have used. In the future, if more work is done, then our results could be improved more. One paper in which VGG16 is used for the WHU-RS-19 dataset but using a different algorithm [49] has 96.5% accuracy and our model has 98% accuracy.

5 Conclusion and Future Scope

The advancement in remote sensing images for scene classification is changing very rapidly. In this paper, the transfer learning is applied and the results achieved are promising, and after comparing them with different papers, our results are better. The use of pre-trained models like VGG16, ResNet50, and Xception network proved useful in the classification purpose of the satellite images. We could use certain more different pre-trained models like GoogleNet, Inception Network, and many more and see the results.

This study could also be proceeded further by a fusion of pre-trained models and GANs. The results could be measured, and more generalized results could be obtained, and also, they could give more promising results than this study. Also, different datasets could be used and results could be compared. As the datasets are huge, measures could be taken into account and the computational cost and time could be optimized as not everyone has access to hardcore machines for deep learning.

References

1. Albawi S, Mohammed TA, Al-Zawi S (2018) 'Understanding of a convolutional neural network'. In: Proceedings of 2017 international conference on engineering and technology, ICET 2017. Institute of electrical and electronics engineers Inc., pp 1–6. https://doi.org/10.1109/ICEngTechnol.2017.8308186
2. Ji M, Jensen JR (1999) Effectiveness of subpixel analysis in detecting and quantifying urban imperviousness from landsat thematic mapper imagery. Geocarto Int 14(4):33–41. https://doi.org/10.1080/10106049908542126
3. Blaschke T (2010) 'Object based image analysis for remote sensing'. ISPRS J Photogrammetry Remote Sens. Elsevier, pp 2–16. https://doi.org/10.1016/j.isprsjprs.2009.06.004
4. Cheng G et al (2015) Effective and efficient midlevel visual elements-oriented land-use classification using VHR remote sensing images. IEEE Trans Geosci Remote Sens 53(8):4238–4249. https://doi.org/10.1109/TGRS.2015.2393857
5. Cheng G, Han J (2016) 'A survey on object detection in optical remote sensing images'. ISPRS J Photogrammetry Remote Sens. Elsevier B.V., 11–28. https://doi.org/10.1016/j.isprsjprs.2016.03.014
6. Tong XY et al (2020) Exploiting deep features for remote sensing image retrieval: a systematic investigation. IEEE Trans Big Data 6(3):507–521. https://doi.org/10.1109/TBDATA.2019.2948924
7. Bank D, Koenigstein N, Giryes R (2020) 'Autoencoders', Machine learning: methods and applications to brain disorders. pp 193–208. Available at: http://arxiv.org/abs/2003.05991 (Accessed: 30 March 2021)
8. Mishra M, Srivastava M (2014) 'A view of artificial neural network'. In: 2014 international conference on advances in engineering and technology research, ICAETR 2014. Institute of electrical and electronics engineers Inc. https://doi.org/10.1109/ICAETR.2014.7012785
9. Sherstinsky A (2018) 'Fundamentals of recurrent neural network (RNN) and long short-term memory (LSTM) network'. Physica D 404. https://doi.org/10.1016/j.physd.2019.132306
10. Kohonen T (1990) The self-organizing map. Proc IEEE 78(9):1464–1480. https://doi.org/10.1109/5.58325
11. Salakhutdinov R, Hinton G (2009) Deep boltzmann machines. PMLR. Available at: http://proceedings.mlr.press/v5/salakhutdinov09a.html (Accessed: 30 March 2021)
12. Goodfellow I et al (2020) Generative adversarial networks. Commun ACM 63(11):139–144. https://doi.org/10.1145/3422622
13. Zhu J-Y et al (2017) 'Unpaired image-to-image translation using cycle-consistent adversarial networks'. Proceedings of the IEEE international conference on computer vision, 2017–October. pp 2242–2251. Available at: http://arxiv.org/abs/1703.10593 (Accessed: 30 March 2021)
14. Deng J, et al (2010) 'ImageNet: a large-scale hierarchical image database'. In: Institute of electrical and electronics engineers (IEEE), pp 248–255. https://doi.org/10.1109/cvpr.2009.5206848
15. Simonyan K, Zisserman A (2015) 'Very deep convolutional networks for large-scale image recognition'. In: 3rd international conference on learning representations, ICLR 2015—conference track proceedings. International conference on learning representations, ICLR. Available at: http://www.robots.ox.ac.uk/ (Accessed: 30 March 2021)

16. Oyelade ON, El A, Ezugwu S (no date) 'Application of a novel and improved VGG-19 network in the detection of workers wearing masks recent citations a state-of-the-art survey on deep learning methods for detection of architectural distortion from digital mammography'. J Phys: Conf Ser Paper—Open access. https://doi.org/10.1088/1742-6596/1518/1/012041

17. He K, et al (2016) 'Deep residual learning for image recognition'. In: Proceedings of the IEEE Computer society conference on computer vision and pattern recognition. IEEE computer society. pp 770–778. https://doi.org/10.1109/CVPR.2016.90

18. Chollet F (2017) 'Xception: deep learning with depthwise separable convolutions'. In: Proceedings—30th IEEE conference on computer vision and pattern recognition, CVPR 2017. Institute of electrical and electronics engineers Inc., pp 1800–1807. https://doi.org/10.1109/CVPR.2017.195

19. Szegedy C, et al (2015) 'Going deeper with convolutions'. In: Proceedings of the IEEE computer society conference on computer vision and pattern recognition. IEEE computer society. pp 1–9. https://doi.org/10.1109/CVPR.2015.7298594

20. Cheng G et al (2020) Remote sensing image scene classification meets deep learning: challenges, methods, benchmarks, and opportunities. IEEE J Sel Top Appl Earth Observations Remote Sens 13:3735–3756. https://doi.org/10.1109/JSTARS.2020.3005403

21. Abdelhack M (2020a) 'A comparison of data augmentation techniques in training deep neural networks for satellite image classification'. https://doi.org/10.5281/zenodo.3268451

22. Abdelhack M (2020b) 'An open-source tool for hyperspectral image augmentation in tensorflow'. Available at: http://arxiv.org/abs/2003.13502 (Accessed 18 Dec 2020)

23. Gu Y, Wang Y, Li Y (2019) 'A survey on deep learning-driven remote sensing image scene understanding: scene classification, scene retrieval and scene-guided object detection'. Appl Sci (Switzerland). MDPI AG. https://doi.org/10.3390/app9102110

24. Cheng G, Han J, Lu X (2017) 'Remote sensing image scene classification: benchmark and state of the art'. Proceedings of the IEEE. Institute of electrical and electronics engineers Inc., pp 1865–1883. https://doi.org/10.1109/JPROC.2017.2675998

25. Zhu Q et al (2016) Bag-of-visual-words scene classifier with local and global features for high spatial resolution remote sensing imagery. IEEE Geosci Remote Sens Lett 13(6):747–751. https://doi.org/10.1109/LGRS.2015.2513443

26. Zhao L, Tang P, Huo L (2016) Feature significance-based multibag-of-visual-words model for remote sensing image scene classification. J Appl Remote Sens 10(3):035004. https://doi.org/10.1117/1.JRS.10.035004

27. Zhao B, Zhong Y, Zhang L et al (2016) The fisher kernel coding framework for high spatial resolution scene classification. Remote Sens 8(2):157. https://doi.org/10.3390/rs8020157

28. Zhao B, Zhong Y, Xia GS et al (2016) Dirichlet-derived multiple topic scene classification model for high spatial resolution remote sensing imagery. IEEE Trans Geosci Remote Sens 54(4):2108–2123. https://doi.org/10.1109/TGRS.2015.2496185

29. Yu H et al (2016) A color-texture-structure descriptor for high-resolution satellite image classification. Remote Sens 8(3):259. https://doi.org/10.3390/rs8030259

30. Yao X et al (2016) Semantic annotation of high-resolution satellite images via weakly supervised learning. IEEE Trans Geosci Remote Sens 54(6):3660–3671. https://doi.org/10.1109/TGRS.2016.2523563

31. Wu H et al (2016) Hierarchical coding vectors for scene level land-use classification. Remote Sens 8(5):436. https://doi.org/10.3390/rs8050436

32. Liu Y et al (2016) Adaptive spatial pooling for image classification. Pattern Recogn 55:58–67. https://doi.org/10.1016/j.patcog.2016.01.030

33. Cui S (2016) Comparison of approximation methods to kullback-leibler divergence between gaussian mixture models for satellite image retrieval. Remote Sens Lett 7(7):651–660. https://doi.org/10.1080/2150704X.2016.1177241

34. Gomez-Chova L et al (2015) Multimodal classification of remote sensing images: a review and future directions. Proc IEEE 103(9):1560–1584. https://doi.org/10.1109/JPROC.2015.2449668

35. Zou Q et al (2015) Deep learning based feature selection for remote sensing scene classification. IEEE Geosci Remote Sens Lett 12(11):2321–2325. https://doi.org/10.1109/LGRS.2015.247 5299

36. Xia G-S et al (2015) Accurate annotation of remote sensing images via active spectral clustering with little expert knowledge. Remote Sens 7(11):15014–15045. https://doi.org/10.3390/rs7111 5014

37. Qi K et al (2015) Land-use scene classification in high-resolution remote sensing images using improved correlatons. IEEE Geosci Remote Sens Lett 12(12):2403–2407. https://doi.org/10. 1109/LGRS.2015.2478966

38. Mekhalfi ML et al (2015) Land-use classification with compressive sensing multifeature fusion. IEEE Geosci Remote Sens Lett 12(10):2155–2159. https://doi.org/10.1109/LGRS.2015.245 3130

39. Cantalloube HMJ, Nahum CE (2013) Airborne SAR-efficient signal processing for very high resolution. Proc IEEE 101(3):784–797. https://doi.org/10.1109/JPROC.2012.2232891

40. Gamba P (2013) Human settlements: a global challenge for EO data processing and interpretation. Proc IEEE 101(3):570–581. https://doi.org/10.1109/JPROC.2012.2189089

41. Cheng G et al (2013) Automatic landslide detection from remote-sensing imagery using a scene classification method based on boVW and pLSA. Int J Remote Sens 34(1):45–59. https://doi. org/10.1080/01431161.2012.705443

42. Plaza A et al (2011) Parallel hyperspectral image and signal processing. IEEE Signal Process Mag 28(3):119–126. https://doi.org/10.1109/MSP.2011.940409

43. Martha TR, et al. (2011) 'Segment optimization and data-driven thresholding for knowledge-based landslide detection by object-based image analysis'. In IEEE transactions on geoscience and remote sensing. pp 4928–4943. https://doi.org/10.1109/TGRS.2011.2151866

44. Stumpf A, Kerle N (2011) Object-oriented mapping of landslides using Random forests. Remote Sens Environ 115(10):2564–2577. https://doi.org/10.1016/j.rse.2011.05.013

45. WHU-RS19-1|Kaggle (no date). Available at: https://www.kaggle.com/sunray2333/whurs191 (Accessed: 30 March 2021)

46. AID: A benchmark dataset for performance evaluation of aerial scene classification (no date). Available at: https://captain-whu.github.io/AID/ (Accessed: 30 March 2021)

47. Brazilian Coffee Scenes Dataset (no date). Available at: http://www.patreo.dcc.ufmg.br/2017/ 11/12/brazilian-coffee-scenes-dataset/ (Accessed: 30 March 2021)

48. Szegedy C, et al (2016) 'Rethinking the inception architecture for computer vision'. In: Proceedings of the IEEE computer society conference on computer vision and pattern recognition. IEEE computer society. pp 2818–2826. https://doi.org/10.1109/CVPR.2016.308

49. Hung S-C, Wu H-C, Tseng M-H (2020) Remote sensing scene classification and explanation using RSSCNet and LIME. Appl Sci 10(18):6151. https://doi.org/10.3390/app10186151

Damage Identification in High-Rise Buildings Using Deep Learning Techniques

Vishal Pandit, Smita Kaloni, Shagun Sharma, and Ghanapriya Singh

Abstract In the last few years, structural damage identification has been considered a strong area of research by the structure engineers community. Vibration-based damage identification using machine learning algorithms has shown a tremendous advantage over other damage identification procedures. Although existing damage detection methods have been adapting machine learning principles, the majority of machine learning-based procedures extract fixed features. Depending on the framework under investigation, their output differs significantly across different data patterns. During the training process, deep learning technique convolution neural networks (CNNs) will fuse and adjust two key sets of an assessment assignment (feature extraction and classification) into a sole learning block that is manually identified in advance. This capability not only improves classification accuracy, but it also improves computational performance. In the proposed work, a damage identification study has been carried out using deep learning CNN algorithms. The performance of two CNN models one-dimensional (1D) CNN and two-dimensional (2D) CNN is discussed. This approach is verified using an analytical model of G+20 storey building modelled in FEM-based software. Probability of damage (POD) is used as the damage indicator in the analysis. It has been witnessed that the presentation of 1D CNN is superior to 2D CNN for damage diagnosis.

Keywords Structural health monitoring (SHM) · Damage identification (DI) · Convolution neural network (CNN) · Machine learning algorithm (MI) · Deep learning algorithms (DL)

V. Pandit · S. Kaloni (✉)
Department of Civil Engineering, National Institute of Technology, Srinagar (Garhwal), Uttarakhand 246174, India

S. Sharma · G. Singh
Department of Electronics Engineering, National Institute of Technology, Srinagar (Garhwal), Uttarakhand 246174, India
e-mail: shagun.ece17@nituk.ac.in

© The Author(s), under exclusive license to Springer Nature Singapore Pte Ltd. 2022   393
D. Gupta et al. (eds.), *Proceedings of Data Analytics and Management*,
Lecture Notes on Data Engineering and Communications Technologies 90,
https://doi.org/10.1007/978-981-16-6289-8_33

1 Introduction

Damage to structures can have an adverse effect on their stiffness and stability, decreasing their life efficiency. Various structural damage detection procedures were proposed during the past decades in order to accomplish an automated structural health monitoring (SHM) [1, 2]. Vibration-based structural damage detection methods, which are divided into parametric and nonparametric methods [3–6], have shown great promise in assessing the state of civil structures. Nonparametric global damage detection methods evaluate the structure's vibration response using statistical methods for damage recognition [7, 8]. Several machine learning processes have recently been employed in the development of more effective parametric and nonparametric global damage identification techniques [9]. Machine learning-based structural damage diagnosis methods basically consider two steps: (1) Feature extraction and (2) classification are the first two steps. Various classifiers models have been developed in past for further implementation of machine learning algorithms [10–13]. In phase 1, the measured acceleration signals are analysed for certain hand-crafted characteristics. The extracted features are then used as inputs in phase 2 to train a classifier. The modal parameters of the controlled structure are derived from the dynamic responses and called features extracted in parametric machine learning techniques. Nonparametric machine learning-based approaches, on the other hand, use techniques like simple statistical analysis (using the signal's mean and variance) to extract features [7–9, 13]. In both parametric and nonparametric machine learning-based methods, several classifiers have been used for the classification stage [14–17]. The performance of machine learning-based structural damage detection techniques is expected to be largely dependent on the identified features as well as the classifier was chosen. As a result, the extracted features should be carefully chosen to capture the analysed signals' main characteristic state. In addition, depending on the type of extracted features, a suitable classifier must be used to properly categorize them.

Recent research has shown that 1D and 2D CNNs can outperform traditional methods [18, 19]. Another distinguishing feature of CNNs is their adaptive architecture, which incorporates feature extraction tasks and damage classification procedures into a single learning block, allowing CNNs to extract and learn the best features from the data simultaneously. In this study, damage identification of civil engineering structures has been carried out using deep learning CNN algorithms. A comparative study is performed on the performance of 1D-CNN, 2D-CNN and further compared with LSTM techniques for damage identification.

The remainder of this paper is as follows: Sect. 2 provides the details of the methodology proposed for damage identification. Section 3 describes the deep learning models used in this study for damage identification purpose. Sect. 4 consists of result and discussion followed by the conclusion in Sect. 5.

2 Methodology

As presented above, the damage diagnosis procedure was established and validated considering a G+20 storey building. In this section, an overview of the analytical building model is provided followed by the type of analysis involves and the details of the dataset created for the damage identification study purpose.

Modelling of G+20 Storey Building

Any damage to civil structures causes loss of life and property. Designing of any such structures is based on Indian standard guidelines following Indian Standard codes IS 456-2000, IS 1893-2016 and IS 13920-2016. In this study, G+20 building is assumed for the damage identification. The building is designed using FEM-based software [20]. The dynamic analyses of structures are performed for different earthquake forces. To consider different damage conditions, earthquake forces are scaled to different labels. The building consists of ground floor, 19 upper floors and one terrace accounting every kind of loading in this and load combination as per I using Indian standard codes. Strong column-weak beam concept was adopted for design purpose. Seismic loads will be considered acting in the two horizontal directions, and the vertical direction is neglected. The building will be used for residential purpose the wall size is 230 mm, and plaster is 35 mm on both sides. The grade of concrete M45 and steel Fe 415 is used. It is also noted that the structure is in seismic zone V. The different types of load cases are considered while performing analysis on structure. The lateral force resistance is normally concentrated in a small portion of the building framing elements in modern style designs. The building's remaining framing, also known as "non-participating" or "gravity-only" framing, is not included in the construction lateral resistance. Despite this, the "gravity-only" framing must be built to remain secure in the face of lateral drifts anticipated for future earthquakes.

To consider the damage cases in the building, the building is analysed using pushover analysis procedure. The analysis is performed as per FEMA-356 [21]. In this modelling, two pushover analysis is performed in both X- and Y-directions on mode 1 and mode 2, respectively, with initial displacement of 5 m given at joint no. 67. It has been observed during pushover analysis (X) hinges are developed at joint 1 and joint 89 in 1058th steps which at support and condition of these hinges are within range of B and IO. However, during pushover analysis (Y) hinges are developed at joint 1 and joint 331 at 103th steps which are at support. The pushover method is basically involving continued and significant use of capacity curve, and the potential of the method is to assist the development of retrofitting strategies. The pushover nonlinear static analysis is done to represent the lateral force-resisting capacity of structures. Figures 1 and 2 describe the formation of X and Y during pushover nonlinear static analysis that is done to represent the lateral force-resisting capacity of structure. The pushover analysis consists of the application of gravity loads and a representative lateral load pattern. The most commonly and precisely used approach for predicting the force and deformation demands at different components

Fig. 1 Push *X* hinges
formation

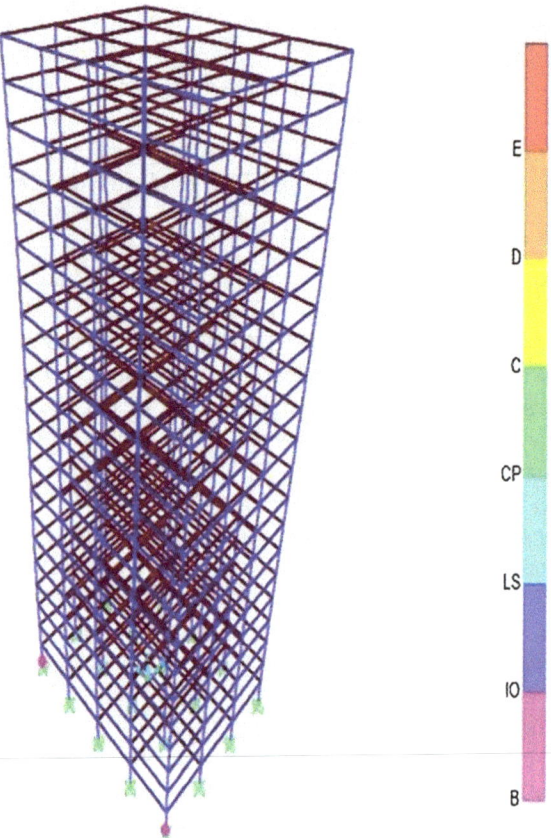

of the system is inelastic time-history analysis. The inelastic time-history analysis
is difficult to use for earthquake performance evaluation due to the calculation time,
the time needed for input planning and the time required to interpret the output.

FEM-based software is used in performing the nonlinear dynamic time-history
analysis on the three-dimensional model of the G+20 building, with the different
earthquake ground motions. There are two methods of nonlinear time-history analysis
which are fast nonlinear analysis and direct integration time history. Different results
of time histories acceleration and displacements features are extracted from different
earthquake analyses.

Data Set

Creation of dataset is one important part in the nonparametric damage identification
techniques. In this study, the training is performed in 60 samples of dataset and
validation is done on 16 samples of dataset. The first set of data is recorded by
measuring the vibration signature from undamaged condition, while the second set
is recorded when the structure is fully damaged condition, the resulting data for fully

Fig. 2 Push Y hinges formation

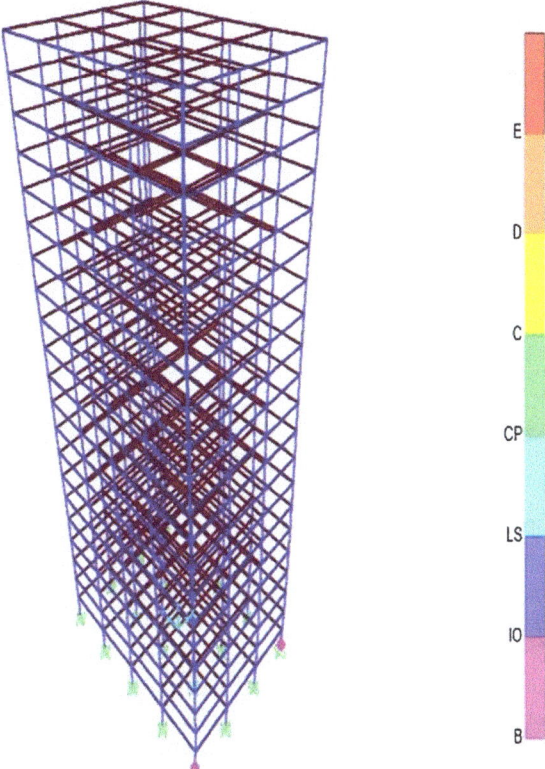

undamaged and damaged cases. The data, after being categorized as undamaged or damaged, is then provided with labels for the same categories to be fed to a supervised learning algorithm. The data was appended from all the cases of earthquakes to form a collective database of damaged and undamaged accelerometer readings of all the joints wherein, accelerometer readings of 4001 timestamps for X- and Y-axes, respectively, were recorded in damaged cases and 440 accelerometer readings of 4001 timestamps for X- and Y-axes, respectively, were recorded in undamaged cases for all the joints. This segregated data was now compiled together to form a pool of readings for a particular joint and then was split into the training and testing sets. This produced randomly shuffled cases of recordings for damaged and undamaged readings in a particular joint and since data of every joint went through the same procedure, the split was done in a similar fashion on every joint, therefore reducing any chance of biased data in any of the joint that might have affected the training algorithm and features extracted from the data for a particular joint. Hence, we ensured that readings for any joint data have similar number of cases for damaged and undamaged recordings. This data was now reshaped as an array of shape (number of samples, 4001, 2). Purpose of choosing this shape was the fact that our data comprised of readings from 2 different axes (X & Y); hence, we fed the data into the 1-D CNN as 2

Table 1 Earthquake details

Location	Magnitude	Hypocentre	PGA iX	PGA iY	PGAi (Vertical)
CHAMOLI	7	21.7	−2.42	2.48	2.89
UTTARKASHI	7	34	−2.37	3.04	−1.93
BERLONGFER	5.9	120.8	−0.885	−0.706	0.452
BAMUGAO	5.9	146.3	0.194	−0.194	0.186
BARAKOT	7	55.8	1.16	1.15	−0.992
CHERAPUNJI	7.2	353	−0.536	−0.511	0.23
KOOMBER	6	231.8	0.36	0.482	0.263
GHANSALI	6.5	39.3	−2.37	3.04	−1.93
SAN FERNANDO	6.6	58.8	0.654	−0.494	0.331
KANGRA	5.3	34.8	−1.45	−1.09	0.708
NORTHRIDGE	6.69	19.48	1.048	1.048	1.048

channels of 4001 readings each to extract features from both the axes simultaneously which is denoted by the third argument of the shape parameter. The data was fed in the form of an array of aforementioned shape to the network whose structure has been defined in Table 1 and was used for each joint independently. Therefore, a total of 440 CNNs were trained for 440 joints. Once the training was over, the predictions from each joint's CNN were evaluated and taken average to get the probability of damage for the overall building. After completion of training, the classifier CNN classifies any input frame measured at the corresponding accelerometer as fully undamaged or damaged. The POD is measure thereafter from n accelerometers. After computing all PoD values corresponding to the n accelerometers, the overall structural score can be obtained simply by computing the average PoD value as given by Eq. (1).

$$PoD_{avg} = \frac{PoD_1 + PoD_2 + PoD_3 + \ldots \ldots \ldots \ldots PoD_n}{n} \qquad (1)$$

3 Deep Learning Algorithms

1. *One-Dimensional Convolution Neural Network (1D-CNN)*

Convolutional neural network (CNN) models were designed primarily for image classification, and they take a two-dimensional input representing an image's pixels as input. This CNN was extended to apply on one-dimensional dataset. The deep learning model CNN extracts feature from sequential dataset and extracts the internal features of the sequence. The activation function is able to account for nonlinear transformations. In comparison to sigmoid and tanh activation rectifier linear unit (ReLU) performs better as sigmoid and tanh activation function tends to saturate.

In this study, different numbers of layers are considered and all neural networks with ReLU activation function in CNNs layers and output layer were experimented. The hidden layer has ReLU (positive linear) function while the output layer has sigmoid (sigm) function. Both activation functions are shown in figures below. Stochastic gradient descent optimizer "Adam" has been used that can handle sparse gradients on noisy problems.

It has been observed that 1D CNN is particularly good at extracting features from a fixed-length segment of a larger dataset, where the location of the feature in the segment is less significant. In vibration-based structural health monitoring, 1D CNNs are a relatively new procedure that promises a reliable and real-time solution.

2. *Two-Dimensional Convolutional Neural Network (CD-CNN)*

In 2D CNN network, first the dataset is to be produced which is in the form of spectrograms that will be used as the input for training purpose. The images produced possess a good resolution of width and height of 432 and 288 pixels, respectively, having an approximate size of 33 KB. These spectrograms generation is the first step for 2D CNN.

The spectrogram datasets are divided into two-axis datasets which is X-axis and Y-axis dataset. Each axis consists of spectrograms of 440 joints, and those joints are further classified under two categories which are damage and undamaged. In training dataset, there are 32 spectrograms in damaged case. In the same training dataset, there are 28 spectrograms in undamaged category. In the validation dataset, there are 5 spectrograms in damage and 11 spectrograms in undamaged case. The input shape of the 2D convolutional neural network is represented in Fig. 3, and the values of input shape are $432 \times 288 \times 3$.

The input layer also has an activation function known as "ReLU" activation function which works more efficiently than other activation functions. The most important task is feature extraction in which multiple features are extracted without losing important information. Once the input is done the next process is pooling its main idea is downsampling so that the complexity is reduced for further layers. Max-pooling is one of the most common types of pooling method, it divides images into subregion rectangles, and it only provides the maximum value output of that sub-region. The size of max-pooling layer is 2×2. When the pooling is done in the top left pink area, it moves 2 and attention on the top right part. This means stride 2 is used for max-pooling as shown in Fig. 4.

Fig. 3 Three-dimensional input representation of 2-D CNN

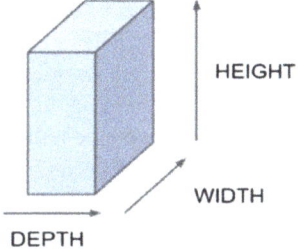

Fig. 4 Max-pooling process

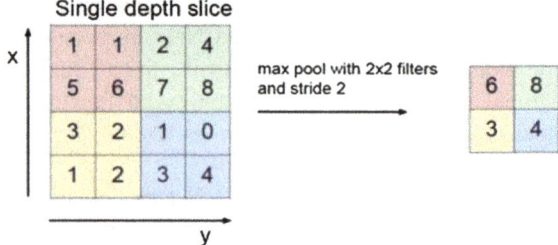

Single depth slice

There is a "flatten" layer between the convolutional and fully linked layers. A two-dimensional matrix of features is flattened into a vector, which may then be input into a fully connected neural network classifier. The main disadvantage of the dense layer (completely connected layer) is that it has a large number of parameters that require complicated computational in order to train for damage detection. Finally, the output dense layer has dimension 1 and uses the sigmoid activation function.

3. *Long Short-Term Memory Network (LSTM)*

Long-term dependency problems can be solved using LSTM networks since they can remember knowledge over long periods of time. The damage detection using LSTM involves the following steps:

(a)   Acquisition of raw acceleration data
(b)   Pre-process the multipoint acceleration data
(c)   Feature extraction
(d)   Training using damage-sensitive features.

Proving test data as input to the system and using the trained LSTM model, it is possible to identify damage patterns of the structure. When the output of the framework is "0", it means there is no damage in the structure. While it indicates damage is present in the structure when the output is "1".

4   Results and Discussion

Despite the fact that 1D CNNs were just launched a few years ago, recent research has demonstrated that compact 1D CNNs can outperform both classic and conventional methodologies with the correct systematic approach. In this work, we concentrate on compact 1D CNNs in particular and provide a detailed overview of their applications in different engineering fields. Compact 1D CNNs have the distinct benefit of being relevant to applications in which marked data for training is very limited, and real-time implementation is required in comparison with 2D CNN. Furthermore, traditional 2D CNNs can only process 2D signals, necessitating an additional 1D to 2D transformation followed by a windowing operation, all of which add time and

Table 2 Damage prediction accuracy of different neural networks

S. No.	Neural network	No of layers	Damage prediction (%)
1	1D CNN	2	77
		3	79
		4	73
2	2D CNN	2	72
3	LSTM	2	68

resources. 1D CNNs are simpler to train and have the least amount of computational complexity while achieving cutting-edge efficiency. Table 2 shows damage prediction accuracy of different neural network.

5 Conclusion

In this paper, study was performed on G+20 storey building considering various earthquake events. The dataset was created for healthy and damaged state of structure for further implementation in deep learning models. In deep learning models, three techniques were experimented, viz. 1D CNN, 2D CNN and LSTM models. It has been found that 1D CNN outperforms 2D CNN and LSTM models. Less efficiency was reported for LSTM as compared to CNN because the LSTM works on the sequential dataset and the dataset (acceleration readings) we have provided is not sequential in nature; hence, the efficiency of LSTM is less than the 1D CNN. To perform 2D CNN, spectrogram of measured response was calculated and provided as input to the deep learning model for damage identification. The accuracy of 2D CNN was lower than 1D CNN. Further, the complexity involved in the computation of 1D CNN is expressively lower than the 2D CNN since forward propagation (FP) and backward propagation (BP) in 1D CNN require simple array operations. Thus, 1D CNN is a more efficient deep learning architecture in comparison with 1D CNN and LSTM models for damage identification of structures.

References

1. Brownjohn JM (2007) Structural health monitoring of civil infrastructure. Philos Trans Royal Soc A: Math, Phys Eng Sci 365(1851):589–622
2. Catbas FN (2009) 1—structural health monitoring: applications and data analysis. Struct Health Monit Civil Infrastruct Syst
3. Chang KC, Kim CW (2016) Modal-parameter identification and vibration-based damage detection of a damaged steel truss bridge. Eng Struct 122:156–173
4. Sohn H, Farrar CR, Hunter NF, Worden K, Structural health monitoring using statistical pattern recognition techniques. J Dyn Syst Meas Control 123

5. Kaloni S, Shrikhande M (2017, Jan 9–13) Damage detection in structural system via blind source separation. In: Proceedings of 16th world conference in earthquake engineering. Santiago Chile
6. Kaloni S, Shrikhande M (2018) Seismic damage detection using blind source separation in 16th symposium on earthquake engineering. Indian Institute of Technology Roorkee, India
7. Figueiredo E, Park G, Farrar CR, Worden K, Figueiras J (2011) Machine learning algorithms for damage detection under operational and environmental variability struct. Health Monit 10:559–572
8. Gul M, Catbas FN (2011) Structural health monitoring and damage assessment using a novel time series analysis methodology with sensor clustering. J Sound Vib
9. Adeli H (2001) Neural networks in civil engineering: 1989–2000. Comput-Aided Civil Infrastruct Eng 16(2):126–142
10. Alzubi JA (2015) Optimal classifier ensemble design based on cooperative game theory. Res J Appl Sci Eng Technol 11(12):1336–1343
11. Alzubi JA, Jain R, Kathuria A, Khandelwal A, Saxena A, Singh A (2020) Paraphrase identification using collaborative adversarial networks. J Intell Fuzzy Syst 39(1):1021–1032
12. Omar AA, Alzubi JA, Mohammed A, Issa Q, Sara Al-S, Manikandan R (2020) An optimal pruning algorithm of classifier ensembles: dynamic programming approach". Neural Comput Appl
13. Acharya UR, Oh SL, Hagiwara Y, Tan JH, Adam M, Gertych A, San Tan R (2017) A deep convolutional neural network model to classify heartbeats. Comput Biol Med 89:389–396
14. Adeli H, Jiang X (2006) Dynamic fuzzy wavelet neural network model for structural system identification. J Struct Eng 132(1):102–111
15. Agarwal N, Sondhi A, Chopra K, Singh G (2021) Transfer learning: survey and classification. In: Smart innovations in communication and computational sciences, pp 145–155. Springer, Singapore
16. Liu Y-Y, Ju Y-F, Duan C-D, Zhao X-F (2011) Structure damage diagnosis using neural net-work and feature fusion. Eng Appl Artif Intell 24:87–92
17. Chun PJ, Yamashita H, Furukawa S (2015) Bridge damage severity quantification using multipoint acceleration measurement and artificial neural networks. Shock Vib
18. Avci O, Abdeljaber O, Kiranyaz S, Inman D (2019, May) Convolutional neural networks for real-time and wireless damage detection. In: Dynamics of civil structures, volume 2: proceedings of the 37th IMAC, a conference and exposition on structural dy-namics. Springer, p 129
19. Avci O, Abdeljaber O, Kiranyaz S, Inman D (2020) Convolutional neural net-works for real-time and wireless damage detection. In: Dynamics of civil structures, vol 2. Springer, Cham, pp 129–136
20. SAP2000 Integrated software for structural analysis and design, computers & structures, Inc., Berkley, CA, USA
21. FEMA-356 (2000) Prestandard and commentary for the seismic rehabilitation of buildings American society of civil engineers

AI-Driven Fraud Detection and Mitigation in e-Commerce Transactions

Iqbal Hasan and SAM Rizvi

Abstract Digital technology has created new channels of digital transaction for e-Commerce and financial services distribution companies. The COVID-19 has intensified the adoption of digital transactions. With increased load, digital transactions have become more vulnerable to frauds and have created a lucrative avenue for fraudsters. Digital transactions pose a significant challenge to enterprises and customers for having secure business and financial transactions. To enhance the ease and convenience of the customers and business enterprises, the e-Commerce and financial service providers are deploying sophisticated tools that employ data analytics, AI and machine learning techniques. However, such tools are susceptible to compromise and challenging to enterprises and customers for having secure business and financial transactions. In this paper, we present the issues and challenges in identifying spurious online payments and detecting fraud in e-Commerce transactions. We also present the AI, and data analytics techniques are employed for the detection, prevention and control of online frauds in e-Commerce and financial transactions.

Keywords Digital transaction · Online payments · Fraud detection · e-Commerce · Fraud mitigation

1 Introduction

Previously, fraud detections were mostly based on the business experience of an individual who had a long history of dealing with anomaly and fraud detection. Later on, rule-based statistical techniques were used to detect anomalies and find frauds. However, this process was lengthy and tedious with an uncertain outcome. Generally, the fraud detection process recognizes previously learned patterns from

I. Hasan (✉) · S. Rizvi
DCS, Jamia Millia Islamia, SIO and Scientist 'F', NIC, New Delhi, India
e-mail: ihasan@nic.in

S. Rizvi
e-mail: sarizvi@jmi.ac.in

the datasets using data mining and machine learning techniques. Large e-Commerce companies such as Amazon and Walmart and financial companies such as MasterCard have integrated machine learning (ML) and artificial intelligence (AI) to track and process variables such as transaction size, time and location, the device used, and purchasing data. The ML-based system assesses buyer's behaviour in each operation and makes a real-time assessment of whether a transaction is suspicious or fraudulent. These systems can detect fraud precisely with substantially reduced reconciliation costs. Business fraud detection systems using data analytics and machine learning can minimize fraud investigation time and improve detection accuracy.

In today's digital era, Internet use is growing drastically with the exchange of online information and financial transactions all around for buying and selling of goods and services [1]. From consumers to e-Commerce industries, everyone is changing their way of financial transactions with the introduction of online payments. The ease and convenience of online payments, lucrative offers, cashback, and discounts by e-Commerce sites have attracted peoples to online transactions. Thus, people use online payment systems for bill payments, buying foods, beverages, electronic equipment, tickets, and essential commodities. Payment gateways, wallets, and net banking nowadays are popular means of online transactions. Countries like India with its Digital India[1] programs and cashless transaction move at the time of demonetization have opted for digital transactions in every sector from the disbursement of relief to financial assistance [2, 3]. However, securing information is a great challenge in today's world as the threats and breaches of security in online transactions are increasing day by day. Though governments and private companies are taking various measures to prevent such security breaches and online frauds, it still remains a major concern. Although consumers are moving towards online payment regularly, it is also observed that the insecurity, fear and threats of fraud are increasing among consumers while performing online transactions. This paper discusses the various security issues, threats and frauds that threaten online payments and presents an AI-driven threat detection and prevention mechanism in e-Commerce transactions.

2 Fraud Detection Techniques

The e-Commerce and financial transaction companies are experiencing losses because of various financial frauds and scams. The digital space has created an environment for fraudsters to trick consumers, companies and financial institutions during online financial transactions. The challenge is to detect such frauds in real time with improved accuracy. Many fraud detection techniques are being used by financial institutions and e-Commerce industries to prevent and detect fraudulent activities.

---

[1] https://www.digitalindia.gov.in/.

Traditionally rule-based systems were used to detect financial fraud in online transactions; however, in recent years the machine learning-based fraud detection has received considerable attention [4].

2.1 Rule-Based Versus Machine Learning Systems in Fraud Detection

Rule-based systems involve algorithms that consider different fraud detection scenarios written by experts or fraud analysts. These systems are simple that apply specific rules to approve transactions along with manual interventions to adjust the transaction scenarios [5, 6]. However, these systems are unable to identify any implicit correlations and process real-time data streams. Since the real-time data streams may contain critical or hidden events that may be vulnerable to fraud, the rule-based systems with their legacy software may not detect such frauds or fraudulent activities. On the other hand, machine learning (ML) systems can process large data sets and can easily detect hidden patterns or correlations between user behaviour and possible fraudulent activities or operations [7, 8]. The ML-based systems process data faster with minimal human efforts. Table 1 presents the comparison between the rule-based systems and machine learning systems for fraud detection. The ML-based systems may be supervised or unsupervised depending upon the training of the system using labelled or unlabelled data. The commonly used supervised fraud detection algorithms are random forests, support vector machine, K-nearest neighbours, logistic regression, artificial neural network, and deep neural network. The unsupervised learning model processes unlabelled data and classifies them in different cluster sets for detecting the hidden relationships among different variables in data sets. The selection of a particular ML technique is based on the type of problem, size of the dataset, processing resource availability, business category, etc.

Table 1 Comparison between rule-based systems and machine learning systems for fraud detection

Rule-based system	Machine learning systems
Based on offline long-term data processing	Online real-time data processing
Catches obvious fraudulent scenarios	Find hidden and implicit correlation in data
Entails a number of manual verification steps giving bad user experience	The verification steps are reduced considerably
Facilitates limited big data analytics options	Makes full use of big data analytics
Substantial manual intervention to enumerate all possible detection rules	Facilitates automatic detection of fraud scenarios

2.2 Big Data and AI Techniques for Fraud Detection

The fraud detection system involves a variety of systems and tools to collect data such as operational systems, legacy systems, back-office systems, social media, mobile communications, and customer relationship management systems. The data could be textual contents of social media posts or the accounting entries from the enterprise resource planning (ERP) systems that require specific data mining techniques for processing and performing analytics [9]. As the data generated by existing systems are huge in volume and diverse in variety constituting big data, these data cannot be processed or analysed for fraud detection using existing techniques because of their internal limitations. With the use of big data analysis technologies, frauds can be detected and prevented. For instance, insurance fraud in the healthcare sector can be detected using business rules, text mining, anomaly detection, social network analysis and database searches [10]. In addition, "big data", which could be heterogeneous, high volume and complex, requires specific data analytics techniques for storing and processing to find out the targeted results for financial frauds [10, 11]. Big data analytics and data mining techniques provide a wide range of solutions that rely on monitoring and identifying suspicious activities based on recognizing the patterns that are indicative of any fraudulent activities such as unusual data entries, unexpected relationships between two or more records, or patterns with similar characteristics of already known frauds. Modelling such fraudulent patterns and unusual transactions from a large amount of functional application data such as supply-chain management, marketing, and social media data could be helpful to identify financial frauds. As the fraudsters are becoming smarter, the AI techniques can be helpful to fight against such fraudsters and prevent any frauds or financial crimes [12]. AI technique such as *anomaly detection* is being used to identify deviation from the norms and helps in predictive and perspective analytics. Such techniques could help to reduce the cost of prevention of financial frauds. Techniques such as big data analytics, data mining and supervised/unsupervised ML can be applied independently or in combination for building advanced anomaly detection systems to ensure fraud detection in real time, improve the credibility of data, analyse the user behaviour and dig out the hidden correlations.

2.3 Fraud Detection in e-Commerce Transactions

In e-Commerce, the transactions are digital, and, in most cases, the provider and seeker are not related. The e-Commerce transactions have electronic data interchange (EDI) transactions as their precursor. The EDI mode of message exchange is based on one-to-one mapping of entities involved; hence, the identity is established before the transaction takes place. But in e-Commerce, the hub and spoke model is where a single hub may cater to million spokes; therefore, chances of false or spurious transactions are very high. Most of the frauds here are related to identity theft and merchant scams. Table 2 presents the online payments systems and their corresponding pros and cons.

3 Types of e-Fraud

3.1 Identity Theft

It is a cybercrime when fraudster captures the personal information of a person to commit fraud and gain advantage. The stolen information can be used in other serious crimes as well as financial frauds.

3.2 Phishing

Phishing steals the confidential personal information of user by coning itself to be as a trusted authority. They especially target the banking details via email attachments and fraudulent hyperlinks. They trick to be a legitimate organization so as to collect sensitive information of user. Upon clicking the hyperlinks, the malware gets placed, and the next online transaction activates it to capture personal financial details. These malwares are Trojans, spyware, worms and other computer viruses. Phishing attacks are presently not limited to just popup, emails, SMS but also are in form of spoof mobile applications and QR codes [13].

3.3 Website Cloning (Spoofing)

Spoofing involves website duplication or hoax website creation for fraud and criminal use. The name, graphics, company logos and other codes of legitimate organizations are used to make illusion for fake sites to be seen real ones. This leads to chat room and trade sites where user compromises his personal financial information and commits fake purchases of a non-existing product and pay for it in real.

3.4 Virtual Casinos (Internet Gambling)

It is online gambling where a person can participate remotely without physical involvement. Internet poker and other active online gambling sites need license but only few of them acquire it genuinely. Rate of online gambling is high as it appears

Table 2 Online payment systems with their pros and cons

System	Pros	Cons
EFT (Electronic Funds Transfer) System	Economical than credit cards and cheques processing, efficient and faster fund transferring method, more secure as eradicates the need of carrying cheques and cash holding bank details. Easily and automated to use as digitally done	Need to be alert while entering the details Sometimes, bank servers might be busy and account may be debited but fails to reach beneficiary's account
NEFT (National Electronic Funds Transfer) System	Near real-time funds transfer to the beneficiary account in a secure manner. Eliminates the use of physical cheque or Demand draft reducing time and effort. Enables fund transfer current location, eliminating unnecessary bank visits. Cost-efficient service, no amount limit on transaction. Credit and debit confirmation receipt is immediately received to remitter and beneficiary via email or SMS	Customers should have balance funds available in account No copy of cancelled cheque will be received
RTGS (Real-Time Gross Settlement) System	Real time, efficient and reliable system for fund transfer. Facilitates online high-value fund transfer. Supports one to one as well as gross transactions by organizations for funds transactions	The gross transfer system of RTGS has the gridlock risk on having finance deficit account On payment completion, it cannot be revoked
ECS (Electronic Clearing Service) System	Eliminates physical actions such as document deposition at the bank for payments, mishandling of paper risks such as cheques lost or misused. Transferring fund to beneficiary is quick and takes a day. Facilitates automatic payments on due dates eliminating the need of remembering due dates. Cost-effective process for making bulk payments	Activation process is tedious and lengthy No complaint cell for resolving issues
NECS (National Electronic Clearing Service) System	Transferring fund to beneficiary is quick and easy. Eliminates risks associated with mishandling of cheques, etc. Facilitates multi-credits of funds to beneficiary accounts for the single debit from account of sponsor bank	Initiation of activation is time-consuming No complaint cell for resolving issues

(continued)

Table 2 (continued)

System	Pros	Cons
RECS (Regional ECS) System	Eliminates physical actions such as document deposition at the bank for payments The overall process is also much convenient for bankers with everything done electronically	No customer care and grievance cell available to settle the disputes arose. Initiation of activation is time-consuming

and disappears daily. It gathers money from losers and does not pay prize money to winners.

3.5 Electronic Cheque Frauds

It can be committed very easily and is a major financial security threat. Fraudster just needs desktop publishing software, printer and a scanner. Electronic cheque frauds involve endorsement forging, counterfeiting the cheques, altering cheques resulting in financial losses.

3.6 Lottery Frauds

It involves scam mails and SMS in form of winning a lottery and seek reply from sender compromising their bank account and personal details for prize money transfer.

3.7 Automated Clearing House (ACH) Frauds

Corporate payments are made using ACH transactions. Increase in ACH transactions results in information frauds. Fraudsters capture route number and account information and use it to steal funds directly from accounts. Government payments like online payments and payrolls face these types of frauds resulting in heavy financial losses.

3.8   Nigerian Advance Fee Fraud (419 Fraud)

This is the most lucrative and popular e-fraud that undercovers as Nigerian law section "419". They send bulk hoax mails to recipients to join business and receive large sum as commission in return. Once initiated, fraudsters ask for some money for opening the bank account or paying some processing fee which is a fraud.

4   Security of e-Payment

4.1   Secure Socket Layer (SSL)

It encrypts data between site server and shopper's computer. The information first gets transformed into packets and then numbered, and lastly an error control system is added. Packets holding information are then transferred on Internet via different routes following the Transmission Control Protocol/Internet Protocol (TCP/IP). TCP/IP reassembles the packets to detect any error and resends the packet holding errors. The SSL guarantees privacy and authentication using digital certificates and PKI [14]. The server is provided with SSL certificate issued by government authority. When shopper requests from his browser to the site server, his browser checks the site for certificate to recognize. When site is not recognized, warning message prompts from browser [15].

4.2   Secure Communication Tunnel for Secure e-Payment System

A secure e-payment system comprises of four divisions. The interconnection between these divisions is made via secure communication tunnels which offer a reliable and private interaction of two or more sections or parties such as customer to payment gateway, and then payment gateway to shop [15, 16].

4.3   Server Firewalls (Software or Hardware Firewalls)

It protects computer, server and network from virus attack and hackers. Several companies implement Kerberos protocol that implements secret key cryptography to confine the access of authorized users. Firewall protects rule analyst and from malignant attack providing secure Internet usage to all. Firewall is placed in between Internet and concealed network. It implements security access rule through link controls between networks [17].

5 Fraud Mitigation and Control Strategies

Fraud mitigation techniques and strategies try to mitigate the fraud frequency or the severity of fraud by incorporating anti-fraud approaches to the financial transaction processes. It requires a robust and effective system that can assess fraud risks and flag the alert in near real time. The fraudsters can use different malware applications and track user behaviour, logs, confidential or sensitive information. They may use different e-fraud techniques discussed in Sect. 3 or some other means to deceive customers. Nevertheless, the frauds can be mitigated and controlled [18]. Table 3 presents some security issues and their mitigation strategies. Despite new tools and techniques, frauds are still perpetrated by fraudsters using new and changed strategies.

The frauds can be controlled using three different strategies including preventive, detective and corrective controls [19, 20]. Preventive control strategies are the steps that are taken before the fraud actually happens such as not have a single point of control for various aspects of processing transactions. Having a robust authentication system for any transactions could help prevent fraud. Detective control strategies are taken during or after the fraudulent events. They are designed to identify irregularities and reconciliation of accounts. Corrective control is strategies that are designed to correct irregularities, errors, or any fraudulent activities as soon as they are detected. These strategies may help enhance the detective and preventive controls and reduce the risk of occurrence of such events in near future.

6 Conclusion

Online transactions have benefited every sector and industry by speeding up the trades and commerce around the world. However, these transactions are risky and may be vulnerable to frauds that need to be detected and prevented in near real time. This paper discussed the mechanisms to detect and prevent online fraud using suitable machine learning and AI techniques and reduce the risks of the existing payment systems. Moreover, the paper presented fraud detection in e-Commerce transactions along with the pros and cons of various online payment systems and types of e-frauds. Finally, it discussed the security challenges associated with online payment systems and various controlling and mitigation strategies to reduce the risks of the existing payment systems. In future, we would like to extend our work by providing a secure framework for fraud detection and mitigation in e-Commerce transactions.

Table 3 Security issues associated with online payment systems and its mitigation strategy

Security issues	Descriptions/Comments	Mitigation strategies
Sophisticated malware	It tracks keystrokes and can learn passwords, infiltrate microphones and cameras of systems. The URL scraping can be done online on user's system. Bot attacks can be incorporated on user's system	Secure attack prevention mechanism against malware and ransomware is to be effectively installed. User must be very vigilant while acting on links, pages that visit and interact online
Application/Middleware Vulnerabilities	Applications of enterprise are proliferated for personal interest by suspicious middle vendors. Attackers detect system vulnerabilities and infiltrate the software applications	A good application security program should be incorporated. Scanning of all the internal applications and review of codes time to time. The security programs need to be updated to latest version time to time. User must remain vigilant from vendors
Service providers	Third-party providers get access to much sensitive and confidential information	Users need to be strict while dealing with third-party service provider and evaluate time to time to keep track record of security
Desensitization caused by media saturation	The first data and security breach in an organization is considered to be very important and taken seriously. If such breach repeatedly breaks, the employees and executives get desensitized to its seriousness	Making data security the priority and repressing the desensitization regarding security breach
Smarter phishing and spear phishing	Phishing can be easily identified due to its message composition, spelling errors, poor grammar. Presently, it upgraded to a targeted messaging system. Spear phishing holds smart language that is very specific and similar to that of intellectual folks with top access and authorization abilities	Users should remain vigilant and should refrain from granting access of payments to those whom they don't recognize

(continued)

Table 3 (continued)

Security issues	Descriptions/Comments	Mitigation strategies
Cloud unpreparedness	Cloud storage is risky in view of hacking and data breaches despite several benefits	Users should be carefully transferring the data and predetermine the things to be stored in clouds, be aware of data controls and decide the time duration for storage carefully
Failed understanding of information securities and cyber risk	Cyber security safeguards the user data against cyber frauds, cybercrimes and law enforcement. Information security prevents data and information breach. These data may be of any form like electronic or physical. Lack of understanding limits the development of effective security system	The technical education and awareness of Information securities and cyber risk should be made priority
Bring your own technology (BYOT) and personal mobiles	BYOT refers to allowing user to use his personal gadgets, mobile, laptops rather than using one officially provided by organization. Majority of employees use at least one personal device at work	Organizations must control and regulate this via implementing systematic comprehensive mobile device management (MDM) strategy, keep check over the employees and keep record of all devices being used within the network

References

1. Khan BUI, Olanrewaju RF, Baba AM, Langoo AA, Assad S (2017) A compendious study of online payment systems: past developments, present impact, and future considerations. Int J Advan Comput Sci Appl 8(8):256–271
2. Kedar MS (2015) Digital India new way of innovating India digitally. Int Res J Multi Stud 1(4):34–49
3. Lahiri A (2020) The great Indian demonetization. J Econ Perspect 34(1):55–74
4. Kou Y, Lu CT, Sirwongwattana S, Huang YP (2004) "Survey of fraud detection techniques." In: IEEE international conference on networking, sensing and control, 2004
5. Gopal RK, Meher SK (2007) A rule-based approach for anomaly detection in subscriber usage pattern. Int J Math, Phys Eng Sci 1(3):396–399
6. Deshmukh A, Talluru L (1998) A rule-based fuzzy reasoning system for assessing the risk of management fraud. Intel Syst Account, Finance Manage 7(4):223–241
7. Yee OS, Sagadevan S, Malim NHAH (2018) Credit card fraud detection using machine learning as data mining technique. J Telecommun, Elect Comput Eng (JTEC) 10(4):23–27
8. Varmedja D, Karanovic M, Sladojevic S, Arsenovic M, Anderla A (2019) "Credit card fraud detection-machine learning methods"
9. Bhowmik R (2008) Data mining techniques in fraud detection. J Digit Forensics, Secur Law 3(2):35–54
10. Bologa AR, Bologa R, Florea A (2013) Big and specific analysis methods for insurance fraud detection. Database Syst J 4(4):30–39
11. Cardenas AA, Manadhata PK, Rajan SP (2013) Big data analytics for security. IEEE Secur Priv 11(6):74–76
12. Ryman-Tubb NF, Krause P, Garn W (2018) "How artificial intelligence and machine learning research impacts payment card fraud detection: a survey and industry benchmark. Eng Appl Artif Intell 76:130–157
13. Alqahtani M (2019) "Phishing websites classification using association classification (PWCAC)". In: 2019 International conference on computer and information sciences (ICCIS)
14. Fei Z, Heping D, Zhengyue L, Zhengfu L (2010) "The analysis of e-commerce online payment status in China". In: 2010 international conference on networking and digital society
15. Mazumder FK, Jahan I, Das UK (2015) Security in electronic payment transaction. Int J Sci Eng Res 6(2):955–960
16. Hassan MA, Shukur Z, Hasan MK (2020) An efficient secure electronic payment system for e-commerce. Computers 9(3):66
17. SenthilKumar P, Muthukumar M (2018) "A study on firewall system, scheduling and routing using pfsense scheme". In: 2018 international conference on intelligent computing and communication for smart world (I2C2SW)
18. Reddy VB, Negi A, Venkataraman S, Venkataraman VR (2019) "A similarity based trust model to mitigate badmouthing attacks in internet of things (IoT)". In 2019 IEEE 5th world forum on internet of things (WF-IoT)
19. Ushad SA, Ramen M (2017) Usage and perceptions of fraud detection and preventive methods evidence from mauritius. Revista Int Adm Finanzas 9(1):87–96
20. Mwangi SW, Ndegwa J (2020) "The influence of fraud risk management on fraud occurrence in Kenyan listed companies". Int J Fin Bank Stud (2147–4486), 9(4):147–160

COVID-19 Vaccination Monitoring Using IoT and Machine Learning

Ayushi Chahal, Preeti Gulia, and Nasib Singh Gill

Abstract World Health Organization (WHO) announced the epidemic situation due to COVID-19 on January 30, 2020. It made drastic alteration of daily routine of every person's life when Coronavirus takes over the globe very quickly and affected every sector of world industry. To maintain the pace of life, one needs a technology which may help in performing daily routine works at different places without being there. IoT is one of the technologies which may get it done. This study explains the role of IoT in COVID-19 pandemic like helping healthcare providers in detection and medication of infected people without directly contacting them. Furthermore, the use of machine learning and deep learning for handling and analyzing COVID-19 data is very well-elaborated. From the last year, different Research & Development (R&D) organizations are continuously working to develop vaccine for coronavirus. Now, major goal for every country is to provide this immunization to their citizens. This study analyzed vaccination distribution with respect to countries, continents and comparison between them. Vaccine vacillation is one of the main concerns for vaccine distribution which is pointed out in the study. For data analysis, datasets are taken from kaggle and github repository. By using three different datasets, this paper presents different vaccines used by different regions of the world according to their popularity/availability. It uses Python's library scikit learn for implementation of linear regression. This supervised machine learning technique is used to predict vaccine provided to people on daily basis. Obtained results of this data analysis are delineated in the form of graphs and bar charts to give a quick review to the readers.

Keywords COVID-19 · Vaccination · Machine learning · IoT · Data analysis

1 Introduction

COVID-19 pandemic has made human race to pay a heavy price in terms of health, deaths, economy, etc. In continuation to this pandemic, a new virus from the family of

A. Chahal (✉) · P. Gulia · N. S. Gill
Department of Computer Science & Applications, Maharshi Dayanand University, Rohtak, Haryana, India

© The Author(s), under exclusive license to Springer Nature Singapore Pte Ltd. 2022
D. Gupta et al. (eds.), *Proceedings of Data Analytics and Management*,
Lecture Notes on Data Engineering and Communications Technologies 90,
https://doi.org/10.1007/978-981-16-6289-8_35

coronavideae called SARS-CoV-2 is spreading disease worldwide [21]. This scenario started from Dec. 2019 when World Health Organization (WHO) warned about several cases of pneumonia in China's city Wuhan, Hubei. Then later on, in Jan. 2020 these cases were identified as coronavirus cases by Chinese authorities. SARS-CoV-2 is considered as genetic sequence of coronavirus only. On Jan 30th, 2020, WHO announced epidemic of new coronavirus named as, COVID-19 [10, 11]. Today, it is been more than one year of this pandemic and every country of every region wants to protect their citizens from this pandemic situation. Research & Development (R&D) team of every country all over the world is working effectively in this area to find out cure of this virus. Some of them succeeded and some are trying hard to come over this problem.

Rigorous efforts of R&D teams and different organizations of healthcare sector resulted in rapid development of various means to avoid to get affected from this virus and vaccines to be taken. Many countries like UK, USA, Canada, and China have developed vaccine which are present in the market till date. Though it is difficult to produce vaccine and make it available to every common person of the world, but not impossible. By the time, an effective vaccine gets distributed among people and one must follow general hygienic manners like using masks, sanitizers, face shields, washing hands regularly, social distancing. Also, there are number of techniques that can be used to detect affected people and handle them with care. Internet of Things (IoT) could be one of the safest possible ways to find out and treat COVID-19 affected people without coming in directly contact with them.

To handle large amount of data related to daily affected people, death rate due to COVID-19, daily vaccination provided to people is also necessary. This helps in analyzing the progress toward conquering this pandemic situation. And to analyze such a large data machine learning techniques can be very useful. It is seen in this study that many researchers have used different machine learning techniques like classification, regression, and clustering to make prediction for this epidemic situation. These predictions help authorities to keep check on their steps in order to handle these rampant circumstances [25].

This paper is organized into four sections as follows: Sect. 1 consists of basic introductory part. Section 2 explains the related research or reviewed work that helps in building understanding of the current scenario of COVID-19, different vaccination being developed and used across the world and how machine learning and IoT contribute to it. Section 3 describes the use of different IoT devices that can help in avoiding direct contact of healthcare providers with coronavirus-infected people. This section also highlighted the use of machine learning and deep learning for this pandemic situation. In Sect. 4, datasets and methodology used for data analysis are described in detail. And then in the last section, i.e., Sect. 5, whole paper is concluded.

2 Related Work

K. Prathyusha et al. have done analysis on COVID-19 in India. Authors have used different regression models like linear regression, polynomial regression models to make future predictions. Dataset downloaded from kaggle is used in this study. The study is done by distributing dataset in multiple sections based upon dates of lockdown period in India. After comparing results of linear regression and polynomial regression, polynomial regression technique is declared to be more affective in case of prediction. This study also presents visualized results of statewise analysis of recovered and death rate of people [5].

Khubchandani et al. [6] have presented a different view of vaccination. However, everyone is focused on developing a vaccine and then exerts to make it reachable to people. No-one has thought of vaccine hesitancy. This study has focused on hesitancy of accepting vaccine for COVID-19 in USA. 1878 adults helped in this study. These adults gave answers to many questions related to COVID-19 vaccine. Data analysis has been done using SPSS 22. Multiple regression technique has applied on vaccine hesitancy data to find out the outcomes. As a result, authors find out that 22% of the participants hesitated to take vaccine. Various factors like mistrust, lack of awareness, former ambiguity, and financial problem create this hesitation among people. Vaccination in the USA is considered to be most cost-effective by a study [23].

COVID-19 vaccination in Asia–Pacific region countries has been studied by N.W.S. Chew et al. In these countries, healthcare workers were placed as frontlines to take immunization. For this study, authors have taken 1720 healthcare employees from six Asian countries, i.e., China, Singapore, Bhutan, India, Indonesia, and Vietnam. This study aims to investigate the willingness of these workers to take COVID-19 vaccine. Authors conducted this study by doing survey based on questionnaire. This questionnaire contains seven questions. Based upon the answers given by the participant un-variate and multi-variate analysis is done. Outcome of this study says that 95% of the healthcare employees are willing to take vaccine without any hesitation [7].

Kelekar et al. [8] have done their study on healthcare providers and the hesitancy they have for taking COVID-19 vaccine. Authors conducted online survey in three dental schools as well as in medical schools. Both medical and dental students have been used as subjects of study to know their behavior and frame of mind toward the vaccine. Medical students were found to be interested in taking vaccine while dental students seem to be in denial of taking vaccine due to illness and allergies, etc. But dental students accepted to take vaccine if it is made compulsory by the government. This study presented various factors for denial of vaccinations and shows that positive recommendation from healthcare service providers will help common people to boost their confidence on vaccination.

Defendi et al. [9] have done analysis on progress of vaccine for COVID-19. Different countries steps for vaccination production and their clinical trials have been studied by the researchers. This study presented various factors on which production of a vaccine depends like technology factor, R&D strategies, financial factors, relations and alliances, manufacturing infrastructure, and many more. They have monitored clinical trials for 10 months, and these trials are divided into four phases. Author also warns regulatory bodies to take care of the usage of the vaccine when it gets into the market to avoid any kind of misuse.

Ashok Kumar Dutta [12] has done a study about vaccinations and their development in third quarter of 2020. This study has divided the vaccine development into different phases like preclinical phase, clinical trial phase I, Phase II, and Phase III. Normal vaccine development process takes 10–15 years but in this pandemic situation due to overlapping of these vaccine development phases reduced this time to 12–15 months. Biggest challenge to develop vaccine and make it available in market is to test it on large number of humans from different geographical regions, different gender, different age group and healthy as well as diseased people.

Sharif et al. [13] have done analysis of COVID-19 in different cities from 5 continents. The cities whose data is used for analysis are London, Maharashtra, Lombardy, Madrid, New York, Hubei, Sydney, Western Cape, Sao Paulo, and Dhaka. For this study, first data was collected from various online means, then statistical analysis (spearman rank correlation test) was done to find correlation of different environmental factors and number of cases in particular region. Results of this study have shown that with decrease in temperature there is possibility that coronavirus will spread more swiftly as compared to high-temperature regions.

Different strategies are discussed by authors to divide population in different groups and then prioritize them to get COVID-19 vaccination. Authors have used real-world data and develop a model to prioritize group of people for getting immunized. This model categorizes people on the basis of their age group and any pre-existing disease and calculates the risk of life. In the first phase, this model is implemented in Italy and then applied on different datasets and tested before actually being used for the purpose [14].

Tuli et al. [24] have used machine learning with cloud computing to predict increasing or decreasing trend of this epidemic. This analysis is done on different countries throughout the world. Authors have used a mathematical model with machine learning to increase accuracy of the model. Prediction of new cases over time is done for 28 countries like Iran, China, Canada, Belgium, Netherlands, India, Switzerland, Portugal, and many more.

Agrawal et al. [29] have analyzed real-time data of different countries to tackle problems of COVID-19. Authors have elaborated the use of machine learning to make classification and prediction of people as infected or not. This study discussed four different approached of machine learning which helps in making predictions for COVID patterns. Harbola et al. [30] also have analyzed and visualized coronavirus outbreak by using long short-term memory (LSTM) model. Authors have used LTSM to predict the growth of COVID-19 infected people weekly.

3   Intelligent IoT in COVID-19

IoT is a platform which is gaining attention from almost all disciplines of line like healthcare, agriculture, industries, infrastructure, and automation. This technology could break a leg in case of identifying affected people from COVID-19 without even being in direct contact with them. This would ensure the health of unaffected people as well as it would be a great help to healthcare providers (like doctors and nurses) also. Different hardware and software related to IoT technologies can help to make a setup for contactless setup for patients and healthcare providers. Moreover, this setup can be made intelligent with the addition of artificial intelligence techniques in it. Machine learning and deep learning are the subsets of artificial intelligence, which are used worldwide for data analytical purpose.

Chinese 'COVID-19 Intelligent Diagnosis and Treatment Assistant Program' (nCapp) is one of these setups which can distinguish patients into different categories of pneumonia [26]. Arogya Setu (Indian), TraceTogeather (Singapore), Civilians (Canada), DetectaChem (USA), Social monitoring (Russia) are some of the applications created by government for their citizens to keep track of COVID-19 pandemic in the country.

Different sensor devices could be used under COVID-19 situations as shown in Fig. 1. [15] These hardware can be used as the primary source of data collection. Raw data is collected directly from persons. Software like smartphone applications, embedded tools in Android/iOS are used to manage this data collected by hardware and then come the data extraction and data analysis part. One can use machine learning or deep learning for decision-making purpose [18].

- In a crowded place, to find out a person whether infected from coronavirus or not is very difficult but this can be done easily with the help of infrared thermometer. This infrared thermometer helps in thermal imaging in crowed. It allows authorities to identify infected persons and also helps to improve social distancing [16].
- To control spread of coronavirus used of drones are perfect these days. Drones can be used for surveillance of infected persons, to provide medicines, to make announcements, for thermal scanning at random places, etc. This helps to protect

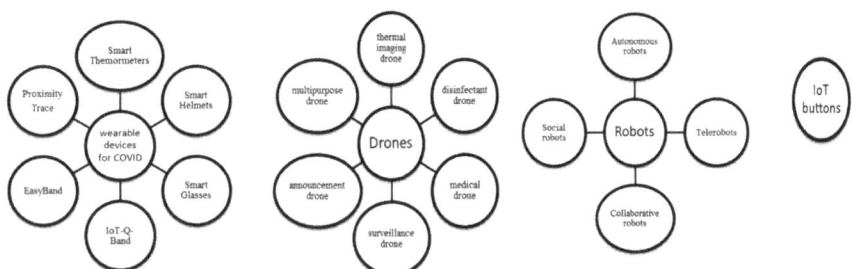

Fig. 1   Different IoT devices for COVID-19

doctor/nurses as well as healthy persons also from further transmission of infection [22].

- Smartphones contains magnetometer. This magnetometer can be used for social distancing which further reduces chances of spreading infection from one person to another [27].
- Robots can be used to detect corona infected symptoms, delivering medical supplies, lowering the use of humans hence reducing chances of getting infected from coronavirus.
- IoT buttons are used to give any particular task like sanitation and medication to be done by any healthcare employee.

In order to provide healthcare services on large scale, these IoT devices can be connected with each other with the help of Internet. These device connections can help clinical staff indeed. These connections may help clinical staff to monitor the patients, provide medicines, give advice, etc. Serious respiratory problem due to SARS-CoV-2 can also be detected with the help of devices like chest X-ray and CT scan. Data collected by these devices can be given as an input to an intelligent system using machine learning or deep learning, and then these systems help in predicting such a severe respiratory problem [17]. Mortality rate, curve, vaccine development, vaccine distribution rate, active virus rate can be predicted with the help of artificial intelligent techniques [19].

Many automated models are in market right now, which use convolutional neural network (CNN), artificial neural network (ANN), deep neural network (DNN) for the classification of CT scan and chest X-ray images of COVID-19 infected persons [20, 28].

4   Data Analysis

This study is focused on the COVID-19 vaccination present in the market, and how well these vaccines are being accepted all over the world. To analyze COVID-19 vaccination, let us get familiarized with datasets first and then move to different results of the analysis. Then to make predictions for future supervised machine learning technique is used called regression.

4.1   Datasets

For this exploratory data analysis, datasets are taken from kaggle. This study uses three datasets, i.e., COVID-19 world vaccination progress, population by country, country mapping, and all these datasets are merged to get the outputs as shown in Fig. 2.

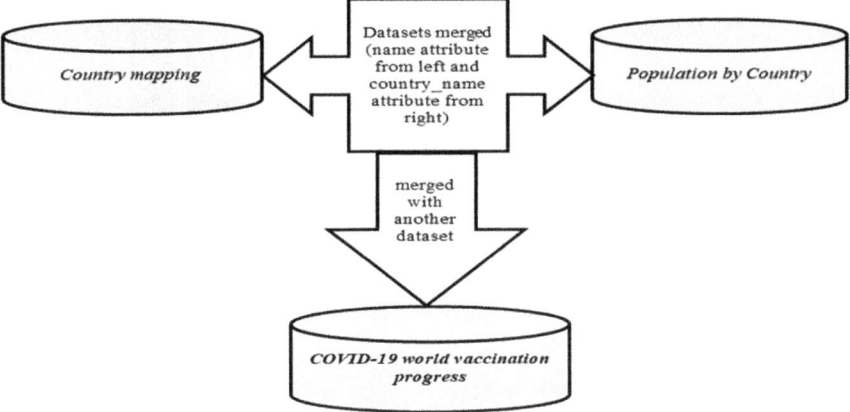

Fig. 2 Datasets used

- *COVID-19 world vaccination progress* dataset is generated by Gabriel Preda by collecting data from 'Our data in world' on daily basis and then merged. This dataset contains 8191 records and 15 attributes (country, iso_code, date, total_vaccinations, people_vaccinated, people_fully_vaccinated, daily_vaccinations_rawdaily_vaccinations, total_vaccinations_per_hundred, people_vaccinated_per_hundred, people_fully_vaccinated_per_hundred, daily_vaccinations_per_million, vaccines, source_name, source_website) [1].
- *Population by country* dataset is generated by Tanu N Prabhu by web-scrapping from the website named 'worldometer'. This dataset contains 235 records and 11 attributes (Country (or dependency), Population (2020) Yearly Change, Net Change, Density (P/KmÂ²), Land Area (KmÂ²), Migrants (net), Fert. Rate, Med. Age, Urban Pop %, World Share). This dataset mainly focuses on countrywise population change [2].
- *Country mapping* dataset describes different countries and their different regional codes. This dataset helps in relating above two datasets. This dataset contains 248 records and 11 attributes (name, alpha-2, alpha-3, country-code, iso_3166-2, region, sub-region, intermediate-region, region-code, sub-region-code, intermediate-region-code) [3].

4.2 Vaccines Analysis

Multiple research and development organizations have given their input to develop vaccines which can reduce the high risk of transmission and cure all the patients who are affected from COVID-19. Through our research, it is seen that majority of vaccines that are being used all over the world are: Covaxin, EpiVacCorona, Johnson and Johnson, Sinopharm/Wuhan, Pfizer/Biontech, Moderna, Sinovac, Sputnik V,

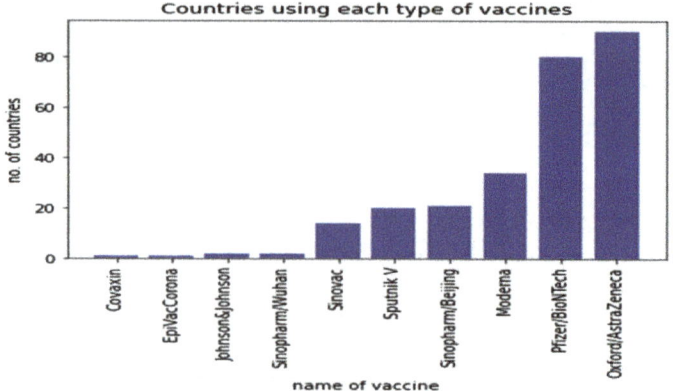

Fig. 3 Analysis of vaccine's popularity with respect to number of countries

Sinopharm/Beijing, Moderma, Oxford/AstraZeneca vaccine. Analysis of most popular vaccine that are being used in majority of countries is shown in Fig. 3.

4.3 Prediction Using Regression

Machine learning has ability to learn from the data and helps in making decisions. It derives predictions from the data provided to any of the machine learning techniques. To get the output in continuous value linear regression technique of supervised machine learning algorithm is used. Linear regression is considered as the easiest method of among all supervised algorithms. It is used to set up a linear relationship between dependent and independent variable(s).

For this study, Python's library scikit learn is used to apply linear regression algorithm to the dataset for which vaccine given to population and date were taken as input. Figure 4 shows the prediction of daily vaccination to the patients (in millions).

4.4 Continentwise Daily Vaccination

After merger of three datasets, a new merged dataset is used to analyze the daily vaccination provided to their population. From the late 2020, severe coronavirus 2 (SARS-CoV-2) is spreading through entire world not only through direct contact but also through non-living objects, etc. People get affected throughout the world at different paces in different regions due to different factors like population density, gender of a person, age of a person, facilities available, environmental factors like temperature, humidity, wind velocity, and UV index ; hence, vaccines named above are used based on these factors in different regions.

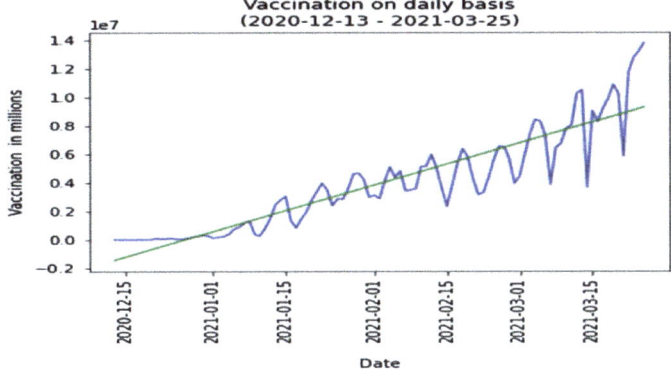

Fig. 4 Prediction of vaccination (given to patients) using linear regression

This study gives an overview on number of vaccination provided to the people of different continents. After analyzing the data, it is found that from 15–12–2020 till present Asian people are getting vaccinated more as compared to Americans, Europeans, Africans, and Oceania people. Asia has hit 6 million people after March 03, 2021, which get vaccinated, Americans crosses 3.5 million, Europeans hit 2 million once and then daily vaccination decreased, Africans have not crossed half a million till March 03, 2020, Oceania people got the last place of getting vaccinated daily. This analysis is being shown in Fig. 5.

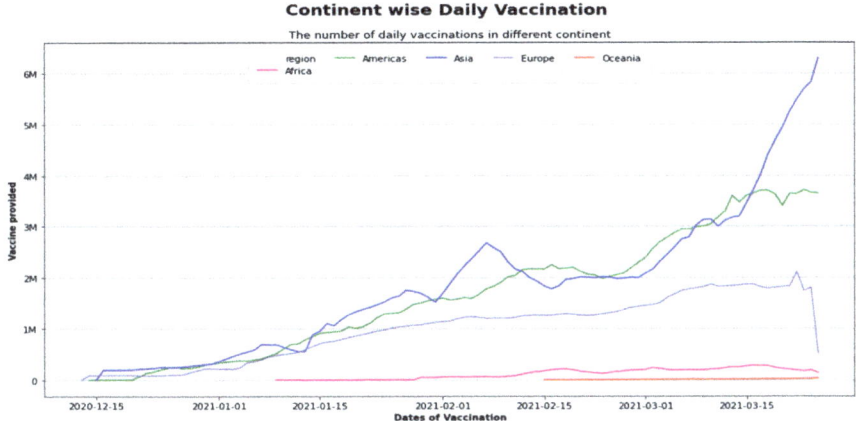

Fig. 5 Daily vaccination in different continents

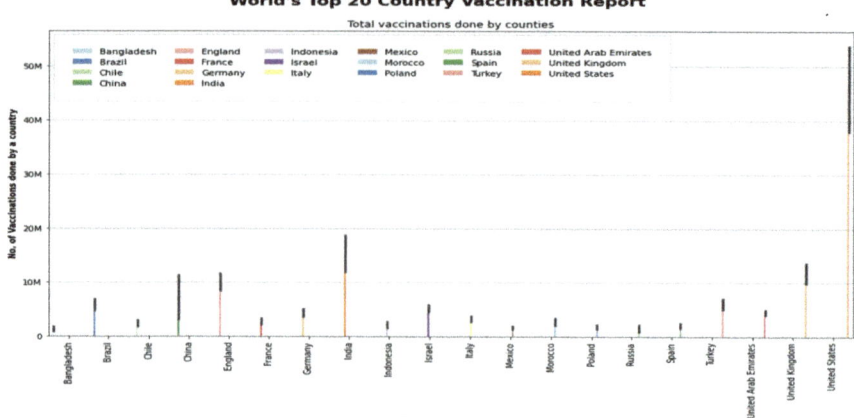

Fig. 6 Total vaccination done by top 20 countries

4.5 Country Wise Vaccination

To do total vaccination analysis over the countries, first top twenty countries are found which have done maximum total number of vaccination for their citizens. Then these twenty counties are compared among themselves in which the USA got first position by providing more than 50 million vaccines to their citizen in total, then India, then China, and so on as you can see in Fig. 6.

4.6 Vaccination in India

Vaccination for COVID-19 in India started in January 2021. Two vaccines named Oxford/AstraZeneca and Covaxine are being given to citizens for immunization. Other vaccines like Moderna, Pfizer are also trying to get clearance but they did not get any place in the market yet [4]. Total COVID-19 vaccination given to citizens of India as on 08 May, 2021, is 16, 73, 46, 544. Total active cases have increased by 17.01% and death rate has also increased by 1.09%. Though daily vaccination rate given to Indians is excellent, since India is the most populated country (density-wise) this increase in number of active case are to be taken seriously (Fig. 7). To immunize Indians, one must find a portion of population with COVID vaccine vacillation and then handle them with care. This would help in keenness and anticipation of Indian citizens toward COVID-19 vaccination.

Total vaccination performance of India is compared with USA as shown in Fig. 8. As the figure shows vaccination in USA had started formerly it started in India; therefore, total vaccination provided by both countries to their citizens has huge

Fig. 7 Daily vaccination rate in India

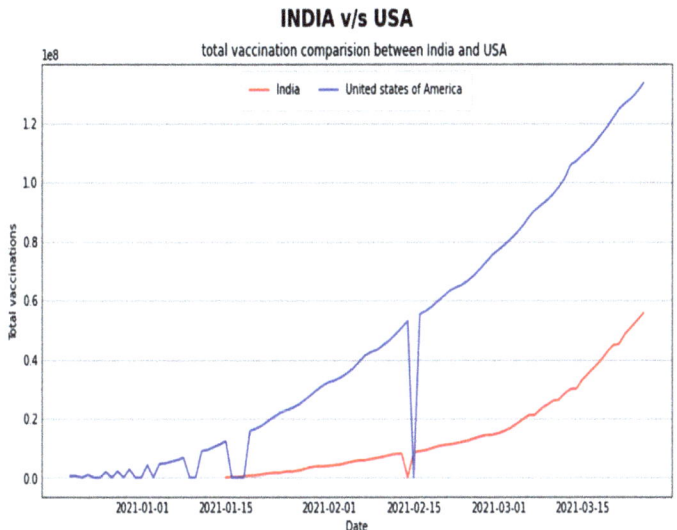

Fig. 8 India versus USA daily vaccination

difference. According to our world data, USA has coved 15.4% of its population for vaccination and India as covered only 0.6%.

5 Conclusion

Readers may find it interesting that how IoT and artificial intelligence collectively removes many of the problems of healthcare employees, data analysts, infected people, healthy people, etc. This study discussed different software applications provided by different countries to their citizens to handle emergency situations and collect data on daily basis. Also, different IoT devices are explained with their usage for COVID-19 pandemic situation. This study helps readers to analyze and visualize current scenario of vaccination and use of IoT+ AI for this COVID epidemic situation.

After covering a stressful long road of COVID-19 pandemic, finally vaccines are available to stabilize the situation. This study presented popular vaccines being used in most of the countries, analyze vaccine distribution in every continent on daily basis, found top 20 countries that are providing vaccines to their citizens. Daily vaccination provided to people is also predicted with the help of a supervised machine learning technique. This prediction is based on the dataset used from different sources. Comparative analysis of providing total vaccination by USA and India is also visualized. A result obtained from this comparison has sown that India is at backfoot in vaccine distribution as compared to the USA. This implies that India needs to take this situation very seriously as it is the densest country in the world. Moreover, it needs to take care of vaccine vacillation.

References

1. COVID-19 world vaccination progress [WWW Document], (n.d.) URL https://kaggle.com/gpr eda/covid-world-vaccination-progress (Accessed 29 March 21)
2. Population by Country—2020 [WWW Document], (n.d.) URL https://kaggle.com/tanuprabhu/ population-by-country-2020 (Accessed 29 March 21)
3. lukes/ISO-3166-countries-with-regional-codes [WWW Document], (n.d.). GitHub. URL https://github.com/lukes/ISO-3166-Countries-with-Regional-Codes (Accessed 29 March 21)
4. Krishnamurthy A, Gopinath KS (2021) The big billion Indian COVID-19 vaccine challenge. Indian J Surg Oncol 12:3–4. https://doi.org/10.1007/s13193-021-01280-1
5. Prathyusha K, Helini K, Raghavendran CV, Kurumeti NK (2021) COVID-19 in India: lockdown analysis and future predictions using regression models. In: 2021 11th international conference on cloud computing, data science engineering (confluence). Presented at the 2021 11th international conference on cloud computing, data science engineering (confluence). pp 899–904 https://doi.org/10.1109/Confluence51648.2021.9377052
6. Khubchandani J, Sharma S, Price JH, Wiblishauser MJ, Sharma M, Webb FJ (2021) COVID-19 vaccination hesitancy in the United States: a rapid national assessment. J Community Health 46:270–277. https://doi.org/10.1007/s10900-020-00958-x
7. Chew NWS, Cheong C, Kong G, Phua K, Ngiam JN, Tan BYQ, Wang B, Hao F, Tan W, Han X, Tran BX, Hoang MT, Pham HQ, Vu GT, Chen Y, Danuaji R, Rn K, Rv M, Talati K, Ho CS, Sharma AK, Ho RC, Sharma VK (2021). An Asia-Pacific study on healthcare worker's perception and willingness to receive COVID-19 vaccination. Int J Infect Dis. https://doi.org/ 10.1016/j.ijid.2021.03.069

8. Kelekar AK, Lucia VC, Afonso NM, Mascarenhas AK (2021) COVID-19 vaccine acceptance and hesitancy among dental and medical students. J Am Dent Assoc. https://doi.org/10.1016/j.adaj.2021.03.006

9. Defendi HGT, da Madeira LS, Borschiver S (2021) Analysis of the COVID-19 vaccine development process: an exploratory study of accelerating factors and innovative environments. J Pharm Innov 1–17. https://doi.org/10.1007/s12247-021-09535-8

10. Coronavirus Disease (COVID-19) pandemic—PAHO/WHO|Pan American health organization [WWW Document], (n.d.) https://www.paho.org/en/topics/coronavirus-infections/coronavirus-disease-covid-19-pandemic (accessed 30 March 21)

11. Le Thanh T, Andreadakis Z, Kumar A, Gómez Román R, Tollefsen S, Saville M, Mayhew S (2020) The COVID-19 vaccine development landscape. Nat Rev Drug Discov 19:305–306. https://doi.org/10.1038/d41573-020-00073-5

12. Dutta AK (2020) Vaccine against Covid-19 disease—present status of development. Indian J Pediatr 87:810–816. https://doi.org/10.1007/s12098-020-03475-w

13. Sharif N, Sarkar MK, Ahmed SN, Ferdous RN, Nobel NU, Parvez AK, Talukder AA, Dey SK (2021). Environmental correlation and epidemiologic analysis of COVID-19 pandemic in ten regions in five continents. Heliyon e06576. https://doi.org/10.1016/j.heliyon.2021.e06576

14. Giampiero Russo A, Decarli A, Grazia Valsecchi M (2021) Strategy to identify priority groups for covid-19 vaccination: a population based cohort study. Vaccine. https://doi.org/10.1016/j.vaccine.2021.03.076

15. Internet of things for current COVID-19 and future pandemics: an exploratory study|SpringerLink [WWW Document], (n.d.) https://doi.org/10.1007/s41666-020-00080-6 (accessed 30 March 21)

16. Kumar K, Kumar N, Shah R (2020) Role of IoT to avoid spreading of COVID-19. Int J Intell Networks 1:32–35. https://doi.org/10.1016/j.ijin.2020.05.002

17. Kumar SS (2020) Emerging technologies and sensors that can be used during the COVID-19 pandemic. In: 2020 international conference on UK-China emerging technologies (UCET). Presented at the 2020 international conference on UK-China emerging technologies (UCET) pp 1–4. https://doi.org/10.1109/UCET51115.2020.9205424

18. Ndiaye M, Oyewobi SS, Abu-Mahfouz AM, Hancke GP, Kurien AM, Djouani K (2020) IoT in the wake of COVID-19: a survey on contributions, challenges and evolution. IEEE Access 8:186821–186839. https://doi.org/10.1109/ACCESS.2020.3030090

19. Ghimire A, Thapa S, Jha AK, Kumar A, Kumar A, Adhikari S (2020) AI and IoT solutions for tackling COVID-19 pandemic. In: 2020 4th international conference on electronics, communication and aerospace technology (ICECA). Presented at the 2020 4th international conference on electronics, communication and aerospace technology (ICECA). pp 1083–1092. https://doi.org/10.1109/ICECA49313.2020.9297454

20. Ozyurt F, Tuncer T, Subasi A (2021) An automated COVID-19 detection based on fused dynamic exemplar pyramid feature extraction and hybrid feature selection using deep learning. Comput Biol Med 104356. https://doi.org/10.1016/j.compbiomed.2021.104356

21. Shereen MA, Khan S, Kazmi A, Bashir N, Siddique R (2020) COVID-19 infection: emergence, transmission, and characteristics of human coronaviruses. J Adv Res 24:91–98. https://doi.org/10.1016/j.jare.2020.03.005

22. Angurala M, Bala M, Bamber SS, Kaur R, Singh P (2020) An internet of things assisted drone based approach to reduce rapid spread of COVID-19. J Saf Sci Resilience 1:31–35. https://doi.org/10.1016/j.jnlssr.2020.06.011

23. COVID-19 vaccination predicted to be cost effective in USA, 2021. PharmacoEcon outcomes news 871, 10–10. https://doi.org/10.1007/s40274-021-7448-y

24. Tuli S, Tuli S, Tuli R, Gill SS (2020) Predicting the growth and trend of COVID-19 pandemic using machine learning and cloud computing. Int Things 11:100222. https://doi.org/10.1016/j.iot.2020.100222

25. Nadeem O, Saeed MS, Tahir MA, Mumtaz R (2020) A survey of artificial intelligence and internet of things (IoT) based approaches against Covid-19. In: 2020 IEEE 17th international conference on smart communities: improving quality of life using ICT, IoT and AI (HONET).

Presented at the 2020 IEEE 17th international conference on smart communities: improving quality of life using ICT, IoT and AI (HONET). pp 214–218. https://doi.org/10.1109/HONET5 0430.2020.9322829

26. Bai L, Yang D, Wang X, Tong L, Zhu X, Zhong N, Bai C, Powell CA, Chen R, Zhou J, Song Y, Zhou X, Zhu H, Han B, Li Q, Shi G, Li S, Wang C, Qiu Z, Zhang Y, Xu Y, Liu J, Zhang D, Wu C, Li J, Yu J, Wang J, Dong C, Wang Y, Wang Q, Zhang L, Zhang M, Ma X, Zhao L, Yu W, Xu T, Jin Y, Wang X, Wang Y, Jiang Y, Chen H, Xiao K, Zhang X, Song Z, Zhang Z, Wu X, Sun J, Shen Y, Ye M, Tu C, Jiang J, Yu H, Tan F (2020) Chinese experts' consensus on the internet of things-aided diagnosis and treatment of coronavirus disease 2019 (COVID-19). Clinical eHealth 3:7–15. https://doi.org/10.1016/j.ceh.2020.03.001
27. Dong Y, Yao Y-D (2021) IoT platform for COVID-19 prevention and control: a survey. IEEE Access 1–1. https://doi.org/10.1109/ACCESS.2021.3068276
28. Gulia AC, Preeti (n.d.) Deep learning: a predictive Iot data analytics method. Int J Eng Trends Technol—IJETT 68:25–33. https://doi.org/10.14445/22315381/IJETT-V68I7P205S
29. Agrawal R, Gupta N (2021) Analysis of COVID-19 data using machine learning techniques. In: Khanna A, Gupta D, Pólkowski Z, Bhattacharyya S, Castillo O (eds) Data analytics and management, lecture notes on data engineering and communications technologies. Springer, Singapore, pp 595–603. https://doi.org/10.1007/978-981-15-8335-3_45
30. Harbola S, Jain P, Gupta D (2021) Analysis, visualization and forecasting of COVID-19 outbreak using LSTM model. In: Khanna A, Gupta D, Pólkowski Z, Bhattacharyya S, Castillo O (eds) Data analytics and management, lecture notes on data engineering and communications technologies. Springer, Singapore, pp 151–164. https://doi.org/10.1007/978-981-15-8335-3_14

An Efficient Feature Extraction Technique for Brain Tumor Detection Using GUI

Faiyaz Ahmad and Tanvir Ahmad

Abstract The brain is the main organ in the human body, answerable for controlling and directing all basic life capacities for the body, and a tumor is a mass of tissue framed by the aggregation of dead cells. MRI imaging techniques is used for brain tumor detection. This paper describes the detection of brain tumor using three important stages: preprocessing, segmentation, and morphological operation. Conversion of RGB image to grayscale image and anisotropic filtering used to remove noise and maintaining edges is applied during preprocessing stage. Thresholding is performed at segmentation stage, and closing operation is done at morphological operation stage. The accuracy of this model was found to be 91.33%. This paper include detection of brain tumors on MRI images by making graphical user interface.

Keywords Preprocessing · Anisotropic · Image segmentation · Thresholding · Morphology operation

1 Introduction

The Brain tumor contains lethal cancerous cell. It has its high effects because it is a special term of the main nervous system of human brain where each small defect can cost a lot. For this reason, it is important to find ways of early detection of anxiety about a brain tumor. This importance comes from the fact that early detection dramatically increases the possibility of treating the disease and saving patient lives. Recently, cancer treatments have evolved considerably, especially in the early stages of infection. The chances of survival are very high for those patients who receive early treatment compared to those who do not have this opportunity in the early stages of the disease. Tumor is either benign or malignant. Benign is non-cancerous cell, whereas malignant is cancerous. Malignant extend quickly which make patient condition worse.

F. Ahmad (✉) · T. Ahmad
Department of Computer Engineering, Faculty of Engineering and Technology, Jamia Millia Islamia, New Delhi, India

© The Author(s), under exclusive license to Springer Nature Singapore Pte Ltd. 2022
D. Gupta et al. (eds.), *Proceedings of Data Analytics and Management*,
Lecture Notes on Data Engineering and Communications Technologies 90,
https://doi.org/10.1007/978-981-16-6289-8_36

Fig. 1 Sample images **a** Normal MRI, **b** tumor infected MRI

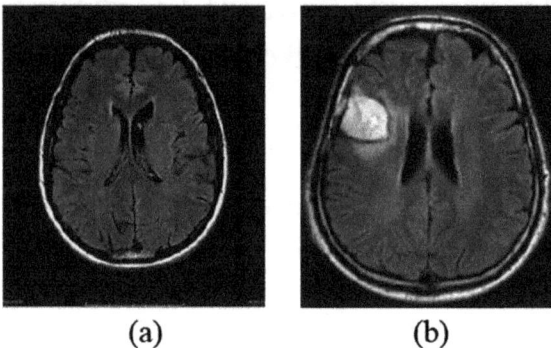

(a)   (b)

There are various imaging techniques on which brain tumor detection is performed like X-ray, ultrasound, PET, MRI and SPECT. But magnetic resonance imaging give better outcome than all other imaging techniques because of its powerful magnetic field component that provides precise picture of tissue and bones inside the brain. Various filtering, segmentation, and morphological operations are used to determine the brain tumor.

Brain tumor image processing concerns the three stages, namely preprocessing, segmentation, and morphological operation. Preprocessing is done by converting the RGB image into grayscale image and after that applying anisotropic filter which helps to remove noise and maintaining the shape of tumor in image. Segmentation is done using thresholding. Next morphological operations is applied, and at last, the brain tumor is detected. This operation is performed on image processing tools such as matrix laboratory (Figs. 1 and 2).

1.1 Problem Statement

Brain tumor are life threatening and should be detected as early as possible. Tumor can be of any size, shape, and intensity that makes it difficult to detect and analyze it. Different pathologist have different expertise and treatment plan. So it may have computational complexity and time consuming. This image processing and ML based project help to overcome all this problem. This research include brain tumor segmentation and classification. Earlier at preprocessing stage median filter were used to remove only salt and pepper noise, now anisotropic filtering is used to remove salt and pepper noise as well as maintaining the shape and edges of the tumor. The motivation of the proposed application is to aid neurosurgeons and radiologists in detecting brain tumors in an inexpensive and non-invasive manner. The main contributions of this work are mentioned below:

- The main purpose of this work is to make an automated system that is used for classification of tumor.

Fig. 2 General architecture
of brain tumor detection

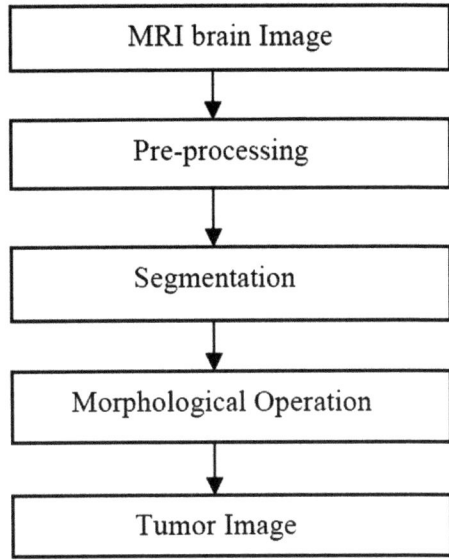

- Goal is to find the tumor from the MRI image in less amount of time with less computational cost.
- Model can be used by neurosurgeons and health care experts easily.
- The problem includes image processing, pattern analysis, and computer recognition strategies that are expected to improve the sensitivity, clarity and efficiency of brain tumor test.
- Model should have user friendly interface and give accurate result.

2 Related Work

Myint and Aung [1]. In his research work titled An Efficient Tumor Segmentation of MRI Brain Images Using Thresholding and Morphology Operation, they used Gaussian high pass filter at preprocessing stage and other image processing techniques like segmentation, thresholding, and morphological operation used to detect tumor in MRI image. The accuracy of this paper was 89.11% on 100 MRI images [1]. Hussain et al. [2]. In this paper titled Brain Tumor Detection and Segmentation Using Anisotropic Filtering for MRI Images, they focused on anisotropic filtering for removing salt and pepper noise, thresholding for segmentation, and morphological operation such as erosion and dilation for extraction of brain tumor. The accuracy of this paper was 84.44% on 90 MRI images [2]. MITS et al. [3]. In his paper titled Efficient Way to Analysis the Textural Features of Brain Tumor MRI Image Using GLCM, they used arithmetic operator for filtering, thresholding and then watershed segmentation and at last morphological operation for detection of tumor. Accuracy of this model was found to be 90.33% on 50 MRI image [3]. Abd Khalid et al. [4].

In his paper titled MRI Brain Tumor Segmentation: A Forthright Image Processing Approach, they used some stages like preprocessing, edge detection, and segmentation. Converting of input image to grayscale were done at preprocessing stage and removing noise if required. Then some enhancement techniques were used. Next, edge detection by Sobel and Canny algorithms. At last, the segmentation is used along with morphological operations to find the area in MRI images. The result was measured by confusion matrix. It provides satisfactory results from the MRI brain images database. The accuracy of this paper was 76% on FLAIR type MRI images in BRATS 2017 dataset [4]. Rajasekhar et al. [5]. In his paper titled Brain Tumor Detection Using Image Processing, they used median filter for removing noise. Sobel operator for edge detection and thresholding was used for segmentation and finally morphological operation for tumor detection. Accuracy of this model was found to be 90% on 50 MRI image [5]. Sarkar et al. [6]. In his paper titled Brain Tumor Segmentation from MRI Images Using Morphological Operation, they used median filter and morphological operation for the detection of brain tumor using python programming. Experiment was performed on 20 MRI images, and all have correctly identified [6]. Deore and Mandawkar [7]. In his paper titled Automated Brain Tumor Detection and Identification using Image Processing, they used different preprocessing methods such as filtering, contrast enhancement, and detection of edges. Post-processing methods such as thresholding, histogram, and morphological operation using image processing tools [7]. Sravan et al. [8]. In his paper titled Magnetic Resonance Images Based Brain Tumor Segmentation-A critical survey, they proposed two models, one using median filter and thresholding and another using high pass filter and K-Means. Then, finally morphological operation is used for extraction of tumor. The accuracy of first model was 89.90%, and second model accuracy was 77% on BRATS 2015 dataset [8]. Suresha et al. [9]. In his paper titled Detection of Brain Tumor Using Image Processing, they proposed a model to check that the MRI has tumor or not by taking combined technique of K Means and support vector machine. The accuracy of this paper was 88.33% on 75 MRI images [9]. Zaman et al. [10]. In his paper titled Image Segmentation of MRI Image for Brain Tumor Detection, there research presents a new technique for brain tumor detection by using the combination of watershed algorithm, Fuzzy K means and Fuzzy C means clustering. The accuracy of this paper was 89.90% on private dataset [10].

3 Proposed Algorithm

Brain MRI image is processed and elegant, well for the finding of tumor by using image processing tools in matrix laboratory using graphical user interface. The methodology engages following steps:

Step 1: Reading the input image
Step 2: Converting the input image into a grayscale image

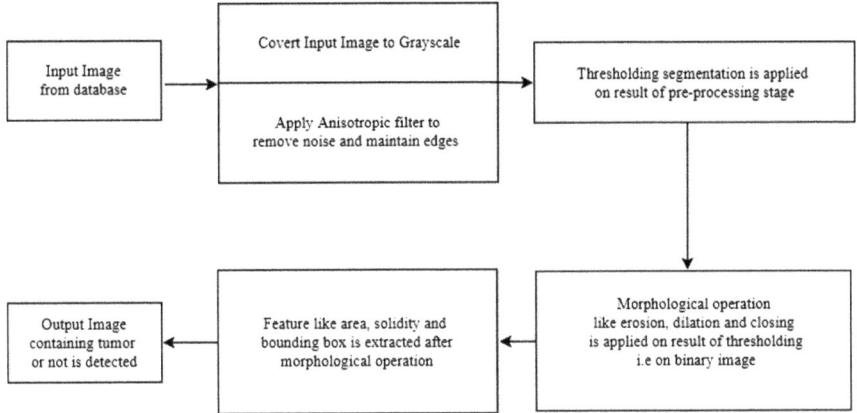

Fig. 3 Block diagram of proposed model

Step 3: Using anisotropic filter for removal of noise and maintaining the edge
Step 4: Applying threshold segmentation
Step 5: Performing morphological closing operation (Fig. 3).

Pseudo Code of Proposed Algorithm

Input: Input Image
Output: Image result based on proposed Brain Tumor Detection
Get Image
//**Data Augmentation**//
Resizing image to 256*256;
//**Create dataset**//
Input set of training and testing Image are separated
Maximum number of iteration to 150
//**Create BTD based model**//
For each iteration
do
{
Convert RGB to grayscale;
Applying filter;
Thresholding for segmentation;
Perform morphological closing operation;
}

Firstly, take an image from the database and performing preprocessing which include converting of grayscale image (0–255) to binary image (0 for black and 1 for white pixels) and applying anisotropic filter on the image. Segmentation is performed using thresholding operation by taking a threshold value that differentiates the background and the object. Erosion and dilation are applied on MRI images during morphological operation. Then, finally the tumor is detected by GUI. Usually, imaging pictures appears like a black and white pictures. The illusion of gray shading during a white and black image is obtained by rendering individual dots determinative the apparent lightness of the gray in their neighborhood.

This work is predicated primarily on thresholding, closing morphological operations, and detection of tumor area for analysis. The experiment has been implemented by making GUI in MATLAB R2016a.

4   Experimental Results

Taking input from database and then preprocessing take place which involve converting the RGB image to grayscale image and then applying anisotropic filtering to get the filtered image. Bounding box is applied after thresholding and morphological operation. Then, separate the foreground from background to get the tumor alone, and then, finally outline the boundary of the tumor. In this way, tumor is detected, and for non-tumor image, it will display popup message saying "No Tumor." This work is performed on MATLAB R2016a by making graphical user interface.

The experiment is performed on database with 150 MRI images out of which 57 MRI images contain tumor and 93 MRI image contain non-tumor. Model correctly identified 137 MRI Images and did not able to identify identified 13 MRI images. Result was measured using confusion matrix. Confusion matrix have four parameter true positive, true negative, false positive, and false negative. True positive (TP) means that the MRI contain tumor, and it is correctly identified as tumor by model. True negative (TN) means that the MRI contain non-tumor, and it is correctly identified as non-tumor by model. False positive (FP) means that the MRI contain non-tumor, but it is incorrectly identified as tumor by model. False negative (FN) means that the MRI contain tumor, but it is incorrectly identified as non-tumor by model. In confusion matrix of this MRI image database, there are 53 TP images, 84 TN image, 9 FP image, and 4 FN image. Final accuracy of this model was found to be 91.33% (Table 1).

The Multimodal Brain Tumor Image Segmentation Benchmark (BRATS) is mostly used for brain tumor detection. It is freely available online [11]. BRATS dataset volumes are arranged in to T1, T2, T1c and Fluid Attenuated Inversion Recovery (FLAIR) sequences. FLAIR and T2 are formed due to collection of water

Table 1 Different type of MRI images in database

Brain multi-modality MRI image	
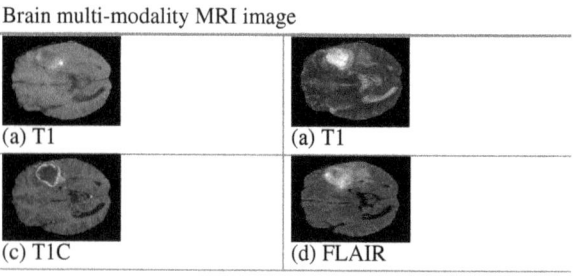	
(a) T1	(a) T1
(c) T1C	(d) FLAIR

Table 2 Outcome of different images

Input	Output				
	Filtered image	Bounding box	Tumor alone	Tumor outline	Detected tumor

or fluid inside the brain. T1 is formed by accumulation of dead cell. T1C is formed by breakdown of blood brain. Image in database is in jpg, jpeg, and png format (Table 2).

Brain tumor detection is performed by Graphical User Interface in MATLAB R2016a. Firstly, take an input image from database, then preprocessing is done by converting the RGB images into grayscale image and then applying anisotropic filter to remove salt and pepper noise and maintaining the edges. Segmentation is done using thresholding. Next, morphological operations is applied and at last the brain tumor is detected if present in the MRI image (Fig. 4).

4.1 Evaluation Matrix

Database with 150 MRI images out of which 57 MRI images contain tumor, and 93 MRI images contain non-tumor images. Model correctly identified 137 MRI images and does not able to correctly identify 13 MRI images. Final accuracy of this MRI image database is 91.33% (Table 3).

$$\text{Accuracy} = (TP + TN)/(TP + TN + FN + FP) \text{ OR}$$
$$\text{Accuracy} = \frac{(\text{Number of correctly classified test samples})}{\text{Total samples}} * 100\%$$
$$= (137/150) * 100\%$$
$$= 91.33\%$$

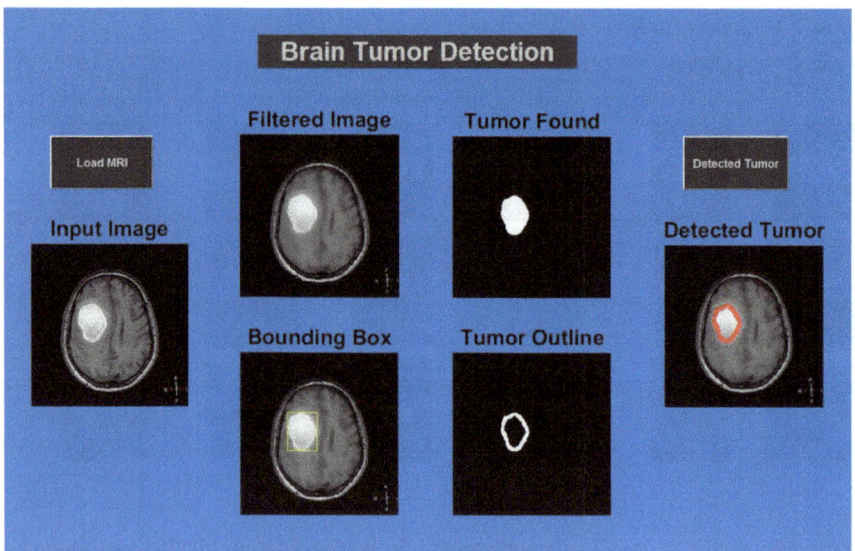

Fig. 4 Graphical user interface for brain tumor detection

Table 3 Table of confusion matrix

		Actual class	
		Yes	No
Predicted class	Yes	TP = 53	FP = 9
	No	FN = 4	TN = 84

5   Conclusion and Future Direction

In medical field, manual identification of brain tumor by doctor referring the MRI images is a very time-consuming task and can be inappropriate for a large amount of data. Instead of manual identification, image processing and machine learning technique can be used to identify the tumor from the images. Therefore, this model helps in understanding the creation of a system that will carry out image processing and identify the brain tumor using machine learning approach. Brain tumor detection include three important stages, preprocessing, segmentation, and morphological operation. Convert of RGB image to grayscale image and apply anisotropic filtering at preprocessing stage. Thresholding is performed at segmentation stage and closing operation at morphological stage. Finally, the area of the tumor in the highlighted in the MRI image. Proposed model have user friendly interface and give 91.33% accurate result on 150 MRI images.

Encouraged by these results, forthcoming work will embrace the progress of classification result and absolute accuracy. With more extensive and diverse dataset, the classification accuracy can be improved by using modern deep learning models.

By increasing the number of hidden layer in neural network, we can improved the classification result by doing this the weights will be better adjusted which help to find better shape, edges and size of tumor, and thus, classification accuracy increases. Fine tuning and transfer learning approaches can also be applied to better tune the model on the basis pretrained models.

References

1. Myint HH, Aung SL (2020) An efficient tumor segmentation of MRI brain images using thresholding and morphology operation. In: Proceedings of the eighteenth international conference on computer applications (ICCA 2020)
2. Hussain CA, Gopi C, Kishore DS, Reddy GG, Sai GC (2020) Brain tumor detection and segmentation using anisotropic filtering for MRI images (JES 2020)
3. MITS G, Wadhwani AK, Wadhwani S (2020) Efficient way to analysis the texural features of brain tumor MRI image using Glcm. Image 5(7)
4. Abd Khalid NE, Ismail MF, Manaf MAA, Fadzil AFA, Ibrahim S (2020) MRI brain tumor segmentation: a forthright image processing approach. Bull Electr Eng Inf 9(3):1024–1031
5. Rajasekhar D, Divya C, Farhana M, Krishnaveni K, Priya AJB (2020) Tumour detection using image processing. Alochana Chakra J
6. Sarkar B, Al Zubaer A, Sen J, Islam MN (2020) Brain tumor segmentation from MRI images using morphological operation. Comput Rev J 7:2581–6640
7. Deore PC, Mandawkar U (2020) Automated brain tumor detection and identification using image processing. J Compos Theory XIII(IV):0731–6755
8. Sravan V, Swaraja K, Meenakshi K, Kora P, Samson M (2020) Magnetic resonance images based brain tumor segmentation—a critical survey. In: 2020 4th international conference on trends in electronics and informatics (ICOEI) (48184). IEEE, pp 1063–1068
9. Suresha D, Jagadisha N, Shrisha HS, Kaushik KS (2020) Detection of brain tumor using image processing. In: 2020 fourth international conference on computing methodologies and communication (ICCMC). IEEE, pp 844–848
10. Zaman A, Ullah K, Ullah R, Imtiaz HH, Yu L (2019) Image segmentation of MRI image for brain tumor detection. Int J Eng Appl Sci Technol 4(8):50–55. ISSN No. 2455-2143
11. Islam R, Imran S, Ashikuzzaman M, Khan MMA (2020) Detection and classification of brain tumor based on multilevel segmentation with convolutional neural network. J Biomed Sci Eng 13(4):45–53

Content-Based Image Retrieval Using Deep Learning

Faiyaz Ahmad and Tanvir Ahmad

Abstract In last few decades, digital images are growing with a rapid pace on and off the Internet, and given the volume of the images, the need of better storage, processing, and retrieval of images has raised. One of the retrieval techniques that is focus of this work is content-based image retrieval (CBIR) in which similar images are searched from a pool of images without manually annotating them; rather, in CBIR, other features of images that discriminate them from other images are used. By finding the better discriminative features of a collection of images, an efficient and generalized CBIR system can be built. The success and the efficiency of such a system depend on the choice of the features of images used to identify them. Generally, the images are stored at a very low level in pixels, but to get better results or in other words better features, we need to store these images at a very high level in order to reduce the semantic gap. There has been very extensive research on CBIR using the traditional methods of image processing. With the advancement of deep learning systems, deep learning can be used for large range of problems. In this work, we investigated the use of deep learning, more precisely auto-encoders, for the feature extraction and representation of images in CBIR, and we reached to the retrieval efficiency of $\approx 80\%$.

Keywords CBIR · Deep learning · Convolutional neural networks · Semantic gap · Auto-encoders

1 Introduction

Various types of multimedia resources (for example images, videos, text, and audio) are quickly developing with an immense advancement of associated technologies, for example, information can be processed and analyzed in mobile phones, media transmission, Web content, 2D/3D applications, and so forth. Last few decades has seen an unrivaled advancement in the amount, complexity, availability, diversity, and

F. Ahmad (✉) · T. Ahmad
Department of Computer Engineering, Faculty of Engineering and Technology, Jamia Millia Islamia, New Delhi, India

D. Gupta et al. (eds.), *Proceedings of Data Analytics and Management*,
Lecture Notes on Data Engineering and Communications Technologies 90,
https://doi.org/10.1007/978-981-16-6289-8_37

significance of images in every domain. Consequently, the analysis and processing of images turns out to be imperative as they play an important role in a wide range of fields and have numerous applications such as medical care, education, advertising, art, social media, Web, entertainment, and art. To retrieve the information from these images, the analysis and processing of these image require immense storage and processing, which makes the case for better indexing methods, better storage techniques, efficient analysis and accurate retrieval methods. In addition, images have associated information, called metadata, which can be employed for better indexing and efficient searching of these images. So, searching images from a huge complex database becomes the major issue given the amount of the data and the underlying complexities.

The need of an efficient image retrieval system has led to the generation of CBIR system. Figure 1 represents the general model of CBIR system.

1.1 Problem Statement

For an input image I, and a collection of images S, the primary goal of our proposed method is to apply search process in S for finding i and return set of closely related or similar images to i according to their contents.

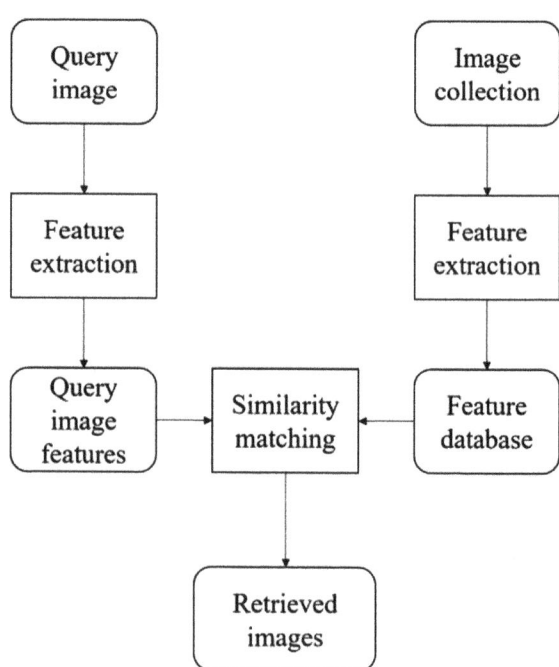

Fig. 1 A general model of CBIR system

Learning similarity measures and effective feature representations are critical to the retrieval performance of a CBIR system [1].

- Rotation variance is a problem that needs to be addressed.
- Real-time deployment of CBIR system is hindered by the exponential time complexity that arises due to similarity comparison of images.

2 Related Work

Learning the portrayal of useful highlights and likeness measures is fundamental for the CBIR framework recuperation measure. Not with standing broad examination endeavors for quite a long time, it stays perhaps the most difficult open issues frustrating the achievement of true CBIR programs. The greatest test originates from the infamous "semantic gap" issue that exists between pixels of high-level pictures taken by machines and excellent semantic perspectives seen by people. Among the different methodologies, AI has been effectively examined as a potential manual for shutting the semantic gap later on.

Rupapara et. al. [2] proposed that auto-encoders can be used to represent features. Further they showed that the nearest-neighbor algorithm can be a good choice for similarity measure. The dataset they used was MNIST [2]. Maji et al. [3] applied many CNN architectures on various datasets for feature extraction and concluded that pretrained deep learning features produces better precision outcomes w.r.t. the features derived by traditional methods, e.g., CCM, wavelet, etc. [3]. Nikolaos et al. [4] proposed regularized discriminative profound measurement learning technique for CBIR, which works by displaying the dormant generative components for each class. It was prepared on numerous datasets and the outcomes got recommended that proposed technique is equipped for learning proficient portrayals that can separate between various classes, displaying both the in-class difference and conceivable subclasses, just as addressing objects from classes that were not seen during the preparation cycle [4]. Zhu [5] in his paper preprocessed images by selecting candidate regions and fed the resulted images to deep learning network. The deep learning architecture they used was ResNet where they replanted Softmax layer with fully connected layer and pooling layer. The algorithm was performed on Caltech 256 and Cifar 100 datasets. The performance was found better by using Minkowski similarity measure [5]. Tya-Shen-Tin et al. [6] analyzed the effect of hyper parameters on accuracy of auto-encoders used in CBIR. In this work, they used three different annealing techniques for optimization of parameters, i.e., random search, tree of Parzen estimator, and annealing. And the result that they concluded was that annealing is not as good as other two techniques for parameter optimization, i.e., random search and tree of Parzen [6]. Shamna et al. [7] in their work focused around medical image CBIR systems proposed T&L based models. In the beginning for topic information formulation, they used other techniques like guided LDA. A position-weighted exactness technique [7] was used to determine the position of the positional grid. Mezzoudj et al. [8] in the work mentioned formulated a new system which was designed to perform retrieval tasks very fast on a huge collection of image databases. The goal of this investigation was to increase the speed of the indexation interaction using a MapReduce dispersed model in the indexation stage. In addition, a memory-driven conveyed stockpiling framework was used to improve the compose activity. At that point, in the subsequent commitment, the speed of picture recovery was expanded with k-NN search strategy [8]. Tzelepi et al. [9] devised a unique model retraining approach for social event information on convolutional representations using deep CNN in CBIR. Three approaches were devised based on the available data. In the event that the data is not accessible, the fully solo retraining technique was used, and the retraining with important data approach will be used within the sight of markings. Additionally, significance input-based retraining was used within the sight of the clients' criticism [9]. Correlated primary visual texton histogram features (CPV-THF) was introduced as a new element descriptor for CBIR by Raza et al. [10]. To incorporate the semantic info as well as the visual substance of the image, the connections between the surface direction, LSSI, shading, and force of a picture were examined. Besides, the case molded primary components were developed in the picture surface investigation based on the texton hypothesis [10]. Dai et al. [11] introduced RS-CBIR, in which the other factors of images like spectral data, just as the spatial data substance of RS pictures, was described with picture portrayal strategy. In RS picture, the sparsity was misused utilizing regulated recovery technique and the spatial substance was displayed utilizing the all-encompassing pack of ghostly qualities descriptors. Shamna et al. presented a machine learning-based unsupervised content-Based medical image retrieval system based on the different visualizations in the spatial domain and then matching them based on features computed [12]. Mistry et al. [13] detailed a hybrid component based effective CBIR framework by using distinctive distance quantifies just as BSIF include descriptors, spatial descriptors, CEDD descriptors, and recurrence descriptors. The exactness was improved utilizing the Gabor wavelet change highlights [13]. Jin et al. [14] proposed a creative CBIR procedure by combining both DML and CSL techniques. The issue of class awkwardness was resolved by categorizing the instances into minority and lion's share classes according to their weight. Furthermore, three important metrics such as edge fluctuation, misclassification penalty cost, and edge mean were considered to improve the suggested model's suitability for CBIR [14]. Unar et al. [15] developed a CBIR approach that was entirely designed with the objective of retrieving similar images with visual and written characteristics. The picture highlights were separated from the picture in the worldwide based picture recovery approach with the help of worldwide descriptors, and the based match similitude between the pictures was assessed in the area-based picture recovery approach with methods for parting the picture into more modest pieces. At that time, the SURF description had eliminated the picture's notable visual features, and the picture's installed text was linked to the MSER computation. Stop-channels were used to distinguish between non-text items and content articles in the image. The strings were created using the neural probabilistic language model [15], exactly like watchwords.

3   Proposed Algorithm

As discussed semantic gap refers to the difference in visualization of images by humans and by computers. This difference hinders the success of any system which need to extract the image features. In our work, we will be using auto-encoder to retrieve the features from images. There are various other techniques by which features of images can be represented and extracted. The next section discusses the feature representation and extraction in images.

Our CBIR system will be composed of two components. The first component will extract the features from each image of the image database, and the second component will compare the features of the query image with the features stored in the database. Feature extraction is discussed in section. Figure 2 illustrates the different components of our design. In the first part of our design, an auto-encoder will be trained and then that trained auto-encoder will be used as feature extractor. After the features are extracted in images from image databases, images will be extracted from the query image by the same auto-encoder and the query image features will be compared with all the images in the feature database. This comparison will be done using Euclidean distance, Manhattan distance, and cosine similarity.

3.1   Feature Extraction Technique

Feature extraction refers to extract and reduce the features given the initial set of features or dimensions of an image. There are various methods of feature extraction

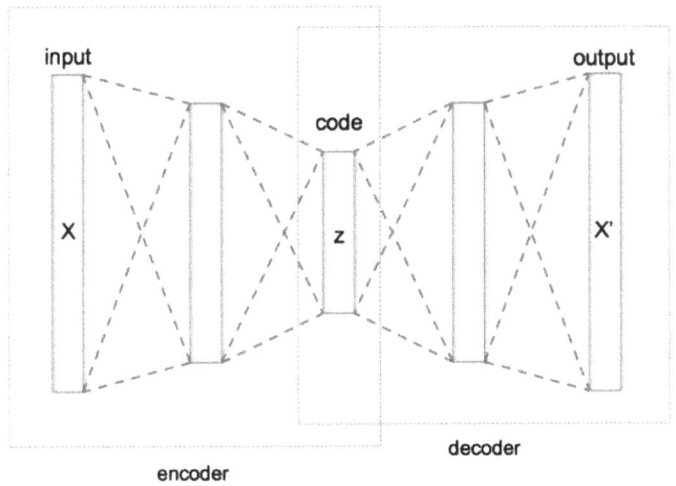

Fig. 2 General model of auto-encoder

techniques such as principle component analysis, latent semantic analysis, independent component analysis, and auto-encoder. In this work, we will be using auto-encoder as a feature extractor. Next section discusses the auto-encoder as a feature extractor. An auto-encoder is a class of neural networks that attempts to recreate the output relative to the input by estimating the identity function. To attenuate the reconstruction error which can be evaluated using loss functions, the model parameters are optimized. Auto-encoder comprises of three main components: an encoder network, a code, and a decoder network. The encoder encodes the input image as a compressed representation in a reduced dimension; the code resembles the latent space encoding, i.e., the compressed input fed to the decoder; the decoder rebuilds input back to the initial dimension from the code.

Pseudocode of the proposed methodology is given in Algorithm 1.

Algorithm 1 Proposed Algorithm

Input: Dataset, D and Query Image Q
Ouput: Set of image similar to image Q

CreateDB: create a dataset D' of size D
for each D **do**
    remove noise
    normalize image
    store $D[i]$ in $D'[i]$
end for

take Q, remove noise, normalize and store it in Q'
$epochs \leftarrow 3$
//Autoencoder training
for epoch in epochs **do**
    for i in D' **do**
        compute loss
        update loss
        $D' \leftarrow$ AE output at Last layer
    end for
end for
$distances \leftarrow$ array[]
//Get similar images to Q
for each image, i, in D' **do**
    compute euclidean distance between i and Q as d
    array[this]$= d$, with index $= j$
end for
Sort $distances$ in ascending order
first k-images are top-k similar images

---

4 Experimental Results

There are two types of dataset which is considered for experiment. First dataset is CIFAR 10 that contains around 60,000 images, and each image is of 32 × 32. These 60,000 images are divided into ten classes, i.e., "airplane," "automobile," "bird," "cat," "deer," "dog," "frog," "horse," "ship," and "truck." Some of the sample images from this dataset are given in Fig. 3.

The second dataset is Modified National Institute of Standard and Technology (MNIST) which is the most used dataset in the deep learning in other fields. It is a small datasets composed of 60,000 images of size 28 × 28 pixels. The images in the MNIST dataset are grayscale, and the images are of handwritten digits from zero to nine. Some of the sample images from this dataset are given in Fig. 4.

4.1 Evaluation Matrix

The metric used to measure the performance of our system is mean average precision (MAP). Mean average precision is one of the popular metric to measure the performance of CBIR systems. The mean average precision for a set of queries is the mean

Fig. 3 CIFAR 10 Input sample images

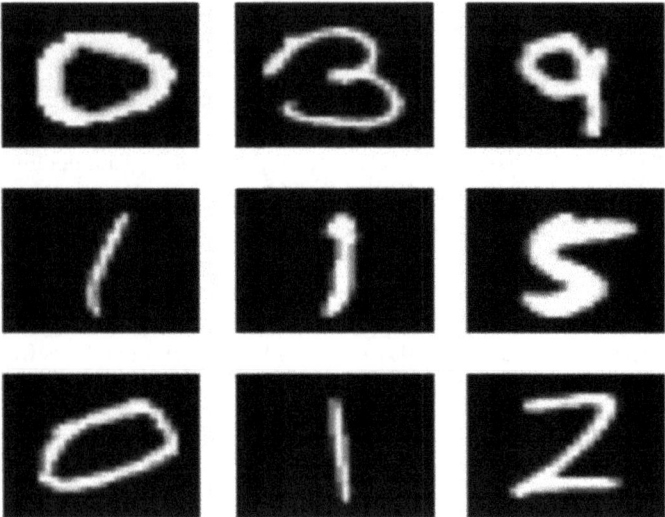

Fig. 4 MNSIT Input sample images

of the average precision scores for each query. It is define as:

$$\text{MAP} = \frac{\sum_{q=1}^{Q} \text{AVe P}(q)}{Q} \tag{1}$$

where Q is the number of queries in the set and $\text{AveP}(q)$ is the average precision (AP) for a given query, q.

4.2 Performance Analysis

Our model was first made to do the classification in order to measure the accuracy, and then, the model was trained on both the datasets, and the accuracy was evaluated for both the retrieval and classification tasks by mean average precision (MAP). By using different approaches and different similarity measurement techniques, the difference in the results of sorting in retrieval tasks was much different. The two datasets that we used for testing and evaluation of our model were CIFAR-10 and MNIST. In order to measure the retrieval accuracy, we had to first compute the classification accuracy. We removed the retrieval part of our design and modified it to classify the images; we achieved some good results which are listed in Table 1.

Table 1 summarizes the TOP-1, TOP-2, and TOP-3 classification accuracy of CIFAR-10 dataset on our model and other standard models. The model 1 (CNN RGB) was used on original images on a CNN without pretrained weights. It achieved the accuracy of 83.4% in 20 epochs. Then our model was modified first by using

Table 1 CIFAR-10: CNN accuracy (classification)

Model	Accuracy		
	Top 1	Top 2	Top 3
Existing models accuracy			
Raw pixels	37.3	–	–
RBM with black propagation	64.8	–	–
Convolutional deep belief network	78.9	–	–
Sparse AE	73.4	–	–
PCANet	78.7	–	–
Our CNN model accuracy in 20 epochs			
CNN RGB	83.4	92.8	94.2
CNN RGB and DA	86.2	93.1	94.8
Pretrain and CNN RGB	85.1	94.4	96.1
Pretrain and CNN RGB and DA	87.8	95.4	97.0

the pretrained weights and then the augmented data (DA), the best TOP-1 accuracy it achieved was 87.8% in 20 epochs. So the using model with pretrained weights performed better than using the model without pretrained weights.

Table 2 summarizes the TOP-1, TOP-2, and TOP-3 classification accuracy of CIFAR-100 dataset on our model and other standard models. On increase in the number of subsets, i.e., classes to identify the results show that our model can still learn the features from this dataset. And pretraining also helped in improving accuracy $\approx 2\%$ more than the random initialized weight network. Table 3 summarize

Table.2 CIFAR-100: CNN accuracy (classification)

Model	Accuracy		
	Top 1	Top 2	Top 3
Enhanced model for 20 epochs			
CNN RGB	50.8	69.4	70.1
CNN RGB and DA	55.2	63.6	70.4
Pretrain and CNN RGB	57.1	62.9	76.2
Pretrain and CNN RGB and DA	58.8	63.7	72.5

Table.3 MNIST: CNN accuracy (classification)

Models	Accuracy
Our enhanced model for 20 epochs	
CNN	96.8
CNN and DA	97.2
Pretrain and CNN	98.1
Pretrain and CNN and DA	99.8

Table.4 CIFAR 10:
minimum average precision
(image retrieval)

Distance metrics	MAP
L_2 Norm	0.871
L_1 Norm	0.894
Cosine metric	0.807

about MNIST data set. In order to measure the retrieval accuracy we had to first compute the classification accuracy and achieve better result.

As mentioned before, the most crucial aspect of better classification or retrieval is to learn better feature representations. In order to estimate how good our representations are, we also predicted the top-3 and top-2 accuracy in both CIFAR-10 and CIFAR-100, and it indicated that the images can be classified 98% accurately in CIFAR 10 when we look at top-3 predicted classes, and 70% accurately when we look at top-3 predicted classes in CIFAR-100 dataset. Now these results indicate that our model has learned the features from both the datasets. Hence, we can consider these features as well learned discriminative representations for our dataset. The conclusion can be drawn that the learned features are well discriminative and can be used for retrieval task. So we used mean average precision (MAP) for evaluating this model for retrieval results. In the test dataset, each image was given to the auto-encoder and auto-encoder in turn generated the latent representations, also called the code, of the query image and that latent representation was stored in the database. We applied different similarity measurements to get the improved results given the query image. Table 4 summarizes the results achieved by applying MAP on the results obtained from the model.

5 Conclusion and Future Direction

In this paper, we discussed the various aspects of content-based image retrieval systems and discussed the techniques and trends in CBIR and also presented a solution to how to employ the deep learning in CBIR systems. The problem in CBIR was identified and a design composed of auto-encoder was deployed to eliminate the problem identified. The problem of semantic gap existed from the beginning of image retrieval systems and different methods were used to tackle the problem, we looked into different solutions that were used to remove the semantic gap. In order to estimate how good our representations are we also predicted the top-3 and top-2 accuracy in both CIFAR-10 and CIFAR-100, and it indicated that the images can be classified 98% accurately in CIFAR 10 when we look at top-3 predicted classes, and 70% accurately when we look at top-3 predicted classes in CIFAR-100 dataset. These results indicate that our model has learned the features from both the datasets. The learned features are well discriminative and can be used for retrieval task. So we used mean average precision (MAP) for evaluating this model for retrieval results. In the test dataset, each image was given to the auto-encoder, and auto-encoder in turn generated the latent representations, also called the code, of the query image and that latent representation was stored in the database. We investigated the use of deep learning, more precisely auto-encoders, for the feature extraction and representation of images in CBIR, and we reached to the retrieval efficiency of $\approx 80\%$.

References

1. Wan J, Wang S, Hoi SCH, Wu P, Zhu J, Zhang Y, Li J (2014) Deep learning for content-based image retrieval: a comprehensive study. In: Proceedings of the 22nd ACM international conference on multimedia, pp 157–166 (2014)
2. Rupapara V, Narra M, Gonda NK, Thipparthy K, Gandhi S (2020) Auto-encoders for content-based image retrieval with its implementation using handwritten dataset. In: Fifth international conference on communication and electronics systems (ICCES), pp 289–294. https://doi.org/10.1109/ICCES48766.2020.9138007
3. Maji S, Bose S (2020) CBIR using features derived by deep learning. *arXiv e-prints,* arXiv: 2002.07877 (2020). arXiv:2002.07877 [cs.IR]
4. Passalis N, Iosifidis A, Gabbouj M, Tefas A (2020) Variance-preserving deep metric learning for content-based image retrieval. Pattern Recogn Lett 131:8–14. https://doi.org/10.1016/j.patrec.2019.11.041
5. Zhu H (2020) Massive-scale image retrieval based on deep visual feature representation. J Vis Commun Image Represent 70:102738
6. Tya-Shen-Tin YN, Razumov AA, Ushenin KS (2019) Hyperparameter optimization for autoencoders that perform content-based image retrieval. In: AIP conference proceedings, vol 2174. AIP Publishing LLC, pp 020260. https://doi.org/10.1063/1.5134411
7. Shamna P, Govindan VK, Abdul Nazeer KA (2019) Content based medical image retrieval using topic and location model. J Biomed Inform 91:103112
8. Mezzoudj S, Behloul A, Seghir R, Saadna Y (2019) A parallel content-based image retrieval system using spark and tachyon frameworks. J King Saud Univ Comput Inf Sci
9. Tzelepi M, Tefas A (2018) Deep convolutional learning for content based image retrieval. Neurocomputing 275:2467–2478
10. Raza A, Dawood H, Dawood H, Shabbir S, Mehboob R, Banjar A (2018) Correlated primary visual texton histogram features for content base image retrieval. IEEE Access 6:46595–46616
11. Dai OE, Demir B, Sankur B, Bruzzone L (2018) A novel system for content-based retrieval of single and multi-label high-dimensional remote sensing images. IEEE J Sel Top Appl Earth Observations Remote Sens 11(7):2473–2490
12. Shamna P, Govindan VK, Abdul Nazeer KA (2018) Content-based medical image retrieval by spatial matching of visual words. J King Saud Univ Comput Inf Sci
13. Mistry Y, Ingole DT, Ingole MD (2018) Content based image retrieval using hybrid features and various distance metric. J Electr Syst Inf Technol 5(3):874–888
14. Jin C, Jin S-W (2018) Content-based image retrieval model based on cost sensitive learning. J Vis Commun Image Represent 55:720–728
15. Unar S, Wang X, Zhang C (2018) Visual and textual information fusion using kernel method for content based image retrieval. Inf Fusion 44:176–187

Diagnosis and Detection of Plant Diseases Using Data Mining Techniques

Harshita Bhati and Monika Rathore

Abstract The destruction of plants by diseases is one of the major problems in the agricultural industries and also suffers in terms of quality and proportion of agricultural production. As agriculture plays a vital role in the economy of India, it is therefore important to increase the agriculture throughput by decreasing such diseases through tracking the plants' life cycle. The traditional oversight followed by farmers is time consuming, and much expertise and adequate monitoring are needed to measure plant situations. Therefore, to improvise the overall process, it is necessary to automate the diagnostic process. Numerous researchers have developed systems based on different image processing techniques. In this paper, we reviewed the most common mechanisms that are utilized in order to diagnose the plant maladies. It includes various processes for diagnosing plant diseases. Later, we proposed the model to diagnosis the plant leaf diseases with the help of a hybrid model of fuzzy C-means with ANN and Bayes classifier, and through this proposed model, we expect better prediction.

Keywords Plant leaf detection · Convolutional neural network · Fuzzy C-means algorithm clustering · Artificial neural network · Bayes classifier

1 Introduction

In India, agriculture plays a major source in the economy. Hence, it is necessary to work on high-quality agriculture production to maintain economic development and crop yielding has turned to be like a nightmare because of numerous diseases affecting the plant [1]. Farmers need to manually monitor and control various environmental factors to achieve high-quality yield [2]. Researchers are working on novel technologies to tackle food shortage which is caused by demographic growth and

H. Bhati (✉)
Rajasthan Technical University, Kota, India

M. Rathore
ISIM Jaipur, Jaipur, India

© The Author(s), under exclusive license to Springer Nature Singapore Pte Ltd. 2022
D. Gupta et al. (eds.), *Proceedings of Data Analytics and Management*,
Lecture Notes on Data Engineering and Communications Technologies 90,
https://doi.org/10.1007/978-981-16-6289-8_38

climate change. But, there are few issues such as plant leaf maladies that affect the yield. Diagnosing the type of diseases on the plants through the manual process is not working all the time. So to resolve this problem, automated system will help to diagnose the disease on time, so that it will be useful for farmers. The paper review intent to explore several approaches for the identification of the plant leaves disease and discuss in terms of various aspects and characteristics. The paper is organized into three sections. The section first explains why plant disease diagnostics are crucial, and the second section explains the ongoing work in this area and focuses on their techniques. The third section includes the baseline methodology. Tracking the development of the plant disease through plant leaf disease detection system is included in this phase. Lastly, Section four will conclude this paper in future directions.

2 Literature Review

Dheeb Al Bashish et al. (2010) they proposed an approach based on image processing, and they opted K-means technique for image segmentation and the second phase, the processed (segmented) image passed via well-trained NN (neural network) technique with a back propagation algorithm [3].

Mrunmayee Dhakate et al. (2016) the main focus on four diseases of the pomegranate plant which is detected and classified by digital image processing and neural network techniques. The results of the experiments describe from the given samples and categories. The healthy fruit and leaf, 100% results from healthy fruit leaf, 87.5% from leaf spot, and 85.71% from bacterial blight, while 83.33% result from fruit spot and rot [4].

Srdjan Sladojevic et al. (2016) in this paper, they utilize deep learning approach to categorize and detect plant diseases from images of leaves. For training procedure, they are using the fine-tuning and deep convolutional neural network. Various tests were performed to examine the progress of the model. They use different techniques like fine-tuning and augmentation for improving accuracy [5].

Vijai Singh et al. (2017) opted genetic algorithm for the image segmentation process in the plant maladies detection and classification. Few images have been utilized for the process (training) and prepare the test sets for f different plants like lemon, banana, beans, tomato, and potato. For the leaves extraction and comparison, they used the co-occurrence feature [1].

Ehsan Kiani et al. (2017) proposed a model for diagnosing plant leaves disease and basically focused on examine infected leaves at strawberry farm especially in outdoor condition using a fuzzy-based classifier [6].

Saradhambal. G. et al. (2018) proposed k-means and Otsu's classifier in an improvised way to predict the contaminated area of the plant leaves by shape and texture extraction. Experiments covered the time of complexity and infection area of the leaves and also classified the plant diseases [7].

Serawork Wallelign et al. (2018) through this paper it elaborates the flexibility and feasibility of CNN for detecting and classifying plant disease, and model is trained in such a way that it takes images from the natural environment, and this model design is based on LeNet architecture to per classification on soya bean plant disease[8].

Santhana Hari et al. (2019) proposed a new architecture to classify the plant disease by new CNN model, this proposed model would learn various properties which are available in the background information, and it will consider the property into different class according to their nature most probably. Thus, it is only focused on plant and their different disease [9].

Abhinav Sagar et al. (2020) focused on multi-class problem (classification) and also compared different CNN models like VGG ResNet169, InceptionV3, etc. They found ResNet50 achieved the best result. For estimation, they used metrics like accuracy, precision, recall, and class-wise confusion metric and achieve accuracy of 0.982, recall of 0.94, and the precision of 0.94 [10].

Junde Chen et al. (2020) proposed approach showed a significant improvement in performance compared to other leading-edge methods. Deep learning CNN architecture is proposed (INC-VGG) for the plant disease detection which was able to correctly identify the disease with 91.83% accuracy on publicly available data and 92% accuracy for images of rice crop disease [11].

3   Related Works

See Table 1.

4   Plant Disease Detection Process

There are four levels to the plant disease detection process, as shown in Fig. 1. The first level consists of acquiring pictures either by mobile phone or digital camera by the Internet. The second level consist image segmentation in which segments the image into clusters through different techniques. Third level is feature extraction and fourth level is classification of the diseases.

4.1   Image Acquisition

In image acquisition, a different image of plant leaves is collected from different devices like good quality camera, mobile phones, or the Internet with desired resolution of size. This construction of a database of images is dependent on the application developer. The database (images) is responsible for the effectiveness of classification at the most recent level of diagnosis [9].

Table 1 Methods and their result

Title	Technique	Accuracy
A framework for detection and classification of plant leaf and stem diseases	K-means technique, neural network classifier along with back propagation	This method was able to identify the plant leaf malady with 93% precision
Deep neural networks based recognition of plant diseases by leaf image classification	Convolutional neural network	This experiment was able to achieve an accuracy of somewhere between 91 and 98%, for separate class tests. The average accuracy achieved is 96.3%
Detection of plant disease by leaf image using convolutional neural network	ANN, deep learning, convolutional neural network	Accuracy is achieved by 86% from CNN new trained and tested model
Detection of plant leaf diseases using image segmentation and soft computing techniques	Genetic algorithm, K-means clustering	This was a two-phase experiment. In the first phase, the author used K-mean combined with minimal distance criterion for clustering with 86.54% accuracy and 93.63% accuracy with its own suggested algorithm for prediction. In the second phase, the support vector machine is used as a classifier with 95.71% accuracy, and the accuracy of the detection after using the SVM classifier with the proposed model is enhanced to 95.71%
Diagnosis of pomegranate plant diseases using neural network	K-means clustering, ANN	They extracted the features from the leaf texture utilizing GLCM method and fed these features to the artificial neural network which was able to reach 90% accuracy
Identification of plant disease infection using soft computing: Application to modern botany	Fuzzy-based classifier,	This experiment was able to identify plant leaf disease with 96% without using neural network and minimal computational load
Plant disease detection and its solution using image classification	K-means clustering, Otsu's classifier Voice navigation,	Implementing improvised K-means with Otsu's algorithm helps to segment and analyzes the infected area of the leaf

(continued)

Table 1 (continued)

Title	Technique	Accuracy
Soybean plant disease identification using convolutional neural network	Convolutional network	CNN is used to diagnose and classify the disease in soya bean plant. Neural network is trained by images which is captured in the natural environment and achieved 99.32%
On using transfer learning for plant disease detection	CNN, transfer learning	Deep learning technique is used for plant ailments detection and also differentiate between healthy and unhealthy plants with 98.2% accuracy
Using deep transfer learning for image-based plant disease identification	CNN, transfer leaning	The author achieves validation accuracy 91.83% on public data set and average accuracy reached 92% for predicting rice disease prediction

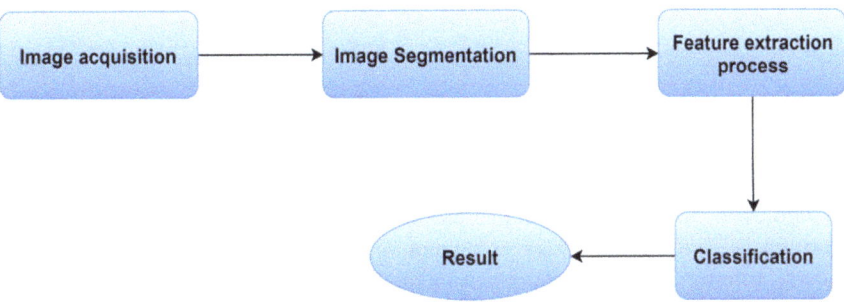

Fig. 1 Plant disease detection process

4.2 Image Segmentation

The main aim is to simplify the presentation of an image, so that it becomes more meaningful and effectively work on analyzing part. As the foundation of the extraction process, this step is also a technique to image processing.

4.3 Feature Extraction Process and Classification

Following the image segmentation process, the result so far achieved is the area of interest. Hence, in this phase, the characteristics of this field of interest need to be extracted. These functions are required to determine the meaning of a sample image.

Features may be based on color, size, form, and texture [2]. After extraction, the classification aims at determining whether the input image corresponds or not.

5 Proposed Model

Our main aim is to merge these algorithms into one hybrid algorithm that will save the whole execution time of the process as well as it improve the efficiency of prediction.

In Fig. 2, at the first level, we will take raw input with different variables, then the raw data transfer for segmentation and feature extraction with the help of fuzzy C-means which iteratively minimized the objective function, and also it is useful for authenticating known or suggesting substructure in unexplored data. Fuzzy c-means algorithm helps to segment and classifying the data efficiently, and then, we will generate a random list of training and testing data set. This will become input to artificial neural network. The data set will be used to train the ANN algorithm for the better selection, and then, a Naïve classifier algorithm will use for probability to predict the class from unknown data set also for better prediction at the final selection method.

6 Discussion

Basically, it describes that the plant diagnosis and detection process has three phases, i.e., segmentation, extraction (feature), and classification. Mostly K-means clustering algorithm is used for the image segmentation process. For the feature extraction process, different color co-occurrence method was applied such as gray level co-occurrence matrix, and for classification different also, applied such as artificial neural network (ANN), support vector machine (SVM), and backpropagation neural network (BPNN), but as comparative study, SVM is simple and robust though every algorithm has own pros and cons. With this inspiration, we expect that our model will perform much better in comparison of other models.

7 Conclusion

This review paper presents synopsis of different techniques for plant diseases diagnosis and detection using image processing, segmentation, etc., which have been used by numerous researchers according to their requirements. The main approaches applied are: K-means clustering, neural network, backpropagation neural network, Otsu's algorithm, fuzzy-based classifier, etc., for image classification and segmentation. This review sum up that leaf disease detection techniques show the accuracy and effectiveness to implement the system developed for diagnosing and detect plant

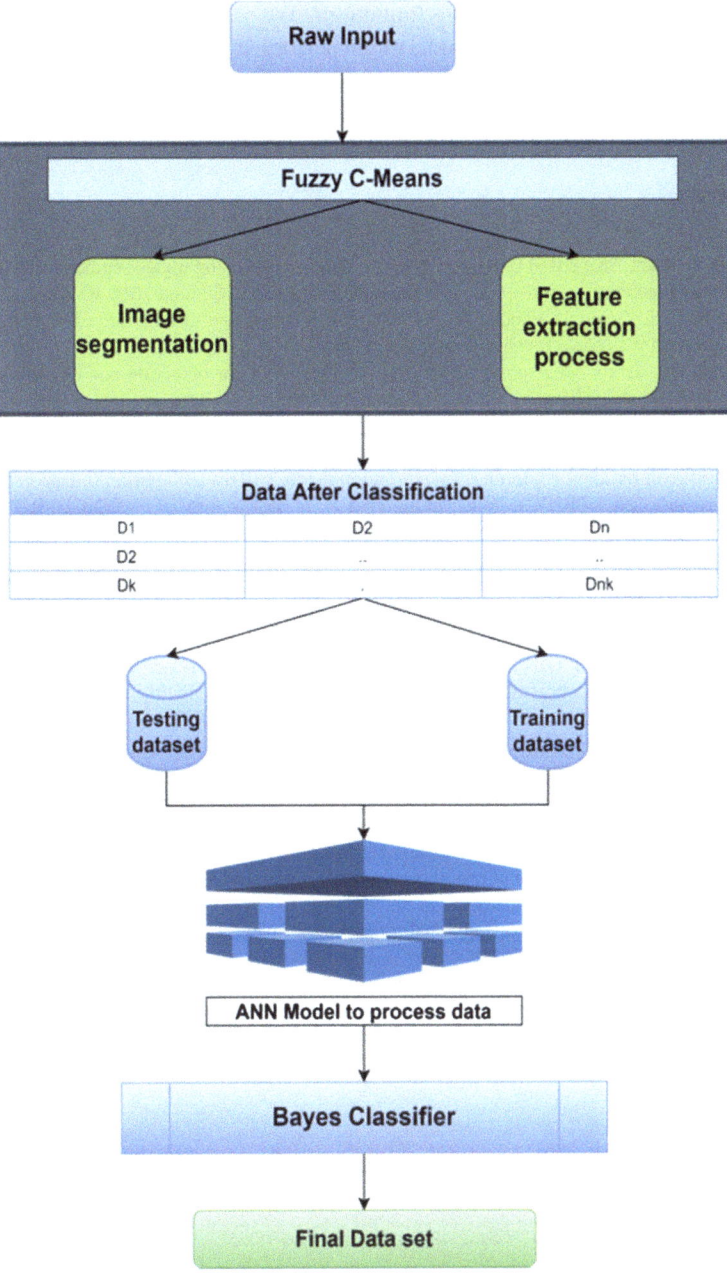

Fig. 2 Hybrid fuzzy C-means-ANN-Bayes classifier for plant leaf disease diagnose and detection

leaf diseases, in addition to having limitations. We also proposed our model after reviewing the paper; this hybrid model might perform effectively as compare to other algorithms.

References

1. Singh V, Misra AK (2016) Detection of plant leaf diseases using image segmentation and soft computing techniques. Info Proc Agri. https://doi.org/10.1016/j.inpa.2016.10.005
2. Chaudhary S, Kumar U, Pandey A (2019) A review: crop plant disease detection using image processing. Int J Innov Technol Explor Eng 8(7)
3. Bashish DA, Braik M, Ahmad SB (2010) A Framework for detection and classification of plant leaf and stem diseases. In: International conference on signal and image processing, pp 113–118
4. Dhakate M, Ingole AB (2015) Diagnosis of pomegranate plant diseases using neural network. In: 2015 fifth national conference on computer vision, pattern recognition, image processing and graphics (NCVPRIPG), Patna, India, pp 1–4. https://doi.org/10.1109/NCVPRIPG.2015.7490056
5. Sladojevic S, Arsenovic M, Anderla A, Culibrk D, Stefanovic D (2016) Deep neural networks based recognition of plant diseases by leaf image classification. In: Hindawi Publishing Corporation computational intelligence and neuroscience, vol 2016, Article ID 3289801, 11 pp
6. Kiani E, Mamedov T (2017) Identification of plant disease infection using soft-computing: application to modern botany. In: 9th international conference on theory and application of soft computing, computing with words and perception, vol 120, pp 893–900
7. Saradhambal G, Dhivya R, Latha S, Rajesh R (2018) Plant disease detection and its Soltion using image classification. Int J Pure Appl Math 119(14):879–884
8. Wallelign S, Polceanu M, Buche C (2018) Soybean plant disease identification using convolutional neural network. In: FLAIRS conference
9. Santhana Hari S, Sivakumar M, Renuga P, Karthikeyan S (2019) Detection of plant disease by leaf image using convolutional neural network. In: International conference on vision towards emerging trends in communication and networking (ViTECoN)
10. Sagar A, Jacob D (2020) On using transfer learning for plant disease detection. https://doi.org/10.13140/RG.2.2.12224.15360/1
11. Chen J, Chen J, Zhang D, Sun Y, Nanehkaran YA (2020) Using deep transfer learning for image-based plant disease identification. Comput Electron Agric 173:105393. ISSN 0168-1699

Artificial Intelligence in Healthcare: Diabetic Retinopathy

Ajay and Manish Pandey

Abstract Artificial Intelligence (AI) has had a major influence on the medical sector. It has the potential to be used for mass screening and may also aid in accurate diagnosis. AI technology is primarily used in optometry and ophthalmology to treat diseases with a high prevalence, such as Retinal Vein Occlusion, Cataract, Age-related Macular Degeneration (AMD), Retinopathy of Prematurity, Diabetic Retinopathy (DR), and Glaucoma. Diabetic Retinopathy is becoming more common among these. It can result in permanent blindness if not diagnosed in a timely manner. As a result, any technologies that can aid in rapid screening, while reducing the need for qualified human resources will likely be beneficial to both patients and ophthalmologists. This paper represents how AI can help with diabetic retinopathy by using *Image Recognition* technique.

Keywords Artificial intelligence · Healthcare · Machine learning · Diabetic retinopathy · Image recognition

1 Introduction

Artificial Intelligence can be defined as the intelligence of machines, as opposed to the intelligence of humans or other living species. AI in simple words means to accomplish a task mainly by a computer or a robot, with none or a minimal involvement of human beings. Simply, AI is a simulation of human intelligence by a software or a machine. It is the ability of a computerized system to show consequent abilities [1].

AI is a hybrid framework that incorporates both software and hardware. Algorithms are at the core of AI in software. A computational framework, known as Artificial Neural Network (ANN), is needed for the execution of algorithms [2]. The artificial intelligence unit ANN is designed to counterfeit the workings of the human

Ajay (✉) · M. Pandey
Department of Computer Science & Engineering, Maulana Azad National Institute of Technology, Bhopal, Madhya Pradesh, India

© The Author(s), under exclusive license to Springer Nature Singapore Pte Ltd. 2022   459
D. Gupta et al. (eds.), *Proceedings of Data Analytics and Management*,
Lecture Notes on Data Engineering and Communications Technologies 90,
https://doi.org/10.1007/978-981-16-6289-8_39

brain. It is the cornerstone of artificial intelligence, and it solves problems that would be impossible or difficult to solve by human or mathematical standards. ANNs have the ability to self-learn, helping them to achieve better results as more data becomes available.

AI systems required a database that would enable them to learn simple targets associated with specific findings or diseases. Learning can be:

1. **Supervised learning**: In supervised learning, the work is to ascertain a function from pair of instances of inputs and outputs that assign an input to an output.
2. **Unsupervised learning**: In unsupervised learning, the aim is to imbibe from unstamped, uncategorized, or unclassified test data to recognize casual visage in the data, rather than reacting to system assessment, to react found on the presence or absence of known casual features in recent data.
3. **Reinforced learning**: In reinforced learning, the job is to operate within the specified constraints to optimize benefits and penalties in some sort of incremental nature [3].

The implementation of neural network (NN) algorithms on a physical data processing device is what AI is all about in hardware. The most straightforward method is to use a multithread or multicore general-purpose Processor to implement a NN algorithm. Graphical Processing Units (GPUs), which are used for convolutional computations, are more efficient than CPUs for extensive NN. Both of these co-processing methods are more effective than relying solely on the CPU [4].

As AI software and hardware innovations advance, AI is increasingly being used in biomedical fields to further improve analysis and patient outcomes, thus improving the overall effectiveness of the healthcare industry. Healthcare's key aim is to become more and more participatory, intimate, predictive, and preventive in nature. AI has made significant contributions in these areas. If we look at the progress made, we can expect that AI will develop and expand as a sovereign means for bio-medicine. Section 2 of the rest of this paper includes a description of information processing and implementation of the algorithm. The third section discusses the prediction and diagnosis of diseases. Case studies of diabetic retinopathy can be found in Sect. 4. Section 5 ends with a review of the results and challenges.

2 Information Processing

Artificial Intelligence (AI) systems needed to be trained before deployment of in healthcare by using data formed by clinical exercises such as treatment assignment, diagnosis, screening, etc., so that they could learn identical groups of subjects, correlations among subject features, as well as desired outcomes. Demographics, electronic recordings of medical equipment, medical documents, physical observations, and clinical laboratory and image evidence are all examples of clinical data [5]. To improve the accuracy of congenital anomaly diagnosis, we need AI techniques to extract phenotypic features from these results.

The majority of AI devices fall into one of two groups:

1. **Machine Learning (ML)**: Structure data, such as genetic, imaging, and EP data, is analyzed by machine learning. In medical systems, machine learning algorithms aim to cluster patients' characteristics or predict disease outcomes [6].
2. **Natural Language Processing (NLP)**: NLP is a collection of techniques for extracting information from unstructured records, such as medical journals and clinical notes, in order to supplement and improve organized medical data. The aim of NLP procedures is to transform the text into machine-understandable structured data that can be analyzed using machine learning technologies [7].

Figure 1 illustrates the journey of clinical decision making from clinical data generation, along with NLP data improvement and machine learning data analysis. The road map, we observe, begins and ends with clinical operations. AI, no matter how powerful it is, must be persuaded by clinical predicament and eventually used to reinforce the clinical practice.

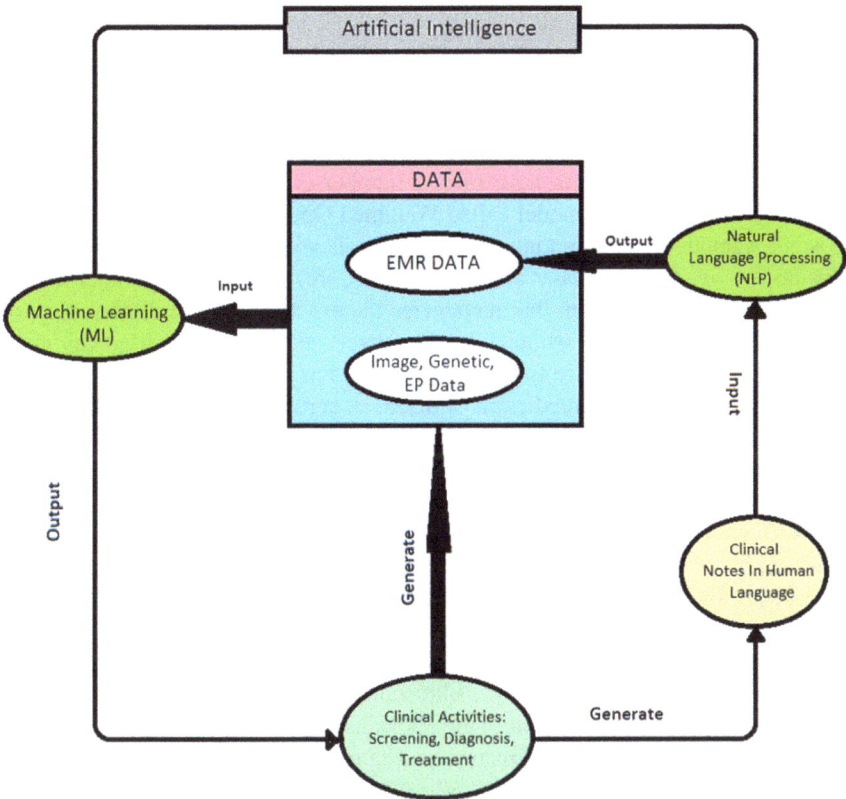

Fig. 1 The blueprint for clinical data creation, data enhancement with NLP, data interpretation with ML, and clinical decision-making. EMR and EP

Essentially, it is a procedure of training a computer to identify particular patterns. It is being used for a variety of abstruse tasks in the past, including precise classification of high-resolution images. Machine learning techniques, natural language processing methods, voice, vision, expert systems, robotics, and other AI devices' methods of operation are broadly divided into the following main categories: [1] machine learning techniques, natural language processing methods, robotics, speech, expert systems, vision, etc. Machine learning capabilities have been used more in ophthalmology so far [8].

The training set and validation set are the two key sections of the machine learning process. This is accomplished by supplying a large amount of training data to the machine/system, such as lacs of retinal scans of waffling grades of DR as the training set [9]. The bulk of the data is pre-labeled with functionality by validated professionals. Machine grasps how to grade diabetic retinopathy (DR) by constructing a model of multifarious network among input data and postulating a performance regulation after being exposed to numerous labeled retinal photos. Other data, known as the validation collection, is also used to validate the entrenched algorithm.

3   Prediction and Diagnosis of Diseases

The most pressing need of Artificial Intelligence in bio-medicine is in disease diagnosis [10]. In this area, there have been a number of notable breakthroughs like an ensemble classification model called Weighted Optimized Neural Network and Boosting is applied for early lung cancer diagnosis with a higher accuracy rate AI helps doctors to provide earlier and more precise treatments for a number of diseases [11], and an AI system that recognizes the exact size of tumor and space of tumor in the brain using an artificial neural network algorithm [12]. Medical imaging and signal processing are two other essential forms of disease diagnostics. In the diagnosis, control, and prediction of diseases, certain methods have been used. AI is being used to derive biomedical signal features like electroencephalography (EEG), electrocardiography(ECG) and electromyography (EMG) for signal processing. Epileptic seizure prediction is an important application of EEG. In later years, AI has been recognized as one of the most critical components of a precise and accurate prediction framework. AI may also play a key role in biomedical image processing-based diagnosis [13]. To improve image quality and analysis performance, AI has been used in thermal imaging, multidimensional imaging, and image segmentation. AI may even be embedded in portable ultrasonic systems, allowing untrained individuals in underdeveloped areas to use ultrasound as an effective tool for diagnosing a variety of illnesses. In accession to the above uses, AI can help standard decision support systems (DSSs) increase diagnostic accuracy and ease disease management, reducing staff workload. The above applications show that AI has the potential to be an advantageous tool for diagnosing, managing, and even predicting diseases and patient conditions early and accurately.

4 Diabetic Retinopathy and Artificial Intelligence

4.1 Diabetic Retinopathy

Diabetic Retinopathy (DR) is an eye disease that causes mild to extreme vision loss and also the main reason for blindness in people of working age who have had diabetes for a long time.

The enormous per capita expense adds to the health burden. With the advent of anti-VEGF drugs, this has risen even further. Sometimes, signs do not appear until the disease has progressed; however, if diagnosed early, vision deterioration can be avoided by early intercession, which also happens to be the most remunerative choice. Given the unprecedented increase in the number of diabetics and the lack of qualified retinal technicians and ophthalmologists, an automated computer-based scrutiny of fundus images will alleviate the pressure on healthcare systems in terms of DR screening and provide a near-ideal system for its management. As a result, screening would be effective at any stage of the disease and will help 90% of patients avoid blindness [14].

4.2 Problem and Solution

Diabetic retinopathy is the most frequent cause of new cases of blindness among adults aged 20–74 years. During the first two decades of disease, nearly all patients with type 1 diabetes and >60% of patients with type 2 diabetes have retinopathy[15].

Irrespective of the nature of diabetes, all people diagnosed with it need regular retinal screening in order to detect diabetic retinopathy early and effectively treat it [16].Traditionally, retinopathy screening has been performed by ophthalmologists performing fundus examinations or by qualified eye technicians or optometrists using color fundus photography with traditional fundus cameras using conventional fundus cameras.

The main concern is the grading of retinal scans by ophthalmologists or other qualified individuals, who are in short supply relative to the number of patients who need to be screened. Second, many patients live in remote areas and are unable to see an eye doctor. Third, since such long-term follow-ups are needed, patients' attitudes and/or behaviors have a detrimental effect on their practice, despite being aware of the consequences.

These problems can be resolved by the availability of an automated imaging device within the easy reach of the patient. As a result, there has been an increase in the interest in the creation of automatic analysis tools for the analysis of retinal images in people with diabetes using AI, thus solving at least a portion of the problem [17]. ?: There are just 11 ophthalmologists per million people in India. In India, nearly 80 million people are suffering from diabetes out of which 55 million peoples are affected by Diabetic retinopathy (DR) with the number expected to rise to 80 million by 2030. So, approximately 65% of diabetes patients are affected by Diabetic retinopathy (DR). A doctor will examine 70–80 people every day, which ensures that only 40–50 patients can be prescribed per day. Giving prescriptions to all patients will take 5 years approximately. However, in the case of Diabetic retinopathy (DR), prompt care is necessary. As a result, patients who do not receive care in a timely manner risk losing their eyesight permanently [18].

The Solution: To address this problem, AI has the potential to be a blessing to humanity. We may use AI to perform screening. For this, image recognition is used. In **Image Recognition**, tagged photographs of objects like cats and dogs are used to train image recognition models. The algorithm learns to recognize new images without the assistance of humans after looking at thousands of examples.

4.3 Algorithm

For our retinopathy project, over 100,000 eye scans were graded by eye doctors. Each eye scan was scored on a scale of one to five, with one being stable and five being diseased. These images were then fed into a machine learning algorithm for training. With training, the AI was able to predict which eyes had disease symptoms.

The algorithm will now be fed the scans of a patient's eyes. Nobody would need to see an ophthalmologist personally because the algorithm will produce the input scans with a grading. If a person is unsure whether or not they have diabetic retinopathy, he or she may go under screening. After screening, they will decide whether or not they need to see an ophthalmologist based on the grading of their scan results.

4.4 Result

Now, the ophthalmologist just has to see patients who are most likely suffering from diabetic retinopathy. Also, ophthalmologist doesn't need to examine the patients. He has to only write prescriptions according to reports generated by the AI system. As a result, instead of just 40–50 patients per day, a doctor will write prescriptions for 250–300 (nearly 6 times of previous numbers).

Before AI system, the time required to tackle all patients = 5 years approximately.
Now, the time required to tackle all patients = 250 days approximately.

5 Challenges and Conclusion

The paper discussed why AI is being used in healthcare shows how AI has evaluated various types of healthcare data, and surveys the most common disease types for which AI has been used. The two main categories of AI products, machine learning and natural language processing, were then explored in depth. An effective AI system must have both a machine learning peripheral for dealing with structured data, and a natural language processing component for prospecting unstructured texts. Until the system will inform about patients with disease diagnosis and treatment decisions, the advanced algorithms must be trained using healthcare data.

5.1 Challenges

The main stumbling block is data transfer. AI systems must be constantly trained by data from clinical trials in order to function properly. However, once an AI system has been implemented and trained with factual data, maintaining data supply is transformed into a critical issue for the system's continued growth and progress.

5.2 Conclusion

It has been addressed why AI is being used in healthcare, the various healthcare data that AI has analyzed, and the major disease categories for which AI has been used. We discovered that while AI technologies are gaining a lot of interest in medical science, real-world implementation is still a challenge. Apart from these, we've seen how AI has miraculously changed healthcare. So, we can say that AI will definitely help doctors in making better clinical decisions or in certain realistic fields of healthcare, artificial intelligence may also replace human judgment.

References

1. Jiang F et al (2017) Artificial intelligence in healthcare: past, present and future. Stroke Vasc Neurol 2(2):230–243
2. Hopfield JJ (1982) Neural networks and physical systems with emergent collective computational abilities. Proc Natl Acad Sci USA 79(8):2554–2558
3. Schmidhuber J (2015) Deep learning in neural networks: an overview. Neural Netw 61:85–117
4. Naveros F et al (2018) Corrigendum: event- and time-driven techniques using parallel CPU-GPU co-processing for spiking neural networks. Front Neuroinform 12:24
5. Administration UFaD (2013) Guidance for industry: electronic source data in clinical investigations
6. Darcy AM et al (2016) Machine learning and the profession of medicine. JAMA 315:551–552
7. Murff HJ et al (2011) Automated identification of postoperative complications within an electronic medical record using natural language processing. JAMA 306:848–855
8. Murff HJ et al (2011) Automated identification of postoperative complications within an electronic medical record using natural language processing. JAMA 306:848–855
9. Lee A et al (2017) Machine learning has arrived. Ophthalmology 124:1726–1728
10. Sajda P (2006) Machine learning for detection and diagnosis of disease. Annu Rev Biomed Eng 8:537–565

11. Alzubi JA et al (2019) Boosted neural network ensemble classification for lung cancer disease diagnosis. Appl Soft Comput 80:579–591. https://doi.org/10.1016/j.asoc.2019.04.031
12. Alzubi JA et al (2019) Efficient approaches for prediction of brain tumor using machine learning techniques. Indian J Public Health Res Dev 10(2)
13. Stoitsis J et al (2006) Computer aided diagnosis based on medical image processing and artificial intelligence methods. Nucl Instrum Methods Phys Res A 569(2):591–595
14. Vashist P et al (2011) Role of early screening for diabetic retinopathy in patients with diabetes mellitus: an overview. Indian J Community Med 36:247–252
15. Fong DS et al (2004) Retinopathy in diabetes. Diabetes Care 27(suppl 1):s84–s87
16. Namperumalswamy P et al (2003) Developing a screening program to detect sight threatening retinopathy in south India. Diabetes Care 26:1831–1835
17. Bhaskaranand M et al (2016) Automated diabetic retinopathy screening and monitoring using retinal fundus image analysis. J Diabetes Sci Technol 10:254–261
18. Zheng Y et al (2012) The worldwide epidemic of diabetic retinopathy. Indian J Ophthalmol 60(5):428–431

Prediction of Heart Disease and Chronic Kidney Disease Based on Internet of Things Using RNN Algorithm

S. Chitra and V. Jayalakshmi

Abstract Nowadays, people are very serious and aware about their health condition, even though people are running towards their work and busy time schedule, they are not taking care of themselves until it shows any kind of symptoms. Weight gain and obesity are the two issues affecting productivity and quality of life throughout the globe. The internet of things (IoT) plays an important role in achieving the shared goal by linking, detecting, recognising and processing data between devices or services. IoT in healthcare provides us the advantages of monitoring, analysing, diagnosing and controlling of the different conditions of overweight and obesity. Also provides solution for the prevention of weight gain and obesity. Since IoT has a limited resources of objects used in it, another alternative method is being introduced for the above mentioned advantages such as machine learning. People who have a high risk of cardiovascular disease may also have a chance of getting kidney diseases, and it can be treated accordingly with the help of historical medical records. But chronic kidney disease (CKD) is an illness that shows no diagnostic signs at all and it is difficult to identify, detect and prevent CKDs and it may sometimes permanently harm the health system, so that the prediction and analysis of therapy are done in machine learning. The main aim of this study is to establish a predictive model for the CKD heart disease data to analyse different open-source Python module and to achieve outcomes, and the 96% prediction and precision machine learning methods may be established by the comparison with various algorithms such as K-nearest neighbours (KNNs) and recurrent neural network (RNN). A data set which is gathered from patient's medical history is predicted by using this algorithm. Based on the amount of potassium in patient's blood, the predicted value provides us the clarification that the person will get chronic kidney disease or not.

Keywords Chronic kidney disease · K-nearest neighbours · IoT · Recurrent neural network · Machine learning

S. Chitra (✉)
Department of Computer Science, Immaculate College for Women, Cuddalore, India

V. Jayalakshmi
Department of Computer Applications, VISTAS, Chennai, India
e-mail: jayasekar.scs@velsuniv.ac.in

© The Author(s), under exclusive license to Springer Nature Singapore Pte Ltd. 2022   467
D. Gupta et al. (eds.), *Proceedings of Data Analytics and Management*,
Lecture Notes on Data Engineering and Communications Technologies 90,
https://doi.org/10.1007/978-981-16-6289-8_40

1 Introduction

Coronary disease is the world's leading cause of mortality these days. Foresight of coronary artery disease is a difficult task since it needs intelligence and cutting-edge knowledge. internet of things (IoT) innovation in the medical care framework has been accepted late for collecting sensor appreciation for the analysis and prediction of cardiovascular diseases. Many experts have examined coronary disease findings but the accuracy of the determination outcomes is poor. An IoT structure for assessing cardiac disease, using a modified deep convolutional neural network (MDCNN), is suggested to solve this problem. The clever and cardiovascular gadget attached to the patient monitors such as the pulse and electrocardiogram (ECG) [1]. Globally, early conclusion and therapy may enhance the prediction of coronary disease. However, current planned coronary disease diagnostic frameworks are hindered by huge necessary information. This paper [2] describes the internet of a clinical device based on subjects for collecting cardiac sensitivities in patients with heart disease.

The latest developments in IoT and innovation detection may be used for online healthcare administrations. The huge quantity of data is formed in the healthcare sector by IoT devices, and distributed computer methods were used to manage the monster measurement of information. In order to gain from the customer using online medical services, a new cloud is developed as an IoT-based healthcare application for screening despite analysing real diseases. The UCI repository data set is utilised to perform an effective system for coronary heart disease, as is the usage of medical care sensors to predict those who suffer heart disease sickness [3]. The fog-based IoT approach may be helpful for patients with cardiovascular infection from remote areas. Normally, a specialised cardiologist is not available in such remote areas.

A few frameworks are available to order coronary heart disease and yet they just take advantage of recommendations in these present frameworks. We present an IoT-based expert local recommendation system for analysing cardiovascular disease and its kind and offering suggested physical and nutritional measures. The first section expects the data from the patient to be collected remotely by using biosensors [4].

IoT is entomb correspondence of the gadgets placed using various organisational advances. IoT invention is excellent to convert later into the next pattern. A framework for checking medical services consisting of ECG sensors. The limits that are important is recognised by the ECG sensors which are essential for the remote control of patients. A flexible application perception is used to constantly check the patient's ECG, and the ECG wave offers various methods for extracting information to correctly predict cardiac disease [5]. IoT and cloud stages are now widely used in many medical care applications. On the cloud stage, instead of subject to restricted capacity and calculating assets, the enormous quantity of information supplied by the IoT devices in the medical field may be analysed. In order to provide strong restorative administrations, an emotionally supportive online clinical choice network for chronic kidney disease (CKD) is provided in this paper [6].

CKD is often seen as a well-being problem, especially in non-industrial countries, where the acceptance of suitable medications is exorbitant. Thus, early CKD expectations that safeguard the kidney and stop the gradual development of CKD have become a major problem for physicians and researchers. In the IoT world-view, easy body sensor and intelligent media clinical devices are used to provide far-reaching observation of kidney work, an important role is assumed especially if clinical considerations are not really accessible to the great majority of people [7, 8]. Cloud processing and IoT have a major role in medical treatment, especially in predicting diseases in bright urban areas.

The internet of medical things (IoMT) has several uses in medical care by inte-grating wellness checks for sensors and clinical equipment in order to remotely notice patient records in order to make the federal medical authorities more smart and clever. In order to help clients using e-wellbeing apps in the finest medical services, an IoT with a cloud-based clinical choice network emotionally supporting awareness and identification of chronic kidney disease (CKD) with its seriousness [9] was included in this paper [10].

The surplus regions are followed by Sect. 2 related works, Sect. 3 proposed methodology, Sect. 4 experimental findings and Sect. 5 concluding.

2 Related Works

As of late, heart disease is the world's leading cause of mortality. The coronary episode is a mind-boggling endeavour for a clinician, since more knowledge and information are needed. Nonetheless, pulse observation is the primary estimate element that affects respiratory failure with other wellnesses such as circulatory strain, serum cholesterol and glucose levels. In times of rapid internet of things (IoT), the pulse check sensors fill patients' accessibility. In this paper [11], I explained the pulsation and other data checking technique design and also revealed how to use an AI method such as a KNN characterisation computation, to predict the coronary episode, using pulse and other wellness associated perimeter information collected. Overall, continuing illnesses such as diabetes, heart disease (HD), malignancy and contin-uous respiratory disease are the major causes of mortality. It is remarkably puzzling to establish that HD has different manifestations or highlights. As the popularity of skilful wearable gadgets expands, there is a chance to provide an internet of things (IoT) arrangement.

Tragically, those with sudden cardiac failures have poor endurance rates [12]. The universal growth of the internet of things (IoT) and its clinical applications has enhanced the suitability of older people or patients with long-term individual consideration in far-off well-being check frameworks. Continuous illnesses such as strokes, coronary heart disease, diabetes, malignant development and continuous respiratory infections are now important causes of mortality in many parts of the globe. We suggest in this article, a patient observation framework for people who are affected by strokes to limit future repeats of something quite similar by disrupting the specialist and guardian of the range of hazardous components of stroke [13].

Coronary disease is probably the world's leading death source today. The cardio-vascular disease forecast is a fundamental clinical information test nearby. AI (ML) has been shown to assist settle decisions and predictions from the huge quantity of information generated by the medical care sector. We have also observed ML tech-niques in late progress in many areas of the internet of things (IoT) [13]. Health is not all that matters, notwithstanding, all the other things is not anything without well-being. At that point and now, individuals are evaluating ways which can expand the lifespan of life. In any case, innovation is far away from accomplishing this objec-tive of diminishing the death rate. Nonetheless, for a beginning, the starting advances have been finished. Propelling innovation and its effect on people groups' life are now prompting sound ways of life. Solid living propensities, schedules, proactive well-being observing and early identification of infections lead to expanded anticipation of life. Today, the world is embracing internet of things in its every day employ-ments. There are different wearable mechanical devices that have been created to screen/measure diverse well-being ascribes. In any case, none of them measures the recorded information to give future well-being help utilising continuous monitoring [14].

At present, medical care frameworks are updated using cutting-edge capabilities such as AI (ML) and information mining as well as computerised thinking to provide people more skilled and masterful medical treatment. This article provides an exten-sive medical care expectation and arrangement framework, particularly density-based feature selection (DFS) with ant colony-based optimisation (ACO) CKD Calculation. The suggested expert approach identifies DFS's unessential or superfluous highlights prior to the construction of the ACO-based classifier [15]. The prevalence of CKD increases in the present exploratory scenario annually. One of the highlights of extra therapy is the CKD prediction, where the process of machine learning is increas-ingly important in clinical findings due to its high accuracy of grouping. In the new past, characterisation precision calculations rely on the proper use of computations to reduce the amount of information. This study is intended to further the detec-tion, division and analysis of continuous renal deception on the internet of medical things (IoMT) platform, using the heterogeneous modified artificial neural network (HMANN) [16].

Advanced technologies such as IoT and fog have mostly affected today's clin-ical control frameworks. Logical measurement by means of different information research methods is obtained from huge clinical data on patients, contributes to remote clinical observation, early disease conclusion, foresees clinical occasions and suggests urgent welfare/clinical directions. Due to the existence of compa-rable well-being/clinical benefits in meaningful terms, appropriate composite well-being/clinical benefits were found as a major concern in the current clinical context [17]. Big information collected enormous information for applications like as advanced mechanics, IoT and healthcare. Although the IoT-based medical care framework plays a vital role in the large information business for certain situations detection may be difficult to predict the precise result.

The suggested framework with computerised reasoning and IoT may greatly improve the implementation of Parkinson's disease. This research clearly defines the share of the robots in Parkinson's disease and how they relate to a massive information review [18]. Associated items are the key for certain canny frames such as direct admission to physical and physiological characteristics and collection of data on the human body. Our examinations will develop non-intrusive methods to predict dialysis patients' risk of end-stage renal disease (ESRD) in a smart home-dependent internet of things (IoT) framework. Nevertheless, the IoT sectors have several additional challenges in collecting even finer granularity data known as biomarkers [19].

3 Proposed Methodology

The proposed system uses KNN architecture and its features of extraction predict the heart disease and CKD by using neural networks. The main aim of this study is to predict whether a patient is suffering from heart disease or not. A medical data of the patient are taken from the public healthcare. The improvement in quality of the patient's healthcare is reduced to low cost by using IoT-based applications. Nowadays, IoT smart devices are used for monitor and record the patient's medical data. And the advanced telecommunication also provides us to use the IoT-based solutions for the problem of overweight and obesity.

Likewise, wearable and smart IoT devices enable us to regularly collect patient data (e.g. stress levels, sleep quality, blood glucose levels, cardiac rate and burnt calories) regardless of their location (whether the patient is at home or in hospital). By using IoT-based applications, the healthcare sectors are highly benefited to find the solutions for different health problems which are connected to IoT-based sensors and smart devices. And also it improves the patient's personal healthcare. The data are given as input to the model which can predict whether the patient who has heart disease is at risk. Likewise, CKD-based previous medical history data are obtained and used to predict the risk chance of the patient who is having heart disease. The whole process is shown in Fig. 1.

Firstly, prehistory medical data of a heart disease patient and normal person are gathered and recorded in the public healthcare database. With the help of monitoring system, both the histopathology image and parameters were gathered. The extraction and selection of features are done by using the KNN. In the process of utilising the pre-trained data, normal or abnormal data are anticipated. The RNN architecture classification technology helps you estimate the risk of heart disease and chronic kidney disease.

3.1 Neural Network Model

Neural network is a novel process of representing the human brain's data processing mechanism. The NN is used for different kinds of applications like pattern recognition, diagnosis of diseases and to classify data with the help of learning process.

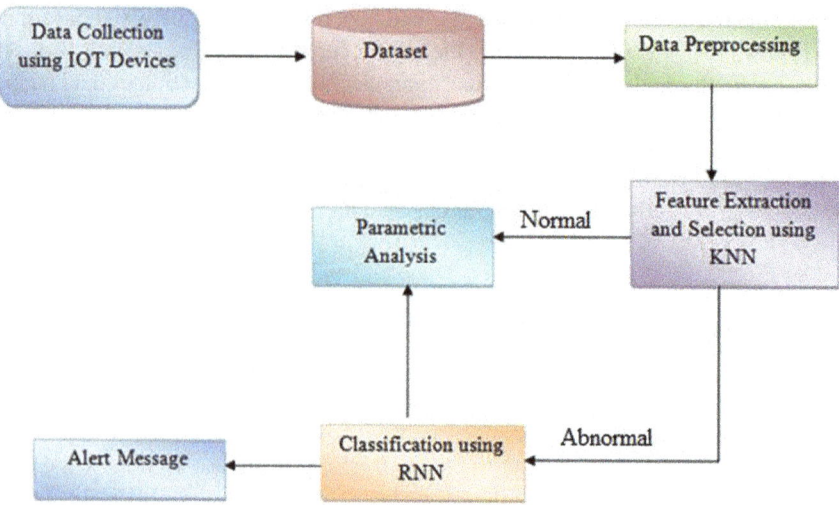

Fig. 1 Architecture for proposed methodology

Figure 2 shows that the NN has different types of networks. They are feedforward network and backpropagation network. And it is made up of different layers.

They are input layers, hidden layers and output layers (decision (CKD or no CKD)) as follows:

- Input layer—In this input units, the raw data are given to network for processing the data.

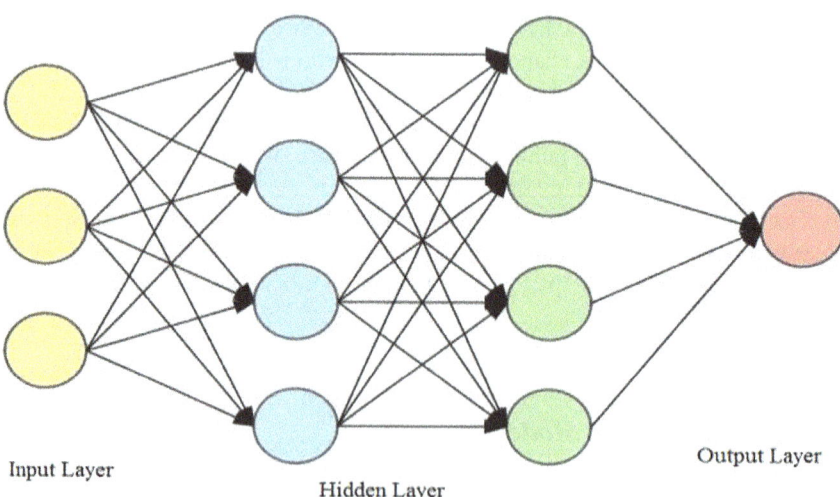

Fig. 2 Neural network model

- Hidden layer—In this layer, the input data and its weights are interconnected.
- Output layer—The result of this unit is based on the action processed in hidden units and weights between output layer and hidden layer.

3.2 K-Nearest Neighbours for Feature Extraction

A Neural network is an artificial neural network which is made up of different layers that combine both process in the extraction and classification and enable it for decision-making. If the prediction is performed, a raw descriptor representation is obtained from hidden layers and it is given as

$$X_1 + 1 = H(W_1 X_1 + B_1), 1 = 1, 2, \ldots, L \tag{1}$$

where H is denoted as activation function, W_1 is denoted as weight matrix, and B_1 is denoted as bias of lth hidden layer, and the parameters are selected according to the rectified linear unit (ReLu). The input which has to be fed into hidden unit is not directly given by training or testing data set. Here, the neural network of three connected layers is used. The neuron of layer t of the hidden layers in layer $t - 1$ is associated with neurons. And with regard to the equation below, each hidden unit generates output.

$$x_{i+1} = \sigma \left(\sum_{i=1}^{n} (w_i x_i + b_i) \right) \tag{2}$$

where w and b are variables of weight and bias, the input is x and the buried layer number of neurons is n. ReLu may be employed with its activation function, $f(x)$ = max, both input and hidden layer $(0, x)$. In the output layer, the sigmoid function is employed. For the loss function, the mean squared error (MSE) is employed and the Adadelta method is used to optimise the MSE loss. Dropout is used to prevent overfitting in the input and hidden layer.

It is a basic algorithm which can store and save all the data, which are available, and classifies the data according to the similar measure. If no unknown data are required, it searches for k more similar data in the complete training data set and simultaneously returns the obtained data with more similar data. A least distance measure is used to find its neighbours. KNN is a simple architecture and mostly for the machine learning processes in which it classifies the pattern recognition and regression. KNN locates its neighbour form the data used in the Euclidean distance.

The value of k finds the similar feature cases with new cases and all finds other new similar cases. Since the 'k' value is very carefully chosen to avoid overfitting problem, if k is loss. The cost of computation becomes high because the training data samples distance from each to 9, KNN algorithm provides a new data point and it is shown in Fig. 3. The classified new point with respect to 'X' is shown as $(0.6, 0.45)$. The two possible classes are existed in the dotted circle in which it has triangle class

Fig. 3 KNN algorithm model

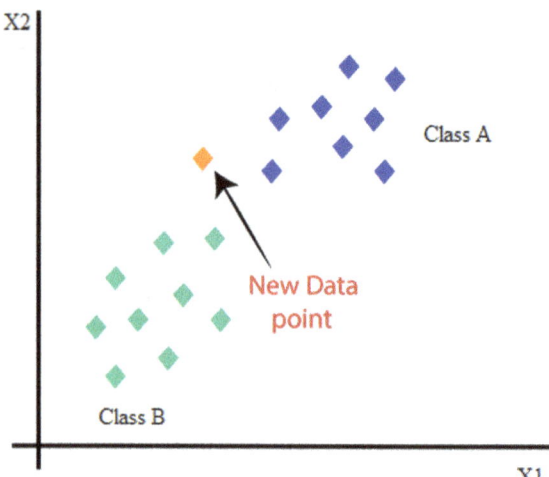

with three objects circle class with five objects. The KNN algorithm which classifies a new point '*X*' to circle class by using the Euclid distance.

3.3   *Recurrent Neural Network (RNN)*

A deep learning is subset of ML which has a numerous non-linear transformations. Different kinds of algorithms are used to learn and view the input data used in many layers with complicated structure. Different types of deep learning architectures are recurrent neural networks (RNNs), deep autoencoders and convolutional neural networks (CNNs). The above algorithms are for natural language processing, speech recognition and medicals field. To store the internal memory RNN is used, in which the data are transferred when it executed from the previous output. The RNN which changes the current forward process nature for adjusting the existing input. Some of the region in the histopathology images are spread across many adjacent slices which gives the output of consecutive similar slices. For solving the complex problems and to classify the numerous data set, RNN algorithm is used and it is made up four individual layers. They are convolution, max pooling, fully connected and output layer. A submedical image of 32×32 of M classes which is classified by a RNN is shown in Fig. 4, where A, B and C are parameters.

The size of the feature maps is reduced and transferred to the further layer when the feature map is evolved by convolution layer and pooling layers, so that the fully connected layer produces the output to predict the class accordingly.

Fig. 4 RNN classification
for proposed system

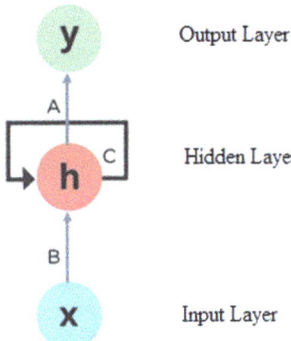

Output Layer

Hidden Layer

Input Layer

Table 1 Attributes of data set

Attribute	Description
Age	Age in years
BP	Mm/Hg
Gravity	1000–10,025
Glucose level	90–120
Albumin	0–5

3.4 Data Set

The data set is used to predict cardiac disease and chronic renal disease using an algorithm for learning from the UCI repository. The collection includes 400 patient records. The data set also includes information on the kidney disease impacted. We have utilised two binary classification classes for CKD and no CKD prediction, and the parameters are shown in Table 1.

3.5 Data Transformation for CKD

In the data transformation process, an additional attribute called "outcome" is added to the current data set. When applying a function to the outcome column, the data set is safe and low. So the values are produced accordingly with the potassium level present in the blood. The main use of this converting process is to apply the numerical data in the machine learning algorithms. Thereafter, 20 attributes are randomly selected form 25 attributes and made into a predictive system model so that if a random values is given for the selected attributes, the system will predict whether the patient has chronic kidney disease or not. If '0' is noted on the display (either by any of the two algorithm, i.e. KNN and RNN), it shows that the patient has not diagnosed with CKD. If '1' is noted on the display, the patient has chronic kidney disease.

```
Confusion matrix :
  [[2 2]
  [1 5]]
Outcome values :
  2 2 1 5
Classification report :
                   precision    recall   f1-score    support

               1       0.67      0.50       0.57          4
               0       0.71      0.83       0.77          6

     micro avg         0.70      0.70       0.70         10
     macro avg         0.69      0.67       0.67         10
  weighted avg         0.70      0.70       0.69         10
```

Fig. 5 Classification output

Then, the confusion matrix error is calculated for the algorithm. The calculated error value gives us the how correctly the model is predicted the CKD in presence of the potassium level in the patient blood.

4 Evaluation Parameters

4.1 Precision and Recall

It is known as positive predictive value, and also it is defined as average probability of relevant retrieval as shown in Fig. 5.

$$\text{Precision} = \text{Number of true positives/Number of true positives} + \text{False positives} \tag{3}$$

$$\text{Recall} = \text{True positives/True positives} + \text{False negative} \tag{4}$$

4.2 Accuracy

As illustrated in Fig. 6, accuracy is defined as the ratio of properly categorised examples to the total number of occurrences in the data set.

$$\text{Accuracy } (A) = \text{No. of classified samples/Total samples} \tag{5}$$

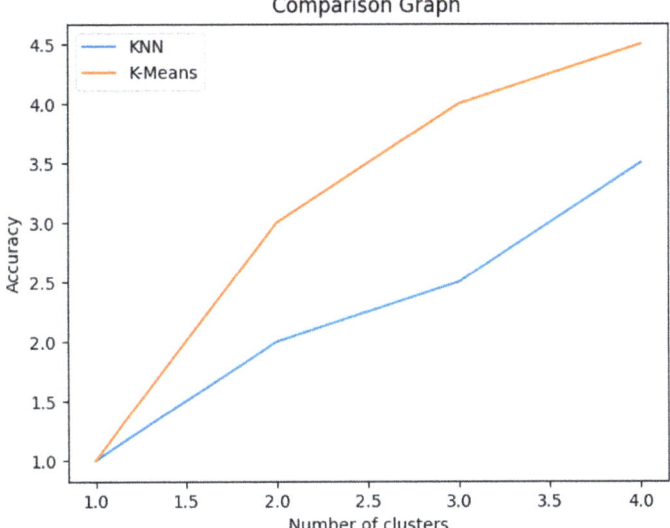

Fig. 6 Comparison graph for accuracy

4.3 Confusion Matrix

The confusion matrix indicates the overall number of accurate and wrong predicted values given by the prediction model when compared to the actual data set classification, and it is represented as an *n*-by-*n* matrix, where *n* represents the number of classes in the data set. The confusion matrix, as shown in Fig. 7, is primarily used to determine the correctness of the classification algorithms which is its primary function.

Fig. 7 Confusion matrix

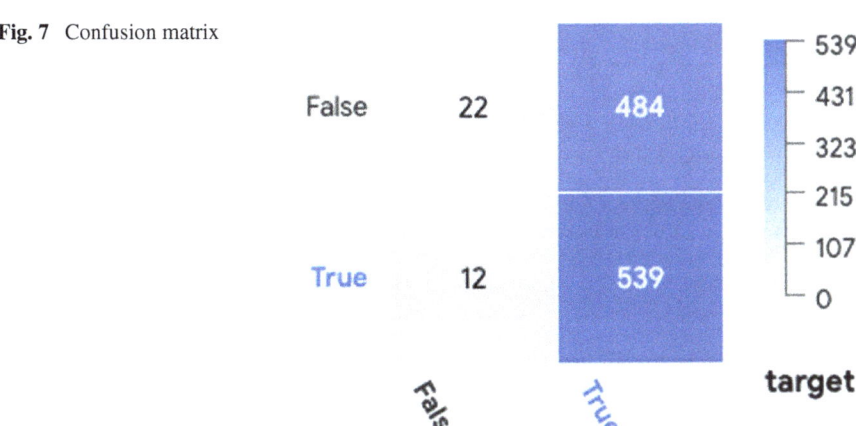

5　Conclusion and Future Work

This research created profiles and modelled chronic kidney disease (CKD) using the EventCKD35, which is considered a goal variable and included gender as an element. There were a total of 12 supervised machine learning algorithms employed in this study. Comparison and assessment with the classification accuracy (CA) and the area under the curve (AUC) are carried out utilising the confusion matrix-based measures. The polynomial SVM was created as a result of this model, and it has a high accuracy and efficiency rate compared to other models. This study began with 253 potential combinations of variables that lead to chronic kidney disease (CKD), particularly in the case of CVDs that are associated with a high risk of developing CKD. Machine learning methods were effectively used in order to investigate CKD, however, some of the data components may need improvement. The Machine learning technique has evolved with the medical sector to provide tools and analyse the data which are related to the diseases. So the machine learning algorithms play an important role to achieve the prior detection of the diseases. This study has given a review of a various machine learning algorithms for prediction of diseases, and its standard data sets are used in several diseases like liver, chronic kidney, breast cancer, heart syndrome, brain tumours and so many diseases. For the detection of the diseases by different ML algorithms, researchers have discovered many outputs and tabulated. Nineteen papers used for the study and compared with various models for predicting the diseases and stated that K-nearest neighbours, and RNN has the best accuracy rate of 96% for prediction. If the factors affect the data sets, feature selection and the number of features, it will also affect the accuracy rate of same algorithm. The other critical points noted and found in this study is that by using various algorithms to obtain the ensemble model, the model's accuracy and performance may get increased.

References

1. Khan MA (2020) An IoT framework for heart disease prediction based on MDCNN classifier. IEEE Access 8:34717–34727
2. Al-Makhadmeh Z, Tolba A (2019) Utilizing IoT wearable medical device for heart disease prediction using higher order Boltzmann model: a classification approach. Measurement 147:106815
3. Ganesan M, Sivakumar N (2019) IoT based heart disease prediction and diagnosis model for healthcare using machine learning models. In: 2019 IEEE international conference on system, computation, automation and networking (ICSCAN). IEEE, pp 1–5
4. Jabeen F, Maqsood M, Ghazanfar MA, Aadil F, Khan S, Khan MF, Mehmood I (2019) An IoT based efficient hybrid recommender system for cardiovascular disease. Peer-to-Peer Netw Appl 12(5):1263–1276
5. Golande A, Sorte P, Suryawanshi V, Yermalkar U, Satpute S (2019) Smart hospital for heart disease prediction using IoT. JOIV: Int J Inf Visual 3(2–2):198–202
6. Arulanthu P, Perumal E (2020) An intelligent IoT with cloud centric medical decision support system for chronic kidney disease prediction. Int J Imaging Syst Technol 30(3):815–827

7. Hosseinzadeh M, Koohpayehzadeh J, Bali AO, Asghari P, Souri A, Mazaherinezhad A, Rawassizadeh R (2020) A diagnostic prediction model for chronic kidney disease in internet of things platform. Multim Tools Appl 1–18

8. Abdelaziz A, Salama AS, Riad AM, Mahmoud AN (2019) A machine learning model for predicting of chronic kidney disease based internet of things and cloud computing in smart cities. In: Security in smart cities: models, applications, and challenges. Springer, Cham, pp 93–114

9. Chimwayi KB, Haris N, Caytiles RD, Iyengar NCS (2017) Risk level prediction of chronic kidney disease using Neuro-fuzzy and hierarchical clustering algorithm(s)

10. Lakshmanaprabu SK, Mohanty SN, Krishnamoorthy S, Uthayakumar J, Shankar K (2019) Online clinical decision support system using optimal deep neural networks. Appl Soft Comput 81:105487

11. Ahmed F (2017) An internet of things (IoT) application for predicting the quantity of future heart attack patients. Int J Comput Appl 164(6)

12. Sarmah SS (2020) An efficient IoT-based patient monitoring and heart disease prediction system using deep learning modified neural network. IEEE Access 8:135784–135797

13. Mohan S, Thirumalai C, Srivastava G (2019) Effective heart disease prediction using hybrid machine learning techniques. IEEE Access 7:81542–81554

14. Gupta A, Yadav S, Shahid S, Venkanna U (2019) HeartCare: IoT based heart disease prediction system. In: 2019 international conference on information technology (ICIT). IEEE, pp 88–93

15. Elhoseny M, Shankar K, Uthayakumar J (2019) Intelligent diagnostic prediction and classification system for chronic kidney disease. Sci Rep 9(1):1–14

16. Ma F, Sun T, Liu L, Jing H (2020) Detection and diagnosis of chronic kidney disease using deep learning-based heterogeneous modified artificial neural network. Futur Gener Comput Syst 111:17–26

17. Asghari P, Rahmani AM, Haj Seyyed Javadi H (2019) A medical monitoring scheme and health-medical service composition model in cloud-based IoT platform. Trans Emerg Telecommun Technol 30(6):e3637

18. Sivaparthipan CB, Muthu BA, Manogaran G, Maram B, Sundarasekar R, Krishnamoorthy S, Chandran K (2019) Innovative and efficient method of robotics for helping the Parkinson's disease patient using IoT in big data analytics. Trans Emerg Telecommun Technol e3838

19. Fki Z, Ammar B, Ayed MB (2018) Machine learning with Internet of Things data for risk prediction: application in ESRD. In: 2018 12th international conference on research challenges in information science (RCIS). IEEE, pp 1–6

Using Big Data Analytics to Analyze Pre- and Post-launch Emotions: A Study of Apple's iPhone 12

Kashish Ara Shakil

Abstract Big data, with its potential to ascertain valued insights, have recently attracted substantial interest from both academics and research practitioners. Big data analytics is increasingly becoming a trending practice in new product development (NPD) using social media. In this article, we analyze the user emotions toward a brands product in particular Apple's iPhone. We compare users' emotions before and after launch of the product, which serve as input for NPD. In this study, we used tweets and comments posted on Twitter and YouTube platforms, respectively, which act as a cheaper, real time, and more accurate alternative to conventional practice of conducting surveys. A total of 20,175 user tweets and comments were collected in two time periods before and after product launch. The user's behavior during the two time periods was compared. It was observed that in pre-launch period anticipation and joy were the common emotions. During the post-launch period, joy and trust were the most used emotions. Thus, the results show that the proposed approach can help the businesses by adjusting their NPD process based upon real-time feedback of consumers expectations and experiences and replace traditional methods.

Keywords Business intelligence · New product development · Big data analytics · Social media · Emotion classification · Emotion analysis

1 Introduction

Social media platforms such as Facebook [1, 2], Twitter [3], Tumblr [4], YouTube [5], and Instagram [6] are considered to be very important source of information in the big data era. It involves constant engagement of people across the globe. Due to this vast user, engagement companies are now focusing toward strengthening their social media community which can lead to better consumer-firm relationship and increased

K. A. Shakil (✉)
Department of Computer Science, College of Computer and Information Sciences, Princess Nourah Bint Abdul Rahman University, Riyadh, Saudi Arabia
e-mail: kashakil@pnu.edu.sa

© The Author(s), under exclusive license to Springer Nature Singapore Pte Ltd. 2022
D. Gupta et al. (eds.), *Proceedings of Data Analytics and Management*,
Lecture Notes on Data Engineering and Communications Technologies 90,
https://doi.org/10.1007/978-981-16-6289-8_41

profits. These platforms are now being utilized for branding and development of new products.

Consumers usually have a lot of expectations and need before making purchasing decisions about any product. Many of these consumers are avid social media platform users. They increasingly share a lot of information about products they want to buy, their needs, expectations, and experiences after purchase. Many people turn to these platforms as a source of information for reviews and feedback from people who have already purchased a product. Thus, information available on these platforms plays a major role in making purchasing decision by consumers. If there is a mismatch between the consumers, expectations from a product and its social media image than the brand can be badly affected. Thus, businesses are constantly trying to align the user expectations with their new products development.

This research contributes to the literature in terms of comparing the user emotions before and after launch of new products. It illustrates the issue how social media can affect new product development NPD. The user emotions are captured from user contents on social media platforms and linked with real-world behavior of users. The literature available for NPD [7] based on social media is very limited with regards to user's emotion analysis. Indeed, most of the existing work focuses on brand sentiments analyzes based on a simple three-point scale, i.e., positive, negative, or neutral. This analysis based upon the user sentiment does not provide much information about user attitude polarity. It does not show the difference in emotions of users before the launch versus after the launch of a new product. These metrics do not help a manager engaged in development of new product to determine which characteristics lead to better sales or pre- and post-sales expectations. Thus, the user analysis should be done beyond the three categories to provide a clear analysis of consumer's expectations.

In this manuscript, given the necessity of monitoring the user emotions toward new product development, to protect a brand and its successive products, we examine user emotions based on tweets and comments posted on two very popular social media platforms, namely Twitter and YouTube, respectively. It highlights that user emotions act as an important source of information in understanding the user's attitude about a product in both its pre- and post-launch phase. This will be a step toward fulfilling the existing gaps in literature and contribute by providing companies with valuable insights about the way people interact about the newly launched product. Thus, the main contribution of this work is to see that how information about user's emotions can be used in new product development. It provides a cost-effective and both a qualitative as well as quantitative mechanism to evaluate customer sentiments.

The rest of the paper is organized in the following manner: Sect. 2 presents theoretical background for this research. Section 3 presents the methodology which has been adopted for this study. Section 4 presents the results obtained and a discussion of the work. Lastly, Sect. 5 synopsizes the paper with conclusion in terms of limitations and future research directions.

2 Related Work

While purchasing any new goods and services, users always try to match their expectations and requirements with their purchases. Several studies in literature have focused on how knowledge of user emotions is related to product innovations. Many consumers make purchasing decisions based on emotional attachment to a product or brand [8].

In [8], the authors have focused on development of new products based upon emotions of its users called as emotion-driven innovation. According to them, the study of user emotions facilitates the designers to come up with new product ideas and features and helps them to identify the associated emotion of consumers linked to a particular category of goods, development of new products based on user emotions. Their work involved direct interaction with the participants in order to know their emotions about products and then these emotions could be translated into final product design. On the contrary, the methodology adopted by us is much more accurate, unbiased, and fast as we have gathered user emotion based upon their social media content. Also, the dataset adopted by us is exhaustive in comparison with the traditional methods.

The choice of social media as an information source for development of new product has been studied in [9]. According to the authors, social media platforms can be used as an informal source of information for collecting consumer preference, product feedback, and current market trends. Even though social media has been used by start-up firms for social media marketing predictions [10]. However, the use of such information sources is still not part of the formal procedures adopted by organizations for NPD.

In [11], the authors have stressed upon how the influence of user's emotions as the main driver for innovations. It explores that unsatisfied and unenthusiastic engagement of consumers has an effect on innovation process. It stresses upon innovation being a by-product of user engagement and leads to improvement in product design and specifications. According to the findings of this research, emotions bring people closer and a study of a community of these users can lead to development of stronger problem driven methods. Thus, in the proposed work, we focus upon study of a community of users interested in a common product, i.e., iPhone 12 and how it affects in new product development.

In [7], the authors have studied the behavior of consumers of three different categories of products pizza, car, and a smartphone. Their work is based on theory of planned behavior [12, 13] which is an influential model for prediction of human behavior in its social environment. According to this theory, user's behavior and their intentions are interrelated with one another and thus users make choices of new features and products based upon their emotional characteristics. This study links the user behavior during pre- and post-launch development of new product and how the user emotions can be used in development of new products. A similar approach has been adopted by us in this work, we have studied human emotions in two temporal phases before and after launch of a product on different social media platforms.

Table 1 Hashtags used and data collected

Launch period	Time of data collection	Hashtags used	Number of tweets and comments collected
Pre-launch	August–October 2020	#iPhone12ProMax #iPhone12Pro #iPhone12 # PRE-ORDER # PRE-launch	10,145
Post-launch	February–March 2021	#iPhone12ProMax #iPhone12Pro #iPhone12	10,030

3 Methodology

This work adopts social media analytics using Twitter and YouTube platform, which have been used as a source of data. It is a cost-effective and time efficient approach for gathering customer requirements and feedback about a product. The usse of social media analytics helps the organizations to come up with product features that are as per the user demands. The following are the steps involved in this study:

3.1 Data Collection

In order to perform this study, data have been collected from two most used social media platforms, i.e., Twitter and YouTube. These platforms enjoy a global following and are frequently referenced by people across the globe to express their opinions and ideas. For extracting tweets, [14] a real-time streaming API provided by Twitter has been used. The user tweets and comments written in English language have been collected and filtered based on the keywords. Table 1 shows a detail of the tweets and comments collected, and keywords are used. The data have been collected in two time periods: The first is the period before the launch of the product, during the months of August to September in the year 2020. The second period is the post-launch period, i.e., after the product has been launched during the months of February to March 2021. A total of 10,145 unique tweets and comments were collected during the pre-launch phase, and around 10,030 comments and tweets were collected during the post-launch phase.

3.2 Data Preprocessing

During this step, all the redundant information is removed from the Twitter tweets and YouTube comments. This is one of the most challenging phases as the English or other language used on Internet is very different from the Standard English. It comprises of slangs, emojis, abbreviations, hashtags, URLs, and spelling mistakes. Therefore, in order to make out some sense from the data, we need to preprocess it. The following are the preprocessing techniques adopted.

1. Removal of non-textual content: Since our analysis is based on user text, we removed non-textual contents such as emojis, etc.
2. Conversion of uppercase characters or entire text to lowercase.
3. Removal of stop words and punctuations.
4. Removing URLs and user mentions.

3.3 Data Analysis

The user emotions reflected in their tweets show the user's state of mind and preferences. An understanding of these users' sentiments and emotions can help firms come up with better product and features depending upon the user's sentiments. It also helps the organizations to improve their marketing strategies as per the market scenario. Users' sentiments such as joy and happiness can be directly linked to the products efficiency.

Thus, in order to assess the sentiments of the users in the pre- and post-product launch era, a lexicon-based [15] method for sentiment analysis has been adopted. This approach relies upon a lexicon or a sentiment-based dictionary, which comprises of words labeled based on their semantics. For this study, we have utilized the NRC emotion intensity lexicon (NRC-EIL) or the NRC affect intensity lexicon (NRC-AIL) [16]. This lexicon comprises of 10,000 English words with scores of intensity corresponding to eight basic emotions proposed by Plutchik [17]. These emotions include anger, anticipation, disgust, fear, joy, sadness, surprise, and trust. However, the intensity scores have not been utilized by us.

4 Results and Discussions

4.1 Pre-launch

In pre-launch analysis phase, all the user tweets and comments before the launch of the product are analyzed. During this phase, we aim at studying the distribution of user along the mood dimensions. After the analysis as shown in Fig. 1, it was observed that anticipation is the most expressed emotion by users (around 34%), expressed by users in form of tweets such as *Mine still says processing but I am anxious to wait, I wanted it by next week* and comments such as *Cannot wait to get mine this weekend*!!." The highest score for anticipation can be attributed to the excitement of users about the launch of new product. Trust (16%) and joy (17%) are

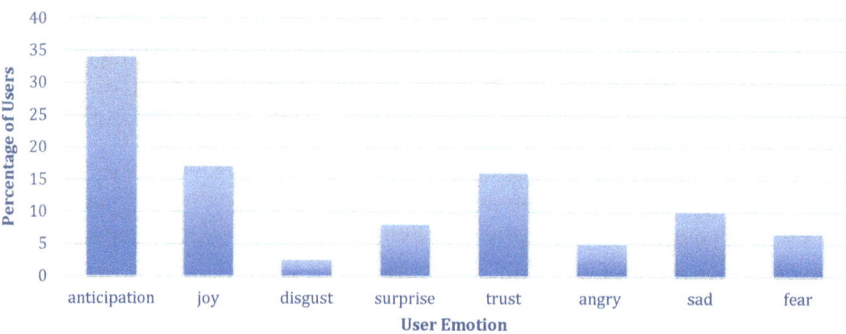

Fig. 1 Pre-launch user emotions

the most expressed emotions followed by anticipation. The users used sentences such as *"You posed 1 angle three times each, and perfect that is great,"* which expresses both the consumer joy as well as excitement before the launch and words such as *"So excited for the new #iPhone12 Pro Max and we bought the case before the phone launch. 18 days to go!!"* expressing the trust of users in the product.

Apart from analyzing the user emotions before the launch of a product, there is also a need to identify the topics used by netizens in each emotion category. To meet this end, a word cloud of the words in each emotion category has been created as shown in Fig. 2. The customers belonging to anticipation category were mostly curious and excited as shown in Fig. 2a. The customers belonging to joy category also expressed their excitement and the pride even after pre-booking of the product through the use of words such as mine, good, and finally. They were eagerly waiting for the product launch and expressed the kind of functions, performance, and features that they are expecting from product. These can be easily analyzed and used for NPD.

4.2   Post-launch

In post-launch analysis phase, all the user tweets and comments after the launch of the product are analyzed. During this phase as shown in Fig. 3, we aim at studying the distribution of user along the mood dimensions. The data that were observed was collected 3–4 months after the launch. On analyzing the post-launch data, it was observed that joy is the most expressed emotion by users (around 28%), expressed by users in form of tweets such as *"I love my @Apple #iPhone12Pro but I hate having that stupid flashlight button on the #iOS14 Lock Screen. I am forever turning it on without wanting to"* and comments such as *"Shifted from #iPhone10 to #iPhone12ProMax. It is too good because of quality make and bigger screen."* The highest score for joy can be attributed to the happiness and satisfaction of users with the new product after their expectations was met. Trust (20%) is the second most expressed emotions followed by anticipation. There is an increase in the trust

Fig. 2 Word categorization of emotions before launch **a** anticipation, **b** joy, **c** disgust, **d** surprise, **e** trust, **f** angry, **g** sad, and **h** fear

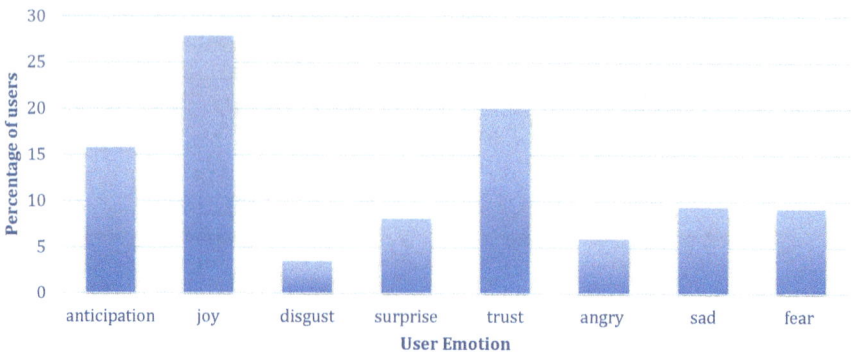

Fig. 3 Post-launch user emotions

percentage (around 5percent) before and after the product launch. It clearly shows that after the launch, the users are satisfied and happy with the product and their trust in the brand and product increases. The users used sentences such as "*I love the shots from my #iPhone12Pro from the north of Denmark,*" which expresses the trust of users in the product. Thus, emotions after the launch of product can be studied by brad managers to observe the impact of the product on their brand value.

An analysis of the topics used under each of the emotion category has been shown in Fig. 4. The users chose words such as beautiful, love, special, and good for expressing their joy. It shows that the users are loving the product and find it beautiful. They trust the product, because they find it as having good timings or speed, is reliable and gives a luxurious feel. They are angry about the battery and wireless issues and also about the high amount of money invested in buying. Thus, the study of emotions after the product launch can give useful insights about the drawbacks and limitations of products. If done within a short period of time after the launch, it can be used for improving the product or coming up with new updates, thus strengthening the trust of the users in the product as well as the associated brands.

Thus, studying user emotions can help the organizations to understand the feeling, expectations, and how much a user feels about a product. These emotions have an important impact on consumers making purchasing decisions. If the users have a positive emotion toward a product such as joy and trust, then they are more likely to buy that good. If the users express negative emotions, they can help the companies by providing as a feedback for improving their quality.

5   Conclusion

Understanding user emotions and sentiments are very important for development of new products and can lead to building up more innovative goods. Netizens share their opinions and comments about different products about to be launched or which

Fig. 4 Word categorization of emotions after launch **a** anticipation, **b** joy, **c** disgust, **d** surprise, **e** trust, **f** angry, **g** sad, and **h** fear

are already available in the market on different social media platforms. These social media posts reflect emotions of users such as joy, trust, surprise, anger, fear, disgust, anticipation, and sadness. An analysis of such user emotions can help organizations in developing new products and deciding what all features should be available in the product. This study shows an analysis of Twitter and YouTube data to help organizations improve their products. It describes the emotions expressed by users before the launch of iPhone 12 mobile phone and after its launch.

Since this method is a cost- and time-effective and much more reliable approach for understanding user emotions. Hence, it can replace traditional methods of capturing consumer's feedback and has a potential to be included as a formal methodology by business developers when strategizing about new product launch. Furthermore, it can be used as a mechanism to generate positive emotions about the product in order to attract consumers and generate more businesses.

The present work analyzes only English language tweets and comments. However, as a future work, we would like to analyze data from multiple languages as we might miss out on important information about a product when we exclude non-English language text.

References

1. Al-Saggaf Y, Simmons P (2015) Social media in Saudi Arabia: exploring its use during two natural disasters. Technol Forecast Soc Chang 95:3–15
2. Alryalat MAA, Rana NP, Sarma HKD, Alzubi JA (2016) An empirical study of facebook adoption among young adults in a Northeastern State of India: validation of extended technology acceptance model (TAM). In: Lecture notes in computer science (including subseries Lecture notes in artificial intelligence and lecture notes in bioinformatics), vol 9844. LNCS, pp 206–218
3. Liu P, Chen W, Ou G, Wang T, Yang D, Lei K (2014) Sarcasm detection in social media based on imbalanced classification. In: Lecture notes in computer science (including subseries Lecture notes in artificial intelligence and lecture notes in bioinformatics), vol 8485. LNCS, pp 459–471
4. Tumblr (2021) "Tumblr." [Online]. Available: https://www.tumblr.com/. Accessed: 31 Mar 2021
5. Kim KH, Beloglazov A, Buyya R (2009) Power-aware provisioning of cloud resources for real-time services. In: Proceedings of the 7th international workshop on middleware for grids, clouds and e-science, MGC'09 held at the ACM/IFIP/USENIX 10th international middleware conference, pp 1–6
6. Worldometers (2020) World population projections—Worldometers. [Online]. Available: https://www.worldometers.info/world-population/world-population-projections/. Accessed: 01 Jan 2020
7. Rathore AK, Ilavarasan PV (2020) Pre- and post-launch emotions in new product development: insights from twitter analytics of three products. Int J Inf Manage 50:111–127
8. Alaniz T, Biazzo S (2019) Emotional design: the development of a process to envision emotion-centric new product ideas. Procedia Comput Sci 158:474–484
9. Bashir N, Papamichail KN, Malik K (2017) Use of social media applications for supporting new product development processes in multinational corporations. Technol Forecast Soc Chang 120:176–183
10. Jung SH, Jeong YJ (2020) Twitter data analytical methodology development for prediction of start-up firms' social media marketing level. Technol Soc 63:101409
11. Brinks V (2020) Fun or frustration? How emotions shape user innovation processes. Emot Space Soc 34:100651
12. Ajzen I (2011) The theory of planned behaviour: reactions and reflections. Psychol Health 26(9):1113–1127
13. Ajzen I, Madden TJ (1986) Prediction of goal-directed behavior: attitudes, intentions, and perceived behavioral control. J Exp Soc Psychol 22(5):453–474
14. "Tweepy." [Online]. Available: https://www.tweepy.org/. Accessed: 01 Apr 2021

15. Taboada M, Brooke J, Tofiloski M, Voll K, Stede M (2011) Lexicon-based methods for sentiment analysis. Comput Linguist 37(2):267–307
16. Mohammad SM (2018) Word affect intensities. In: Proceedings of the 11th edition of the language resources and evaluation conference (LREC-2018)
17. Plutchik R (1984) Emotions: a general psychoevolutionary theory. Approach Emot 1984:197–219

A Deep Learning-Based Feature Extraction Model for Classification Brain Tumor

Astha Jain, Manish Pandey, and Santosh Sahu

Abstract The computer-aided diagnostic-based that supports deep learning (DL) algorithms consists of several processing layers, which symbolize data with several stages of the construct. In current years, deep learning has increased speedily in almost all areas, especially in medical imaging, medical image investigation, or bioinformatics. Therefore, deep learning has effectively untouched or enhanced the methods of recognition, calculation, or diagnosis in many medical and health areas such as pathology, brain tumors, lung cancer, stomach, heart, or retina. As we know there are many applications of deep learning, the purpose of this paper is to appraise the most important deep learning perception related to tumor analysis detection and classification. In recent pretrained models, usually, features are taken from last layers that are different for each dataset from natural to plants to medical images. GLCM feature extraction and ResNet-50 techniques are used for feature extraction and support vector machine (SVM) are used for brain tumor detection and classification to overcome this difficulty in the proposed method. A practical and efficient deep learning model is proposed, in which backpropagation neural network feature is used to predict brain stroke through CT/MRI scan images. The performance and accuracy of proposed model are evaluated and compared with preexisting models, and checked whether it produces high sensitivity, specificity, precision, and accuracy.

Keywords Feature extraction · Deep learning · Magnetic character imaging · GLCM · Support vector machine · Brain tumor classification · Pretrained model

1 Introduction

Generally there are two types of tumors, benign and malignant, out of which the malignant tumor is known as cancer. Brain tumor is referred as the abnormal growth of cells inside the brain. The two general groups of brain tumors are primary and secondary brain tumor. The primary ones starts in the brain and usually stays there.

A. Jain (✉) · M. Pandey · S. Sahu
Department of Computer Science and Engineering, Maulana Azad National Institute of Technology, Bhopal, India

© The Author(s), under exclusive license to Springer Nature Singapore Pte Ltd. 2022   493
D. Gupta et al. (eds.), *Proceedings of Data Analytics and Management*,
Lecture Notes on Data Engineering and Communications Technologies 90,
https://doi.org/10.1007/978-981-16-6289-8_42

The secondary ones are more common; they start somewhere else in the body but travels to the brain. The reason for cause of brain tumors is still unknown. Few of the probable reasons can be several conditions like neurofibromatosis, exposure to chemical vinyl chloride, EpsteinBarr virus, and ionizing radiation. There is no solid evidence but the use of mobile phones is also measured as one of the risk factors. According to the World Health Organization and American Brain Tumor Organization [1–3], for the most part, the common grading system is used which has scale from grade I to grade IV for classification of benign and malignant tumor. According to the scale, benign tumors fall under grade I and II glioma while malignant tumors are of category grade III and IV glioma. The grade I and II glioma are also called low-grade tumor as they possess a slow growth, and the grade III and IV are called high-grade tumor types as they have rapid expansion of tumors. In case the low-grade brain tumor is not treated in time, it is likely to develop into a high-grade malignant brain tumor. Patients with grade II gliomas need to be monitored serially and observed by magnetic timbre imaging (MRI) or computed tomography (CT) scan every 6–12 months. Furthermore, glioblastoma is the most malignant form of astrocytoma that is the highest form of glioma. Unlike other tumor class grades, in glioblastomas, there is abnormal fast growth of blood vessels and necrosis (dead cells) around the tumor. Grade IV tumor class, glioblastoma, is a rapidly growing and compared to other class of tumors, it is a highly malignant form of tumor.

The process of analyzing and manipulating an image in order to perform operation to extract the information from it is known as image processing. The aim of Check-up imaging is to disclose internal structure which is hidden by skin and bones leading to disclosure of disease properly. It establishes a file of normal anatomy and physiology so by comparing it is possible to identify abnormalities. Brain tumor is one the many reasons for rise in mortality in today's world. According to few researches done before [4, 5], the abnormal or uncontrolled growth of cell inside individual skull is called a brain tumor. They say these tumors grow inside the skull, affecting regular brain activity. A brain tumor is a severe existence frightening disease. Brain tumors are classified into three varieties called benign, malignant, and premalignant by few, out of which it is said the malignant tumor leads to cancer. Treatment of brain tumors depends on many factors such as diagnosis on the right time as well as on the different factors like the type of tumor, location, size, and stage of development. In early stages of the tumor appears to be bodily but with the assist of observation of image by professionals from time to time it can be diagnosed [6, 7].

In the last few years, the advancement in biomedicine and technology has made it possible for human to overcome many diseases, but as cancer is unpredictable in nature, it still exists. Therefore, this disease is still a major difficulty for humanity and brain tumor is one of the major serious diseases. In USA, nearly 23,000 patients have been diagnosed with brain tumor in 2015 [4]. Effective treatment of the disease is very important, which depends on its timely or correct recognition. The treatment's specificity depends on the tumor's size at the time of examination, the nature of pathology, or tumor's type. The brain is mainly a complex and critical part of a person anatomy. It controls tissues or nerve cells to synchronize the body's key actions, such as breathing, our senses, and muscle function. The formation of abnormal cell populations in or near the brain leads to the initiation of brain tumors. Abnormal cells will intensify brain function and affect patients' health [2]. Brain imaging, diagnosis, and conduct using medical imaging method are research focus of researchers, radiologists, and clinical experts [5]. The examination of brain imagery is a top priority because brain diseases called brain tumors in developed countries are fatal and cause a great number of deaths. For example, according to data from National Brain Tumor Foundation (NBTF), 29,000 people in USA (USA) are diagnosed with brain tumors, and 13,000 patients die each year [8]. Many superior MRI systems embrace diffusion tensor imaging magnetic resonance spectroscopy (MRS) or perfusion MRI to analyze brain tumors by MRI [6, 7, 9]. Brain tumors separate into two types: cancerous tumors (called malignant tumors) and noncancerous tumors (called benign tumors). The World Health Organization (WHO) further divides malignant tumors into grade I–IV [3]. The first-class tumor is called malignant astrocytoma, the second-class tumor is a low-grade astrocytoma, the third-grade tumor is anaplastic astrocytoma, or the fourth-class tumor is glioblastoma. Primary tumors or secondary tumors are less aggressive semi-malignant tumors. Grade III or IV are harmful tumors that have a major impact on patients' health and can lead to tumor patients' death [10]. Many ultrasound practices or processes have been used to diagnose and treat brain tumors. Dissection is a basic step in the imaging technique used to remove brain cells' affected area from MRI [11]. The tumor area division plays an important role in cancer identification, treatment, and evaluation of treatment response. Several semi-automatic or automatic division method and practice are used in tumor division [12]. MRI has a variety of sequencing techniques, including $T1$-weighted (TI) and $T1$-weighted contradictory enhancement ($T1c$) for tumor distribution in the brain, $T2$ weight, and $T2$-weighted change-in-conversion (FLAIR) technology. Some cells usually develop; some cells fall into function, stop growing, and then become abnormal. Such a large number of irregular cell populations produce tissues described as tumors. Therefore, brain tumors are sovereign or unbalanced reproduction of brain cells [2, 5]. For neurologists using computer-aided diagnosis (CAD) as a support tool for medical surgeries, brain tumor analysis, organization, or identification are crucial issues. There are three different types of brain tumors: malignant, pituitary tumors, and gliomas. Accurate or suitable analysis of brain cancer is essential for reasonable conduct of the disease. The resolution to treat depends on the type of pathology, the stage of examination, or the tumor's degree. CAD arrangement has helped neurologists in many ways. The use of CAD in neurology supports tumor classification, organization, and recognition [8]. Brain tumors are one of mainly severe cancers in children or adults. Provisional approval, classification, and analysis of brain tumors is mainly important for the adequate treatment of tumors. Recently, several CAD systems have been introduced in medical imaging which helps clinicians and radiologists to diagnose different categories of diseases or health-related problems [6]. Segmentation is the most used method for brain tumor classification. Regrettably, the significance of function extraction and classification problems is not high, which is mainly imperative step and can also recover computer-aided medical diagnosis presentation. Therefore, researchers focus on categorization and assignments using deep learning procedure. Lately, some study has used deep learning to recover presentation of computer-assisted medical diagnosis to study intellect tumor growth. Deep learning strategies play a significant role in the medical field or have proven to be useful tools for many significant infections such as lung cancer discovery [9] or breast cancer picture analysis [3].

Previously, machine learning technology (ML) was considered to take over classification or mining. Newly, the low precision of predictive models or the critical environment of medical data analysis has forced researchers to seek new processes for detecting brain tumors to improve organization and accuracy. Therefore, deep learning (DL) is a subfield of machine learning and has come into attention because it can provide effective predictable models by using large amounts of data such as images or text. Also, deep learning can predict models on large datasets to provide conclusions. Deep learning is mostly used in medical imaging to recognize smashed parts of anything, such as the affected part of the lungs, or it is also helpful to organize object images or predictable mold.

Today in various areas, such as medical analysis, object appreciation systems, and object recognition, deep learning is widely used [10, 13]. In order to classify dissimilar patterns in cell images, advancement throughout deep learning has important rewards. The precision of predictable models or data investigation through DL technology depends largely on data samples or their education because it requires more precise data to achieve better results. To surmount shortcomings of the training data examples, cross-learning can be used to ensure better presentation. Transfer learning is a deep learning technique where the trained big data functions can be implemented in small datasets known as target dataset, where big data is described as the source dataset or the basic data set [13]. The two main schemas for transfer erudition are coordinated ConvNet and frozen ConvNet. To continue backpropagation, the pretrained ConvNet can be preserved and replaced on a small dataset (objective dataset). The target dataset is then proceeded forward according to fully connected layers.

The idea of usage of pretrained deep learning models aims to save time because fallout can be achieved using smaller dataset. These division into smaller sets also extract random features and functions from the images for categorization. The starting layer extracts basic physical features at lower levels, such as touch, color, and edges. The deeper layer of the model extracts minute and high-level features, such as such as objects or curve. In the literature, a preformed model from the bottom layer is usually used to extract functions because the functions on the top layer of the preformed model are roughly identical in natural images or medical images. The underlying skin texture varies from natural images to medical imagery. The main aim is to remove features from different levels of the pretrained model qualified in our planned datasets. These functions are combined or cascaded to extract in sequence on multiple scales from the input image to enhance the classification model's performance model further. According to the literature review, information on multilevel functions is removing from dissimilar bottom layers in the pretraining model, the most important contribution from this paper, while other researchers use extraction with sole layer function.

Deep Features of Brain Tumors: Exploring and representing deep features is an important task in predicting and diagnosing brain tumors through radiological MRI. Extract deep features from MRI images for oncological diagnosis, treatment, or prognosis. The radiological assets of images are clearly linked to meaningful biological features and provide qualitative information known to radiologists. Once the network is pretrained as a feature extractor, the deep degradation neural network can achieve the latest prediction and classification. Deep feature extractor technique or practice is more suitable for predicting patients' overall survival time [7]. The deep convolutional neural network (CNN) establishment technique is used to remove skin texture from ImageNet to train the CNN for organization or segmentation. The CNN activation feature technique uses a variety of techniques, including feature selection, function merging, or figures enhancement algorithms [8]. To condense varying concentration of different average filters in the image, function selection, function removal, or fusion are executed. Gabor wavelet function technology can be used to attain texture information of the image, which includes the tumor's location, direction, and frequency. Core principal component analysis (KPCA) decides on a small portion of the function condenses redundancy by mounting correlation of function. The Gaussian radial basis function (GRBF) provides function differentiation information from many sets of functions to function fusion [6]. Function extraction based on fine-tuning use in the pretrained CNN method.

2   Literature Review

Machine learning methods are widely used in various fields such as clinical diagnostics and preventive medicine. However, limited research has focused on the identification of brain tumors, mainly using magnetic resonance imaging (MRI). In general, machine learning schemes train or test traditional machine learning algorithms with MRI data. More freshly, some methods have used DL to define brain tumors. Rehman et al. [1] proposed a system that uses space to describe a three-dimensional convolutional neural network (CNN to classify the types of tumors (GoogLeNet, AlexNet, and VGGNet). This organization includes the types of pituitary tumors, gliomas, and malignant. The algorithm described above cuts the MRI brain to detect the area of interest. The data was also well organized and converted into further classifications. The authors also measured the data amplification system to obtain the accuracy of the results. Using the VGG16 model to recover classification and detection, this study's exactness reached 98.69%.

Deepak et al. [6] also used deep removal to classify the images or used the same source of information discussed there [5]. Remove the functions from the image, and use these functions in the experimental and classification models. At the patient level, using a classification model of 5, the author's accuracy was reached 98%. The research concluded that automatic classification can help in the organization of areas and is better than the classification of urban areas.

Another study by Afshar et al. [7] classifies CapsNets-based brain tumors as a capsule network model. The research improves the degree of accuracy by incorporating CapsNet mapping modifications to certain convolutional layers. The study revealed that 86.50% accuracy can be achieved by using CapsNet on the convolutional layer. This installation is achieved by using 64 maps to improve the accurate metrics.

Another study by Abiwinanda et al. [9] took a CNN-based in-depth study model and applied it to image classification of brain tumors. Although the study uses five classification models, conclusion 2 is the best approach to classify the images. The final building consists of a RELU layer or a topmost pool layer. This site has 64 latent neurons in a covered structure. The lessons revealed that it would achieve 98.5% proficiency in training or 84.19% in certification. The authors in [14] used a two-dimensional contrast pedestal on wavelet and Gabor filters to explore brain MRI's efficacy. By using the above system with NN backpropagation, the study achieved 91.9% accurate measurements.

Pashaei et al. [10] created a feature mining planning based on CNN. They also calculated a five-layer architecture, with all layers as learning layers, and with a special three-layer layout. The revision said it would reach an 81% accuracy rate or further improve CNN's standard classification model's accuracy based on extreme learning machine (ELM). The revision found that in examining classification, the differences between the pituitary gland and malignant images were limited by discrimination.

Sajjad et al. [3] planned an arrangement based on neural network organization, which further helped provide a clear picture (separating the tumor area from the data). Also, the research uses noise suppression techniques through the use of concepts of variability and invariance. CNN's analysis can correct the accuracy of the predictions to predict the magnitude of the swelling. For the truth of the prophecy, the data is transmitted to CNN modified. Some experiments performed on both radioactive and brain tumors. This study uses innovative data or improved data to determine the system's accuracy, which is 90.67%. Classifying tumors by different numbers of imaging data can be very useful in clinical performance, as it can also speed up conduct plan of these.

Anaraki et al. [15] projected a CNN or genetic algorithm (GA)-based strategy to organize different types of glioma imagery using MRI data. The proposed system uses a genetic algorithm to select the CNN system automatically. They obtained a 90.9% exactness in predicting images of three glioma types. The accuracy of this study in the organization of glioma, malignant, or pituitary gland reached 94.2%. Zhou et al. [16] proposed a way to directly use 3D general graphics. First, the overall 3D image is then converted into a 2D segment, and then DenseNet is used to extract features from each 2D segment. Subsequently, repetitive neural networks are linked to 2D networks that use long- or short-term memory for classification. They conducted conduct test on public and property data. They also installed a pure convolutional neural complex in DenseNet as a convolutional autoencoder for illustration study. Therefore, they used long- or short-term DenseNet sensing and DenseNet convolutional neural system for tumor detection of classification of tumor types. Their system uses DenseNet-LSTM to achieve 92.13% accuracy. The existing method cannot remove features from the bottom layer of the predesigned model, which are unusual from the natural image to the clinical image. To conquer this problem, a multifeature exploration method is proposed that improves the model's ability to classify brain tumors (Fig. 1).

The rapid development of in-depth learning and practice has provided inspiration for automatic illustration and attribute learning. So, we use ResNet for attribute

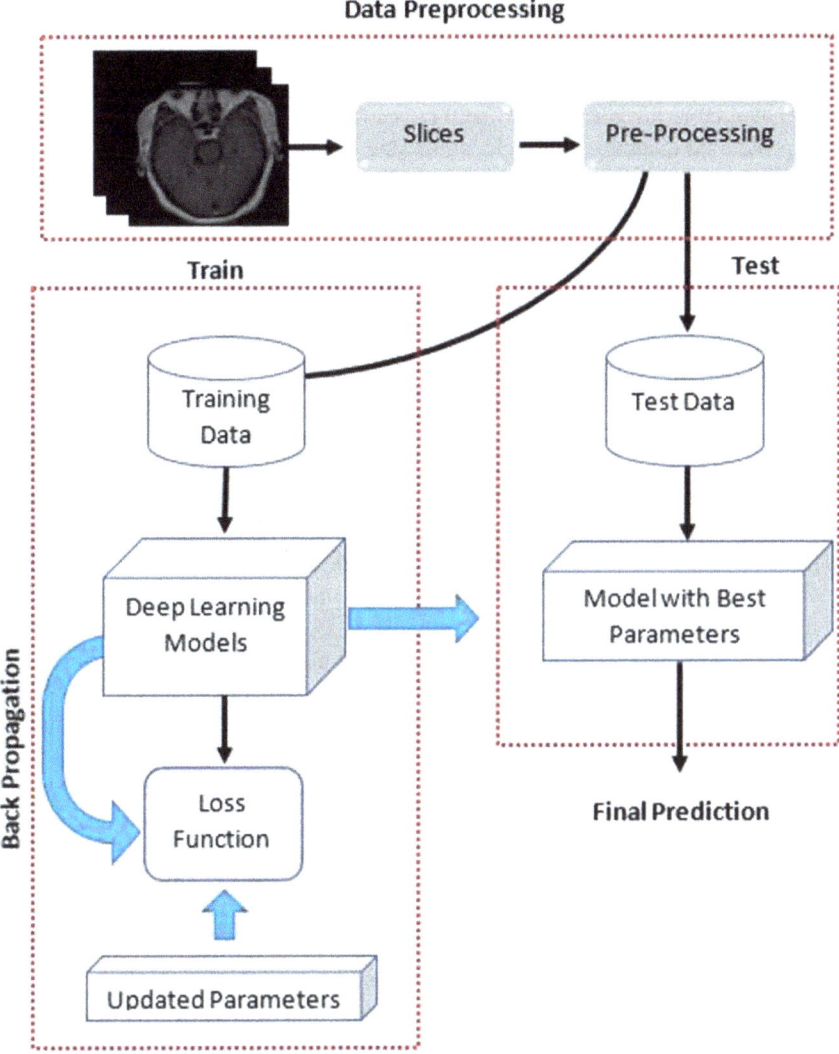

Fig. 1 Complete process of the existing deep learning model

mining, which is a deep CNN system. CNN feels that learning from the lowest to the highest levels is automatic through change and construction. The highlighted images are sent to the neural network for cataloging training. The speed of stochastic neural networks is faster than traditional BPNN because it is based on ramp ancestry. CBA is used to optimize parameters of random neural complex to recover their organization function further.

3 Problem Identification

Radiologists traditionally do the segmentation and feature extraction of brain tumor MRI images manually knowing that the process is time consuming as well as can cause unavoidable mistakes. So, to help radiologist, proper segmentation of tumor region is required to identify tumor location, size, and brain structure [15]. For appropriate and timely treatment, this information is very essential. One of the key subjects of radiology departments is to do the correct assessment of tumors by the use of imaging modalities. Brain tumor can affect people of any age and it is the main cause of death by cancer. Brain tumor exists in brain and surrounded by cancerous cell, which results in growth of defect. We have studied and extract different features for classification of tumor type. In this work, we extracted features; features are selected for the classification of the tumor type using optimization technique. To improve the classification accuracy, the feature extraction and selection are performed for accurate diagnosis analysis. The brain tumor detection and then deciding the right therapy is a long process, and once it acquaint, then time to time evaluation and its progress is extremely important. The purpose of this research work is to extract relevant information from the deep learning-based feature extraction technique and classify healthy and infected tumor tissues for a large database of medical images. The results of this research are helpful in classifying benign and malignant tumors, fast, and accurately and thus, improving the diagnosis of tumor slices [3, 12] Traditional human-based imaging methods have been used to diagnose and classify MRI brain tumors, and rely on the radiologist's ability to examine or analyze the imaging elements. For large amounts of data, the operator-assisted classification method is irreversible or inefficient, as manual processing of large amounts of data is a time-consuming procedure. To trounce these troubles, computer-assisted testing tools are needed to effectively procedure great amounts of data. In practice, the classification of brain tumors can be divided into two types: (1) MRI is divided into normal or abnormal tumors; (2) abnormal brain tumors are classified as dissimilar nature of tumors. Compared with the categorization of tumors for tumors (normal and abnormal), it is somewhat difficult to classify the brain's tumors into various pathological types automatically.

In general, the method of exploiting traditional features used in machine learning depends on high-level operations and low-level operations to get idea about hands-on approach which is also a major problem in analysis of tumors by machine learning algorithms. The structure of the head is seen more clearly in order to know the structure of the tumor with adjacent cells or edema, in CE-MRI. Gliomas come in a variety of shapes and are often surrounded by edema. Malignant tumors are usually located near the skull or in cerebrospinal cord. The pituitary gland is in near to the sphenoid sinus or chic optic nerve. Thus, the discriminant function or most applicable information associated with brain tumors is related to the location, shape, size, or end of the tumor area in all MR images. The shape, size, or severity of brain tumors vary greatly. Therefore, manual work based on conventional machine learning techniques may not be a very suitable resolution to the complexity of the in turn. More lately, automated feature exploration and classification methods based on in-depth studies have been well established, highlighting the work of computer-assisted therapy. In-depth studies have involved a great deal of work in self-study methods and require minimal data processing for most of the process.

Significant extraction and classification: A major problem with image organization and MR recognition is to meet the gap between high-quality data required by human evaluators and the low-level data obtained by MRI machines. This project's success on the categorization difficulty is to remove the best skin texture that symbolize low or high quality in turn by disregard hand-crafted tasks. CNN's deep learning model uses a hierarchical learning approach to extract key features, proving that the deep learning model can provide better results. The in-depth study-based model exploits simple structural information in the previous layer (especially edges, shapes, etc.), and the last layer incorporates specific shapes or constructs abstract interpretations. While the model based on in-depth learning is compared to manual feature extraction, it can provide a very efficient feature extraction mechanism, thereby reducing the need for field knowledge to improve performance of the system. When compared to the latest proposal for study-based in-depth feature exploration, an arrangement of high-level and low-level models can produce very good results.

4   Dataset

The brain dataset examined in this revise consists of 233 3064 T1-weighted MRI images with contrast [5]. There are three types of tumors in this dataset: malignant, gliomas, and pituitary tumors. An image with dimension of 512×512 with a voxel spacing size of 0.49×0.49 mm^2 has been used in this dataset. The image resolution consists of axial (lateral planes), coronal (frontal planes), or sagittal (lateral) planes. The axial planar allotment based on number of categories consists of 708, 1426, and 930 specimens of glioma, meningiomas, or pituitary tumors.

5   Proposed Model

This research proposes a new approach for brain recognition structure based on deep leaning feature extraction. First, use ResNet as an attribute extractor, which is a well-known neurological network construction. Then we used three random neural complexes, namely Schmidt neural complex, chance vector function link network, or extreme learning machine. The weights or biases in the three networks are taught by chaotic bat algorithm. The three planned techniques attain comparable consequences based on five runs, and they produced comparable presentation compared to the latest method. The proposed models with dissimilar numeral of layers and pretrained models are described in follow segment. Figure 2 illustrate pretreatment, training, testing, or calculation process of brain tumors. The planned model uses pretreatment using deep learning. Various hyperparameters are used for training and loss function is used to optimize these parameters during the training process. Model learning through the loss utility is an efficient way of evaluating how an algorithm models the delivered data. Gradually, loss occupation learns to use some optimization functions to minimize the prediction error. We used pretrained ResNet-50 model to extract the brain's MRI functions and after that SVM classifier used to classify the types of tumor (Fig. 3).

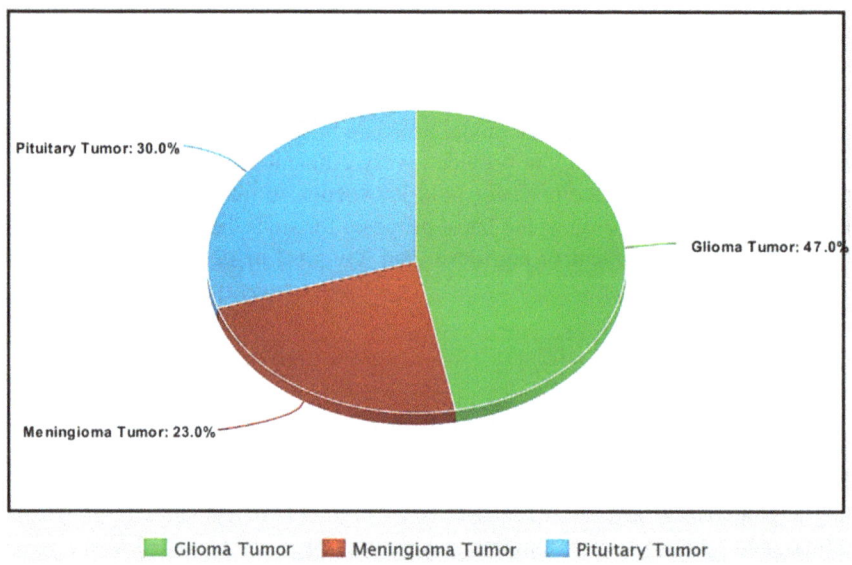

Fig. 2 Sample distribution of tumor's types in dataset

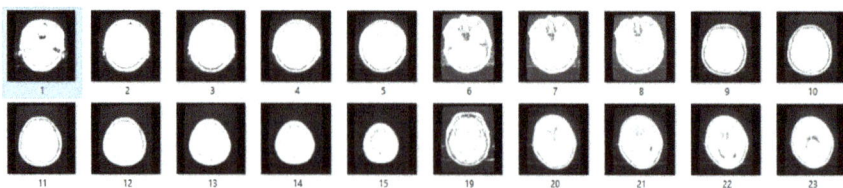

Fig. 3 3064 T1-weighted MRI images

Architecture of ResNet-50 Now, we will discuss building of ResNet50. The architecture of ResNet50 is divided into four phases, as shown in figure below. The network can use height or width of the input image as multiples of 32 and 3 of the channel width. For sake of illustration, we treat the input size as $224 \times 224 \times 3$. Each ResNet architecture uses 7×7 and 3×3 core sizes to perform initial folding and maximum pooling, respectively. Then the first phase of the complex starts. It has three residual blocks and each residual block contains three layers. The core sizes used to perform folding operations in all three layers of the block in step 1 are 64, 64 and 128, correspondingly. The curved arrow refers to individuality association. The dashed projectile indicates that the difficulty process in residual block is executed in step 2. Therefore, the input size is reduced by half in terms of height or width, but channel width is twofold. As we go from one step to another, channel width doubles or input size is condensed by half. For deep learning system like ResNet50, ResNet152, etc., need to use bottleneck design. For each residual utility F, three layers are stacked on top of each other. These three layers are $1 \times 1, 3 \times 3, 1 \times 1$ folding. The 1×1 folding layer is answerable for shrinking and then restoring size. The 3×3 layer is still the bottleneck in the smaller input / output size (Figs. 4 and 5).

Fig. 4 Architecture of ResNet-50

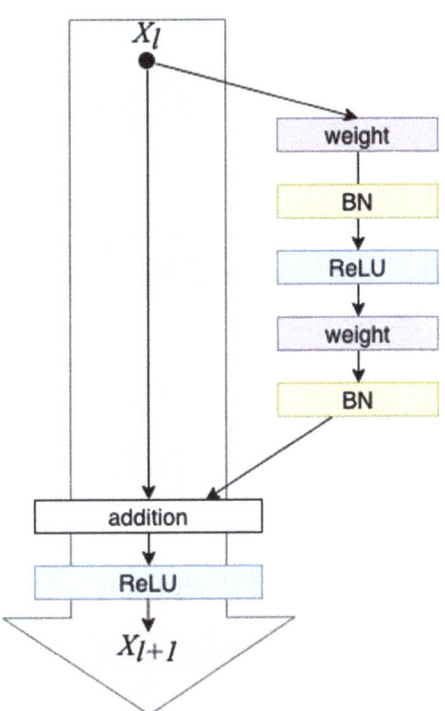

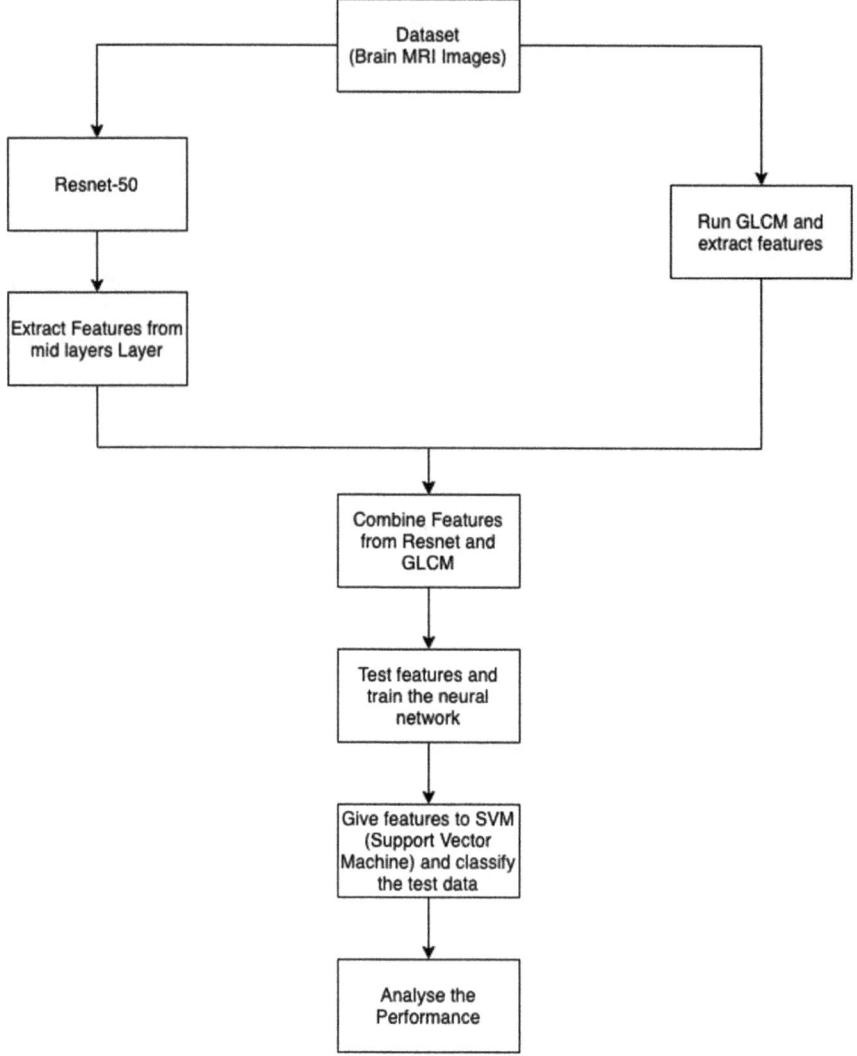

Fig. 5 Proposed flow diagram

6 Results and Discussion

In this work, GLCM and ResNet-50 methods used to improve the presentation of the model but also facilitate the examination of the results. Finally, SVM is used to classify the descriptions into normal or abnormal type tumor (Fig. 6).

Preprocessing constitutes of two sub divisions. One is filtering and the second is the image-enhancing section. The main objectives of preprocessing are the reduction of

Fig. 6 Input image

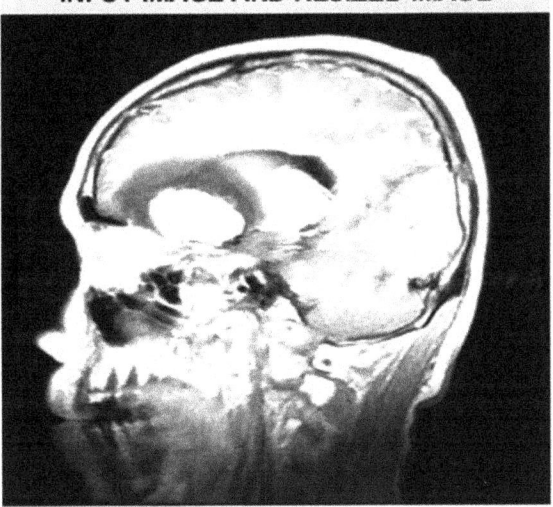

redundant and irrelevant data before processing or giving them into the model [7]. For smoothing and removing the noise from the image, median filter is used. This filter preserves the image edges while reduces the image noise. The process of image adjustment images so that it is more suitable is known as image enhancement. To enhance the contrast, we used adaptive histogram equalization.

Support Vector Machine SVM was originally developed by Vapnik used earlier just for ninary classification in order to minimize structural risk. It is a supervised ML technique used for one-class to multiple-class classification problem [12]. These are also kernel machines, as kernel trick has the main impact on transformation of feature vector into space kernel so that it can fit the maximum margin hyperplane.

Performance Metrics The mathematical notation for each concert indicator is shown below: Process performance is calculated in terms of presentation indicators (such as accuracy, sensitivity, specificity—TP is a total number of dataset confidential suitably (true positive).

TN—is a total number of dataset that are not well classified (true negative number).

UN—is a total number of false rejections representing the numeral of false pixels classified as background pixels (false negatives).

FP—is a total number of false positives, which means that pixel is imperfectly confidential as foreground (false positive). Calculate representation value for each image of the input video according to the cost index (Fig. 7 and Table 1).

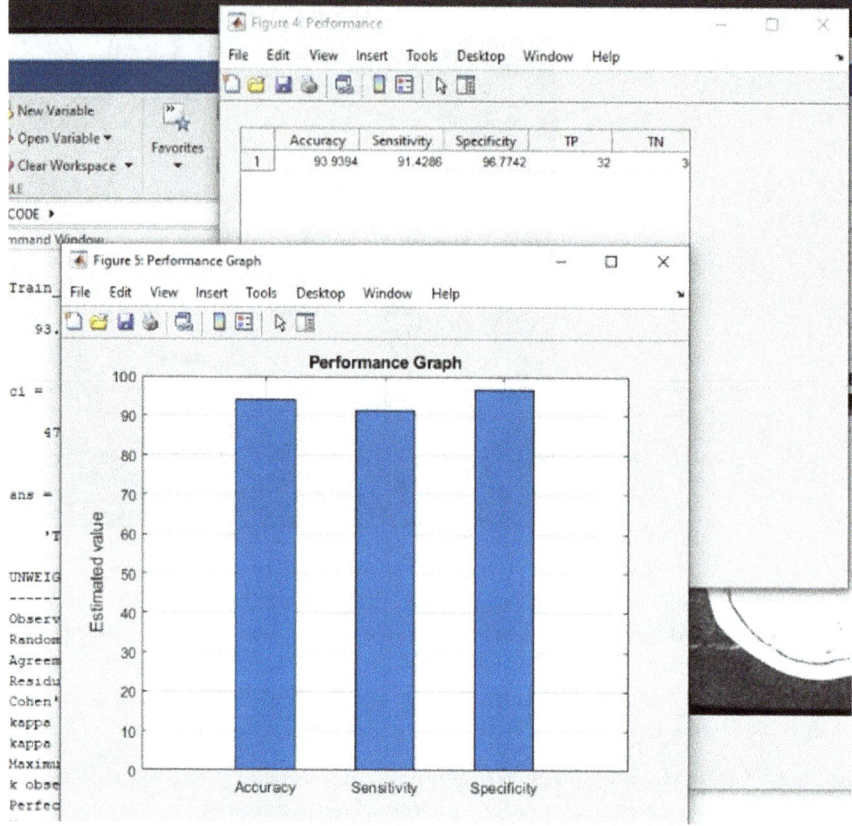

Fig. 7 Performance results of proposed work

Table 1 Comparison on different techniques

Reference	Feature extracted	Technique	Accuracy
[4]	CNN	CNN	91.0%
[17]	Morphological operation	Naïve Bayes classifier	85.0%
[1]	GLCM	ROI (Identification)	87.5
Proposed work	GLCM feature and ResNet-50	SVM	93.93%

7 Conclusion

This paper converse functions of deep learning models for the feature extraction in classification process of brain tumors. First, use the qualified ResNet50 deep learning model and remove functions from different ResNet50 blocks. These attributes are then linked and transferred to the softmax category to organize brain tumors. Connect them and then transfer them to softmax for the brain tumor category. The algorithm can significantly reduce the computational load and increase classification accuracy. The main motive of this work is to provide an aid to our clinicians or a radiologist by giving them an efficient and cost-effective tool for classification of benign and malignant tumors. Abnormal growth of cells in the brain leads to formation mass known as brain tumor. It may affect anyone of any and can have different shapes and sizes; can be present at any location in different image intensities. Brain tumors can be of any type, benign or malignant. Benign tumors, like low-grade gliomas and malignant are quite common. In this paper, two dissimilar scenarios were taken into consideration. First, use the qualified ResNet50 deep learning model and remove functions from different ResNet50 blocks. These attributes are then linked and transferred to the softmax category to organize brain tumors which are then connected and then transferred to softmax for the brain tumor category. Both cases were evaluated using three brain tumor datasets. Therefore, evaluating the existing explore technique for brain tumors' organization, the cascade block-based ensemble method's effectiveness using the DensNet201 pretraining model is much better. The proposed method yielded 93.93% of the test example or realized the highest increase in brain tumors' detection. In the future, we will discover or pertain ne-tune technology to overtrained models and may classify hair-based models and data-enhancing methods to classify brain tumors. We will also examine the company methods (fusion of output classifier) based on ne-tune or floor-based operations derived from the in-depth learning model. In the future, we will assemble more brain MRIs to expand the data. We will also try to use the transmission of studies to model the brain MRI's deep network accurately. Visualization of the neural complex is another exploration course that can help humans comprehend how complex works in a throughway. We shall apply our technique to notice another ailment like Alzheimer's disease, hearing loss, etc.

References

1. Shahriar Sazzad TM, Ahmmed KMT, Hoque MU, Rahman M (2019) Development of automated brain tumor identification using MRI images. In: 2019 international conference on electrical, computer and communication engineering (ECCE). IEEE. 978-1-5386-9111-3/19/$31.00 ©2019
2. Siegel R, Miller CR, Jamal A (2017) Cancer statistics. Cancer J Clin 67(1):7–30. Brain Tumor Statistics, American Brain Tumor Association. Accessed: 26 Oct 2019. [Online]. Available: http://abta.pub30.convio.net/
3. Sajjad M, Khan S, Muhammad K, Wu W, Ullah A, Baik SW (2019) Multi-grade brain tumor classification using deep CNN with extensive data augmentation. J Comput Sci 30:174–182. https://doi.org/10.1016/j.jocs.2018.12.003
4. Divyamary D, Gopika S, Pradeeba S (2020) Brain tumor detection from MRI images using Naive classifier brain tumor detection from MRI images using Naive classifier. IEEE, pp 620–622. 978-1-72815197-7/20/$31.00 ©2020
5. Razzak MI, Imran M, Xu G (2019) Efficient brain tumor segmentation with multiscale two-pathway-group conventional neural networks. IEEE Biomed Health Inform 23(5):1911 1919. https://doi.org/10.1109/JBHI.2018.2874033

6. Deepak S, Ameer PM (2019) Brain tumor classification using deep CNN features via transfer learning. Comput Biol Med 111(Art. no. 103345). https://doi.org/10.1016/j.compbiomed.2019. 103345
7. Afshar P, Mohammadi A, Plataniotis KN (2018) Brain tumor type classification via capsule networks. In: Proceedings of 25th IEEE international conference on image processing (ICIP), pp 3129–3133. https://doi.org/10.1109/ICIP.2018.8451379
8. Razzak MI, Akram F, Imran M (2019) A deep learning-based framework for automatic brain tumors classification using transfer learning. Circ Syst Signal Process 39(2):757–775. https:// doi.org/10.1007/s00034-019-01246-3
9. Abiwinanda N, Hanif M, Hesaputra ST, Handayani A, Mengko R (2019) Brain tumor classification using convolutional neural network. In: Proceedings of world congress on medical physics and biomedical engineering. Springer, Singapore, pp 183–189 [Online]. Available: https://www.springerprofessional.de/en/brain-tumor-classification-using-convoluti onalneural-network/15802612
10. Pashaei A, Sajedi H, Jazayeri N (2020) Brain tumor classification via convolutional neural network and extreme learning machines. In: Proceedings on 8th international conference on computer and knowledge engineering (ICCKE), vol 8, pp 314–319
11. Cheng J (2019) Brain tumor dataset. Figshare. Dataset. Accessed: 19 Sept 2019 [Online]. Available: https://doi.org/10.6084/m9.gshare.1512427.v5
12. Gu Y, Lu X, Yang L, Zhang B, Yu D, Zhao Y, Zhou T (2018) Automatic lung nodule detection using a 3D deep convolutional neural network combined with a multi-scale prediction strategy in chest CTs. Comput Biol Med 103:220–231
13. Shao L, Zhu F, Li X (2018) Transfer learning for visual categorization: a survey. IEEE Trans Neural Netw Learn Syst 26(5):1019–1034
14. Tan C, Sun F, Kong T, Zhang W, Yang C, Liu C (2018) A survey on deep transfer learning. In: Proceedings of international conference on artificial neural network. Springer, Cham, Switzerland, pp 270–279 [Online]. Available: https://doi.org/10.1007/978-3-030-01418-6, https://doi. org/10.1007/978-3-030-01424-7_27
15. Rani AM, Naz S, Razzak MI, Imran M, Xu G (2019) Rening Parkinson's neurological disorder identification through deep transfer learning. Neural Comput Appl 32:839–854. https://doi. org/10.1007/s00521-019-04069-0
16. Zuo H, Fan H, Blasch E, Ling H (2017) Combining convolutional and recurrent neural networks for human skin detection. IEEE Signal Process Lett 24(3):289–293
17. Hemanth G, Janardhan M, Sujihelen L (2019) Design and implementing brain tumor detection using machine learning approach. In: Proceedings of the third international conference on trends in electronics and informatics (ICOEI 2019). IEEE Xplore Part Number: CFP19J32-ART, pp 1289–1294. ISBN: 978-1-5386-9439-8

Normalized Feature Plane Alteration for Dental Caries Recognition

Shashikant Patil, Smita Nirkhi, Suresh Kurumbanshi, Mayank Kothari, and Sachin Sonawane

Abstract *Objectives*: Dentists and other professionals are extensively using digital dental explorer for dental treatment and assessment of dental disease. Digital imaging and image processing are widely used in dentistry for recognition, identification and assessment of various lesions in oral treatment. Acquiring the images and extraction useful from it for correct diagnosis is widely acclaimed by usage of advanced image processing tools. *Methodology*: This paper is an attempt to propose an outline of numerous techniques used in oral radiology and for teeth-related complications. Here, a unique and novel approach are used to repair the cracks and stretch of RCT as well as sternness using x-ray image. Tools and methods: This approach includes the series of the measures which are used in preprocessing method using contrast stretching process, parting system by means of active shape prototypical and other practises to devise the region of root canal and weakening progression to determine midline of the root canal region and sternness of cavities. The multilinear subspace learning approach which comprises of multilinear principal component analysis (MPCA) and multilinear discriminant analysis (MLDA) are used here for better outcomes. *Results*: The results are proving better than conventional methods used in dentistry in terms of statistical measures like accuracy and others. It will have better scope in the field of dentistry and can be used as secondary opinion for improving diagnostic efficacy.

S. Patil (✉) · S. Kurumbanshi · M. Kothari · S. Sonawane
EXTC Department, MPSTME, SVKMs NMIMS, Shirpur, Maharashtra, India
e-mail: sspatil@ieee.org

S. Kurumbanshi
e-mail: suresh.kurumbanshi@nmims.edu

M. Kothari
e-mail: mayank.kothari@nmims.edu

S. Sonawane
e-mail: sachin.sonawane@nmims.edu

S. Nirkhi
AI Department, GHRIET, Nagpur, Maharashtra, India

D. Gupta et al. (eds.), *Proceedings of Data Analytics and Management*,
Lecture Notes on Data Engineering and Communications Technologies 90,
https://doi.org/10.1007/978-981-16-6289-8_43

Keywords Root canal treatment (RCT) · Active shape model (ASM) · Image processing · Cavities · Features

1 Introduction

Medical image processing is a proven discipline with social relevance. Among diverse processing stages and fields, caries detection from x-ray images is highly required to aid diagnosis [1]. The high-tech methods concentrate on diagnosing the caries using perceptron and rule-based systems. In dentistry imaging and using image processing refined radiograph of exteriors display pictorial indication of deterioration in external portion of enamel, nonetheless, it does not confirm if deterioration cross the threshold in the dentin and if exteriors want to be reinstated and has entered the dentin for explicit surface and deterioration has arrived the dentin of interaction exterior [2, 3]. But the contemporary refurbishment instrument in which image taken all through refurbishment displays deterioration has pierced the coating (snowy quantifiable due to decalcification) and arrived the dentin (brunet adverts) of both exteriors. Recent tools have capability to photograph mistrustful lacerations through it, and it will disappointment you how it finds deterioration, and upon grounding of the tooth, it is deep-rooted every solitary interval [4]. Imageries of an outmoded dental radiographic image bitewing of tooth show no deterioration. Afterwards, for the patient's subsequent appointment, the copy evidently displays enormous deterioration. The patient may not have uneasiness with that tooth, but continued subsequently seeing the copy and the validation with these contemporary utensils [5–7]. Upon groundwork, widespread deterioration stayed, noted and confirmed. Also imaging and image processing practise aids in teeth impersonations, i.e. digital impressions.

There are imaging apparatuses which are established to support dental expert in the hard-hitting job of spotting dental radiographic images for cavities in teeth. It excerpts image structures and relates them with a catalogue of recognized cavity glitches. This instrument is shared with the cardinal radiography arrangement and its efficacy displays improved fallouts [8]. Efficacy was assessed by conniving three trials of recital are type I parameters like $F1$ Score, Matthews correlation coefficient (MCC), recall, precision along with sensitivity, specificity and accuracy for cavity finding by each dental experts both afore and later by means of this technique. Sensitivity amongst all dentists formerly using this instrument was around seventy (70.3%) and subsequently was almost ninety (90.5%), an enhancement of nearby twenty (20.2%). Radiologist' specificity parameter was 88.6% beforehand using the sophisticated technique and 88.3% consequently, with a modification of 0.3%. Radiologist' correctness was 75.6% afore using the process and 88.3% subsequently, with an upgradation of 12.7% [9, 10]. Tools like this permits radiologist to find 20% additional cases of caries piercing into dental area than they can to find devoid of it, whereas not instigating them to mistreat any supplementary vigorous teeth. With the aid of imaging techniques and tools, dentist can predictably diagnose caries at the previous stage and treat with minimal loss of tooth structure [11]. Even more prominently, it retains from cutting into tooth where the decay is not yet into the dentin.

Dental imaging sharpened radiograph of surfaces shows graphical indication of deterioration in outer part of coating, but it is not vibrant if deterioration move in the dentin and if exteriors need to be reinstated, modern IP tools assist the dental experts for perceiving if deterioration has moved in the dentin of connection surface. The up-to-date refurbishment apparatus will have capability in which image taken in the course of refurbishment displays deterioration has infiltrated the enamel and moved in the dentin (brown spots) of both the exteriors [12, 13]. Our aim is to devise a mechanism which will have facility to scan suspicious lesions with it and finding a decay. Images of an out-of-date radio graphical images having bitewing of tooth shows no deterioration. Afterwards, at the patient's subsequent appointment to prior procedures, the image clearly shows enormous deterioration [14, 15]. The patient may not have uneasiness with that teeth, but progressed after seeing the appearance. Using the image processing techniques, extensive decay can be noted and confirmed.

2 Motivation and Related Work

Majority of the works reported in the literature has concentrated more on segmenting and processing the images than diagnosing caries. Few caries detection approaches such as perceptron-based and rule-based caries detection systems require considerable improvement. The state-of-the-art caries detection methods still suffer from serious challenges such as lack of accuracy and computational complexity. The methods are highly sensitive to the image artefacts and acquisition noises.

To develop a procedure which will support dental experts in the challenging task of detecting radiographs for critical caries [16–18]. It abstracts image topographies and associates them with a databank of identified caries complications. The usefulness of the image processing procedure might stretch enhanced outcomes. Efficacy can be evaluated by conniving three events of enactment viz. Type I and type II errors like sensitivity, specificity and accuracy for caries diagnosis both afore and later using this procedure. This will help dentist in many cases of caries penetrating into dentin will preventing from mistreatment of any additional healthy teeth [19–21]. With the aid of imaging techniques and tools, dentist can probably analyze caries at an earlier stage and treats with trifling damage of tooth construction. Even more significantly, it retains dentist from cutting into teeth where the deterioration is not yet into the dentin [22].

Medical image processing is a proven discipline with social relevance. Among diverse processing stages and fields, caries detection from x-ray images is highly required to aid diagnosis. The state-of-the-art methods concentrate on diagnosing the caries using perceptron and rule-based systems.

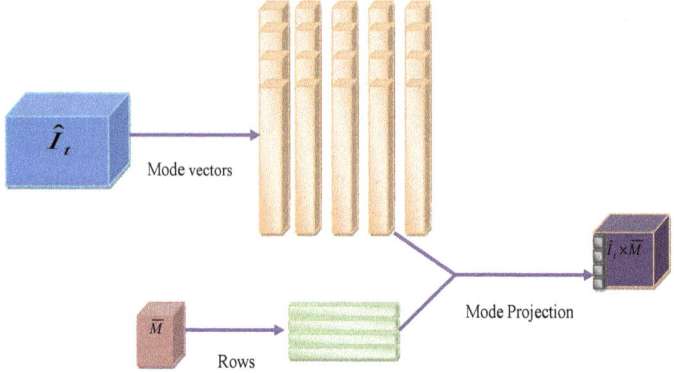

Fig. 1 Graphical representation of multilinear subspace learning

Majority of the works reported in the literature has concentrated more on segmenting and processing the images than diagnosing caries. Few caries detection approaches such as perceptron-based and rule-based caries detection systems require considerable improvement. One can use structural image enrichment procedures which will be deployed to assimilated images, comprising endodontic, periodontics, and dentin-enamel junction and a comprehensible sharpness filter with dynamic slider bar sorts it stress-free to see disparity vicissitudes in actual [23–25].

One can recommend an approach to create it at ease and quicker to quantify the bone mineral density of graft zone by assimilating the image processing and oral therapeutic picture (Fig. 1).

3 Methodology

The proposed methodology focuses on enhancing the features in such a way that it can distinguish the caries and the normal portions of the tooth [26]. Despite the multidimensional projection distinguishes the features, the intention of reducing the dimension often tends to deviate from retaining the margin. So a nonlinear correlation-based function will be derived so that the features of caries and non-caries portions can be transformed into a plane of high margin. Secondly, a regularization function will be developed to ensure the lack of over-fitting with the unknown and uncertain test data [27–29] (Fig. 2).

To devise a technique, which will support clinical experts in the problematic job of spotting radiographs for critical caries. It abstracts image topographies and relates them with a database of well-known caries hitches [30–33].

The helpfulness of the image processing method could give superior consequences. Usefulness can be assessed by scheming various performance parameters [34–36].

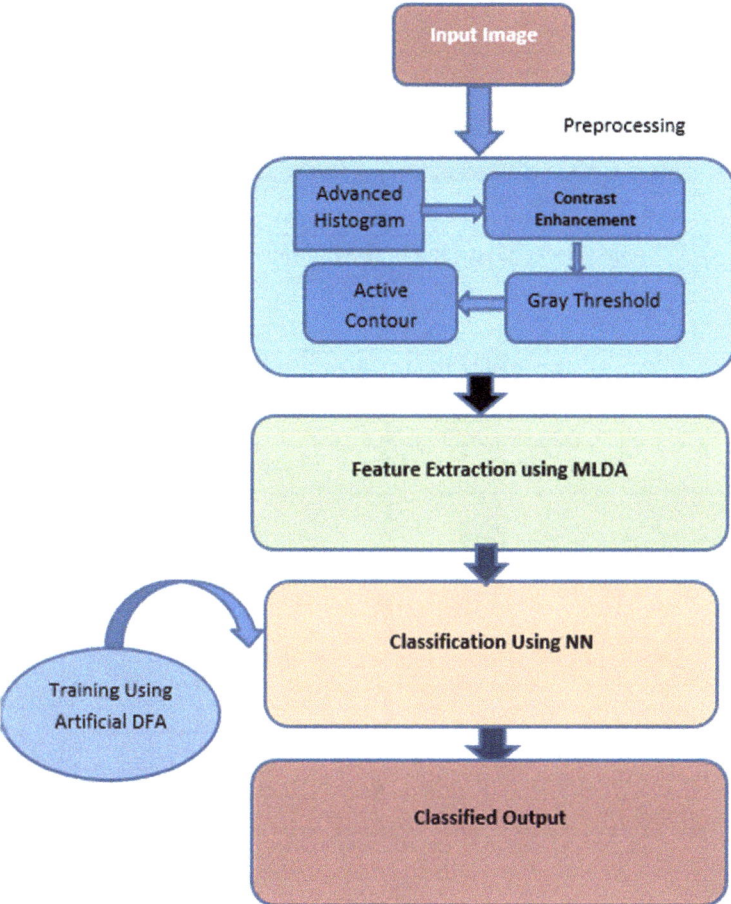

Fig. 2 Proposed model

The low sensitivity indicates that it is not good at detecting caries. The high specificity indicates that it is good at ruling out caries. Along with positive predictive value, negative predictive value can be also included.

The steps involved in the methodology are given below [37–39] (Fig. 3).

4 Preprocessing Results

See Fig. 4.

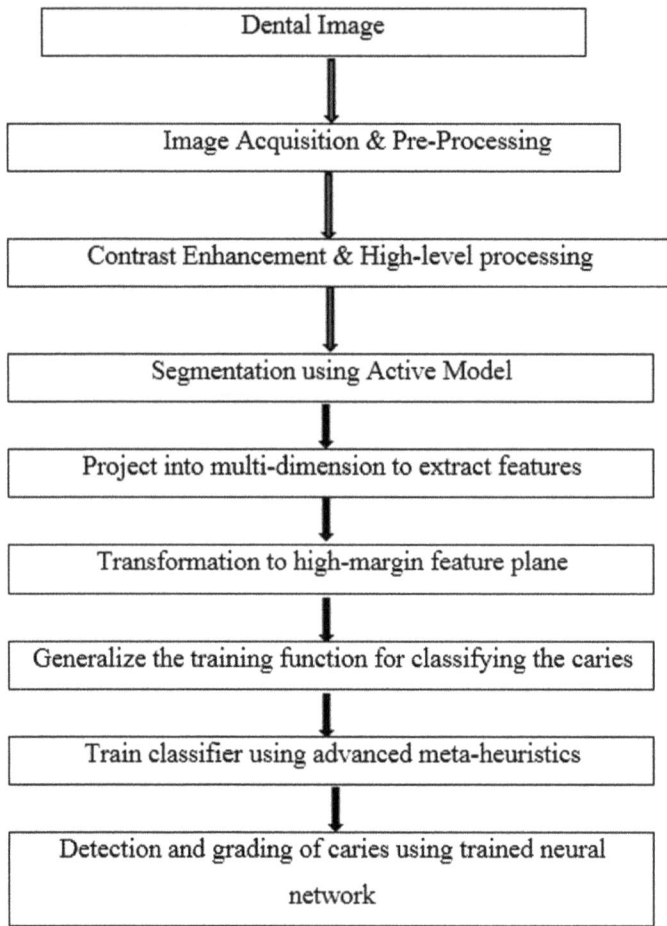

Fig. 3 Flow diagram of MPCA-MLDA method

5 Results

The investigation was conceded out in MATLAB 2021a. The databank was downloaded from https://mynotebook.labarchives.com/share/Vahab/MjAuOHw4N Tc2Mi8xNi9UcmVlTm9kZS83NzM5OTk2MDZ8NTIuOA==, which was splitted into two sets (database 1, database 2) arbitrarily for the scrutiny tenacity. Such test case separations are essential to confirm fair investigation. For each set, it contains of 60 caries imageries. Subsequently, the performance of the projected prototypical was explored in the regard with ten statistical parameters like type I and type II for proving our model's superiorities like precision, recall, accuracy, sensitivity, specificity, FPR, FNR, NPV, FDR, F_1 score and MCC (Tables 1 and 2).

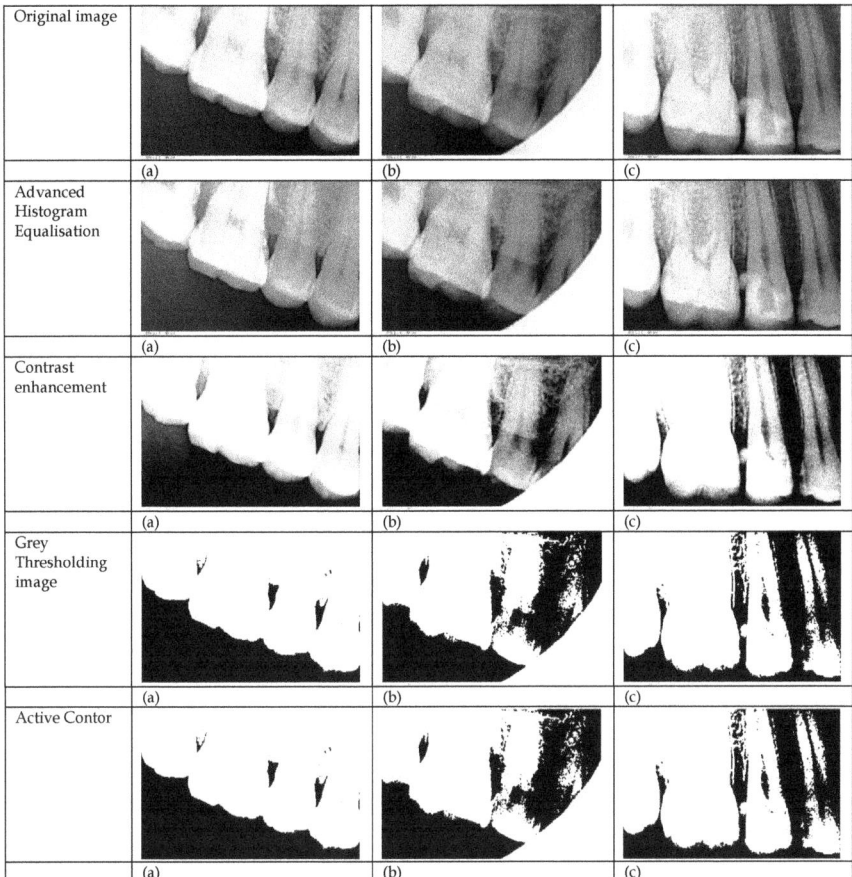

Fig. 4 Results of preprocessing stage

Table 1 Performance exploration of projected model over conservative approaches for database 1

Methods	PCA	LDA	ICA	MPCA-MLDA
Accuracy	0.5	0.5	0.58	0.82
Sensitivity	0.6	0.2	0.71	0.83
Specificity	0.2	0.6	0.44	0.83
Precision	0.43	0.33	0.5	0.81
FPR	0.9	0.4	0.56	0.19
FNR	0.4	0.8	0.29	0.18
NPV	0.2	0.6	0.45	0.7
FDR	0.57	0.67	0.5	0.19
$F1$ score	0.5	0.25	0.59	0.79
MCC	-0.22	-0.22	0.16	0.59

Table 2 Performance exploration of projected model over conservative approaches for database 2

Methods	PCA [40]	LDA [23]	ICA [24]	MPCA-MLDA
Accuracy	0.4	0	0.56	0.81
Sensitivity	0.33	0	0.57	0.79
Specificity	1	0	0.5	1
Precision	1	0	0.89	1
FPR	0	1	0.5	0
FNR	0.67	1	0.43	0.21
NPV	1	0	0.5	0.9
FDR	0	1	0.11	0
$F1$ score	0.5	0	0.7	0.87
MCC	0.22	− 1	0.05	0.52

6 Conclusion

Whilst summing up, the recommended and proposed scheme affords enhanced sophisticated model with images through exterminating noise and inaccuracies which can be used for the more precise elucidation of oral radiographic images. Moreover, by deploying the distinction in approaches, it classes managing technique more correct and delivers efficacious patient analysis. The procedure has been advanced in MATLAB 2021a, and the investigational studies are conceded out for the x-ray images. The standard x-ray images with ground truth information are used for the study. The portrayal of the anticipated procedure is compared against the conservative methods using the standard measures such as type I and II error functions. Our approach is better than other procedures in terms of accuracy and other statistical measures of classifiers existing.

References

1. Liang X et al (2010) A comparative evaluation of cone beam computed tomography (CBCT) and multi-slice CT (MSCT): part I. On subjective image quality. Eur J Radiol 75:265–269
2. Liang X, Lambrichts I, Sun Y, Denis K, Hassan B, Li L, Pauwels R, Jacobs R (2010) A comparative evaluation of cone beam computed tomography (CBCT) and multi-slice CT (MSCT). Part II: On 3D model accuracy
3. Gaia BF, de Sales MAO, Perrella A, Fenyo-Pereira M, Paraíso Cavalcanti MG (2011) Comparison between cone-beam and multislice computed tomography for identification of simulated bone lesions
4. Liang X et al (2009) A comparative evaluation of cone beam computed tomography (CBCT) and multi-slice CT (MSCT). Part II: On 3D model accuracy. Eur J Radiol. https://doi.org/10.1016/j.ejrad.2009.04.016
5. Ahmad M, Jenny J, Downie M (2012) Application of cone beam computed tomography in oral and maxillofacial surgery

6. Huang Y, Van Dessel J, Depypere M, EzEldeen M, Iliescu AA, Dos Santos E, Lambrichts I, Liang X, Jacobs R (2014) Validating cone-beam computed tomography for peri-implant bone morphometric analysis

7. Geraets WGM et al (2007) Prediction of bone mineral density with dental radiographs. Bone 40: 1217–1221

8. Chiang Y-Y, Wang S-L (2011) Using cone beam CT with image processing in detecting the bone mineral density of jaw

9. Lamichane M, Anderson NK, Rigali PH, Seldin EB, Will LA (2009) Accuracy of reconstructed images from cone-beam computed tomography scans. Am J Orthod Dentofacial Orthop 136:151–157

10. Chen J (2011) A new annotation method for 3D cephalometric landmark in CBCT

11. Automatic dental CT image segmentation using mean shift algorithm by 2013. In: 8th Iranian conference on machine vision and image processing (MVIP)

12. Behere RR, Lele SM (2011) Reliability of Logicon caries detector in the detection and depth assessment of dental caries: an in-vitro study. Indian J Dent Res 22:362

13. Shakhnarovich G, Moghaddam B (2004) Face recognition in subspaces. In: Li SZ, Jain AK (eds) Handbook of face recognition. Springer, New York, NY, USA, pp 141–168. ISBN 038740595X

14. Zhang J, Li SZ, Wang J (2004) Manifold learning and applications in recognition. In: Tan YP, Yap KH, Wang L (eds) Intelligent multimedia processing with soft computing, vol 168. Springer, Berlin, Germany, pp 281–300. ISBN 354023053X

15. Law MHC, Jain AK (2006) Incremental nonlinear dimensionality reduction by manifold learning. IEEE Trans Pattern Anal Mach Int 28:377–391

16. Jolliffe IT (2002) Principal component analysis, 2nd edn. Springer, New York, NY, USA, pp 1–488. ISBN 9780387954424

17. Lu H, Plataniotis KN, Venetsanopoulos AN (2008) MPCA: Multilinear principal component analysis of tensor objects. IEEE Trans Neural Netw 19:18–39

18. Lu H, Plataniotis KN, Venetsanopoulos AN (2013) Multilinear subspace learning: dimensionality reduction of multidimensional data, 1st edn. CRC Press, London, UK, pp 1–296. ISBN 9781439857243

19. Sahambi HS, Khorasani K (2003) A neural-network appearance-based 3-D object recognition using independent component analysis. IEEE Trans Neural Netw 14:138–149

20. Li N, Liu C, Pfeifer N, Yin JF, Liao ZY, Zhou Y (2016) Tensor modeling based for airborne LiDAR data classification. In: Proceedings of the congress of 23rd ISPRS, Prague, Czech Republic, pp 283–287

21. Ye J (2005) Generalized low rank approximations of matrices. Mach Learn 61:167–191

22. Chen J (2014) Gait correlation analysis based human identification. Sci World J 2014:1–8

23. Patil S, Kulkarni V, Bhise A (2018) Caries detection using multidimensional projection and neural network. Int J Knowl Intell Eng Syst 22(3):155–166

24. Patil S, Kulkarni V, Bhise A (2018) Intelligent system with dragonfly optimisation for caries detection. IET Image Process 13(3):429–439

25. Shashikant P, Vaishali K, Archana B (2018) Caries detection with the aid of multi-linear component analysis and neural network. In: Second international conference on green computing and internet of things (ICGCIoT 2018), IEEE conference; IEEE Record No.:#44090; IEEE. ISBN: 978-1-5386-5657-0

26. Patil S, Kulkarni V, Bhise A (2019) Algorithmic analysis for dental caries detection using an adaptive neural network architecture. Heliyon 5(5):e01579

27. Byeon Y-H, Lee J-N, Pan S-B, Kwak K-C (2018) Multilinear Eigen ECGs and Fisher ECGs for individual identification from information obtained by an electrocardiogram sensor. Symmetry

28. Chiang Y-Y, Wang S-L, Liu S, Wu F (2011) Using cone beam CT with image processing in detecting the bone mineral density of jaw. In: 2011 4th international conference on biomedical engineering and informatics (BMEI)

29. Darekar RV, Dhande AP (2018) Emotion recognition from Marathi speech database using adaptive artificial neural network. In: Biologically inspired cognitive architectures

30. Chen J, Cheng E, Gabler B, Yang J (2011) A new annotation method for 3D cephalometric landmark in CBCT. In: 2011 4th international congress on image and signal processing
31. Sun M-L, Liu Y, Liu G-M, Cui D, Heidari AA, Jia W-Y, Ji X, Chen H-L, Luo Y-G (2020) Application of 23 machine learning to stomatology: a comprehensive review. IEEE Access (2020)
32. Bouchahma M, Hammouda SB, Kouki S, Alshemaili M, Samara K (2019) An automatic dental decay treatment prediction using a deep convolutional neural network on X-ray images. In: 2019 IEEE/ACS 16th international conference on computer systems and applications (AICCSA)
33. Li M, Xu X, Punithakumar K, Le LH, Kaipatur N, Shi B (2020) Automated integration of facial and intra-oral images of anterior teeth. Comput Biol Med
34. Wu X, Lai J (2010) Tensor-based projection using ridge regression and its application to action classification. In: IET image processing
35. Alzubi JA, Kumar A, Alzubi OA, Manikandan R (2019) Efficient approaches for prediction of brain tumor using machine learning techniques. Indian J Public Health Res Dev. https://doi.org/10.5958/0976-5506.2019.00298.5
36. Alzubi OA, Alzubi JA, Tedmori S, Rashaideh H, Almomani O (2018) Consensus-based combining method for classifier ensembles. Int Arab J Inf Technol
37. Alweshah OA, Alzubi JA, Alzubi SAM (2016) Solving attribute reduction problem using wrapper genetic programming. Int J Comput Sci Netw Secur
38. Alzubi OA, Alzubi JA, Alweshah M, Qiqieh I, Al-Shami S, Ramachandran M (2020) An optimal pruning algorithm of classifier ensembles: dynamic programming approach. Neural Comput Appl. https://doi.org/10.1007/s00521-020-04761-6
39. Gupta D, Rodrigues JJPC, Sundaram S, Khanna A, Korotaev V, Albuquerque VHC (2018) Usability feature extraction using modified crow search algorithm: a novel approach. In: Neural computing and applications. Springer, Berlin. https://doi.org/10.1007/s00521-018-3688-6
40. Bowyer KW, Chang K, Flynn P (2006) A survey of approaches and challenges in 3D and multi-modal 3D + 2D face recognition. Comput Vis Image Underst 101:1–15

Performance Analysis of Smart System with Algorithmic Optimization for Cavities Detection

Shashikant Patil, Smita Nirkhi, Suresh Kurumbanshi, Mayank Kothari, and Sachin Sonawane

Abstract *Objectives*: Various interpretation and analytical approaches of oral and dental cavities have significantly got enough attention and consideration in imaging and imaging analysis as it led to numerous carious lesions. Prevailing cavities and caries finding approaches are not able to verify and scrutinize the carious lesions at the beginning level. It is an attempt which proposes to develop a unique and novel cavities identification approach which comprises three segments, namely (a) pre-processing, (b) feature extraction and (c) classification. *Methodology*: In the beginning, the quality of the image is enriched by numerous pre-processing techniques such as contrast improvement, grey thresholding and active contour. Subsequently, the features are found with the aid of "multi-linear principal component analysis (MPCA)". Lastly, classification is done using neural network (NN), which is trained by the adaptive dragonfly algorithm (DA) algorithm. *Results*: The proposed MPCA model nonlinear programming with adaptive DA (MNP-ADA) yields the correct and scrutinized as well as classified outcome. It is compared with prevailing approaches in terms of numerous statistical measures, and it is proved superior over it. It can be used as an additional approach for enriching the diagnostic efficiency.

Keywords Dental caries · Features · Classification · BEASF · MPCA · Neural network

S. Patil (✉) · S. Kurumbanshi · M. Kothari · S. Sonawane
EXTC Department, MPSTME, SVKMs NMIMS, Shirpur, Maharashtra, India
e-mail: sspatil@ieee.org

S. Kurumbanshi
e-mail: suresh.kurumbanshi@nmims.edu

M. Kothari
e-mail: mayank.kothari@nmims.edu

S. Sonawane
e-mail: sachin.sonawane@nmims.edu

S. Nirkhi
AI Department, GHRIET, Nagpur, Maharashtra, India

© The Author(s), under exclusive license to Springer Nature Singapore Pte Ltd. 2022   519
D. Gupta et al. (eds.), *Proceedings of Data Analytics and Management*,
Lecture Notes on Data Engineering and Communications Technologies 90,
https://doi.org/10.1007/978-981-16-6289-8_44

Nomenclature

Abbreviation	Description
MPCA	Multi-linear principal component analysis
NN	Neural network
DCNN	Deep convolutional NN
DMFT	Decayed, missing and filled teeth
MLR	Multiple logistic regressions
NILTI	Near-infrared light trans illumination
AUC	Area under curve
LM	Levenberg–Marquardt
FNR	False negative rate
MNP-ADA	MPCA model nonlinear programming with adaptive DA
FDR	False discovery rate
NPV	Net present value
MCC	Matthews correlation coefficient
DA	Dragonfly algorithm
FPR	False positive rate
FOR	False omission rate
BM	Bookmaker informedness
MK	Markedness

1   Introduction

"Dental caries is a disease, defined as the process of progressive demineralization of the inorganic component of the tooth accompanied by the disintegration of the organic portion. It can also be defined as a dynamic disease process, in which early lesions undergo many demineralization and remineralization cycles before being expressed clinically" [1–3]. As a result, it is necessary to identify the symptoms of caries at an earlier stage rather than searching for cavities [4, 5]. The accurate treatment prior to cavitations would allow targeted preventative diagnoses like fissure, pit and fluoride sealants, thus considerably enhancing dental health. This minimizes the requirement for extensive filling and drilling [6–8].

A clinician needs skill, ability and knowledge to exploit the right diagnostic technique and to handle them. Visual assessment by means of radiography, mouth mirrors and conventional probes was the analytical techniques that were usually used in previous days [9]. The noticeable variations within tooth composition affect the microporosity of enamel that consecutively affects the transmission of light via the enamel [10]. The outcomes of numerous studies point out that the deployment of the probe has restricted value in detecting caries and is also found to interrupt remineralisation [6, 11, 12]. Diagnosing caries exists to be a challenging task for dental experts [13–15]. Moreover, the automatic diagnosis of caries is done through machine learning algorithms.

The main contribution of the paper is as follows:

1. At the initial phase, the quality of the image is improved by pre-processing schemes like contrast enhancement, grey thresholding and active contour.
2. Moreover, the features get extracted using MPCA, and subsequently classification is carried out using optimized NN, where ADA model is proposed for the training process.
3. Finally, a parametric analysis is performed to verify the performance of the proposed model.

The paper is arranged as shown: Sect. 2 shows the reviews on dental caries detection. Section 3 delineates the adopted caries recognition model, and Sect. 4 depicts the computation of distance measure. Moreover, Sect. 5 elaborates the proposed adaptive dragonfly algorithm for solving the optimization problems. Sects. 6 and 7 portray the outcomes and conclusion, and literature review.

1.1 Related Works

In 2018, Lee et al. [1] have adopted a method for evaluating the efficiency of DCNN approaches for diagnosis and detection of dental caries on "periapical radiographs". Accordingly, this analysis focused on the potential effectiveness of DCNN framework for the diagnosis and detection of dental caries. From the analysis, DCNN framework has offered significant performance in recognizing dental caries in "periapical radiographs".

In 2019, Yue et al. [2] have carried out an analysis on detecting dental caries on three eighty-six kids residing in Mexico town. Here, "graphite-furnace atomic-absorption spectroscopy" was used for quantifying the Pb levels of blood. Accordingly, the existence of dental caries was computed by means of DMFT scores. Furthermore, the residual approach was exploited in this work for determining the total energy produced in the children based on the consumption of sweets and beverages.

In 2019, Cácia et al. [3] have analysed how the risk factors of patients influenced operative diagnostic decisions in a dental-oriented system at the Netherlands. In this work, the data were gathered from eleven dental practices, and the patients attended the practice regularly throughout the observation time. Consequently, a descriptive study was carried out after performing the MLR process.

In 2019, Ayşe et al. [4] have presented a vivo study for confirming the recognition of proximal caries by means of NILTI. Moreover, the diagnostic performance of the device was compared over other caries recognition techniques, together with visual assessment. Accordingly, here a total of nine seventy-four proximal surfaces of stable posterior teeth from thirty-four patients were taken into account. The data were examined with statistical analysis, and the AUC, specificity and sensitivity were computed.

In 2019, Darshan et al. [5] have computed the relationship amongst suscepti-bility of dental caries progression risk and ENAM gene polymorphisms. The imple-mented analysis was performed on one sixty-eight children from South India, and kids affected by dental caries were also taken into account. "Preliminary insilico analysis" has revealed that variations in "rs7671281 (Ile648Thr) amino acid" lead to the functional and structural changes in the ENAM.

2 Materials and Methods

Figure 1 demonstrates the architecture of the adopted dental caries recognition model. The implemented scheme includes three major stages: "(i) pre-processing, (ii) feature extraction and (iii) classification". At first, the input image Im is subjected to pre-processing, which involves four major image improvement features such as contrast improvement, grey thresholding and active contour. From the pre-processed image I_{pre}, the features are extracted by means of the MPCA model. These extracted features are then subjected to classification using NN classifier that offers the classified output (non-caries or caries).

2.1 Pre-processing

The image Im is improved by carrying out the below processes.

2.1.1 Contrast Enhancement

The contrasting of the resized input image Im^g is enhanced here. The particular procedure controls the image intensity [15, 16], and thus the image resolution is developed via the brightness and darkness of Im^g, as given by Eq. (1), in which V refers to the contrast improvement of image. Therefore, the current Im^g transforms into a grey image Im^g_{new}.

$$V = (((\text{Im} - \text{low\_in})/(\text{high\_in} - \text{low\_in}))\hat{g}\text{amma})$$
$$*(\text{high\_out} - \text{low\_out})) + \text{low\_out} \tag{1}$$

Fig. 1 Pictorial representation of the adopted scheme

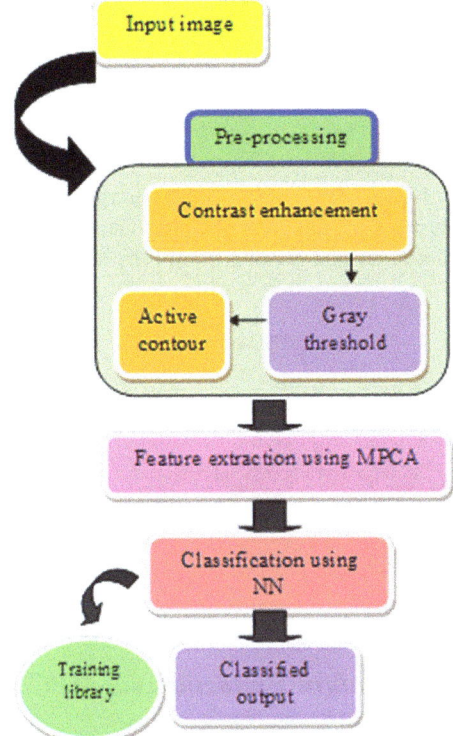

2.1.2 Grey Thresholding

The Otus's oriented grey thresholding [19] method portrays the threshold of the image, which is exploited for converting the grey pixel to either black or white. This is performed depending on the grey intensity.

2.1.3 Active Contour [18]

Here, two types of driven forces, namely external and internal energy, are exploited. This framework gets smoothed via internal forces, and it is reallocated in the direction through the external energy. Therefore, the contour $G(n)$ is formed by the coordinate sets such as $l(n)$ and $k(n)$ as given in Eq. (2), where (k, l) indicates the contour coordinates and n denotes the normalized index of the control point.

$$G(n) = (k(n)l(n)) \, ; \, G(n) \in \text{Im}_{\text{new}}^{C}(k, l) \tag{2}$$

Equation (3) shows the total energy of deformed design, where $\text{Im}^{g\text{int}}$ indicates the internal energy of the curve, $\text{Im}^{g\text{con}}$ denotes the exterior restriction, and Im denotes the energy of the image.

$$FO^* = \int_0^1 (FO^{intl}G(n) + FO^{im}G(n) + FO^{con}G(n))dn \tag{3}$$

In addition, the bending energy and elastic energy are summed up to form the internal energy as specified in Eq. (4), where $\alpha(n)$, $\beta(n)$ indicate the varying parameter that denotes continuity and contour curving, respectively.

$$FO^{intl} = FO^{elastic} + FO^{bend} = \alpha(n)\left|\frac{du}{dn}\right|^2 + \beta(n)\left|\frac{d^2u}{dn^2}\right|^2 \tag{4}$$

$$FO^{elastic} = \alpha(G(n) - G(n-1))^2 dn \tag{5}$$

$$FO^{bend} = \beta(G(n-1) - G(n) + (G(n+1))^2 dn \tag{6}$$

Finally, the pre-processed image Im_{pre} is determined from the initial stage.

2.2 Feature Extraction via MPCA

The pre-processed image Im_{pre} is then subjected to the MPCA [16, 17] model for extracting the features. MPCA is an extended version of PCA, and it is also known as "data tensor". MPCA includes image reconstruction, by which the image is reorganized into 3D tensor as $Im_{te} \in K^{Im_{pre}^1 \times Im_{pre}^2 \times Im_{pre}^3}$ in which Im_{pre}^1, Im_{pre}^2, and Im_{pre}^3 refer to the height, width and count of the images. The explanation of the MPCA model is given below:

(1) Calculate the mean matrix $\overline{O} = \left(1/Im_{pre}^3\right)\sum_{i=1}^{Im_{pre}^3} E_i$.

(2) The tensor $I\hat{m}_{te} = \left[E_1 - \overline{O}, E_2 - \overline{O}, \ldots, E_{Im_{pre}^3} - \overline{O}\right]$ is centred.

(3) $I\hat{m}_{te}$ is unfolded into a matrix. The components of the mode md unfolding matrix are portrayed as $\left(I\hat{m}_{te}^{(md)}\right)_{(index)} = b_{i_1 i_2 i_3}$, in which index $= i_{md}, \sum_{o=3}^{md+1}(i_o - 1) + \sum_{o=md-1}(i_o - 1)$.

(4) By covariance matrix, the eigenvectors are determined by mode-md that computes the matrix K, $C^{(m)} = \hat{K}_{(md)}\hat{K}_{(md)}^K$, and assume the eigenvectors as $B_{(md)} = [B_1, B_2, \ldots, B_{LA_{(md)}}]$ in which $LA_{(md)}$ refers to the highest eigenvectors.

(5) The feature selection is of two kinds: (a) by the sample $S_{i(md)} = (E_i - \overline{O}) \times B_{(md)}^K$ for diverse modes and (b) by $S_i = B_1^P(E_i - \overline{O})B_{(2)}$. There will be various percentages captured for each node.

(6)   The classification is performed on any one of these feature sets.

2.3 Classification

This work exploits NN [17] for recognizing caries. The input feature set is given by Eq. (7), in which N_D denotes the count of elected features [18–20].

$$FE^{\text{weight}} = [F_1, F_2, F_3, F_4, \ldots, F_{N_D}] \tag{7}$$

The weight WE of the network model is portrayed by LM framework. Equation (8) portrays the NN framework, in which the resultant output from ith node of jth layer is given by $ou_l^{(j)}$. The input is signified by $FE_i^{\text{weight } j}$, $af(\cdot)$ indicates the activation function, the entire count of input to jth layer is given by $nu^{(j)}$, bi_i symbolizes the input bias to jth layer, and c and d denote the weight coefficient of WE as specified in Eq. (9). The predicted network output $\hat{P}$ is given by Eq. (10), in which w^0 signifies the bias weight and $w^{(h)}$ defines the hidden neuron weight.

$$ou_l^{(j)} = af \left[c_l^{(j)} bi_j + \sum_{i=1}^{nu^{(j)}} FE_i^{'\text{weight}(j)} d_{il}^{(j)} \right] \tag{8}$$

$$WE = [c; d] \tag{9}$$

$$\hat{P} = w^0 + \sum_{i=1}^{nu^{(j)}} ou_l^{(j)} w_i^{(h)} WE \tag{10}$$

So as to train the network, the network weight WE^* is optimally chosen with the determination of objective function as in Eq. (11), where P indicates the actual output.

$$WE^* = \arg \min[WE] \| P - \hat{P} \| \tag{11}$$

Thus, the classifier classifies the input image (non-caries or caries image).

2.4 Computation of Distance Measure

For enhancing the adopted scheme, the extracted features FE are multiplied with the weight W given in Eq. (12). The weight size should be equal to the attained feature size. The distance D amongst the attained features FE^{weight} is given by Eq. (13).

$$W = \left[W_1, \ldots, W_{N_D} \right] \tag{12}$$

$$FE^{\text{weight}} = FE \times W$$
$$= FE_1^{\text{weight}}, \ldots, FE_{N_D}^{\text{weight}} \tag{13}$$

The distance di_o, in which $o = 1, 2, 3, \ldots, N_D$ amongst attained f^{weight}, is computed by "nonlinear programming optimization model". The objective function Obj is given by Eq. (14).

$$\text{Obj} = \max(di_o) : o = 1, 2, \ldots, N_D \tag{14}$$

2.5 Nonlinear Programming Optimization

The issues regarding the nonlinear programme are given in Eq. (15), in which $\hat{h}(\hat{x})$, $\hat{i}(\hat{x})$ and $\hat{j}(\hat{x})$ are portrayed as "deferential functions".

$$\min_{\hat{y}} \hat{h}(\hat{x}) = 0 \tag{15}$$

So that

$$\hat{i}(\hat{x}) = 0$$
$$\hat{j}(\hat{x}) = 0 \tag{16}$$

The substitution of Eq. (15) is done by a sequence of barrier sub-issues as specified in Eq. (17), in which $\hat{l} > 0$ points out the vector of slack parameters and $\hat{k} = (\hat{x}, \hat{l})$ and $\mu > 0$ denotes the barrier constraint.

$$\min_{\hat{k}} \phi_\mu(\hat{k}) \equiv \hat{h}(\hat{x}) - \mu \sum_{\hat{o}}^{\hat{n}} \ln \hat{l}_{\hat{o}} \tag{17}$$

So that

$$\hat{i}(\hat{y}) = 0$$
$$\hat{j}(\hat{x}) + \hat{l} = 0 \tag{18}$$

The Lagrangian function associated with Eq. (17) is specified in Eq. (19), in which $\zeta_{\hat{i}}, \zeta_{\hat{a}}$ indicates the "Lagrange multipliers" and $\zeta = (\zeta_{\hat{i}}, \zeta_{\hat{a}})$.

$$\aleph(\hat{k}, \zeta; \mu) = \phi_\mu(\hat{k}) + \zeta_{\hat{i}}^{\hat{v}} \hat{i}(\hat{x}) + \zeta_{\hat{a}}^{\hat{v}}(\hat{a}(\hat{x}) + \hat{l}) \tag{19}$$

The optimality states in Eq. (17) could be specified as per Eq. (20), in which $\hat{l}$ and $\zeta_{\hat{a}}$ are nonnegative, $\hat{Y}_{\hat{i}}$ and $\hat{Y}_{\hat{a}}$ refer to Jacobian matrices, and $\hat{D}$ and $\Gamma_{\hat{a}}$ point out the diagonal matrices.

$$\begin{bmatrix} \nabla \hat{h}(\hat{x}) + \hat{Y}_{\hat{i}}(\hat{x})^{\hat{v}} \zeta_{\hat{i}} + \hat{Y}_{\hat{a}}(\hat{x})^{\hat{v}} \zeta_{\hat{a}} \\ \hat{D}\Gamma_{\hat{a}}\hat{e} - \mu\hat{e} \end{bmatrix} = \begin{bmatrix} 0 \\ 0 \end{bmatrix} \tag{20}$$

Further, the current iterate $\left(\hat{k}, \zeta\right)$ outcomes in the primal–dual system are given by Eq. (21), in which $\hat{z}_{\hat{k}} = \begin{bmatrix} \hat{z}_{\hat{x}} \\ \hat{z}_{\hat{l}} \end{bmatrix}$, $\hat{z}_{\zeta} = \begin{bmatrix} \hat{z}_{\hat{i}} \\ \hat{z}_{\hat{a}} \end{bmatrix}$, $\hat{c}\left(\hat{k}\right) = \begin{bmatrix} \hat{i}(\hat{x}) \\ \hat{j}(\hat{x}) + \hat{l} \end{bmatrix}$, $\hat{Y}(\hat{x}) = \begin{bmatrix} \hat{Y}_{\hat{i}}(\hat{x}) & o \\ \hat{Y}_{\hat{a}}(\hat{x}) & 1 \end{bmatrix}$ and $\hat{R}\left(\hat{k}, \zeta; \mu\right) = \begin{bmatrix} \nabla^2_{\hat{x}\hat{x}}\aleph\left(\hat{k}, \zeta; \mu\right) & 0 \\ 0 & \hat{D}^{-1}\Gamma_{\hat{a}} \end{bmatrix}$

$$\begin{bmatrix} \hat{R}\left(\hat{k}, \zeta; \mu\right) & \hat{Y}(\hat{x})^{\hat{v}} \\ \hat{Y}(\hat{x}) & 0 \end{bmatrix} \begin{bmatrix} \hat{z}_{\hat{k}} \\ \hat{z}_{\zeta} \end{bmatrix} = - \begin{bmatrix} \nabla_{\hat{k}}\aleph\left(\hat{k}, \zeta; \mu\right) \\ \hat{c}\left(\hat{k}\right) \end{bmatrix} \tag{21}$$

The novel iteration is specified as given in Eq. (22), in which $\alpha_{\hat{k}}$ and α_{ζ} denote the step lengths that are modelled as per Eq. (23).

$$\hat{k}^+ = \hat{k} + \alpha_{\hat{k}}\hat{z}_{\hat{k}}, \ \zeta^+ = \zeta + \alpha_{\zeta}\hat{z}_{\zeta} \tag{22}$$

$$\alpha_{\hat{k}}^{\max} = \max[\alpha \in (0, 1)] : \hat{l} + \alpha\hat{z}_{\hat{i}} \geq (1 - \tau)\hat{l}$$
$$\alpha_{\zeta}^{\max} = \max[\alpha \in (0, 1)] : \zeta_{\hat{a}} + \alpha\hat{z}_{\hat{a}} \geq (1 - \tau)\zeta_{\hat{a}} \tag{23}$$

2.6 Theory/Calculation

In this work, modified ADA is implemented for training the NN classifier. The DA model [21, 22] concerns on five factors for updating the location of the dragonfly. They are "(i) control cohesion, (ii) alignment, (iii) separation, (iv) attraction and (iv) distraction". The separation of rth dragonfly, M_r is calculated by Eq. (24), and here A denotes the current dragonfly position, A'_s refers to the location of sth neighbouring dragonfly, and H' denotes the count of neighbouring dragonflies.

$$M_r = \sum_{s=1}^{H'} \left(A' - A'_s\right) \tag{24}$$

The alignment and cohesion are computed by Eqs. (25) and (26). In Eq. (25), Q'_s refers to the velocity of sth neighbour dragonfly.

$$J_r = \frac{\sum_{s=1}^{H'} Q'_s}{H'} \tag{25}$$

$$V_r = \frac{\sum_{s=1}^{H'} A'_s}{H'} - A \tag{26}$$

Attraction towards food and distraction to the enemy is illustrated in Eq. (27) and Eq. (28). In Eq. (27), Fo refers to the food position, and in Eq. (28), ene denotes the enemy position.

$$W_r = \text{Fo} - A' \tag{27}$$

$$Z_r = \text{ene} + A' \tag{28}$$

The vectors such as position A' and $\Delta A'$ step are considered here for updating the position of the dragonfly. The step vector $\Delta A'$ denotes the moving direction of dragonflies as given in Eq. (29), in which q', t', v', u', z' and δ refer the weights for separation, alignment, cohesion, food factor, enemy factor and inertia, respectively, and l denotes to the iteration count[24–26].

$$\Delta A'_{l+1} = \left(q'M_r + t'J_r + v'V_r + u'W_r + z'Z_r \right) + \delta \cdot \Delta A'_l \tag{29}$$

The position vector is computed by Eq. (30), in which l denotes the present iteration, L^{best} and G^{best} denote the local and global best solutions, respectively, and τ indicates the variation of fitness as given by Eq. (31).

$$A'_{l+1} = A'_l + \Delta A'_{l+1} + \left(L^{\text{best}} + G^{\text{best}} \right) \times \tau \tag{30}$$

$$\tau = \frac{A(l-1) - A(l)}{A(l)} \tag{31}$$

If there are no neighbourhood solutions, the position is updated as per Eq. (32), and here d refers to the size of position vectors, y_1 and y_2 refer to the two arbitrary integers in [0, 1], and $\wp$ signifies a stable parameter. With the increase in iteration, the positions and steps of all dragon flies are updated as per Eqs. (32)–(34).

$$A'_{l+1} = A'_l + levy(d) \times A'_l \tag{32}$$

$$levy(d) = 0.01 \times \frac{y_1 \times \Phi}{|y_2|^{\frac{1}{d}}} \tag{33}$$

$$\Phi = \left(\frac{\Psi(1 + \wp) \times \sin\left(\frac{\pi \wp}{2}\right)}{\psi\left(\frac{1+\wp}{2}\right) \times \wp \times 2^{\left(\frac{\wp-1}{2}\right)}} \right)$$

(34)

$$\psi(m) = (m - 1) \, !$$

At last, the algorithm provides the optimized weight that is multiplied with the features that are extracted, which is then classified using NN for detecting caries [27–29].

3 Results

The analysis on dental carried recognition was simulated using MATLAB. The data set was attained from "https://mynotebook.labarchives.com/share/Vahab/MjA uOHw4NTc2Mi8xNi9UcmVlTm9kZS83NzM5NzM5OTk2MDZ8NTIuOA==" that was defined into three test cases, which were necessary to offer better outcomes. Every set includes 40 caries images. Accordingly, the algorithmic analysis was carried out with respect to "accuracy, sensitivity, specificity, FPR, FNR, FDR, NPV, FOR, BM, MK, MCC and F_1-score" by varying the percentage of variation (T) captured for each node in MPCA. For analysis purpose, T was varied from 95, 97, 98, 99 and 100 with respect to cohesion weight v' that was varied from 0.2, 0.4, 0.6, 0.7 and 0.9 [30–33].

3.1 Performance Analysis

The performance analysis of the adopted model with respect to varied values of T is given by Figs. 2, 3, 4 and 5 for accuracy, sensitivity, specificity and F_1-score, respectively. For instance, from Fig. 2a, accuracy of T at 97 is high, which is 3.06, 3.06, 8.16 and 6.12% better than T at 95, 98, 99 and 100 when v' is 0.2. From Fig. 2b, the accuracy of the adopted model when $T = 95$ is high, which is 8.16, 13.27, 8.16 and 16.33% better than T at 97, 98, 99 and 100 when v' is 0.4. On considering Fig. 2c, the accuracy at $T = 95$ is high, which is 7.53, 3.23, 3.23 and 3.23% better than T at 97, 98, 99 and 100 when v' is 0.2. Likewise, from Fig. 3a, the sensitivity of the adopted scheme when $T = 97$ is higher, which is 1.08, 2.15, 1.08 and 16.13% better than T at 95, 98, 99 and 100 when v' is 0.9. Also, from Fig. 3b, the sensitivity at $T = 97$ is more, which is 7.22, 12.37, 7.22 and 6.19% better than T at 95, 98, 99 and 100 when v' is 0.7. Moreover, Fig. 4 shows the specificity of the adopted model, which revealed better results for all the test cases.

From Fig. 4a, the specificity of the presented model at $T = 95$ is high, which is 3.23, 8.6, 8.6 and 8.6% better than T at 97, 98, 99 and 100 when v' is 0.7. From

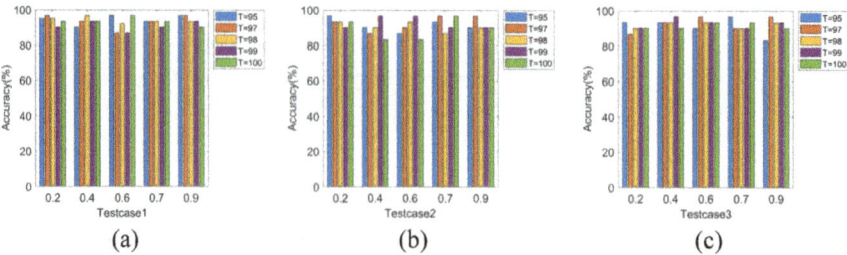

Fig. 2 Accuracy analysis of the adopted model by varying *T* **a** test case 1, **b** test case 2 and **c** test case 3

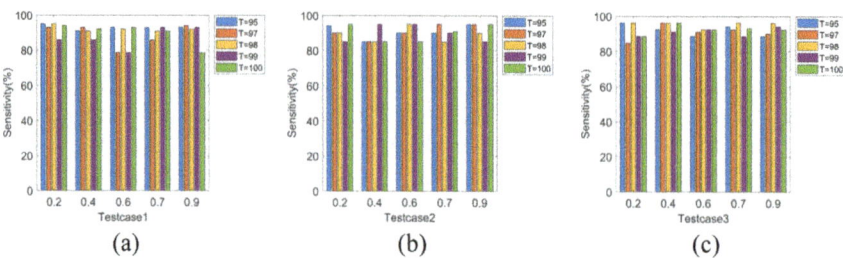

Fig. 3 Sensitivity analysis of the adopted model by varying *T* for **a** test case 1, **b** test case 2 and **c** test case 3

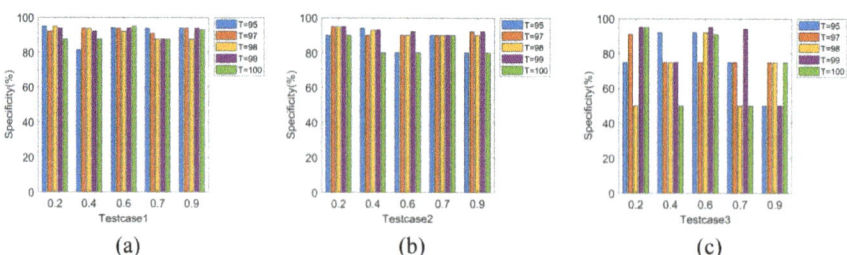

Fig. 4 Specificity analysis of the adopted model by varying *T* **a** test case 1, **b** test case 2 and **c** test case 3

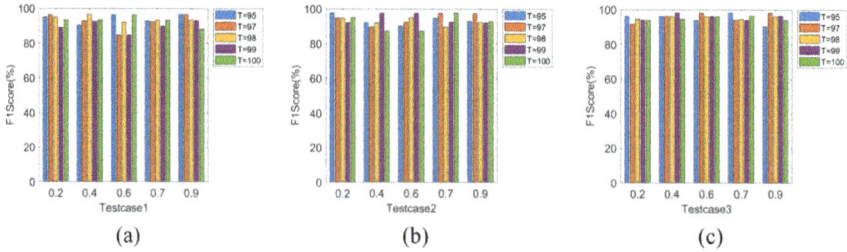

Fig. 5 *F*1-score analysis of the adopted model by varying *T* **a** test case 1, **b** test case 2 and **c** test case 3

Fig. 4b, the specificity of the presented model at $T = 99$ is high, which is 13.04, 2.17, 2.17 and 13.04% better than T at 95, 97, 98 and 100 when v' is 0.6. From Fig. 4c, the specificity when $T = 99$ is high, which is 21.05, 21.05, 47.37 and 47.37% better than T at 95, 97, 98 and 100 when v' is 0.7. The F1-score of the adopted model is revealed by Fig. 5, which shows betterment for all values of T. From Fig. 5a, the $F1$-score of the implemented model at $T = 95$ is high, which is 3.23, 8.6, 8.6 and 8.6% better than T at 97, 98, 99 and 100 when v' is 0.4. From Fig. 5b, the $F1$-score at $T = 99$ is high, which is 3.23, 8.6, 8.6 and 8.6% better than T at 95, 97, 98 and 100 when v' is 0.4. Thus, the betterment of the adopted scheme has been validated effectively.

4 Discussion

The performance analysis under different variation of learning percentage for three test cases is given in Tables 1, 2 and 3, respectively, for varied values. Here, the learning percentage is varied from 10, 25, 50 and 100% for the three test cases (Tables 4, 5 and 6).

5 Conclusion

This manuscript has designed a novel caries detection scheme that includes three phases. At the initial phase, the quality of the image was improved by pre-processing schemes like contrast enhancement, grey thresholding and active contour. Moreover, the features get extracted using MPCA, and subsequently classification was carried out using optimized NN, where ADA model was proposed for the training process. Finally, parametric analysis was performed to verify the performance of the proposed model. From the analysis, the accuracy of the adopted model when $= 95$ was high, which was 8.16, 13.27, 8.16 and 16.33% better than at 97, 98, 99 and 100 when is 0.4. Moreover, the accuracy at $= 95$ was high, which was 7.53, 3.23, 3.23 and 3.23% better than at 97, 98, 99 and 100 when was 0.2. Thus, the enhancement of the adopted scheme has been verified effectively from the attained outcomes. It has better performance in terms of accuracy as well other statistical measures. It is very much efficient in terms of dimensionality reductions and preserving original structure of image using multi-linear subspace learning approaches.

Table 1 Performance analysis by varying with respect to learning percentage of 10 and 25% for database 1

Training percentage	10%					25%				
Varying values of T	95	97	98	99	100	95	97	98	99	100
Sensitivity	0.88889	0.88889	0.94444	0.94444	0.94444	0.95	0.91	0.92857	0.92857	0.93
Accuracy	0.94444	0.91667	0.97222	0.91667	0.94444	0.95	0.9	0.96667	0.93333	0.96667
Precision	0.95	0.94118	0.94	0.89474	0.94444	0.95	0.82353	0.94	0.92857	0.93333
Specificity	0.95	0.94444	0.94	0.88889	0.94444	0.95	0.8125	0.94	0.9375	0.9375
FNR	0.11111	0.11111	0.055556	0.055556	0.055556	0.1254	0.1254	0.071429	0.071429	0.1254
FPR	0.1254	0.055556	0.1254	0.11111	0.055556	0.1254	0.1875	0.1254	0.0625	0.0625
FDR	0.1254	0.058824	0.1254	0.10526	0.055556	0.1254	0.17647	0.1254	0.071429	0.066667
NPV	0.95	0.94444	0.94	0.88889	0.94444	0.95	0.8125	0.94	0.9375	0.9375
$F1$-score	0.94118	0.91429	0.97143	0.91892	0.94444	0.95	0.90323	0.96296	0.92857	0.96552
MCC	0.89443	0.83462	0.94591	0.83462	0.88889	0.95	0.818	0.93485	0.86607	0.93541
FOR	0.1254	0.055556	0.1254	0.11111	0.055556	0.1254	0.1875	0.1254	0.0625	0.0625
BM	0.88889	0.83333	0.94444	0.83333	0.88889	0.95	0.8125	0.92857	0.86607	0.9375
MK	0.95	0.88562	0.94	0.78363	0.88889	0.95	0.63603	0.94	0.86607	0.87083

Table 2 Performance analysis by varying with respect to learning percentage of 50 and 75% for database 1

Training percentage	50%					75%				
Varying values of T	95	97	98	99	100	95	97	98	99	100
Sensitivity	0.9	0.9	0.91	0.95	0.9	0.8	0.9	0.94	0.6	0.9
Accuracy	0.9	0.95	0.91	0.85	0.9	0.8	0.9	0.9	0.7	0.9
Precision	0.9	0.94	0.91	0.76923	0.83333	0.8	0.9	0.83333	0.75	0.83333
Specificity	0.9	0.94	0.91	0.7	0.8	0.8	0.9	0.8	0.8	0.8
FNR	0.1	0.1	0.1254	0.1254	0.1254	0.2	0.1254	0.1254	0.4	0.1254
FPR	0.1	0.1254	0.1254	0.3	0.2	0.2	0.1254	0.2	0.2	0.2
FDR	0.1	0.1254	0.1254	0.23077	0.16667	0.2	0.1254	0.16667	0.25	0.16667
NPV	0.9	0.94	0.91	0.7	0.8	0.8	0.9	0.8	0.8	0.8
$F1$-score	0.9	0.94737	0.91	0.86957	0.90909	0.8	0.9	0.90909	0.66667	0.90909
MCC	0.8	0.90453	0.91	0.7338	0.8165	0.6	0.9	0.8165	0.40825	0.8165
FOR	0.1	0.1254	0.1254	0.3	0.2	0.2	0.1254	0.2	0.2	0.2
BM	0.8	0.9	0.91	0.7	0.8	0.6	0.9	0.8	0.4	0.8
MK	0.8	0.94	0.91	0.46923	0.63333	0.6	0.9	0.63333	0.55	0.63333

534 S. Patil et al.

Table 3 Performance analysis by varying with respect to learning percentage of 10 and 25% for database 1

Training percentage	10%					25%				
Varying values of T	95	97	98	99	100	95	97	98	99	100
Sensitivity	0.86957	0.86957	0.86957	0.86957	0.86957	0.94	0.85	0.9	0.9	0.95
Accuracy	0.91667	0.91667	0.91667	0.91667	0.91667	0.96667	0.9	0.86667	0.93333	0.9
Specificity	0.94	0.91	0.91	0.93	0.94	0.9	0.94	0.8	0.9	0.8
FPR	0.1254	0.1254	0.1254	0.1254	0.1254	0.1	0.1254	0.2	0.1254	0.2
Precision	0.94	0.91	0.91	0.93	0.94	0.95238	0.94	0.9	0.9	0.90476
NPV	0.94	0.91	0.91	0.93	0.94	0.9	0.94	0.8	0.9	0.8
FNR	0.13043	0.13043	0.13043	0.13043	0.13043	0.1254	0.15	0.1	0.1	0.05
FDR	0.1254	0.1254	0.1254	0.1254	0.1254	0.047619	0.1254	0.1	0.1254	0.095238
F1-score	0.93023	0.93023	0.93023	0.93023	0.93023	0.97561	0.91892	0.9	0.94737	0.92683
MCC	0.84055	0.84055	0.84055	0.84055	0.84055	0.92582	0.80861	0.7	0.86603	0.77152
FOR	0.1254	0.1254	0.1254	0.1254	0.1254	0.1	0.1254	0.2	0.1254	0.2
BM	0.86957	0.86957	0.86957	0.86957	0.86957	0.9	0.85	0.7	0.9	0.75
MK	0.94	0.91	0.91	0.93	0.94	0.85238	0.94	0.7	0.9	0.70476

Table 4 Performance analysis by varying with respect to learning percentage of 50 and 75% for database 2

Training percentage	50%					75%				
Varying values of T	95	97	98	99	100	95	97	98	99	100
Sensitivity	0.92	0.92	0.75	0.9375	0.95	0.88889	0.88889	0.9	0.88889	0.94
Accuracy	0.95	0.92	0.8	0.9	0.95	0.9	0.9	0.9	0.9	0.94
Precision	0.94118	0.92	0.92	0.9375	0.95	0.9	0.95	0.9	0.92	0.94
Specificity	0.75	0.92	0.92	0.75	0.95	0.9	0.95	0.1254	0.92	0.94
FNR	0.1254	0.1254	0.25	0.0625	0.1254	0.11111	0.11111	0.1254	0.11111	0.1254
FPR	0.25	0.1254	0.1254	0.25	0.1254	0.1254	0.1254	0.9	0.1254	0.1254
NPV	0.75	0.92	0.92	0.75	0.95	0.9	0.95	0.1254	0.92	0.94
FDR	0.058824	0.1254	0.1254	0.0625	0.1254	0.1254	0.1254	0.1	0.1254	0.1254
$F1$-score	0.9697	0.92	0.85714	0.9375	0.95	0.94118	0.94118	0.94737	0.94118	0.94
MCC	0.84017	0.92	0.61237	0.6875	0.95	0.66667	0.66667	0.4576	0.66667	0.94
FOR	0.25	0.1254	0.1254	0.25	0.1254	0.1254	0.1254	0.9	0.1254	0.1254
BM	0.75	0.92	0.75	0.6875	0.95	0.88889	0.88889	0.1254	0.88889	0.94
MK	0.69118	0.92	0.92	0.6875	0.95	0.9	0.95	-0.1	0.92	0.94

Table 5 Performance analysis by varying with respect to learning percentage of 10 and 25% for database 2

Training percentage	10%					25%				
Varying values of T	95	97	98	99	100	95	97	98	99	100
Sensitivity	0.92857	0.91	0.92857	0.92857	0.96429	0.96154	0.92308	0.88462	0.94	0.88462
Accuracy	0.88889	0.94444	0.94444	0.86111	0.88889	0.93333	0.93333	0.9	0.96667	0.83333
Specificity	0.75	0.75	0.91	0.625	0.625	0.75	0.92	0.92	0.75	0.5
Precision	0.92857	0.93333	0.91	0.89655	0.9	0.96154	0.92	0.92	0.96296	0.92
FNR	0.071429	0.1254	0.071429	0.071429	0.035714	0.038462	0.076923	0.11538	0.1254	0.11538
FPR	0.25	0.25	0.1254	0.375	0.375	0.25	0.1254	0.1254	0.25	0.5
FDR	0.071429	0.066667	0.1254	0.10345	0.1	0.038462	0.1254	0.1254	0.037037	0.08
NPV	0.75	0.75	0.91	0.625	0.625	0.75	0.92	0.92	0.75	0.5
$F1$-score	0.92857	0.96552	0.96296	0.91228	0.93103	0.96154	0.96	0.93878	0.98113	0.90196
MCC	0.67857	0.83666	0.86189	0.5815	0.65738	0.71154	0.78446	0.71098	0.84984	0.35082
FOR	0.25	0.25	0.1254	0.375	0.375	0.25	0.1254	0.1254	0.25	0.5
BM	0.67857	0.75	0.92857	0.55357	0.58929	0.71154	0.92308	0.88462	0.75	0.38462
MK	0.67857	0.68333	0.91	0.52155	0.525	0.71154	0.92	0.92	0.71296	0.42

Table 6 Performance analysis by varying with respect to learning percentage of 50 and 75% for database 2

Training percentage	50%					75%				
Varying values of T	95	97	98	99	100	95	97	98	99	100
Sensitivity	0.94444	0.94444	0.88889	0.9	0.91	0.9	0.91	0.93	0.93	0.9
Accuracy	0.95	0.9	0.9	0.95	0.91	0.9	0.91	0.93	0.93	0.9
Specificity	0.91	0.5	0.95	0.5	0.91	0.4576	0.4576	0.4576	0.4576	0.4576
FPR	0.1254	0.5	0.1254	0.5	0.1254	0.4576	0.4576	0.4576	0.4576	0.4576
Precision	0.91	0.94444	0.95	0.94737	0.91	0.91	0.91	0.93	0.93	0.91
NPV	0.91	0.5	0.95	0.5	0.91	0.4576	0.4576	0.4576	0.4576	0.4576
FNR	0.055556	0.055556	0.11111	0.1254	0.1254	0.1	0.1254	0.1254	0.1254	0.1
FDR	0.1254	0.055556	0.1254	0.052632	0.1254	0.1254	0.1254	0.1254	0.1254	0.1254
$F1$-score	0.97143	0.94444	0.94118	0.97297	0.91	0.94737	0.91	0.93	0.93	0.94737
MCC	0.79349	0.44444	0.66667	0.68825	0.91	0.4576	0.4576	0.4576	0.4576	0.4576
FOR	0.1254	0.5	0.1254	0.5	0.1254	0.4576	0.4576	0.4576	0.4576	0.4576
BM	0.94444	0.44444	0.88889	0.5	0.91	0.4576	0.4576	0.4576	0.4576	0.4576
MK	0.91	0.44444	0.95	0.44737	0.91	0.4576	0.4576	0.4576	0.4576	0.4576

References

1. Lee J-H, Kim D-H, Jeong S-N, Choi S-H (2018) Detection and diagnosis of dental caries using a deep learning-based convolutional neural network algorithm. J Dent 77:106–111
2. Wu Y, Jansen EC, Peterson KE, Foxman B, Martinez-Mier EA (2019) The associations between lead exposure at multiple sensitive life periods and dental caries risks in permanent teeth. Sci Total Environ 654:1048–1055
3. Signori C, Laske M, Bronkhorst EM, Huysmans M-CDNJM, Opdam NJM (2019) Impact of individual-risk factors on caries treatment performed by general dental practitioners. J Dent 81:85–90
4. Dündar A, Çiftçi ME, İşman O, Aktan AM (2019) In vivo performance of near-infrared light transillumination for dentine proximal caries detection in permanent teeth. Saudi Dental J In press, corrected proof, Available online 28 Aug 2019
5. Divakar DD, Alanazi SAS, Assiri MYA, Halawani SM, Mustafa M (2019) Association between ENAM polymorphisms and dental caries in children. Saudi J Biol Sci 26(4):730–735
6. Kang H, Jiao JJ, Lee C, Le MH, Darling CL, Fried D (2010) Nondestructive assessment of early tooth demineralization using cross-polarization optical coherence tomography. IEEE J Select Top Quant Electron 16(4):870–876
7. Sampathkumar A, Hughes DA, Kirk KJ, Otten W, Longbottom C (2014) All-optical photoacoustic imaging and detection of early-stage dental caries. In: 2014 IEEE international ultrasonics symposium
8. Joris P et al (2018) Preprocessing of heteroscedastic medical images. IEEE Access 6:26047–26058
9. Sultana J, Islam MS, Islam MR, Abbott D (2018) High numerical aperture, highly birefringent novel photonic crystal fibre for medical imaging applications. Electron Lett 54(2):61–62
10. Angelino K, Edlund DA, Shah P (2017) Near-infrared imaging for detecting caries and structural deformities in teeth. IEEE J Trans Eng Health Med 5:1–7
11. Lee RC, Staninec M, Le O, Fried D (2016) Infrared methods for assessment of the activity of natural enamel caries lesions. IEEE J Select Top Quant Electron 22(3):102–110
12. Lashgari M, Shahmoradi M, Rabbani H, Swain M (2018) Missing surface estimation based on modified Tikhonov regularization: application for destructed dental tissue. IEEE Trans Image Process 27(5):2433–2446
13. Winstone B, Melhuish C, Pipe T, Callaway M, Dogramadzi S (2017) Toward bio-inspired tactile sensing capsule endoscopy for detection of submucosal tumors. IEEE Sens J 17(3):848–857
14. Top CB, Tafreshi AK, Gençer NG (2016) Microwave sensing of acoustically induced local harmonic motion: experimental and simulation studies on breast tumor detection. IEEE Trans Microw Theory Tech 64(11):3974–3986
15. Kaur R, Kaur S (2016) Comparison of contrast enhancement techniques for medical image. In: 2016 conference on emerging devices and smart systems (ICEDSS), Namakkal, pp 155–159
16. Lu H, Plataniotis KN, Venetsanopoulos AN (2008) MPCA: multilinear principal component analysis of tensor objects. IEEE Trans Neural Netw 19(1)
17. Mohan Y, Chee SS, Xin DKP, Foong LP (2016) Artificial neural network for classification of depressive and normal in EEG. In: 2016 IEEE EMBS conference on biomedical engineering and sciences (IECBES)
18. Bakoš M (2007) 5th Slovakian-Hungarian joint symposium on applied machine intelligence and informatics
19. Vala MHJ, Baxi A (2013) A review on otsu image segmentation algorithm. Int J Adv Res Comput Eng Technol (IJARCET) 2(2)
20. Jafari M, Chaleshtari MHB (2017) Using dragonfly algorithm for optimization of orthotropic infinite plates with a quasi-triangular cut-out. Eur J Mech A/Solids 6:1–146
21. Nirkhi S, Patil S (2020) Comprehensive assessment of imbalanced data classification. Int J Eng Adv Technol 9(4):1426–1431. https://doi.org/10.35940/ijeat.d7349.049420
22. Shashikant Patil SN (2020) Deep learning techniques for oral diagnosis and cavity recognition: a systematic approach. Int J Adv Sci Technol 29(9):192–199

23. Patil S, Kulkarni V, Bhise A (2019) Algorithmic analysis for dental caries detection using an adaptive neural network architecture. Heliyon 5(5). https://doi.org/10.1016/j.heliyon.2019.e01579
24. Patil S, Kulkarni V, Bhise A (2019) BEASF-based image enhancement for caries detection using multidimensional projection and neural network. Int J Artif Life Res 8(2):47–66. https://doi.org/10.4018/ijalr.2018070103
25. Patil S, Kulkarni V, Bhise A (2018) Caries detection with the aid of multilinear principal component analysis and neural network. In: Proceedings of the 2nd international conference on green computing and internet of things, ICGCIoT 2018. Institute of Electrical and Electronics Engineers Inc., pp 272–277. https://doi.org/10.1109/ICGCIoT.2018.8753002
26. Patil S, Kulkarni V, Bhise A (2018) Caries detection using multidimensional projection and neural network. Int J Knowl Based Intel Eng Syst 22(3):155–166. https://doi.org/10.3233/KES-180381
27. Patil S, Kulkarni V, Bhise A (2020) Comprehensive assessment of dental cone beam computed tomography (CBCT): a systematic approach. Test Eng Manag 83:16243–16256. ISSN: 0193-4120. Publication Issue: May–June 2020
28. Alzubi JA, Kumar A, Alzubi OA, Manikandan R (2019) Efficient approaches for prediction of brain tumor using machine learning techniques. Indian J Public Health Res Dev. https://doi.org/10.5958/0976-5506.2019.00298.5
29. Alzubi OA, Alzubi JA, Tedmori S, Rashaideh H, Almomani O (2018) Consensus-based combining method for classifier ensembles. Int Arab J Inf Technol
30. Alweshah OA, Alzubi JA, Alzubi SAM (2016) Solving attribute reduction problem using wrapper genetic programming. Int J Comput Sci Netw Secur
31. Alzubi OA, Alzubi JA, Alweshah M, Qiqieh I, Al-Shami S, Ramachandran M (2020) An optimal pruning algorithm of classifier ensembles: dynamic programming approach. Neural Comput Appl. https://doi.org/10.1007/s00521-020-04761-6
32. Chen TM, Blasco J, Alzubi J, Alzubi O (2014) Intrusion detection. IET Publishing. https://doi.org/10.1049/etr.2014.0007
33. Gupta D, Rodrigues JJPC, Sundaram S, Khanna A, Korotaev V, Albuquerque VHC (2018) Usability feature extraction using modified crow search algorithm: a novel approach. Neural Comput Appl. https://doi.org/10.1007/s00521-018-3688-6

Image Classification in Python Using Keras

Akhil Kumar, Kartik P. Singh, Sawan Kumar, and L. Vetrivendan

Abstract Image classification is a confounding cycle that may be affected by utilizing different various components. This paper dissects current strategies, issues, and parts of image classification. The supplement in this manner is put on the theoretical of a huge headway advanced method of collection close and the procedures utilized for improving how precise classification is. Additionally, several huge variables impacting classification execution are analyzed. This composing review recommends that arranging a sensible image-processing system is a basis for a compelling classification of remotely distinguished data into a well-defined thematic map. Incredible usage of different features of indirectly recognized data and the assurance of a sensible order procedure are especially immense for improving characterization precision. Classifiers that are non-parametric, for instance, (DT) Decision tree classifier, neural networks, and knowledge-based gathering have logically become critical procedures for multisource data characterization. Blend of geographical information systems (GIS), distant detecting, and expert systems emerges as another examination frontier. More assessment, regardless, is required to perceive and diminish weaknesses in the image processing chain to improve characterization precision. Deep Learning and Computer Vision (CV) is stressed over the programmed extraction, examination, and perception of accommodating information from a single picture or a classification of pictures. We've used Convolutional Neural Networks (CNN) in an automated image classification framework. Generally speaking, we utilize the features/attributes from the top layer of the CNN for arrangement; regardless, those highlights may not contain enough accommodating information to predict an image precisely. Occasionally, highlights from the lower layer pass on more discriminative power than those from the top. Thus, applying highlights from a specific layer just to the arrangement is apparently a cycle that doesn't utilize CNN's potential discriminant ability to its full degree. Because of this property we are requiring a blend of features from different layers. We need to make a model with various layers

A. Kumar (✉) · K. P. Singh · S. Kumar · L. Vetrivendan
SCSE, Galgotias University, Greater Noida, India

L. Vetrivendan
e-mail: vetrivendan.l@galgotiasuniversity.edu.in

D. Gupta et al. (eds.), *Proceedings of Data Analytics and Management*,
Lecture Notes on Data Engineering and Communications Technologies 90,
https://doi.org/10.1007/978-981-16-6289-8_45

that will have the alternative to see and group the pictures. It needs to complete our model by using the thoughts of CNN and the CIFAR10 dataset. Keras is used as an independent API to support the build environment and help with error analysis. Vast documentation of Keras also provides an extensive support to the study. Further, for optimizing the prediction model, we will utilize one of the most widely used Keras optimizers—stochastic gradient descent (SGD). Moreover, we will show how Convolution-2D (Conv2D) can be used to execute our model with Central Processing Unit (CPU) planning similarly as less getting ready time. The objective of our project is to learn and in every practical sense, apply the thoughts of Convolutional Neural Network. Google Colab is used as the platform for building, training and testing the prediction model. However, the methodology and techniques used are platform independent and are ready to be implemented as long as the appropriate libraries are compatible with the environment. Pillow python and NumPy library is utilized for image attribute modifications and array calculations. After a successful training and optimizing period of prediction model, it'll then be tested by predicting random images picked from the internet which in theory; will be classified correctly in one of the preset labels/classes.

Keywords CIFAR10 dataset · Convolution · Convolutional neural networks · Deep learning · Image classification · Keras · Machine learning · Max pooling · Rectified linear unit · Stochastic gradient descent

1 Introduction

Present day's web is stacked up with an abundance of pictures and chronicles, which is enabling the headway of search execution that can review a [1] semantic analysis of picture with animation/videos; giving every customer, an improved inquiry content as well as their diagram. In the past, there were huge disclosures in the picture captioning, object Identification (ID), scene grouping [2, 3], and territories report from different analysts from all around the world. This prompt makes it very feasible to design approaches that leans to the concerns of the detection of objects and scene characterization issues. Since artificial neural networks have always been a pioneer in showing a presentation accomplishment in the domain of item identification and scene grouping, especially convolutional neural organizations (CNN) [4], this paper/work targets to attain the best networks thus [5, 6]. Highlight/Features extraction is an imperative advance of such calculations. Feature-extraction from the picture incorporates removing an insignificant arrangement of features containing a high proportion of item or scene information right from a low-level picture pixel regards, subsequently, getting the qualification among the object classes included. A bit of the conventional feature extracting systems used on the picture is scale invariant feature transform, also abbreviated as (SIFT) as in [7], Content-Based Image Retrieval (CBIR) [8], Local Binary Patterns (LBP) [9], a histogram-graph of arranged slants [10], etc. Once features are removed their grouping is done subject to object availability in a picture. A couple of examples of efficient classifiers taken into account are support vector machines (SVM), Decision Random Forest, Tree Logistic Regression, etc.

Image Classification is a major issue in machine learning and computer vision for many years, in the case of people the image processing, and classification is done extremely simple undertaking, however, if there should arise an occurrence of computers, it is an exceptionally extensive assignment [11]. When all is said and done, each image is made out of a set of pixels and every pixel is spoken to with various qualities. Thus, to store a picture, the computer system must need more space to store information. To characterize pictures, it must perform a higher number of computations. For this, it requires frameworks with higher setup and additional computing power. In genuine to make choices based on the information is preposterous on the grounds that it requires some investment for playing out these numerous calculations to give a result.

In the following study, we're going to discuss related literatures of existing methodologies, and use a CIFAR-10 dataset to implement what we know as Convolutional Neural Network architecture for building a prediction model for classifying random images in predefined classes. The model is optimized by the SGD optimizer provided by Keras. The study leans towards the practical usage of the model and is intended to be ready to classify images in real time.

2 Related Work

In [12], we found Neural Network (NN) as a proficient strategy for image classification. The structure contains a mix between imitates of two sets' normal eye and assortment gathering auto-encoding. It incorporated various confusing pictures anyway during the time spent this assessment; the system steadily improved the MNIST, an acronym of [Modified National Institute of Standards and Technology] models. This is one of many public-accessed databases used as the planning set. It is tested with Street View House Numbers dataset and the resulting yield was revised considering the way that even the characteristic eyes can't remember it. Given a deep thought on the study of [13], the discussion relies over a picture collection function based on an architecture of neural networks known as Convolutional Neural Network (CNN). Preparation was performed with the ultimate objective that a reasonable number of pictures having a face and pictures without one, were taken into account for planning by getting more faces from the face pictures data. This function uses the bi-scale model of deep neural networks with 120 arranged data with pre-staged preparing achieves an approximating 80% distinguishing proof rate and with Face Detection Data Set and Benchmark (FDDB) having only 5–6 counterfeit positives where the current status of the craftsmanship attained about 80% area rate and had a sham positives count of 50. Making the note of [3] the exploration used Decision Tree (DT) as the strategies in picture characterization. There are a large number of datasets under DT that are arranged under all of the stratified classifiers.

This seems necessary in order to figure enrollment for every given labels or class. DT allowed some excusal of the labels on the go-between stages. There were certain 3 areas, from where this procedure was expected from [14], one of them being to find terminal hubs, other being in the arrangements of class inside it. The very last area had a specific job of parceling of the hubs. This procedure is seen as a particularly direct and high movement of capability. In [15], a specific machine learning algorithm in discussed. This multi utility algorithm is Support Vector Machine (SVM) and the dynamic finding was successfully and becoming a beacon in the eyes of researchers. It furthermore contemplated some noteworthy ideas by merging spatial information from a progressive cycle in the primer cycle with ghostly. It requires a combination of 3-way technique. The first technique is the Euclidean distance. It decided a segment of the preparation tests from the major bit of spatial. The ensuing approach relies upon the Parzen window methodology ultimately, it joins spatial entropy. The result demonstrated that two of the photos have a significant standard with respect to the sufficiency of regularly. Considering the journal [16], it proposed a fast picture portrayal by promoting the Fuzzy Classifiers. This turned out to be an essential strategy to isolate among established and obscure classification. The strategy is basically all about promoting Metadata where neighborhood trademarks are commonly found. A gigantic data of renders was provided for its attempt and differentiated and the sack-of-features picture model. This yielded a much superior request precision, it being a difficult cycle that gave a jiff of time where it made 30% more restricted and stood out from the previous one.

3   Literature Review

The developing recognizing system of Machine learning plays a significant function in a wide scope of generic utilizations, for example, information boring, common language preparing, picture recognition, and master frameworks. And since Machine learning provides a possible arrangement in all such territories, it definitely is supposed to be a mainstay of forthcoming development. Artificial Intelligence (AI) contains a variety of building calculations the computer can gain from the information and settling on information-driven choices just as expectations. The fast development of AI from the previous few years welcomes a huge impact on our everyday life with such instances of AI for insolvency expectation, precipitation determining, climate estimating, self-driving frameworks and optical character recognition and so forth. By consolidating AI approaches with counterfeit knowledge creates a superior outcome [17]. Despite the fact that machine learning shows an excellent presentation, it isn't productive in human data handling frameworks, for example, discourse acknowledgment, and Computer Vision. Now this can be overwhelmed by the genuine forefront of Machine learning as in Deep Learning. Image processing inculcates some constitutional activities for a given picture rebuilding, picture upgrade, image classification, pictures combination, and other such operations. Image classification builds up a consequential piece of picture management. Thus, the ambition of picture characterization is the lineup assignment of the picture to vital classes [18].

Since 2006, deep structured learning, alias deep learning/progressive learning, has risen as a new region of AI development and research [19]. A middling definitions are accessible for Deep Learning; housing one of the populous definitions from Deep Learning is type casted as: A class of AI methods that abuse numerous panels of nonlinear data handling for managed or solo highlight extraction and change and for design investigation furthermore, grouping. Computational models built upon neural networks have been near for quite a while; the very pioneer model proffered was by McCulloch and Pitts given in [20]. Neural networks are very generally comprised of various layers with each layer associated with different layers shaping the network. A Feed-Forward Neural Network alias FFNN can be deduced in functions of neural activation and the quality of the associations between every pair of significant neurons [21]

4 Implemented Method

CNNs are significant Artificial Neural Networks (ANNs). We use them to describe pictures, pack them by comparability (photo search), and also to perform object detection/recognition inside scenes. CNNs might be used to perceive faces, specific individuals, platypuses, tumors, street signs, and various pieces of visual data. The Convolution Layer (Conv layer) is the middle structure-square of a CNN. The parameters in layers include a lot of learnable filters also known as kernels that have a little responsive field yet connect through the full significance of the data volume. While running the forward pass, each and every kernel is convolved over the height and width of the data volume calculating the dot-product and conveying a two dimensional (2D) activation map of that kernel. Accordingly, the network finds out about the similarities and working of kernels. The kernels institute when they see some specific kind of highlight at some spatial circumstance in the data. By then, the actuation maps are dealt with into a downsampling-layer, and like Conv's, this system is applied on one fix for every turn. Convolutional Neural Networks have in like manner a totally associated layer that gatherings yield with one mark for every hub.

5 Dataset

The dataset used and tested in this publication is the infamous CIFAR10. CIFAR10 dataset consists of images categorized into 10 classes with over a database of 6000 images. These categories contain images of airplanes, frogs, cars, trucks, birds, cats, dogs, horses, deer, ships [22].

6 Neural Networks

Neural Network gets information and goes it through various concealed/hidden layers. Each layer consists of a carriage of neurons, with each and every single neuron completely associated with all the rest neurons in the preceding layer. Each layer in a solitary layer works freely. The terminating layer in a neural network is known to be the 'output layer,' which basically categorizes/groups to the class to which information belongs.

7 Layers of Convolutional Neural Networks

Convolution Layer

The Convolutional layer (Conv layer) is the nucleus piece of a CNN. Pictures are ordinarily static in nature. That insinuate the development of one piece of the picture has a similar resemblance as some other part. Along these lines, a component learnt in one vicinity can coordinate equipollent examples in another region. In a huge picture, we go through one little area at a time and pass that through all the points in the enormous picture (input). While passing anytime, it is then suggested to convolve them into a solitary position (Output). Every slight part of the picture that passes over the huge picture can be labeled as a filter (Kernel). The filters are later arranged depending on the back propagation strategy. Figure shows commonplace convolutional activity (Fig. 1).

Sub Sampling or Pooling Layer

Pooling can be easily understood as a system of down sampling/examining a picture. It takes minute blocks of the convolutional output as information and sub-examples it to create a solitary output. Many distinctive pooling procedures are there to use, for

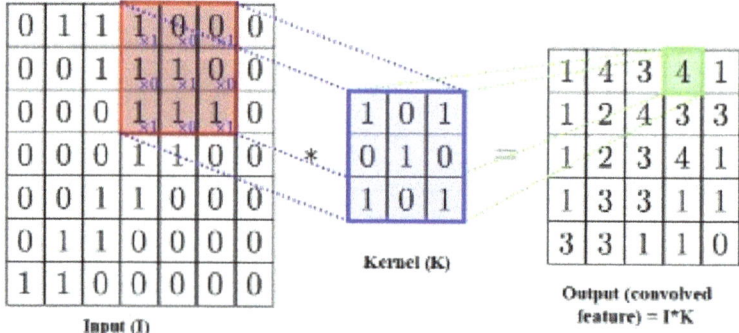

Fig. 1 Convolutional layer

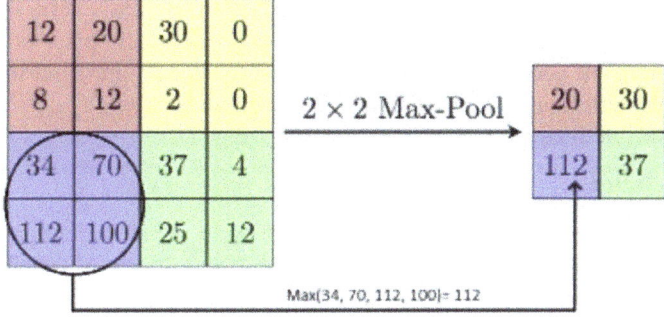

Fig. 2 Max-pooling operation

example, max as well as mean pooling, average pooling, and many others like these. Max-pooling yield the maximum of all the pixel values of an area as presented in Fig. 2. Pooling subside the count of parameters to be figured yet makes the network steady to perception fit like a shape, size and scale.

ReLU Layer

The Rectified Linear Unit, or ReLU, is just another segment of the convolutional neural network's cycle.

It is a benign footstep to the convolution activity that we scrutinized in the past discussion. There have been many educators and creators who examine the two stages independently, yet for our locus, we will believe the two of them to be segments of the initial phase in our cycle.

In the event that you are finished with the related works on artificial neural networks, at that point you should be acquainted with the rectifier work that you find in the picture underneath (Fig. 3).

The answer to why we apply rectifier function, is to fatten the nonlinearity in our renders. We may simply put it to account that images and renders are usually nonlinear. If and when a render is observed closely, what can be seen that it contains a lot of nonlinear features (for example, the transitions between pixels, the outskirts, the shadings, and so on). The rectifier serves to isolate the linearity significantly

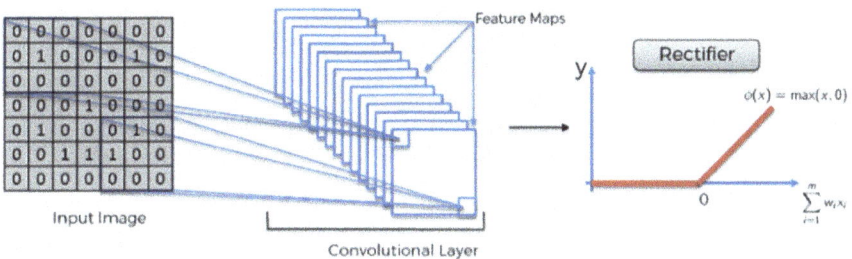

Fig. 3 ReLU layer activation

further to recoup for the linearity that we may impose a picture whenever it is passed through the convolution operation. To perceive how that really plays out, we can take a gander at the accompanying picture and witness the progressions that to it as it goes through the convolution activity followed by correction.

This grayscale image is the initial information (Fig. 4).

Feature Detector

By getting the picture through the convolution cycle, or at the end of the day, by applying it to a feature detector, the outcome is the thing that you find in the accompanying picture (Fig. 5).

As you see, the whole picture is currently made out of pixels that fluctuate from white to dark with numerous shades of gray in the middle.

Fig. 4 Initial grayscale image

Fig. 5 Processed Image with pixel range from white to dark with multiple gray shades

Fig. 6 Processed image
with only positive values

Rectification

What the rectifier works never really picture like this is eliminating all the dark components from it, keeping just those conveying a positive worth (the dim and white tones).

The fundamental contrast between the non-corrected adaptation of the picture and the amended one is the movement of colors. In the event that you take a gander at the first, you will discover parts where a white streak is trailed by a dim one and afterward a dark one.

After we correct the picture, you will discover the tones changing all the more unexpectedly. The steady change is no longer there that shows the linearity has been discarded (Fig. 6).

You need to endure at the top of the priority list that the route by which we just inspected this model just gives an essential non-specialized comprehension of the idea of the amendment. The numerical ideas driving the cycle are superfluous here and would be pretty intricate now.

Fully Connected Layer (FC Layer)

Last part of CNN are fundamentally altogether connected layers as portrayed in the figure beneath. This layer in the CNN model takes contributions from every significant neuron in the past layer and performs activity with singular neurons in the prevailing layer to create output (Fig. 7).

ReLU Activation Function

ReLU $F(x) = \max(x, 0)$ is the most widely used deep learning activation function, used in subsequent hidden layers. This rectified linear unit gives '0' if the input is less than '0' or the bias/threshold value, and raw output 'otherwise'. ReLU is the most modest and easiest nonlinear activation function. Research shows that the Rectified Linear Unit result is much more efficient and quicker for large networks training. There are many frameworks, Keras for instance make it convenient to use ReLU on the intermediate layers/hidden layers (Fig. 8).

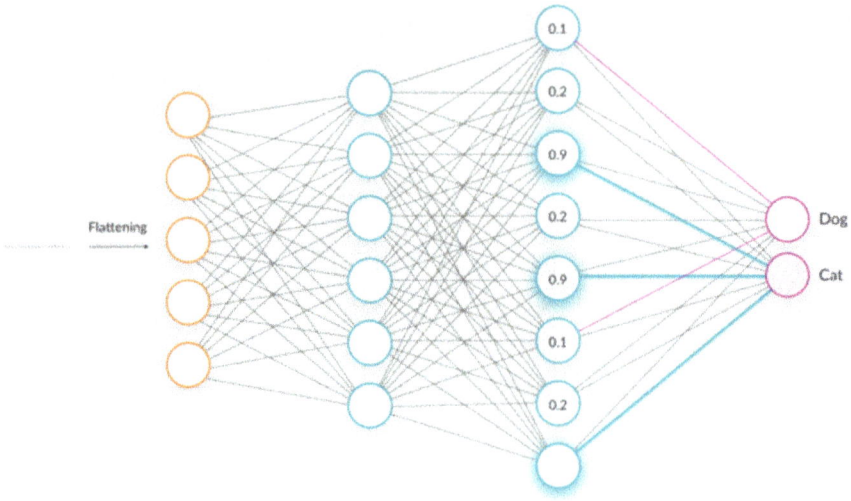

Fig. 7 CNN layers segmenting to FC layer

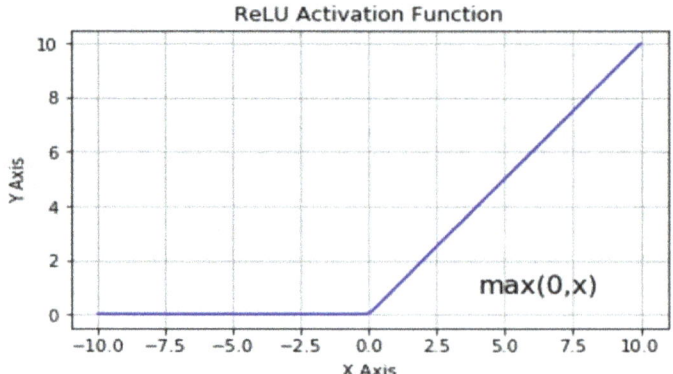

Fig. 8 ReLU activation function

Softmax Function

The softmax classifier

$$\left[F(x) = e^{xi} \left(\sum_{j=0}^{k} e^{xj} \right) \right]$$

distort the resultant outputs of each unit to be in the range of 0 and 1, similar to what sigmoid classifier has to offer. Also, it carve each output in a way that the resulting summation of all the outputs results to be 1. The yield of the softmax classifier is comparable to a categorical likelihood distribution; it discloses to you the likelihood that any of the classes are true. Softmax classifiers can be brought into use for multiple data classification.

8 Preparing Database

Usually, an actual image is provided as the fundamental information. These render are then switched over to grayscale mode because it's not the color data that matters to the network, but the information data. Furthermore, these renders are resized to a size of 32 × 32 because it's what a principle dimension taken into convention. Information structure like picture pyramid is the structures intended to help proficient scaled convolution by miniature picture representation. It comprises an arrangement of duplicates of an original picture where both resolution and sample density are decreased in ordinary steps [23].

9 Explanation of the Model

A convolutional network is the ordering of layers. The layers change one volume of activation to another by using a differentiable function. We use three standard kinds of layers to create the architecture of the model. These layers are the Convolutional (Conv) layer, pooling layer, and Fully Connected (FC) layer. We'll stack these three layers of network architecture. We will go into more nuances under (Fig. 9).

Figure presented above speaks to the visualization of CNN model architecture. From the beginning, we require some pre-handling care on the photos like normalizing the pixel, resizing pictures, and so forth. After the significant pre-handling with data, it is an appropriate condition to be given to the model to train.

First Layer (Layer-1) involves the convolutional layer along with the Rectified Linear Unit (ReLU) activation function which we see as the first convolutional layer in our CNN architecture. This layer acquires the pre-processed picture as the committed

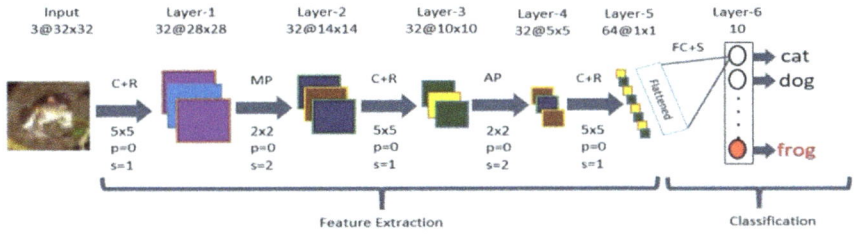

Fig. 9 Architecture of proposed CNN

size of $n * n = 32 * 32$. The convolutional kernel size $(f * f)$ is $5 * 5$, step (s) is one, padding (p) is zero (around all the edges of an image), and there are filters/kernels. After this Conv activity, we get condition $((n = 2p - f)/s) + 1 = ((32 + 2 * 0 - 5)/1) + 1 = 28$ which is equalizes to the feature map of size 32 @ 28 * 28 where 32 is the quantity of the feature maps. By then, the ReLU activation function has been applied to every feature map.

Layer-2 is known to be the max-pooling layer. Max-pooling layer gets the input-data of the picture in the size of 32 @ 28 * 28 from the last layer. The size of the pooling is physically set to 2 * 2; rowing is set to zero, the stride being two. Once this maximum max-pooling is done, a feature map is acquired of size 32 @ 14 * 14. Max-pooling is performed in each and every feature map uninhibitedly, so we acquire a similar measure of feature maps as the last layer, and an additional 14 is received from a comparable equation $((n + 2p - f)/s) + 1$. This doesn't require an activation function.

Layer-3 is known to be the 2nd Conv layer that utilized the Rectified Linear Unit as an activation function. This gets the data of size 32 @ 14 * 14 from the last layer. The size of the kernel is 5 * 5; however, padding isn't available, the stride is set to one, and there are 32 kernels. After this convolutional movement, we acquire feature maps of size 32 @ 10 * 10. By then, ReLU is applied to each and every feature map.

Layer-4 or alternatively called as average pooling layer is then applied. The information given to this layer is in the dimensions of 32 @ 10 * 10 from the last layer. The dimensions of pooling are 2 * 2. There is no padding included and the stride is set to be 1. After this max pooling, we acquire a feature map of dimensions 32 @ 5 * 5.

Layer-5 is the third Conv layer having Rectified Linear Unit as an activation function. This Conv layer gets the Input of dimensions 32 @ 5 * 5 from the last layer. The size of this kernel is 4 * 4; without any padding present, the step is 1, and there are 64 kernels. After this Conv action, we get a feature map of dimensions 64 @ 1 * 1. This layer goes probably as a fully connected layer and conveys a 1-D vector size 64 by flattening.

Layer-6 is the end layer of our neural network. This layer is a fully connected layer. This layer will enlist the class scores, achieving a vector size of which is 10, where all of the ten classes of the CIFAR-10 dataset. For conclusive yields, we use the SoftMax as an activation function.

Accordingly, Convolutional Neural Networks changes the fundamental picture layer by layer from the fundamental pixel esteems to the last class scores. It is to be noted that where few layers contain params (parameters), different layers don't. In particular, the Conv/fully connected layers, perform changes that are a segment of the activations in the information volume just as of the params (the heaps and biases of the neurons). On the other hand, the ReLU/pooling layer will execute a specific limit. We then train the params in the Conv/fully connected layer along with SGD. Through this cycle, we set up the readied model which will be probably utilized to perceive the image present in test data. In this manner, we can characterize the photos as class-planes, feathered creatures, felines, deer, ponies, ships, vehicles, canines, frogs, and trucks.

```
| model.summary()
|⊳ Model: "sequential_2"

   Layer (type)                   Output Shape              Param #
   =================================================================
   conv2d_1 (Conv2D)              (None, 32, 32, 32)        896
   _____
   dropout_1 (Dropout)            (None, 32, 32, 32)        0
   _____
   conv2d_2 (Conv2D)              (None, 32, 32, 32)        9248
   _____
   max_pooling2d_1 (MaxPooling2   (None, 16, 16, 32)        0
   _____
   flatten_1 (Flatten)            (None, 8192)              0
   _____
   dense_1 (Dense)                (None, 512)               4194816
   _____
   dropout_2 (Dropout)            (None, 512)               0
   _____
   dense_2 (Dense)                (None, 10)                5130
   =================================================================
   Total params: 4,210,090
   Trainable params: 4,210,090
   Non-trainable params: 0
   _____

[ model.fit(train_X,train_Y,validation_data=(test_X,test_Y),epochs=10,batch_size=32)
```

Fig. 10 Summary of CNN model architecture

Figure represents the CNN model (Fig. 10).

10 Optimizing Model with SGD

An ideal functioning of gradient descent algorithm is to assess and update parameters of the $J(\theta)$ as,

$$\theta = \theta - a\nabla\theta E[J(\theta)]$$

where the desire in the given equation can be calculated by approximations and by assessing the expense and gradient over the complete set of data given for training. Thus, this optimizer known as Stochastic Gradient Descent (SGD) just gets rid of the desire in the update and figures the gradient of the parameters utilizing just a single or a couple of training models. The new update is given by,

$$\theta = \theta - \alpha\nabla\theta J(\theta; x(i), y(i))\theta = \theta - \alpha\nabla\theta J(\theta, x(i), y(i))$$

with a couple $(x(i), y(i))$ $(x(i), y(i))$ within our training set.

By and large every update in the parameters done by SGD is figured w.r.t a little amount of training models or a minibatch instead of a solitary model. There are two primary reasons for this: first this decreases the change in the parameter update and can helps in a steadier convergence. The other reason being that this permits the calculation to exploit profoundly highly optimized matrix tasks that should be utilized in a well vectorized calculation of the expense and gradient. A common minibatch size is 256, in spite of the fact that the ideal size of the minibatch can shift for various applications and designs.

In SGD, the learning rate $\alpha\alpha$ is regularly a lot more modest than a comparing learning rate in batch gradient descent on the grounds that there is considerably more change in the update. Picking the best possible learning rate and schedule (for example, changing the estimation of the learning rate as learning advances) can be genuinely troublesome. One standard strategy that functions admirably practically speaking is to utilize a little enough consistent learning rate, which in return provides a stable assembly in the underlying epoch. An epoch is a full pass through the dataset dedicated for training, or two of training and afterward divide the estimation of the learning rate as intermingling eases back down. A far better methodology is that after every epoch, it is assessed and strengthen the learning rate in the time when adjustment in target between epochs is under a little threshold. This will in general give great chance to achieve a local optima by its convergence. Another ordinarily utilized schedule can be to temper the learning rate at every iteration tt as [ab + tab + t] where aa and bb direct the underlying learning rate and when the toughening starts separately. More modern strategies incorporate utilizing a backtracking line search to locate the ideal update.

One last yet significant point with respect to SGD is the request where we present the information to the algorithm. On the off chance that the information is provided in some important request, this can bias the gradient and lead to improper convergence. For the most part a decent strategy to keep away from this is to arbitrarily rearrange the information preceding every epoch of training.

11 Training Model

For training the model, we require:

- Training images from CIFAR10 dataset and their corresponding image labels
- Validation images from CIFAR10 dataset and their corresponding image labels (we will not use these labels until the validation of model).

We also declare the number of epochs for training the model. For starters, we will run the model for 25 epochs and batch size 32 (you can change these numbers later).

12 Result and Interpretation

An exactness of 70.75%! This demonstrates that it is essential to pick the correct learning rates and correct number of epochs to locate the best model for your dataset. As expressed before, this is in all likelihood because of the way that the model is taking a gander at every pixel individually. This is the place where convolution networks

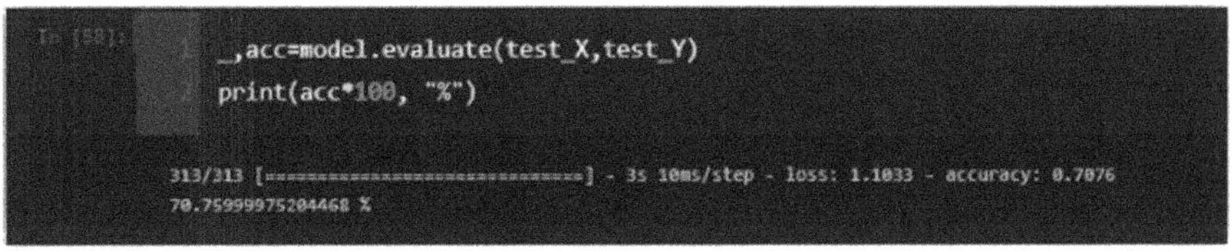

Fig. 11 Evaluated accuracy of trained model

proves their effectivity. Networks like these are deep neural networks that utilize convolution layers. These layers incorporated with convolutional filter, measure and deliver pictures. Kind of how people can improve thought of what a picture is the point at which we take a gander at the full picture or parts of an image, convolution networks can take a gander at a segment of an image, permitting it to hold more data about a picture instead of taking a gander at a pixel (Fig. 11).

13 Conclusion

Here, we demonstrated a neural network model capable of recognizing and classifying the image. Further, this model was able to be extended for real-time valued object recognition, and character recognition. As per the observations and the results from the study, CNN turns out to be so much better than many other researched traditional classifiers. The results, however, can be made closer to precise by increasing the number of Conv layers and hidden neurons. Ability to recognize the object from blurred images by utilizing this model was provided with an accuracy of 70.75% even with the minimum hardware resources and limited Conv layers and training epochs. This represents the future capabilities of this model when to be used with sufficient resources. In the near future, this study can be used for a focused planning to develop and deploy a real-time valued image recognition system and much more.

References

1. Kou F, Du J, He Y, Ye L (2016) Social network search based on semantic analysis and learning. In: CAAI transactions on intelligence technology
2. Garcia-Garcia A, Orts-Escolano S, Oprea S, VillenaMartinez V, Garcia-Rodriguez J (2017) A review on deep learning techniques applied to semantic segmentation
3. Li LJ, Su H, Lim Y, Li FF (2010) Objects as attributes for scene classification." ECCV workshops, pp 57–69
4. Srinivas S, Sarvadevabhatla RK, Mopuri KR, Prabhu N, Kruthiventi SS, Babu RV (2016) A taxonomy of deep convolutional neural nets for computer version
5. Zhou B, Khosla A, Lapedriza A, Oliva A, Torralba A (2014) Objects detectors emerge in deep scene cnns
6. Wang Y, Wu Y (2014) Scene classification with deep convolutional neural networks

7. Lowe DG (2004) Distinctive image features from scale-invariant keypoints. Int J Comput Vis 60(2)
8. Khan SMH, Hussain A, Alshaikhli IFT (2012) Comparative study on content based image retrieval (CBIR). In: 2012 international conference advance computer science applications and technologies (ACSAT)
9. Cheung YM, Deng J (2014) Ultra local binary pattern for image texture analysis. In: 2014 international conference security pattern analysis, and cybernetics (SPAC)
10. Dalal N, Triggs B (2005) Histograms of oriented gradients for human detection. In: Computer vision and pattern recognition. CVPR 2005
11. Barboza F, Kimura H, Altman E (2017) Machine learning models and bankruptcy prediction. Expert Syst Appl 83:405–417
12. Gregor K, Danihelka I, Graves A, Rezende DJ, Wierstra D (2015) DRAW: a recurrent neural network for image generation. https://doi.org/10.1038/nature14236
13. Rastegari M, Ordonez V, Redmon J, Farhadi A (2016) XNOR-net: imagenet classification using binary convolutional neural networks. In: Lecture notes in computer science (including subseries lecture notes in artificial intelligence and lecture notes in bioinformatics), 9908 LNCS, pp 525–542. https://doi.org/10.1007/978-3-319-46493-0_32
14. Kamavisdar P, Saluja S, Agrawal S (2013) A survey on image classification approaches and techniques. Int J Adv Res Comput Commun Eng 2(1):1005–1009. https://doi.org/10.23883/IJRTER.2017.3033.XTS7Z
15. Pasolli E, Melgani F, Tuia D, Pacifici F, Emery WJ (2014) SVM active learning approach for image classification using spatial information. IEEE Trans Geosci Remote Sens 52(4):2217–2223
16. Korytkowski M, Rutkowski L, Scherer R (2016) Fast image classification by boosting fuzzy classifiers. Inf Sci 327(175):182. https://doi.org/10.1016/j.ins.2015.08.030
17. Ling Z-H et al (2015) Deep learning for acoustic modeling in parametric speech generation: a systematic review of existing techniques and future trends. IEEE Signal Process Mag 32(3):35–52
18. Lillesand TM, Kiefer RW, Chipman JW (2004) Remote sensing and image interpretation, 5th edn. Wiley
19. Deng L, Yu D (2014) Deep learning: methods and applications. Microsoft research [Online]. Available at: http://research.microsoft.com/pubs/209355/NOWBookRevisedFeb2014-online.pdf
20. McCulloch W, Pitts W (1943) A logical calculus of ideas immanent in nervous activity. Bull Math Biophys 5(4):115–133
21. An introduction to convolutional neural networks [Online]. Available at: http://white.stanford.edu/teach/index.php/An_Introduction_to_Convolutional_Neural_Networks
22. CIFAR 10 and CIFAR 100 Datasets. https://www.cs.toronto.edu/~kriz/cifar.html
23. Adelson EH, Anderson CH, Bergen JR, Burt PJ, Ogden JM (1984) Pyramid methods in image processing. RCA Eng 29(6):33–41

Predictive Analytics on E-commerce Annual Sales

Sandhya Makkar and Sneha Jaiswal

Abstract Sales forecasting is essential for any organization or industry, but it comes with challenges and problems. Spotting pattern in the past data is one of the approaches of prediction when the past data is known in advance. If the same pattern is followed by the data frequently, we can say that there is a relationship between them and same can be used to predict the future sales. In this paper, we have used the Autoregressive Integrated Moving Average (ARIMA) model in R for predicting the next year sales of fruits and vegetables through ecommerce platform. We have calculated the possible ARIMA(p, d, q) values and selected the best models out of it and predicted the sales of next one year using the selected model. For this purpose, we are using a time series data of past 3 years (2012–14) and using these data, we are predicting the sales of next 1 year.

Keywords Sales prediction · ARIMA · Time series

1 Introduction

Forecasting is a technique of estimating the future of some variables based on the past data. The variables are demand, supply or price. Sales prediction is the most important component of any organization. It helps in decision making for other components or units of the business. Sales prediction guides the company to cover expenses, decide salaries and wages, in marketing, inventory planning, etc. Sales forecasting helps the sales team to focus on the area where they are lacking and to retain the existing customers by fulfilling their demand on time. In an organization, other than sales department, it also provides insights to other business functions such as, business planning, budgeting, inventory management, etc. It helps the decision makers to efficiently allocate the resources to achieve growth. Ultimately, we can say that sales forecasting is very important when it comes to organization, be it small, big, old or new. Sales forecasting is beneficial as it prepares the company for any

S. Makkar (✉) · S. Jaiswal
Lal Bahadur Shastri Institute of Management, Delhi 110075, India

© The Author(s), under exclusive license to Springer Nature Singapore Pte Ltd. 2022   557
D. Gupta et al. (eds.), *Proceedings of Data Analytics and Management*,
Lecture Notes on Data Engineering and Communications Technologies 90,
https://doi.org/10.1007/978-981-16-6289-8_46

unknown or unexpected surge or decrease in demand in the market. In recent years, e-commerce has developed rapidly in the world. The forecasting of e-commerce sales can greatly affect the order quantity, inventory, and logistics strategies, as it is very important for e-commerce companies to predict the sales accurately. When it comes to the brands or products, a customer only wants that whenever they need a product, it can be delivered to them as soon as possible. So, in online purchasing, customers compare the products on different platforms to check who will deliver the product early. No company can speed up the product delivery if they don't have the sufficient products in their hands. So, sales forecasting helps businesses predict the future sales so that they can manufacture the sufficient products in advance and make their customers happy. Also, there is a substantial fluctuation in e-commerce sales volume, and the e-commerce sales which is highly dependent on the seasons, shows a linear trend of increase or decrease over a period of time, but there can be non-linear fluctuations also because of the various uncertainties in the market. All these observations make it highly recommendable to forecast the e-commerce sales to find the best prediction model which is suitable for the mixed characteristics of both the linear and nonlinear changes.

Sales forecasting is basically the prediction of sales which will be done by the company. The models of sales forecasting are Autoregressive (AR), Moving Average (MA), Autoregressive Moving Average (ARMA) and Autoregressive Integrated Moving Average (ARIMA). Compared with all other models, ARIMA is more accurate in predicting the future demand or sales. The ARIMA analysis is based on the observations of time series data for generating a perfect model which can show the process-generating mechanism accurately. Time series data is the data according to the chronological order of time. Mathematical techniques based on the historical data has been used by the time series forecasting models to forecast the demand or sales of the organization.

2   Literature Review

The very first operational sales prediction model consisted of only one layer and because of it, it could model only the slight variation of the mean vertical structure of the sales [1]. Sana Prasanth Shakti et al., 2018 have mentioned about the benefits of forecasting by predicting sales for the "Mahindra Tractors." The historical data for the period of 10 years (2003–2014) has been used to forecast the sales of next 5 years using Autoregressive Integrated Moving Average (ARIMA). Historical sales data of the tractors has been utilized to develop the ARIMA models by using Box-Jenkins time series procedure. In ARIMA, Autoregressive (AR), Moving Average (MA) and Stationary Series (Integrated) processes has been used to generate the data. At last, through ACF and PACF plots, the best fit model has been decided.

In today's organization, which are highly subjected to abrupt changes, the forecast is becoming very crucial. In food industry, forecasting is important as the food is perishable and we need to know as accurate as possible about the consumption so that we can reduce the wastage. In their research, Fattah et al. [2] have mentioned about the forecasting of demand for a Moroccan food company. The time series approach has been used to forecast the demand for the next 10 months. Historical demand data has been utilized to develop the several ARIMA models by using Box-Jenkins time series procedure and the best one has been chosen. The large and consistent demand data from January 2010 to December 2015 has been used to develop model using Autoregressive Integrated Moving Average (ARIMA). The implementation of the model has been done using IBM-SPSS. The selection of the best fit model was based on the 4 criteria: Akaike criterion, Schwarz Bayesian criterion, maximum likelihood, and standard error [2]. The selected model has also been tested using the previous demand information and based on the results, the selected model has been provided to the company for forecasting their demand. The model which they selected as best fit is ARIMA (1, 0, 1) and it minimizes the four previously mentioned criteria, and this model was suggested to the food company owner to predict their future demand.

3   Research Methodology

For predicting the demand of next 1 year, the time series data of sales of ecommerce of year 2012–2014 (monthly data) has been taken from Kaggle [3]. The data included the sales of all the products available on the website of the ecommerce. The sales data of fruits and vegetables has been filtered out from all the available data. Total data available for predicting the sales of next year is 13359. For predicting the sales, ARIMA model has been used [4, 5].

For understanding the concepts of ARIMA models, research papers based on the same concept have been studied.

4   ARIMA (Auto Regressive Integrated Moving Average)

ARIMA model is applied on time series data. Time series data is the data or observation collected from a process with equally spaced periods of time.

Examples: Daily/Monthly data of sales.

ARIMA is also known as Box-Jenkins approach. To build a time series model using ARIMA, the value of p, d, q should be known.

$p =$ Auto Regressive (Auto Correlation)

$d =$ Integrated (Stationary)

$q =$ Moving Average.

- **Identification**: Determine the appropriate values of p, d, q using Autocorrelation Function (ACF), Partial Autocorrelation Function (PACF), and unit root tests.
- **Estimation**: Estimate the ARIMA model using values of p, d, q which is appropriate for predicting the sales.

- **Forecasting**: Using the same model, predict the sales of next year.

Figure 1 is the line graph of monthly sales of fruits and vegetables at ecommerce online platform. The graph has been drawn using SPSS platform.

The monthly sales of fruits and vegetables of the year 2012, 2013 and 2014 has been shown in Fig. 1. Using this data, sales of next year will be predicted using ARIMA model. Here, we can see that for the 2012 and 2013, the sales have been increasing but after the month of July 2014, it started decreasing.

Figure 2 is the pie chart for the sales of the 3 years. If we talk about the overall year sales, sales have been increasing from 2012 to 2013. We can see a slight decrease in the year 2014 from 2013.

In Fig. 3 we can see that the sales of each month of the 3 years, 2012, 2013 and 2014. By above graph, we can say that the sales have been continuously increasing from 2012 to 2014 but for the month of Oct, Nov and Dec of the year 2014, the sales have been decreasing. But, if we talk about the overall sales, the trendline has been increasing.

Summary of data has been computed in *R*. The values of mean, median, 1st and 3rd quartile, minimum and maximum values have been shown in the Table 1. We can see the difference between the minimum and maximum value is very high. This

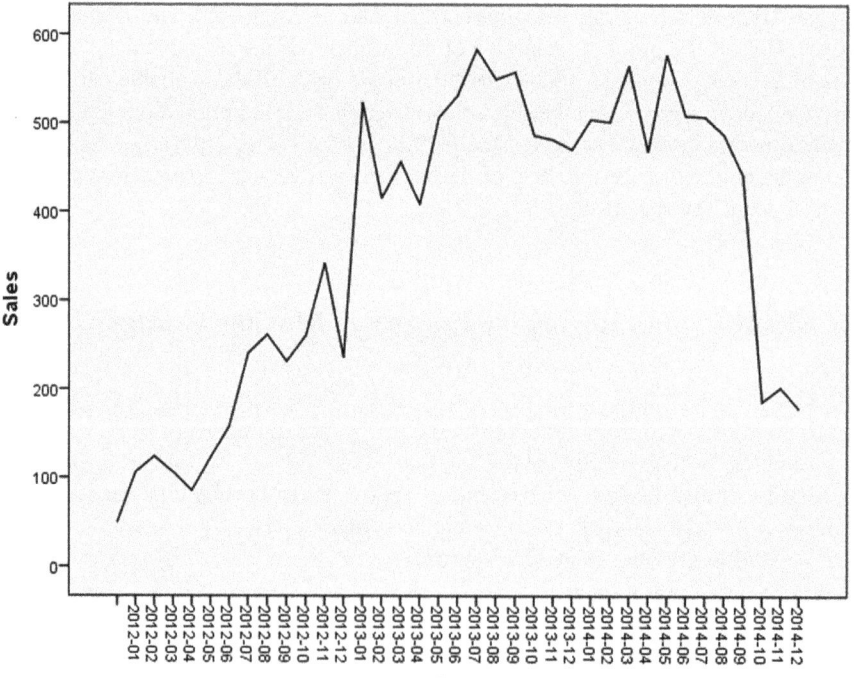

Fig. 1 Line graph of the monthly sales

Fig. 2 Pie chart showing the yearly sales of the ecommerce company

Fig. 3 Bar chart showing the monthly sales

Table 1 Summary of the sales data

Minimum	1st quartile	Median	Mean	3rd quartile	Maximum
85	223.8	448.5	371.1	506.2	583

shows that there are lots of variations in the sales of the fruits and vegetables through the online platform.

Figure 4 is the graph showing the regular counts, modified moving average and weighted moving average. These plots are used to identify the trends in the data. Using these, we can find out the small shifts or trends in the available data. Modified moving average values are the sum after subtracting the last average and adding the new data to the previous final sum. Weighted moving average is the value in which more weightage is given to the recent data.

In Fig. 5, we can see the data decomposed in three parts, trend-cycle, seasonality and remainder. In the seasonal panel, the seasonality data has been extracted and shown in the graph. The seasonal component of the data changes slowly and that is why we see the similar pattern for any two consecutive years. When we remove the seasonality and trend from the data, we are left with the remainder. All the three plots have been shown in Fig. 5 computed from *R* programming.

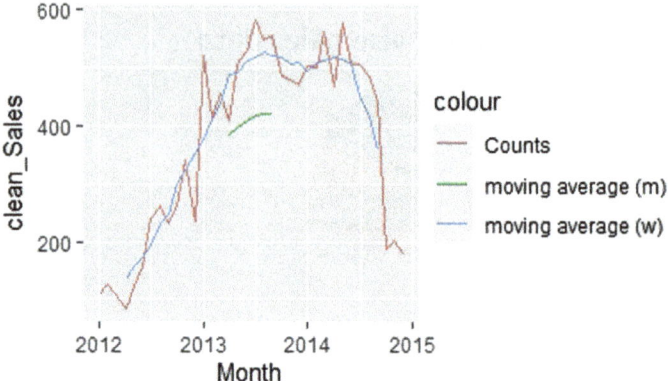

Fig. 4 Moving average graph

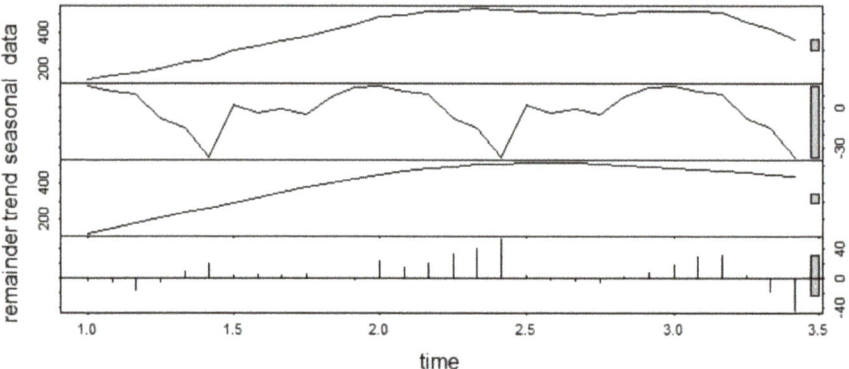

Fig. 5 Seasonality, trend and remainder shown in the graph

4.1 Stationary Test (Augmented Dickey-Fuller (ADF) Test)

The Augmented Dickey Fuller Test (ADF) is a stationarity test [6]. If the data is non-stationary, it can cause unpredictable results in your time series analysis.

A stationary time series data is the data which is not dependent on the time at which the series is observed. So, before going for the forecasting steps, there is need to check that the data must be stationary.

The hypotheses for the test:

- Null hypothesis: The time series is not stationary.
- Alternate hypothesis: The time series is stationary (or trend-stationary).

The results of the stationarity test (ADF test) have been shown in the Table 2. We can see that the p-value is greater than 0.05 in the sales data. This shows that the data is not stationary. The remedy for this problem is simple differentiation. In

Table 2 Output of the stationarity test at different orders

Data	Dickey fuller	Lag order	p-value
Sales	0.53086	3	0.99
1st order diff	− 2.525	3	0.3713
2nd order diff	− 3.9902	3	0.02315

the output of 1st differentiation, we can see that the p value is still greater than 0.05, this shows that the data is still not stationary. So, we will differentiate the data again (2nd order). After the 2nd order differentiation, the p value is less than 0.05, so we can say that the time series data is stationary and we can proceed with the further prediction steps.

The next step in the process is to find out the values of ARIMA(p, d, q). From the above stationarity test, we get the value $d = 2$. Now, to find out the values of p and q, we need to plot the ACF and partial ACF graphs of the differenced series.

Figure 6 shows the (Auto-Correlation Function (ACF) and PACF Partial Auto-Correlation Function (PACF) graphs. In the graph, we can see the values which are going out of the significance level. Using the graph, we can compute the possible p and q values to make the ARIMA models.

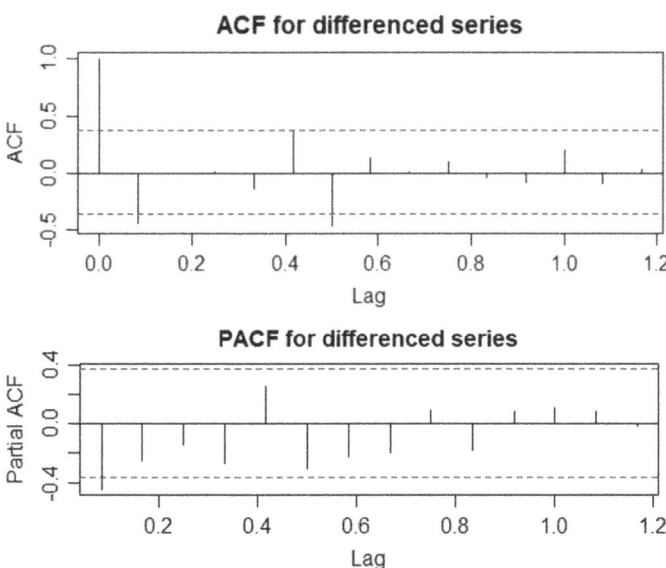

Fig. 6 ACF and PACF plots

5 Forecasting

When autoarima function has been run on the R-programming with the data, the system has returned the output and the model suggested by the system is ARIMA (0, 2, 1).

The Table 3 contains the sigma sqr value, AIC and AICc values.

The Akaike Information Criteria (AIC) values are used to measure the statistical model. It is generally used to quantify the goodness of fit, and the simplicity of any model into a single statistic. The lowest the value of AIC, the better the model is.

But if we see the ACF and PACF graph (Fig. 6), there are values which are going out of the significance level. So, we have come with all the possible values of p and q and made the ARIMA models according to the p, d, q values. So, here we get the possible values of p, d and q.

$p = 0, 1, 5$ (from the ACF plot)

$d = 2$ (Stationarity)

$q = 1$ (from PACF plot).

So, the three possible models are ARIMA(0, 2, 1), ARIMA(1, 2, 1) and ARIMA(5, 2, 1). On applying autoarima on all the possible models, the coefficient values, Sigma sqr, log likelihood and AIC of the different models have been compared in the Table 4.

The values of AIC, RMSE, MAPE and ME of the three models have been compared in Table 5. In the table, we can see that the AIC value of the model ARIMA(0, 2, 1) is the minimum of all the models and MAPE and ME value are minimum for model ARIMA (0, 2, 1) and ARIMA(1, 2, 1).

The best model is ARIMA(0, 2, 1) which has maximum minimum values. The best models can be selected by comparing the AIC, RMSE, MAPE and ME parameters of the models. The lowest the values are, best the model is. So, we will look for the models having the maximum number of lowest values of the given parameters. We can see the parameters with the lowest value for ARIMA(0, 2, 1) is 3 and for ARIMA(1, 2, 1) is 2 and for ARIMA (5, 2, 1) is 1. So, the best model is ARIMA(0, 2, 1). After selecting the best model, the forecast of next 12 months have been done through R programming.

Next 12 months sales of the ecommerce company has been forecasted using the ARIMA(0, 2, 1) model. The graph (Fig. 7) is showing the forecasted value of next 12 months.

We can see that the value of next 1 year is decreasing. As we have seen in the past data analysis that the sales of the fruits and vegetables for the year 2014 has been decreasing, so, the forecasting of next 12 months has taken the same trend and

Table 3 Summary of the differenced data

Coefficients	sigma^2	log likelihood	AIC	AICc	BIC
ma1 = − 0.5408	303.6	− 119.42	242.83	243.31	245.5

Table 4 Summary of the selected models

Models	Coefficients						Sigma^2	Log likelihood	AIC
	ma1	ar1	ar2	ar3	ar4	ar5			
ARIMA(0, 2, 1)	− 0.5408	Null	Null	Null	Null	Null	303.6	− 119.42	242.83
ARIMA(1, 2, 1)	− 0.5408	0	Null	Null	Null	Null	292.8	− 119.42	244.83
ARIMA(5, 2, 1)	0.855	− 1.1273	− 0.6198	− 0.2891	− 0.1213	0.2027	216	− 116.08	246.17

Table 5 Parameters to select the best model

	AIC	RMSE	MAPE	ME
ARIMA(0, 2, 1)	242.83	16.531	2.6282	− 3.956
ARIMA(1, 2, 1)	244.83	16.531	2.6282	− 3.956
ARIMA(5, 2, 1)	246.17	14.1981	2.71	− 3.19

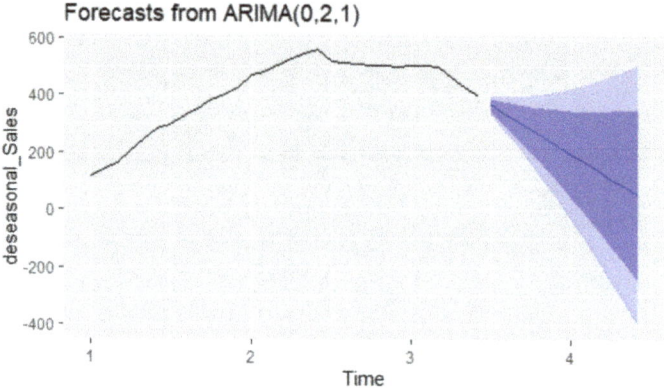

Fig. 7 Forecasted value of next 12 months

showing the decreasing sales for next 1 year. We can see the sales for December 2015 has been predicted somewhere near to 50.

6 Conclusion

Sales forecasting is very important for any organization. It helps the decision makers to take the decision and make the strategies for the benefit of the company. One of the most important application of sales forecasting is how it integrate with other areas of any business and makes it easy to take decisions for other functions as well. We have successfully developed a model of ARIMA for sales forecasting of a product line (fruits and vegetables) of an ecommerce company. The past sales data (2012, 2013 and 2014) were used here to develop all the possible models. In the next step, the best model was selected on the basis of AIC, RMSE, MAPE and ME values. The best model which fulfills all the criteria and which was used for further processing is ARIMA (0, 2, 1). The details of results can be found in Table 5. In our conclusion, we can say that the sales forecast from this model can provide insightful details to the managers which will further be used in taking important decisions for the organization.

References

1. Shakti SP, Hassan MK, Zhenning Y, Caytiles RD, Iyenger SN (2017) Annual automobile sales prediction using ARIMA model. Int J Hybrid Inf Technol 10(6):13–22
2. Fattah J, Ezzine L, Aman Z, El Moussami H, Lachhab A (2018) Forecasting of demand using ARIMA model. Int J Eng Bus Manag 29(10):1847979018808673
3. https://www.kaggle.com/shikhnu/big-basket-sales-transaction-records-20112014
4. https://datascienceplus.com/time-series-analysis-using-arima-model-in-r/#:~:text=arima()%20function%20in%20R,chosen%20by%20minimizing%20the%20AICc
5. https://machinelearningmastery.com/arima-for-time-series-forecasting-with-python/
6. https://www.machinelearningplus.com/time-series/augmented-dickey-fuller-test/

Proposed Method to Identify Oil Seed Leaf Diseases by Deep Learning Techniques

Abhilasha, Vaibhav Vyas, Vijay Singh Rathore, and Neelam Chaplot

Abstract In oilseed production, India has fourth rank after three countries which are United States of America, China and Brazil. Soybeans, groundnut and rapeseed/mustard are three main oilseeds. Disease can be caused by fungus, virus and bacteria in oilseed leaves. Diseases cause production to go down affecting economy. If we can detect diseases at early stage, then we can save crops. Purpose of this paper is to describe a method to identify the oilseeds leaf diseases using multi-layered deep learning technique. We will be using a convolution neural network (CNN) to get reliable and authenticate results. The images needed to be acquired for this process should be collected from online sources and an agriculture research center.

Keywords Oilseeds leaf · Machine learning · Deep learning · Convolution neural network · Leaf diseases identification · Image processing · Supervised learning

1 Introduction

Deep learning is extension to machine learning. It works on neural network concept. Deep learning has characteristics such as robustness, universal learning approach, scalability and generalization. Deep learning is applied in systems where large amount of data needs to be processed and systems such as automatic voice recognition, image reorganization or medical image progressing. DL mechanism can be categorized in three different learning methods which are SL, i.e., supervised learning; UL, i.e., unsupervised learning and RL, i.e., reinforcement learning. These learning

Abhilasha (✉) · V. Vyas
Banasthali Vidhyapeeth, Jaipur, Rajasthan, India

V. Vyas
e-mail: vvaibhav@banasthali.in

V. S. Rathore
IIS University, Jaipur, Rajasthan, India

N. Chaplot
Poornima College of Engineering, Jaipur, Rajasthan, India

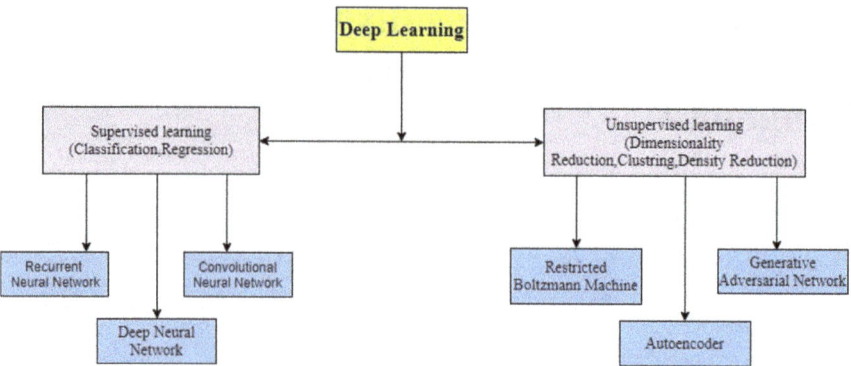

Fig. 1 Mechanism of deep learning [1]

methods have different working features and characteristics. We will discuss all of them here. Figure 1 is about mechanisms of deep learning.

Convolution Neural Networks This neural network basically depends on feedforward learning method. One input layer, one output layer and many hidden layer form a CNN. Hidden layers depend on working requirements. Convolution layer, pooling layer, full connected layer, Relu and Softmax layers are usually part of hidden layers. Neural networks work in similar manner as brain functions with combination of neurons and synapses. CNN is used in natural language processing, speech signal and image processing .AlexNet, Google Net, VGG Net, ResNet and Inception V4 are CNN architectures. This CNN architecture has many hidden layers and is in use by many researchers for deep learning purpose [2]. Phases for disease prediction using CNN are shown in Fig. 2.

Deep learning is used in many areas with image processing. Here, our concern is plant diseases identification with the help of convolution neural network model in oilseeds crops. As we discussed, there are multiple neural network available, but we would prefer a CNN model which has advantages with image processing.

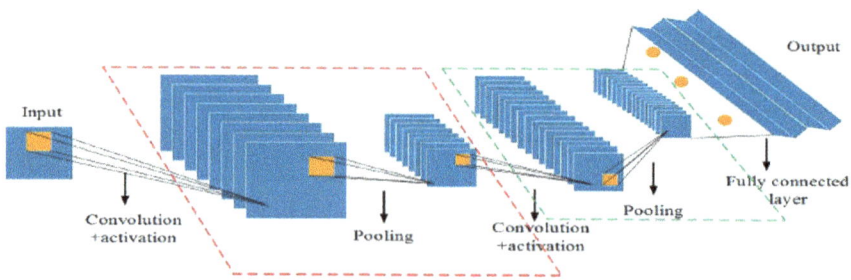

Fig. 2 Phases for disease prediction using CNN [2]

Agriculture is a major segment of Indian economy. If there are ways to figure out diseases in early state of infection, then mitigation mechanism can be used by farmers for safety of crops. Fungus, bacteria and virus can possibly attack a leaf, and these are harmful for crops and damage a plant and spread from one plant to other. The major aid of this paper is: We compared related work according to different leaves and technology of deep learning. The next section explains proposed work and its methodology with the help of flowchart and algorithm. In brief, we give the overview of tools and technology which is used for implementing deep learning CNN model. In the end of the paper, we will acheive the expected output with the help of proposed work.

2 Related Work

Many researchers have developed different models or modified already developed model to identify diseases in leaves by using deep learning methods with help of image processing. Researchers have used deep learning techniques with CNN for diseases identification at early stage in banana, potato, tomato, maize, soybeans, apple and many others crops. Table 1 is the list of distinct papers which we have surveyed.

3 Proposed Work and Methodology

The main objective of the current research work is to identify disease in oilseeds crops leaves using deep learning techniques. The above flowchart depicts the steps we are going to follow in complete process. At initial stage, number of infected and healthy images should be collected from an agriculture research center of oilseeds for training, testing and validation. We can also collect high quality images using camera DSLAR. This initial stage has major role in whole process because having low quality images can hinder the process of disease identification or can make it slow or less accurate. After collecting images, we will apply image augmentation on images to increase the size of dataset. With the help of augmentation, we can rotate, change the angles of images, at times resulting in multiple images from a single image. Image preprocessing methods can be used to resize and adjust images according to requirement. Different methods to do image preprocessing are available for our use. After applying preprocessing method, feature extraction will be used. Then after, we do train, test and validate our data. Classification is done by using deep learning model. If accuracy of developed model is good from our model, then disease identification can be generated otherwise we will have to adjust parameter optimization. With the help of developed model, we can input new images for classification and to identify diseases in leaves (Fig. 3).

Steps Performed

Table 1 Analysis of different crops with accuracy and methodology

S. No.	Leaves	Methodology	Accuracy (%)	Reference
1	Cotton	Pattern recognition system	85	[3]
2	Apple	Fine-tuned VGG16 model trained with transfer learning	90.4	[4]
3	Apple	Deep CNN model using Alex net CNN model	97.62	[5]
4	Rice	Deep CNN model with the use of transfer learning	90.9	[6]
5	Mango	Introduced CNN model	96.67	[7]
6	Maize	Improvement in developed deep CNN GoogleNet and ciffer10	98.8	[8]
7	Tomato	Deep CNN model using algorithm of stochastic gradient descent and Adam optimization method	96.51	[9]
8	Rice	Developed deep CNN model	94	[10]
9	Maize	Optimized VGG 16 and VGG 19 with the OLPSO algorithm	98.2	[11]
10	Rice	Deep CNN model	99	[12]
11	Potato	Deep CNN model	100	[13]
12	Soybean	Modified AlexNet and GoogleNet models	98.75 and 96.25	[14]
13	Tomato	Introduced CNN model	91.2	[15]
14	Tomato	Developed attention-based residual CNN model	98	[16]
15	Rice	Deep CNN model	92.46	[17]
16	lady finger	Introduced CNN model	96	[18]
17	Potato	Modified VGG 16 and VGG 19	91	[19]
18	Oilseeds	Hybrid ensemble	94.73	[20]
19	Guava	Deep CNN model	70	[21]
20	Strawberry	Using ResNet50	99.60	[22]
21	Cassava	Novel deep residual convolution neural network	96.75	[23]

A. **Image Augmentation**: Image augmentation is used to increase the data set size because more images are required for accuracy of any process. In image augmentation rotation, zoom, shear, width shift, height shift, horizontal flip, fill mode, intensity transformation techniques are applied for expand number of images in datset. By augmentation, it will reduce the over fitting problem [7].

B. **Image Preprocessing**: Image preprocessing is used on RGB input images of oilseed leaf. In this preprocessing step, few major steps are applied which are image resize, image enhancement, color space conversion and image normalization [25].

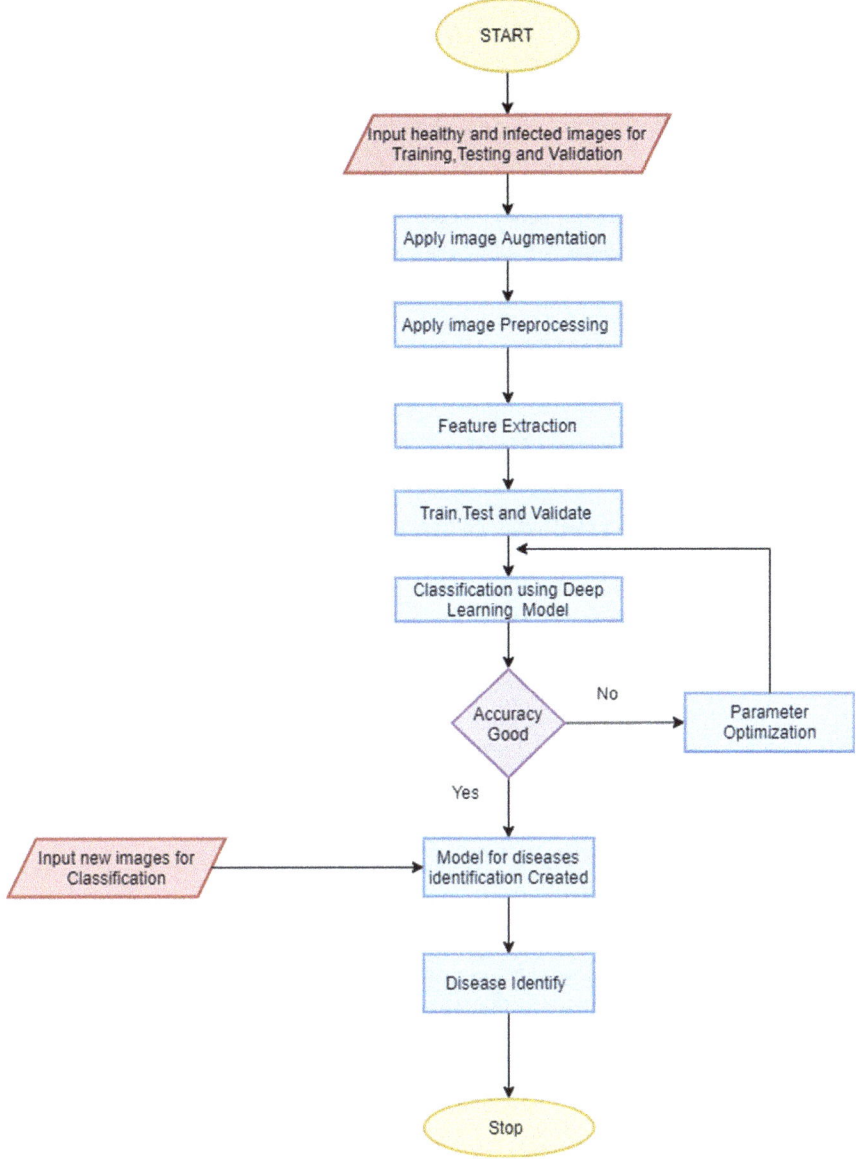

Fig. 3 Process flow diagram of processed system

C. **Feature Extraction**: Feature extraction is used to extract more significant features from infected leaf. In feature extraction color, edge and texture are extended by using grey-level co-occurrence matrix (GLCM) and district wavelet transform (DWT) [26].

ALGORITHM for Proposed System

Step1: Take RGB image of oilseeds leaf as input dataset.
Step2: Apply image augmentation techniques to input images dataset.
Step3: Apply image preprocessing techniques to input images dataset.
Step4: Extract the feature of the infected part of oilseeds leaf dataset.
Step5: Train, test and validate the dataset.
Step6: Apply classification using deep learning techniques.
Step7: If good accuracy is achieved, then model is created.
Step8: Else change the parameter, and repeat step 6.
Step9: Input new images for classified model.
Step10: Disease identified.

4 Tools and Technology Used

AlexNet [14], GoogLeNet [14], VGGNet 19 and VGG 16 [19], ResNet [22] different deep learning architecture of convolution neural network Model for disease identification in leaves available. Many platforms are also required such as Tensor Flow [27], Torch, DeepLearning4J, MXNet, Keras [28], Pytorch, Theano, Open CV [29], Microsoft Cognitive Toolkit, Caffe and Caffe2 [30] for executing different convolutional neural network model. So, there are many tools and technology available for leaf diseases identification. With the help of transfer learning, we will find which CNN model will give better result and develop a new convolutional neural network model which has less hidden layer.

5 Expected Outcome

As we know, agriculture is a major contributor to Indian economy. Healthy crops result in improved income for farmers, while crops' diseases negatively impact farmer's income as well economy. Deep learning techniques are used for diseases identification in number of crops. Early detection of diseases through leaves can be very beneficial. With the help of early detection, farmers can apply timely treatment or preventive intervention to infected crops. In our work, we are planning to design a deep learning model for early disease identification in oilseeds crops leaves. Our work would be helpful for farmers as it will be help them to quickly and accurately identify diseases in oilseeds crops leaves. Proposed deep learning model should be easily accessible, time-consuming and cost-effective.

6   Conclusion

In this paper, we proposed architecture for disease identification in oilseeds using convolutional neural network which we will apply on oilseeds leaf. Deep learning-based approach can perpetually bring out the disease in leaf. In future, we apply this model on different and large dataset of oilseeds crops leaf and real images for more accurate result. We will collect dataset from agriculture institute and agriculture field. Many optimization techniques available are like gradient descent, Nesterov accelerated gradient (NAG), Adam, stochastic gradient descent (SGD), etc. As we study many optimization techniques for good accuracy, try to find which is better for disease identification of oilseeds leaf in proposed approach.

References

1. Hatcher WG, Yu W (2018) A survey of deep learning: platforms, applications and emerging research trends. IEEE Access 6:24411–24432
2. Xiong J, Yu D, Liu S, Shu L, Wang X, Liu Z (2021) A review of plant phenotypic image recognition technology based on deep learning. MDPI (2021)
3. Rothe PR, Kshirsagar RV (2015) Cotton leaf disease identification using pattern recognition techniques. In: International conference on pervasive computing (ICPC), Apr 2015
4. Wang G, Sun Y, Wang J (2017) Automatic image-based plant disease severity estimation using deep learning. In: Computational Intelligence and Neuroscience, vol. 2017
5. Liu B, Zhang Y, He D, Li Y (2017) Identification of apple leaf diseases based on deep convolutional neural networks. Symmetry 10(1)
6. Mique EL, Palaoag TD (2018) Rice pest and disease detection using convolutional neural network. In: Proceedings of the 2018 international conference on information science and system, pp 147–151, Apr 2018
7. Arivazhagan S, Ligi SV (2018) Mango leaf diseases identification using convolutional neural network. Int J Pure Appl Math 120(6):11067–11079
8. Zhang X, Qiao Y, Meng F, Fan C, Zhang M (2018) Identification of maize leaf diseases using improved deep convolutional neural networks. IEEE Access 6:30370–30377
9. Zhang K, Wu Q, Liu A, Meng X (2018) Can deep learning identify tomato leaf disease? In: Hindawi advances in multimedia, vol 2018
10. Bhattacharya S, Mukherjee A, Phadikar S (2020) A deep learning approach for the classification of rice leaf diseases. In: Advances in intelligent systems and computing. Springer
11. Darwish A, Ezzat D, Hassanien AE (2020) An optimized model based on convolutional neural networks and orthogonal learning particle swarm optimization algorithm for plant diseases diagnosis. In: Swarm and evolutionary computation, vol 52. Elsevier
12. Desai S, Nayak R, Patel R (2019) Identification of plant diseases using convolutional neural networks. In: Recent Advances in Communication Infrastructure, vol 618. Springer, pp 95–104
13. Dasgupta SR, Rakshit S, Mondal D, Kole DK (2019) Detection of diseases in potato leaves using transfer learning. In: Computational intelligence in pattern recognition. Springer, pp 675–684
14. Jadhav SB, Udupi VR, Patil SB (2020) Identification of plant diseases using convolutional neural networks. Int J Inf Technol
15. Agarwal M, Singh A, Arjaria S, Sinha A, Gupta S (2020) ToLeD: tomato leaf disease detection using convolution neural network. In: International conference on computational intelligence and data science Elsevier, Procedia Computer Science, vol 167, pp 293–301

16. Karthik R, Hariharan M, Anand S, Mathikshara JPA, Menaka R (2019) Attention embedded residual CNN for disease detection in tomato leaves. Appl Soft Comput J
17. Ghosal S, Sarkar K (2020) Rice leaf diseases classification using CNN with transfer learning. In: Proceedings of 2020 IEEE Calcutta Conference (CALCON)
18. Selvam L, Kavitha P (2020) Classification of lady finger plant leaves using deep learning. J Ambient Intell Humanized Comput
19. Sholihati RA, Sulistijono IA, Risnumawan A, Kusumawati E (2020) Potato leaf disease classification using deep learning approach. In: International electronics symposium, IEEE Explore, pp 392–397
20. Chaudhary A, Kolhe S, Kamal R (2016) A hybrid ensemble for classification in multiclass datasets: an application to oilseed disease dataset. Elsevier, pp 0168–1699
21. Srinivas B, Satheesh P, Rama Santosh Naidu P, Neelima U (2021) Prediction of guava plant diseases using deep learning. Springer Nature Singapore Pte Ltd., pp 1495–1505
22. Xiao J-R, Chung P-C, Wu H-Y, Phan Q-H, Andrew Yeh J-L, Hou MT-K (2021) Detection of strawberry diseases using a convolutional neural plants. MDPI, Journal/Plants
23. Oyewola DO, Dada EG, Misra S, Damaševičius R (2021) Detecting cassava mosaic disease using a deep residual convolutional neural network with distinct block processing. PeerJ Computer Science
24. Wang G, Sun Y, Wang J (2017) Automatic image-based plant disease severity estimation using deep learning. In: Hindawi computational intelligence and neuroscience
25. Ashok S, Kishore G, Rajesh V, Suchitra S, Sophia SGG, Pavithra B (2020) Tomato leaf disease detection using deep learning techniques. In: Proceedings of the fifth international conference on communication and electronics systems
26. Abadi M, Barham P, Chen J, Chen Z, Davis A, Dean J et al (2016) TensorFlow: a system for large-scale machine learning, In: Proceedings of the 12th USENIX symposium on operating systems design and implementation, Savannah, GA, USA
27. Rampasek L, Goldenberg A (2016) Tensorflow: biology's gateway to deep learning? Cell Systems 2(1):12–14
28. Jia Y, Shelhamer E, Donahue J, Karayev S, Long J, Girshick R, Gudarrama S, Darrell T (2014) Caffe: convolutional architecture for fast feature embedding. In: Proceedings of the 22nd ACM international conference on multimedia, pp 675–678 (2014)
29. Keras (2017) The python deep learning library. https://keras.io/
30. https://opencv.org/

Heart Disease Detection Using Feature Selection Based KNN Classifier

Rajendrani Mukherjee, Srestha Sadhu, and Aurghyadip Kundu

Abstract Heart disease is the leading cause of death across world. With an advent of several clinical decision support (CDS) mechanisms, it has become possible to detect the disease early and also to predict the prognosis. In this paper, an attempt has been made to apply machine learning classifier KNN for heart disease prediction. Several benchmark datasets were chosen, and the KNN classifier was applied with different values of k. In the second phase of experimentation, the KNN classifier was augmented with Least Absolute Shrinkage and Selection Operator (LASSO) feature selection mechanism. Results indicated KNN coupled with LASSO mechanism yielded greater accuracy than KNN without feature selection. Results also indicated significant decrease in error rate with negligible increase in execution time. It is hoped usage of LASSO aided KNN classifier will help in building a robust clinical decision support (CDS) system for treating heart disease efficiently.

Keywords Feature · Classifier · Dataset · Accuracy · Disease · Prediction

1 Introduction

Cardiovascular disease is a common cause of death in elder population nowadays, and it even sometimes affects young adults and infants also. Underlying health conditions (diabetes, blood pressure, etc.), stress, life style disorders (smoking, tobacco usage, lack of exercise, etc.), pollution are provoking the disease and putting the human population in risk at large. In 2015, 17.9 million deaths occurred due to cardiovascular disease (CVD). CVD mainly involves heart or blood vessels. However, it has been observed that 90% of CVD is preventable. Toward paving this attempt of prevention,

R. Mukherjee (✉) · S. Sadhu
Department of Computer Science and Engineering, University of Engineering and Management, Kolkata, India

A. Kundu
Department of Computer Science and Engineering, Brainware University, Kolkata, India

© The Author(s), under exclusive license to Springer Nature Singapore Pte Ltd. 2022
D. Gupta et al. (eds.), *Proceedings of Data Analytics and Management*,
Lecture Notes on Data Engineering and Communications Technologies 90,
https://doi.org/10.1007/978-981-16-6289-8_48

this paper focuses on building a machine learning classifier which will prompt us to detect and avert heart diseases early.

Machine learning has seen widespread application is building intelligent clinical decision support (CDS) system. A survey conducted by Ramalingam et al. in 2018, has showed that ML classifiers can increase the accuracy in predicting heart disease to great extent [7]. Throughout the years, researchers have tried to improve and redesign several intelligent techniques in order to optimize the performance of ML classifiers. Application of feature selection methods is one of those improvement mechanisms. While several ACO based [2, 3] or GA based [4] feature selection techniques are gaining momentum, LASSO is also another robust but less exploratory FS technique to improve the performance of ML classifiers. This paper addresses this research gap.

The contribution of the paper is as follows:

RQ1: How KNN classifier can be used for predicting heart disease?

The paper applied KNN classifier for different k values on three subject datasets and measured the accuracy value from generated confusion matrix. The error rate, execution time parameters were also noted down to check the implementation potential.

RQ2: What is the contribution of feature selection methods along with KNN classifier in terms of the performance of ML classifier in accurately predicting heart disease?

Usage of feature selection methods to improve the performance of ML classifiers is explored through this research question. LASSO technique is a robust feature selection (FS) mechanism and this paper leverages that.

The organization of the paper is as follows. Section 2 surveys the literature while Sect. 3 describes the subject datasets. Section 4 elaborates the implementation details. Section 5 narrates the results and findings. The work is concluded in Sect. 6 with future scope.

2   Related Work

Several researchers [4, 10, 11] have performed many elaborate studies to show the effectiveness of AI in decision making, prediction and detection of heart disease, kidney disease, breast cancer detection, neurological screening, etc. Of late, several studies are conducted using ANN and deep learning methodologies to analyze the prognosis of COVID-19 pandemic.

A survey conducted by Ramalingam et al. in 2018 indicated the popularity of ML classifiers for heart disease analysis [7]. In 2008, Parthiban and Subramanian used fuzzy logic for intelligent heart disease diagnosis [5]. In 2013, Jabbar et al. performed a study to accurately predict heart disease using KNN and genetic algorithm [4]. In 2016, Song et al. also found KNN helpful for predicting heart disease regarding big datasets [11]. The study leveraged the MapReduce technique for big data analysis. Shinde et al. utilized Naïve Bayes classifier along with k-means clustering for categorizing heart disease in 2015 [6].

The usage of feature selection mechanism [1–3, 12] for intelligent data analysis (IDA) is also very popular. In 2016, Li et al. steered a study with EEG graphs and showed that optimal performance is obtained using correlation features selection (CFS) and KNN [8]. Remeseiro et al. took two real-life case studies and gaged the benefit of using feature selection techniques [9].

Being motivated by all these prior studies, our experimentation explored the benefit of ML classifiers along with feature selection mechanism.

3 Subject Datasets

As part of this experimentation, three subject datasets were chosen from UCI (University of California, Irvine) heart disease dataset repository.

The first dataset is of Cleveland heart disease dataset. Even though the original dataset has 76 attributes, 14 of those attributes are used by all state-of-the art experimentations. In the dataset, absence of heart disease is indicated by 0 and presence of heart disease is indicated by 1 to 4. The second dataset is Hungarian dataset which is having almost similar attributes like Cleveland dataset. The third dataset is from Switzerland and it has 14 attributes. Some of the attributes from 14 attributes are—age, chest pain, cholesterol level, blood sugar, maximum heart rate, etc. Table 1 enlists the 14 attributes.

Table 1 List of 14 attributes and explanation

Name of attribute	Description of attribute
Age	Age in years
Sex	Gender of patient
Chol	Cholesterol level
CP	Chest pain
Fbs	Fasting blood sugar
Trestbps	Resting blood pressure
Restecg	Resting ECG result
Ca	Number of vessels colored while testing (indicated by 0–3)
Num	Presence (1–4)/Absence (0) of disease
Thal	Type of defect
Slope	Slope of peak exercise point (flat/up/down, etc.)
Exang	Exercise-induced angina
Thalach	Achieved maximum heart rate
Oldpeak	ST depression prompted by exercise comparative to rest

4 Implementation Mechanism

4.1 Data Preprocessing

In the very first step, the datasets were cleaned by removing duplicate records. Very few records were discarded from each dataset because of junk values in some fields. On the other hand, in very few cases, some fields with NULL values in some records were updated with average value of that particular field. It is hoped these data cleaning measures will make KNN implementation more accurate. The input datasets were split into training and testing data. For each datasets, 80% of records were chosen as train data, while rest 20% was considered as test data.

4.2 Application of KNN

K nearest neighbor (KNN) is a prominent machine learning algorithm which can be utilized for both classification and regression. It is a supervised learning method which does not presume any data distribution. KNN is nonparametric in nature, and k is number of nearest neighbors. If the k value becomes 1, then it is termed as the nearest neighbor algorithm.

In this experimentation, for all three subject datasets, several values of k were applied. For example, for the Cleveland heart disease dataset, k was chosen as 9, 13, and 15. As the KNN classifier was run, the confusion matrix got generated. The accuracy values were obtained from the confusion matrix. Table 2 tabulates all the obtained accuracy values. For each run, the error rate was also determined, and the time taken to complete the execution was also noted down. Comparison of all these

Table 2 Performance parameters of KNN run on subject datasets

Dataset	Size (N) (number of instances)	k value	Accuracy (in %)	Precision (in %)	Error rate (in %)	Time taken (s)
Cleveland dataset	303	9	65.4	87.8	7.5	3.2
		13	89.6	85.6	7.9	3.8
		15	76.8	84.2	8.2	4.2
Hungarian dataset	294	7	78.9	83.1	6.5	2.8
		9	84.5	84.7	5.6	2.9
		11	91.2	84.3	6.6	2.95
Switzerland dataset	123	5	65.2	79.8	6.5	1.75
		9	87.5	84.4	7.1	1.8
		11	89.4	83.2	6.8	2.0

parameters will help us to evaluate the implementation potential of KNN classifier in terms of heart disease.

4.3 Application of LASSO

In this paper, the LASSO feature selection mechanism is used to select features. Feature selection is a critical task, and there are many available feature selection mechanisms. In case of feature selection mechanism (FSM), the attributes which control the response variable mostly are kept. In 1996, least absolute shrinkage and selection operator (LASSO) was first used by Robert Tibshirani. With an attempt to test the LASSO method efficacy for heart disease, our experimentation applied it on all three subject datasets.

The LASSO method applies a shrinkage mechanism in which the coefficients of some variables are reduced to almost 0. Ultimately, the variables with nonzero coefficients are chosen for usage. The tuning parameter (λ) of LASSO model actually determines how many coefficients will be reduced to 0.

The regression equation relating dependent and independent variables can be represented as

$$Y_i = \beta_0 + x_{i1}\beta_1 + \cdots + x_{im}\beta_m \tag{1}$$

where $i = 1, 2, 3 \ldots, n$

m is number of independent variables.

β_0, β_1 etc. are the coefficients.

Y is the response.

In our case, m is 14 (from Table 1). If the tuning parameter (λ) is of high value, then more number of coefficients will be shrinking toward 0. So, in order to avoid this scenario, we have kept $\lambda = 0.35$ in this experimentation. From the 14 attributes as mentioned in Table 1, after running LASSO using R software with $\lambda = 0.35$ on the Cleveland, Hungarian, and Switzerland heart disease dataset, the attributes which appeared most significant are—age (Age in years), chol (Cholesterol level), ca (Number of vessels colored while fluoroscopy), thal (type of defect), and trestbps (resting blood pressure).

All the three datasets were rerun using KNN with these five features. As new confusion matrix got regenerated, accuracy values were recalculated. The time required for the entire LASSO-KNN execution was also noted down. This will promote fair comparison to judge usage of KNN over KNN with LASSO method.

5   Results and Findings

As the KNN classifier was run on the three subject datasets, accuracy and precision values were calculated from the generated confusion matrix. Table 2 enlists all these performance indicators.

- While the Cleveland dataset produced highest accuracy of 89.6% at $k = 13$, the Hungarian dataset yielded highest accuracy of 91.2% at $k = 11$. On the other hand, the Switzerland dataset generated highest accuracy of 89.4% at $k = 11$. In all three cases, the obtained accuracies are significant and the highest accuracies are achieved at different values of k for different datasets.
- At $k = 15$, for the Cleveland dataset, the error rate increased to 8.2% with a reduced accuracy to 76.8% (Fig. 1). The time taken to execute the classifier run also increased to 4.2 s.
- For the Switzerland dataset, at $k = 11$, the accuracy is high (89.4%) and the error rate is also low (6.8%). At $k = 9$, an accuracy of 87.5% was obtained with an increased error rate of 7.1%. However, there is very negligible time difference in both execution (2.9 s and 2.95 s).
- At $k = 9$, for Hungarian dataset, the error rate reduced to 5.6% with an accuracy value of 84.5%. For the same dataset, an accuracy of 91.2% was obtained at $k = 11$, but the error rate slightly increased to 6.6% (Fig. 2).

After this application of KNN, feature selection technique least absolute shrinkage and selection operator (LASSO) with tuning parameter 0.35 was applied on each dataset with 14 attributes using R software. R has built in packages and libraries to support LASSO. Five attributes (age, cholesterol, resting blood pressure, defect type,

Fig. 1 Accuracy value obtained after KNN application on Cleveland Dataset

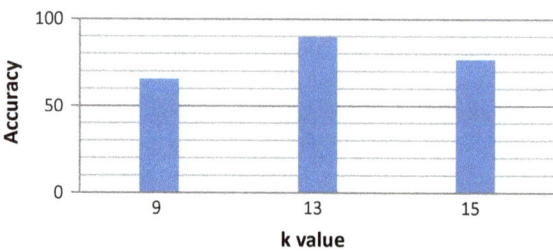

Fig. 2 Accuracy value obtained after KNN application on Hungarian Dataset

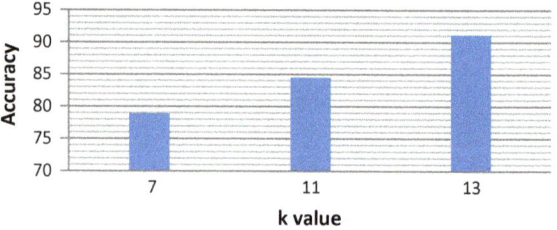

and number of vessels affected) got selected from these 14 attributes. As the datasets changed to new dimensionality, KNN was reapplied on the subjects with previous k values (for fair comparison). Table 3 tabulates recalculated performance parameter values and compares it with KNN without feature selection result.

According to Table 3, it can be observed that LASSO based KNN is highly effective in terms of increasing accuracy values with very nominal increment in execution time. It is also effective in terms of decreasing the error rate. Following results highlights these facts

- In all the cases, the accuracy value increased prominently. For example, while ordinary KNN was producing an accuracy of 89.6% at $k = 13$ for Cleveland dataset, LASSO-KNN produced an accuracy of 92.3% for the same k value.
- In almost all the cases, the error rate also decreased. For example, for the Hungarian dataset, at $k = 11$, the error reduced to 4.2% from 6.6%. In very few cases, the error rate remained same (Switzerland dataset, at $k = 5$) (Cleveland dataset at $k = 13$).
- The increment in time (in seconds) for implementing LASSO-KNN over ordinary KNN is negligible according to the last two columns of Table 3. For example, for the Switzerland dataset at $k = 11$, the execution time increased to 2.2 s from 2.0 s. There are very few cases where the time increased by moderate amount ($k = 7$ for Hungarian dataset).

Thus, the results and findings bolster the application of LASSO based ML classifier for detecting heart disease accurately.

6 Conclusion

Heart disease is becoming rampant nowadays, and it has become a major cause of death across the population. Several underlying health issues along with lifestyle changes contribute a lot toward this disease. paper tried to gage the disease using several standard datasets and conducted an experimentation to apply the K nearest neighbor (KNN) classifier with promising accuracy. As the LASSO feature selection method was applied along with KNN classifier, it was observed that the accuracy increased to great extent. Because of usage of feature selection methodology, increase in execution time was negligible.

In future, we would like to apply our model for larger datasets. The number of features might be greater than the number of observations in case of datasets of larger dimensionality and future work should address that issue. Experimentations to choose optimal value of k for KNN classifier are also other avenues of future work.

Table 3 Performance parameters of LASSO-KNN run on subject datasets

Dataset	Size (N) (number of instances)	k value	Accuracy (in %) for ordinary KNN	Accuracy (in %) for LASSO-KNN	Error rate (in %) for ordinary KNN	Error rate (in %) for LASSO-KNN	Time taken (s) for ordinary KNN	Time taken (s) for LASSO-KNN
Cleveland dataset	303	9	65.4	76.4	7.5	6.1	3.2	3.6
		13	89.6	92.3	7.9	7.9	3.8	4.5
		15	76.8	87.6	8.2	7.6	4.2	4.7
Hungarian dataset	294	7	78.9	82.3	6.5	7.4	2.8	4.1
		9	84.5	86.5	5.6	4.3	2.9	3.5
		11	91.2	95.7	6.6	4.2	2.95	3.76
Switzerland dataset	123	5	65.2	79.0	6.5	6.5	1.75	1.75
		9	87.5	89.8	7.1	5.4	1.8	2.1
		11	89.4	93.4	6.8	5.3	2.0	2.2

References

1. Bolón-Canedo V, Rego-Fernández D, Peteiro-Barral D, Alonso-Betanzos A, Guijarro-Berdiñas B, Sánchez-Maroño N (2018) On the scalability of feature selection methods on high-dimensional data. Knowl Inf Syst 56(2):395–442. https://doi.org/10.1007/s10115-017-1140-3
2. Kanan HR, Faez K, Taheri SM (2007) Feature selection using Ant Colony Optimization (ACO): a new method and comparative study in the application of face recognition system. Lecture Notes in Computer Science (Including Subseries Lecture Notes in Artificial Intelligence and Lecture Notes in Bioinformatics), 4597 LNCS, 63–76. https://doi.org/10.1007/978-3-540-734 35-2_6
3. Menghour K, Souici-Meslati L (2016) Hybrid ACO-PSO based approaches for feature selection. Int J Intell Eng Syst 9(3):65–79. https://doi.org/10.22266/ijies2016.0930.07
4. Jabbar MA, Deekshatulu BL, Chandra P (2013) Classification of heart disease using k-nearest neighbor and genetic algorithm. Procedia Technol 10:85–94. https://doi.org/10.1016/j.protcy.2013.12.340
5. Parthiban L, Subramanian R (2008) Intelligent Heart disease prediction system using CANFIS and genetic algorithm. Int J Biol Med Sci 3(3):157–160
6. Shinde R, Arjun S, Patil P, Waghmare PJ (2015) An Intelligent heart disease prediction system using k-means clustering and Naïve Bayes algorithm. Int J Comput Sci Inf Technol 6(1):637–639
7. Ramalingam VV, Dandapath A, Karthik Raja M (2018) Heart disease prediction using machine learning techniques: a survey. Int J Eng Technol (UAE) 7(2.8 Special Issue 8):684–687. https://doi.org/10.14419/ijet.v7i2.8.10557
8. Li X, Hu B, Sun S, Cai H (2016) EEG-based mild depressive detection using feature selection methods and classifiers. Comput Methods Programs Biomed 136:151–161. https://doi.org/10.1016/j.cmpb.2016.08.010
9. Remeseiro B, Bolon-Canedo V (2019) A review of feature selection methods in medical applications. Comput Biol Med 112(February):103375. https://doi.org/10.1016/j.compbiomed.2019.103375
10. Sahu SK, Kumar P, Singh AP (2018) Modified K-NN algorithm for classification problems with improved accuracy. Int J Inf Technol (Singapore) 10(1):65–70. https://doi.org/10.1007/s41870-017-0058-z
11. Song G, Rochas J, El Beze LE, Huet F, Magoulès F (2016) K nearest neighbour joins for big data on MapReduce: a theoretical and experimental analysis. IEEE Trans Knowl Data Eng 28(9):2376–2392. https://doi.org/10.1109/TKDE.2016.2562627
12. Wang L, Wang Y, Chang Q (2016) Feature selection methods for big data bioinformatics: a survey from the search perspective. Methods 111(August):21–31. https://doi.org/10.1016/j.ymeth.2016.08.014

Clustering-Based Hybrid Approach for Wind Speed Forecasting

Vendra Akhil, Rajesh Wadhvani, Manasi Gyanchandani, and Anil Kumar Kushwah

Abstract A variety of forecasting techniques for wind speed are available to forecast the wind's uncertainty, which is important for predicting the grid's availability of wind power generation. More attention is being gained by the recent creation of the smart grid, which challenges wind power integration into the grid. In order to provide wind speed prediction, many methods have been proposed. There has been a lot of study in recent years to estimate wind speed using many mathematical models. This paper proposes a hybrid model for wind speed forecasting that combines statistical model and segment slope clustering algorithm. In the proposed approach, hybrid models are used, namely clustering ARIMA and clustering SARIMA. Firstly, the wind time series data is divided into different clusters using slope-based clustering. Thereafter, the ARIMA and SARIMA models are applied to each cluster. This work is measured using the metrics: absolute mean error and root-mean-squared error. The experimental result in this work suggests that the proposed hybrid approach performs well as compared to statistical model used alone.

Keywords Wind time series · ARIMA · SARIMA · Slope-based clustering · Hybrid model

1 Introduction

The increase in the prices of crude oil and other non-renewable energy sources highlights their exploitation. The wind is a great alternative to this problem, due to its low pollution and high performance. However, the output power produced by the wind

V. Akhil (✉) · R. Wadhvani · M. Gyanchandani · A. K. Kushwah
Department of Computer Science and Engineering, Maulana Azad National Institute of Technology, Bhopal, Madhya Pradesh, India

R. Wadhvani
e-mail: rajeshwadhvani@manit.ac.in

M. Gyanchandani
e-mail: mansigyanchandani@manit.ac.in

© The Author(s), under exclusive license to Springer Nature Singapore Pte Ltd. 2022   587
D. Gupta et al. (eds.), *Proceedings of Data Analytics and Management*,
Lecture Notes on Data Engineering and Communications Technologies 90,
https://doi.org/10.1007/978-981-16-6289-8_49

energy conversion systems (WECS) is highly dependent on wind speed and atmospheric meteorology [1]. This change in meteorology, which is completely unpredictable, may have variations which are its biggest challenge. These unpredictable variations in output power would result in unstable electricity. They would force the systems to use their energy reserves to suppress these variations, which would raise running costs and an alternate option to resolve the energy crisis. Global climate change has recently drawn considerable attention to wind energy. Wind power has evolved extensively and rapidly around the world. The forecast of wind speed is an important technique in wind power generation. Detailed information on wind speed can offer guidance for wind power regulation, scheduling, and maintenance. Several variables determine the wind speed; usually, with time, it fluctuates wildly, and it becomes difficult to render forecasting of the wind accurately [2]. Therefore, we can develop wind energy conservation systems that dynamically control wind turbines and schedule the power systems by predicting these changes. In the place of direct wind power, forecasting in terms of wind speed is more accurate to account for the spatial correlation of wind. Therefore, there is a high demand for this energy production, so a great deal of research has been invested in it. There is a great need for accurate wind forecasting to reduce the use of the primary reserves of energy and for better wind power penetration. Therefore, the need for forecasting is very imminent, but there are some challenges that come with it. Nonlinear behavior without typical patterns due to differences in high rates of change and dependence on a number of factors such as temperature, terrain, atmospheric pressure, and elevation, accurate and reliable wind speed forecasting is not an easy challenge, resulting in significant variations in wind speeds [3].

There are three categories of wind speed forecasting models: physical models, statistical models, and neural network-based models [4]. The physical prediction techniques or model requires meteorological data and various physical factors affecting the wind speed [5]. This technique is considered efficient for long-term prediction. However, the limitations with these models are that it is location-specific and requires a physical presence to record the changes [6]. Research reported in [7, 8], and [9] presented a broad view of various statistical methods. In these methods, data based on history is applied to build a prediction model. Statistical models like autoregressive (AR), moving average (MA), autoregressive integrated moving average (ARIMA), and ARIMA with exogenous variable (ARIMAX) are applicable when the data can be linearly approximated. Various nonlinear statistical modeling methods, such as generalized autoregressive score (GAS) and vector autoregression (VAR), have also been proposed to deal with nonlinearity in time series data. [10–12]. Each of the mentioned approaches has its pros and cons. A hybrid approach can overcome the limitations of simple forecasting mechanisms and result in accurate wind speed predictions. The time series clustering-based prediction model, on the other hand, is a hybrid model that blends clustering and statistical models.

Clustering methods break data points into many groupings. It is an unsupervised method for applying a range of approaches to both sequential and non-sequential data points. Due to variations in sequential and non-sequential data features, techniques employed for non-sequential time series data do not yield appropriate results. When applied to non-sequential data, clustering techniques such as k-mean clustering yield a minimum number of clusters based on the distance metric. [13]. Time series data, however, has the feature of a serial correlation between subsequent observations. Without affecting serial correlation, the distance metric is unable to combine related data into classes. Wind power's inherent randomness, intermittency, and fluctuation, in particular, have resulted in a slew of issues, including power supply reliability, balancing management and reserve capacities, and wind turbine harm. The risk of the power system being harmed by the uncertainties of wind power is normally mitigated by accurate and reliable wind power prediction. In [14], the time series data, trend, and seasonality are generally used to describe similar patterns and generate the clusters accordingly.

This paper describes a wind time series clustering technique for identifying segments of time series data with the same slope, and a different number of clusters of the similar slope are generated using the clustering technique. After that, the statistical models for time series forecasting, ARIMA, and SARIMA are applied to model each cluster. The hybrid model is proposed for wind speed forecasting, namely clustering ARIMA (C-ARIMA) and clustering SARIMA (C-SARIMA). The performance of proposed hybrid models is measured using the criteria of mean absolute error (MAE) and root-mean-square error (RMSE).

2 Time Series Data Clustering

Clustering and external detection are often treated as separate problems. However, both problems are connected. For example, outliers may have inconsistent effects on the size of groups, which may mask outspoken outliers [15]. KNN is computationally costly as it calculated the distance between the nearest neighbors for a point. KNN requires large memory to store all the data points (Fig. 1).

The prediction platform is very expensive. Sensitive to outliers, accuracy is affected by noise or irrelevant data. The Mahalanobis distance interval is assumed to spherically assign the sampling points around the center of mass where the distribution is obviously non-spherical and uniformly connected, normalizing the distance of the characteristics to the squares. If the feature is affected by noise, a single feature may have such a large value that it covers the information provided by other features and leads to a miscarriage due to the increasing time of distances. Here, we have done clustering of series based on the slope and applied ARIMA. Therefore, the result may vary depending on the complexity of the slope and the redundancy of the slope values.

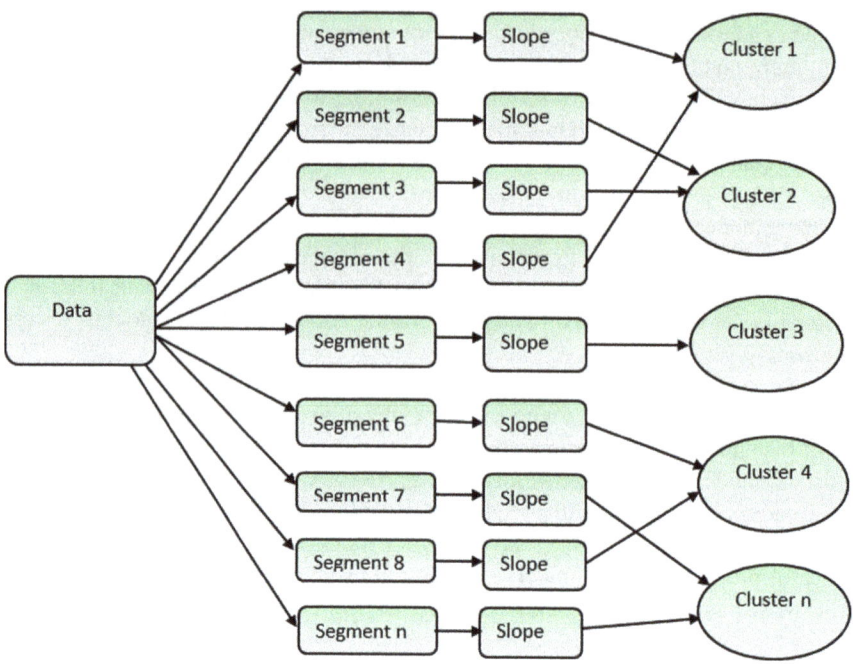

Fig. 1 Slope-based clustering of dataset

3   Statistical Models for Wind Speed Forecasting

Statistical models are used to forecast time series data in terms of data, horizon, and precision. The direction of the wind, wind speed, air density, temperature, and other factors are used to forecast wind speed. The predicted wind speed values are primarily included in the forecasted data from the wind time series. Using the following statistical models, the wind speed is forecasted: ARIMA and SARIMA.

3.1   Autoregressive Integrated Moving Average Model (ARIMA)

The autoregressive integrated moving average model (ARIMA) collects univariate time series analyzing models that use their past values of the variable to forecast future values of the variable. A white noise time series data cannot be used to apply the ARIMA model [16]. An ARIMA model consists of three terms: p, d, and q. p gives us the number of past variable values to take for the AR term, q gives us the number of past error terms for the MA term, and d is the order of differencing required to make the data stationary. ARIMA is made of two models: autoregressive

(AR) and moving average (MA). ARIMA model can be expressed as:

$$y_t = \alpha_0 + \alpha_1 y_{t-1} + \cdots + \alpha_p y_{t-p} + \varepsilon_t + \beta_1 \varepsilon_{t-1} + \cdots + \beta_q \varepsilon_{t-q} \tag{1}$$

where y_t is a variable of interest that is lagged up to pth value, $\alpha_0, \alpha_1 \ldots \alpha_p$ are the AR coefficients, $\beta_1 \ldots \beta_q$ are the MA coefficients, and ε_t is assumed to be white noise. An ARIMA would require the data to have no seasonal components. After making the data stationary and acquiring the number of lags to be included, the linear regression function is then carried out then on the variables to get the predicted values [4]. A value of 0 for the order of p, q, d would suggest that the particular parameter of the model is not needed. Thus, the ARIMA model can be used to call a standalone AR, I, or MA model. An ARIMA model can only be applied on those datasets that generate the observations following an ARIMA process. Therefore, we can be sure that the properties needed for the ARIMA model are met in the raw observations and the residual errors of the forecasts.

3.2 Seasonal Autoregressive Integrated Moving Average (SARIMA)

The basic concept behind the ARIMA model is to treat the prediction target's data sequence over time as a random sequence. SARIMA is an ARIMA extension that works with univariate time series data. With a seasonal variable specifically and for both the trend and seasonal components, configuring SARIMA involves the collection of hyperparameters.

A seasonal autoregressive integrated moving average (SARIMA) model is a step up from an ARIMA model that uses seasonal trends as a notion. Seasonal effects are common in many time series datasets. Take, for instance, the average temperature in a four-season location. On a yearly basis, there will be a seasonal impact, and the temperature in this season will almost certainly have a strong correlation with the temperature measured the previous year in the same season.

The non-seasonal ARIMA (p, d, q) approach is used to create the pure seasonal model of SARIMA $(p, d, q) \times (P, D, Q)_s$, where the first term (p, d, q) determines the non-seasonal component and the second part $(P, D, Q)_s$ represents the seasonal component of the model [11]. Mathematically, it is represented as

$$\phi_p(B)\Phi_P\left(B^s\right)W_t = \theta_q(B)\Theta_Q\left(B^s\right)\omega_t \tag{2}$$

Equation (2) is explained as follows: p, d, and q are represented in previous Eq. (1), p stands for the order of the seasonal AR Model, d stands for the amount of seasonal differencing, q stands for the order of the seasonal MA, and s stands for the duration of the season (periodicity), and ω_t and B stand for the white noise value at time period t and the backward shift operator, respectively.

Equation (2) represents the seasonal components of the SARIMA model, which can be described as after we substitute the value of

$$W_t = \nabla^d(B)\nabla_s^D(B)X_t$$

$$\phi_p(B)\Phi_p(B^s)(1 - B)^d(1 - B^s)^D X_t = \theta_q(B)\Theta_Q(B^s)\omega_t \tag{3}$$

The components of seasonal ARIMA can be defined as

- Non-seasonal AR

$$\phi_p(B) = 1 - \phi_1 B - \phi_2 B^2 - \phi_3 B^3 - \cdots - \phi_p B^p$$

- Non-seasonal MA

$$\theta_q(B) = 1 - \theta_1 B - \phi_2 B^2 - \theta_3 B^3 - \cdots - \theta_q q$$

- Seasonal AR

$$\Phi_P(B^s) = 1 - \Phi_1 B^s - \Phi_2 B^{2s} - \Phi_3 B^{3s} - \cdots - \Phi_p B^{ps}$$

- Seasonal MA

$$\Theta_Q(B^s) = 1 - \Theta_1 B^s - \Theta_2 B^{2s} - \Theta_3 B^{3s} - \cdots - \Theta_Q B^{Qs}$$

And,

- $B^s X_t = X_{t-s}$.
- $\nabla_s X_t = \nabla_s(B)X_t = (1 - B^s)X_t = X_t - B^s X_t = X_t - X_{t-s}$.
- $\nabla^d(B)X_t = (1 - B)^d X_t$.
- $\nabla_s^D(B)X_t = (1 - B^s)^D X_t$.

4 Proposed Hybrid Methodology

The workflow that will be carried forth in this paper is as shown in Fig. 2. Modeling is done on the data using ARIMA and SARIMA, after which prediction of future values is made. The data is broken down into clusters of similar properties of segment slope. This gives us more stationary data. The prerequisite of most forecasting models is stationary data. After clustering, we apply the model on each cluster iteratively and

Fig. 2 Flowchart for prediction model

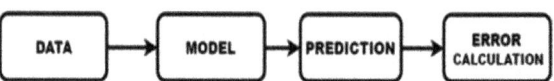

observe the results. We take a specific window of data and calculate its slope. If this slope has been already calculated previously and exists, we include this data with the time series that is formed using the previous windows of data. In the alternate case, if the window of data chosen for processing has not been used previously, we create a new time series function with the given slope. This process is applied iteratively on all datasets. Consequently, clusters are formed on the basis of similar slope values. Thus, windows of data will be added to the same time series function for that slope.

5   Research Design and Analysis

This section presents the experimental parts of the research, including the description of the dataset, performance measuring parameters, and results analysis.

5.1   *Dataset*

For the experimentation, the dataset is collected from the National Renewable Energy Laboratory (NREL) with site ID 38,382. This site ID is having a geographical location with a longitude of $-100.841°$ and a latitude of $37.978°$. From January 1st to December 31st, 2012, the data was gathered every five minutes at a mean wind speed of 7.587 m/s. In this work, six years of datasets are taken from the NREL site from 2007 to 2012.

5.2   *Performance Measuring Criteria*

Statistical and hybrid model efficiency is assessed by acceptable criteria that evaluate the potential of the models. Wind speed forecasting efficiency is calculated using absolute mean error (MAE) and root-mean-squared error (RMSE) for our experimental study. The measuring criteria like MAE and RMSE are represented as follows:

$$\text{MAE} = \frac{1}{N} \sum_{i=1}^{N} y(i) - x(i) \tag{4}$$

$$\text{RMSE} = \sqrt{\frac{1}{N} \sum_{i=1}^{N} (y(i) - x(i))^2} \tag{5}$$

where N denotes the total number of input data, y denotes the expected variable, x denotes the input variable, and $\bar{x}$ is the input variable average. The model with the lowest MAE and RMSE values outperforms the others. The autoregressive integrated moving average model (ARIMA), seasonal autoregressive integrated moving average (SARIMA), and hybrid model are implemented in Python 3.6.

5.3 Result Analysis

Prior to the model being applied on the complete dataset, data has been split into train and test datasets to avoid overfitting and increase accuracy to observe the validation of the prediction on real-time data. We have split the complete data into five datasets with 2,000 records each. We have taken the training dataset of each of the six datasets as 90% of the data, which is 1,800 records. The testing dataset is the remaining 10% which is 200 records. We have applied the ARIMA and SARIMA models in two ways:

- The ARIMA model and SARIMA model are applied to the training dataset, and the prediction is compared with the remaining test dataset.
- The training data is clustered, and the ARIMA model and SARIMA model are applied on each cluster, and the trained ARIMA and SARIMA are applied to each cluster of the test data to make predictions.

In this paper, the statistical models are applied to six datasets. Future prediction of the datasets is forecasted using the ARIMA model and SARIMA model. Each dataset consists of 2000 values. The model is trained in prior with 1,800 datasets. The result of the model plotted against the test data for a simple ARIMA model without clustering for each dataset is shown in Fig. 3, and data for a simple SARIMA model without clustering for each dataset is shown in Fig. 5.

The second method of ARIMA and SARIMA was applied after clustering. The clusters generated on each dataset are shown in Figs. 4 and 6. The predicted results of the ARIMA and SARIMA models in terms of root-mean-squared error (RMSE) and absolute mean error (MAE) are compared in Table 1. Table 1 demonstrates the experimental results for the ARIMA and SARIMA variants in terms of MAE and

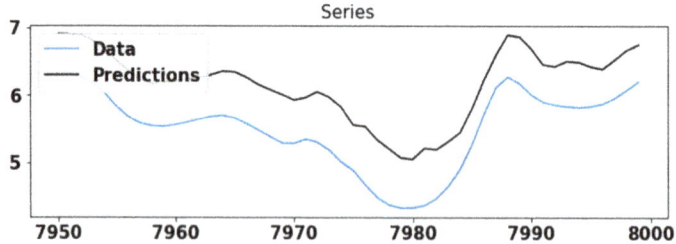

Fig. 3 Prediction plot of the ARIMA model on dataset #1

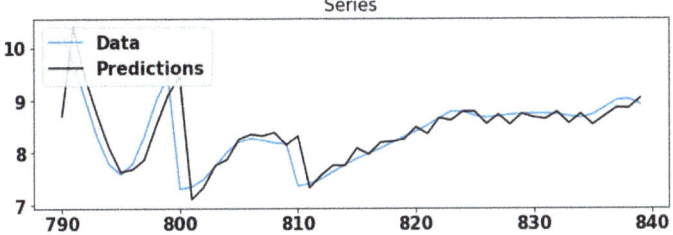

Fig. 4 Prediction plot of C-ARIMA on dataset #1

Fig. 5 Prediction plot of SARIMA model on dataset #1

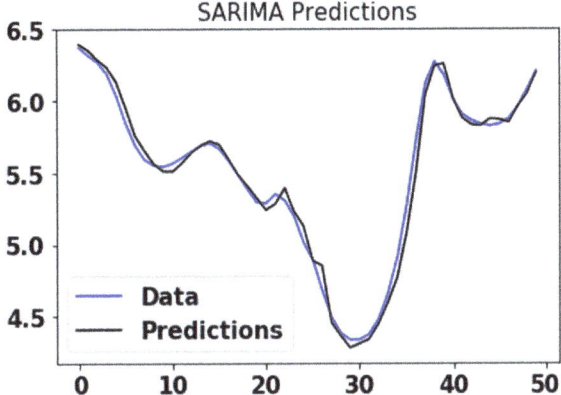

RMSE values. From all the clusters tested, we conclude from the result that Cluster 1 has the lowest RMSE and MAE values for dataset #1, Cluster 2 has the lowest RMSE and MAE values for dataset #3, Cluster 4 has the lowest RMSE and MAE values for dataset #4, and the ARIMA without clustering model had the lowest RMSE and MAE values for datasets #2 and #5. Thus, ARIMA with clustering models provides us with better results for a majority of the datasets. Therefore, we conclude that ARIMA with clustering gives us a better result than ARIMA without clustering.

6 Conclusion

The statistical wind speed prediction model is important in the creation of the best wind energy framework. For wind speed forecasting, statistical models, as well as hybrid models, are used. The current research focuses on determining the best wind speed prediction model. This paper suggested a hybrid model, C-ARIMA, C-SARIMA, for precise and robust wind speed forecasting. For each dataset in the proposed hybrid approaches, the original series was clustered into various clusters using the clustering technique. The ARIMA and SARIMA models are then applied to each cluster. Finally, the predicted wind speeds of each cluster are added together

Fig. 6 Prediction model of
C-SARIMA on dataset #1

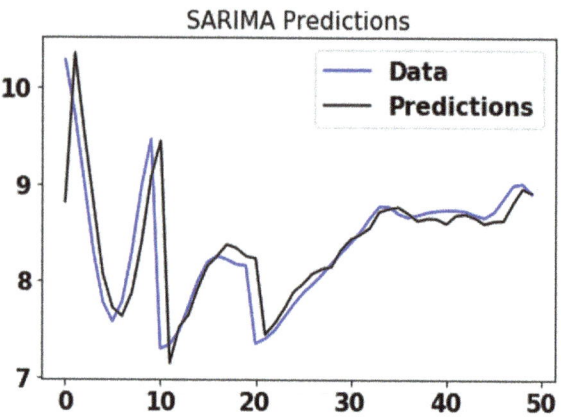

Table 1 MAE and RMSE values using the ARIMA, SARIMA, and hybrid models (C-ARIMA and C-SARIMA)

Dataset	ARIMA	C-ARIMA	SARIMA	C-SARIMA
	MAE RMSE	MAE RMSE	MAE RMSE	MAE RMSE
#1	6.511 6.901	2.642 2.869	**1.827** **2.251**	2.641 2.869
#2	3.325 3.912	**3.186** **3.382**	4.151 4.964	3.381 3.977
#3	5.165 5.312	1.331 1.569	2.813 3.041	**1.327** **1.568**
#4	2.192 2.646	2.201 2.633	4.956 5.631	**2.043** **2.693**
#5	3.708 4.262	**3.619** **4.433**	3.621 4.244	3.627 4.459
#6	2.248 2.802	3.026 3.360	2.184 2.698	**2.091** **2.591**

to produce the overall predicted series. When compared to the ARIMA and SARIMA models currently in use, the hybrid model outperforms them in our tests.

References

1. Morshedizadeh M, Kordestani M, Carriveau R (2017) Improved power curve monitoring of wind turbines. Wind Eng 41:260–327
2. Thapar V, Agnihotri G, Sethi V (2011) Critical analysis of methods for mathematical modelling of wind turbines. Renewable Energy 36:3166–3177
3. Liu H, Tian HQ, Li YF (2015) Comparison of four Adaboost algorithm based artificial neural networks in wind speed predictions. Energy Convers Manage 92:67–81

4. Lei M, Shiyan L, Chuanwen J (2009) A review on the forecasting of wind speed and generated power. Renew Sustain Energy Rev 13:915–920
5. Carvalho D, Rocha A, Gomez-Gesteira M, Santos C (2012) A sensitivity study of the WRF model in wind simulation for an area of high wind energy. Environ Model Softw 33:23–34
6. Buhan S, Cadirci I (2015) Multistage wind-electric power forecast by using a combination of advanced statistical methods. IEEE Trans Industr Inf 11:12–31
7. Chen DS, Li PQ, Li XR, Xu CH, Xiao YY, Lei B (2011) Short-term wind speed forecasting considering heteroscedasticity. In: APAP 2011—Proceedings: 2011 international conference on advanced power system automation and protection, vol 2, pp 884–888
8. Kavasseri RG, Seetharaman K (2009) Day-ahead wind speed forecasting using f-ARIMA models. Renewable Energy 34:1388–1393
9. Lydia M, Suresh Kumar S, Immanuel Selvakumar A, Edwin PremKumar G (2015) Wind resource estimation using wind speed and power curve models. Renewable Energy 83:425–434 (2015)
10. Xu J, Zhang Z, Zhao L, Ai D (2011) The application review of GARCH model. In: 2011 International Conference on Multimedia Technology, ICMT 2011, pp. 2658–2662 (2011)
11. Xie CS (2008) Dynamic vs static autoregressive models for forecasting time series (2008)
12. Creal D, Koopman SJ, Lucas A (2013) J Appl Economet 28(5):777
13. Zhu E, Zhang Y, Wen P, Liu F (2019) Fast and stable clustering analysis based on Grid-mapping K-means algorithm and new clustering validity index. Neurocomputing 363:149–170
14. Lim BY, Wang JG, Yao Y (2018) Time-series momentum in nearly 100 years of stock returns. J Bank Finance 97:283–296
15. Wang X, Smith K, Hyndman R (2006) Characteristic-based clustering for time series data. Data Min Knowl Disc 13:335–364
16. AitMaatallah O, Achuthan A, Janoyan K, Marzocca P (2015) Recursive wind speed forecasting based on Hammerstein Auto-Regressive model. Appl Energy 145:191–197
17. Harvey AC (2013) Dynamic Models for Volatility and Heavy Tails: With Applications to Financial and Economic Time Series. Cambridge University Press, New York, USA
18. Chang WY (2013) An RBF neural network combined with OLS algorithm and genetic algorithm for short-term wind power forecasting. J Appl Math Article ID: 971389, 9 p
19. Sideratos G, Hatziargyriou ND (2007) An advanced statistical method for wind power forecasting. IEEE Trans Power Syst 22:258–265. https://doi.org/10.1109/TPWRS.2006.889078
20. Lange M, Focken U (2008) New developments in wind energy forecasting. In: Proceedings of the 2008 IEEE Power and Energy Society General Meeting, Pittsburgh, 20–24 July 2008, pp 1–8
21. Wang XC, Guo P, Huang XB (2011) A review of wind power forecasting models. Energy Procedia 12:770–778. https://doi.org/10.1016/j.egypro.2011.10.103
22. Zhao DM, Zhu YC, Zhang X (2011) Research on wind power forecasting in wind farms. In: Proceedings of the 2011 IEEE Power Engineering and Automation Conference, Wuhan, 8–9 September 2011, pp 175–178. https://doi.org/10.1109/PEAM.2011.6134829
23. Zhao X, Wang SX, Li T (2011) Review of evaluation criteria and main methods of wind power forecasting. Energy Procedia 12:761–769. https://doi.org/10.1016/j.egypro.2011.10.102
24. Wu YK, Hon JS (2007) A literature review of wind forecasting technology in the world. In: Proceedings of the IEEE Conference on Power Tech, Lausanne, 1-5 July 2007, pp 504–509
25. Soman SS, Zareipour H, Malik O, Mandal P (2010) A review of wind power and wind speed forecasting methods with different time horizons. In: Proceedings of the 2010 North American Power Symposium, Arlington, 26–28 September 2010, pp 1–8. https://doi.org/10.1109/NAPS.2010.5619586
26. Catalao J, Pousinho H, Mendes V (2011) Short-term wind power ˜ forecasting in portugal by neural networks and wavelet transform. Renewable Energy 36(4):1245–1251. [Online]. Available: https://doi.org/10.1016/j.renene.2010.09.016
27. Conejo A, Plazas M, Espinola R, Molina A (2005) Dayahead electricity price forecasting using the wavelet transform and ARIMA models. IEEE Trans Power Syst 20(2):1035–1042. [Online]. Available: https://doi.org/10.1109/tpwrs.2005.846054

28. Mohandes M, Halawani T, Rehman S, Hussain AA (2004) Support vector machines for wind speed prediction. Renewable Energy 29(6):939–947
29. Liu D, Niu D, Wang H, Fan L (2014) Short-term wind speed forecasting using wavelet transform and support vector machines optimized by genetic algorithm. Renewable Energy 62:592–597. [Online]. Available: https://doi.org/10.1016/j.renene.2013.08.011
30. Pousinho H, Mendes V, Catalao J (2011) A hybrid PSO–ANFIS approach for short-term wind power prediction in Portugal. Energy Convers Manag 52(1):397–402. [Online]. Available: https://doi.org/10.1016/j.enconman.2010.07.015
31. Wei Fei S, He Y (2015) Wind speed prediction using the hybrid model of wavelet decomposition and artificial bee colony algorithm-based relevance vector machine. Int J Electr Power Energy Syst 73:625–631. [Online]. Available: https://doi.org/10.1016/j.ijepes.2015.04.019
32. Salcedo-Sanz S, Pérez-Bellido AM, Ortiz-García EG, PortillaFigueras A, Prieto L, Paredes D (2009) Hybridizing the fifth generation mesoscale model with artificial neural networks for short-term wind speed prediction. Renewable Energy 34(6):1451–1457. [Online]. Available: https://doi.org/10.1016/j.renene.2008.10.017
33. Croonenbroeck C, Ambach D (2015) A selection of time series models for short- to medium-term wind power forecasting. J Wind Eng Ind Aerodyn 136(Supplement C):201–210. [Online]. Available: http://www.sciencedirect.com/science/article/pii/S016761051400244X
34. Lydia M, Kumar SS, Selvakumar AI, Kumar GEP (2016) Linear and non-linear autoregressive models for shortterm wind speed forecasting. Energy Convers Manage 112:115–124. [Online]. Available: https://doi.org/10.1016/j.enconman.2016.01.007

Data Analytics and Visualization to Aid Mental Health Care

Faizul Aqtab and Suraiya Parveen

Abstract Mental state of a person is an indication of emotive, psychological and social welfare. Various parameters like pressure, strain, social fretfulness, dejection, obsessive compulsive disorder, drug obsession and personality disorders leads to mental illness. The purpose of the research was to analyse, predict and make better mental health solutions for the individuals using data visualization tool on selected datasets in metropolitan and semi-urban population. The study was undertaken among wide age range group from less than 18 to 50 plus of all around India. A 747 samples were collected using Google Form, out of which 425 were male (56.09%) and female were 319 (42.07%). The figure of 307 people of total population showed prevalence of diagnosable mental disorders. Hence, there is a great need to upsurge the prevention policies both at initial and tributary levels to overcome the mental distress and economic loss to society due to mental disorders.

Keywords Mental health · Data analysis · Data visualization · Tableau · COVID-19

1 Introduction

Mental welfare of a person is the mindset of the personal, influence his/her overall nature. Mental disorder is an outcome of inequities in brain interaction. Psychological health is most significant in all phases of life, from childhood and youth through adulthood. It refers to a wide array of mental fitness–illnesses that influence the mood, thinking and behaviour [1]. The illness conditions include depression, anxiety, schizophrenia, food consumption disorders and addictive behaviours. The mental health care service applies to the treatment of psychological infirmities and the improvement of mental condition in people with disorders or problems. Mental health is often related with a lot of stigma and discrimination, leading to most critical assessment to understand and suggests treatment to be given to patients with

F. Aqtab (✉) · S. Parveen
Department of Computer Science, SEST, Jamia Hamdard, New Delhi 110062, India

D. Gupta et al. (eds.), *Proceedings of Data Analytics and Management*,
Lecture Notes on Data Engineering and Communications Technologies 90,
https://doi.org/10.1007/978-981-16-6289-8_50

diverged mental behaviour [2]. There is very small comprehensive data available on the widespread of psychological disorders around the world.

Patient mindset is a vital parameter to diagnose his/her mental state, as its action is highly reflected in the social community. Initially, people should refer to a psychologist to get analysed by the type of mental state if they endure from anomalous behaviour. The diagnosis is based on face-to-face interaction and trailed by psychoanalytic sessions. Through the advent of sophisticated technology in today's time, there have been developed various modalities which can envisage the state of mental health. The poor diagnosis and remedies provided have been discussed by WHO [3].

An administrative board of World Health Organization (WHO) prophesied that by the year 2030, foremost reasons of global disease burden will be depression. In the upcoming future, mental health profile of a patient will be radically shifted by health care professional, and it will be made compulsory to deliver better treatment and to contribute in quicker recovery [4]. The various scientists have discussed about different ways of medical predictive analytics to transform the health care field internationally. Visual analytics plays a vital part in efficiently examining mental health care data. Analytics comprises of tools that conjoins both automated analysis procedures through visualizations and analytical abilities of users to benefit from intricate data to convalesce indulgence and decision-making [5, 6]. The estimates suggested that one in seven people on the planet have mental or substance-use disorders and 4.0% have anxiety disorders. At present, over 36 million people around the world ever affected by COVID-19 infection, which further had impact on mental health [7].

Worldwide, young people (20%) encounter mental illness. In India, solitary 7.30% of its 365 million adolescents reported to have such issues. Although civic stigma related with mental sickness mainly affects help-seeking adults, the magnitude of stigma is still unidentified in India [8]. The degree and indexes of communal stigma and synthesized indication of commendations to decrease mental health-related stigma among adults in India need to be explored.

The notable cause for India to be affected by mental health is the dearth of responsiveness and sensitivity regarding the disorder. There is immense humiliation around people misery, leading to a nasty cycle of embarrassment, sorrow and seclusion of the patients from the surrounding [7]. According to WHO, mental illness comprises of around 15% of the overall illness conditions all around the biosphere. The same evaluation suggested that largest populations of India has been shook from mental disorder. Because of this, WHO described India as the world's 'most depressing country'. Also, the mental state of Indian people around 1990 to 2017 showed one in seven people has experienced mental disorder ranging from depression, anxiety to severe conditions such as schizophrenia, according to a study. Considering the situation, it is suggested that the country is under a mental health epidemic [7].

In India, the states like Odisha, Telangana, Andhra Pradesh, Tamil Nadu and Goa diagnosed with most cases of depression, whereas Manipur has anxiety disorders, but Kerala has both depression and anxiety disorder as shown in Fig. 1. Similarly, Madhya Pradesh and Assam have idiopathic developmental intellectual disability,

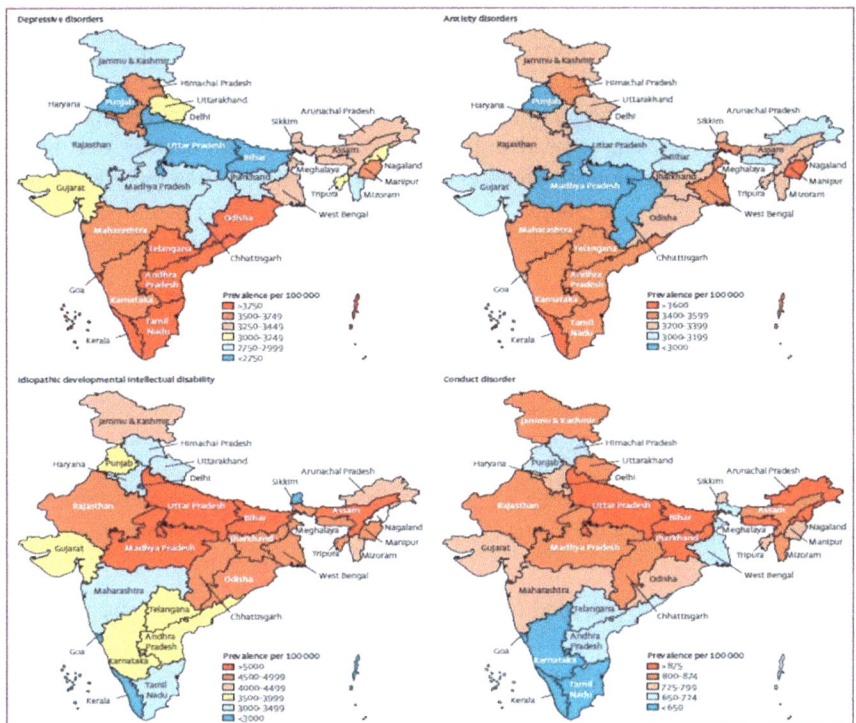

Fig. 1 Incidence of mental health in different parts of India

and Jharkhand, Arunachal Pradesh, Meghalaya, and Nagaland have conduct disorder with Uttar Pradesh and Bihar have shown both [4, 7].

Around 365 million population are young in the world [9] and have huge burden of untreated mental issues, whereas in India, younger generation faces challenges in accomplishing their social and economic capability. In the years 2015–2016, India's national mental health survey was done which stated that statistics on such mental health-related stigma were limited [10].

2 Material and Methods

The study was performed on targeted population of wide age group who were mentally distressed. These individuals required special care and devotion in order to mentally balance the community. To know the status of individuals, the basic criteria was to collect and rate the responses in the population for common benchmarked questions mentioned in Google Form. This research was model work for further core study about effect of mental health in their day-to-day routine life. The target population were from different communities ranging from high school/college-going students to working professionals to unemployed from diverse organizations. The questionnaire for the focussed group was designed in discussion with a medical professional. As per the professional's recommendation, we prepared the questionnaire to achieve the purpose of identifying deviations in mental health. A set of 20 questionnaire was developed, which comprised of multiple choice questions. The survey was managed via Google forms platform, which required individual's to sign in to their own email account to take part in the survey. The circulation of the questionnaire was operated by means of social media platforms, email and standard messaging services.

2.1 Study Design

A web-based survey was held through the means of Google online platforms from 5th February 2021 to 17th February 2021. The online survey questionnaire was as follows:

(a) Participants were enquired to define their general demographics, such as age, gender, the region of residence.
(b) Report on their impact on work due to mental illness or physical health, mental disorder diagnosis, examination, treatment and medication was asked.
(c) Personal regime like sleeping hours, quality of sleep and habits like smoking or drinking were also asked to assess their mental health.

The aim of this survey is to scrutinize the impact of the mental health lifestyle of individuals from different age groups and the relations between various parameters.

2.2 Evaluation of Dataset/Survey

Export Survey Data

For the survey, Google Form was used which was free, and data was downloaded and exported in a CSV file. Various software like Survey Monkey are paid plan for exporting the survey responses. However, the Google Form survey response procedure was simpler for the study.

Data Cleaning

To clean and to arrange the data, Google Form survey was inspected in CSV file, where the first column was timestamp, which was created by Google Form. The 'timestamp' column was an add-on column which was deleted with other questions or columns which was not required to analyse. The questions of interest were kept in the survey; others were deleted to avoid the confusion.

Analysing the Dataset

The data was analysed, and the appropriate parameters were chosen. The chosen data was further analysed using the visualization tool. Data examination permitted the probability of uncovering relations, patterns and trends that were formerly neither expected nor hypothesized [11, 12].

Data Visualization and Interpretation

The parameters selected for visualization were association graphs such as bar, curve and scatter plot or variation graph that included bar, line and radar plots. Different comparisons were done on the visualization that led to better policy, funding and quality of services for people with mental health care [9].

Data Visualization Tool

For our data visualization, we had used Tableau. Tableau is a prevalent, also fleetest emergent data visualization tool used in the business intelligence industry. It aids in streamlining data to a simple comprehensible format. Tableau helped in creating the information which will be valued by professionals at any level in a corporation. It connects and extricates the information stockpiled in various places. It can take out data from any platform thinkable, an easy database like an Excel, PDF, a database within the cloud like Amazon webs services, Microsoft Azure SQL database, Google Cloud SQL, to a posh database like Oracle, etc. The worksheet presentation is broadly utilized for controlling the data, while Tableau is an ideal visualization tool used for scrutiny. Tableau software in this survey helped seamlessly in understanding data to analyse support system for mental health and guided to amend further.

2.3 Ethical Consideration

The ensuing survey was completed in a well-notified set up, and permission for the participation was taken from the people. No person was enforced against their wish, and no recognizing evidence was gathered.

3 Result and Discussion

3.1 General Data Statistics

The survey was conducted on 747 people of the different age groups in the month of February 2021 as shown in Table 1. It was observed that out of 747 people, around 425 individuals were male (56.09%) and 319 were female (42.07%), respectively. The individuals belonging to the age group of 19–30 were avid participants comprising of total 568 responds, i.e. around 76.03% of the total respondents, whereas individuals

Table 1 General statistics of dataset obtained by doing the survey in 2021

Gender distribution	Male	Female	Other
	425 (56.09%)	319 (42.07%)	3.0 (0.40%)
Age distribution	Less than 18	111 (14.85%)	
	19–30	568 (76.03%)	
	31–50	60 (08.03%)	
	50 +	8.0 (01.07%)	
People with mental illness 307 (41.09%)			
Total respondent 747			

belonging to age of less than 18 years were 111 (14.85% of total respondent), followed by the age group of 31–50 (08.03% of total respondent) and individuals of age above 50+ (01.07%), respectively.

3.2 Mental Health Disorder Distribution

After carrying out the survey, out of 747 respondent, it was found that around 307 individuals (41.09% of the total respondent) were suffering from mental illness that could be anxiety disorder, mood disorder, eating disorder, depression, etc. Figure 2 shows a distribution of different mental health disorders among the survey data participants.

It was accessed that 47 people had complained about anxiety disorder, panic disorder and obsessive–compulsive disorder, whereas around 56 people were suffering from depression, bipolar or mood disorder. Twenty-seven individuals had

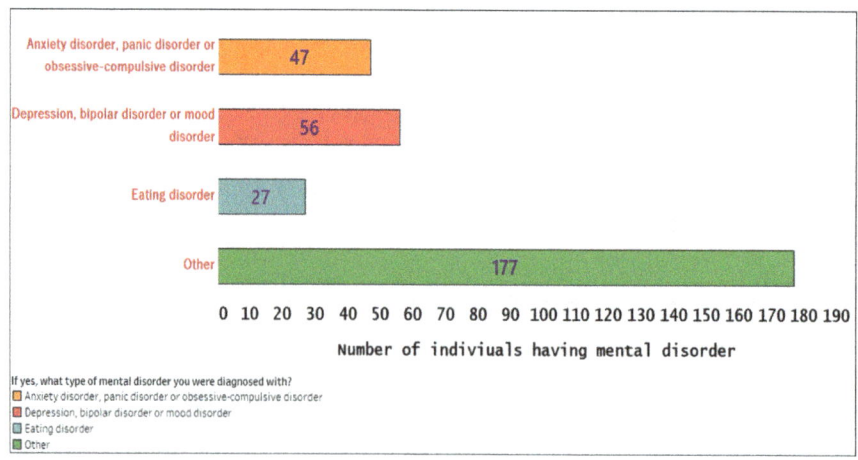

Fig. 2 Mental health disorder distribution

eating disorder, and the most 177 individuals could be suffering from other mental disorder like personality disorders, post-traumatic stress disorder, psychotic diseases, including schizophrenia.

3.3 COVID-19 Impact on Mental Health

The pandemic has disrupted the mental health for many. Isolation and lack of physical contact with other family members, friends and colleagues, loss of income and fear are triggering mental health conditions or aggravating existing ones. The care providers, older adults and people having underlying health conditions are more concerned and worried as the coronavirus pandemic is sweeping at high pace around the globe arising the level of fright in population at large [13]. Adapting to lifestyle changed due to the existing pandemic, managing the fear of catching the virus and worrying about individuals close with us are predominantly vulnerable challenging for all of us.

Home quarantine was not the blessing for many persons, specifically individuals being in abusive relationships. National Commission of Women had given statistical data that the total grievances from females ascended in the last week of March 2020 as out of 214 disputes, 58 complaints were of domestic violence [14].

The upsurge has been because of the abusers being kept in their house with no means of release of their rage or frustration and the vulnerability of the victims to disclose their misery or to move to their relatives due to isolation from public. The Government of India's has received beyond 92,000 distress calls on child abuse and fierceness on their 'ChildLine' helpline number in 11 days lockdown period asking for protection, and this duration becomes long-lasting detention for children as they were being house captive with their abusers [15]. Meanwhile, COVID-19 itself has led to neurological and mental complications, such as delirium, agitation and stroke. [16].

A survey conducted on psychological impact of COVID-19 on residents of 194 cities in China during January–February 2020, 54% of respondents considered COVID-19 outbreak had affected there psychology moderately and severely, while 29% showed moderate to severe anxiety symptom, and 17% people responded moderately to severe depressive symptoms [17].

According to the survey conducted, it was found that 480 individuals (64.26% of the total) had no effect on their mental health due to COVID-19, whereas around 123 people (16.47%) were uncertain, if COVID-19 pandemic had affected their mental health condition in any way, and around 144 individuals (19.28%) were aware of their mental well-being as shown in Fig. 3.

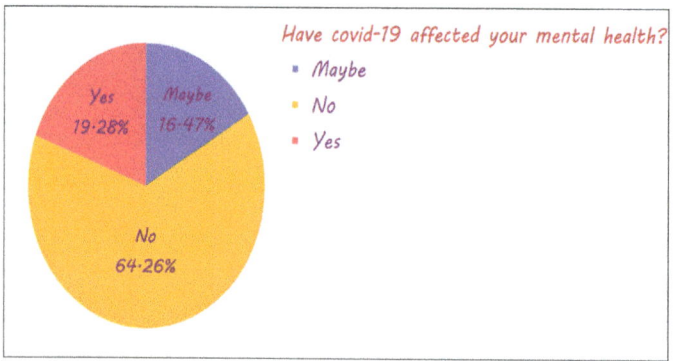

Fig. 3 Number of individual's mental health affected by COVID-19

3.4 Mental Illness Versus Marital Status

One of the most consistent findings in research on social aspect of illness revealed there were more frequent results showed co-relation between marital status and mental illness [18].

Spousal status is well-recognized as a key demographic changeable parameter linked equally with mental and physical health. Spousal status is the distinct possibilities that describe a relationship of a person with significant other that could be single, married, divorced, widowed, etc. Moreover, marriage relates with lower depressive affect, lower anxiety, lower suicidal ideation and less substance abuse on an average [19]. On average, marriage relates with lower anxiety, lower depressive affect, lesser suicidal ideation and less substantial abuse. In various surveys, it is found that married people have shown longer life expectancy, reduced death rates and enhanced psychological well-being. The dissolved marital relationship leads to stressful affair with a succeeding adjustment period, but aftermaths hang on the context of the divorce. Hence, in case of divorced individuals, the after effect of dissolved marital relationship may vary from person to person as short-term depressive symptoms usually occur in women and in men, there were sign of short-term increase in drug abuse. The separated and divorced individual displays poorer health compared to married, single and often to widowed people [20].

The connection of marital status and mental illness was examined and, hence, found that 236 single people are frequently suffering from mental illness, followed by married (48 respondent), separated (18 respondent), widowed (2 respondent) and divorced (3 respondent) people, respectively. The majority of the people could be suffering from personality disorders, post-traumatic stress disorder, psychotic disorders, including schizophrenia (other) and depression, bipolar disorder or mood disorder [21] (Fig. 4).

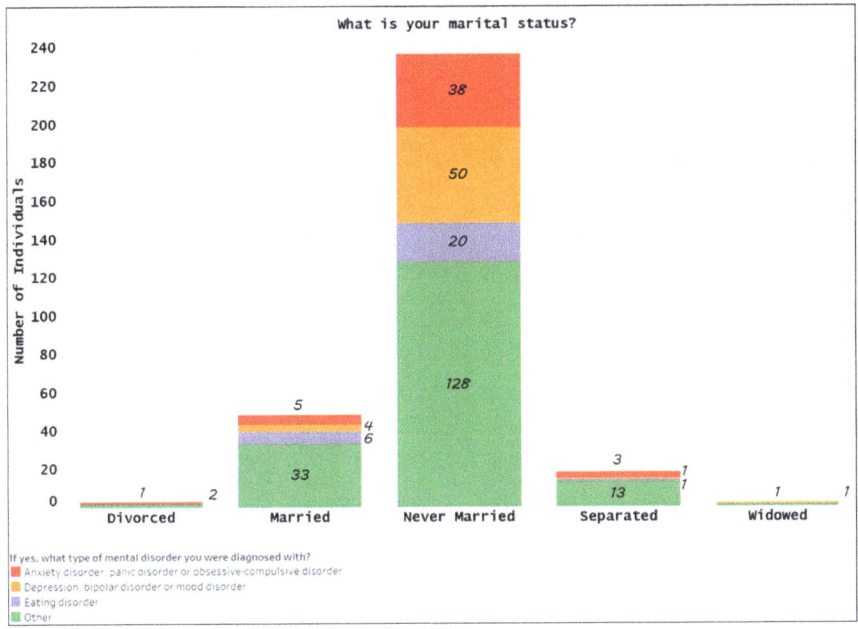

Fig. 4 Effect of marital status on individual's mental health

3.5 *Illness Family History Versus Willingness to Discuss It with Other*

There is inherited risk of developing mental illness for some individual where mental health issues were commonly proliferating among families, for example mental depression. Ten per cent of people in the USA at some point in their lifespan had experienced depression according to Stanford university school of medicine. In case of presence of family, history of mental illness, risk of getting depression or other mental disorder in individual was increased by two to threefold, but severity of symptom varied among family members as mental disease does not follow typical pattern of inheritance [22, 23].

Younger people in comparison to adults do not ask for help for mental health issues because of distinctive fears of death of confidentiality, peer pressure, longing to be self-reliant [24] and lack of knowledge and awareness to acknowledge mental health problems and related services [25].

Over the years, it has been seen that people have now become more aware of mental illness and are open to talk about it. Individuals having history of mental disorder in their family are more willingly to discuss about their mental health with people then those who do not have any family history of mental disorder.

According to the conducted survey, majority of the people (around 261 individual) were open to discussion about their mental disorder, whereas 258 individuals were not

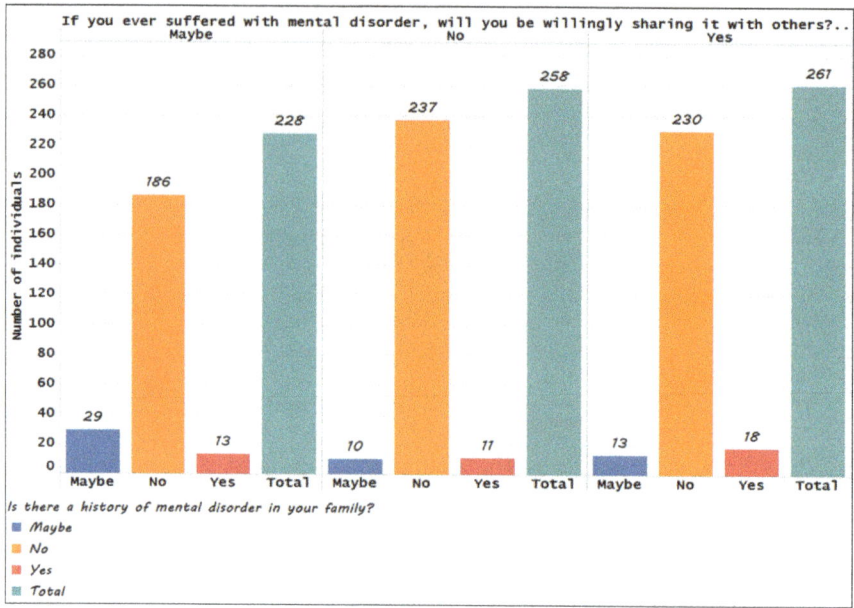

Fig. 5 Number of individual open to talk about their mental health

whether they had a history of mental disorder in family or not, and 228 individuals were unsure to discuss about it. It was also found that individuals having family history of mental disorder were more open to discuss about it (Fig. 5).

3.6 Occupation Versus Mental Illness

Occupation and mental health are related in many ways. A person's occupation (life role, kind of work) not only provides him with a livelihood but also symbolizes his/her degree of success. When an individual is out of work, in the wrong field, misplaced or dissatisfied with his job regarding basic requirements, seek to satisfy in other ways. Sometimes he may successfully meet these needs; at other times, he may fail, find himself thrown off balance, and become emotionally ill.

The high prestige attached to certain occupations leads parents to select the future careers for children and causes people to strive for occupational roles beyond their abilities. Such fruitless striving produces failure and rejection feelings and results in withdrawal, defeatism and other states of disequilibrium [26].

India is placed third in world's youth population. Mental health difficulties are likely to adversely influence the efficiency and capabilities of India's youth. Occupation has a straight effect on mental health of the individual. Secure job, good

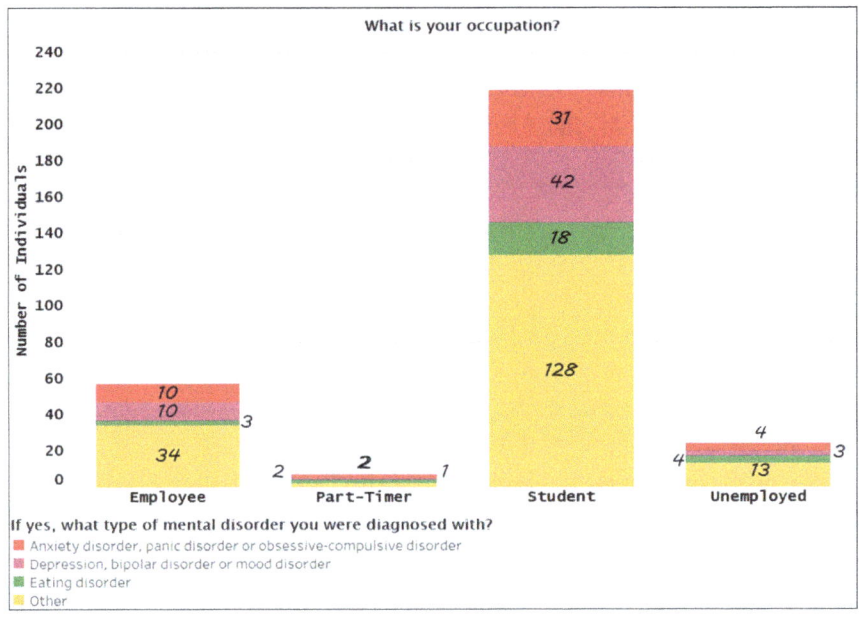

Fig. 6 Effect of occupation and their effect of mental health

money, healthy lifestyle are directly proportional to the good mental health. Individuals lacking in any of those are in stress, tensed and sometimes anxious which affect their state of mind. Putting effort in what they do and still not getting result will affect them and in a long term will cause depression.

Occupation and mental health have affected the individual in a following manner: student > employee > unemployed > part-timer. Students are mostly affected than employee, unemployed and part timers due to arising competition in education field, workload, insecurities of future, anxiety and panic disorder, depression, eating disorder, obsessive compulsive disorder, job stress and lack of performance (Fig. 6).

A large number of adults identify mental health problems potentially only when it becomes acute. Selective facets of conventional Indian culture, like significance of marriage, is accountable for specific indexes of stigma. Scholastic intrusions to lower stigma related with mental well-being can upsurge help-seeking behaviours by evading the usage of psychiatric tags that are not understandable, else it openly discusses various mental health problems with varying cruelty emphasizing on symptomatic vignettes. Intercession content that clearly and interactively converses young brains mental health-related stigma responses and age-appropriate social roles, other than getting targeted for forthcoming responsibility such as wedding, will assist in achieving timely detection of mental health diseases among younger generation.

Flexible work environment with flexible work scheduling and break timings should be allowed also work from home facility and flexible place of work allowed by employers. Day-to-day feedback and guidance should be given for nurturing employees' well-being. Employees should also have a knowledge of health benefits offered by their organization and allowed to participate in any wellness programmes. Proper feedback should be provided to employee when they resign from their job will play important role in improving their health.

3.7 Mental Health Versus Physical Health

A clear distinction is usually made between 'mind' and 'body'. But when considering psychological state and physical health, the two should not be thought of as separate. Poor physical health can cause an increased risk of developing psychological state problems. Similarly, poor psychological state can negatively impact on physical health, resulting in an increased risk of some conditions.

Several studies have been conducted and have revealed that the individual suffering from diabetic, arthritis, cardiac and asthma shows sign of mental illness, namely depression [27], panic, phobia [28], anxiety and mood disorder [29, 31], while obese individual showed eating disorder and sign of clinical depression [32, 33].

The 242 individuals had shown effect on their daily life cause of mental health. It has impacted them due to various mind disorders like sensing miserable all the time, dropping interest in crucial stage of life, oscillating between extreme pleasure and extreme sorrow. Out of 747 people, 501 individuals had shown no effect on daily life due to physical health (Fig. 7). Our research survey discovered that psychological illness had displayed disturbance in daily routine of life; moreover, living standard shown no affect by any physical health problem such as fatigue, agony, laziness until unless it is a serious health issue.

4   Conclusion

In many parts of the world, research is conducted by using big data of mental health and for many different purposes. Data science is briskly surfacing that deals with many treasured uses to mental health research. This study validates the data complexity, the type of study, visualizations and the impact of visual analytics. The survey was done to analyse and predict and make better mental health solutions for the individuals using data visualization tool, i.e. Tableau on selected datasets in urban and semi-urban population. The various parameters chosen were marital status, family history, willingness to discuss with others, occupation, mental health or physical health and mental illness, etc. The figure of 307 people of entire population showed prevalence of diagnosable mental disorders. Hence, there is a great need to upsurge the prevention policies at both primary and secondary stages to defeat the mental distress and economic shortfall to organization due to mental disorders. Nationalized mental state policies should be made, and psychological healthy state mind campaigning should be mainstreamed into non-governmental and

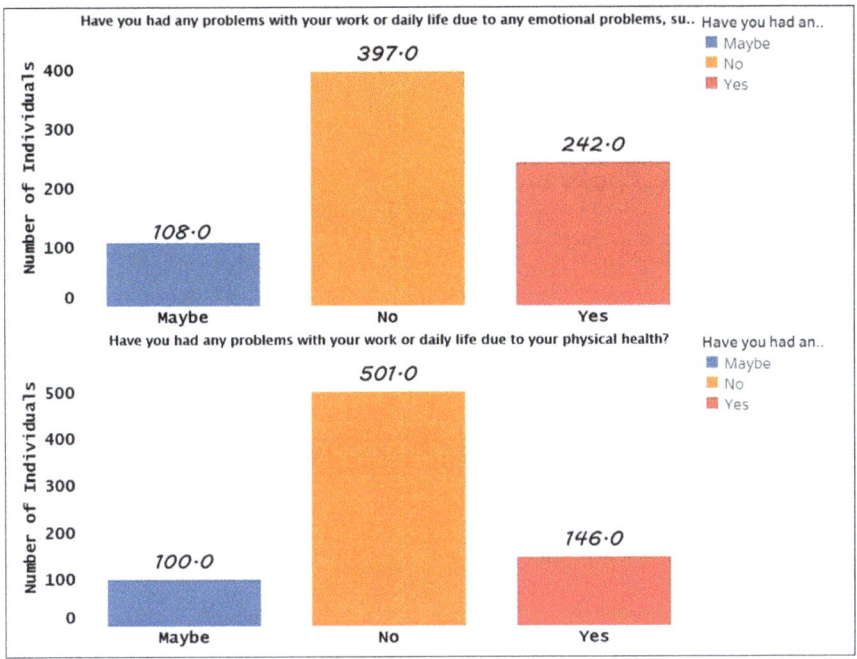

Fig. 7 Effect of physical health with respect to mental health

governmental policies. Ongoing developed graphic analytical tactics and involving domain experts in better understanding the responsibility of visual analytics and safeguarding their genuine impact in better decision-making are the key next steps for this emerging research area.

References

1. Srividya M, Mohanavalli S, Bhalaji N (2018) Behavioral modeling for mental health using machine learning algorithms. J Med Syst. https://doi.org/10.1007/s10916-018-0934-5
2. World Health Organization (2001) Mental Health: A Call for Action by World Health Ministers. World Health Organization, Department of Mental Health and Substance Dependence, Geneva
3. Koh HC, Tan G (2005) Data mining applications in healthcare. J Healthc Manag 19(2):64–72
4. Miner L et al (2014) Practical Predictive Analytics and Decisioning Systems for Medicine: Informatics Accuracy and Cost-Effectiveness for Healthcare Administration and Delivery Including Medical Research. Academic Press, Cambridge
5. Ruppert T (2018) Visual Analytics to Support Evidence-Based Decision Making. Technische Universität
6. Torous J, Baker JT (2016) Why psychiatry needs data science and data science needs psychiatry: connecting with technology. JAMA Psych 73(1):3–4
7. Sagar R, Dandona R, Gururaj G, Dhaliwal RS, Singh A et al (2020) The burden of mental disorders across the states of India: The Global Burden of Disease Study 1990–2017. Lancet Psychiatry 7(2):148–161

8. Parcesepe AM, Cabassa LJ (2013) Public stigma of mental illness in the United States: a systematic literature review. Adm Policy Ment Health Ment Health Serv Res 40(5):384–399

9. Gururaj G, Varghese M, Benegal V, Rao GN, Pathak K, Singh LK et al (2016) National Mental Health Survey of India, 2015–16 (NIMHANS, editor). NIMHANS Publication, Bengaluru

10. Reddy MV, Chandrashekar C (1998) Prevalence of mental and behavioural disorders in India: a meta-analysis. Indian J Psy 40(2):149–157

11. Koivisto J, Hamari J (2019) The rise of motivational information systems: a review of gamification research. Int J Inf Manage 45:191–210

12. https://timesofindia.indiatimes.com/on-international-womens-day-dr-batras-shines-a-spotli ght-on-womens-health/articleshow/81343206.cms

13. https://www.euro.who.int/en/health-topics/health-emergencies/coronavirus-covid-19/public ations-and-technical-guidance/mental-health-and-covid-19

14. The Economic Times (2020) Govt helpline receives 92,000 calls on abuse and violence in 11 days. 8 April 2020. https://economictimes.indiatimes.com/news/politics-and-nation/govt-hel pline-receives-92000-calls-on-abuseand-violence-in11days/articleshow/75044722.cms?utm_ source=contentofinterest&utm_medium=text&utm_campaign=cppst

15. World Health Organization (2020a) Coronavirus disease 2019 (COVID-19) situation report – 35. https://www.who.int/docs/defaultsource/coronaviruse/situation-reports/20200224-sitrep-35-covid-19.pdf?sfvrsn=1ac4218d_2

16. World Health Organization (2020b) Mental health and psychosocial considerations during the COVID-19. https://www.who.int/docs/default-source/coronaviruse/mental-health-consid erations.pdf

17. https://academic.oup.com/qjmed/article/113/5/311/5813733?login=true

18. Rushing WA (1979) Marital status and mental disorder: evidence in favor of a behavioral model. Soc Forces 58(2):540–556. https://doi.org/10.1093/sf/58.2.540

19. Spiker RL (2014) Mental Health and Marital Status. The Wiley Blackwell Encyclopedia of Health, Illness, Behavior, and Society, pp 1485–1489 (2014). https://doi.org/10.1002/978111 8410868.wbehibs256

20. Shankardass MK (2018) Mental health issues in India: concerns and response. Indian J Psy Nsg 15(1):58–60. http://www.ijpn.in/text.asp?2018/15/1/58/262509

21. Gulliver A, Griffiths KM, Christensen H (2010) Perceived barriers and facilitators to mental health help-seeking in young people: a systematic review. BMC Psychiatr 10:113

22. https://ct.counseling.org/2019/08/challenging-the-inevitability-of-inherited-mental-illness/

23. https://www.healthychildren.org/English/health-issues/conditions/emotional-problems/ Pages/Inheriting-Mental-Disorders.aspx

24. Kelly CM, Jorm AF, Wright A (2007) Improving mental health literacy as a strategy to facilitate early intervention for mental disorders. Med J Aust 187(S7):S26–309

25. Clement S, Schauman O, Graham T, Maggioni F, Evans-Lacko S, Bezborodovs N et al (2015) What is the impact of mental health-related stigma on help-seeking? A systematic review of quantitative and qualitative studies. Psychol Med 45(1):11–27

26. Wolford JA (1964) Mental health and occupation. Public Health Rep 79(11):979–984

27. Ducat L, Philipson LH, Anderson BJ (2014) The mental health comorbidities of diabetes. JAMA 312(7):691–692

28. Nicassio PM (2010) Arthritis and psychiatric disorders: disentangling the relationship. J Psychosom Res 68(2):183–185

29. Kubzansky LD, Kawachi I (2000) Going to the heart of the matter: do negative emotions cause coronary heart disease? J Psychosom Res 48:323–337

30. Kawachi I, Colditz GA, Ascherio A, Rimm EB, Giovannucci E, Stampfer MJ et al (1994) Coronary heart disease/myocardial infraction: prospective study of phobic anxiety and risk of coronary heart disease in men. Circulation 89:1992–1997

31. Weissman MM, Markowitz JS, Ouellette R, Greenwald S, Kahn JP (1990) Panic disorder and cardiovascular/cerebrovascular problems: results for a community survey. Am J Psychiatry 147:1504–1508

32. Vila G, Nollet-Clemencon C, deBlic J, Mouren-Simeoni MC, Scheinmann P (2000) Prevalence of DSM-IV anxiety and affective disorders in a pediatric population of asthmatic children and adolescents. J Affect Disord 58:223–231
33. Collingwood J (2016) Obesity and mental health. Psych Central. Retrieved on September 25, 2018 from https://psychcentral.com/lib/obesity-and-mental-health/

Baseline Evaluation of COVID-19 Impact on Developing Countries Workforce by Machine Learning

Forhad An Naim

Abstract A skilled workforce is the ultimate strength of the economy. COVID-19 has affected the continuity of the regular workforce remarkably across the world. A large number of the workforce had been splashed from operations, and a significant number had been continued operations from remotely. The aim of this research is to analyze the COVID-19 effects on the developing countries workforce like Bangladesh. In this paper, we have performed the statistical analysis and developed a machine learning model based on Bangladeshi IT company daily activities. The dataset has been separated into three parts based on periods such as pre-COVID-19, lockdown and post-lockdown periods. We have used several supervised learning algorithms such as SVM, LR, RF and DT for both classification and regression problem. From statistical analysis, it has been observed that regular operations of pre-COVID-19 have been disrupted during the lockdown period, and to minimize the damages of the lockdown period, company workforces are putting extra effort during the post-lockdown period. From machine learning analysis, it has been observed that random forest (RF) has performed better than other classifiers for both classification and regression problem.

Keywords Workforce · COVID-19 effect · Machine learning · Regression · Baseline evaluation

1 Introduction

Bangladesh has fulfilled eligibility criteria of least developer country (LDC) in 2018 and declared as a developing country and is on track for graduation in 2024 [1]. Bangladesh must have to set at the three-year average of the level of GNI per capita and maintain the three gross national income (GNI), human assets index (HAI) and economic and environmental vulnerability index (EVI) thresholds guided by the World Bank [2]. Now, Bangladesh economy is the 35th largest in the world in nominal

F. An Naim (✉)
Department of Computer Science and Engineering, United International University, Dhaka 1212, Bangladesh

terms. According to the Asian Development Bank (ADB), the GDP of Bangladesh had been projected 8% for the year 2019–2020 [3].

Early 2020, COVID-19 had been spread out across the world. World Health Organization (WHO) had declared it as a pandemic. People have to stay home and maintain social distance to stay safe. As a result, across the world, many organizations, industries, mills, factories, etc., had suspended their regular operations, and many had completely shut off, and some had tried to maintain their operations from home. Eventually, all countries' economy has suffered from massive degradation. Especially, newly developing countries had suffered a lot because they had to maintain World Bank thresholds. Experts had been suggested a significant impact of COVID-19 on developing countries, with a particularly large influence on South Asian economies. Because, the forecast was saying the world economy could be shrunk by 0.9% in 2020 instead of growing a projected 2.5% [4]. Another research said, the consequences of the COVID-19 pandemic would slow down the development, jobless in many sectors and create messes in the workforce that could lead to a major economic crisis [5].

The impact of the COVID-19 on the workforce is directly related to the economic growth of any country. So, the impact of the COVID-19 on the workforce is a well-timed research topic. A few pieces of researches have already conducted on workforce management based on machine learning. But, to the best of our knowledge, no paper is available researching the impact of the COVID-19 on the workforce based on machine learning. The ICT sector of Bangladesh is growing day by day. The ICT sector is playing a vital role in building Digital Bangladesh. The Government of Bangladesh has aimed to earn 5 billion dollars from ICT export by 2021 [6]. In the paper, we have researched the impact of COVID-19 on the workforce of a Bangladeshi IT firm based on their daily activities. We have selected three different periods such as pre-COVID-19, lockdown and post-lockdown to research the daily activities of the respective period. This research has selected the three important reports of the activities of the employees by the employer as classification and regression targets. Supervised learning can deal with both classification and regression problem. We have tried different supervised classifiers, e.g., SVM, RF, LR and DT. For a regression problem, we have used SLR, RFR and SVR. During the experimental period, we have fed the classifiers by labeled data and then we have provided unobserved data to predict the output label. We have evaluated the performance of the proposed model by some evaluation matrices, e.g., precision, recall, F1-score and area under curve.

2   Related Works

There are some related works from which we founded the idea of this topic. To the best of our knowledge, no paper is available researching the impact of the COVID-19 on the workforce based on machine learning. Kniffin et al. [7] have researched the implications, issues and insights on COVID-19 and the workplace. Researchers focused on: (i) emerging changes in work practices and (ii) economic and social-psychological impacts. They also illustrated the benefits of team science and provided an integrative approach for considering the implications of COVID-19 for work and organizations. Blundell and Machin [8] have researched self-employment in the COVID-19 crisis. Researchers have reported results from the LSE-CEP survey of UK self-employment. The result claimed that self-employed workers have felt that their health was at risk while working during the coronavirus crisis than those who work in app-based jobs. On average, self-employed workers were prepared to sacrifice 10% of their income. Fana et al. [9] have researched the employment impact of the COVID-19 crisis based on short-term effects on long-term prospects. Their research was based on confinement decrees of three European countries (Germany, Spain and Italy). The analysis has done to assess the implications of the COVID-19 crisis on labor markets and also to speculate on mid- and long-term developments. Akter. S. provided a comprehensive and demonstrative review of the observed data and the potential impact of unemployment that will arise after the lockdown. The study found that massive job losses could have happened in all of the sectors in Bangladesh, e.g., RMG, remittance, export and import, transportation, tourism, banking and insurance, education [10]. Hamid et al. researched 'new normal' rules for organizations to reduce the impact of coronavirus imposed by the Government of Bangladesh. The study asked some online question to 199 employees randomly from different organizations. The research found that COVID-19 has impacted employee's performance negatively and 'new normal' affected employee's work concentration negatively [11].

3   Dataset Description

COVID-19 pandemic has imposed an adverse effect on socio-economic growth and development across the world. Especially, the developing countries economy is suffering significantly. According to the World Bank, Bangladesh's economy has significantly impacted by the COVID-19 by creating unemployment and deepening poverty [12]. The majority workforce of Bangladesh stayed away from work during the lockdown period and rest minority part worked from home. After the lockdown period, still, a part of the workforce is working from home, and others are working from office premises. So, a new change has introduced in the management of workforce in Bangladesh. This factor encouraged us to build the dataset for analyzing the impacts on Bangladeshi workforce by COVID-19.

3.1   Dataset Collection

We have collected more than 35,000 employee activities of an IT company as the dataset. Each data point of the dataset represents an activity of a single employee.

Employees of the company have recorded every day activities by workforce management software. We have separated the dataset into three periods such as pre-COVID-19, lockdown and post-lockdown. The pre-COVID-19 dataset consists of one-year activities of the company before COVID-19. Bangladesh government has announced the lockdown across the country from 26 March 2020 to 30 May 2020. The lockdown dataset consists of activities of the lockdown period, and the post-lockdown dataset consists of six-month activities after the lockdown. The pre-COVID-19 period dataset has 18,502 activities as data points, the lockdown period has 3515 data points, and the post-lockdown period has 13,209 data points.

3.2 Features Description

Features are the descriptive attributes. Each data point consists of six features or independent variables. There are two types of data points called as activity such as personal and check-in. The personal activity can be recorded to back date. But, the check-in activity must be pre-scheduled at a future time and performed checked-in in scheduled time; otherwise, the activity will be marked as a missed activity. SCHEDULE_STATUS (SS) represents the status of the activity. Personal activity marked as nine (9), checked-in activity as seven (7), missed activity as zero (0) and scheduled ready to check-in as one (1). CATEGORY_TYPE (CATT) is another feature of activities or data points. CATT has two states such as normal (1) and priority (2). Priority represents that activities must have stakeholder-related activities, and for normal, the stakeholder is optional. CATEGORY_ID (CATI) is another feature. Each activity or data point belongs to one category. Each CATI represents a unique meeting category. The dataset has thirty unique categories. Table 1 shows the summary of CATT, SS, and Table 2 shows the summary of the CATI, and Table 3 shows the summary of the IG, EI and CI for pre-COVID-19, lockdown and post-lockdown periods. IS_GROUPSCHDEULE (IG) shows whether the activity belongs to more than one person or single? IG $==$ 2 represents the activity is single, and IG $==$ 1 represents the activity is group schedule. EMPLOYEE_ID (EI) represents the

Table 1 Summary of the CATEGORY_TYPE (CATT) and SCHEDULE_STATUS (SS) for pre-COVID-19, lockdown and post-lockdown periods

Feature	Pre-COVID-19	Lockdown	Post-lockdown
CATT	Normal → 13046 (70.5%) Priority → 5456 (29.5%)	Normal → 3482 (99%) Priority → 33 (1%)	Normal → 12199 (92.3%) Priority → 1010 (7.7%)
SS	Checked → 9416 (51%) Personal → 8744 (47.2%) Missed → 342 (1.8%)	Checked → 2122 (60.4%) Personal → 1276 (36.4%) Missed → 117 (3.2%)	Checked → 8128 (61.6%) Personal → 4685 (35.5%) Missed → 396 (2.9%)

Table 2 Summary of the CATEGORY_ID (CATI) for pre-COVID-19, lockdown and post-lockdown periods

Pre-COVID-19	Lockdown	Post-lockdown
Divisional work → 7671 (41.5%)	Remote work → 1421 (40.4%)	Desk job → 4981 (37.7%)
Stakeholder visit → 4850 (26.3%)	Divisional work → 804 (22.9%)	Divisional work → 1621 (12.3%)
Proposal → 1053 (5.7%)	Internal meeting → 560 (15.9%)	Remote work → 1122 (8.5%)
Internal meeting → 1013 (5.4%)	Virtual meeting → 254 (7.2%)	Team building → 1017 (7.7%)
Travel → 952 (5.15%)	Training → 111 (3.1%)	Virtual meeting → 949 (7.2%)
Pre-sales → 498 (2.7%)	External meeting → 100 (2.8%)	Stakeholder visit → 591 (4.5%)
Rest categories → 2465 (13.25%)	Rest categories → 265 (7.7%)	Rest Categories → 2410 (22.1%)

Table 3 Summary of IS_GROUPSCHEDULE (IG), EMPLOYEE_ID (EI) and CUSTOMER_ID (CI) for three periods

Feature	Pre-COVID-19	Lockdown	Post-lockdown
IG	Group → 40 (0.7%) Single → 18,462 (99.3%)	Group → 26 (0.7%) Single → 3489 (99.3%)	Group → 1010 (7.7%) Single → 12,199 (92.3%)
Unique EI	52	53	58
Unique CI	942	25	332

employee who owns the activity. CUSTOMER_ID (CI) represents the customer or stakeholder related to any activity.

3.3 Targets Description

Targets are the final outputs that machine learning trying to predict based on features. It is also known as labels. This research has selected the three important reports of the activities of the employees by the employer as targets. TOTAL_TIME (TT) is the total minute elapsed for each activity. TOTAL_TIME (TT) indicates the total recorded time for each activity based on the minute. Previously, we explained that stakeholder is mandatory for priority category and optional for normal. IS_CUSTOMER (IC) shows whether the activity related to any stakeholder or not? Table 4 shows the summary of the IS_CUSTOMER (IC), SCHEDULE_STATUS (SS) and TOTAL_TIME (TT) as targets for pre-COVID-19, lockdown and post-lockdown periods. Only SCHEDULE_STATUS (SS) has also been used as a target

Table 4 Summary of the SCHEDULE_STATUS (SS), IS_CUSTOMER (IC) and TOTAL_TIME (TT) as targets for pre-COVID-19, lockdown and post-lockdown periods

Target	Pre-COVID-19	Lockdown	Post-lockdown
SS	Checked → 9416 (51%) Personal → 8744(47.2%) Missed → 342 (1.8%)	Checked → 2122 (60.4%) Personal → 1276 (36.4%) Missed → 117 (3.2%)	Checked → 8128 (61.6%) Personal → 4685 (35.5%) Missed → 396 (2.9%)
IC	No → 12003 (64.8%) Has → 6499 (35.2%)	No → 3377 (96% %) Has → 138 (4%)	No → 10047 (76%) Has → 3162 (24%)
Average TT	241 min	379 min	170 min

and feature to predict customer-related activities and status of activities because of the employer demand.

4   Feature Selection

Feature selection is another important technique in machine learning that selects the most important features correlated to targets. Feature selection is the process of selecting highly relevant features automatically or manually to predict the targets. The reason behind this while predicting the targets, all features do not contribute equally [13]. Highly relevant features provide an effective way to improve learning accuracy and facilitate a better understanding of the learning model. In this research, we have used different features for predicting different target variables based on univariate selection and feature importance technique. In univariate selection, a statistical test (chi-squared) can be used to select the most relevant features that have the strongest association with the targets. While predicting the targets, feature importance assigns a score for each feature, and the higher the score indicates more relevant features [14]. We have listed all of the highly correlated unique features based on univariate selection and feature importance. As the number of features is six and proposed target is three, while predicting the targets, we have used a maximum of four (4) and a minimum of three (3) features that are highly correlated to targets. From Table 5, we can see that both feature selection techniques identify the same relevant features such as CATI, CI and SS to predict IC. But, to predict SS, both selection techniques identify the same relevant features except IG and EI. As the maximum threshold value (4) did not exceed, we have selected all the identified features to predict SS. We can also see that both feature selection techniques identify the same relevant features such as CATI, CI, EI and SS to predict TT. Figure 1 shows the identified features by feature importance technique to predict IC, SS and TT.

Table 5 Selected features and most correlated features with the score of both techniques for IC, SS and TT

Targets	Univariate selection	Feature importance	Selected features
IS_CUSTOMER (IC)	CATI (17,240) CI (12,780) SS (4,446)	CI (0.4696) CATI (0.2441) SS (0.1376)	CATI CI SS
SCHEDULE_STATUS (SS)	CATI (3,228) CI (2300) IG (358.9)	CI (0.370) CATI (0.357) EI (0.267)	CI CATI EI IG
TOTAL_TIME (TT)	CATI (7.92) SS (5.346) CI (3.111) EI (1.523)	CI (0.487) EI (0.31) CATI (0.16) SS (0.02)	CI CATI EI SS

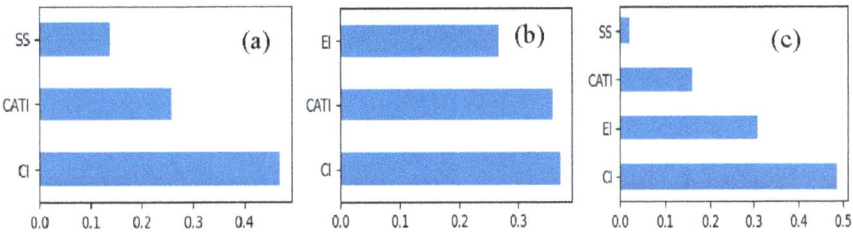

Fig. 1 Selected relevant features by feature importance: **a** IS_CUSTOMER (IC); **b** SCHEDULE_STATUS (SS); **c** TOTAL_TIME (TT)

5 Methodology

5.1 Dataset Preparation

Dataset preparation is one of the crucial steps in machine learning areas. The reason behind this is each dataset is different and highly specific to projects. Data preparation step transforms that raw data into a transformed form that can be used in modeling [15]. The proposed method has three periods of data such as pre-COVID-19, lockdown and post-lockdown. For the pre-COVID-19 dataset, we have used 80% of the dataset as the training dataset consists of 14,802 data points and 20% of the dataset as the test dataset consists of 3700 data points. Similarly, the lockdown dataset had 2812 training data points and 703 test data points. The post-lockdown dataset had 10,567 training data points and 2642 test data points.

5.2 Classification and Regression

Classification is the process of classifying the discrete or nominal class of given data points or observations. The output is categorical for classification [16]. Classification is the task of approximating a mapping function (f) from input variables (X) to discrete output variables (y). The mapping function predicts the labels or targets for a given observation [17]. We have used several supervised classification classifiers such as support vector machine (SVM), logistic regression (LR), decision tree (DT) and random forest (RF) to evaluate the proposed model. We have tried different hyperparameters to maximize model performance. Regression is the process of predicting numeric or continuous values. It is the task of approximating a mapping function (f) from input variables (X) to continuous output variables (y) [18]. In other words, it is the process of establishing the correlations between the dependent or target variables and independent or input variables. Real-life problems can be modeled as regression problems [19]. We have used several supervised regression classifiers such as simple linear regression (SLR), support vector regression (SVR) and random forest regression (RFR) to evaluate the proposed model.

5.3 Proposed Model

In this research, we have used two discrete or nominal targets along with one continuous target variables. We have split the original dataset into three different datasets based on periods, e.g., pre-COVID-19, lockdown and post-lockdown. We have selected highly relevant feature from the three datasets using the univariate selection and feature importance technique. After that, we have split three datasets into training and test dataset which was used in modeling. During the training phase, we have fed the algorithms by labeled data. During the testing phase, we have provided unobserved data to predict the output label. The base of classification is to predict discrete or nominal values, but in contrast, regression is to predict continuous or numeric values. Supervised learning can deal with both classification and regression problem. We have tried different supervised classifiers, e.g., SVM, RF, LR and DT. For a regression problem, we have used SLR, RFR and SVR. We have evaluated the performance of the proposed model by some evaluation matrices, e.g., precision, recall, F1-score and area under curve. The proposed model is shown in Fig. 2.

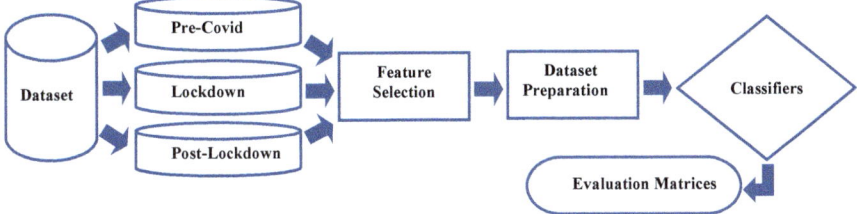

Fig. 2 Proposed model

6 Experimental Result and Analysis

6.1 Experimental Environment

Experimental environment is an important factor in the field of machine learning when dealing with large datasets. In this research, we have used Intel Core i7 Processor, 16 GB ram and 250 GB SSD. We have used Python 3.8, Java 11 and Google map.

6.2 Statistical Data Analysis

Statistical data analysis is the process of performing various statistical operations to understand the outcome of the study. It is a kind of quantitative research that seeks to quantify the data, describes the size of samples, data distribution and compares data from a different group of observations. Table 6 shows the statistics of total activities, working hours, working days and no. of employees of the pre-COVID-19, lockdown and post-lockdown periods. The total number of activities of the three periods such as pre-COVID-19, lockdown and post-lockdown are 18,502, 3515 and 13,209, respectively, as shown in Table 6. The total working days of the three periods

Table 6 Statistics of total activities, working hours, working days, visited companies and no. of employees of three periods

Report	Pre-COVID-19	Lockdown	Post-lockdown
Total activities	18,502	3515	13,209
Total working hours	74,314 h	22,237 h	37,356 h
Total working days	361 day	63 day	113 day
Total employees	52	53	58
Visited companies	942	25	332

are 361, 63 and 113, respectively. The total working hours of the three periods are 74,314, 22,237 and 37,356, respectively. The total visited companies of the three periods are 942, 25 and 332, respectively. The total employees involved in activities of the three periods are 52, 53 and 58, respectively.

Table 7 shows the statistics based on per day and employee. Per day activities of post-lockdown period have increased twice than other two periods as shown in Table 7. Company has spent more per day working hours on lockdown period than other two periods. The per day working hours of the three periods such as pre-COVID-19, lockdown and post-lockdown are 205.85, 352.96 and 330.58, respectively. Per employee has recorded more activities and spent more working hours on pre-COVID-19 period than two others. The number of activities per employee of the three periods is 358.0, 67.0 and 228. The number of per day per employee activities of the three periods is 0.98, 1.05 and 2.01, respectively. During lockdown period, per day per employee has spent more working hours than two others. Per day per employee elapsed working hours of the three periods is 3.95, 6.65 and 5.69.

Table 8 shows the statistics of the visited customers and per customer activities. During post-lockdown period, the company employees have monthly visited more customer than two other periods. The per month customer visits of the three periods are 219.13, 27.8 and 351.44. The company has spent 1.5 times more activity on each customer at post-lockdown period than others. The numbers of customer visits per month by per employee of the three periods are 4.21, 0.524 and 6.05. The number of working hours per customer of the three periods is 23.81, 29.9 and 19.23.

The company from where the dataset has been collected is located at Dhaka city, Bangladesh. Government of Bangladesh had declared public holiday across the country during the lockdown period. Most of the private companies like this IT

Table 7 Statistics based on per day and per employee

Report	Pre-COVID-19	Lockdown	Post-lockdown
Per day activity	51	56	117
Per day working hours	205.85 h	352.96 h	330.58 h
Per employee activity	358	67	228
Per employee working hours	1429.75 h	419.11 h	644.0 h
Per day per employee activity	1	2/3	2
Per day per employee working hours	3.95 h	6.65 h	5.69 h

Table 8 Statistics of the visited customers and per customer activities

Report	Pre-COVID-19	Lockdown	Post-lockdown
Per month customer visits	219	28	351
Per customer activity	6	5	9
Per customer working hours	23.81 h	29.9 h	19.23 h
Per month per employee customer visit	4	1/2	6

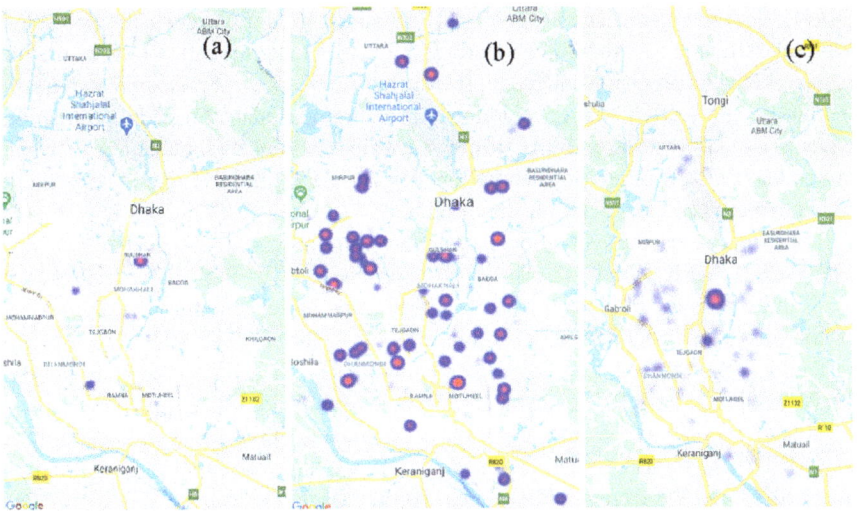

Fig. 3 Heat map of activities location: **a** Pre-COVID-19; **b** Lockdown; **c** Post-lockdown

company had started home office during the lockdown to safe their business from the adversity of the pandemic. Figure 3 shows the location of the activities along with the activities intensity of the pre-COVID-19, lockdown and post-lockdown periods.

From Fig. 3, the dark color indicates more activities than the light color. During the pre-COVID-19 period, employees have done most of their activities from the physical office located at Dhaka and some from other locations as shown in Fig. 3a. But, during the lockdown period, employees have done most of their activities from different location of Dhaka city as shown in Fig. 3b. Now, in post-lockdown period, the company is trying to return their normal office routine. Now, most of the employees have done their activities from office premises, but some still doing home office. From Fig. 3c, we can see that most activities indicated by dark color are from the office premises and some portion of the activities still from different locations indicated by slight light color shades.

6.3 Machine Learning Results and Analysis

Measurement of model performance is indispensable in machine learning. Different evaluation matrices are widely used to identify how one classifier is better than other classifiers. In this research, we have presented precision, recall, F1-score and area under curve (AUC) as performance evaluation matrices for classification problems. Evaluation matrices are measured based on four parameters such as true positive (TP), false positive (FP), false negative (FN) and true negative (TN). The proportion of positively identifications correctly predicted by classification methods for a particular class is called precision. The proportion of actual identifications correctly predicted by the classification methods for a particular class is called recall. F1-score makes balances between precision and recall. It is the harmonic mean of precision and recall. Area under curve (AUC) is a performance measurement matrix based on threshold values. AUC indicates the ability of classifiers in terms of distinguishing between classes known as a degree of separation. Higher AUC indicates good separation efficiency between classes by the classifier.

Table 9 represents the evaluation matrices analysis of stakeholder-related activities as IS_CUSTOMER (IC). It is a classification problem. From Table 9, we can see that for all three periods, random forest (RF) performed better among other classifiers. For the pre-COVID-19 period, RF achieved an F1-score of 0.95 and AUC of 0.97. For the lockdown period, RF achieved an F1-score of 0.71 and AUC of 0.93. For the post-lockdown period, RF achieved an F1-score of 0.67 and AUC of 0.90.

Table 10 represents the evaluation matrices analysis of the status of the activities as SCHEDULE_STATUS (SS). From Table 10, we have seen that three classifiers except SVM and RF were unable to separate the target classes.

Regression models are used to predict a continuous value. In this research, we have presented mean squared error (MSE), root-mean-squared error (RMSE) and R-squared error as performance evaluation matrices for regression problems. MSE is the average squared difference between the predicted or observed value and actual

Table 9 Representation of evaluation matrices analysis of IS_CUSTOMER (IC)

Classifier	Pre-COVID-19	Lockdown	Post-lockdown
LR	Precision → 0.88 Recall → 0.88 F1-score → 0.88 AUC → 0.87	Precision → 0.50 Recall → 0.48 F1-score → 0.49 AUC → 0	Precision → 0.50 Recall → 0.38 F1-score → 0.43 AUC → 0
SVM	Precision → 0.71 Recall → 0.73 F1-score → 0.65 AUC → 0.73	Precision → 0.40 Recall → 0.48 F1-score → 0.49 AUC → 0	Precision → 0.48 Recall → 0.48 F1-score → 0.32 AUC → 0.48
DT	Precision → 0.90 Recall → 0.95 F1-score → 0.92 AUC → 0.94	Precision → 0.66 Recall → 0.79 F1-score → 0.70 AUC → 0.90	Precision → 0.60 Recall → 0.87 F1-score → 0.61 AUC → 0.87
RF	Precision → 0.94 Recall → 0.97 F1-score → 0.95 AUC → 0.97	Precision → 0.65 Recall → 0.93 F1-score → 0.71 AUC → 0.93	Precision → 0.64 Recall → 0.91 F1-score → 0.67 AUC → 0.90

Table 10 Representation of evaluation matrices analysis of SCHEDULE_STATUS (SS)

Classifier	Pre-COVID-19	Lockdown	Post-lockdown
LR	Precision → 0.49	Precision → 0.33	Precision → 0.35
	Recall → 0.48	Recall → 0.21	Recall → 0.39
	F1-score → 0.48	F1-score → 0.25	F1-score → 0.31
	AUC → 0	AUC → 0	AUC → 0
SVM	Precision → 0.42	Precision → 0.38	Precision → 0.40
	Recall → 0.55	Recall → 0.38	Recall → 0.36
	F1-score → 0.38	F1-score → 0.36	F1-score → 0.24
	AUC → 0.69	AUC → 0.53	AUC → 0.51
DT	Precision → 0.56	Precision → 0.50	Precision → 0.44
	Recall → 0.57	Recall → 0.53	Recall → 0.45
	F1-score → 0.56	F1-score → 0.50	F1-score → 0.44
	AUC → 0	AUC → 0	AUC → 0
RF	Precision → 0.60	Precision → 0.54	Precision → 0.62
	Recall → 0.62	Recall → 0.58	Recall → 0.57
	F1-score → 0.60	F1-score → 0.53	F1-score → 0.55
	AUC → 0.75	AUC → 0.70	AUC → 0.70

value. This is the absolute measure of the model. The lower the MSE, the better the model is. RMSE is the squared root of the average squared difference between the predicted or observed value and actual value. The lower the RMSE, the better the model is. R-squared error compares the proposed model with a baseline model to measure method efficiency. The baseline model is chosen by taking the mean of the data points. R-squared is calculated by the MSE of proposed model divided by MSE of the baseline model. R-squared will always be one (1) or less than one. One (1) indicates a better fit, and less than one indicates the worst fit.

Table 11 represents the evaluation matrices analysis of the total time elapsed of activities as total time (TT). From Table 11, it has been seen that random forest regression (RFR) has performed better among other regression methods for all periods. RFR has achieved the lowest MSE and RMSE score for all three periods. RFR has achieved better R-squared value for three periods such as 0.43, 0.48 and 0.47.

7 Summary of the Findings

- The research has been done based on three different periods such as pre-COVID-19, lockdown and post-lockdown periods. The pre-COVID-19 period consists of

Table 11 Representation of evaluation matrices analysis of TOTAL_TIME (TT)

Classifier	Pre-COVID-19	Lockdown	Post-lockdown
SLR	MSE → 33,668.41 RMSE → 183.48 R-squared → 0.129	MSE → 56,626.27 RMSE → 237.96 R-squared → 0.082	MSE → 28,925.87 RMSE → 170.07 R-squared → 0.049
SVR	MSE → 47,095.90 RMSE → 217.01 R-squared → -0.09	MSE → 68,786.51 RMSE → 262.27 R-squared → -0.10	MSE → 35,027.44 RMSE → 187.15 R-squared → -0.106
RFR	MSE → 25,008.43 RMSE → 158.14 R-squared → 0.43	MSE → 32,088.83 RMSE → 179.13 R-squared → 0.48	MSE → 16,804.92 RMSE → 129.63 R-squared → 0.47

the previous one-year activities before lockdown, the lockdown period consists of two-month activities, and the post-lockdown period consists of six-month activities after the lockdown.

- The datasets consist of a total of six feature variables and three target variables. SCHEDULE_STATUS (SS) has been used as both the feature and the target. The model has evaluated based on two classification and one regression problem.
- Feature selection technique identifies the features that are highly relevant to the target variables. Highly relevant features provide an effective way to improve learning accuracy and facilitate a better understanding of the learning model.
- From statistical analysis, it has been observed that during the post-lockdown period, daily activities have increased twice than pre-COVID-19 and lockdown periods. In the lockdown period, per day working hours was greater than the other two periods. In the lockdown, customer visit by employees had decreased by 90% than the pre-COVID-19 period. In contrast, customer visit had increased by 60% than the pre-COVID-19 period. In the post-lockdown period, per customer activities had increased by 50% than two other periods. In the pre-COVID-19 period, employees have conducted their operations from office premises as shown in Fig. 3a. In the lockdown, employees have conducted their operations remotely from different locations as shown in Fig. 3b. In the post-lockdown, employees have conducted most of their operations from office premises but still few conducting operations remotely as shown in Fig. 3c.
- Random forest (RF) has performed better among other classifiers for two classification problems. Random forest regression (RFR) has performed better among other regression methods for TOTAL_TIME (TT) that is shown as residual errors. The residual error is the difference between the actual value and the prediction

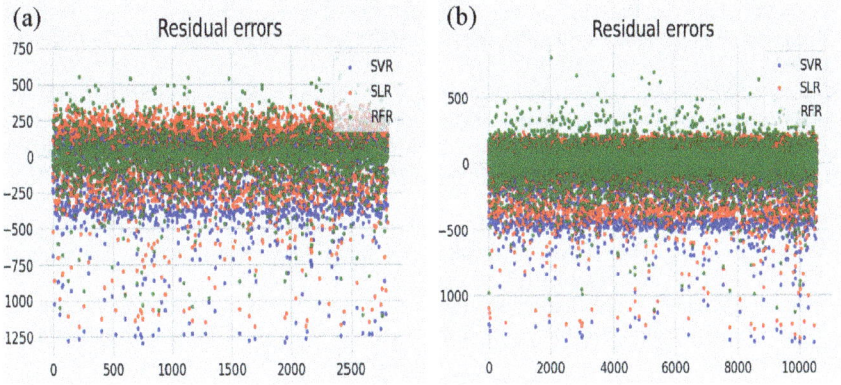

Fig. 4 Representation of residual error for **a** Lockdown and **b** Post-lockdown periods

value. A residual error near zero indicates a better prediction result. Residual error of SLR, SVR and RFR is shown in Fig. 4 for lockdown and post-lockdown periods.

8 Conclusion

In this research, we have presented a statistical and machine learning analysis on three different periods, e.g., pre-COVID-19, lockdown and post-lockdown for evaluating the COVID-19 impact on the workforce. The study has provided the baseline because to the best of our knowledge, no research has conducted researching the impact of the COVID-19 on the workforce based on machine learning. We have used only the three most important employer reports as targets, but others are ignored. In the future, we will research the rest reports. Because of our limitation, we only researched employee's daily activities. In the future, we will research employee's attendance, targets and sales based on the three periods, e.g., pre-COVID-19, lockdown and post-lockdown.

References

1. Least Developed Country Category: Bangladesh Profile. [Online]. Available: https://www.un. org/development/desa/dpad/least-developed-country-category-bangladesh.html/. Accessed: 10 Jan 2021
2. LDC Identification Criteria & Indicators. [Online]. Available: https://www.un.org/develo pment/desa/dpad/least-developed-country-category/ldc-criteria.html/. Accessed: 10 Jan 2021
3. WB projects 7.2 pc GDP growth for Bangladesh in 2019–20 FY. [Online]. Available: https://www.thedailystar.net/country/news/wb-projects-72-pc-gdp-growth-bangladesh-2019-20-fy-1852267/. Accessed: 10 Jan 2021

4. Islam MM, Jannat A, Rafi DAA, Aruga K (2020) Potential economic impacts of the COVID-19 Pandemic on South Asian economies: a review. MDPI Journal. https://doi.org/10.3390/world1 030020
5. Sarwar ASM, Tarafder S, Rahman MM, Razzak KSB, Bushra A, Rahman S (2020) COVID 19 outbreaks and impact on developing countries like Bangladesh. Asian J Res Infectious Diseases
6. Bangladesh capable of $5bn ICT export by 2021. [Online]. Available: https://www.dhakat ribune.com/bangladesh/development/2016/10/21/the-road-to-5-billion-ict-export-by-2021/. Accessed: 16 Jan 2021
7. Kniffin KM, Narayanan J, Anseel F, Antonakis J, Ashford et al (2020) COVID-19 and the workplace: implications, issues, and insights for future research and action. Am Psychol 76(1):63–77. https://doi.org/10.1037/amp0000716
8. Blundell J, Machin S (2020) Self-employment in the Covid-19 crisis. A CEP Covid-19 analysis, Paper No. 003, Centre for Economic Performance: London School of Economics and Political Science
9. Fana M, Pérez S, Macias E (2020) Employment impact of Covid-19 crisis: from short term effects to long terms prospects. J Ind Bus Econ 47:391–410
10. Akter S (2020) Covid-19 and Bangladesh: threat of unemployment in the economy. North Am Acad Res 3:79–104. https://doi.org/10.5281/zenodo.3986907
11. Hamid M, Wahab SA, Hosna AU, Hasanat MW (2020) Impact of coronavirus (COVID-19) and employees' reaction to changes on employee performance of Bangladesh. Int J Bus Manage 8:34–43. https://doi.org/10.24940/theijbm/2020/v8/i8/BM2008-013
12. Bangladesh Must Ramp Up COVID-19 Action to Protect its People, Revive Economy. [Online]. Available: https://www.worldbank.org/en/news/press-release/2020/04/12/bangla desh-must-act-now-to-lessen-covid-19-health-impacts/. Accessed: 16 Jan 2021
13. Rawat T, Khemchandani V (2019) Feature engineering (FE) tools and techniques for better classification performance. Int J Innov Eng Technol (IJIET). https://doi.org/10.21172/ijiet. 82.024
14. Feature Selection Techniques in Machine Learning with Python. [Online]. Available: https://towardsdatascience.com/feature-selection-techniques-in-machine-learning-with-python-f24e7da3f36e/. Accessed: 20 Jan 2021
15. What Is Data Preparation in a Machine Learning Project. [Online]. Available: https://machin elearningmastery.com/what-is-data-preparation-in-machine-learning/. Accessed: 30 Jan 2021
16. Kim SW, Gil JM (2019) Research paper classification systems based on TF-IDF and LDA schemes. Hum Cent Comput Inf Sci 9:30. https://doi.org/10.1186/s13673-019-0192-7
17. Difference Between Classification and Regression in Machine Learning. [Online]. Available: https://machinelearningmastery.com/classification-versus-regression-in-machine-learning/. Accessed: 20 Jan 2021
18. Doan T, Kalita J (2015) Selecting machine learning algorithms using regression models. IEEE International Conference on Data Mining Workshop (ICDMW). https://doi.org/10.1109/ ICDMW.2015.43
19. Iyer SJ, Pawar AD (2019) Machine learning model for predicting price of processors using multivariate linear regression. International Conference on Smart Systems and Inventive Technology (ICSSIT). https://doi.org/10.1109/ICSSIT46314.2019.8987936

A Survey on Deep Learning Methods in Image Analytics

Pramod Kumar Vishwakarma and Nitin Jain

Abstract The domain of machine learning (ML) is at bright phase as deep learning (DL) gradually turns into pioneer in this sphere. DL utilizes numerous covers to address the reflections of information to construct computational prototypes. The main empowering agent DL algorithms, for example, generative adversarial networks, convolutional neural networks, and model exchanges have totally changed our impression of data handling. Notwithstanding, there happens a gap of comprehension behind this colossally speedy domain, since it was never recently addressed from a multi-scope viewpoint. The absence of center arrangement delivers these incredible techniques as discovery machines that repress improvement at an essential level. In addition, DL has repeatedly been seen as a silver slug to all hindrances in ML, which is a long way from reality. This paper presents a thorough review of authentic and late condition state-of-the art approaches in visual, sound, and text preparing; informal community examination; and common language handling, trailed by the top to bottom investigation on turning and pivotal improvements in DL claims. Likewise, attempted to audit the issues looked in deep learning, for example, unsupervised learning, discovery models, and Internet learning and to delineate how these difficulties can be changed into productive future exploration roads.

Keywords Deep learning · Artificial intelligence · Machine learning · Frameworks · Algorithms · Deep neural networks · Convolutional neural networks · Recurrent neural networks · Short-term memory · Gated recurrent units

1 Introduction

Since the 1950s, a little subset of artificial intelligence (AI), regularly called machine learning (ML), has reformed a few fields over the most recent couple of many years.

P. K. Vishwakarma (✉) · N. Jain
Computer Science & Engineering, Chandigarh University, Gharuan, Mohali, Punjab 140413, India

N. Jain
e-mail: nitin.e8466@cumail.in

© The Author(s), under exclusive license to Springer Nature Singapore Pte Ltd. 2022   631
D. Gupta et al. (eds.), *Proceedings of Data Analytics and Management*,
Lecture Notes on Data Engineering and Communications Technologies 90,
https://doi.org/10.1007/978-981-16-6289-8_52

Neural networks (NN) are a subfield of ML, and it was this subfield that brought forth deep learning (DL). Since, its inception DL has been making ever-bigger interruptions, indicating remarkable achievement in pretty much every application domain [1, 2].

Learning is a methodology comprising of assessing the model boundaries, so the learned model may play out a particular assignment. For instance, in artificial neural networks (ANN), the boundaries weight grids. DL, then again, comprises of a few layers in the middle of the information and yield layer which considers numerous phases of non-direct data handling units with various leveled structures to be available that are misused for include learning and example characterization [2, 3]. Learning strategies dependent on portrayals of information can likewise be characterized as portrayal learning [4].

Late writing positions and DL-created portrayal learning include a chain of command of highlights or ideas, where elevated glassy ideas can be characterized from the low-level ones and low-level ideas can be characterized from significant-level ones. In certain articles, DL has been depicted as a general learning approach that can tackle practically a wide range of issues in various application domains. All in all, DL is not task explicit [4–6].

Supervised learning is a learning method that utilizations named information. On account of supervised DL draws near, the climate has a bunch of data sources and comparing yields. For instance, if for input, the insightful specialist predicts the specialist will get misfortune esteem. The specialist will at that point iteratively alter the organization boundaries for a superior guess of the ideal yields. After fruitful training, the specialist will have the option to find the right solutions to inquiries from the climate. There are distinctive supervised learning approaches for deep inclining, including deep neural networks (DNN), convolutional neural networks (CNN), recurrent neural networks (RNN), including long short-term memory (LSTM), and gated recurrent units (GRU). These networks will be portrayed in details in the particular segments. Semi-supervised learning will be learning that happens dependent on somewhat marked datasets. At times, DRL and generative adversarial networks (GAN) are utilized as semi-supervised learning procedures [7–9].

Unsupervised learning frameworks are able to perform with much data. For this situation, the specialist learns the inner portrayal or significant highlights to find obscure connections or construction inside the info information. Frequently, bunching, dimensionality decrease, multiplicative methods are measured as unsupervised learning draws near. Few individuals from the deep learning family that are acceptable at bunching and non-straight dimensionality decrease, including autoencoders (AE), restricted Boltzmann machines (RBM), and the as of late created GAN. What is more, RNNs, for example, LSTM and RL, are likewise utilized for unsupervised learning in numerous application domains. Areas 6 and 7 examine RNNs and LSTMs in detail [10–12].

Once in a while, this methodology is called semi-supervised learning also. There are numerous semi-supervised and unsupervised procedures that have been actualized dependent on this idea (in Sect. 8). In RL, we do not have a straightforward misfortune work, in this manner making learning more enthusiastically contrasted with conventional supervised methodologies [10, 13].

The crucial contrasts among RL and supervised learning are: First, you do not have full admittance to the capacity you are attempting to enhance; you should question them through collaboration, and second, you are interfacing with a state-based climate: Input relies upon past activities. Contingent on the difficult extension or space, one can choose which kind of RL should be applied for addressing an assignment. On the off chance that the issue has a ton of boundaries to be streamlined, DRL is the most ideal approach. On the off chance that the issue has less boundaries for advancement, a deduction-free RL approach is acceptable. An illustration of this is annealing, cross-entropy techniques, and SPSA [3, 5, 14] (Fig. 1).

Artificial intelligence is one of the parts of software engineering and information technology that makes PCs or machines that carry on as keen as people. John McCarthy, who is the dad of artificial intelligence, said, "The Science and designing of making savvy machines especially artificial intelligence endeavors to mimic intelligence of people and results with another smart machine that will have the option to deliver data as human mindfulness and conduct". Artificial intelligence has the application in different fields, for example, normal language handling, neural networks, mechanical technology, and picture preparing. The core of artificial intelligence is machine learning and the essential methodology for planning scholarly frameworks. Machine learning remembers a numerous practices for insights and likelihood hypothesis, curved examination, and estimation just as algorithm complexities. The ideas of machine learning basically receive instinct and joining, to make the contemporary learning by duplicating the human instinct and execution, then rebuilds the new information to improve the presentation routinely. Machine learning has

Fig. 1 Categories of deep learning approaches

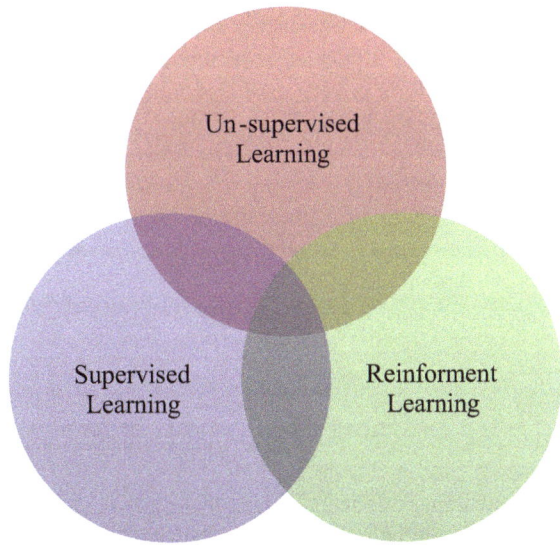

been widely applied in different territories, for example, space offices, automation, industry finance, industry health care, government organization, marketing [6, 15].

The idea of deep learning has been proposed with the deepening of the artificial neural organization. Deep learning utilizes deep neural networks, and it is a product that reproduces human brain neural organization. Thus, deep learning is a development of machine learning which is bringing machine learning and artificial intelligence closer to make innovative applications. The human brain is a relationship of organization of neurons; this plan has led to the development of artificial neural networks [16, 17].

The artificial neural organization has a disparity with a human brain, which the human brain neurons network produces information, connections, and discrete layer spread course. In 2011, the original learning frameworks disbelieve were dispatched by Google. Google had the option to examine a huge number of their information from different server forms. Google is broadly utilizing deep learning in preparing its items, for example, Google searches and Google photographs. One of the significant perspectives in picture preparing and examination is highlight portrayal. Coming up next are the two favorable circumstances that deep learning yields: (1) Given a dataset, deep learning discovers include normally for an unmistakable application. Normally, semi-automatic learning technique is a traditional element extraction strategy that utilizes a piece of earlier information to extricate the highlights. (2) Deep learning can discover new highlights that are pertinent to the reasonable applications. As indicated by customary extraction strategies, they are constantly characterized by past information, which just gathers certain highlights related to a proper application. Also, there are two different segments that change picture handling results. They are picture procurement and translation. Image Acquisition: It is exhibited tentatively by specialists that the higher the picture quality, the more the outcomes obtained in picture preparation and investigation. Regardless, picture quality relies upon picture procurement. Picture translation: It is a strategy of looking into an aerial picture or a computerized far-off detecting picture and physically determines picture highlights. This methodology is very trusted, commendable, and expansive highlights can be gathered from a picture. Some of them can be riparian highlights, anthropogenic highlights, and so on. However, it requires some investment, and just master individuals like picture experts having sound information can do it. In spite of the fact that picture translation works best on the implicit, ascribes of pictures are made out of seven perspectives, for example, surface, size, affiliation, tone, example, and shape. These viewpoints are utilized to know the data about the articles present in the picture [18–21].

2 Deep Learning Applications in Image Analytics

There are numerous deep learning techniques that have been proposed for application in image examination and image handling zones, similar to image characterization, image enrollment, image division, and image recognition. The issues in the previously mentioned zones can be tackled utilizing highlight portrayal as a team with deep learning procedures. In this bit of the paper, we give an investigation on the progression of placing deep learning engineering in all the above zones referenced [21, 22].

2.1 Use of Deep Learning in Feature Extraction and Image Segmentation

One of the significant and intriguing ideas with regards to picture handling is picture division; the significant classes of PC vision are: (1) picture classification, (2) picture detection, and (3) picture division; the characterization of pictures by recognizing a presence of a picture in an image is known as picture arrangement. Picture location follows picture grouping and limitation. The idea confinement finds the territory if a given item is found. A limit is attracted around it to feature the district. The troublesome and the most advantageous sort among the over three is picture division. This is identified with order each pixel into a gathering of articles in a given picture [23].

2.2 Image Classification Using Deep Neural Networks

Picture order is a cycle of drawing out the data of classes from a multiband raster picture. Here, the PC is educated to distinguish pictures and sort into a portion of the given classes. In this model, a dataset CIFAR-10 having pictures aligned 60,000 in number every one of size 32×32 pixels is taken. This dataset is ordered into ten classes that do not happen all the while going with each class that has pictures in number 6000. The pictures in this dataset have no clamor, little in size and are unquestionably named [24].

3 Deep Learning Historical Analysis

Constructing a mechanism that can mimic humanoid intelligences had been a fantasy of mentors for quite a long time. The absolute starting point of DL can be followed back to 300 B.C. at the point when Aristotle proposed the model that began the historical backdrop of people's aspiration in attempting to comprehend the brain, since such a thought requires the researchers to comprehend the component of human acknowledgment frameworks. The cutting edge history of deep learning began in 1943 when the McCulloch-Pitts (MCP) model was presented and got known as the model of artificial neural models. They made a PC model dependent on the neural networks practically imitating neo-cortex in human brains. The mix of the algorithms and arithmetic called "edge rationale" was utilized in their model to imitate the human manner of thinking yet not to learn. From that point forward, deep learning has advanced consistently with a couple of huge achievements in its turn of events. After the MCP model, the Hebbian hypothesis, initially utilized for the organic frameworks in the regular habitat, was actualized. From that point onward, the primary electronic gadget called "perceptron" inside the setting of the cognizance framework was presented in 1958; however, it is not quite the same as average perceptrons these days. The perception exceptionally looks like the advanced ones that have the ability to validate associations. Toward the finish of the primary AI winter, the rise of "back propagandists" turned into another achievement. Werbos presented back propagation, the utilization of mistakes in training deep learning models, which opened the door to current neural networks. In 1980, "neocogitron," which roused the convolutional neural organization, was presented, while recurrent neural networks (RNNs) were proposed in 1986. Then, LeNet made the deep neural networks (DNNs) work essentially during the 1990s; notwithstanding, it did not get exceptionally perceived. Because of the equipment limit, the construction of LeNet is very naive and cannot be applied to huge datasets. Around 2006, deep conviction networks (DBNs) along with a layer-wise pre-training structure were created. Its main thought was to train a basic two-layer unsupervised model like restricted Boltzmann machines (RBMs), freeze all the boundaries, stick another layer on top, and train simply the boundaries for the new layer. Scientists had the option to train neural networks that were a lot deeper than the past endeavors utilizing such a strategy, which incited a re-branding of neural networks to deep learning. Initially, from artificial neural networks (ANNs) and following quite a while of improvement, deep learning currently is, perhaps, the most proficient devices contrasted with other machine learning algorithms with extraordinary execution. We have seen a couple of deep learning strategies established from the underlying ANNs, including DBNs, RBMs, RNNs, and convolutional neural networks (CNNs). While illustrations preparing units (GPUs) are notable for their presentation in processing huge scope grids in organization structures on a solitary machine, various disseminated deep learning systems have been created to accelerate the training of deep learning models. Since the immense measures of information come without names or with loud names, some examination considers center more around improving clamor power of training modules utilizing unsupervised or semi-supervised deep learning methods. Since the majority of the current deep learning models just spotlight on a solitary methodology, this prompts a restricted portrayal of genuine information. Specialists are currently focusing closer on a cross-methodology structure, which may yield a tremendous advance forward in deep learning [25–28].

4 Methods

Distinctive deep learning algorithms help improve the learning execution, expand the extents of uses, and disentangle the computation cycle. In any case, the very long training season of the deep learning models remains a significant issue for the specialists. Besides, the characterization exactness can be radically upgraded by expanding the size of training information and model boundaries. To quicken the deep learning preparing, a few progressed methods are proposed in the writing. Deep learning systems consolidate the execution of modularization deep learning algorithms, enhancement methods, circulation procedures, and backing to frameworks. They are created to streamline the usage cycle and lift the framework level turn of events and examination. In this segment, a portion of these agent methods and structures are presented [29, 30].

4.1 Unsupervised Learning

In opposition to the huge measure of work done in supervised deep learning, not many investigations have tended to the unsupervised learning issue in deep learning. Be that as it may, as of late, the advantage of learning reusable highlights utilizing unsupervised strategies has indicated promising outcomes in various applications. In the most recent decade, having a self-educated learning structure has been broadly examined in the writing. In late couple of years, generative models, for example, GANs and VAEs have become predominant strategies for unsupervised deep learning. For example, GANs are trained and reused as a fixed element extractor for supervised errands in. This organization depends on CNNs and has indicated its matchless quality as unsupervised learning in visual information examination. In another work, a deep scanty auto-encoder is trained on an exceptionally huge scope picture datasets to learn highlights. This organization creates an undeniable-level component extractor from unlabeled information, which can be utilized for face location in an unsupervised manner. The produced highlights are likewise discriminating enough to recognize other undeniable-level items like creature appearances or human bodies [30, 31].

4.2 Web-Based Learning

Normally, the organization geographies and models in deep learning are time static (i.e., they are predefined before the learning begins) and are likewise time invariant. This limitation on time unpredictability represents a genuine test when the information is streamed on the web. Internet learning recently came into mainstream research, yet just an unobtrusive headway has been seen in online deep learning. Expectedly, DNNs are based upon the stochastic inclination plunge (SGD) approach in which the training tests are utilized exclusively to refresh the model boundaries with a known name. The need is that instead of the successive handling of each example, the updates ought to be applied as clump preparing. One methodology was introduced in where the examples in each group are treated as autonomous and indistinguishably appropriated (IID). The cluster handling approach relatively balances the figuring assets and execution time. Another test that piles up on the issue of web-based learning is high-speed information with time-differing dissemination. This test addresses the retail and banking information pipelines that hold huge business esteems. The current reason is that the information is generally close so as to securely expect piece savvy fixed, hence, having a comparable conveyance. This suspicion portrays information with a certain level of relationship and builds up the models appropriately, as talked about in. Shockingly, these non-fixed information streams are not IID and are regularly longitudinal information streams. In addition, Internet learning is regularly memory delimited, is more diligently to parallelized, and requires a straight learning rate on each information test. Creating techniques that are equipped for Internet learning from non-IID information would be a major jump forward for large information deep learning [32, 33].

4.3   Optimization Approaches

Training a DNN is an improvement cycle, i.e., finding the boundaries in the organization that limit the misfortune work. Practically speaking, the SGD strategy is a major algorithm applied to deep learning, which iteratively changes the boundaries dependent on the slope for each training test. The computational multifaceted nature of SGD is lower than that of the first angle plummet strategy, in which the entire informational index is viewed as each time the boundaries are refreshed. In the learning cycle, the refreshing rate is constrained by the hyper boundary learning rate. Lower learning rates will at last prompt an ideal state after a long time, while higher learning rates rot the misfortune quicker yet may cause variances during the training. To control the swaying of SGD, utilizing energy is presented. Enlivened by Newton's first law of movement, this procedure gets a quicker combination and an appropriate energy that can improve the enhancement aftereffects of SGD [34, 35].

4.4   Usage of Distributed Systems in Deep Learning

The productivity of model training is restricted to a solitary machine framework, and the appropriated deep learning strategies have been created to additionally quicken the training cycle. There are two main ways to deal with training the model in an appropriated framework, to be specific, information parallelism and model parallelism. For information parallelism, the model is repeated to all the computational hubs, and each model is trained with the doled out subset of information. After a certain timeframe, the weight update should be synchronized among the hubs. Relatively, for model parallelism, all the information is prepared with one model where every hub is answerable for the incomplete assessment of the boundaries in the model. Then again, a model parallelism approach parts the training venture across various GPUs. In a straightforward model-equal procedure, each GPU processes just a subset of the model. For instance, for a model with two LSTM layers, the framework with two GPUs can utilize every one of them to figure one LSTM layer [36–39]. The benefit of the model-equal procedure is that it makes training and forecasts with gigantic deep neural networks conceivable. For example, the bunks HPC framework trained a neural organization with in excess of 11 billion boundaries, which needs about 82 GB of memory. It is difficult to fit a particularly model into one machine, and in this way, it should be apportioned utilizing model-equal procedures. Notwithstanding, since the model is parceled across hubs, one disadvantage of model parallelism is that every hub can just process a subset of results, and synchronization is consequently expected to get the full outcomes. The synchronization misfortune and correspondence overhead of model-equal procedures are more than those of information equal methodologies since every hub in the previous should synchronize the two slopes and boundary esteems on each update step. All in all, the versatility of model parallelism is second rate. To handle this issue, Google has proposed an automated gadget position system dependent on deep reinforcement learning to locate the best plan of the model segment and situation. The structure takes the inserting portrayal of every activity, puts the gathered tasks to various gadgets, and shows a 60% presentation improvement contrasted with the human specialists [40–42].

4.5 Frameworks

Table 1 records a sprinkling of famous deep learning systems for engineering plans, for example, Caffe, deep learning 4j (DL4j), light, neon, Theano, MX Net, TensorStream, and Microsoft Psychological Tool compartment (CNTK). In Table 2, the permit, center language, upheld interface language, and system backing of CNN, RNN, and DBN are likewise recorded. It very well may be seen from Table 2 that C++ is normally utilized for execution of deep learning systems since it quickens the speed of training. Since GPU is essentially useful to accelerate the lattice calculation, the vast majority of the previously mentioned structures likewise uphold GPU through the interface given by CuDNN. In the interim, Python has become the most widely recognized language for deep learning engineering plan since it can make the programming more productive and simpler by rearranging the programming cycle [43–45]. Likewise, appropriated computation gets regular in some as of late delivered structures, for example, TensorStream, MX Net, and CNTK. The objective is to additionally improve the computation productivity for deep learning. In addition, TensorFlow likewise incorporates uphold for the tweaked deep learning application-explicit coordinated circuit (ASIC), called tensor preparing unit (TPU), to help increment the productivity and reduction the force utilization. Caffe, executed

Table 1 Popular deep learning methods

Paper	NLP tasks	Architecture	Datasets
Socher et al. [60]	Sentimental analysis	RNTN	SST
Kim [61]	Sentimental analysis General classification	CNN	SST
Wehrmann et al. [62]	Sentimental analysis	Conv-Char-S	MTD
Bahdanau et al. [63]	Translation	Bidir RNN Encoder-decoder	WMT-14-EF
Cho et al. [64]	Translation	RNN Encoder-decoder	WMT-14-EF
Wu et al. [65]	Translation	GNMT	WMT-14-EF WMT-14-EG
Socher et al. [66]	Paraphrase identification	Unfolding RAE	MSRP
Yin et al. [67]	Paraphrase identification Question and answer	ABCNN	Wiki QA MSRP
Kageback et al. [68]	Summarization	Unfolding RAE	OD
Dong et al. [69]	Question and answer	MCCNN	WQ
Feng et al. [46]	Question and answer	CNN	IQA

by Berkeley Vision and Learning Center, is quite possibly the most generally utilized structures. It bolsters the most generally utilized layers for both CNN and RNN yet does not straightforwardly empower the utilization of DBN. Clients of Caffe plan their engineering by announcing the design of a calculation diagram, for example, convolutional layers. There are pre-trained models available for a wide scope of neural networks, for example, AlexNet, GoogLeNet, and ResNet. Moreover, Caffe is a solitary machine structure [46, 47]. At the end of the day, it does not uphold multi-hub execution, while the multi-GPU computation is upheld when there are outer contributions like Caffe on Sparkle by Hurray that coordinate Caffe with a major information motor like Flash. DL4j is the most famous structure executed in Java, created and maintained by Sky mind since 2014. Helping out Hadoop and Flash, DL4j is fit for disseminated calculation too. Notwithstanding, this system is accounted for to possess a longer training energy for comparative structures bench-marked with different structures. Light was first delivered in 2002 and broadened its deep learning highlight in 2011. Joined with Facebook's deep learning CUDA library (fbcunn), light can work model- and information-level equal calculation. In contrast to different structures, light is assembled dependent on a powerful diagram portrayal rather than a static chart. The powerful chart permits the client to refresh the computational diagram (i.e., to change the model design) during runtime, while the static chart utilizes certain capacities to characterize the charts ahead of time [48, 49]. As of late, light delivered its Python interface, PyTorch, and the utilization of this system has significantly expanded because of its adaptability. Neon and Theano are two structures created in Python by Intel and the College of Montreal, individ-ually. The two of them perform code enhancements in the framework and portion

Table 2 Popular visual datasets for deep learning

Dataset	Data type	Num. of instances	Num. of classes	Ground truth	Applications
ImageNet [70]	Images	14 M	1,000	Yes	Image classification, object localization, object detection, etc.
CIFAR10/100 [71]	Images	60 K	10/100	Yes	Image classification
Pascal VOC [72]	Images	46 K	20	Yes	Image classification, object detection, semantic segmentation
Microsoft COCO [73]	Images	2 M	80	Yes	Object detection, semantic segmentation
MNIST [74]	Images	70 K	10	Yes	Hand written digit classification
YFCC100M [75]	Images, videos	100 M	8 M	Partially	Video and image understanding
YouTube-8M [76]	Videos	8 M	4,716	Automatic	Videos classification
Trecvid [77]	Videos	Varies	Varies	Partially	Video search, event detection, localization, etc.
UCF-101 [78]	Videos	13 k	101	Yes	Human action detection
Kinetics [79]	Videos	306 k	400	Yes	Human action detection

level. In this manner, their training speeds normally beat different systems. In any case, albeit just parallelism and multi-GPU are upheld, and the multi-hub count is not planned in these structures. MX Net backings a few interfaces, including C++, Python, R, Scala, Perl, MATLAB, Java content, Go, and Julia. It upholds both calculation diagram announcements and basic calculation revelations for engineering plan. MX Net backings information and model parallelism as well as follow boundary worker plans to help circulated computation too. MX Net has the most thorough usefulness; however, the presentation is not streamlined as much as other cutting edge systems [50–53]. TensorStream is executed by Google and gives a progression of inside capacities to help actualize any deep neural organization dependent on the static computational diagram. As of late, Keras began to help TensorFlow by means of a significant-level interface and permitted clients to plan the engineering without stressing over the interior plan. The structure gives various degrees of equal and dispersed activities and very much planned lethal resistance. The power of its plan draws in a ton of clients, and it has gotten quite possibly the most well-known deep learning structures since its delivery [54, 55]. CNTK, planned by Microsoft, has a particular significant-level content language, brain content, for neural organization execution. CNTK models the neural organization as a coordinated diagram. Every hub in the diagram addresses an activity or a channel, and each edge alludes to the information stream. Rather than the boundary worker model, the message passing interface is applied for dispersed count upholds [30, 40, 42, 56–59].

5 Dataset in Deep Learning

Video investigation has pulled in extensive consideration in the PC vision local area and is considered as a difficult errand since it incorporates both spatial and transient data. In an early work, huge scope YouTube recordings containing 487 game classes are utilized to train a CNN model. The model incorporates a multi-goal engineering that uses the neighborhood movement data in recordings and incorporates setting stream (for low-goal picture demonstrating) and fovea stream (for high-goal picture handling) modules to order recordings. An occasion location from sport recordings utilizing deep learning is introduced in. In that work, both spatial and fleeting data are encoded utilizing CNNs and highlight combination by means of regularized auto-encoders. As of late, another procedure called recurrent convolution networks (RCNs) were presented for video preparing [28, 57, 80–82].

It applies CNNs on video outlines for visual arrangement and afterward takes care of the casings to RNNs for investigating worldly data in recordings. Another RCN model proposed in utilizes RNN on the intermediate layers of CNNs. Likewise, a gated recurrent unit is utilized to use the sparsity and area in the RNN modules. This model is approved on the UCF-101 and YouTube2Text datasets. The critical headways in picture and video preparing not just depend on the advancement of new learning algorithms and use of incredible equipment yet additionally urgently rely upon exceptionally enormous scope public datasets. A few huge scope visual datasets used to train deep learning algorithms are recorded in Table 2. PictureNet can be considered as the most significant and powerful informational index in deep learning. It is utilized to train all well-known networks, for example, AlexNet, GoogLeNet, VGGNet, and ResNet because of its enormous scope marked picture assortments. A more limited size picture informational collection that is used in many explorations examines is CIFAR10/100. This informational index is additionally utilized for assessing numerous DNNs in the picture arrangement task. As referenced before, PASCAL VOC and Microsoft COCO are utilized for different item identification and semantic division errands. At long last, YouTube-8M is a moderately new informational collection produced by Google to assume a similar part as PictureNet for video preparing. It very well may be used as a benchmark informational collection for different video examinations, including occasion identification, comprehension, and characterization [83–88].

6 Key Challenges in Deep Learning

With the intense advancement in deep learning and its exploration settings being in the spotlight, deep learning has gained uncommon force in discourse, language, and visual discovery frameworks. In any case, a few domains are essentially still immaculate by DNNs because of either their difficult nature or their absence of information availability for the overall population. This makes huge chances and fruitful ground for compensating future exploration roads. In this part, these domains, key experiences into their difficulties, and likely future headings of significant deep learning strategies are talked about. There is a waiting discovery impression of DNNs, implying that deep learning models can be evaluated dependent on their last yields without the comprehension of how they get to these choices [89–92]. This power-less factual decipher capacity has likewise been distinguished in, particularly in applications where the information is created not by an actual appearance. Mama et al. explain neural networks utilizing cell science from the sub-atomic scale up. They planned the layers of a neural organization to the segments of a yeast cell, beginning with the minuscule nucleotides that make up its DNA, moving upward to bigger constructions, for example, ribosomes (which take guidelines from the DNA and make proteins), lastly moving to organelles like the mitochondrion and core (which runs the cell activities). Since it is a noticeable neural organization, they could undoubtedly notice the progressions in cell instruments when the DNA was changed. Another milestone challenge looked by deep learning strategies is the decrease of dimensionality without losing basic data required for order. In clinical applications like malignancy RNA sequencing examination, usually the quantity of tests in each name is definitely not exactly the quantity of highlights. In current deep learning models, this causes serious over fitting issues and represses appropriate order of untrained cases. Barely, any strategies attempt to observationally conclude vari-able consistency and lessen the list of capabilities in a supervised manner; however, this regularly brings about the deficiency of goal and details [93–96]. Comparable difficulties are confronted while dissecting clinical pictures on the grounds that the training information is enormously expensive and tedious to obtain. A couple central papers have endeavored to construct the models that require an insignificant number of tests during learning, where stands apart as a pioneer distribution in applying CNNs to bosom and prostate disease location. One of the developing pains of deep learning identifies with the issue of computational productivity, i.e., accomplishing the most extreme throughput while devouring minimal measure of assets. Current deep learning structures require extensive measures of computational assets to move toward the best in class exhibitions. One technique endeavors to defeat this test by utilizing repository processing. Another option is to utilize the steady methodologies that abuse medium and huge datasets on disconnected training. In ebb and flow years, numerous scientists have moved concentration to fabricate equal and adaptable deep learning systems. Recently, the center has been moved to move the learning cycle on GPUs. Be that as it may, GPUs are famous for their spillage flows and this modified works any conceivable acknowledgment of the deep learning models on compact gadgets. One arrangement is to utilize field-programmable door exhibits (FPGAs) as deep learning quickening agents to improve the information access pipelines to accomplish altogether better outcomes [97–100].

7   Future Scope

This article examines the difficulties and gives a few existing answers for these difficulties. In any case, there are as yet a few issues that should be tended to later on for deep learning. A few discoveries of this article and conceivable future work are summed up beneath:

- Albeit deep learning can retain an enormous measure of information and data, and its feeble thinking and comprehension of the information makes it a discovery answer for some applications. The interpretability of deep learning ought to be investigated later on.
- Deep learning actually experiences issues in displaying different complex information modalities simultaneously. Multimodal deep learning is another well-known bearing in ongoing deep learning research.
- In contrast to human brains, deep learning needs broad datasets (ideally marked information) for training the machine and foreseeing the inconspicuous information. This issue turns out to be all the more overwhelming when the available datasets are little (e.g., medical care information) or when the information should be prepared progressively. One-shot learning and zero-shot learning have been concentrated in the new couple of years to reduce this issue.
- Most of the current deep learning executions are supervised algorithms, while machine learning is step-by-step moving to unsupervised and semi-supervised learning to handle genuine information without manual human marks.
- Disregarding all the deep learning progressions lately, numerous applications are as yet immaculate by deep learning or are in the beginning phases of utilizing the deep learning methods (e.g., calamity data the board, money, or clinical information examination). With everything taken into account, deep learning, another and quickly developing strategy, gives various difficulties just as promising circumstances and arrangements in an assortment of utilizations. All the more significantly, it moves machine learning to its new stage, in particular, the "more brilliant AI."

8 Conclusion

Deep learning, another and interesting issue in machine learning, can be characterized as a course of layers performing nonlinear preparing to gain proficiency with various degrees of information portrayals. For quite a long time, machine learning specialists have attempted to find the examples and information portrayals from the crude information. This strategy is called portrayal learning. Dissimilar to ordinary machine learning and information mining procedures, deep learning can produce undeniable-level information portrayals from huge volumes of crude information. Along these lines, it has given an answer for some genuine applications. The overview of this article reflects cutting edge algorithms and procedures in deep learning. It begins with a background marked by artificial neural networks since 1940 and moves to late deep learning algorithms and significant advancements in various applications. At that point, the key algorithms and systems around there, just as well-known strategies in deep learning, are introduced. It first momentarily presents the customary neural networks and a few supervised deep learning algorithms, including recurrent, recursive, and convolutional neural networks, just as the deep conviction networks and Boltzmann machines. From that point, further developed deep learning approaches, for example, unsupervised and web-based learning are examined. In addition, a few advancement methods have additionally been given. Well-known structures here incorporate TensorFlow, Caffe, and Theano. Likewise, to handle enormous information challenges, the circulated procedures in deep learning are momentarily examined. From that point, this article audits the best deep learning techniques in different applications, including NLP, visual information preparing, discourse and sound handling, and interpersonal organization investigation.

References

1. Oord Avd, Kalchbrenner N, Kavukcuoglu K (2016) Pixel recurrent neural networks. *arXiv* 2016, arXiv:1601.06759
2. Xue W, Nachum IB, Pandey S, Warrington J, Leung S, Li S (2017) Direct estimation of regional wall thicknesses via residual recurrent neural network. In: International conference on information processing in medical imaging. Springer: Cham, Switzerland, pp 505–516
3. Tjandra A, Sakti S, Manurung R, Adriani M, Nakamura S (2016) Gated recurrent neural tensor network. In: Proceedings of the 2016 international joint conference on neural networks (IJCNN), Vancouver, BC, Canada, 24–29 July 2016, pp 448–455
4. Wang S, Jing J (2015) Learning natural language inference with LSTM. *arXiv* 2015, arXiv: 1512.08849
5. Sutskever I, Vinyals O, Le QV (2014) Sequence to sequence learning with neural networks. In: Advances in neural information processing systems (NIPS). MIT Press, Cambridge, MA, USA, pp 3104–3112
6. Akhani VA, Mahadev R (2016) Multi-language identification using convolutional recurrent neural network. *arXiv* 2016, arXiv:1611.04010
7. Längkvist M, Karlsson L, Loutfi A (2014) A review of unsupervised feature learning and deep learning for time-series modeling. Pattern Recognit. Lett. 42:11–24

8. Zhu BB, Swanson MD, Tewfik AH (2004) When seeing isn't believing [multimedia authentication technologies]. IEEE Signal Process Mag 21:40–49
9. Farid H (2006) Digital doctoring: how to tell the real from the fake. Significance 3:162–166
10. Cao YJ, Jia LL, Chen YX, Lin N, Yang C, Zhang B, Liu Z, Li XX, Dai HH (2019) Recent advances of generative adversarial networks in computer vision. IEEE Access 7:14985–15006
11. Yin W, Schutze H, Xiang B, Zhou B (2015) ABCNN: attention-based convolutional neural network for modeling sentence pairs. *CoRR* abs/1512.05193 (2015). arxiv:1512.05193. Retrieved from http://arxiv.org/abs/1512.05193
12. Yu D, Eversole A, Seltzer ML, Yao K, Guenter B, Kuchaiev O, Seide F, Wang H, Droppo J, Huang Z, Zweig G, Rossbach CJ, Currey J (2014) An introduction to computational networks and the computational network toolkit. In: The 15th annual conference of the International Speech Communication Association. ISCA.
13. Yu D, Kolbak M, Tan Z-H, Jensen J (2017) Permutation invariant training of deep models for speaker-independent multi-talker speech separation. In: IEEE international conference on acoustics, speech and signal processing. IEEE, pp 241–245
14. Yu Q, Wang C, Ma X, Li X, Zhou X (2015) A deep learning prediction process accelerator based FPGA. In: 15th IEEE/ACM international symposium on cluster, cloud and grid computing. IEEE, pp 1159–1162
15. Zeiler MD (2012) ADADELTA: an adaptive learning rate method. *CoRR* abs/1212.5701. Retrieved from http://arxiv.org/abs/1212.5701
16. Zhang C, Li P, Sun G, Guan, Bingjun Xiao, and Jason Cong. 2015. Optimizing FPGA-based accelerator design for deep convolutional neural networks. In *ACM/SIGDA International Symposium on Field- Programmable Gate Arrays*. ACM, 161–170.
17. Zhang X, Wang DeLiang (2017) Deep learning based binaural speech separation in reverberant environments. IEEE/ACM Transactions on Audio, Speech, and Language Processing 25(2017):1075–1084
18. Beridze I (2019) Butcher, J When seeing is no longer believing. Nat Mach Intell 1:332–334
19. Yang P, Ni R, Zhao Y (2016) Recapture image forensics based on Laplacian convolutional neural networks. In international workshop on digital watermarking. Springer, Berlin, Germany, pp 119–128
20. Choi HY, Jang HU, Son J, Kim D, Lee HK (2017) Content recapture detection based on convolutional neural networks. In: International conference on information science and applications. Springer, Berlin, Germany, pp 339–346
21. Piva A (2013) An overview on image forensics. ISRN Signal Process
22. Guan H, Kozak M, Robertson E, Lee Y, Yates AN, Delgado A, Zhou D, Kheyrkhah T, Smith J, Fiscus J (2019) MFC datasets: large-scale benchmark datasets for media forensic challenge evaluation. In: Proceedings of the 2019 IEEE winter applications of computer vision workshops (WACVW), Waikoloa Village, HI, USA, 7–11 January 2019, pp 63–72
23. IEEE (2018) Signal processing society—camera model identification. IEEE, Piscataway, NJ, USA
24. Kovalev V, Kalinovsky A, Kovalev S (2016) Deep learning with Theano, Torch, Caffe, Tensorflow, and Deeplearning4J: Which one is the best in speed and accuracy? In: The 13th international conference on pattern recognition and information processing
25. Krause J, Sapp B, Howard A, Zhou H, Toshev A, Duerig T, Philbin J, Fei-Fei L (2016) The unreasonable effectiveness of noisy data for fine-grained recognition. In: European conference on computer vision. Springer, pp 301–320
26. Krizhevsky A, Sutskever I, Hinton GE (2012) ImageNet classification with deep convolutional neural networks. In: Pereira F, Burges CJC, Bottou L, Weinberger KQ (eds) Advances in neural information processing systems 25. Curran Associates, pp 1097–1105
27. Lang M, Kotthaus H, Marwedel P, Weihs C, Rahnenfuhrer J, Bischl B (2015) Automatic model selection for high-dimensional survival analysis. J Stat Comput Simul 85(1):62–76
28. Le Quoc V (2013) Building high-level features using large scale unsupervised learning. In: IEEE international conference on acoustics, speech and signal processing. IEEE, pp 8595–8598

29. LeCun Y, Bengio Y (1995) Convolutional networks for images, speech, and time series. Handbook of Brain Theory and Neural Networks 3361(10):255–257
30. LeCun Y, Bengio Y, Hinton G (2015) Deep learning. Nature 521(7553):436–444
31. Hinton GE, Osindero S, Teh YW (2006) A fast learning algorithm for deep belief nets. Neural Comput 18:1527–1554
32. Amerini I, Uricchio T, Caldelli R (2017) Tracing images back to their social network of origin: A CNN-based approach. In: Proceedings of the 2017 IEEE workshop on information forensics and security (WIFS), Rennes, France, 4–7 December 2017, pp 1–6
33. Caldelli R, Amerini I, Li CT (2018) PRNU-based image classification of origin social network with CNN. In: Proceedings of the 2018 26th European signal processing conference (EUSIPCO), Rome, Italy, 3–7 Sept 2018, pp 1357–1361
34. Hinton GE, Srivastava N, Krizhevsky A, Sutskever I, Salakhutdinov RR (2012) Improving neural networks by preventing co-adaptation of feature detectors. *arXiv* 2012, arXiv:1207. 0580
35. Srivastava N, Hinton G, Krizhevsky A, Sutskever I, Salakhutdinov R (2014) Dropout: a simple way to prevent neural networks from overfitting. J Mach Learn Res 15:1929–1958
36. Ronneberger O, Fischer P, Brox T (2015) U-net: convolutional networks for biomedical image segmentation. In: International conference on medical image computing and computer-assisted intervention. Springer, Cham, Switzerland, pp 234–241
37. Khan A, Sohail A, Zahoora U, Qureshi AS (2019) A survey of the recent architectures of deep convolutional neural networks. arXiv 2019, arXiv:1901.06032
38. Krizhevsky A, Sutskever I, Hinton GE (2012) Advances in neural information processing systems. MIT Press, Cambridge, MA, USA, pp 1097–1105
39. Simonyan K, Zisserman A (2014) Very deep convolutional networks for large-scale image recognition. arXiv 2014, arXiv:1409.1556
40. Wan L, Zeiler M, Zhang S, le Cun Y, Fergus R (2013) Regularization of neural networks using dropconnect. In: Proceedings of the international conference on machine learning, Atlanta, GA, USA, 16–21 June 2013, pp 1058–1066
41. Bulò SR, Porzi L, Kontschieder P (2016) Dropout distillation. In: Proceedings of the international conference on machine learning, New York, NY, USA, 20–22 June 2016, pp. 99–107
42. Ruder S (2016) An overview of gradient descent optimization algorithms. *arXiv* 2016, arXiv: 1609.04747
43. Chen C, McCloskey S, Yu J (2018) Focus manipulation detection via photometric histogram analysis. In: Proceedings of the IEEE conference on computer vision and pattern recognition, Salt Lake City, UT, USA, 18–23 June 2018, pp 1674–1682
44. Marra F, Gragnaniello D, Verdoliva L, Poggi G (2018) Do GANs leave artificial fingerprints? arXiv 2018, arXiv:1812.11842
45. Chen L-C, Papandreou G, Kokkinos I, Murphy K, Yuille AL (2014) Semantic image segmentation with deep convolutional nets and fully connected CRFs. *arXiv* 2014, arXiv:1412. 7062.
46. Feng M, Xiang B, Glass MR, Wang L, Zhou B (2015) Applying deep learning to answer selection: a study and an open task. In: IEEE workshop on automatic speech recognition and understanding. IEEE, pp 813–820
47. Alzubi JA, Kumar A, Alzubi OA, Manikandan R (2019) Efficient approaches for prediction of brain tumor using machine learning techniques. Indian J Public Health Res Dev. https://doi.org/10.5958/0976-5506.2019.00298.5
48. Alzubi OA, Alzubi JA, Tedmori S, Rashaideh H, Almomani O (2018) Consensus-based combining method for classifier ensembles. Int Arab J Inf Technol (2018)
49. Alweshah, Alzubi OA, Alzubi JA, Mohammed SA (2016) Solving attribute reduction problem using wrapper genetic programming. Int J Comput Sci Netw Secur
50. Badrinarayanan V, Kendall A, Cipolla R (2015) Segnet: a deep convolutional encoder-decoder architecture for image segmentation. *arXiv* 2015, arXiv:1511.00561.

51. Lin G, Milan A, Shen C, Reid I (2017) Refinenet: Multi-path refinement networks for high-resolution semantic segmentation. In: Proceedings of the 2017 IEEE conference on computer vision and pattern recognition (CVPR), Honolulu, HI, USA, 21–26 July 2017, pp 5168–5177

52. Zhao H, Shi J, Qi X, Wang X, Jia J (2017) Pyramid scene parsing network. In Proceedings of the IEEE conference on computer vision and pattern recognition (CVPR), Honolulu, HI, USA, 21–26 July 2017, pp 2881–2890

53. Chen L-C, Papandreou G, Kokkinos I, Murphy K, Yuille AL (2018) Deeplab: Semantic image segmentation with deep convolutional nets, atrous convolution, and fully connected CRFs. IEEE Trans Pattern Anal Mach Intell 40:834–848

54. Alzubi OA, Alzubi JA, Alweshah M, Qiqieh I, Al-Shami S, Ramachandran M (2020) An optimal pruning algorithm of classifier ensembles: dynamic programming approach. Neural Comput. Appl. https://doi.org/10.1007/s00521-020-04761-6

55. Gupta D, Rodrigues JJPC, Sundaram S, Khanna A, Korotaev V, Albuquerque VHC (2018) Usability feature extraction using modified crow search algorithm: a novel approach. Neural Comput. Appl. (Springer). https://doi.org/10.1007/s00521-018-3688-6

56. Le QV, Ngiam J, Coates A, Lahiri A, Prochnow B, Ng AY (2011) On optimization methods for deep learning. In Proceedings of the 28th international conference on international conference on machine learning, Bellevue, WA, USA, 28 June–2 July 2011, pp 265–272

57. Koushik J, Hayashi H (2016) Improving stochastic gradient descent with feedback. *arXiv* 2016, arXiv:1611.01505

58. LeCun Y, Bengio Y, Hinton G (2015) Deep learning. Nature 521:436

59. Goodfellow I, Bengio Y, Courville A (2016) Deep learning. MIT Press, Cambridge, MA, USA

60. Socher R, Perelygin A, Wu JY, Chuang J, Manning CD, Ng AY, Potts C (2013) Recursive deep models for semantic compositionality over a sentiment treebank. In: Conference on empirical methods in natural language processing. Citeseer, Association for Computational Linguistics, pp 1631–1642

61. Kim Y (2014) Convolutional neural networks for sentence classification. *CoRR* abs/1408.5882 (2014). Retrieved from http://arxiv.org/abs/1408.5882

62. Wehrmann J, Becker W, Cagnini HEL, Barros RC (2017) A character-based convolutional neural network for language-agnostic Twitter sentiment analysis. In: International joint conference on neural networks. IEEE, pp 2384–2391

63. Bahdanau D, Cho K, Bengio Y (2014) Neural machine translation by jointly learning to align and translate. *CoRR* abs/1409.0473 (2014). Retrieved from http://arxiv.org/abs/1409.0473

64. Cho K, van Merrienboer B, Gulcehre C, Bahdanau D, Bougares F, Schwenk H, Bengio Y (2014) Learning phrase representations using RNN encoder-decoder for statistical machine translation. In: The conference on empirical methods in natural language processing, pp 1724–1734

65. Wu Y, Schuster M, Chen Z, Le QV, Norouzi M, Macherey W, Krikun M, Cao Y, Gao Q, Macherey K, Klingner J, Shah A, Johnson M, Liu X, Kaiser L, Gouws S, Kato Y, Kudo T, Kazawa H, Stevens K, Kurian G, Patil N, Wang W, Young C, Smith J, Riesa J, Rudnick A, Vinyals O, Corrado G, Hughes M, Dean J (2016) Google's neural machine translation system: bridging the gap between human and machine translation. *CoRR* abs/1609.08144(2016). arxiv:1609.08144. Retrieved from http://arxiv.org/abs/1609.08144

66. Socher R, Huang EH, Pennington J, Ng AY, Manning CD (2011) Dynamic pooling and unfolding recursive auto encoders for paraphrase detection. In: Advances in neural information processing systems, vol 24. Neural Information Processing Systems Foundation, pp 801–809

67. Yin W, Schutze H, Xiang B, Zhou B (2015) ABCNN: attention-based convolutional neural network for modeling sentence pairs. *CoRR* abs/1512.05193(2015). arxiv:1512.05193. Retrieved from http://arxiv.org/abs/1512.05193

68. Kageback M, Mogren O, Tahmasebi N, Dubhashi D (2014) Extractive summarization using continuous vector space models. In: 2nd workshop on continuous vector space models and their compositionality. Citeseer, Association for Computational Linguistics, pp 31–39

69. Dong L, Wei F, Zhou M, Xu K (2015) Question answering over freebase with multicolumn convolutional neural networks. In: 53rd annual meeting of the association for computational linguistics, vol 1. Association for Computational Linguistics, pp 260–269
70. Image Net (2017) Retrieved from http://image-Net.org. Accessed 18 April 2017
71. CIFAR (2009) CIFAR-10 and CIFAR-100 datasets. Retrieved from https://www.cs.toronto.edu/~kriz/cifar.html. Accessed 18 April 2017
72. Pascal VOC (2012) The PASCAL visual object classes. Retrieved from http://host.robots.ox.ac.uk/pascal/VOC/. Accessed 18 April 2017
73. Lin T-Y, Maire M, Belongie S, Hays J, Perona P, Ramanan D, Dollar P, Lawrence Zitnick C (2014) Microsoft COCO: common objects in context. In: European conference on computer vision. Springer, pp 740–755
74. MNIST (2017) The MNIST database of handwritten digits. Retrieved from http://yann.lecun.com/exdb/mnist/. Accessed 18 April 2017
75. Thomee B, Shamma DA, Friedland G, Elizalde B, Ni K, Poland D, Borth D, Li L-J (2016) YFCC100M: The new data in multimedia research. Commun ACM 59(2):64–73
76. Abu-El-Haija S, Kothari N, Lee J, Natsev P, Toderici G, Varadarajan B, Vijaya-narasimhan S (2016) YouTube-8M: a large scale video classification benchmark. CoRRabs/1609.08675(2016). Retrieved from http://arxiv.org/abs/1609.08675
77. Trecvid (2017) TREC video retrieval evaluation. Retrieved from http://trecvid.nist.gov. Accessed 18 April 2017
78. Soomro K, Zamir AR, Shah M (2012) UCF101: a dataset of 101 human actions classes from videos in the wild. CoRR abs/1212.0402 (2012). Retrieved from http://arxiv.org/abs/1212.0402
79. Kay W, Carreira J, Simonyan K, Zhang B, Hillier C, Vijayanarasimhan S, Viola F, Green T, Back T, Natsev P, Suleyman M, Zisserman A (2017) The kinetics human action video dataset. CoRR abs/1705.06950 (2017). Retrieved from http://arxiv.org/abs/1705.06950
80. Sathasivam S, Abdullah WA (2008) Logic learning in Hopfield networks. arXiv 2008, arXiv:0804.4075
81. De Rezende ER, Ruppert GC, Theóphilo A, Tokuda EK, Carvalho T (2018) Exposing computer generated images by using deep convolutional neural networks. Signal Process Image Commun 66:113–126
82. Ranzato M, Poultney C, Chopra S, LeCun Y (2007) Efficient learning of sparse representations with an energy-based model. In: Advances in neural information processing systems. MIT Press: Cambridge, MA, USA, pp 1137–1144
83. Elman JL (1990) Finding structure in time. Cogn Sci 14:179–211
84. Jordan MI (1997) Serial order: a parallel distributed processing approach. Adv Psychol 121:471–495
85. Hochreiter S, Bengio Y, Frasconi P, Schmidhuber J (2001) Gradient flow in recurrent nets: the difficulty of learning long-term dependencies. IEEE Press, New York, NY, USA
86. Schmidhuber J (1993) Habilitation thesis: Netzwerkarchitekturen, Zielfunktionen und Ketten-regel (Network architectures, objective functions, and chain rule), PhD, Technische Universität München, 15 April 1993
87. Bengio Y, Lamblin P, Popovici D, Larochelle H (2007) Greedy layer-wise training of deep networks. In: Advances in neural information processing systems. MIT Press, Cambridge, MA, USA, pp. 153–160
88. Nguyen HH, Tieu T, Nguyen-Son HQ, Nozick V, Yamagishi J, Echizen I (2018) Modular convolutional neural network for discriminating between computer-generated images and photographic images. In: Proceedings of the 13th international conference on availability, reliability and security, Hamburg, Germany, 27–30 Aug 2018, p 1
89. Lu J, Young S, Arel I, Holleman J (2015) A1 TOPS/W analog deep machine-learning engine with floating-gate storage in 0.13 μm CMOS. IEEE J Solid-State Circuits 50(1):270–281
90. Ma J, Ku Yu M, Fong S, Ono K, Sage E, Demchak B, Sharan R, Ideker T (2018) Using deep learning to model the hierarchical structure and function of a cell. Nature Methods 15(4):290–298

650 P. K. Vishwakarma and N. Jain

91. Ma X, Yu H, Wang Y, Wang Y (2015) Large-scale transportation network congestion evolution prediction using deep learning theory. PLoS ONE 10(3):e0119044

92. Manning C (2016) Understanding human language: can NLP and deep learning help? In: The 39th international ACM SIGIR conference on research and development in information retrieval. ACM, pp 1–1

93. McCulloch WS, Pitts W (1943) A logical calculus of the ideas immanent in nervous activity. Bulletin of Mathematical Biophysics 5(4):115–133

94. Schöttle P, Schlögl A, Pasquini C, Böhme R (2018) Detecting adversarial examples—a lesson from multimedia security. In Proceedings of the 26th European signal processing conference (EUSIPCO), Rome, Italy, 3–7 Sept 2018, pp 947–951

95. Carlini N, Wagner D (2017) Adversarial examples are not easily detected: bypassing ten detection methods. In: Proceedings of the 10th ACM workshop on artificial intelligence and security, ACM, Dallas, TX, USA, 30 October–3 November 2017, pp 3–14

96. Du C, Zhu J, Zhang B (2018) Learning deep generative models with doubly stochastic gradient MCMC. IEEE Trans Neural Networks Learn. Syst. 29:3084–3096

97. Hoang Q, Nguyen TD, Le T, Phung D (2017) Multi-generator gernerative adversarial nets. *arXiv* 2017, arXiv:1708.02556

98. Bousmalis K, Silberman N, Dohan D, Erhan D, Krishnan D (2017) Unsupervised pixel-level domain adaptation with generative adversarial networks. In: Proceedings of the IEEE conference on computer vision and pattern recognition (CVPR), Honolulu, HI, USA, 21–26 July 2017, vol 1, p 7

99. Kansky K, Silver T, Mély DA, Eldawy M, Lázaro-Gredilla M, Lou X, Dorfman N, Sidor S, Phoenix S, George D (2017) Schema networks: zero-shot transfer with a generative causal model of intuitive physics. *arXiv* 2017, arXiv:1706.04317

100. Ledig C, Theis L, Huszár F, Caballero J, Cunningham A, Acosta A, Aitken A, Tejani A, Totz J, Wang Z et al (2016) Photo-realistic single image super-resolution using a generative adversarial network. *arXiv* 2016, arXiv:1609.04802

An Estimation of PCA Feature Extraction in EEG-based Emotion Prediction with Support Vector Machines

M. Malathi⊙, **G. Aloy Anuja Mary, J. Senthil Kumar, P. Sinthia, and M. Nalini**

Abstract Emotion recognition from the electroencephalogram (EEG) is growing recent field, which has challenges to acquire the emotions from the extracted features using various classification strategies. For these procedures, it is hard to get the high performance. The implemented technique helps to achieve emotions like happy, sad and angry. Further, principal component analysis (PCA) is other technique which was used to get features from the EEG signal. The proposed research work uses EEG data which have been collected for 15 healthy subjects using three EEG channels, namely Fp1, Fp2 and bipolar channel of F3 and F4 positions of system. Further, the extracted feature undergone for two different classifiers to identify the nature of emotion. The removed features applied as an input to train a network, which classify the type of emotions of the patients. Here, we use support vector machine (SVM) and multiclass support vector machine (MSVM) to categorize the type of emotion. Finally, the performance of the both classifier can be evaluated based on the parameters like accuracy, mean square error (MSE), time and peak signal to noise ratio (PSNR). Overall performance for this proposed system provides good accuracy over the SVM. The SVM and multiclass SVM give us the accuracy of about 89.51% and 96.71% for the EEG signal.

M. Malathi (✉)
Rajalakshmi Institute of Technology, Chennai, India

G. Aloy Anuja Mary
Vel Tech Rangarajan Dr Sagunthala R&D Institute of Science and Technology, Chennai, India
e-mail: draloyanujamary@veltech.edu.in

J. Senthil Kumar
Mepco Schlenk Engineering College, Sivakasi, India
e-mail: senthilkumarj@mepcoeng.ac.in

P. Sinthia
Saveetha Engineering College, Chennai, India

M. Nalini
Saveetha School of Engineering, Saveetha Institute of Medical and Technical Sciences, Chennai, India
e-mail: nalinim.sse@saveetha.com

© The Author(s), under exclusive license to Springer Nature Singapore Pte Ltd. 2022   651
D. Gupta et al. (eds.), *Proceedings of Data Analytics and Management*,
Lecture Notes on Data Engineering and Communications Technologies 90,
https://doi.org/10.1007/978-981-16-6289-8_53

Keywords EEG signals · Emotion recognition · Support vector machine (SVM) ·
Multiclass SVM (MSVM) · Principal component analysis (PCA)

1 Introduction

In this highly technically developed era, most of the machines are trained specifically
to do many works which reduce the requirement of human resources as such. While
these have been efficiently deployed in many of the situations, there is a huge need to
make it trained to recognize and identify the expressions as well as the gestures of the
human beings [1]. As said, only ten percent of a human's life is on birth and death,
and the rest ninety percent rely on the reactions and the expressions conveyed on this.
Therefore, it has become highly necessary and at the same time important to design
a machine that can clearly identify and recognize the type of facial expressions that
are conveyed and reply based on the stored expression. There are many expressions
like happy, sad. So far this, there are many techniques developed. In this, there are
many difficulties for identification of the facial expressions.

Initially, in most of the cases, the image obtained is blurry, and this makes it
difficult for the perfect detection of the expression of the face [2]. Secondly, we
can see that there is no availability of great differences between any two obtained
images of any facial expressions. This makes it much more difficult to identify the
kind of facial expression it is, and this stands to be a great drawback. The last and
the most important difficulty is that unlike the humans, the system does not possess
a capability of understanding a facial expression into emotions of them. These pose
a great difficulty in the identification of the emotions.

To overcome these and provide a successful module, many solutions have been
posed out of which many have also proved to be a good solution too. One of the
best solutions is by obtaining the image using the electroencephalogram which has
proved to be one of the best prevailing solutions in the field of this identification of the
facial expressions is the detection of emotions. Many other classification techniques
like the k-nearest neighbor (KNN) [3].

Many other neural networks are also trained for this purpose, and also many
machine learning based solutions are also provided. From these, the best solution is
identified by researching all the available research which has already been carried
out. With these being supposed to be a ground work, a more deep work is carried out
on the basis of these works. These are carried out in order to increase the efficiency of
the human–machine interaction which has already been in hike since the start of the
twenty-first century. By this, over the period, it has made it completely effective. This
has been very effective in cases of health care functionalities where the identification
of the gestures of the face has been seen as one of the most reliable sources for the
further continuation. Even after this, there are drawbacks like the user must always
focus on the camera for better reception and identification, and even for the audio,
the surrounding noise must be nullified. These disadvantages are taken aback by the
aforementioned technique or technology [4].

These drawbacks can be easily overtaken by the operation of the EEG technique in which it uses HOC-EC where EC stands for emotion controller. The output of this emotion controller is directly sourced to the EEG in which the datasets available in the EEG which relate to the healthy patients are compared with the recently obtained data.

By this, the difference between these two sets is identified, and the correct emotion can be detected and recognized and at the same time can also be applied for the further identifications. When this method is compared with the other existing techniques, this has proved to be a better method or technique and also at the same time has shown a better percentage of perfection [5].

With these as its advantages, a thorough research is made on this technique which is designed for the recognition of the face expression or the face gestures which leads to the emotions like being sad, happy, anger.

Some of the technical contributions of the proposed research work are listed as follows:

- Datasets are preprocessed and collected into a new big dataset.
- Work flow and methodology are clearly illustrated
- Based on standard literature survey, identified a most excellent feature extraction method
- Accuracy of classification algorithm was compared for emotion recognition
- Investigation and visualizations of dataset "emotion analysis using EEG, physiological and video signals (DEAP) dataset".

2   Related Work

The following survey deals the different kinds of feature extraction techniques from EEG signal.

A paper [6] focuses on the recognition of the emotions which are observed in the faces of the humans or by their facial gestures. For this, they have used the logic called the fuzzy logic. Using this, different kinds of expressions are differentiated separately based on the frames obtained. By this, largely different expressions are identified and stored, and it is by that extracted and given for the process of evaluation. There is also a technique provided for holding control over these emotions. Therefore, this technique proves to be an easy and efficient method for this process.

In another reference [7], it makes use of the crossings which are relatively of a higher order. It is based on the electroencephalogram which is used for the recognition of emotions. This has been seen to be a brand new field. This at the same time also gives faster results than the other methods. In this, it is seen that it uses the concept of the k-nearest neighbor. By using this, it is seen that the obtained expression is processed through the dataset which is already present by using the abovementioned principle. By this, it identifies the emotions. This method is combined with the classification method called the higher order crossing emotion classifier which is abbreviated as HOC-EC.

In a similar reference [8], it is also mentioned the use of the electroencephalogram or the EEG which has been employed for the purpose of the emotion detection in a continuous manner. In this, the features which are registered with EEG and the expressions or the gestures of the face which are obtained by the continuous valence. At last, the result of such correlation is obtained as the result. With this, we can get or obtain a continuous detection of various emotions.

In this [9], the process of deep learning is made in EEG signals for the process of the recognitions of the emotions. This makes it easier for the computers to understand the expressions given by the humans. This is made possible by using an interface called the brain computer interface. In this, the emotion models are trained based on subject-wise. For this, a three-layered way is implemented on the restricted Boltzmann machines. By this, back propagation method can be finally applied for the tuning by subject-wise. By this, a relatively higher accuracy can be assured.

Similarly in [10], a survey on the emotion recognition by using the EEG signal is provided. In this, they have brought to the notice that the method of the EEG is found to be comparatively cheaper as well as simple for the identification of the emotions. At the same time, it also has given a study of the research that is performed using PCA feature extraction and T-statistical approach to recognize emotion from EEG signals. The recommended technique uses four classifiers—support vector machine (SVM), artificial neural network (ANN), linear discriminates analysis (LDA) and k-nearest neighbor (KNN) scheme were used to categorize the emotional states.

In [11], they have introduced an electrode device which has been used for the recognition of the emotions that are obtained through EEG signals. They have also mentioned its uses in the medical as well as the non-medical fields. In this, they have proposed a primary method in which they use a single-channel along with the brain computer interface for the identification and the detection of the emotions. By this, using a machine learning methods, it obtained a success of about a percentage of eighty-seven percent.

In [12], author proposed a research which is performed on the estimation of emotion using the signals of the EEG along with the brain computer interfaces. In this, they have mentioned that weak electric signals are produced by the brain which can be in turn measured at the skull. It is said over here that the results are produced and processed by the brain computer interface. In this, they have also mentioned the recognition of voice which is achieved through the recognition with the mimics. By this, they tend to identify the person's feelings. Over here, the latency or the disadvantages of this method are also made as a part of this paper.

Again in [13], it provides the method for the identification of facial emotion. This is taken a step forward because of the increasing demand in the market for its applications. This also makes use of the support vector machine and the deep Boltzmann machine. This uses the data for testing and training that are provided in the FERA 2015. By this, it focuses on constructing a perfect module for the identification of the facial emotions.

In [14], it proposes a method for the recognition or the identification of the emotions in the cases of a social interaction which is simulated. This is a challenge that has to be addressed because of its high requirement. It compares the avatars in the trained set and the one which is obtained during any conversation and then it is manipulated. By this, they have developed a module which is trained to speak with all the facial gestures during its conversation. Finally, it was seen that same was not a part of its facial recognition which was later sustained with suitable reasons.

3 Materials and Methods

3.1 Existing Method

The existing method uses STFT and SVM to extract emotions from EEG signals. Since STFT has limited frequency resolution, it also depends on quantity of samples used, for processing nonlinear and transient signals. Different time frequency technologies are utilized to separate features from EEG signals. The STFT analyzes the non-stationary signals. Many frames can be removed from the original EEG signal. Further, it can be processed to get time-shifted signal. By taking the discrete Fourier transform of the frames, STFT estimates time-varying spectrum.

Next, magnitude squared value of the spectrogram was obtained through STFT. Consider $x\,(n)$ is non-stationary EEG signal when applying STFT to the EEG signal given by the equation

$$X(\mu, w) = \sum_{n=-\infty}^{\infty} x(n)\varphi(n - \mu)e^{-jwn} \tag{1}$$

A spectrogram is defined by equation

$$Sx(\mu, w) = |X(\mu, w)|^2 \tag{2}$$

Hence, the suggested method uses PCA along with MSVM classifier which is mainly based on linear transformation to reduce the dimensionality of data and also helps to extract the useful information from the large dataset to analyze the variable structures. The resolution of EEG signal is poor while using STFT, because the entire channels are considered for feature extraction which leads complexity. Hence, the proposed research work uses PCA in order to detect the significant variations from the input higher dimensions.

3.2 Proposed Method

Figure 1 is the process flow diagram of emotion recognition from EEG signal.

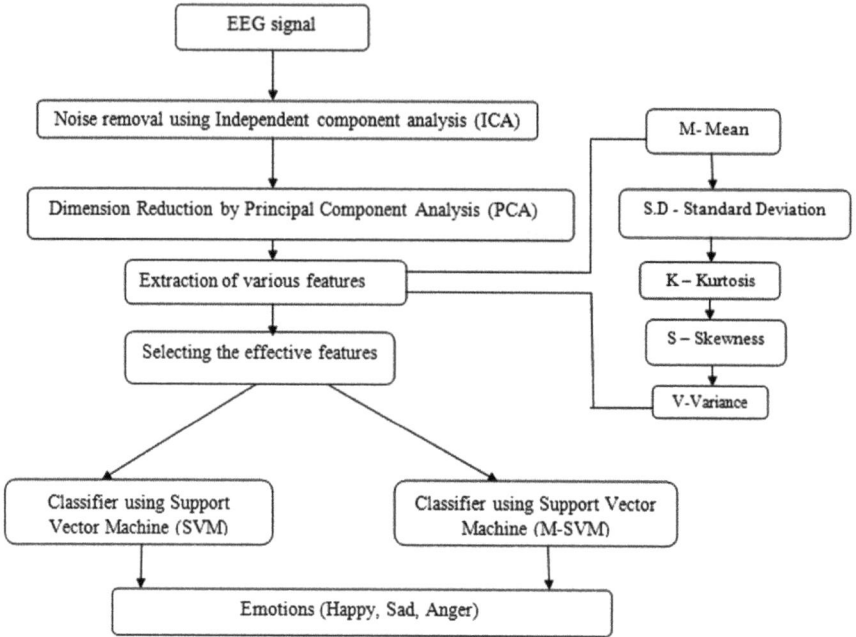

Fig. 1 Flow diagram of proposed model

The suggested method of emotion recognition operation accomplishes by performing the two main processes like feature extraction using PCA and classification by MSVM.

Various stages for emotion recognitions

- Acquire EEG signal from database.
- ICA helps to remove unwanted artifacts present in the recorded EEG signal.
- Convert an image into grayscale
- Feature extraction using PCA
- The separated features are compared with training dataset using SVM and MSVM.
- Types of emotions are recognized like happy, anger, sad.

3.2.1 Data Base

The suggested method utilizes a benchmark multimodal dataset called as DEAP dataset. The dataset was prepared by recording the EEG signals of 32 members while watching various music videos. The data were acquired with the help of 32 channels bio-semi-acquisition system. Before using this signal into MATLAB, it was downsampled to the frequency of 128 Hz.

Fig. 2 Preprocessed EEG
signal

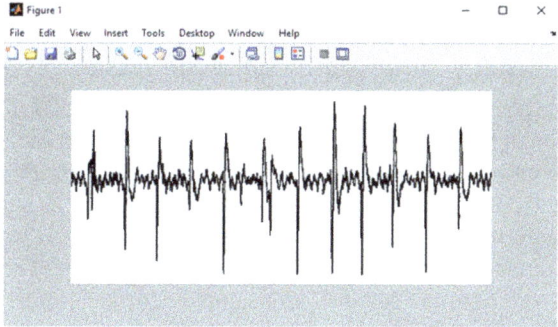

Fig. 2 Preprocessed EEG signal

3.3 Preprocessing

Independent component analysis is one of the technologies, which transforms a collection of mixed signals into a set of independent components. It is one of the best technologies to remove artifacts in EEG signal. It can act as a control signal in brain computer interfaces. After preprocessing, some EEG data reflect as original signal, and remaining act as artifacts shown in Fig. 2.

3.4 Feature Extraction Using Principal Component Analysis

The principal component analysis (PCA) is a well-known technique for the reduction of dimensionality, and this has been used in the applications of machine learning. This principal component analysis obtains information which is available in a large set of variables into a cut short or a comparatively fewer variables. This process is done by the application frequency transformation onto them. This transformation is applied in such a kind that linearly correlated variables get transformed into totally uncorrelated variables. This kind of correlation gives us information that there is a presence of redundancy of the information, and if this redundancy is in case reduced, then the information that has to be processed has to be compressed. For instance, in case, there are two variables in a variable set whose values are highly correlated, then there is no extra gain of any related information which is obtained by retaining both the variables because one can be nearly expressed as the linear combination of the other. In these cases, the principal component analysis transfers the variance of the second variable onto the first variable by the translation as well as the rotation of the original axes and simultaneously projecting data onto new axis. This projection's direction is evaluated using the eigen values and the eigenvectors. Therefore, the initial few transformed features which are termed as principal components are highly rich in information, but still, the last few features contain predominantly noise with very negligible information in them. The retaining of the initial principal components is allowed by the transferability which reduces the number of variables with minimum loss of information significantly.

An image is a matrix of pixels whose brightness represents the reflectance of a surface feature within that pixel. The reflectance value ranges between 0 and 255 for an 8-bit integer image. So the operation goes as the pixels with zero reflectance would appear as black, pixels with the value of 255 would appear as pure white and the pixels with the value in between appear as a gray tone. Over here, even the large images that are captured are initially resized to a smaller scale to reduce the computational load on the central processing unit. The images set totally consists of seven band images that are captured across the blue, green, red, near-infrared and the mid-infrared range of the electromagnetic spectrum.

The proposed method utilizes principal component analysis in order to obtain the absolute features, which helps to train the network using SVM, MSVM. Finally, the performance of the classifier was analyzed.

List of extracted features is mean, standard deviation, kurtosis, skewness, variance. The features can be described as follows.

Mean

It can be calculated by taking the ratio of the sum of all entire pixel values over an image to the total number of pixels in an image.

$$\mu = \frac{1}{MN} \sum_{i=1}^{M} \sum_{j=1}^{N} P(i, j) \tag{3}$$

$P(i, j)$ is the pixel value at that point (i, j).

Standard deviation

It is the measure of probability distribution of in homogeneity distribution of a certain object. It provides the large intensity and high-contrast edges of an image.

$$\sigma = \sqrt{\frac{1}{MN} \sum_{i=1}^{M} \sum_{j=1}^{N} (P(i, j) - \mu)^2} \tag{4}$$

Skewness

It is the nature of degree of the distribution of particular window is called skewness.

$$S = \frac{1}{MN} \sum_{i=1}^{M} \sum_{j=1}^{N} \left(\frac{P(i, j) - \mu)}{\sigma} \right)^3 \tag{5}$$

$P(i, j)$ is the pixel value at that point (i, j) and μ—mean value.

Variance

It is the square root of the median

$$\text{var} = \sqrt{\text{SD}} \tag{6}$$

SD—standard deviation.

Mean square error

The relationship between the two images is defined as mean square error. It results a numerical value.

$$\text{MSE} = \frac{1}{MN} \sum_{i=0}^{m-1} \sum_{j-0}^{n-1} [I(i, j) - k(i, j)]^2 \tag{7}$$

Table 1 shows few samples and its feature values. The extracted features fed as input to SVM and MSVM classifier to classify the type of emotions from the EEG signal.

Table 1 Feature extraction table

S. No.	EEG signal	M	S.D	K	V	S	Status
1		248	27.83	35.13	774.7	−5.55	Happy
2		240.5	38.11	14.6	$1.4e^{+3}$	−3.3	Angry
3		243.7	32.34	16.97	$1.04e^{+3}$	−3.6	Sad
4		245.8	30.66	21.64	940	−4.20	Happy

M—mean, S.D.—standard deviation, K—kurtosis, S—skewness, V—variance

4   Discussion

Emotion recognition has been accomplished, and it can be classified using the following two classifier.

4.1   Support Vector Machine

It is an unsupervised learning model used to examine the data for classification or regression analysis. The interclass separation is accomplished by hyperplane. The RBF kernel function used for mapping of EEG data in the hyperplane. SVM is used only for linear classification. It can effectively achieve the nonlinear grouping with the help of trick. During this process, the inputs are mapped into higher dimensional feature space. It builds a hyperplane in a high-dimensional space, which is used to perform classification and regression analysis. To obtain optimum hyperplane which helps to separate various classes of data inputs, a minimum amount of the observations that locate on the edge of separation called support vectors (SV) are used

$$\text{Let } W_0 Z + b_0 = 0 \tag{8}$$

Equation (8) represents an optimal hyperplane in feature space. To obtain optimal hyperplane, the weights W_0 can be written as a linear combination of support vectors

$$W_0 = \sum \alpha_i z_i \tag{9}$$

The optimal hyperplane for linear and nonlinear SVM is shown in Figs. 3 and 4.

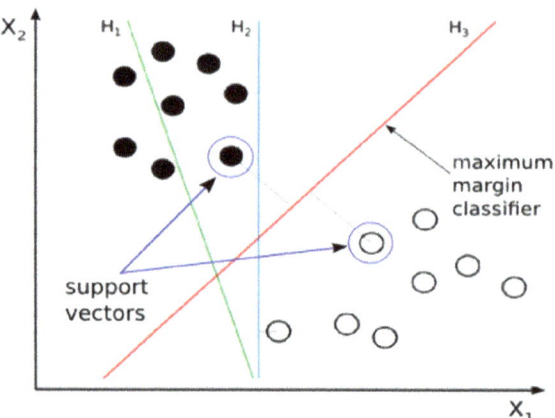

Fig. 3 Linear support vector machine

Fig. 4 Nonlinear support
vector machine

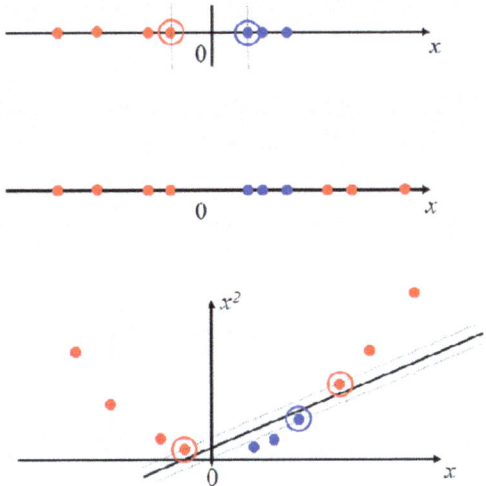

$W_0 Z + b_0 = 0$ is Eq. (8) which separates the training dataset exactly with the maximal margin. The computation of data points separation is based on kernel function. There are linear polynomial, Gaussian radar basis function (RBF) and sigmoid. This function helps to find the smoothness and efficiency of the class separation. The recommended method uses sigmoid function.

4.2 Multiclass Support Vector Machine

SVM supports binary classification and classifies data points into two classes. But, it does not support multiclass classification. In order to map data points to high-dimensional space, it is compulsory to perform linear separation between every two classes. It is represented as one-to-one approach. The technique breaks down the multiclass problem into multiple binary classification problems. It assigns a binary classifier per each pair of class. The classifier uses $m(m - 1)/2$. Another approach is called as one-to-rest approach, in which classifier can use MSVMs; each SVM estimates the membership in one of the m-classes. MSVM is a technique used in classification phase. Two types are one-against all (OAA) or one-against one (OAO). In the case of SVM, N number of independent class is built for N-class pattern recognition. Each class can be separated from others through training process. Consider a sample which belongs to class N1, SVM trained to separate class N1 from the others. Next technique is called as one-against one. It is used to separate one class from the other class. Finally, decision is taken against testing sample via voting of the results of these SVM. The research work uses the OAO method for emotion recognition.

4.3 Performance Analysis

The overall performance of the emotion classification from EEG signal as discussed follows (Fig. 5 and Table 2).

Peak Signal to Noise Ratio (PSNR)

It is the degree of amount of reconstruction of a processed image.

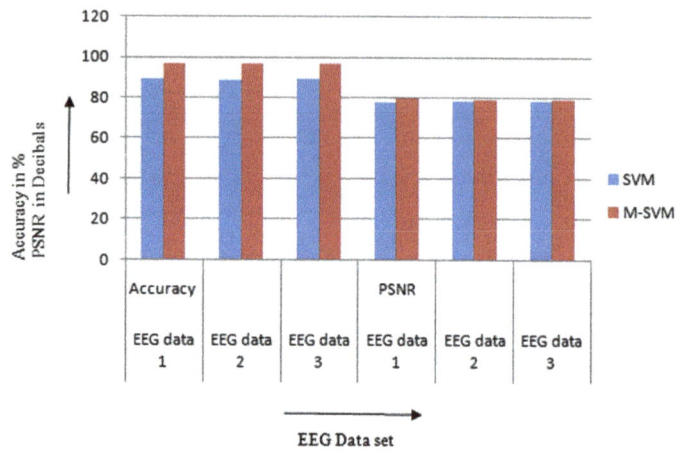

Fig. 5 Comparison chart of SVM and MSVM

Table 2 Performance analysis table

EEG signal	Classifier	A	PSNR	MSE	T
	SVM	89.5	78	$6.40e^{-0.4}$	41.21
	MSVM	96.7	79.9	$6.56e^{-0.4}$	35.6
	SVM	89	78.67	$6.50e^{-0.4}$	20.23
	MSVM	96.9	79.5	$6.61e^{-0.4}$	15.9
	SVM	89.5	78.4	$6.80e^{-0.4}$	22.3
	MSVM	97	79.6	$7.02e^{-0.4}$	15.5

A—accuracy, PSNR—peak signal to noise ratio, MSE—mean square error, T—time

$$\text{PSNR} = 10 \log_{10} \frac{\text{MAX}^2}{\text{MSE}}$$
$$= 20 \log_{10} \frac{\text{MAX}}{\sqrt{\text{MSE}}}$$
$$\text{PSNR} = 20 \log_{10}(\text{MAX}) - 10 \log_{10}[\text{MSE}] \tag{10}$$

The value varies from 0 to ∞. The PSNR is stated in dB. If the PSNR value is 100%, then the processed image is reconstructed accurately.

Elapsed time

The total amount of time required to reconstruct the partitioned image from the input image.

Accuracy

It is a decimal or binary digits. Measure of exactness of the true value.

$$\text{Accuracy} = \frac{\text{TP} + \text{TN}}{\text{TP} + \text{FP} + \text{TN}} \tag{11}$$

TP—true positive; TN—true negative; FP—false positive

5 Conclusion

The neurophysiologic diseases and abnormalities can be detected by using EEG signals, which measure the brain activity. The offered research work provides an efficient feature extraction technique from the multi-channel EEG signal that uses PCA approach. We identified that the recommended technique has good classification performance by reducing the feature size to the optimum level. The spatial and temporal characteristics of the EEG signals are analyzed by using the proposed model. The emotional states of DEAP dataset for each participant were analyzed using multiple support vector machine and SVM. The recommended procedure gives greater results than the existing scheme with the same dataset. The performance of the SVM and MSVM classifiers was evaluated using accuracy, PSNR, mean square error and time. The SVM and multiclass SVM give us the accuracy of about 89.51% and 96.71% for the EEG signal. The proposed work helps to promptly extract the emotions from EEG signals.

References

1. Kim M-K, Kim M, Oh E, Kim S-P (2013) A review on the computational methods for emotional state estimation from the human EEG. Comput Math Methods Med 2013:1–13

2. Mohammadi Z, Frounchi J, Amiri M (2016) Wavelet-based emotion recognition system using EEG signal. Neural Comput Appl 28:1–6
3. Alarcão SM, Fonseca MJ (2017) Emotions recognition using EEG signals: a survey. IEEE Trans Affective Comput, 99
4. Li M, Xu H, Liu X, Lu S (2018) Emotion recognition from multichannel EEG signals using K-nearest neighbor classification. Technol Health Care 26:S509–S519
5. Chakraborty A, Konar A, Chakraborty UK, Chatterjee A (2009) Emotion recognition from facial expressions and its control using fuzzy logic. IEEE Trans Syst Man Cybern-Part A: Syst Humans 39(4):726–743
6. Petrantonakis PC, Hadjileontiadis LJ (2010) Emotion recognition From EEG using higher order crossings. IEEE Trans Inf Technol Biomed 14(2):186–197
7. Mert A, Akan A (2016) Emotion recognition from EEG signals by using multivariate empirical mode decomposition. Pattern Anal Appl, 1–9 (2016)
8. Gaol Y, Lee HJ, Mehmood RM (2015) Deep learning of EEG signals for emotion recognition. In: 2015 IEEE International Conference on Multimedia & Expo Workshops (ICMEW), 29 June–3 July 2015, Turin, Italy. ISBN: 978-1-4799-7079-7
9. Wang XW, Nie D, Lu BL (2011) EEG-based emotion recognition using frequency domain features and support vector machines. In: Proceeding of the International Conference on Neural Information Processing, pp 734–743, Guangzhou, China
10. Rahman MA, Hossain MF, Hossain M, Ahmmed R (2020) Employing PCA and t-statistical approach for feature extraction and classification of emotion from multichannel EEG signal. Egypt Inf J 21(1):23–35
11. Doma V, Pirouz M (2020) A comparative analysis of machine learning methods for emotion recognition using EEG and peripheral physiological signals. J Big Data 7(18):1–21
12. Asghar MA, Khan MJ, Fawad, Amin Y, Rizwan M et al (2019) EEG-based multi-modal emotion recognition using bag of deep features: an optimal feature selection approach. Sensors 19:1–16
13. Mumenthaler C, Sander D, Manstead ASR (2020) Emotion recognition in simulated social interactions. IEEE Trans Affect Comput 11(2):308–312
14. Zhuang N, Zeng Y, Zhang LTC, Zhang H, Yan B (2017) Emotion recognition from EEG signals using multidimensional information in MD domain. BioMed Res Int 2017:1–9

Real-Time Anomaly Detection Surveillance System

Preeti Nagrath, Dharana, Shivansh Dwivedi, Raunak Negi, and Narendra Singh

Abstract Video surveillance plays an important role in today's generation. Due to its relationship with image understanding and video analysis, it has been attracting a lot of research in recent times. With the advancement in the technologies when artificial intelligence, deep learning and machine learning are introduced into the system, more powerful tools which are able to extract high semantic features are introduced to deal with the traditional architecture problems. Majority of anomaly detection applications are concerned with indoor surveillance with monitoring activities which helps to identify anomalous behaviour by tracking live stream footage. The main one of them is human behaviour detection. It is the most unpredictable behaviour, and it is very difficult to decide whether it is suspicious or normal. This paper demonstrates a deep learning approach to detect anomalous behaviour in a campus of school or college. The monitoring is performed through consecutive frames of cameras which are extracted from a video. Our model will be implemented on the extracted camera frames which detect the activities as abnormal. The feed is saved from the time of detection of an anomaly which sends an alert message to the respective authority. So we only need to save that portion of the video at which the anomaly is happening rather than saving the entire feed. The entire system is divided into two parts. In the first part, features will be computed from the live camera feed, and in the second part, based on the features extracted, the classifier will predict the anomaly. Proposed system detects the anomalies with a loss of 4.795.

Keywords LSTM · CNN · Autoencoders · Reconstruction error · Regularity score

P. Nagrath · Dharana (✉) · S. Dwivedi · R. Negi · N. Singh
Department of Computer Science and Engineering, Bharati Vidyapeeth's College of Engineering, New Delhi, India

P. Nagrath
e-mail: preeti.nagrath@bharatividyapeeth.edu

665
D. Gupta et al. (eds.), *Proceedings of Data Analytics and Management*,
Lecture Notes on Data Engineering and Communications Technologies 90,
https://doi.org/10.1007/978-981-16-6289-8_54

1 Introduction

There are about 350 million surveillance cameras worldwide as of 2016, and according to studies, this number is going to reach about 1 billion by 2021. Leaving some developed countries which use modern surveillance techniques with vast networks of cameras and artificial intelligence systems, all other simple and small surveillance systems require human personnel to monitor a number of feeds or camera footage continuously creating a repetitive and prone to error task. The automation in monitoring these camera feeds can get rid of the errors as well as decrease the human dependency for these systems. The aim of this project is to create a fully edged human-independent surveillance system for a campus which can detect anomalous activities and behaviours inside the campus and warn the authority about these suspected behaviours.

Human behaviour detection [1] has plenty of applications, especially in indoor and outdoor places. E-surveillance becomes an important agenda in Digital India as security cameras become an important part of life. Its advantages include effective monitoring, cost efficiency and less manpower requirement. Since, nowadays, monitoring is done by humans which takes a lot of time and a huge amount of space to store the video, this problem is solved by automation in video surveillance [2]. The video will start recording only when the anomaly will be detected and be saved in the database. This solves the problem of space and manpower requirements. It too makes it easier to end the exact event rather than seeing the entire footage of cameras.

The main objectives of the study are listed below:

- Preparing a model using various [3–6] convolutional, long short-term memory (LSTM) layers to detect and monitor human behaviour during campus and warning when any suspicious event occurs, like smoking, drinking or fighting.
- Training of the model on [7–9] deep learning [10] networks like LSTM [11] architecture.
- Setting up an alarm system [12] using ESP32 microcontroller in case any suspicious activity is detected by the model.

After performing a thorough literature survey and finding out the research gaps, the paper goes on to explain the methodology used by us to perform the aforementioned task of detecting suspicious human behaviour. The methodology uses a supervised and unsupervised approach, with different datasets used for each approach. This is followed by results obtained from the prepared model and its related discussions, which includes sample clips from training sets and accuracy figures yielded by the model. The paper winds up with the conclusion section summarizing the paper and finishes with the references.

2 Literature Survey

The anomaly can be detected in many problem statements like bank frauds, [13] video surveillance, etc. This paper mainly focuses on anomaly detection in video surveillance. There has been various researches about detecting anomalies in [14] surveillance videos, considering some recent years papers for our literature survey.

In 2002 [15], a unique study came out in which the researchers presented the model which recognized all the normal activities and then detected the anomaly present in the video surveillance by differentiating between the normal and anomalous activity, and they have done this using the hidden Markov model. In this current paper, the same type of procedure is followed but with different techniques.

In 2008 [16], a group of researchers presented a novel approach in which the framework will learn the movement patterns and object sizes in static camera surveillance. The probability density function of pixel level is used for modelling of the past, and the novel approach contains the modelling of object speed and size to pixel level pdfs. PDFs of motion of the object at that location are modelled as a multivariate Gaussian mixture model. After implementing this tracking method, the output of the teaching method is used for EM-based learning of every GMM (unsupervised learning). They develop this PDF technique to overcome failures of most object path modelling approaches.

In the same year, another study came out in which [17] the person is trying to detect anomalies in sea surveillance, and he also proposed an unsupervised approach in which he is clustering the normal vessel traffic patterns and implementing the same. Like the previous research, the Gaussian mixture model (GMM) is used for cluster modelling, and the algorithm which is used for clustering is the expectation–maximization (EM) algorithm. Training data set contains real videos of sea traffic. His proposed model has potential to detect moving objects in other domains also, and a more complex model for more complex object detection is yet to be proposed.

In 2010, a novel approach came in, [18] which is divided into three parts: the rst part consists of dense motion eld and motion statistics method, the second part consists of SVM classifier for one-class classification, and the last parts consist of principal component analysis (PCA) for feature dimensionality reduction. Their main motive is to take a surveillance video and reduce its feature dimensionality and then using SVM to predict its class. According to them, this method works very well in detecting anomalies in surveillance with low false alarm rates, and along with that, it works well in complex situations in which most of the tracking and detection modules will not work.

LSTMs are the 1st choice for the researchers in this particular topic, [19] in 2016, another study came out in which the researchers are using the LSTM but this time, it is being used in the form of multi-scale, which result in a better accuracy than any conventional machine learning models, it can improve accuracy by 10%, and at the end, the best accuracy that is achieved by this model is 99.5%.

Now, take a look at recent papers, and [20] in 2018, research came out which gave us the replacement of annotating the videos. In their approach, they considered the videos as bags, normal videos as one bag and anomalous videos as another bag, and a segment of the videos as multiple instances learning (MIL), and it will automatically learn to predict high accuracy scores for anomaly detection. Along with this research, they also produced a large data set which contains 128 h of videos, and it also contains 13 activities which can be labelled as anomaly like, accident, burglary, etc.

In 2019 [21], a study came out which is based on sparse coding, for better results many different techniques are used such as parallel computing, a new algorithm is developed known as iterative hard-thresholding algorithm (ISTA) for learning of sparse representation and dictionary, and also the LSTM model is used in this sparse coding-based anomaly detection.

In 2020 [22], a study came out which presents the hybrid model which contains pre-trained CNN model and bidirectional LSTM model, the CNN model is used for extracting the features belonging to both space and time and Bi-LSTM used to detect ongoing anomalous activity, and the final result came out to be an increase in the accuracy by 3.41 and 8.09%.

All the previous studies show different methodologies by which the anomaly detection can be done. Many studies consist of the ensemble methods of two or more techniques, they are capable of producing better results than the normal techniques, and it is also found that there is a new method in which there is no way to annotate the video parts. The paper includes as many things as possible for a better outcome, and it starts with the deep learning methods but trying to involve more methods and ensemble them to produce better results.

2.1 Research Gaps

As you see, consideration of some recent research papers is taken, going through each one thoroughly, it is found that most of the study includes only the unsupervised approach, and there are very few studies about the supervised approach. It is also found that [23] 1–2 papers on supervised approach but they are capable of doing only one-class classification, unlike unsupervised approaches which are able to do multi-class classification.

3 Research Methodology

Our proposed system uses footage from CCTV cameras set up around the campus. The footage is passed through our model, which predicts the anomaly if present and generates the message accordingly. The methodology can be divided into three parts:

- Perception: Normal wireless cameras that transmit video feed wirelessly through radio bands [24] are set up around the campus to cover all the areas and provide a continuous feed for the processing by the models.

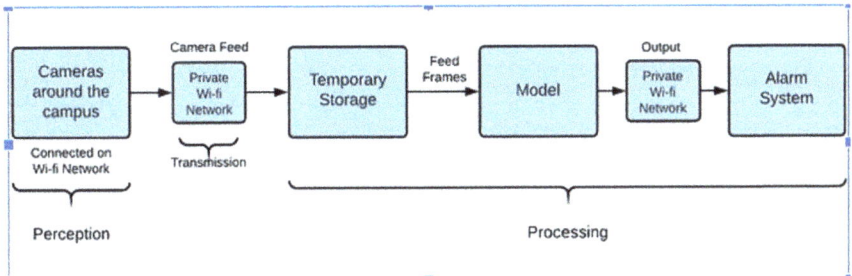

Fig. 1 Methodology

- Transmission: Wireless local area network [25] can be set up over the underlying wired network of the campus to provide for the reception of the video signals transmitted by the wireless cameras. These video signals are processed, the frames of the footage are stored temporarily to be processed by the processing layers, and on the basis of the nature of the output of the processing, the frames are saved or discarded.
- Processing: The temporarily saved frames are fed into the model over the machine where they are saved temporarily, and the output of these models predicts if there are anomalous activities or behaviour inside the footage. According to the prediction, the authority is either warned or not (Fig. 1).

Our model contains various types of layers like convolution, max pooling, input, flatten and dense layers, each of them is doing a different task than another and all the layers are connected to each other directly or indirectly, the model contains two types of activation functions which are ReLu and softmax and it contains a dropout rate of 0.5, and in the compilation of the model, categorical cross-entropy loss function is used and Adam optimizer with learning rate 0.01 is used. Given Fig. 2 represents which layers are present and how they are connected to each other.

Figure 3 represents how the training took place.

3.1 Data Set

We will use the UCSD anomaly data set [26, 27]. The clips were recorded using a stationary camera filming a pedestrian walkway having around 200 frames. Common anomalies that occur in these videos are bikers, skaters, small carts and people walking across a walkway or in the grass that surrounds it. These anomalies, when recorded, occurred naturally. The data set was split into two sets, each for a different scene. These video frames are divided into temporal sequences of size 10, followed by resizing of images to 256 × 256.

These pre-processed images act as inputs for the auto-encoders, which try to reconstruct [28] these sequences and update themselves accordingly.

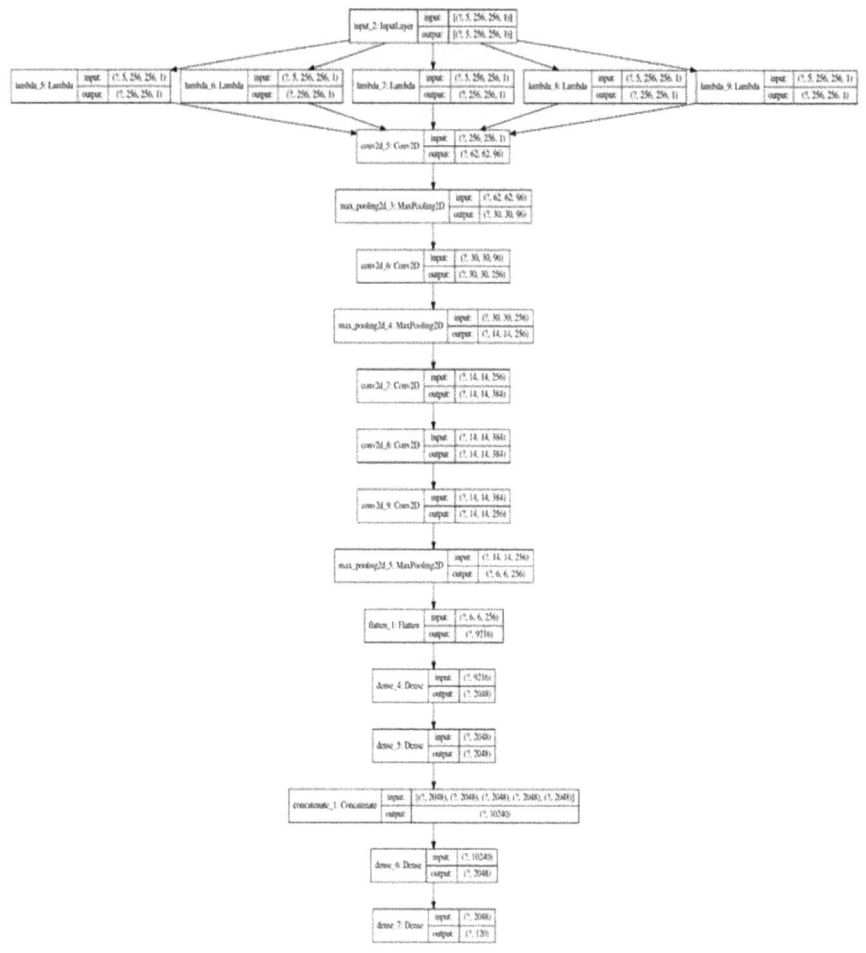

Fig. 2 Information about model layers and how they are connected

3.2 Alarm System

In order to set up an alarm system, we are using the ESP32 microcontroller with a piezo buzzer which can generate simple beeps and tones. If a very loud alarm is needed, a siren with high wattage can be used which can be controlled with MOSFET switch. For the communication between the ESP32 and the Python script, socket communication is used. It can be simply implemented with the use of Python's socket library. For this to work, the IP address of the machine and port number of the socket created on the Python should be known to the ESP32 (Fig. 4).

As the alarm requires a running 24/7 Wi-Fi connection to receive the data and setup the alarm, FreeRTOS is used to create a task on ESP32 that checks the Wi-Fi connection every 10 s. FreeRTOS is an operating system kernel which comes with

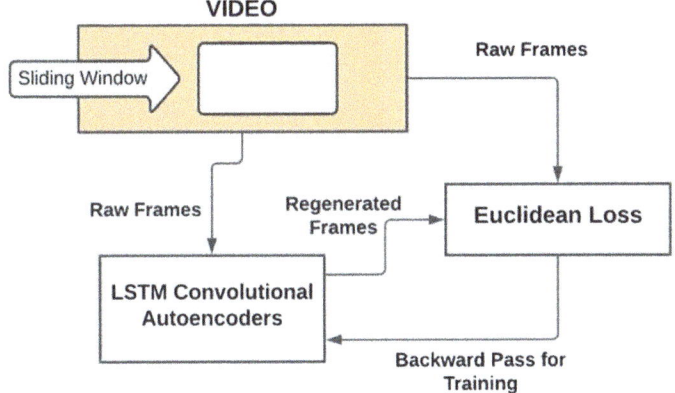

Fig. 3 Algorithm applied for the training

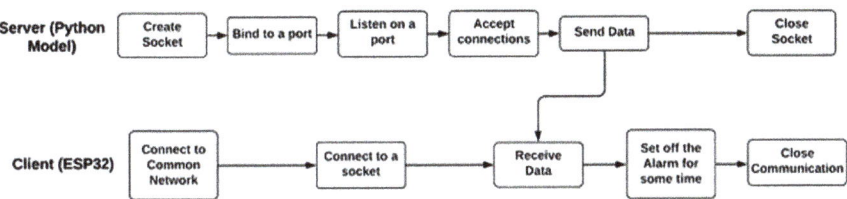

Fig. 4 Socket communication flow chart

the Espressif Internet Development Framework (ESP-IDF) which is generally used to program the ESP32 microcontroller (Fig. 5).

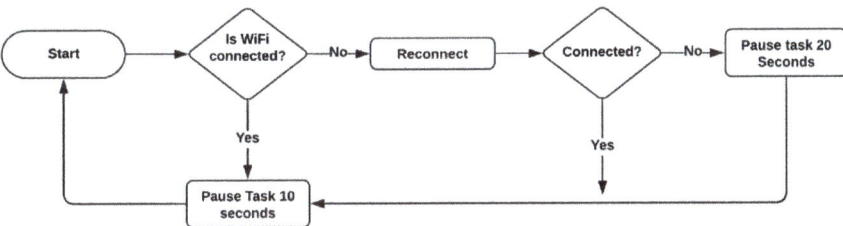

Fig. 5 Flow chart of free RTOS task

3.3 Future Scope

3.3.1 Supervised Approach

This approach classifies the anomaly (if present) into one of these three categories:

- Assault
- Using mobile phones
- Firearms.

3.3.2 Data Set

For supervised learning, we have used KTH data set [29], which is a standard action recognition data set having six types of human actions, namely running, walking, hand waving, clapping, jogging and boxing. Apart from jogging and boxing, the rest of the actions are utilized to train the model for normal or non-suspicious activities.

There are 25 subjects performing the above-mentioned activities in four scenarios: outdoor, outdoor with scale variation, indoors and outdoors with different clothes. This makes it a total of 600 videos, and each video is shot at 25 fps. To train the model for suspicious activities, we will use the CAVIAR [30] data set. The recorded videos were filmed in the entrance lobby of the INRIA laboratories and in a shopping centre in Lisbon. Each video has a resolution of 384×288 pixels with 25 frames per second. Each video has a le size of around 6–12 MB, few of them going to 21 MB. After the collection of video data, these are pre-processed by converting each video into 7000 frames of images. These images were resized to $224 \times 224 \times 3$ pixels via OpenCV library.

VGG16 [31] is a very popular deep network model, as shown in the figure. The network has a total of 16 layers, with convolutional layers of size 3×3 and 1×1 with ReLu activation, max pooling of size 2×2 and three fully connected layers having 4096, 4096 and 1000 units. For feature extraction, the fully connected layers are discarded, and pre-trained convolutional and max pooling layers are used. The extracted features are then passed through LSTM [32] networks, which are RNN networks having the capacity to retain information and detect sequence (Fig. 6).

3.3.3 Unsupervised Approach

Due to the rarity and diversity of abnormal events, supervised learning can lead to biases in the model and fails to produce optimum results. Using unsupervised or semi-supervised learning on unlabelled images produces realistic outcomes. Auto-encoders are one of the many unsupervised learning techniques to learn patterns and regularities in the given data.

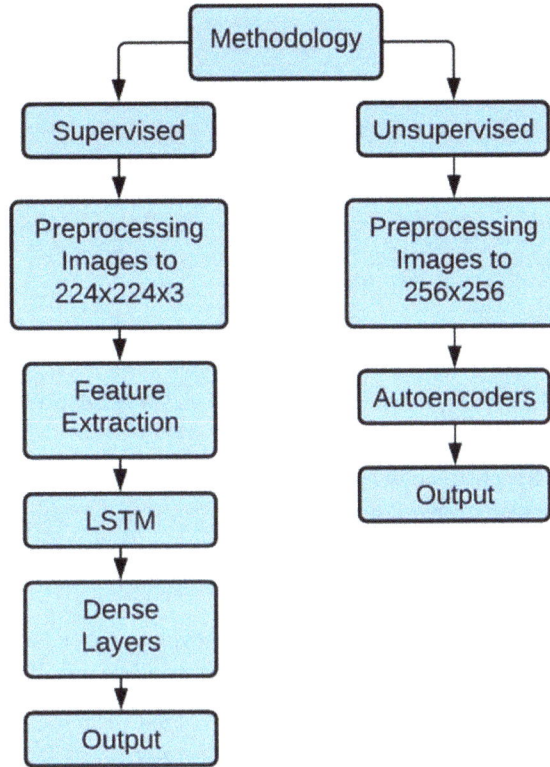

Fig. 6 Work flow in supervised and unsupervised

4   Results and Discussions

This section is for the introduction of all the results and grades obtained from the training and testing of the anomaly detection model.

Technique used in this paper is [33–36] LSTM convolutional [37–40] auto-encoder model on UCSD anomaly data set, in which we trained the model with epoch size as 3 in which each epoch contains 50 sub-steps and batch size as 4. The approach that we have used is of unsupervised type, we have not used the supervised methods because these types of methods need labelled data, collecting labelled data of anomaly is difficult because these activities are very rare, these activities are of many types, and labelling them will take much more time. These all are the reasons which impede the use of supervised techniques, there are other unsupervised and semi-supervised techniques also which only require the abnormal activities' video data which are easily available in real world.

The main approach is all about the reconstruction error, in this approach we first train the auto-encoder, and then that trained auto-encoder will reconstruct only the regular video with low error. Figure 7 thoroughly explains the algorithm behind the reconstruction error.

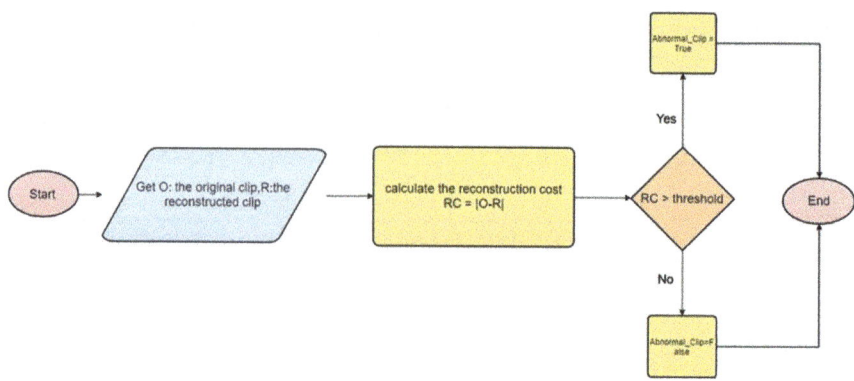

Fig. 7 Algorithm explaining the concept of reconstruction error

Reconstruction error of a pixel's intensity value I at the location (x, y) in frame t of video using L2 norm:

$$e(x; y; t) = j j_1(x; y; t) \ f w(I(x; y; t)) j j_2 \tag{1}$$

Here, fw is the learned LSTM convolutional encode, and the reconstruction error of frame t is calculated by summing up all the pixel's error:

$$e(t) = \sum_{(x,y)} e(x, y, t) \tag{2}$$

Let us suppose we have to find out reconstruction cost which is starting at t can be calculated as follows:

$$\text{Sequence Reconstruction Cost } (t) = t' = \sum_{t=t}^{t+10} e(t') \tag{3}$$

Now, the abnormality score is calculated

$$sa(t) = \frac{\text{Sequence Reconstruction Cost } (t) - \text{Sequence Reconstruction Cost } (t)_{\min}}{\text{Sequence Reconstruction Cost } (t)_{\max}} \tag{4}$$

After this regularity score is calculated for the plotting of graph and finding the abnormal events:

$$s_r(t) = 1 - s_a(t) \tag{5}$$

Each frame of the videos is resized to 256×256 so that the input images will have the same resolution, scaling the pixel size between 0 and 1 by dividing them by 256. After the training is over, the loss comes out to be 4.795 and validation loss comes out to be 4.7935. Given Fig. 8 contains a sample clip of the data set.

The auto-encoder consists of encoder and decoder in which the encoder takes the input video as a sequence of frames, and encoder itself consists of two things which are spatial encoder which encodes the features and temporal encoder which is used for motion encoding.

For obtaining the results, the model is applied on a test set in which a video is chosen for the testing of the model and then each clip of that video is converted into tensors, and these tensors represent the activity happening in the testing video. Figure 9 represents the activity happening in the form of tensors with respect to the time.

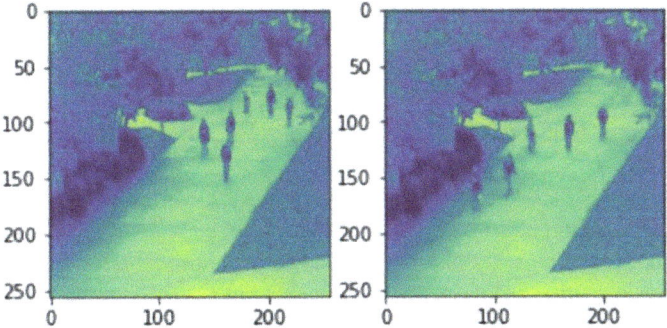

Fig. 8 Clips from training set

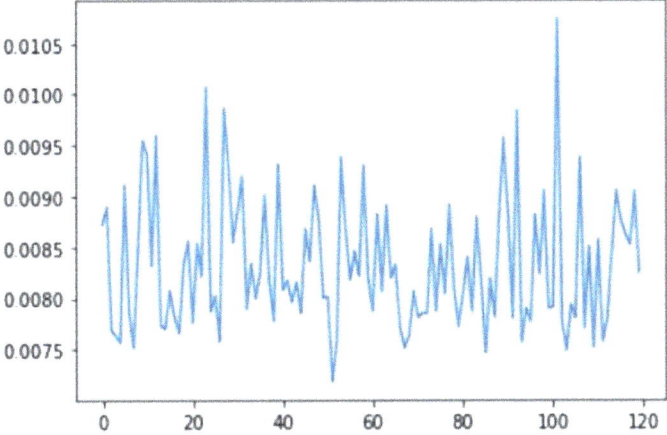

Fig. 9 Representation of activities in the form of tensors, the *x*-axis is the time and *y*-axis is tensors

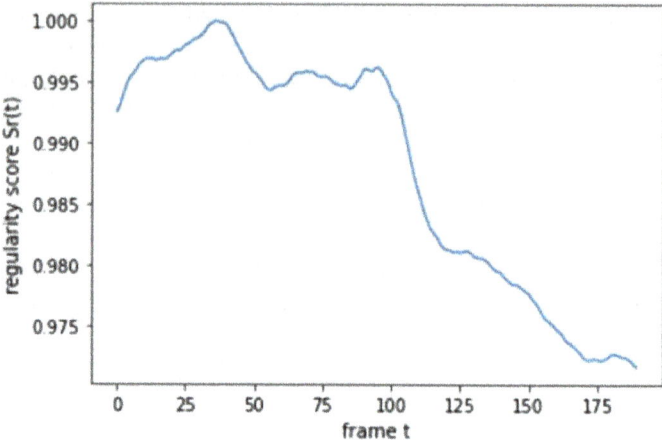

Fig. 10 Detection of abnormal activity through regularity score

In this model, convolutional LSTM is used instead of fully connected LSTM, because the fully connected LSTM layers cannot keep their data because of its full connections.

Now the anomaly is detected using regularity score, given below Fig. 10 represents the anomaly detection through the regularity score of a single test video.

In this figure, the high regularity score indicates that the particular frame at that time contains no type of abnormal or anomalous behaviour, and the low regularity score indicates that the frame at that particular time contains some type of abnormal or anomalous behaviour.

5 Conclusion

This paper is mainly focused on detecting some type of anomaly happening in the real world, and for this purpose, we have used the convolutional LSTM auto-encoders model which is an unsupervised approach. In the end, we found out a method in auto-encoders known as regularity score which changes along with the change in behaviour of the particular scene or video. The proposed study will be applied in the academic grounds in which it can detect all the anomalies happening around the academic environment.

This paper contains the successful implementation of anomaly detection on UCSD anomaly detection data set with a loss of approximately 4.7. This study will help in detecting further anomalies happening around us.

References

1. Proceedings of the Second International Conference on Innovative Mechanisms for Industry Applications (ICIMIA 2020)
2. Sreenu G, Saleem Durai MA (2019) Intelligent video surveillance: a review through deep learning techniques for crowd analysis. J. Big Data
3. Kim T-Y, Cho S-B (2018) Web traffic anomaly detection using C-LSTM neural networks. Expert Syst Appl
4. Fang F et al (2019) Anomaly detection in ad-hoc networks based on deep learning model: a plug and play devices. Ad Hoc Networks 84:82–89
5. Khan MA, Karim M, Kim Y (2019) A scalable and hybrid intrusion detection system based on the convolutional-LSTM network. Symmetry 11(4):583
6. Cao K et al (2018) CNN-LSTM coupled model for prediction of waterworks operation data. J Inform Process Syst 14(6):1508–1520
7. Zhong Z et al (2019) A deep learning approach to anomaly detection in geological carbon sequestration sites using pressure measurements. J Hydrol 573:885–894
8. Xie Guo et al (2020) Motion trajectory prediction based on a CNN-LSTM sequential model. Sci China Inform Sci 63(11):1–21
9. Naseer S et al (2018) Enhanced network anomaly detection based on deep neural networks. IEEE Access 6:48231–48246
10. Chalapathy R, Chawla S (2019) Deep learning for anomaly detection: a survey. arXiv preprint arXiv:1901.03407
11. Abellan-Abenza J, Garcia-Garcia A, Oprea S, Ivorra-Piqueres D, Garcia-Rodriguez J (2017) Classifying behaviours in videos with recurrent neural networks. Int J Computer Vision Image Process
12. Liu Y et al (2020) Deep anomaly detection for time-series data in industrial IoT: a communication-efficient on-device federated learning approach. IEEE Internet Things J
13. Kumaran SK, Dogra DP, Roy PP (2019) Anomaly detection in road traffic using visual surveillance: a survey. arXiv preprint arXiv:1901.08292
14. Hu W et al (2004) A survey on visual surveillance of object motion and behaviors. IEEE Trans Syst Man Cybern Part C (Appl Rev) 34(3):334–352
15. Clark VNJJ (2002) Automated visual surveillance using hidden Markov models. In: International conference on vision interface
16. Basharat A, Gritai A, Shah M (2008) Learning object motion patterns for anomaly detection and improved object detection. In: IEEE conference on computer vision and pattern recognition 2008
17. Laxhammar R (2008) Anomaly detection for sea surveillance. In: 11th International conference on information fusion 2008
18. Liu C, Wang G, Ning W, Lin X, Zi Z, Liu Z (2010) Anomaly detection in surveillance video using motion direction statistics. In: IEEE International conference on image processing
19. Cheng M, Xu Q, L.V. J, Liu W, Li Q, Wang J (2016) MS-LSTM: a multi-scale LSTM model for BGP anomaly detection. In: 2016 IEEE 24th international conference on network protocols (ICNP), Singapore
20. Bhagya Divya P, Shalini S, Deepa R, Reddy BS (2017) Inspection of suspicious human activity in the crowdsourced areas captured in surveillance cameras. Int Res J Eng Technol (IRJET)
21. Zhou, JT, Du J, Zhu H, Peng X, Liu Y, Goh RSM (2019) AnomalyNet: an anomaly detection network for video surveillance. IEEE Trans Inform Forens Secur
22. Ullah W et al (2020) CNN features with bi-directional LSTM for real-time anomaly detection in surveillance networks. Multimedia Tools Appl
23. Kiran BR, Thomas DM, Parakkal R (2018) An overview of deep learning based methods for unsupervised and semi-supervised anomaly detection in videos. J Ima 4(2):36
24. Zhou Z, Yu H, Shi H (2020) Optimization of wireless video surveillance system for smart campus based on internet of things. IEEE Access 8:136434–136448

25. Yin C et al (2020) Anomaly detection based on convolutional recurrent autoencoder for IoT time series. IEEE Trans Syst Man Cybern Syst
26. UCSD Anomaly Detection Dataset. http://svcl.ucsd.edu/projects/anomaly/dataset.htm
27. Khaleghi A, Moin S (2018) Improved anomaly detection in surveillance videos based on a deep learning method. In: 2018 8th Conference of AI robotics and 10th RoboCup Iran open international symposium (IRANOPEN) 2018
28. Zhou Y, Dong H, El Saddik A (2020) Deep learning in next-frame prediction: a benchmark review. IEEE Access 8:69273–69283
29. Schuldt C, Laptev I, Caputo B (2004) recognizing human actions: a local SVM approach. In: Proceedings ICPR 04, Cambridge, UK
30. Brdiczka O, Maisonnasse J, Reignier P (2005) Automatic detection of interaction groups. In: ICMI'05, 2005 international conference on multimodal interaction, Trento, Italy, Oct 2005
31. Simonyan K, Zisserman A (2015) Very deep convolutional networks for large-scale image recognition. In: ICLR 2015
32. Hochreiter S, Schmidhuber J (1997) Long short-term memory. Neural Comput 9(8):1735–1780
33. Luo W, Liu W, Gao S (2017) Remembering history with convolutional LSTM for anomaly detection. In: 2017 IEEE international conference on multimedia and expo (ICME), Hong Kong
34. Javed AR et al (2020) Anomaly detection in automated vehicles using multistage attention-based convolutional neural networks. IEEE Trans Intell Transport Syst
35. Heryadi Y, Warnars HLHS (2017) Learning temporal representation of transaction amount for fraudulent transaction recognition using cnn, stacked LSTM, and CNN-LSTM. In: 2017 IEEE International conference on cybernetics and computational intelligence (CyberneticsCom). IEEE
36. Swapna G, Kp S, Vinayakumar R (2018) Automated detection of diabetes using CNN and CNN-LSTM network and heart rate signals. Procedia Computer Sci 132:1253–1262
37. Zhao Y et al (2017) Spatio-temporal autoencoder for video anomaly detection. In: Proceedings of the 25th ACM international conference on multimedia
38. Sakurada M, Yairi T (2014) Anomaly detection using autoencoders with nonlinear dimensionality reduction. In: Proceedings of the MLSDA 2014 2nd workshop on machine learning for sensory data analysis
39. Chen Z et al (2018) Autoencoder-based network anomaly detection. In: 2018 wireless telecommunications symposium (WTS). IEEE
40. Cozzolino D, Verdoliva L (2016) Single-image splicing localization through autoencoder-based anomaly detection. In: 2016 IEEE international workshop on information forensics and security (WIFS). IEEE

Design of an iOS App Architecture for Cotton Plant Disease Detection Using Artificial Intelligence and Machine Learning Techniques

Sandeep Kumar, Rajeev Ratan, and J. V. Desai

Abstract Identifying the correct crop disease is very crucial to timely control of the disease. The mobile app will be one of the easy ways to identify crop diseases using the latest machine learning and artificial intelligence techniques. This paper proposed the architecture design of the iOS app which will be used to identify cotton plant diseases. The proposed architecture is designed with the latest iOS design pattern, the latest tech stack, and compliance with SDLC. The proposed architecture is based on scalability, high performance, and usability. This proposed architecture will be a blueprint of the actual development of an iOS mobile app. The proposed architecture also supports the usage of ML model in iOS app. The proposed architecture is also supporting the latest machine learning and artificial techniques that are used in identifying the cotton plant disease. Based on the development of this app, the future enhancement can be done by using the server-side implementation when network available in device. The performance of the app is measured by properly doing the memory management using the instruments and profiling of the source code.

Keywords iOS · Machine learning · Artificial intelligence · Cotton leaf · Disease detection · Apple · Architecture · Designing · Software architecture VIPER · CoreML · CreateML · SDLC · ML model · Mobile app · App development

1 Introduction

The major population of India is still dependent on agriculture. Nearly, 70% of the Indian population depends on agriculture [1]. In rural India, the main source of income of people is farming. Almost, three main regional crops are taken by the farmers in a season. Cotton is one of the most commercial crops in India where farmers get good capital. Different species of cotton are used in India at a different

S. Kumar (✉) · R. Ratan · J. V. Desai
Electronics and Communication Engineering, MVN University, Palwal, India

R. Ratan
e-mail: rajeev.arora@mvn.edu.in

© The Author(s), under exclusive license to Springer Nature Singapore Pte Ltd. 2022 679
D. Gupta et al. (eds.), *Proceedings of Data Analytics and Management*,
Lecture Notes on Data Engineering and Communications Technologies 90,
https://doi.org/10.1007/978-981-16-6289-8_55

part of the country at a different time. In northern India, the cotton crop is mainly done from April to November month, while on the southern side of India, the cotton crop is done in the winter season. The four main species of cotton are used in India. These are *Gossypium arboreum, Gossypium herbaceum, Gossypium hirsutum,* and *Gossypium barbadense.* The *G. hirsutum* is the dominant species and contributes 90% of total cotton production. Even cotton is one of most commercial crops, but the disease plays an important role in the production of cotton which impacts the income of the farmer. The cotton crops very easily get affected by the diseases [2–7].

To increase the farmer's income, crop diseases must be minimized. The early identification of diseases will help farmers to mitigate the diseases in an early stage with minimum impact. The Government of India and the agriculture department are committed to double the income of the farmer. By minimizing the disease, productivity and crop production can be increased which is required as the food demand is increasing every year.

We are moving toward digitization in retail shopping and other sectors. Change is happening very fast in every sector. It is also very important to adopt the changes and change ourselves accordingly. Agriculture is India's most vast and wide area where technology is still not used significantly. The rapid growth of urbanization will lead to a reduction in the workforce in rural areas. This challenge can be addressed by using innovative technologies using the cognitive system which will help in easing the farmer's work, removing the need for large numbers of people to work in the land.

This paper presents the proposed iOS mobile app architecture which lays out the design pattern and usage of different components in the app. This architecture will be used in the development of an actual iOS mobile app for cotton plant disease detection [8, 9].

The success of this system will depend on how accurately the system carries the image classification and ML techniques. The huge real image data sets will be required. The significance of this paper is important as it helps in identifying the correct ML and latest libraries that can be used easily in the iOS app. In this paper, while proposing the architecture, the scalability, performance of an app, memory management, and usability are kept in mind [10, 11].

The rest of this paper is structured as follows. The related works of this paper are summarized in Sect. 2. The methodology of this proposed architecture is presented in Sect. 3. The performance analysis is discussed in Sect. 4. Section 5 evaluates the result of the paper, and finally, we conclude the paper in Sect. 6.

2 Related Work

Hase et al. [12] present the android mobile application to diagnose and identify the infected leaf. The system works on the plants that are infected by diseases like fungi and viruses which detect and classify them by using image processing techniques.

The author used the OpenCV open-source framework for image processing and to identify the disease pattern.

Ranjith et al. [13] present a novel smart irrigation system that can control irrigation automatically using the android mobile application. This system is based on cloud back-end integration where image analysis has been done and the result has been shown on android application. The cloud-based back-end system also keeps into consideration of soil context, texture, climatic factors, etc.

Bodhe [6] has proposed a development of a prototype of an android mobile application using the template matching technique. The proposed mobile application is to diagnose and identify the pattern of cotton leaf disease. This proposed android app is based on Internet connection and back-end implementation for disease detection.

Echalar and Subion [14] have developed an android mobile application to detect the rice health condition utilizing image processing and pigment analysis. The author used binary pattern histogram, CNN, and color segmentation algorithms for automatic detection. In the presented paper, android app is evaluated as per ISO standards for software quality standards.

Elhassouny and Smarandache [15] proposed an efficient smart mobile application model based on CNN to recognize tomato leaf diseases. The model is trained from the MobileNet CNN model and used to recognize the 10 most common types of tomato leaf diseases. The author used thousands of sample images to increase the performance and accuracy of the model.

3 Methodology of Proposed Architecture

3.1 System Architecture—Overview

The iOS mobile app architecture is a blueprint for the iOS mobile app system. The system architecture is generally a combination of structural elements and an individual set of interfaces. The system is also composed in addition to the framework behavior of all the structural elements. The mobile app architecture pattern consists of how the data would move, the UI/UX of the app, choice of the platform, tech stack, etc., as shown in Fig. 1.

Presentation Layer: The responsibility of this layer is to present the app to end users. It defines the flow and how to present the app to the end users. When designing this layer, the intended infrastructure is always kept in mind. Additionally, the other deployment restrictions must also be kept in mind. Another necessity is to protect the app from invalid entry by using the robust data validation mechanism and selecting the proper data format. Good architecture also focuses on decoupling business logic from the presentation layer. This layer represents the UI logic.

Business Layer: This layer looks into elements on the business front. In other words, this layer defines the way to present the app to the end users. This consists of workflow,

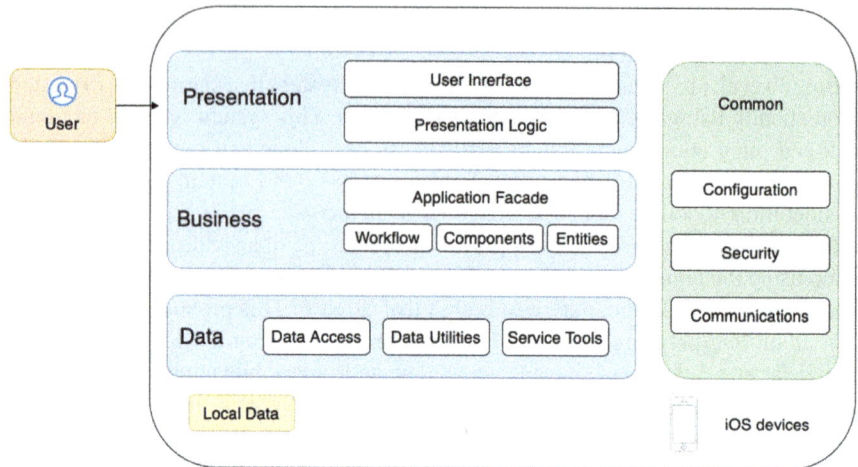

Fig. 1 A high-level overview of mobile app architecture

business components, and the entities under the other sublayers like domain and service.

The responsibility of the service layer is to look into the definition of the end users common application functions, while the domain model layer checks the knowledge and expertise of the linked specific problem areas.

This layer is sandwiched between the presentation layer and data layer, and all the business logic of the application is handled in this layer. This layer must be decoupled with the presentation layer and data layer. The business layer receives the data request from the UI layer based on the user interaction in the application, and then, it requests the data from the data layer. Once the data is received, then based on the requirement from the UI layer, the data is manipulated, all the algorithms, logic, encryption and decryption, etc., are done in this layer, and the final data will be sent to the UI layer to display to the user. Mostly, the following logic is handled by this layer:

- Caching
- Logging
- Authentication
- Exception management
- Security.

Data Layer: The data access layer should help in offering efficient and secure data transactions which meet the application requirements. This layer is kept flexible for the maintenance side of the application while ensuring that this layer can be modified easily with the changing business requirements.

Access components, utilities, helpers, and service agents are part of this data-specific layer. This layer consists of the following two sub-layers to handle the data:

Persistence layer: This layer is responsible to handle the data which is stored locally in the mobile app. The detail of the database management is discussed in Sect. 4. When the app has the requirement to read/retrieve the data locally within session or when the app is not connected to the Internet, then the data is retrieved from the persistence layer. Similarly, the data is saved in a local database using the persistence layer.

Network layer: When the data is retrieved from the remote server and the app is connected to the Internet, then the network layer will fetch the data from a remote server. Once the data is fetched from the remote server, then if needed the data will be cached locally or saved in a local database, and then data will be given to the next layer for further processing. There are a lot of third-party open-source libraries available to fetch the remote data using the APIs.

Now, once the data is fetched from either the persistence layer or network layer, the data will be parsed and converted into the respective model. The models will be passed to the respective layer where data is requested.

3.2 Architecture Principles

The mobile app architecture is a set of patterns and techniques, which follow some architecture principles. The proposed architecture complies with the following architecture principles:

Portability: The system can react to the changing environment. When the requirement is changing slightly or at a large scale, then the app must support the changes with minimum changes in system design. The proposed architecture ensures that the system is portable enough to absorb the changes in app requirements.

Maintainability: Every app or mobile product app starts with minimum numbers of functionality which are essentials for the app's core business logic, and the user can perform some basic functionality. Now, after some time based on the user's feedback some extra and more features will be added to the application. The proposed architecture is highly supporting the maintainability of the app with minimum efforts to keep the app up and running. It reduces the efforts and complexity of the change's implementation in less time and cost.

Reusability: The aim of good architecture is faster and easy development, for which the components must be designed in such a way that the components will be reused across the app. The reusability of the code will be highly supported by the proposed architecture.

Security: The proposed architecture will be designed to support the basic NFRs of the mobile app. The proposed architecture will be robust enough for securing the data which is consumed by the app. The locally stored data must be encrypted.

Performance: The mobile must be good in performance and user experience such as easy navigation and good UX followed by Apple's human guideline. The proposed architecture will be designed in such a way that it met every single user's expectation. The layers will be designed in such a way that they are not tightly coupled, and each layer will take less time to process the data; overall, it will take less time to display the data to the user. The proposed architecture will be designed to consume less memory, and there should be no memory warning or memory-related issue in the application.

Testability: The proposed architecture will be designed in such a way that each layer will be loosely coupled, and each layer will be separately testable. The data will be mocked whenever needed. The proposed architecture will be proposing the writing of the unit test cases for the business layer with high code coverage.

3.3 Design Pattern

VIPER goes a step further by separating the view logic from the data model. Only, the presenter talks to the view, and only the interactor talks to an entity (model). The presenter and interactor coordinate with each other. The presenter is concerned with display and user action, and the interactor is concerned with manipulating the data.

View Layer: The view layer is consisting of view and controller classes and responsible for rendering the UI to the user. The view layer will receive the action from the user and send the action to the presenter layer where the presenter layer will decide the next action for that particular user gesture. The main responsibility of the view layer is to display the data to a user in a specified UI and receive the user gesture when the user interacts with the application. The view layer will not validate any data before showing it to the user (Fig. 2).

Fig. 2 VIPER iOS design pattern

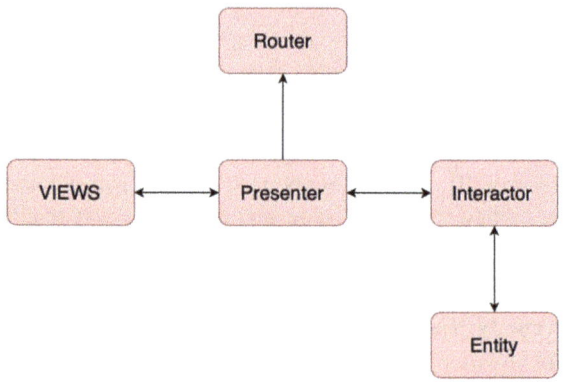

Presenter Layer: The presenter layer is a central layer between view, router, and interactor layers. The main responsibility of the presenter is to decide what to do with the received user gesture from the view layer. For example, if the user clicks on any navigation button, then the presenter layer will interact with the router to go to the next screen, while if the user clicks on to display more data, then might be presenter layer interact with the interactor layer to get the required data. So, the presenter layer is the main deciding layer to further process the app. All the UI-related business logic will be handled by the presenter layer before showing on the UI to the user.

Router Layer: The router layer will receive the command from the presenter layer. The main responsibility of this layer is navigating the user to the next screen. All the navigation animation and navigation styles like push, present, or any specific animation will be implemented by the router layer.

Interactor Layer: The interactor is a very important layer in the VIPER architecture design pattern. All the business logic of the app will be handled by this layer. All the algorithms, encryption, etc., will be done by this layer. In the proposed architecture, artificial intelligence and machine learning algorithms will be written in this layer. The logic of creating the ML model will be written in this layer. The main responsibility of this layer is to handle the app data received from the entity layer and then manipulate the data based on the requirement. The data manipulation will be like process the image data using the ML model, implement artificial intelligence algorithms, and ML techniques will be done in this layer.

Entity Layer: This is also called a data layer. This layer will receive the request from the interactor to give the data. This layer will retrieve the data from either the network layer or persistence layer based on what kind of data requested from the interactor layer. In the proposed architecture, only local/cached or persistence will be used as the app will work offline, so the network layer will not be used.

3.4 Dependency Injection

Dependency injection (DI) is a technique that allows populating a class with objects, rather than relying on the class to create the objects themselves.

The basic idea of DI is to have a separate object, an assembler, that populates a field in the class with an appropriate implementation for the interface. There are mainly three types of DI techniques: initializer injection, property injection, and interface injection.

In the proposed architecture, the initializer injection type of DI will be used where all the required data or model or dependencies will be passed in the init function of the class and that class should not create a new object of the dependency.

3.5   Database—Core Data

The database is used in mobile applications to store the information locally inside the app. Only that specific app-only access the data stored for that specific application and no other application can get any other app's data. Core data is a majorly used modern database.

Overview: Core data is one of the most popular frameworks provided by Apple for iOS. Core data is used to manage the model layer object in the application. Core data is used to save, modify, and filter the data within iOS apps. For more details, refer to [16]. As depicted in Fig. 3, the code data is a middle layer between the iOS mobile app and DB. The app will use the core data for permanent data offline, cache temporary data, and add undo functionality to an app on a single device.

Persistence: To make the saved data easy, core data abstracts the details of mapping objects to a store.

The responsibility of the core data's undo manager is to track the changes and roll them back individually, in groups, or all at once, making the undo and redo support easy in the mobile app. Background data tasks perform the UI-blocking tasks like API call, parsing the JSON data into objects in a background thread. The app can then store the data locally to reduce the API calls.

View Synchronization: Core data also helps keep your views and data synchronized by providing data sources for table and collection views.

Versioning and Migration: Core data includes mechanisms for versioning the app data model and migrating the user data.

The core data supports the lightweight and heavyweight data migration based on the changes in the user model. The lightweight data is easy and automatic like when adding a new property to the model or removing any existing property from the user model. On the other side, the heavyweight data migration requires a lot of

Fig. 3 Core data structure

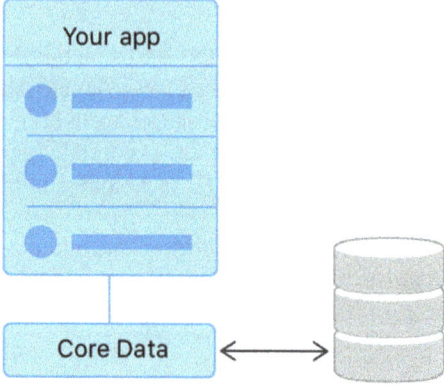

data changes in the model and adding a new model in the database. This kind of data migration requires a lot of effort and code to migrate the user model.

3.6 Programming

To set up the environment, some hardware and software will be required and are as follows:

The iOS mobile application—Tech Stack: To develop a mobile application, the following hardware has been used—MacBook Pro and iOS device. The following software and frameworks will be required: Xcode IDE, CreateML mac application [17], ML framework for iOS, and instrument for memory and performance analysis. Swift 5.x version language will be used for programming. A real iOS device will be used for testing the application.

The leaf disease detection system: The real-time capture image source data will be used to create the ML model [18]. The CreateML mac application will prepare the ML model from the image data source. Approximately 30% image data sources will be used as testing data.

4 Performance Analysis

The performance analysis is a very important phase in app development. It helps in analyzing the different aspects of the app like memory usage, slowness of the screen's navigation, etc. The behavior of the app will change if the performance of the app is slow that results in bad user experience. On a high level, the performance of the app at the code level can be analyzed via the following two ways.

4.1 Static Code Analyzer

As shown in Fig. 4, the static code analyzer is to analyze the static code. It reports about object life cycle and if the allocated object is not deallocated. It reports memory

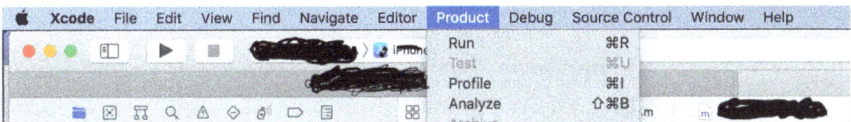

Fig. 4 Static code analyzer

leaks and if any UI-related task is done in a background thread. The Xcode static analyzer parses the project source code and identifies these types of problems:

- Logic flow, such as accessing uninitialized variables and dereferencing null pointers.
- Memory management flaws, such as leaking allocated memory.
- Dead stores (unused variables) flaws.
- API usage flaws result from not following the policies required by the frameworks and libraries the project is using.

The static analyzer reports problems in the issue navigator.

4.2 Instruments

Figure 5 shows all options for instruments to profile the project source code. Now, based on the requirement the option can be selected.

Instruments are part of the Xcode toolset which is a powerful, flexible performance analysis and testing tool. It is used to profile the iOS apps, processes, and devices to better understand and optimize their behavior and performance. Incorporating instruments in the app workflow from the beginning of the development process can

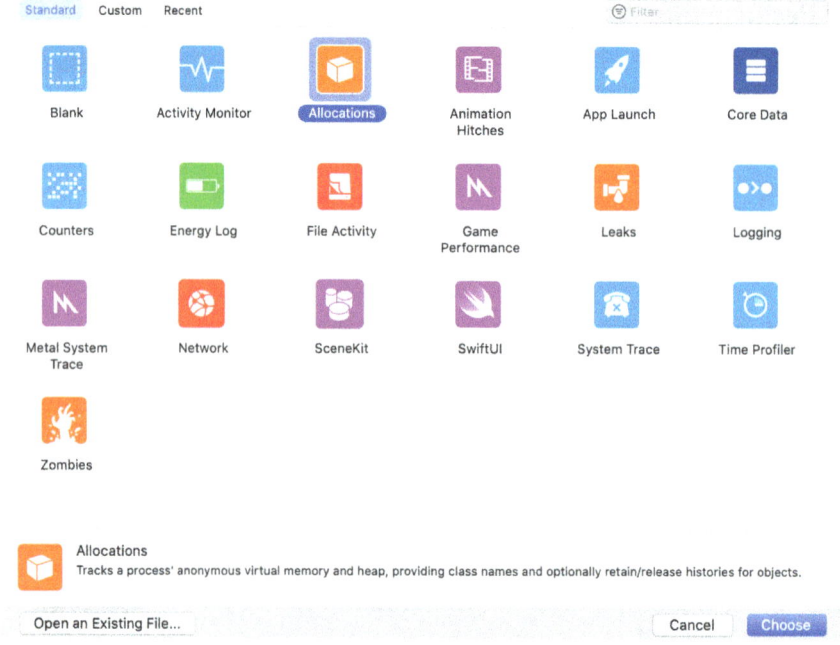

Fig. 5 Instruments and profiling

save time by helping in finding the issues in the early stage of the app development cycle. In instruments, specialized tools are used, known as instruments, to trace different aspects of the apps, processes, and devices over time. Instruments collect data as it profiles and presents the results in detail for analysis.

By using instruments effectively, it can:

- Examine the behavior of one or more apps or processes.
- Examine device-specific features, such as Wi-Fi and Bluetooth.
- Perform profiling in a simulator or on a physical device.
- Track down problems in the source code.
- Conduct performance analysis on the app.
- Find memory problems in the app, such as leaks, abandoned memory, and zombies.
- Identify ways to optimize the app for greater power efficiency.
- Perform general system-level troubleshooting.
- Save instrument configurations as templates.

5   Results

A flexible, scalable iOS app architecture is proposed which can be used in developing an iOS app for cotton plant disease detection. The architecture supports the usage of machine learning and artificial intelligence algorithms and the usage of the ML model. Based on the proposed architecture, the app can be further scalable to enhance the features in the app. The proposed architecture covers all the aspects of software design standards and guidelines. It also compliance the Apple human guideline and app review guidelines.

6   Conclusion

The proposed mobile app architecture is very much scalable for future work. Based on the proposed mobile iOS app architecture, the developed app can successfully detect the cotton plant disease with high accuracy and without Internet support. However, further work can be carried out to integrate the server-based implementation also. In this future work, if the Internet is available in the device, then plant detection can be done via server-based back-end implementation, and if the device is offline, then detection can be done without Internet with a local embedded ML user model. This kind of implementation will give huge flexibility to the user, and a very high level of accuracy can be obtained.

Acknowledgements I would like to thank Prof. Dr. Rajeev Ratan and Senior Prof. Dr. J.V Desai, MVN University, Palwal, India, for their guidance and help with my research.

Also, I would like to thank Dr. Devvrat Akheriya for the support throughout this work.

References

1. Food and Agriculture Organization in India. http://www.fao.org/india/fao-in-india/india-at-a-glance/en/
2. Punn M, Bhalla N (2013) Classification of wheat grain using machine algorithms. Int J Sci Res (IJSR) 2(8):363–366
3. Pooja V, Das R, Kanchana V (2017) Identification of plant leaf diseases using image processing techniques. In: IEEE international conference on technological innovations in ICT for agriculture and rural development (TIAR), pp 130–133
4. Sarangdhar AA, Pawar VR (2017) Machine learning regression technique for cotton leaf disease detection and controlling using IoT. In: International conference on electronics,communication and aerospace technology (ICECA). IEEE, pp 449–454
5. Singh KK (2018) An artificial intelligence and cloud based collaborative platform for plant disease identification, tracking and forecasting for farmers. In: 2018, IEEE International conference on cloud computing in emerging markets (CCEM), pp 49–56
6. Bodhe KD, Taiwade HV, Yadav VP, Aote NK (2018) Implementation of prototype for detection and diagnosis of cotton leaf disease using rule based system for farmers. IEEE, pp 165–169. ISBN 978-1-5386-4765-3/18/$31.00 ©2018
7. Kodana T, Hata Y (2018) Development of classification system of rice disease using artificial intelligence. In: IEEE International conference on systems, man, and cybernetics. IEEE, pp 3699–3702
8. Merchant M, Paradkar V, Khanna M, Gokhale S (2018) Mango leaf deficiency detection using digital image processing and machine learning. In: 3rd International conference for convergence in technology (I2CT), Pune India. IEEE, pp 1–3
9. Chanda M, Biswas M (2019) Plant disease identification and classification using back-propagation neural network with particle swarm optimization. In: Proceedings of the third international conference on trends in electronics and informatics (ICOEI 2019). IEEE Xplore Part Number: CFP19J32- ART; 2019, pp 1029–1036
10. Hasan Md.Z, Ahamed Md.s, Rakshit A, Hasan KMZ (2019) Recognition of jute diseases by leaf image classification using convolutional neural network. In: IEEE 45670, 10th ICCCNT 2019, 6–8 July 2019, IIT, Kanpur, Kanpur, India
11. Howlader Md.R, Habiba U, Faisal RH, Rahman Md.M (2019) Automatic recognition of guava leaf diseases using deep convolution neural network. In: 2019 International conference on electrical, computer and communication engineering (ECCE)
12. Hase AK, Aher PS, Hase SK (2017) Detection, categorization and suggestion to cure infected plants of tomato and grapes by using OpenCV framework for android environment. In: 2017 2nd International conference for convergence in technology, pp 956–959
13. Ranjith, Anas S, Badhusha I, Zaheema OT, Faseela K, Shelly M (2017) Cloud based automated irrigation and plant leaf disease detection system using an android application. In: 2017, International conference on electronics, communication and aerospace technology (ICECA) India, pp 211–214
14. Echalar LS, Subion MAT (2018) PaLife: a mobile application for palay (Rice) health condition classification utilizing image processing and pigment analysis towards sustainability of palay production. In: International seminar on research of information technology and intelligent system, pp 443–448
15. Elhassouny A, Smarandache F (2019) Smart mobile application to recognize tomato leaf disease using convolutional neural network. In: IEEE/ICCSRE2019, pp 1–4
16. Documentation Core Data. https://developer.apple.com/documentation/coredata
17. Framework Create ML. https://developer.apple.com/documentation/createml
18. Framework Core ML. https://developer.apple.com/documentation/coreml

Unknown Attack Detection in Cloud Based on Correlation Analysis

B. Kiranmai and S. V. Vasantha

Abstract Cloud computing is our present and future. Most of us knowingly and unknowingly depend on cloud for everyday life for storing, retrieving, and accessing. As the cloud is vulnerable to attacks, we do not know what kind of attacks is encountering which might collapse our security. Unknown attacks are the one that has not perceived yet; hence, we focused on detecting unknown attacks in the cloud. We proposed a novel approach for detecting attacks based on correlation coefficients which will give casual relationships between pair of attributes, calculated Euclidean distance, and applied hierarchical agglomerative clustering algorithm for building clusters which separate clusters of normal and abnormal users. Our approach has shown better results compared with other approaches.

Keywords Cloud · Machine learning · Correlation coefficients · Unknown attack · Hierarchical clustering

1 Introduction

This work attempts to classify between the normal and the abnormal data for predicting the happening of attack. This is made possible by analysing the traffic data, and the anomalies are detected. The anomalies can be detected by many techniques such as data mining, machine learning, soft computing-based techniques, knowledge, and statistical-based approaches.

Data mining consists of advanced set of techniques, which essentially takes a set of data as input and detects the patterns and deviations through data analysis. Thus, it becomes natural choice not only to detect anomalies, but also to construct the profiles of normal traffic [1].

Machine learning is an algorithmic method wherein an application automatically learns from the input and the feedbacks to improve its performance over time. Unlike statistical methods, which aim at determining the deviations in traffic features,

B. Kiranmai (✉) · S. V. Vasantha
Department of Computer Science and Engineering, KMIT, Narayanguda, Hyderabad, India

© The Author(s), under exclusive license to Springer Nature Singapore Pte Ltd. 2022
D. Gupta et al. (eds.), *Proceedings of Data Analytics and Management*,
Lecture Notes on Data Engineering and Communications Technologies 90,
https://doi.org/10.1007/978-981-16-6289-8_56

machine learning-based methods aim at detecting anomalies using some mechanism, and then, based upon false positive rates, the improvisations can be done [2].

A large number of statistical schemes assume that the presence of anomalies results in the deviation of certain traffic characteristics from normal, in terms of the volume [1]. In knowledge-based approaches, network events are checked against the predefined rules or patterns of attack [3]. In [4], the consensus-based combining method (CCM) is a new method for combining an ensemble of classifiers that is proposed and evaluated [5]. The outputs of multiple classifiers are weighted and added together in a single final classification judgement, as in the most other mixture methods. CCM, on the other hand, changes the weights iteratively after evaluating all of the classifiers' outputs, unlike the other approaches [6]. Artificial intelligence methods are often employed in computing for the purposes of preparation, forecasting, and assessment. Artificial neural networks (ANNs) are one of the most commonly used techniques for designing prediction models. For training neural networks, ANNs use a variety of meta-heuristic algorithms, including approximation methods [7]. In this field, artificial neural networks (ANNs) play a significant role and can be useful in determining the neural network input coefficient [8–10].

In these approaches, general representations of known attacks are formulated to identify actual occurrences of attacks. Examples of knowledge-based approaches include expert systems, signature analysis, self-organizing maps, and state transition analysis [5, 11].

The organization of paper is done as follows: Sect. 1 is about introduction; Sect. 2 is about the proposed unknown-attack detection scheme for cloud; Sect. 3 is about results and discussion, and finally, results are given in Sect. 4.

2 Proposed Unknown Attack Detection Scheme for Cloud

In order to detect unknown attacks in cloud, this work implements an innovative method, which is divided into five stages. The outcome of each stage is passed as an input to the following stage.

The initial stage prepares the data to be processed, such that the data are suitable for identifying the attacks. The second stage detects the relationship between the two different attributes for detecting the key attributes or the combination of attributes that can help in identification of attacks.

The third stage accepts the outcome of the second stage, where the correlation matrix is passed as input and the correlation coefficients are converted to distances. The fourth stage applies a hierarchical clustering algorithm to form clusters, and attack detection scheme is modelled based on distance.

Hence, the proposed anomaly detection algorithm for cloud relies on four stages, and they are data pre-processing, relationship analysis, distance computation, and hierarchical clustering (Fig. 1).

Fig. 1 Overall flow of the work

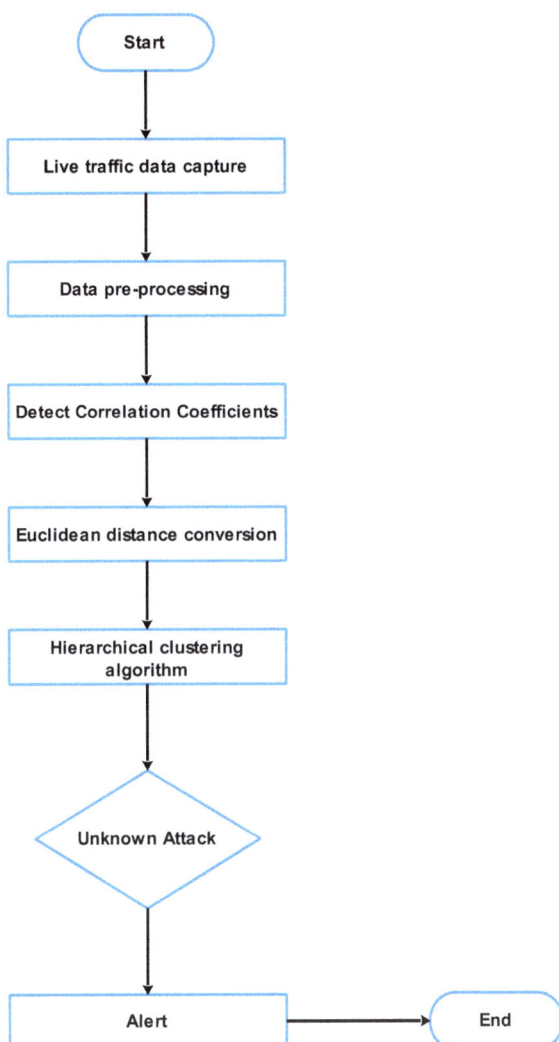

2.1 Data Pre-processing

Cloud log files are utilized for detecting abnormalities with respect to attack detection. Certain important attributes of the cloud log file are given in Table 1.

Most of the attributes in cloud log file are of type string. However, data analysis is possible only when the data can be performed correlation analysis. This is achieved by transforming the string data to the type integer. Similarly, the data in integer format are represented in the same way. The data are set to perform correlation analysis as described below.

Table 1 Sample cloud log file

Field	Type
C-IP	String
CS-method	String
CS-URI	String
SC-status	Integer
CS-bytes	Integer
CS-host	String

2.2 Correlation Coefficient Computation

Correlation coefficients are used in statistics to measure the strength of a relationship between a pair of features. This analysis is carried out on the cloud data. The values of correlation coefficient (CC) range between -1 and 1 [12].

$$(\forall) - 1 \leq CC \leq 1 \tag{1}$$

When the value of CC is -1, a strong negative bond is determined for the attribute pairs. Similarly, when the CC value is 1, a strong positive bond is observed between the attribute pairs. In case, when the CC value is 0, no correlation is observed between the attributes.

Let $D(x)$ be the cloud data to be processed; $F(x)$ denotes the features available in the dataset, and the correlation is indicated by $C(x)$. The correlation between the attributes is computed by

$$X_1 X_2 \ldots X_n = [X_1 \ldots X_n] x_i \varepsilon x_{i \forall i} \varepsilon [1 \ldots n] \tag{2}$$

Correlation coefficient determines the fitting of data in either line or curve. This work employs the Pearson correlation coefficient for detecting the casual relationship between a pair of attributes, which is a measure of the linear correlation between two variables.

$$\rho = \frac{\text{cov}(xy)}{\sigma_x \sigma_y} \tag{3}$$

where

$$\text{cov}(xy) = \frac{\sum_{i=1}^{n} \left(X_i - X^-\right)\left(Y_i - Y^-\right)}{n - 1} \tag{4}$$

In the above equation, X is an independent variable and Y is a dependent variable. The total number of data points in the sample is denoted by n. X^- and Y^- are the mean of independent variable X and dependent variable Y. The covariance is then represented by Eq. (4).

$$Cov(D(x)) = \begin{pmatrix} \cos x_{11} & \cos x_{12} & \cos x_{13} & \cos x_{14} & \dots & \cos x_{1n} \\ \cos x_{21} & \cos x_{22} & \cos x_{23} & \cos x_{24} & \dots & \cos x_{2n} \\ \vdots & \vdots & \vdots & \vdots & \vdots & \vdots \\ \cos x_{m1} & \cos x_{m2} & \cos x_{m3} & \cos x_{m4} & \dots & \cos x_{mn} \end{pmatrix} \tag{5}$$

The following section presents the distance computation.

2.3 Correlation to Euclidean Distance Conversion

The geometric and the scalar products of variables are given by the following equation.

$$d^2 = 1 - |r| \tag{6}$$

The computed geometric distances are employed for constructing clusters as follows.

2.4 Hierarchical Clustering

The clustering is performed to group the related entities together, and hence, the distance between the entities present in a single cluster is minimal [13, 14]. On the other hand, the entities present in different clusters are greater. The clustering algorithm employed to cluster the data is as follows:

Hierarchical Clustering Algorithm

Set cluster level $L(0) = 0$ and sequence number $m = 0$
Begin
Detect the least dissimilar cluster pairs (r) and (s) by Eq. (6)
Increment m;
Merge (r) and (s) into single cluster to proceed further and set $L(m) = d[(r), (s)]$
Update the proximity matrix D by modifying the existing matrix;
Compute proximity of the existing and modified matrices by Eq. (7);
If all the entities are in a cluster;
Stop the process;
Else
Repeat the process;
End;

$$d[(r), (s)] = \min d[(i), (j)] \tag{7}$$

In the above equation, the keyword min denotes all the pairs of clusters in the technique. The proximity matrix is updated by removing the rows and columns with respect to the clusters (r) and (s) and adding a row and column for the newly created cluster.

The proximity between the new and the existing cluster is computed by the following equation.

$$d[(k), (r, s)] = \min d[(k), (r)], d[(k), (s)] \tag{8}$$

The clustering operation is carried out by this way, and the following section presents the scheme for attack detection.

3 Results and Discussion

A dendrogram is a helpful technique to envision the bunching comes about. It is a tree diagram that is utilized to analyse the shapes of groups in progressive bunch investigation. The vertical hub demonstrates a separation or uniqueness measure. The stature of a hub speaks to the separation of the two bunches that the hub joins the bigger the tallness the more unique the groups.

This work employed hierarchical agglomerative clustering algorithm for unlabelled data. The data are not trained but can efficiently identify the entities of a particular group. Clearly, there are two groups of clusters, and they are attack cluster and normal cluster. The attained results clearly separate the attackers and the normal users.

Correlation and linear regression analysis are the statistical techniques to quantify the associations between an independent and dependent variables. The clusters are evaluated by using the sum of squares and silhouette coefficient method. The cluster dendrogram is presented as follows (Fig. 2).

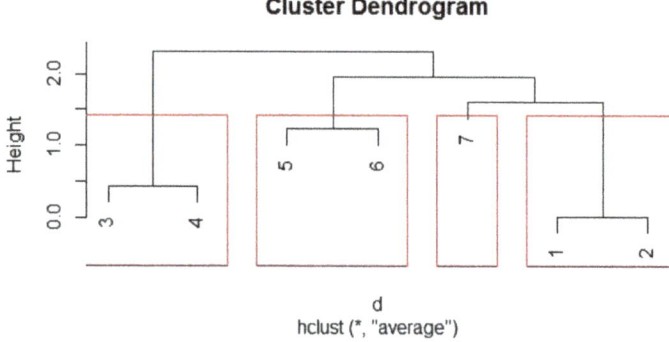

Fig. 2 Cluster dendrogram

The sum of squares estimation indicates the presence of related entities in groups, the littler the esteem, all the more firmly related items are inside the bunch and is computed by

$$\sum_{i \in c} (X_i - X^-)^2 \tag{9}$$

Silhouette coefficient is an estimation that considers how firmly related articles are inside the group and how bunches are isolated from each other. The outline esteem more often than not runs from 0 to 1 an esteem nearly to one indicates the information is better grouped.

The value says that the smaller the value the more closely related objects are in cluster. As per our results, hierarchical agglomerative with single linkage has performed well.

The average silhouette width value says that the value closer to one is better clustered. Agglomerative complete linkage has done well when compared with agglomerative single and k-means clustering. The following figure shows the cluster validation results (Fig. 3).

The performance comparison results with other clustering techniques are presented in Table 2.

Hence, the performance of the proposed approach is better in terms of SSE and Sil.

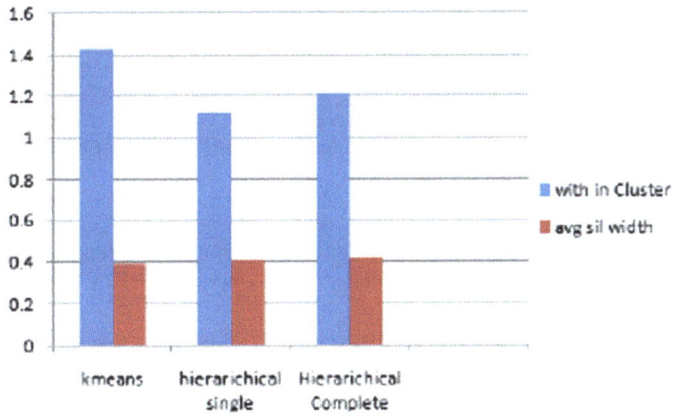

Fig. 3 Cluster validation results

Table 2 Results comparison with existing approaches

SSE	1.428	1.12	1.124	0.8
Sil	0.39	0.4	0.41	0.76

4 Summary

This paper presents a system for unknown attack detection in cloud environment with the help of five important phases. Initially, the collected traffic data are pre-processed and the correlation coefficients are computed. The Euclidean distance is then found out, and the hierarchical clustering algorithm is employed to group similar entities. Based on the clustering results, the unknown attack is detected and the alarm is generated to safeguard the system. Results are compared with existing methods and proven our method has better results.

References

1. Ghanem TF, Elkilani WS, Abdul-Kader HM (2015) A hybrid approach for efficient anomaly detection using metaheuristic methods. J Adv Res 6(4):609–619
2. Tan Z, Jamdagni A, He X, Nanda P, Liu RP (2011) Denial-of-service attack detection based on multivariate correlation analysis. In: International conference on neural information processing. Springer, pp 756–765
3. Alzubi OA, Alzubi JAA, Tedmori S, Rashaideh H, Almomani O (2018) Consensus-based combining method for classifier ensembles. Int Arab J Inf Technol 15(1):76–86
4. Movassagh AA, Alzubi JA, Gheisari M, Rahimi M, Mohan S, Abbasi AA, Nabipour N (2021) Artificial neural networks training algorithm integrating invasive weed optimization with differential evolutionary model. J Am Intell Hum Comput
5. Alhomoud A, Munir R, Disso JP, Awan I, Al-Dhelaan A (2011) Performance evaluation study of intrusion detection systems. Proc Comput Sci 5:173–180
6. Geva M, Herzberg A, Gev Y (2014) Bandwidth distributed denial of service: attacks and defenses. IEEE Secur Priv 12(1):54–61
7. Bhuyan, M. H., Kashyap, H. J., Bhattacharyya, D. K., & Kalita, J. K. (2014). Detecting distributed denial of service attacks: methods, tools and future directions. Comput J 57(4):537–556
8. Kumar, S. (2007). Survey of current network intrusion detection techniques. Washington Univ. in St. Louis, 1–18
9. Aljawarneh S, Aldwairi M, Yassein MB (2018) Anomaly-based intrusion detection system through feature selection analysis and building hybrid efficient model. J Comput Sci 25:152–160
10. Kiranmai B, Damodaram A (2016) Extenuate DDoS attacks in cloud. In: 2016 2nd international conference on applied and theoretical computing and communication technology (iCATccT) IEEE, pp 235–238
11. Ashfaq RAR, Wang XZ, Huang JZ, Abbas H, He Y-L (2017) Fuzziness based semi-supervised learning approach for intrusion detection system. Inf Sci 378:484–497
12. http://www.statisticshowto.com/
13. Chiba Z, Abghour N, Moussaid K, El Omri A, Rida M (2018) A novel architecture combined with optimal parameters for back propagation neural networks applied to anomaly network intrusion detection. Comput Secur 75:36–58
14. Jeyanthi N, Barde U, Sravani M, Tiwari V, Sriman Narayana Iyengar NC (2013) Detection of distributed denial of service attacks in cloud computing by identifying spoofed IP. Int J Commun Netw Distrib Syst 11(3):262–279
15. Idowu RK, Maroosi A, Muniyandi RC, Othman ZA (2013) An application of membrane computing to anomaly-based intrusion detection system. Proc Technol 11:585–592 (2013)

16. Mazzariello C, Bifulco R, Canonico R (2010) Integrating a network IDs into an open source cloud computing environment. In: 2010 sixth international conference on information assurance and security (IAS). IEEE, pp 265–270

Song/Music Recommendation Using Convolutional Neural Network and Keylogger

Kshitiz Badola, Deepesh Sengar, and Pooja Mudgil

Abstract Computers usage in combination with applications is the most important part of today's life. Computer users work on different computing applications such as data entry, freelance transcription, medical transcription to carry out daily routine tasks related to their work. These jobs are very time-consuming which lead to different mood swings of the user. So, a novel framework is proposed which will use convolutional neural network (CNN)-layered model for predicting facial emotions for each second of the user. Furthermore, to track the activity of user's typing speed, a keylogger is implemented which will tell typing speed of user for every 2-min interval. Then, results are computed using convolutional neural network (CNN) model in combination with keylogger and predict the user's mood in seven categories which are angry, disgust, sad, neutral, happy, surprise and fear and three typing speeds which are bad, good and very good. Then, using these variations in moods and typing speeds along with usage of Selenium library, recommended songs/music are auto-played using different playlists from YouTube platform (in background), just to uplift their mood (if tired, unhappy, etc.) or to make sure their mood remains uplifted (if happy, peaceful, etc.) to ensure their better performance in work.

Keywords Machine learning · Convolutional neural network · Selenium · Keylogger · Deep learning · Artificial intelligence

1 Introduction

It was only a few decades back, when machine learning was introduced. With its growing opportunities and its stretching fields all over the world, it builds intelligent

K. Badola (✉) · P. Mudgil
Guru Gobind Singh Indraprastha University, New Delhi, India

D. Sengar
Riga Technical University, Riga, Latvia

© The Author(s), under exclusive license to Springer Nature Singapore Pte Ltd. 2022   701
D. Gupta et al. (eds.), *Proceedings of Data Analytics and Management*,
Lecture Notes on Data Engineering and Communications Technologies 90,
https://doi.org/10.1007/978-981-16-6289-8_57

computer programs that could discover, learn, predict, and even revise its improvements with time to time. Machine learning also allowed us to train computer, not only to copy behavior of human being but also to perform targets and tasks that could have otherwise required considerable human efforts. Thus, machine learning lowers the human effort as much as possible to a larger extent. In our paper, all authors present a solution for a particular problem, i.e., as we have seen that there are many jobs that require continuous amount of typing for continuous hours, with that much of work, user/employee gets various kind of mood swings while working like happy, sad, angry, disgust, fear, surprise and neutral. All these emotions are generally common and normal for any user during any work. Authors in this project provide a framework which will combine these seven emotions with the typing speed of a user anywhere in the operating system and will result in generating a song/music from YouTube platform in the background (of operating system), such that our user maintains his/her work with good speed. The purpose of targeting song/music industry for this project was behind the logic of human psychology, i.e., user generally when gets tired, sad, angry, etc., he/she tries to uplift their mood by listening songs or music. We here in our project to improvise results, and we already categorized our various playlists matching with their moods (emotions) and typing speed. Our presented project can lead many companies and organizations to grow in their different fields of work, if any company using this project and can help their employees in working fast as well as completing work in short time. Companies with this project will be able to extract more amounts of work and efficiency from a particular employee. Also on other hand, it will help employees to stay relax while working through the medium of songs and music.

In the presented paper, our work has been described under few sections like, after the introduction part we will discuss about targeting milestones Sect. 1.1 which will tell us about all the targets and approach to the targets in our complete framework. After that, we will like to discuss on literature review Sect. 1.2 part which will tell us about the previous related research on our topic. After that, we will discuss on the important section of the paper which is facial emotion detector Sect. 2 which will tell us about the workflow of our created model and its results, and then, on the next section, another important topic will be introduced which is typing speed detector (keylogger) Sect. 2.1 which will tell us about the working and designing of keylogger for our framework. And then, one more important section called song/music generation Sect. 2.2 will be introduced which will discuss on how song/music will be generated for the user. After all this work, results and discussions Sect. 3 will be presented for revealing all the results from our framework and work done along with the complete flowchart of our work. Conclusion Sect. 4 is after results and discussions section which will tell us on future work for the project and also limitation in our project. And on last section, references of our previous related research work have been presented.

1.1 Targeting Milestones

We use facial emotion detector to classify seven common emotions to predict from a particular user. Emotions are like sad, angry, disgust, surprise, neutral, happy and fear; then, these results get combined with the typing speed of a user. Typing speed in this project will be measured in three categories bad, good and very good. These three categories will be revised for every two minutes, so that we can track our user's current mood time to time. With combination of results, from emotion detector and typing speed detector, we categorize our auto-playing songs from different playlist in background of desktop from YouTube so that our user should not face any kind of problem while typing in his/her work. Songs are also categorized with happy emotions, so that our user could complete his/her work with more speed in less time and with the pleasant mood till the time user is done with his/her work.

1.2 Literature Review

In paper [1], different approach called electroencephalography (EEG) was used in emotion recognition which results in average accuracy rate of 98%. Paper [2] proposed an emotion detection technique from images using convolutional neural networks (CNN)-based deep learning architecture. In paper [3], the proposed LMDS technique efficiently identifies malicious user behavior with minimum computational overhead (CO) and computational time (CT), and the proposed LMDS technique's performance is analyzed in terms of CO, CT and authentication accuracy. The purpose of the paper [4] is to create a deep learning model to generate captions for a given image by decoding the information available in the image. In paper [5], a novel algorithm for building ensembles called dynamic programming-based ensemble design algorithm (DPED) is introduced and studied in detail. In paper [6], a new method for combining an ensemble of classifiers, called consensus-based combining method (CCM), is proposed and evaluated, and experiments are then conducted on 14 public datasets and on a blog spam dataset created by the authors. In paper [7] approach of an affective cross-platform music player, EMP is presented which helps in recommendation of music based on the real-time mood of the user and provides an accuracy of 90.23%. In paper [8], building of emotion aware graph (EAG) on the basis of user, music, emotion has been done which is then used for music recommendation. In paper [9], results of classified facial emotions show the feasibility in education, consequently, which can help teachers to modify their presentation according to the student's emotions. In paper [10], authors proposed a hardware–software co-design for an FPGA-based real-time video processing system to convert video in standard phase alternating line (PAL) 576i format to standard video of video graphics array (VGA)/super video graphics array (SVGA) format with little utilization of resources, and they achieved results for performing efficient conversion with very less resource utilization compared to the existing solutions. In paper [11], presentation of new optimal mathematical model for the prediction of the degree of stakeholder satisfaction is presented in which optimal models validate the real data using the relationship impacts of various quality attributes which will give the maximum and minimum values for stakeholder satisfaction. The main goal of paper [12] is to train the neural network using meta-heuristic approaches and to enhance the perceptron neural network precision, and authors used an integrated algorithm to determine the neural network input coefficients; later, the proposed algorithm was compared with other algorithms such as invasive weed optimization and ant colony for performance evaluation. In paper [13], two independent methods were used to detect emotions, one was auto-encoders to construct a unique representation of each emotion, and the other was 8-layer convolutional neural network (CNN), and they were then tested on random images. In research [14], reviewing is done in three areas, which are the communication and perception of emotion in music, the emotional consequences of music listening and predictors of music preferences.

2 Facial Emotion Detector

Authors, here, use TensorFlow to train the model, and the dataset used in our model is from Kaggle and trained by several competitors from the competition in 2013 conducted by Kaggle only. In our model, there were 35,000 datasets in which we trained 28,000 successfully, and rest dataset were used as testing. On 25 epoch, we predict our accuracy to be 92.10% and with the validation loss of 21.96%. We, basically with the help of deep neutral network (DNN), used dataset named FER2013, and we design our own neural network with maxpool, averagepool and conv2D layers, which are depicted in Fig. 1. Authors named this model as Model 25.h5; now, this model will take live input image using webcam and then it will predict output as classified between seven categories, which are angry, sad, neutral, surprise, fear,

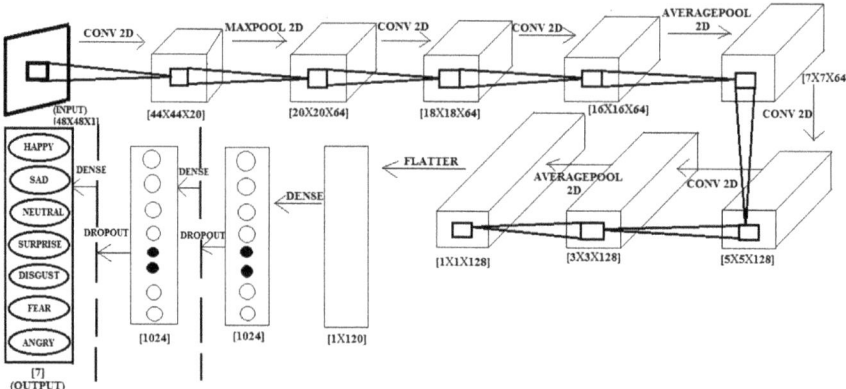

Fig. 1 Layers of CNN model

happy and disgust. These seven categories are generally common for any user while working.

In Fig. 1, we provided an input image (live feed using webcam) of size 48 × 48 × 1 in our depicted model, then after passing it from first layer of conv2D, the input image reforms with the size of 44 × 44 × 20, then after maxpool2D action on our input image, size reforms it to 20 × 20 × 64, and after this layer, two conv2D layers performed their action on image and size changes to 18 × 18 × 64 and 16 × 16 × 64 sequentially. Then, a new layer called averagepool2D layer comes into action which reforms the image size into 7 × 7 × 64. Till now, the image size was partially reformed and half of the moods were detected by these mentioned layers, and for further, we provide two more conv2D layers which again reformed the size of image by 5 × 5 × 128 and 3 × 3 × 128 sequentially. Again with this obtained size, there was a need of action of averagepool2D layer to reform it with the size of 1 × 1 × 128. Now, the new layer called flatter layer comes into action which changes three-dimension size of input image into two dimension, that is, 1 × 120. Now, the input image could be classified into seven emotions, but before that, the input size was passed through the action of dense and dropout layers which converted the two-dimension input size of the image into one dimension 1024 in the last step, which was then easily used to classify our input image into its seven emotions. Also, input image must be a having some character in it; otherwise, model cannot predict any emotion from the image.

2.1 Typing Speed Detector (Keylogger)

Our next milestone was to predict the typing speed of our user, and for that, we designed keylogger for our project. Keylogger was designed in way such that it will count number of words we type anywhere in the system, and the purpose of keylogger was to categorize the typing speed of a user into three types bad, good and very good. These three categories were selected on the basis of average typing speed of human being, and to ensure that, we took survey of 15 different people and we make them to type for about 15 min. With that survey, we found that average typing speed of a particular human being is 45 words per minute, so we categorized it as a "good" category of tying speed, then few of the survey takers were quite slow, so we recorded their typing speed and considered less than 35 words per minute as to be "bad" typing speed category, and few of the survey takers were very fast in typing, so their analyzed speed was more than 45 words per minute. So, this was considered to be as "very good" typing speed. Authors in this feature set use pynput keyboard library to run our keylogger in background for any application and for any work, so that it would store all the predicted typing data into working directory and count them and return counted word per minute as a result which then will be used for further progress in our project. Also with the help of time library, we set a timer for every 2 min while using keylogger so that keylogger keeps on refreshing its revisions and could select the current typing speed of user, and we did this for a reason to predict our user's current mood swing and time-to time with perfect accuracy.

2.2　Song/Music Generation

After considering both results together from facial emotion detector and typing speed detector (keylogger), we found 21 combinations of them (seven from emotion detector and three from keylogger). Authors, here, used Selenium library to work on chrome driver (browser) such that YouTube could be targeted for generating/playing songs according to those 21 combinations, on the basis of current mood (emotion). After that, we manually provided songs/music playlist for those final 21 combinations in our code work/project (e.g., if our user is in a combination set of sad mood and bad typing speed, we predicted motivational songs, for motivation). We also used Selenium keys to advance our project and to set four fixed pixels of a system screen, such that our YouTube screen which automatically launched in the system with playing song gets automatically minimized leaving, song running in background for the user, we did this thing, so that user need not to select song and minimize screen by him/her self, and this was only to save our user's time while working. And so, we are having 21 combinations playlist which can automatically play songs in background of work, and these playlists are designed with both user's positive and negative mood. So, if an employee working for any typing company or any user, while typing a larger data, once he/she lands in any one of those 21 combinations [like sad-bad, angry-bad, neutral-good, disgust-very good (emotions-typing speed)], our proposed playlist according to their mood will be played as a result, in the background of a system.

3　Results and Discussions

As for the final evaluation of our proposed framework, we tested our model on 25 different people with the time interval of 50–60 min each and provide them with their typing contents. And as for the results, our model successfully predicted all kind of mood songs for 23 people, according to their typing speed as well as their emotions. Also, our framework helped them in completing work in less time as the data were for 50–60 min without using this framework, and with using this feature set, they completed the task in 40–45 min. Thus, our model works successfully for any work related to typing, and it will help in avoiding all the tiredness and laziness of any user/employee and will help in uplifting of mood time to time with the help of various mood songs from YouTube. Now, considering Fig. 2 which depicts the model loss and accuracy on each epoch while running the complete model of CNN layers, the validation loss is presented by blue dots which get a drastic decrease on each increasing epoch, and on observation, we can also see the validation accuracy

Fig. 2 Graph for validation loss and accuracy

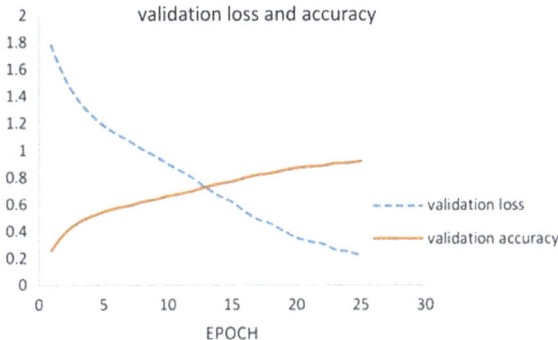

presented by red line which gradually increases on increasing epoch. And for our working framework, we also present a block diagram of complete model which is depicted by Fig. 3.

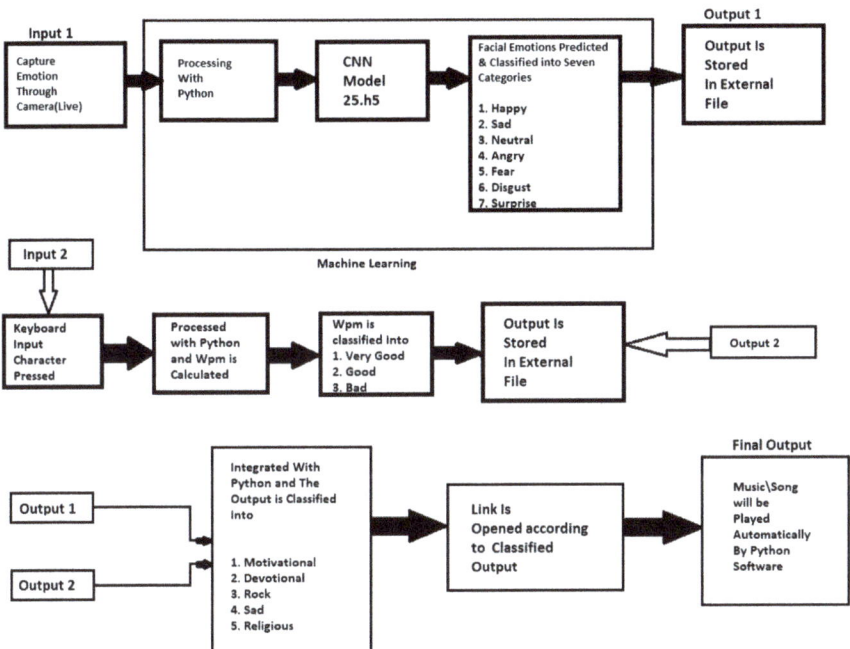

Fig. 3 Flowchart of complete project

4 Conclusion

In current framework, we proposed song recommendation system using emotions classified by CNN layers and typing speed of a particular user. For our future work, we would like to combine all the feature set and make it work for the real-time database system with more kind of emotions to understand users mood in an automatic fashion which will provide a better performance in our model; also, we would like to add a feature which will track user's time of typing, and for after 5–6 h of work, we will recommend sleeping songs so that user will be made to leave his/her desk for a break. This is only to ensure user's better health. One limitation in our feature set is needing of a high-processor PCs to run this framework as it is using machine learning algorithms in its procedure.

References

1. Gómez A et al (2016) An approach to emotion recognition in single-channel EEG signals: a mother child interaction. J Phys Conf Ser 705(1):012051
2. Jaiswal A, Raju AK, Deb S (2020) Facial emotion detection using deep learning. In: 2020 international conference for emerging technology (INCET). IEEE, pp 1–5
3. Alzubi JA (2021) Blockchain-based Lamport Merkle digital signature: authentication tool in IoT healthcare. Comput Commun 170:200–208
4. Alzubi JA et al (2021) Deep image captioning using an ensemble of CNN and LSTM based deep neural networks. J Intell Fuzzy Syst. Preprint: 1–9
5. Alzubi OA et al (2020) An optimal pruning algorithm of classifier ensembles: dynamic programming approach. Neural Comput Appl 32(20):16091–16107
6. Alzubi OA et al (2018) Consensus-based combining method for classifier ensembles. Int Arab J Inf Technol 15(1):76–86 (2018)
7. Gossi D, Gunes MH (2016) Lyric-based music recommendation. In: Complex networks VII. Springer, pp 301–310
8. Wang D, Deng S, Xu G (2016) Gemrec: a graph-based emotion-aware music recommendation approach. In: International conference on web information systems engineering. Springer, pp 92–106
9. Lasri I, Solh AR, El Belkacemi M (2019) Facial emotion recognition of students using convolutional neural network. In: 2019 third international conference on intelligent computing in data sciences (ICDS). IEEE, pp 1–6
10. Jain DK et al (2019) An efficient and adaptable multimedia system for converting PAL to VGA in real-time video processing. J Real-Time Image Process 1–13
11. Gheisari M et al (2019) An optimization model for software quality prediction with case study analysis using MATLAB. IEEE Access 7:85123–85138. https://doi.org/10.1109/ACC ESS.2019.2920879
12. Movassagh AA et al (2021) Artificial neural networks training algorithm integrating invasive weed optimization with differential evolutionary model. J Ambient Intell Hum Comput 1–9
13. Dachapally PR (2017) Facial emotion detection using convolutional neural networks and representational autoencoder units. arXiv preprint arXiv:1706.01509
14. Swaminathan S, Schellenberg, EG (2015) Current emotion research in music psychology. Emot. Rev. 7(2):189–197 (2015)

Journey of Letters to Vectors Through Neural Networks

Tathagat Banerjee, Amandeep Sharma, K. Charvi, Sruti Raman, Rahul Ganesh Regalla, and Taduru Sindhupriya

Abstract Getting an accurate image description has been one of the most discussed topics in the field of Artificial Intelligence. Numerous models/techniques have been developed in the past few years which makes it difficult to trace the exact path of Image Captioning models. This paper attempts its best to give the reader a clear idea of the evolution in the field of image captioning research elaborating on both traditional procedures and the advancements made with the aid of deep learning. This paper is aimed to discuss methods in detail and understand the very essence of depth and logic behind. It also relies upon the ravishing brilliance proceeded by the forthcoming authors. Further it shreds luminance on how the idea can grow in the near and long future.

Keywords Transfer learning · NLP · Reinforcement learning · Neural network · Image captioning · LSTM

T. Banerjee · A. Sharma · K. Charvi (✉) · S. Raman · R. G. Regalla · T. Sindhupriya
Computer Science and Engineering, VIT, Amaravati, AP, India
e-mail: charvi.19bce7002@vitap.ac.in

T. Banerjee
e-mail: banerjee.tathagat@vitap.ac.in

A. Sharma
e-mail: amandeep.18bcd7018@vitap.ac.in

S. Raman
e-mail: sruti.18bce7102@vitap.ac.in

R. G. Regalla
e-mail: rahul.18bec7045@vitap.ac.in

T. Sindhupriya
e-mail: sindhupriya.18bce7343@vitap.ac.in

1   Introduction

The field of artificial intelligence, or AI, attempts not just to understand but also to build intelligent systems. Ever since the research in the field of deep learning, which is an overlapping subset of AI, has taken pace, this advanced community has been trying its best to make computers smarter. Deep learning is trying to replicate the biological mechanisms of a human brain to achieve similar purposes through computational units which are connected to one another through weights. They take in data and train themselves to recognize the patterns and then predict the outputs for a new set of similar data.

Now, a question arises: Can machines be trained to interpret visuals and produce a logical description of an image in a similar manner?

The answer is yes to the very extent, researchers first attempted to make computers understand the handwritten digits. LeNet *(discussed in one of the sections)* paved the path for stronger beliefs in people to replicate the nature of human brains in computers. There were several models proposed after LeNet like Belu1-4 *(2002)*, Rouge *(2004)*, Space Mapping *(2011)*, n-gram *(2011)*, baby talk *(2013)*, meteor *(2014)*, etc. Models prior to 2015 were not really powerful but back in 2015 CIDEr proposed a model which uses CNN and LSTM to generate image descriptions which proved out to be one of the best models and set a benchmark. Our aim would be to discuss the latest model in detail so that the reader gets a clear idea about the advancement in the field of Image Captioning.

To generate captions for an image we need to extract the feature of the image and most commonly used feature extractors are ResNet, AlexNet, VGGNet, ImageNet, etc. Generation of accurate and well described image captions provided a considerable variation of advanced vocabulary remains a challenge to be solved. Yet there are a wide range of datasets which exploit the difficulties to a large extent, such as Flickr 8k, Flickr 30k, MS-COCO, etc. Automated machine translation of image description has great potential and importance when it comes to real-world scenarios as well.

The paper is broadly divided into different sections. Initially discussing the image captioning techniques that do not use deep learning procedures. Further segments will be a detailed review and explanation of image captioning techniques that highlights the extensive usage of deep learning standards and in the last section we will discuss the results and conclude our literary review.

2   Preliminary Exertion

An elaborated study on the initial attempts for Image Captioning (IC) which do not include deep learning facilities is described in this section. Traditional approaches synthesize the sentences from the most similar images to the query image.

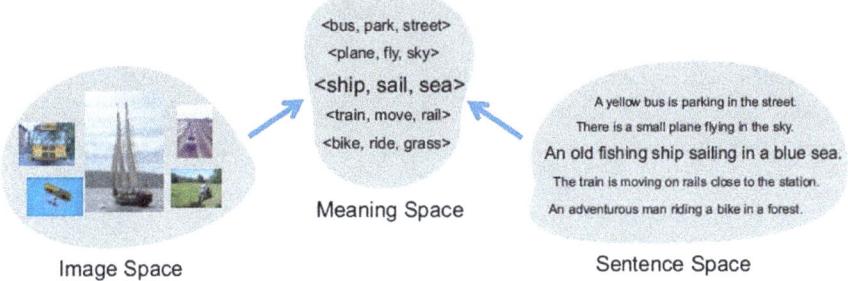

Fig. 1 Intermediate meaning space retrieved from Image and sentence space [9]

- **Retrieval Based**

The main aim is to develop a system which automatically verbalizes the information present on the source image to a target language. Also referred to as transfer-based processing this approach is performed in two classified steps. During the former step, all the candidate images which satisfy the similarities with a given query image are retrieved using content-based retrieval techniques.

The latter part, re-ranking procedures [10] are performed on the retrieved candidate images, computed using various methods. The global image features are used in the process and the captions are accessed with the existing set of descriptions. Finally, the captions of the top ranked images are used for composing a new caption for the query instance. There are cases of considering a top 'n' related image, for example consider a set of 10 images which satisfy the criteria of the objects present in the input image, with all the similar properties. Given this, we may consider the captions of maybe 3 samples to generate the description of our target image.

Furthermore, an intermediate *meaning space*, as shown in Figure 1 [9] consisting of three features: *(object, action, scene)* [20] are retrieved from the intersection of Image *space* and Sentence *space*. There is a scoring strategy to compare, estimate and decide whether the retrieved images match with input query's features. Mapping to the *meaning space* as described in the triplet uses a Markov Random field which is a graphical model of a joint probability distribution satisfying the Markov properties.

This entire process of recapturing or extraction strategies constitute our early works in IC without using deep learning techniques in retrieval-based image captions.

- **Template Based**

This method involves a prior execution of a detection algorithm, which generates a pre-defined language model trained over the Gigaword corpus [11] which fixes a template classifying and describing the scene. Unlike the previous retrieval method, this approach imposes these constraint mechanisms to compose the sentence.

A very famous method called the Conditional Random Field (CRF) [12], a statistical modeling for pattern recognition, which resulted in corresponding objects of the image being represented as nodes in a graph as shown in Figure 2 [12], showing

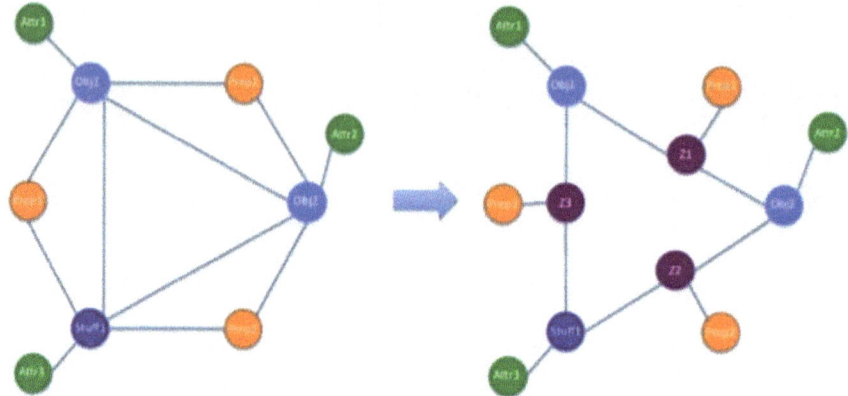

Fig. 2 CRF example image with left showing original CRF with trinary potentials and right showing CRF reduced to pairwise potentials by introducing z variables whose domains are all possible triples of the original 3-clique [12]

a clear relationship between interrelated objects. These are fairly helpful in building the bridge between various templates formed like the (noun, verb, scene, proposition) quadruplet [21] based approach used in analyzing the query image.

These works mainly focused on producing a new sentence by splitting each word and then combining based on templates like grammar semantics. For further enhancement of the process at a later stage, Common Subspace for Model and Similarity (CoSMos) [18] came into picture, wherein it considered phrases instead of individual words. This may improve the relevance in resulting captions but lacks the commonness like the human written descriptions and increases triviality.

These early works mentioned assumed captions from the existing data and failed to provide descriptions to unseen images. The task at hand was often maximizing correlation among the filtered species. It falls into a category of ranking more than generating captions which shows poor performance on both handling larger sets and optimizing space used for storing the same. This paved the way for neural networks and marked the beginning in advancements.

3   Model Analysis Involving Deep Learning Techniques

- **Encoder and Decoder Mechanism**

LeNet (1998): It was developed by Yann LeCun in 1998 which marked the beginning of a new era. It was first of its own kind and later was used as reference for many deep learning algorithms. Average pooling was used back in the late 90s *(max pooling is used extensively now),* and padding was also not exceedingly popular, which makes retaining the image size difficult. Tanh & Sigmoid functions were used instead of

ReLu (Rectified Linear) functions. The total parameter used in the architecture was close to 60K.

LeNet solved the extremely basic problem of recognizing the handwritten digit which laid the foundation for advanced architecture like VGGNet, Inception & ResNet-50 to solve more complex problems. LeNet did not provide any captions but it classified handwritten digits which form the base for the later advancements in the field of Image Captioning. The model architecture is shown in Fig. 3 [3].

The ground was all set and later with the use of an existing CNN model, a deep CNN model was developed which won the 2010 ImageNet-LSVRC *(Large Scale Visual Recognition Challenge)* [4]. After the successful intervention of CNN based model, soon researchers combined the Computer Vision (CV) and Natural Language Processing (NLP) to generate the captions. A simple architecture is followed to generate captions where we feed the vector generated from the feature extractor to the language models in a sequence and the output was again transformed from vector to sentence *(caption)*.

The comparison table [5] shows how the model performed in the ImageNet-LSVRC challenge (Fig. 4; Table 1).

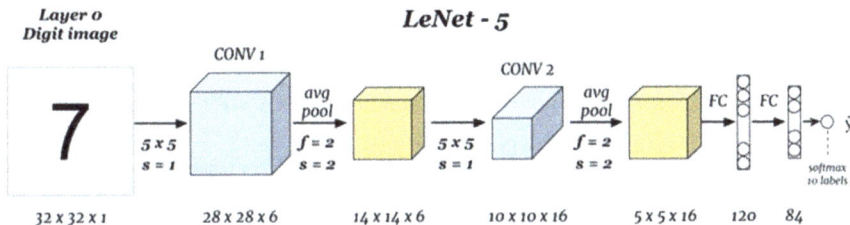

Fig. 3 Concise architecture of LeNet 5

Comparison					
Network	Year	Salient Feature	top5 accuracy	Parameters	FLOP
AlexNet	2012	Deeper	84.70%	62M	1.5B
VGGNet	2014	Fixed-size kernels	92.30%	138M	19.6B
Inception	2014	Wider - Parallel kernels	93.30%	6.4M	2B
ResNet-152	2015	Shortcut connections	95.51%	60.3M	11B

Fig. 4 Comparison of model performances in ImageNet challenge

Table 1 Some common feature extractors and language models

Feature extractors	Language model
VGGNet	RNN
AlexNet	CNN
Inception	TPGN
ResNet	LSTM

The encoder and decoder architecture were used extensively to improve accuracy during the challenge by making neural network deeper.

- **Visual Attention-Based Model**

This model was inspired by the recent advancement in the field of object detection and machine translation. It uses standard backpropagation technique to assign weights to the vectors produced by the feature extractor. Visual attention-based model is nothing but an enhanced version of encoder & decoder where the output of feature vectors is first passed through a specially calculated layer *(attention layer)* which assign more weightage to vectors, which are of more importance. The output of the attention layer is then fed to the Language models which create a corresponding text.

The attention layer shown in Fig. 5 is nothing but a weighted layer. To understand the mathematics behind the calculation of weighted layer better, refer to [6]. The model attempted to replicate human behavior, as humans tend to focus more on a specific region in the photo but in the previous mechanism discussed the weight factor was missing. Visual attention-based models added weight to make the model understand which region is of more importance.

Figure 6 [7] shows exactly how the model attended the correct object, and the word underlined shows the object in white spot. The model focuses particularly on objects that are of higher importance rather than considering every part of the image equally important. Another argument came as the model was developed, that every part of an image is equally important but to generate the caption we need to focus

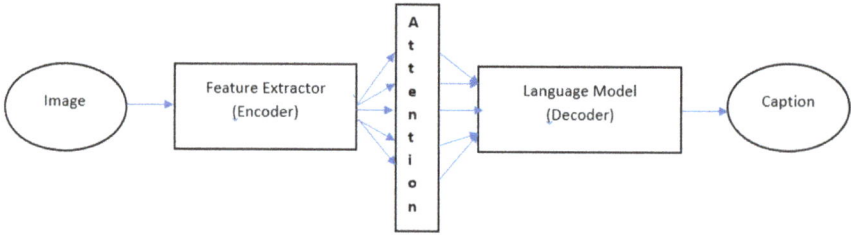

Fig. 5 Feature Vectors are fed to a weighted layer which act as input to the language model to get output after going through attention layer (weighted)

A woman is throwing a <u>frisbee</u> in a park.   A <u>dog</u> is standing on a hardwood floor.

Fig. 6 Object recognition along the white spots

Dataset	Model	BLEU				METEOR
		BLEU-1	BLEU-2	BLEU-3	BLEU-4	
Flickr8k	Google NIC(Vinyals et al., 2014)[†Σ]	63	41	27	—	—
	Log Bilinear (Kiros et al., 2014a)[∘]	65.6	42.4	27.7	17.7	17.31
	Soft-Attention	**67**	44.8	29.9	19.5	18.93
	Hard-Attention	**67**	**45.7**	**31.4**	**21.3**	**20.30**
Flickr30k	Google NIC[†∘Σ]	66.3	42.3	27.7	18.3	—
	Log Bilinear	60.0	38	25.4	17.1	16.88
	Soft-Attention	66.7	43.4	28.8	19.1	**18.49**
	Hard-Attention	**66.9**	**43.9**	**29.6**	**19.9**	18.46
COCO	CMU/MS Research (Chen & Zitnick, 2014)[a]	—	—	—	—	20.41
	MS Research (Fang et al., 2014)[†a]	—	—	—	—	20.71
	BRNN (Karpathy & Li, 2014)[∘]	64.2	45.1	30.4	20.3	—
	Google NIC[†∘Σ]	66.6	46.1	32.9	24.6	—
	Log Bilinear[∘]	70.8	48.9	34.4	24.3	20.03
	Soft-Attention	70.7	49.2	34.4	24.3	**23.90**
	Hard-Attention	**71.8**	**50.4**	**35.7**	**25.0**	23.04

Fig. 7 Model performance on different parameters

more on the important objects present to create a caption rather than focusing on every part.

The result table [6] below shows the model performance on different parameters. The major parameters that were considered to determine how good the model was BLEU-1,2,3,4, Meteor (Fig. 7).

- **Reinforcement Based Learning**

In complex domains, reinforcement learning has proved to be an efficient feasible solution to train a model and perform at high levels. The model can be provided with information to learn an evaluation function that gives a reasonably accurate estimate of the probability of winning. Given the definition, we can now explore the possibility of including the same learning paradigms in the image captioning process.

Although the Encoder and Decoder and Visual Attention-based models were good enough to give an image description, recent research has proved that Encoder-Decoder and Attention-based models suffer exposure bias problems as it uses maximum likelihood estimation [15]. To address such problems Reinforcement based models were introduced, which is developed on the top of reinforcement learning.

The existing models usually use a single decoder to produce descriptions as multi decoders can suffer from vanishing gradient problems. To deal with this intermediary problem, researchers came up with a new objective function [17]. To put in simple words, reinforcement learning uses output of every intermediary and preceding decoder to normalize reward *(Evaluation metrics score received by network)*.

Interesting models proposed by Rennie et al. [16] have shown a significant amount of improvement in optimizing the test metrics in their MS-COCO task, with the realization that reinforcement learning can be used to train deep end to end systems. This new optimization approach as they call it Self-Critical Sequence Training (SCST) has provided remarkable results.

Figure 8 [19] clearly shows that reinforcement learning and CNN-RNN based are much better than CNN-CNN.

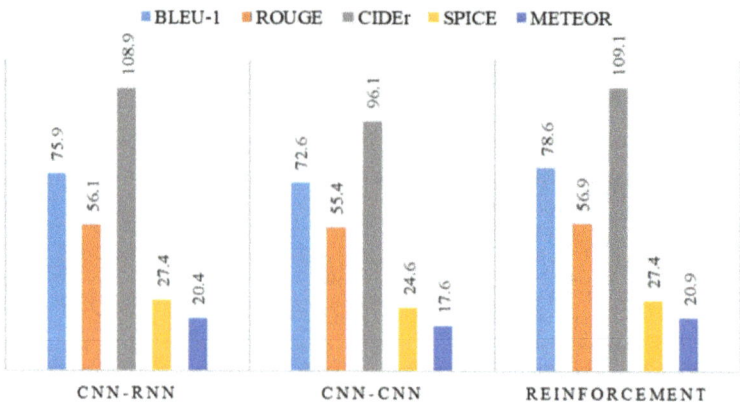

Fig. 8 Evaluation index for different models in the three methods [19]

4 Conclusion

This paper depicts the journey of Image Captioning Research. The early works reviewed in the paper, give a description about the initial approaches in Image Captioning. These techniques mainly focused on ranking the related content rather than generating new captions which led the researchers toward deep learning. Their performance was affected even due to the storage space. Handling a larger dataset had been challenging. Also, on a note the ranking based techniques showed a difficulty in correlating to human judgements and grammatical relationships.

Eventually usage of these conventional methods involving simple technology were not immensely powerful, and hence came an era where Dr Yann LeCun classified the handwritten digits using neural networks and thus paved the path for machines to understand images efficiently. Present technologies have really been developed in a magnificent manner. The work has been mostly done using publicly available datasets like FLICKR and COCO. Deep learning algorithms have proved to increase performance with increasing data.

Reinforcement Based Learning came into existence to solve the exposure bias problem and loss-evaluation mismatch which was a major drawback of Encoder-Decoder and Visual Attention Model.

It was able to achieve equally better results as existing models along with solving the major drawbacks. Although technologies have got really advanced at an enormous scale in the past two decades, there is still a considerable amount of work undone. Researchers can predict from the objects inside image to give a general description but still it lacks the natural compositionality. Work has been going on to fill this gap using generative adversarial networks (GANs) [14]. Researchers are now trying to make GAN's learn deep features to generate more diverse and distinct captions.

References

1. Krizhevsky A, Sutskever I, Hinton GE (2012) ImageNet classification with deep convolutional neural networks. In: NIPS, pp 1106–1114
2. Alom MZ, Taha TM, Yakopcic C, Westberg S, Sidike P, Nasrin MS, Van Essen BC, Awwal AAS, Asari VK Topbots: the history began from AlexNet: a comprehensive survey on deep learning approaches
3. Topbots.com/important-cnn-architectures/
4. Image-net.org/challenges/LSVRC/
5. Towardsdatascience.com/the-w3h-of-alexnet-vggnet-resnet-and-inception
6. Xu K, Ba JL, Kiros R, Cho K, Courville A Show, attend and tell: neural image caption generation with visual attention
7. Medium.com/image-captioning-using-attention-mechanism
8. Bergstra J, Breuleux O, Bastien F, Lamblin P, Pascanu R, Desjardins G, Turian J, Warde-Farley D, Bengio, Y (2010) Theano: a CPU and GPU math expression compiler. In: Proceedings of the Python for scientific computing conference (SciPy), 2010
9. Farhadi A, Hejrati M, Sadeghi MA, Young P, Rashtchian C, Hockenmaier J, Forsyth D (2010) Every picture tells a story: generating sentences from images. Springer, Berlin, Heidelberg
10. Hodosh M, Young P, Hockenmaier J (2013) Framing image description as a ranking task: data, models and evaluation metrics. J Artif Intell Res
11. Graff D, Cieri C (2011) English gigaword: linguistic data consortium. LDC2003T05
12. Kulkarni G, Premraj V, Dhar S, Li S, Choi Y, Berg AC, Berg TL (2011) Baby talk: understanding and generating simple image descriptions. In: 24th IEEE conference on CVPR 2011, vol 18, Colorado Springs, USA. IEEE, pp 1601–1608
13. Bernardi R, Cakici R, Elliott D, Erdem A, Erdem E, Ikizler N, Keller F, Muscat A, Plank B (2016) Automatic description generation from images: a survey of models, datasets, and evaluation measures. J Artif Intell Res
14. Dognin P, Melnyk I, Mroueh Y, Ross J, Sercu T Adversarial semantic alignment for improved image captions. IBM Research, Yorktown Heights, NY
15. Myung IJ (2003) Tutorial on maximum likelihood estimation. J Math Psychol 47:90–100
16. Rennie SJ, Marcheret E, Mrouch Y, Ross J, Goel V (2017). Self-critical sequence training for image captioning. In: 2017 IEEE conference on computer vision and pattern cognition (CVPR), Honolulu, USA. IEEE, pp 1179–1195
17. Gu J, Cai J, Chen T Stack-captioning: coarse-to-fine learning for image captioning. Nanyang Technological University, Gang Wang Alibaba AI Labs
18. Ushiku Y, Yamaguchi M, Mukuta Y, Harada T (2015) Common subspace for model and similarity: phrase learning for caption generation from images. In: 2015 IEEE international conference on computer vision (ICCV). IEEE, pp 2668–2676
19. Liu S, Bai L, Hu Y, Wang H Image captioning based on deep neural networks. College of Systems Engineering, National University of Defense Technology, Changsha, China
20. Mitchell M, Dodge J, Goyal A, Yamaguchi K, Stratos K, Han X, Mensch A, Berg AC, Berg TL, Daume III H (2012) Midge: generating image descriptions from computer vision detections. In: Proceedings of the 13th conference of the European chapter of the Association for Computational Linguistics, pp 747–756
21. Li S, Kulkarni G, Berg T, Berg A, Choi Y (2011) Composing simple image descriptions using web-scale n-grams. In: Proceedings of the 15th conference on computational natural language learning (CONLL 2011), Portland, USA. Association for Computational Linguistics, pp 220–228

A Dynamic Approach of Eye Disease Classification Using Deep Learning and Machine Learning Model

Rahul Pahuja, Udit Sisodia, Abhishek Tiwari, Siddharth Sharma, and Preeti Nagrath

Abstract In today's world, machine learning and deep learning models are applied in the diagnosis of eye diseases in an automated manner. Cataract, one of the eye diseases, can be especially dangerous, increasing the risk of eye damage. Early detection of Cataract is a requirement for Cataract patients to take the same medications at the same time. We suggest the use of in-depth reading to detect the presence of eye disease. In this research, Convolution Neural Network (CNN) and Support Vector Machine (SVM) were used for cataract detection on a dataset containing normal eye images and images of eye with cataract. Various techniques like data augmentation as a preprocessing step, label encoding, feature extraction have been applied in the research as a part of the model building process. SVM model has provided an accuracy score of 87.5% with an F1 score of 91.3% and CNN model has given 87.08% training accuracy and 85.42% validation accuracy.

Keywords Convolution neural network · Data augmentation · Support vector machine · Label encoding · Feature extraction · Preprocessing · Training · Accuracy

1 Introduction

A Cataract is an eye disease or issue majorly caused by aging, injury, or diabetes. It can cause permanent blindness if it is not controlled. Both eyes are affected usually by this disease. There are 3 type of this disease: nuclear cataracts, cortical cataracts, and posterior cataracts.

R. Pahuja · U. Sisodia · A. Tiwari · S. Sharma (✉) · P. Nagrath
Department of C.S.E, Bharati Vidyapeeth College of Engineering, New Delhi 110063, India

P. Nagrath
e-mail: preeti.nagrath@bharatividyapeeth.edu

© The Author(s), under exclusive license to Springer Nature Singapore Pte Ltd. 2022   719
D. Gupta et al. (eds.), *Proceedings of Data Analytics and Management*,
Lecture Notes on Data Engineering and Communications Technologies 90,
https://doi.org/10.1007/978-981-16-6289-8_59

Therefore, it has become a necessity to detect this disease in patients as early as possible in order to avoid its serious impact on the very sensitive organ, that is, eye. If a person remains ignorant and unaware of cataract for a longer period of time, cataract develop serious results and become more dangerous with time.

Abnormal variation in blood sugar level causes the onset of diabetes, and diabetic patients are more prone to cataract disease. Initially, in this disease, there may be absence of symptoms but with time, several symptoms develop like blurred vision, spots, impaired colored vision, and vision loss. Cataract can be prevented if a checkup of the eye is done at proper time.

In this paper, the eye disease detection model was build using deep learning and machine learning techniques, and finally accuracy was compared. Various works have been done in this field using various classifiers like SVM, neuro-fuzzy classifier, and ANN have been used. The accuracy obtained was although good, but when compared to hybrid approaches, they are less efficient. We have put effort to develop a binary classification model using the following two techniques:

CNN: A convolutional neural network (CNN) is a type of neural implant network used for image detection and operation designed to process pixel data. CNN uses a multilayer perceptron system designed to reduce processing needs. CNN layers consist of an input layer, an output layer, and a hidden layer that incorporates multiple convolutional layers, composite layers, fully connected layers, and standard layers.

SVM: Classification problems can be easily solved using a machine learning algorithm named as SVM. N dimensional space is used to plot every data point where the number of features is denoted by N. It consists of a hyper plane to differentiate and perform classification. We can choose different hyperplanes for dividing data points into two categories. A plane having maximum margin or we can say the plane having the largest distance among both classes data points, has to be chosen because it will ensure confidence in the classification of data points. This technique is also beneficial for higher-dimensional data.

1.1 Motivation

In recent scenarios, the eye disease detection has become a necessity to prevent the permanent loss of vision. Normally, colored fundus images are investigated by skilled experts, and this manual work takes a lot of time and may have some errors. So automated detection systems using deep learning models can be made to check the presence of particular complications in the person. Deep learning is advancing at a great pace so effective use of algorithms of deep learning can provide an insight to the disease.

1.2   Contribution

The model is built by employing two algorithms—first model is CNN which is a deep learning model and second is SVM, being a machine learning algorithm. In the initial stages, the binary classification of the disease can be done, and later on, a more advanced model using different deep learning algorithms can be developed to have multi-classification of different diseases. Model scans images to classify patient images into two categories, that with and without cataract. Our main motive in this research is to perform the classification of images into two categories to distinguish a normal eye from an eye having an issue such that extra work of segmentation and extraction can be avoided. The model will have below mentioned objectives:

- Surveying various eye diseases and finding out various datasets.
- Analyzing the techniques previously used for detection.
- Implementation and comparison of the accuracies obtained while performing disease detection.
- Creating an ensemble approach for effective detection.
- Combining the best approaches to achieve a new hybrid approach.

1.3   Uniqueness of the Paper

In this paper, we have used SVM for cataract detection, which is an effective model in high dimensional space, and a stable model, and CNN which helps us to automatically detect relevant and crucial features without much effort.

1.4   Organization

The organization of the remaining portion of the paper is in the following manner—the related works that have been done for disease detection using deep learning and machine learning models are contained in Sect. 2. Proposed approach is described in Sect. 3. Performance evaluation of the proposed approach is indicated in Sect. 4 and Sect. 5 is the concluding part of the paper.

2   Related Works

In 2002, Osareh et al. [1] used the c means clustering technique for retinal image segmentation and then classification into two classes—exudates and non-exudates were performed using different classifiers. Linear Delta Rule (LDR), K-Nearest Neighbor (KNN), Quadratic Gaussian classifier (QG), and Neural Networks (NN) were used for the purpose of classification and the highest accuracy was obtained by NN, that is, 90.1%.

Fuzzy classification, neuro-fuzzy classification, and ANN were used in another work by Acharya et al. [2] in 2006. They distinguished three types of diseases (cataract, iridocyclitis, and corneal haze). The accuracy obtained by the neuro-fuzzy classifier was high compared to the other two algorithms namely 85%.

In 2011, Karegowda et al. [3] identified that model with feature extraction performed before application of the final prediction algorithm is much better compared to the results obtained with the algorithm used on all input data. In this model, decision tree extracted the key features and the back propagation network was used to obtain the final detection of exudates with an accuracy of 98.45%.

The strategies of SVM and PNN (Probabilistic Neural Network) were used in a work conducted by Kandan and Aruna [4] in 2012. They divided patients into three general groups, nonproliferative eye diseases (NPDR), and proliferative eye disease (PDR). SVM has a high accuracy of 97.6% compared to the accuracy obtained by PNN. However, SVM produces ineffective results in large datasets.

It was proposed in 2013 by Yang et al. [5] that the use of an ANN can effectively differentiate the eye disease severity in an automated manner. High-quality carpet conversion and three-dimensional filter were used to improve image quality. Separately, a patient's cataract is classified as normal, mild, moderate, or severe. The accuracy obtained was 85%.

In a 2014 paper by Zheng et al. [6], they proposed a method of classifying the image of a fundus using cataract classification using pattern recognition. They consider the brightness of the lens of the eye as a low-resolution filter, which filters out the details of the eye fundus. This includes the two-dimensional discrete Fourier transform of the fundus images, in which the frequency of the images is obtained. Latent Dirichlet allocation (LDA) has been used as a base classifier and AdaBoost improves its performance. Accuracy of 95.22% and 81.52% was obtained for two and four categories classification.

In 2015, Salam et al. [7] has obtained texture elements from the fundus image, and the KNN model was used to distinguish the image and the feature extracted from the model were co-occurrence matrix, angular moment, sum average, sum variance, sum entropy correlation, and variance.

The 2015 paper of Guo et al. [8] proposes an integrated learning approach that incorporates a variety of different learning models for more accurate predictions of cataract classification. After that, mass ballot and multi-approval were adopted to cover the many foundations of the fundus image development fund raiser. The accuracy obtained was 92.8%.

In 2015, Chen et al. [9] and a team introduced a DL framework for glaucoma detection based on deep CNN, capable of capturing discriminatory features that better reveal hidden glaucoma-related patterns.

In 2015, Gao et al. [10] has stated a new methodology in which local filters, recursive neural networks, and support vector regression (SVR) were used to detect extent of cataract. Image patches are clustered and used to obtain learned filters, which are provided as input to neural networks, which provided output as extracted features, and finally, SVR was applied to be able to tell the severity.

The 2016 paper by Yang et al. [11] proposes an integrated learning approach that incorporates a variety of different learning models for more accurate predictions of cataract classification. After that, multiple base classifiers were assembled by employing stacking and majority voting for providing improved image categorization. The accuracy obtained using this method was 93.2%.

In 2016, Labhade et al. [12] used various subdivisions such as Adaboost, random forest, SVM, gradient boost, naive Bayes to diagnose eye diseases. The accuracy obtained by SVM is 88.71%, which is the highest among all strategies.

In the 2016 edition of the International Research Journal of Engineering and Technology, Umesh et al. [13] and the team used a novel approach to the automatic categorization of fundus images. This approach uses pre-image and data processing techniques to improve the performance of machine learning classifiers (Naive Bayes).

In 2017 Abbas [14] used a softmax linear classifier to find differences between these two common categories normal and images of glaucoma ROI.

In 2017 Burlina et al. [15] and the team developed a deep convolutional neural network that provided accuracy (SD) of between 88.4% (0.5%) and 91.6% (0.1%).

In the 2017, Xiong et al. [16] has provided a four-step model for grading the cataract which includes preprocessing, improving the image quality using two-dimensional Gaussian filter, extracting important features and finally by applying decision tree, they were able to grade the cataract with an accuracy of 92.8%.

The detection and grading of cataract were performed in a work done by Zhang et al. [17] in 2017. Authors used Deep CNN on 5620 images 0.93.52% accuracy was obtained for detection and 86.69% accuracy was obtained for cataract grading.

A 2017 paper made by Qiao et al. [18] proposed a new classifier called to support a vector machine based on the genetic algorithm for measuring features. High accuracy was observed as compared to the previous classifier and was considered an efficient approach. A major step forward is being made in the use of advanced divisions. The algorithm skips the time and effort consuming steps of a traditional algorithm depending on processing of fundus images and instead extracted and calculating features from the original images directly.

In 2018 Jain et al. [19] and the team developed a structure called LCDNet using CNN, which has successfully made the classification in category of two. They found between 96.5 and 99.7% accuracy in those datasets. Their model can be developed to study specific diseases and label appropriate diagnostic images. This will involve obtaining the largest and most diverse database and training the appropriate model for several labels.

Alzubi et al. [20] in 2018 has proposed a consensus based combining method for integrating various ensembles in which comparison of results of classifiers was done initially and then adjustment of weight was done in an iterative manner accordingly. CCM was found to be an efficient and accurate technique than standard product and average method. For providing efficient usage of system resources, Jain et al. [21] has developed a system for video conversion present in phase alternating line format to video graphic array form with additional advantages like allowing switching to different streams and detection of color of skin.

In 2019, a paper published by Ganguly et al. [22] reported an in-depth study method for the automatic diagnosis of eye melanoma using convolutional neural networks. Although the proposed method requires extensive calculation, a high accuracy rate of 91.76% is obtained by surpassing previous ANN work.

Gheisari et al. [23] in 2020 has provided a new mathematical model for forecast of participant satisfaction degree. Various parameters for quality measurement were calculated mathematically and relationship among them was studied for obtaining the optimal value of satisfaction degree. With the use of MATLAB optimal satisfaction degree, that is, Q was obtained to be 22.7 while idle being equals to 30.

ImageNet model was used to classify dataset of 100 images into two categories—having cataract and normal by Yusuf et al. [24] in 2019. A new model was built by training it with Transfer learning. 78% accuracy was obtained through this model while making classification.

Neural networks with three different activation functions (sigmoid, Gaussian, and ArcTan) used by Zaheer et al. [25] in 2019 to separate 16 retinal infections. Sigmoid work gave 92% accuracy, gaussian gave 90% accuracy while very little 46% accuracy was obtained with ArcTan.

In 2019, Prasad et al. [26] and team also proposed a system that could be used to diagnose eye disease and Glaucoma. This program serves as a starting point for referral, advising the patient to consult a retinal specialist when they are properly diagnosed. The pre-trained model is tested using a test set and real-time images. The accuracy obtained was 80%. This accuracy can be further enhanced by performing parameter adjustments and adopting methods such as cross validation.

Pratap et al. [27] in 2019 has proposed another work in which feature extraction was done through CNN and then machine learning model, SVM was applied for final classification of cataract into four stages. Accuracy obtained through this model was 92.91%.

Another work of Qummar et al. [28] presented a comprehensive integrated learning model by incorporating 5 CNN models—Resnet 50, Inceptionv3, Xception, Dense121, and Dense169 to place DR in 5 categories (normal, mild, moderate, severe, and PDR (Infectious Eye Disease) in 2019. The accuracy obtained by the model was 80.8%.

DCNN and SVM algorithms were used in the work done by Li et al. [29] in 2019. Fractional max-pooling layers were used in DCNN instead of max-pooling layers. SVM finally performed the classification of images into five levels of eye disease with an accuracy of 86%.

In another work, an integrated algorithm was proposed to find coefficients of neural networks with the main goal for minimizing error in prediction by Movassagh et al. [30] in 2020 and it was found that coefficients were effectively determined. Weights of the neural network were produced by using Algorithm named meta-heuristic and thus it was a heuristic approach.

In September 2020, Bhandari et al. [31] at Vivekanand Education Society's Institute of Technology, Mumbai, India performed cataract detection using RESNET.

Gadekallu et al. [32] has proposed a work in 2020 in which significant features were extracted using principle component analysis and standard scalar. Dimensionality reduction was done using firefly algorithm and then classification was obtained using a deep neural network model and machine learning models. The highest accuracy was obtained by a hybrid model of PCA-Firefly-DNN, that is, 97%.

Production of image description with the use of two deep learning models, that is, CNN and RNN have been proposed by Alzubi et al. [33] in 2020. Semantic tagging model was incorporated for generating semantic features, classification task was performed by CNN and thereby producing visual features and finally to generate captions RNN was used.

In 2020, Nazir et al. [34] and the FRCNN model could produce in-depth features and results showing that the proposed solution achieved an IoU mean of 0.95 and a mAP value of more than 0.94. A hybrid model based on dynamic programming was suggested in the work done by Alzubi et al. [35] in 2020. For reducing ensemble size, two phases were incorporated in which first phase involved game theory and second phases incorporated the approach of dynamic programming. Accuracy of the model was higher when compared with the classical hybrid model, that is, 96.5%.

For providing security to patients' data, a system for IoT, based on blockchain was suggested by Alzubi et al. [36] in 2021. A tree was created for purpose of verification where patient-sensitive data was stored in leaf nodes containing hash functions.

3 Proposed Approach

We propose the use of deep learning and machine learning to detect the presence of disease in the eye. There are various techniques that can be applied for cataract detection. We came across different approaches to solve the problem statement and used the CNN and SVM model to get a better result. The accuracy of models was calculated and inserted in tabular form in the table.

Steps for implementation of CNN model are stated below.

3.1 Dataset Description

A publicly accessible dataset of retina images was used by us in this research to train two models. Dataset used by us contains a total of 601 images, out of which we have used 399 images in the model building process. 299 images are of normal patients and 100 images are of patients having cataract. The dataset is used for building a CNN model to classify images into two classes, that is, cataract and normal. The source for the dataset is provided in [37]. Hence, early detection of cataract can be done when proper image analysis algorithms are made and evaluated.

3.2   Scaling of Images

Data preprocessing is a very crucial step while designing the model so that inconsistency is removed from the dataset. Biasness in classification is produced if we train the model with imbalanced data. Various preprocessing steps are incorporated in different models according to the need. In this model, we have used scaling as a preprocessing step.

We have used image data augmentation to generate randomly augmented image data. ImageDataGenerator class of Keras library is used in the model through which batch of images are taken and then each image of batch undergoes series of random operations like rotation and then CNN is applied on this randomly generated data instead of original data. There is a parameter rescale used in ImageDataGenerator class which is used to multiply every pixel of the image by the specified value. In our model rescale value is 1/255 because pixel value varies from 0 to 255 and we need value in the range of 0–1.

3.3   Label Encoding of Columns

Label encoder is applied to specific columns of the dataset so as to convert the two classes' names which are cataract and non-cataract in this model into numeric classes in the form of 0 and 1. Label encoder is present in sklearn preprocessing class and can be easily imported.

3.4   Splitting the Dataset into Training Set and Testing Set

Data arrays can be easily divided into two sets: train set and test set, using a function named as train_test_split() present in sklearn model selection. The test set is employed for predicting the accuracy of the designed model, and training of the model is performed using the train set. We have split dataset with the test size being equal to 20%.

For the dataset, first twenty-five images of training and testing data are plotted as shown in Figs. 1 and 2, respectively, along with the appropriate title, that is, whether it belongs cataract class or of normal class.

3.5   Building of Convolution Neural Network Model

First neural network consists of four 3 * 3 conv2D layers with activation function relu applied on it, four 2 * 2 max-pool layers, four drop out layers with dropout rate being

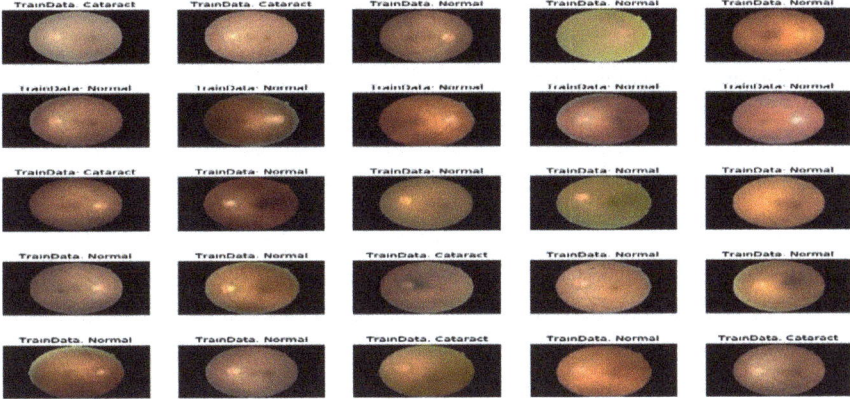

Fig. 1 Training images

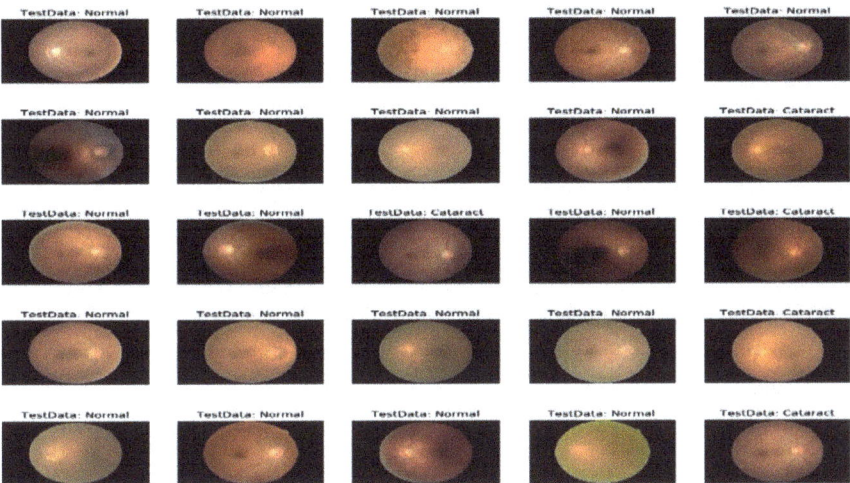

Fig. 2 Testing images

equals to 0.2, one flatten layer and four dense layers. The first conv2D layer consists of 64 filters, second and third conv2D layer consist of same number of filters, that is, 128 and fourth conv2D layer consist of 256 filters. A number of units in the dense layers are 256, 64, 16, 2, respectively. Among the four dense layers, three layers are with activation function relu applied on it and the last dense layer softmax activation function is applied.

The most widely used activation function in the neural networks is Relu function. It maps the function $F(x)$ to two values either zero for x less than zero and x if the value of x equal to or more than zero.

Softmax activation function:

The output layer of our model consists of the softmax activation function which is used for multi-classification problems. The mathematical formula involved in the softmax activation function is given as:

$$\frac{e^{zi}}{\sum_{j=1}^{K} e^{zj}} \tag{1}$$

In above equation, that is, Eq. 1, it is stated that on every element of output layer, an exponential function is applied, after that values are normalized by dividing with exponentials sum.

3.6 Configuring of CNN Model

After developing the desired model, it was configured with compile function of the model class present in the Keras library in which categorical_crossentropy was used as loss function, optimizer used was adam optimizer and accuracy was the desired metrics used in the model.

Adam optimizer is an adaptive learning optimizer that has gained wide popularity due to its efficiency. It is basically observed as a blend of stochastic gradient descent (SGD) with momentum and RMSprop. Learning rate are scaled using squared gradients in a similar way as RMSprop and then moving average of gradients are used to obtain the momentum benefit in a similar way as that of SGD with momentum.

The first moment is also called as mean denoted by m_t and the second moment is known as an uncentered variance of the gradients denoted by v_t. Final corrected estimators, that is, $\hat{m}_t$ and $\hat{v}_t$ are calculated by removing the bias as shown in Eqs. 2 and 3.

$$\hat{m}_t = \frac{mt}{1 - \beta 1}t \tag{2}$$

$$\hat{v}_t = \frac{vt}{1 - \beta 2}t \tag{3}$$

Updating of parameter requires following equation

$$\theta_{t+1} = \theta_t = \frac{\eta * \widehat{mt}}{\sqrt{\widehat{vt}} + \varepsilon} \tag{4}$$

here, θ_{t+1}—this is the weight at instant $t + 1$ or new weight.

θ_t—weight at instant t or previous weight.

η—it is the step size.

$\beta 1$—This is used for decaying the running average of the gradient (0.9).

α—Step size parameter (0.001).

ε—parameter to prevent Division from zero error (10^{-8}).

$\beta 2$—This is used for decaying the running average of the square of gradient.

3.7 CNN Training and Evaluation of Accuracy

Models were then trained on the training data by employing fit function of model class present in the Keras library. In the CNN model, when trained on provided dataset, number of epochs used for training are 24 while steps per epoch being equals to 8. Accuracy for training the model was evaluated to be 87.08% with validation accuracy being observed as 85.42%. Architecture of CNN model is shown in Fig. 3.

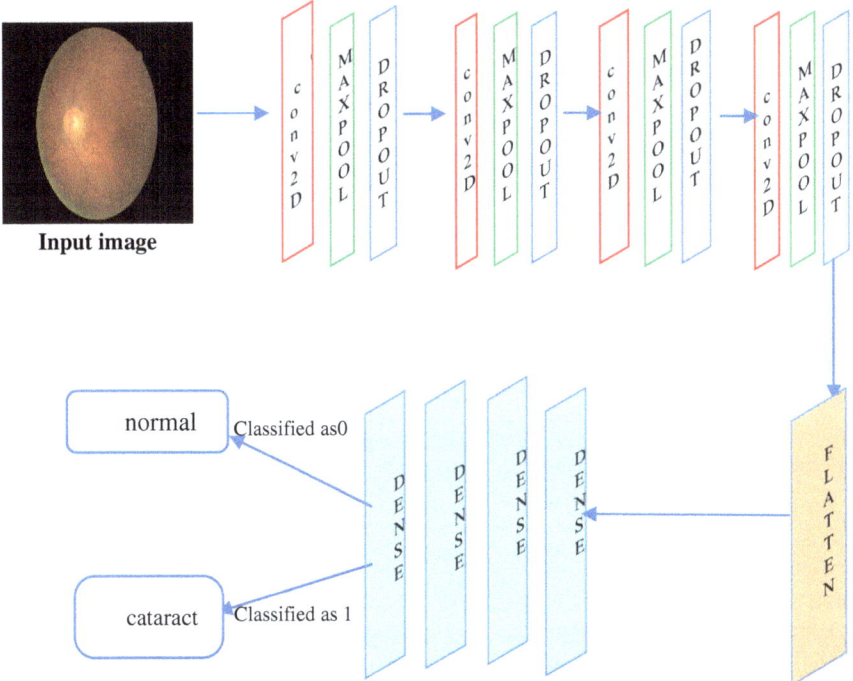

Fig. 3 Architecture of CNN model for eye disease classification

3.8 SVM

Steps of implementation of SVM are as follows:

(1) We read the image in the dataset.
(2) Images are rescaled for better visualization and to obtain efficient predictions.
(3) Hog feature descriptor was used which helped in extracting features from the input eye images.
(4) rgb images are converted into grayscale format using rgb2gray function.
(5) Then label encoding was done as described in Sect. 3.3.
(6) Then splitting of data into training and testing set in the ratio of 70 and 30 was performed respectively as described in Sect. 3.4.
(7) Training of model was done and accuracy scores were calculated as described in Sect. 3.8.1.

3.8.1 SVM Training and Its Evaluation

The training data was fed to the sklearn.svm. SVC classifier using the fit method.

Test data was fed to predict function of the classifier and predicted values are determined. Cross validation score was obtained using cross_val_score method present in sklearn model selection. Various accuracy scores are imported using sklearn metrics.

Accuracy score is defined as:

$$\text{Accuracy score} = \frac{\text{True positive} + \text{True negative}}{\text{True positive} + \text{True negative} + \text{False positive} + \text{False negative}}$$

F1 score is defined as:

$$\text{F1 score} = \frac{\text{True positive}}{\text{True positive} + \frac{1}{2}*(\text{False positive} + \text{False negative})}$$

Accuracy score achieved with SVM was 87.5% and F1 score being equals to 91.3%.

Figure 4 depicts the flow of steps involved in model building. Initially, the image dataset is read in step 1. Then in step 2 preprocessing of the dataset is done. Important features were extracted in steps 3. In step 4, we split data into two sets- train and test. Then the model is trained on training image set in step 5. Then the model was tested using various deep learning models and accuracy was obtained for different models.

Fig. 4 Flow chart for model building process

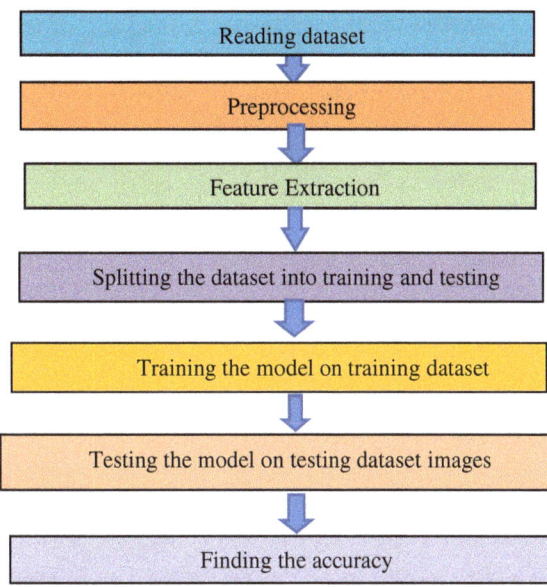

4 Performance Evaluation

4.1 Discussion on Results

Patients' images have been classified into two classes using a CNN and SVM model. Label Encoder converted the column name into two numeric classes. Dataset imported was split into two parts so that training of model was done and then made evaluation with test data size of 20% of whole data. Test data was fitted to the developed model and accuracy was calculated. The accuracy of the model is a crucial parameter for the evaluation of any model. The plot of accuracy and loss on training and validation dataset with respect to the number of training epochs has been plotted for CNN when tested on our dataset as shown in Figs. 5 and 6, respectively.

Accuracy obtained using CNN model:

Training	Testing	Validation
87.08	86.42	85.42

Accuracy obtained using SVM:

Accuracy score	F1 score
87.5	91.37

Fig. 5 Accuracy versus epochs

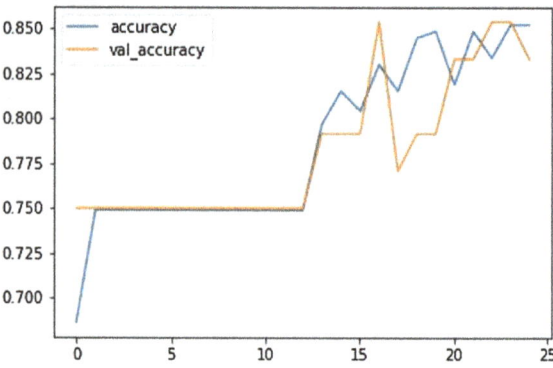

Fig. 6 Loss versus epochs

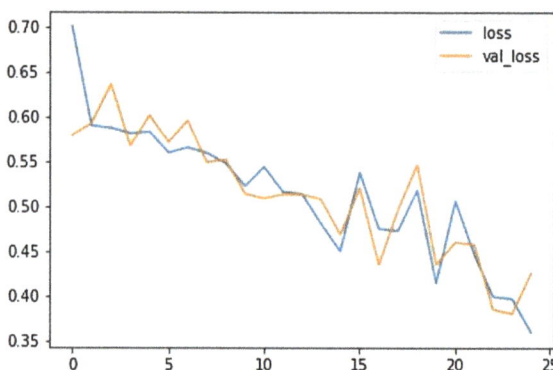

The heat map is the representation of matrix data in a graphical form. It is used to represent the confusion matrix or any two-dimensional data in form of color.

Figure 7 depicts heat map for CNN in which it has identified 57 true positives, 8 false positives, 12 true negatives along with 3 false negatives.

The Heat map for SVM is depicted in Fig. 8. It has successfully been able to identify 53 true positives, 3 false positives, 17 true negatives, and 7 false negatives.

Table 1 depicts comparative analysis of various previous works without model.

The accuracy can be improved by designing hybrid models in the future, which involve combining two or more algorithms together to form an efficient model. The

Fig. 7 Heat map for CNN model

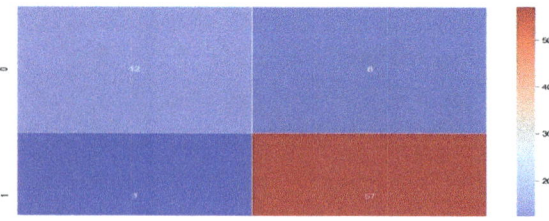

Fig. 8 Heat map for SVM

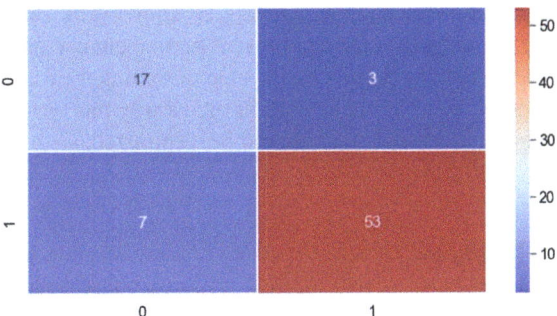

Table 1 Comparative analysis of various works

References	Year	Algorithm used	Accuracy (%)
[1]	2002	(1) LDR (2) BP (3) QG (4) Scaled conjugate gradient (5) KNN	81.8 89.6 81.3 90.1 87.1
[12]	2016	(1) SVM (2) Random Forest (3) GaussianNB (4) AdaBoost (5) GradientBoost	88 83 37 54 83
[16]	2017	Decision Tree	92
[19]	2018	CNN	97
[24]	2019	ImageNet Model	78
[31]	2020	Resnet	92.5
Our model	2021	(1) CNN (2) SVM	85.42 87.5

hybrid approach gives better results as compared to single deep learning or machine learning algorithm.

5 Conclusion

This paper has justified the need of detecting cataract disease so that permanent loss of vision can be prevented. Simple deep learning model, that is, convolution neural network (CNN) and a machine learning model, that is, SVM was used to classify patient images into two categories, that is, with cataract and without cataract. SVM has provided high accuracy as compared to CNN while making prediction of cataract. This is the primary need in every hospital to detect cataract with least possible effort in an automated manner. Patients, by knowing the disease, can follow certain precautions at an early stage can prevent permanent eye damage. The current models built have involved a preprocessing step of scaling the images using ImageDataGenerator class. Models should be built in such a way that we can better result in the least time and preprocessing steps should be minimized.

References

1. Osareh A, Mirmehdi M, Thomas B, Markham R (2002) Classification and localisation of diabetic-related eye disease. In: Heyden A, Sparr G, Nielsen M, Johansen P (eds) Computer vision—ECCV 2002. ECCV 2002. Lecture notes in computer science, vol 2353. Springer, Berlin, Heidelberg. https://doi.org/10.1007/3-540-47979-1_34
2. Acharya UR, Kannathal N, Ng EYK, Min LC, Suri JS (2006) Computer-based classification of eye diseases. In: 2006 international conference of the IEEE Engineering in Medicine and Biology Society, New York, NY, 2006, pp 6121–6124.https://doi.org/10.1109/IEMBS.2006. 260211
3. Karegowda A, Nasiha A, Jayaram MA, Manjunath A (2011) Exudates detection in retinal images using back propagation neural network. Int J Comput Appl 25. https://doi.org/10.5120/3011-4062
4. Kandan RP, Aruna P (2012) SVM and neural network-based diagnosis of diabetic retinopathy. Int J Comput Appl 41:6–12. https://doi.org/10.5120/5503-7503
5. Yang M, Yang J-J, Zhang Q, Niu Y, Li J (2013) Classification of retinal image for automatic cataract detection. In: IEEE international conference on e-health networking, applications & services, 2013, pp 674–679
6. Zheng J, Guo L, Peng L, Li J, Yang J, Liang Q (2014) Fundus image based cataract classification. In: IEEE international conference on imaging systems and techniques, pp 90–94
7. Salam AA, Akram MU, Wazir K, Anwar SM (2015) A review analysis on early glaucoma detection using structural features. In: 2015 IEEE international conference on imaging systems and techniques (IST), Macau, 2015, pp 1–6. https://doi.org/10.1109/IST.2015.7294516
8. Guo L, Yang J-J, Peng L, Li J, Liang Q (2015) A computer-aided healthcare system for cataract classification and grading based on fundus image analysis. Comput Ind 69:72–80
9. Chen X, Xu Y, Kee Wong DW, Wong TY, Liu J (2015) Glaucoma detection based on deep convolutional neural network. In: 2015 37th annual international conference of the IEEE Engineering in Medicine and Biology Society (EMBC), Milan, 2015, pp 715–718.https://doi.org/10.1109/EMBC.2015.7318462
10. Gao X, Lin S, Wong TY (2015) Automatic feature learning to grade nuclear cataracts based on deep learning. IEEE Trans Biomed Eng 62(11):2693–2701. https://doi.org/10.1109/TBME.2015.2444389
11. Yang J-J, Li J, Shen R, Zeng Y, He J, Bi J, Li Y, Zhang Q, Peng L, Wang Q (2016) Exploiting ensemble learning for automatic cataract detection and grading. Comput Methods Prog Biomed 124:45–57
12. Labhade JD, Chouthmol LK, Deshmukh S (2016) Diabetic retinopathy detection using soft computing techniques. In: 2016 international conference on automatic control and dynamic optimization techniques (ICACDOT), pp 175–178
13. Umesh M, Mrunalini MM, Shinde DS (2016) Review of image processing and machine learning techniques for eye disease detection and classification
14. Abbas Q (2017) Glaucoma-Deep: detection of glaucoma eye disease on retinal fundus images using deep learning. Int J Adv Comput Sci Appl 8. https://doi.org/10.14569/IJACSA.2017. 080606

15. Burlina PM, Joshi N, Pekala M, Pacheco KD, Freund DE, Bressler NM (2017) Automated grading of age-related macular degeneration from color fundus images using deep convolutional neural networks. JAMA Ophthalmol 135(11):1170–1176
16. Xiong L, Li H, Xu L (2017) An approach to evaluate blurriness in retinal images with vitreous opacity for cataract diagnosis. J Healthc Eng 2017:Article ID: 5645498
17. Zhang L et al (2017) Automatic cataract detection and grading using Deep Convolutional Neural Network. In: 2017 IEEE 14th international conference on networking, sensing and control (ICNSC), Calabria, 2017, pp 60–65. https://doi.org/10.1109/ICNSC.2017.8000068
18. Qiao Z, Zhang Q, Dong Y, Yang J-J (2017) Application of SVM based on genetic algorithm in classification of cataract fundus images. In: IEEE international conference on imaging systems and techniques, pp 1–5
19. Jain L, Murthy HVS, Patel C, Bansal D (2018) Retinal eye disease detection using deep learning. In: 2018 fourteenth international conference on information processing (ICINPRO), Bangalore, India, 2018, pp 1–6. https://doi.org/10.1109/ICINPRO43533.2018.9096838
20. Alzubi OA, Alzubi JA, Tedmori S, Rashaideh H, Almomani O (2018) Consensus-based combining method for classifier ensembles. Int Arab J Inf Technol 15(1)
21. Jain DK, Jacob S, Alzubi J, Menon V (2019) An efficient and adaptable multimedia system for converting PAL to VGA in real-time video processing. J Real-Time Image Process. https://doi.org/10.1007/s11554-019-00889-4
22. Ganguly B, Biswas S, Ghosh S, Maiti S, Bodhak S (2019) A deep learning framework for eye melanoma detection employing convolutional neural network. In: 2019 international conference on computer, electrical & communication engineering (ICCECE). https://doi.org/10.1109/iccece44727.2019.9001858
23. Gheisari M, Panwar D, Tomar P, Harsh H, Zhang X, Solanki A, Nayyar A, Alzubi JA (2019) An optimization model for software quality prediction with case study analysis using MATLAB. IEEE Access 7
24. Yusuf M, Theophilous S, Adejoke J, Hassan AB (2019) Web-based cataract detection system using deep convolutional neural network. In: 2019 2nd international conference of the IEEE Nigeria computer chapter (NigeriaComputConf), Zaria, Nigeria, 2019, pp 1–7.https://doi.org/10.1109/NigeriaComputConf45974.2019.8949636
25. Zaheer N, Shehzaad A, Gilani SO, Aslam J, Zaidi SA (2019) Automated classification of retinal diseases in STARE database using neural network approach. In: 2019 IEEE Canadian conference of electrical and computer engineering (CCECE), Edmonton, AB, Canada, 2019, pp 1–5.https://doi.org/10.1109/CCECE.2019.8861588
26. Prasad K, Sajith PS, Neema M, Madhu L, Priya PN (2019) Multiple eye disease detection using Deep Neural Network. In: TENCON 2019—2019 IEEE Region 10 conference (TENCON). https://doi.org/10.1109/tencon.2019.8929666
27. Pratap T, Kokil P (2019) Computer-aided diagnosis of cataract using deep transfer learning. Biomed Signal Process Control 53:101533. https://doi.org/10.1016/j.bspc.2019.04.010
28. Qummar et al (2019) A deep learning ensemble approach for diabetic retinopathy detection. IEEE Access 7:150530–150539. https://doi.org/10.1109/ACCESS.2019.2947484
29. Li Y-H, Yeh N-N, Chen S-J, Chung Y-C (2019) Computer-assisted diagnosis for diabetic retinopathy based on fundus images using deep convolutional neural network. Mob Inf Syst 2019:Article ID 6142839, 14 pp. https://doi.org/10.1155/2019/6142839
30. Movassagh AA, Alzubi JA, Gheisari M, Rahimi M, Mohan SK, Abbasi AA, Nabipour N (2020) Artificial neural networks training algorithm integrating invasive weed optimization with differential evolutionary model. J Ambient Intell Hum Comput. https://doi.org/10.1007/s12652-020-02623-6
31. Bhandari AA (2020) Eye disease detection using RESNET. Int Res J Eng Technol (IRJET) 07
32. Gadekallu TR, Khare N, Bhattacharya S, Singh S, Maddikunta PKR, Ra I-H, Alazab M (2020) Early detection of diabetic retinopathy using PCA-firefly based deep learning model. Electronics 9:274
33. Alzubi JA, Jain R, Nagrath P, Satapathy S, Taneja S, Gupta P (2020) Deep image captioning using an ensemble of CNN and LSTM based deep neural networks. J Intell Fuzzy Syst. https://doi.org/10.3233/JIFS-189415

34. Nazir T, Irtaza A, Javed A, Malik H, Hussain D, Naqvi RA (2020) Retinal image analysis for diabetes-based eye disease detection using deep learning. Appl Sci 10:6185
35. Alzubi OA, Alzubi JA, Alweshah M, Qiqieh I, Al-Shami S, Ramachandran M (2020) An optimal pruning algorithm of classifier ensembles: dynamic programming approach. Neural Comput Appl
36. Alzubi JA (2021) Blockchain-based Lamport Merkle digital signature: authentication tool in IoT healthcare. Comput Commun 170:200–208
37. Source of dataset. https://www.kaggle.com/jr2ngb/cataractdataset

Feature and Decision Fusion for Breast Cancer Detection

Rohit Yadav, Richa Sharma, and Pushpendra Kumar Pateriya

Abstract With the advancement of technology and its implementation in the medical field, a lot of new methods and techniques are created to solve complex real-life problems. Breast cancer is a major issue in women and corresponds to a quarter of deaths related to it. In this paper, we have used the MIAS dataset which consists of 322 mammograms (161 pairs). Mammograms are cheap and easy to understand. For contrast enhancement, we have used contrast limited adaptive histogram equalization (CLAHE) with a clip size of 0.2. We have compared the PSNR of histogram equalization (HE), CLAHE, minimum mean brightness error bi-histogram equalization (MMBEBHE), and recursive mean-separate histogram equalization (RMSHE). We found that CLAHE and RMSHE perform better than MMBHBHE and HE. A convolutional neural network (CNN) architecture is created for feature extraction, and a total of 2115 features is provided for classification. SVM, decision tree, and random forest are used as classifiers. The accuracy achieved by SVM, decision tree, and random forest is 92.30, 94.03, and 95.05%, while using decision fusion (SVM, decision tree, and random forest), the highest accuracy of 96.12% is achieved.

Keywords Breast cancer · CLAHE · Image fusion · Machine learning

1 Introduction

Image fusion is a way of merging several pictures of information from the same platform or from separate spectroscopic platforms to generate a single output image using various approaches and procedures. For machine perception or human comprehension, the resultant image (known as the fused image) includes more detailed

R. Yadav (✉) · R. Sharma · P. K. Pateriya
SCSE, Lovely Professional University, Jalandhar, India

R. Sharma
e-mail: richa.18364@lpu.co.in

P. K. Pateriya
e-mail: pushpendra.14623@lpu.co.in

and valuable or predictable information [1]. A general process of the images fusion process is shown in Fig. 1 [2, 3]. Many of the scientific fields use these techniques for real-life application use like in satellite photography, medical fields, and professional photography. To carry out these applications, we need to understand the complete picture of the problem. Different imaging modalities like ultrasound, X-ray, magnetic field are used to capture the different segments of the image, but to understand the final structure, these images are fused. This process helps us to predict or perform the operation more precisely. However, the implementation of the fusion process will not always be the last step. The fused image created after the fusion process can either be directly given as output to the expert, or it can be used as input data for machine learning algorithms for further process.

Breast cancer, lung cancer, etc., cases are on the decline from 1991 to 2018 due to improvements in the medical field and better living conditions which resulted in 3.2 million fewer deaths due to cancer [4]. With new technologies and a lot of digital data, we now started experimenting and implementing machine learning and deep learning algorithms for processing and computing complex information. Researchers have used SVM, random forest, decision tree, KNN, different variants of artificial neural network (CNN, BPNN, RBFNN) for the classification of cancer legions [5–8]. Improvements are achieved by using image fusion techniques in their machine learning process. Figure 2 shows the stage of using pixel, feature, and decision-level fusion in the digital image process.

This section gives an overview of the types of the image fusion process and discusses their disadvantages and issues. Image fusion is of three types, pixel-level fusion, feature-level fusion, and decision-level fusion. First, pixel-level image fusion

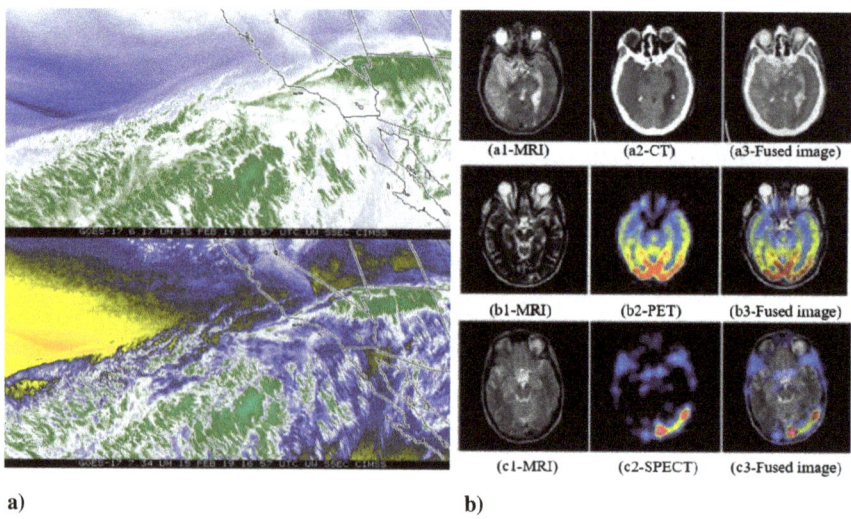

a)   b)

Fig. 1 a Data fusion imagery created using the GOES-17 ABI band 13 (10.3 μm) imagery. b Multi-modal medical image fusion of MRI-CT, MRI-PET, MRI-SPECT

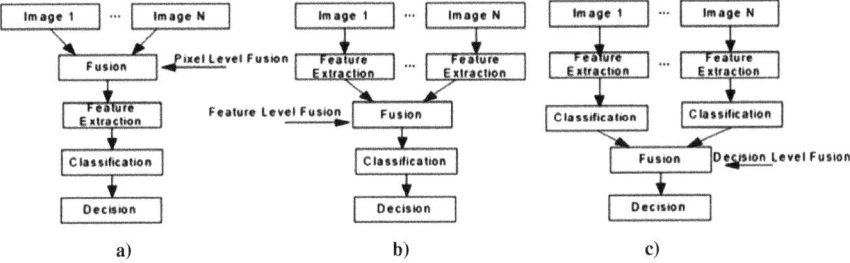

Fig. 2 a Pixel-level fusion. **b** Feature-level fusion. **c** Decision-level fusion in digital image processing

is a technique that depends on simple pixel-by-pixel operations. Arithmetic, logical, and probabilistic are few mathematical techniques that are used in this process. The image value includes pixel-gray, feature map values, and decision map labels. Even though many better techniques exist, pixel fusion is still widely used in many applications [9]. Second, in feature-level fusion, various features in the image are extracted from different image inputs and combined to create feature maps. These features include lines, edges, corners, textures. Preprocessing for image segmentation requires these features. Change detection also uses this method and works with features [1]. Finally, when the classification phase is done, we utilize the fusion procedure in decision-level fusion. To obtain the final image, this method often employs a mix of algorithms. Soft fusion occurs when confidences are employed instead of decisions. Otherwise, it is referred to as hard fusion.

To combine multiple images, all images should be of the same area, and different angled photographs result in the complexity of the fusion process. Before using a fusion method in applications, we have to understand the quality metrics of the algorithm. To say one method is better than the other, it should preserve the useful/relevant information in different images. Second, no artifacts should be introduced while processing which results in misleading information for human perception or any further image processing steps. Finally, it should be robust. Algorithms which take days to process will be less prioritized than faster methods. The fused image might contain more information than a single image, but it should be noted that no salient information should be discarded from any input image. Since the number of images to process for further steps decreases, it significantly reduces the data and processing speed [10]. Table 1 lists the quality parameters which are used by authors to check the quality of fused images corresponding to different algorithms [11].

Section 2 reviews the already published work for breast cancer detection. Multiple authors have presented novel algorithm, framework, and CADx systems for cancer classification. Different image modalities, algorithms are used in the processing of digital images. Section 3 proposes our work in classification technique using image fusion technique. Finally, in Sects. 4 and 5, we publish our results and concluded our paper.

Table 1 Image quality parameters as listed are used for analyzing the fused images

Signal to noise ratio (SNR)	Peak signal to noise ratio (PSNR)	Root-mean-square error (RMSE)	Normalized cross-correlation (NCC)
Mutual information (MI)	Universal image quality index (UIQI)	Fusion factor (FF)	Spatial frequency (SF)
Average gradient (AG)	Fusion symmetry (FS)	Edge information preservation (QAB/F)	Entropy (E)

2 Literature Review

This section reviews the related work to feature and decision-level fusion and different techniques implemented in their experiments.

To detect breast cancer, different image modalities can be used [12]. Authors of the paper [11] have used the MIAS dataset which consists of mammograms while [13] implemented their experiment on MIAS and TMCH. DDSM is also a popular mammograms dataset consisting of 2620 images and used by multiple authors [14, 15]. Different modalities like MRI, CT, PET, SPECT are also used for research, but this paper focuses on mammograms as they are cheap, low complexity, and easily available in countries. In the fusion process, many researchers have also combined different modalities like PET/MRI, MRI/CT, MRI/SPECT [3].

Medical images are special and require some extra steps before processing [16]. Due to low contrast and high noise in digital medical images, it makes them difficult to understand by humans or in further processing steps. MRI [17], mammogram [18], computer topography [19] are a few of the medical processes whose output images suffer from this same pitfall. De-noising, enhancement of structure, and enhancement of contrast are performed on digital images in preprocessing step, and different techniques are used like contrast limited adaptive histogram equalization (CLAHE). Many variant algorithms have been derived from histogram equalization (HE) technique such as adaptive histogram equalization (AHE) [20], CLAHE [21], BBHE, RMSHE, and MMBEBHE which address some or other issues in histogram equalization. This paper focuses on CLAHE and its working. These preprocessing techniques either can be applied to a particular region or can be applied globally on the image. Figure 3 shows the way of distributing excess intensities above threshold which will be clipped and added to the new bin.

Fig. 3 Distribution of excess intensities among all histogram bins in CLAHE

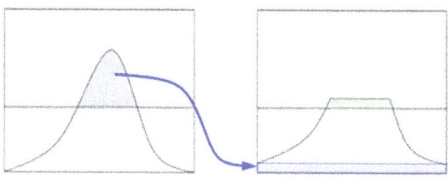

CLAHE is AHE-based algorithm. While AHE improves the contrast, it also amplifies the noise. To reduce this issue, CLAHE is developed. It clips the histogram at a value (clip limit) and distributes the excess intensities above it among different histogram bins. Clip limit and block size are the two main parameters of this algorithm.

Makandar and Halalli [22] implemented CLAHE for contrast enhancement in mammograms. They used 20 images from the mini-MIAS dataset. Median filter, min–max filter, and Wiener filter were used for noise removal. Their paper showed that using Wiener filter with CLAHE showed the best results with RMSE value between 0.586 and 0.7294 and PSNR between 49.5352 and 50.8783 [23], while working on different medical images presented a combination of CLAHE with 2D discrete wavelet-based fusion technique for efficient results. To compute, the performance calculated the SNR and entropy of their finding showed that entropy was increased due to CLAHE and neared maximum and average entropy. Sajeev et al. [24] working on the DDSM dataset presented a self-adjusted mammogram solution as adaptive clip limit CLAHE (ACL-CLAHE). While comparing their work with basic CLAHE, their method presented better performance. Bhat and Tarun [25] proposed a two-step process in which first they applied the CLAHE using estimated clip size and secondly chose the optimum clip size based on AMBE and PSNR values. They used STARE, DDSM, and OASIS for their analysis, and the SSIM index showed better results than traditional CLAHE [26], while working on mammograms from the MIAS dataset purposed a novel fuzzy clipped CLAHE which automates the clip limit size for enhancement. They presented a fuzzy rule-based flexible system that updates the control parameters. They compared the algorithm performance on CII, DE, AMBC, PSNR and showed improved results than traditional algorithms.

Shi et al. [27] presented an automated pipeline of image processing steps for the mammogram. They presented a novel texture-based filter for classification. Their findings on the MIAS dataset showed 97.08% and 96.15% segmentation and classification accuracy. They were able to achieve a bit higher accuracy for full-field digital mammography image datasets. They indicated that comparison in the test of estimates of Jaccard and Dice indices between segmented breast regions and ground realities, which may offer useful guidance for clinical mammogram research [28]. In his review paper, he used the ImageNet dataset pre-trained on VGG19 model for CNN feature extraction and was able to achieve 79.8 sensitivity and 78.9% specificity. Along with CNN, they also evaluated fusion-based classifiers and were able to achieve 80–86.2% sensitivity and 81.6–87.7% specificity on four fusion classifiers. Merati et al. [8] proposed a new triplet CNN where they used DDSM database and 4000 augmented images. Their method was able to achieve 96% sensitivity, 90.25% specificity [29]. In his paper, he presented a method based on linear support vector machine (LSVM) and CNN on thermal images. With this method, they were able to achieve 90.06% sensitivity and 91.8% specificity. For faster and better results, they used application-specific hardware. Sarosa et al. [30] presented that using GLMC and BPNN methods can produce 90.16% sensitivity, 95.57% specificity, and 94.06% accuracy for classification. They used MIAS database. By decreasing the input neurons, they were able to reduce the BPNN learning time drastically [31]. In their paper, they compared

RBFNN and BPNN. Based on GLCM texture-based features, they performed the classification methods on the MIAS dataset. In their experiment, they were able to achieve 93.98% accuracy for RBFNN, which is higher than BPNN, while benign and malignant accuracy is reported to 94.29%, which is again higher than BPNN. Viswanath and Guachi-guachi [7] presented a CAD system for classification. The experiment included SVM, KNN, and random forest (RF) for final decision-making in the classification step. The experiment showed that preprocessing improved the classification effectively for mammograms [32].

3 Materials and Methods

3.1 Datasets

There are multiple free datasets of mammograms available. Mammographic image analysis society (MIAS) database is a popular dataset that is used by many authors for their research. This dataset contains 322 digitalized films which are available in 50-micron pixel edge and 200-micron pixel edge. Another one is digital database for screening mammography (DDSM), and this dataset contains 2500 studies. Each study contains two images of each breast along with patient info. This dataset is a collaboration of hospitals and multiple universities. However, many authors collect their datasets (mammograms) from hospitals and universities. There are more datasets like, B-SCREEN, AMDI, which are present at [33], and this page contains information regarding both old datasets and upcoming ones.

3.2 Preprocessing

Low contrast in mammography pictures makes it harder to identify lumps, and studies have revealed a high proportion of false-positive (regular change as malignant) and false-negative instances (actual abnormality not detected). The CLAHE enhancement approach has been used to overcome this problem. To check the effect of different algorithms, we have calculated the PSNR of some of MIAS images (mdb03, mdb06, mdb09, mdb12, mdb15) as shown in Table 2. CLAHE is a variant of AHE in which the intensities are clipped at a certain threshold level. This experiment was developed in Python and uses a clip limit of 0.2.

Table 2 PSNR value of HE, CLAHE, MMBEBHE, RMSHE

Image	HE	CLAHE	MMBEBHE	RMSHE
mdb03	6.54	23.64	30.48	26.93
mdb06	8.63	22.96	24.92	24.71
mdb09	8.14	24.53	32.34	23.96
mdb012	9.76	24.76	32.08	27.10
mdb015	4.39	24.88	32.16	24.11

3.3 Feature Extraction, Selection, and Feature Fusion

It is a very important step in the detection of diseases and system overall performance. We will select the features which we require for our processing and extract them. Different types of features can be extracted such as shape feature extraction, statistical feature extraction, wavelet-based feature extraction [34]. They can be from features extracted directly from mammograms such as parameter, area, background. Features are extracted from co-occurrence matrix, gray-level run length (GLRL) matrix, gray-level difference (GLD) matrix [35], and features extracted from CNN.

In this paper, we have manually extracted features [36] like area, density, area and combined them with features extracted from CNN. A CNN architecture is created from which 1821 high level and 288 low level were extracted. A total of 2115 features (1821 high level, 288 low level, 6 manually extracted) was used by the classification algorithm. Our CNN model architecture is shown in Fig. 4.

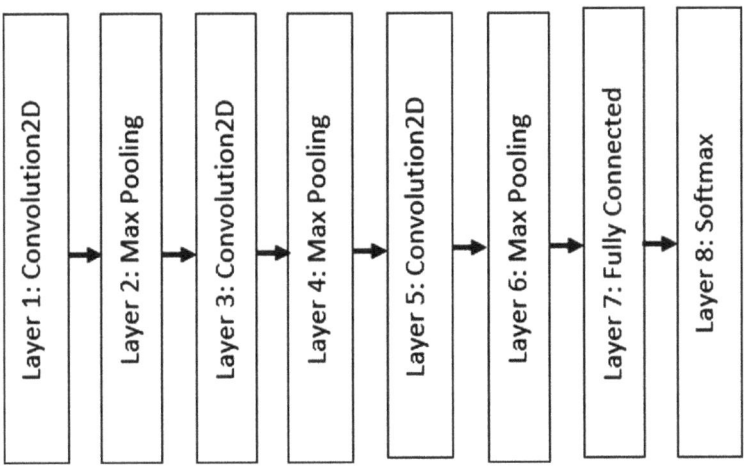

Fig. 4 CNN model architecture for feature extraction

3.4   Classification and Decision Fusion

We implemented SVM, decision tree, and random forest classifiers in this stage. SVM is a machine learning method that may be used to predict and classify data. While the decision tree creates a representation tree, that is easy to understand. However, it sometimes creates a biased tree based on output class resulting in large variance, which can be corrected by bagging and boosting techniques. Advantages and disadvantages are of these algorithms are mentioned in Table 3. Finally, a voting classifier is used for making a decision based on these three input classifiers, and the final output is generated. Figure 5 shows the work flow of the process which we have used for our experiment.

Table 3 Advantages and disadvantages of classification algorithms (SVM, random forest, decision tree)

Method	Advantages	Disadvantages
SVM	• Memory efficient • Works better in high-dimensionality spaces	• Not suitable for large datasets • Prone to overfitting
Decision tree	• Less effort for data preprocessing • Easy to visualize	• High time to train model • Change in data can drastically change the tree structure
Random forest	• Can handle large dataset with higher dimensionality • Feature scaling is not required	• Requires high computation power • Longer training period

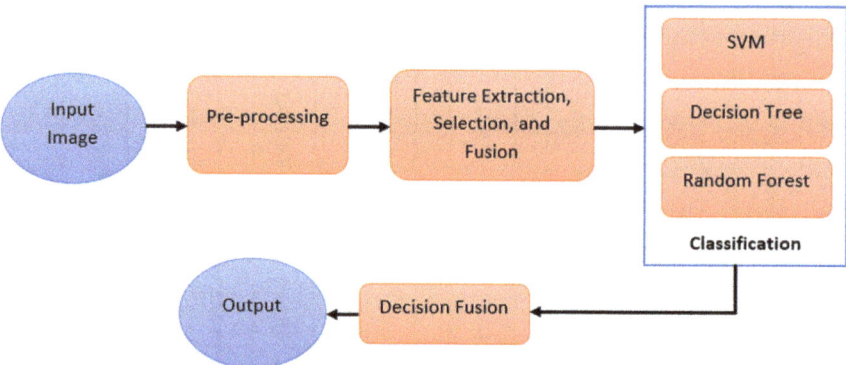

Fig. 5 Work flow of the process for breast cancer detection

4 Results

In this experiment, we have used the MIAS dataset which consists of 322 images (161 pairs). For contrast enhancement in preprocessing step, we have implemented HE, CLAHE, MMBEBHE, and RMSHE and found that CLAHE performs better than others in mammograms. A clip size of 0.2 is used for this experiment. A CNN is used for feature extraction, and six manual features are used. A total of 2115 features (1821 high level, 288 low level, 6 manually extracted) was used by the classification algorithm. Standalone, SVM achieved 92.3% accuracy, 87.8% sensitivity, and 93% specificity, while the decision tree achieved 94.03% accuracy, and the random forest achieved 95.05%. After combining the three classification techniques using a voting classifier, we were able to achieve 96.12% accuracy.

5 Conclusion

Breast cancer is one of the most common illnesses among women. We discovered that a CAD system appears to be a good option for real-life application by radiologist after examining various approaches and methodologies in this article. By enhancing the picture and selecting the ROI, the radiologist's own knowledge, as well as a second and helpful opinion from the CAD system, the accuracy of diagnosis may be improved. To enhance the outcomes, a technique involving pixel fusion, feature fusion, and decision fusion can be used.

References

1. Suthakar J (2014) Study of image fusion-techniques, method and applications. Int J Comput Sci Mob Comput 3(11):469–476
2. GOES-17 Data fusion: an example, and where to find the data « CIMSS Satellite Blog » [Online]. Available at: https://cimss.ssec.wisc.edu/satellite-blog/archives/31865. Accessed 22 Feb 2021
3. Du J, Li W, Lu K, Xiao B (2016) An overview of multi-modal medical image fusion. Neurocomputing 215:3–20
4. Siegel RL, Miller KD, Fuchs HE, Jemal A (2021) Cancer statistics, 2021. CA Cancer J Clin 71(1):7–33
5. Král P, Lenc L (2016) LBP features for breast cancer detection. In: 2016 IEEE international conference on image processing (ICIP), pp 2643–2647
6. Dong M, Lu X, Ma Y, Guo Y, Ma Y, Wang K (2015) An efficient approach for automated mass segmentation and classification in mammograms. J Digit Imaging 28(5):613–625
7. Viswanath VH, Guachi-guachi L (2019) Breast cancer detection using image processing techniques and classification algorithms
8. Merati M, Mahmoudi S, Chenine A, Chikh MA (2019) A new triplet convolutional neural network for classification of lesions on mammograms. Egypt J Radiol Nucl Med 33(3):213–217
9. Mitchell HB Image fusion theories, techniques and applications

10. Ardeshir Goshtasby A, Nikolov S (2007) Image fusion: advances in the state of the art. Inf Fusion 8(2):114–118
11. Kumar MP, Svecw A, Pradesh A (2015) Pixel level weighted averaging technique for enhanced image fusion in mammography, vol 3, pp 10–15
12. Iranmakani S et al (2020) A review of various modalities in breast imaging: technical aspects and clinical outcomes. Egypt J Radiol Nucl Med
13. Pawar MM, Talbar SN (2018) Local entropy maximization based image fusion for contrast enhancement of mammogram. J King Saud Univ Comput Inf Sci
14. Shanmugam S, Shanmugam AK, Muthusamy E (2019) Analyses of statistical feature fusion techniques in breast cancer detection, vol 17, pp 311–316
15. Sert E, Ertekin S, Halici U (2017) Ensemble of convolutional neural networks for classification of breast microcalcification from mammograms. In: Proceedings—annual international conference of the IEEE Engineering in Medicine and Biology Society. EMBS, pp 689–692
16. Okada DR, Blankstein R (2009) Digital image processing for medical applications. Perspect Biol Med
17. Jalalian A, Mashohor SBT, Mahmud HR, Saripan MIB, Ramli ARB, Karasfi B (2013) Computer-aided detection/diagnosis of breast cancer in mammography and ultrasound: a review. Clin Imaging 37(3):420–426
18. Sree SV (2011) Breast imaging: a survey. World J Clin Oncol 2(4):171
19. Chen B, Ning R (2002) Cone-beam volume CT breast imaging: feasibility study. Med Phys 29(5):755–770
20. Pizer SM et al (1987) Adaptive histogram equalization and its variations Comput Vis Graph Image Process
21. Zuiderveld K (1994) Contrast limited adaptive histogram equalization. In: Graphics gems
22. Makandar A, Halalli B (2015) Breast cancer image enhancement using median filter and CLAHE. Int J Sci Eng Res 6(4):462–465
23. Bhan B, Patel S (2017) Efficient medical image enhancement using CLAHE enhancement and wavelet fusion. Int J Comput Appl 167(5):1–5
24. Sajeev S, Bajger M, Lee G (2015) Segmentation of breast masses in local dense background using adaptive clip limit-CLAHE. In: 2015 international conference on digital image computing: techniques and applications, DICTA 2015
25. Bhat M, Tarun PMS (2015) Adaptive clip limit for contrast limited adaptive histogram equalization (CLAHE) of medical images using least mean square algorithm. In: Proceedings of 2014 IEEE international conference on advanced communication, control and computing technologies. ICACCCT 2014, vol 978, pp 1259–1263
26. Jenifer S, Parasuraman S, Kadirvelu A (2016) Contrast enhancement and brightness preserving of digital mammograms using fuzzy clipped contrast-limited adaptive histogram equalization algorithm. Appl Soft Comput J
27. Shi P, Zhong J, Rampun A, Wang H (2018) A hierarchical pipeline for breast boundary segmentation and calcification detection in mammograms. Comput Biol Med 96:178–188
28. Whitney H, Li H, Yu J, Liu P, Giger ML (2019) Comparison of breast MRI tumor classification using radiomics, transfer learning from deep convolutional neural networks, and fusion methods, pp 1–15
29. Iqbal HT, Majeed B, Khan U, Awais M, Altaf B (2019) An infrared high classification accuracy hand-held machine learning based breast-cancer detection system. In: 2019 IEEE biomedical circuits and systems conference (BioCAS), pp 7–10
30. Sarosa SJA, Utaminingrum F, Bachtiar FA (2019) Breast cancer classification using GLCM and BPNN. Int J Adv Soft Comput Appl 11(3)
31. Pratiwi M, Alexander, Harefa J, Nanda S (2015) Mammograms classification using gray-level co-occurrence matrix and radial basis function neural network. Procedia Comput Sci 59:83–91
32. Verma B, McLeod P, Klevansky A (2010) Classification of benign and malignant patterns in digital mammograms for the diagnosis of breast cancer. Expert Syst Appl 37(4):3344–3351
33. Mammographic Image Analysis Homepage—Databases [Online]. Available at: https://www.mammoimage.org/databases/. Accessed 23 Sept 2020

34. Nithya R, Santhi B (2011) Comparative study on feature extraction method for breast cancer classification. J Theor Appl Inf Technol 33(2):220–226
35. Cheng HD, Cai X, Chen X, Hu L, Lou X (2003) Computer-aided detection and classification of microcalcifications in mammograms: a survey. Pattern Recognit 36(12):2967–2991
36. Munisami T, Ramsurn M, Kishnah S, Pudaruth S (2015) Plant leaf recognition using shape features and colour histogram with K-nearest neighbour classifiers. Procedia Comput Sci 58:740–747

Hybridization of Harmony and Cuckoo Search for Managing the Task Scheduling in Cloud Environment

Krishan Tuli and Amanpreet Kaur

Abstract Cloud computing plays a vital role in gathering the physical and virtual resources given to the client on-demand and on a pay as per uses basis using the internet. In today's world, cloud computing has been considered as the best concept for the virtualization of various resources. There are various approaches that have been available for improvising the load balancing and also to improvise the job scheduling in the concept of cloud. But there are two most important thing in cloud computing, that is, task scheduling and allocation of resources. So, to achieve better performance of resources, task scheduling and allocation of resources must be organized in a more optimized way and hence cloud computing provides great performance in less maintenance cost. In this paper, cuckoo search and harmony search is hybrid and proposed cuckoo harmony search algorithm (CHSA) for enhancing the scheduling process and make it more optimized. As per proposed CHSA, a new hybrid function is developed based on cost, memory, energy consumption, credit, and penalty and the proposed algorithm is compared with cuckoo search and harmony search on all these parameters. After the analysis of proposed CHSA, comparing to cuckoo search and harmony search, it has been find out that, it attains minimum cost, memory usage, energy consumption, penalty, and maximum credit.

Keywords Cuckoo search · Harmony search · Task scheduling · Cuckoo harmony search algorithm (CHSA) · Cloud computing · Quality of service (QoS)

1 Introduction

Basically, distributed computing focused on the classical framework, which works on the model of on-request, and it is a very advantageous service, and it organizes the resource pool, and it can compute and manage the resources like servers, storage,

K. Tuli (✉) · A. Kaur
Chandigarh University, Mohali, India

A. Kaur
e-mail: amannpreetkaur.uic@cumail.in

and other application, etc. The other purpose of distributed computing is to quickly provide the administrative and various service provider interactions. In other word, we can say that when we are talking about cloud computing, here cloud means a service provider that provides various services and resources to all the users. Cloud computing is a distributed computing model which are interconnected to various users and it virtualizes many personal computers that are connected with each other. The purpose is to provide at least one consolidated algorithm for the various resources which manages the various tasks without violating the service level agreement. The SLA is between the various consumers and the service providers. There are few difficulties with distributed computing like security issues, execution issues, reliability, and so on. The distributed computing helps in spreading the computational workload among various other servers [1]. So, the distributed computing is an execution environment where the various other resource can meet with the requirements of other applications. The requirement includes:

(a) **Security**: Security is the most important requirement in the data and processing of data. Security remains with the data centers of various enterprises which one public cloud is attached the sharing of data and resources.
(b) **Location**: Location is the most important requirement because it is used with some of the applications to make it more responsive and for making the performance of the application better.
(c) **Redundancy**: The next requirement in cloud computing is redundancy. Redundancy helps to mitigate the various other applications and other large scale enterprises.
(d) **Regulations**: The most important requirement of distributed cloud is the regulations. So that data cannot leave the geographical area. It is shared only on the location where it is permitted to.

Hence, the service providers must ensure for the end to end management for the placement of data, interconnections, and various other computing processes and all these must appear as a single solution from cloud computing point of view. Here, content delivery network (CDN) is the most common example of distributed cloud environment where storage is done based on the geographical diverse regions. The purpose is to reduce the latency of delivery to various resources.

The first very common resource issue is the managing of various tasks. In cloud computing, task scheduling refers to allocation of various task to the virtual machines by physical machines to increase the processing of tasks and it will also increase the utilization of resources. So to get better performance from the cloud, task scheduling plays a vital role. There are numerous tasks that might get scheduled to increase the performance of the system and for minimizing migration. Thus, for this work, task management plays vital role to have improved reliability and flexibility of the system in cloud computing. Most of the algorithms are rule-based to perform the implementation, but rule-based task scheduling performed not very well to solve the complex problems of task scheduling. Another point in task scheduling is the allocation and scheduling of resources related to quality of service (QOS) which will affect the cloud computing service providers. So another important issue here is the resource scheduling, which is a critical issue in cloud based environment. Distributed computing must need to scale with the various numbers of assignments and various customers and must have a scheduling algorithm and that can spread various assets [2]. They form key point in the investigation and must follow the different methods of probability. Examples are the Ant Colony Optimization (ACO), Particle Swarm Optimization (PSO), and so on.

In this paper, author proposed a model of task scheduling which is based on various objectives and used them with combination of harmony search optimization and cuckoo search algorithm. On the basis of these multi-objectives functions, one can have obtained the task scheduling from the set and also it is used for the memory usage, energy consumption, and most importantly the cost. Following contributions are made in the research for the process of task scheduling [3].

- CHSA is done for the process of task scheduling and it has the major advantage of converging and recognizing the various processes of task scheduling. The result of this is that scheduling process will get the optimal solution than harmony search and cuckoo search algorithms.
- Another function is the multi-objectives which are based on the problem of task scheduling are mainly focuses on the cost minimization, proper utilization of memory, energy consumption, etc.

2 Related Work

There are various techniques for the process of scheduling process in cloud environment. In this section, author will discuss about the various review of multi-scheduling method that need to be managed in the proper environment.

Zhu et al. [4] had designed the scheduling process in cloud environment. In that, they have calculated the various rules for bidding value for both the phases of forward as well as regressive announcement phase. Further, they have also developed the periodic tasks in cloud environment based on a dynamic scheduling based on the agents. They have given the name ANGEL. This scheduling algorithm mainly focuses on the issues called schedule-ability, scalability, and real-time in cloud based environment. The drawback of this method is that, they have not considered the communication and dispatching of various resource utilization. Another challenging problem that are there in cloud is that the challenge of workflow scheduling which means that satisfying of quality of service of the particular user and on a minimal cost.

To overcome this issue, Zhong et al. [5] has been explained the algorithm of workflow based on the cost awareness in cloud based environment. The cost aware algorithm signifies the various difficulties on the scheduling of workflow and they are viewed and focused on the quality of service (QOS), design, and framework functionality that will result in taxonomy set.

Wang et al. [6] had proposed algorithm based on energy efficient scheme which is particularly based on time constraint tasks particularly in cloud computing. In this paper, author has developed the mathematical notations of various mobile computing devices. Author develops a probabilistic algorithm for scheduling. In cloud computing, the main focus is how efficiently to use the MCC to enable the real-time complex problems while maintaining the energy efficiency at advanced end. To overcome this problem, the algorithm has developed which is based on multi-objective task.

Jacob et al. [7] had developed multi-objective algorithm particularly for task scheduling and resource allocation in cloud environment. Author made two objectives, first is to make span and second is to minimize the total cost. This problem of multi-objective is solved by genetic algorithm. This algorithm is best for solving the problems that are based on execution time of a process at much faster rates.

From the literature survey above, we have clearly understood that all the work is well-organized for the process of task scheduling. But there is some limitation such as higher cost, Quality of Service (QoS), execution time of the process and so on. Moreover, most of the work uses a single objective functions which are based on the various scheduling processes but the problem with single objective functions is that they are not cost effective for all task scheduling method. So to overwhelm this problem, author proposed one multi-objective task scheduling using hybridization of algorithms.

3   Background Work

In this section, author explains hybridization of algorithm and then focuses on the proposed multi-objective task scheduling.

Cuckoo Search

Cuckoo search algorithm is a swarm algorithm which works on natural behavior of cuckoo bird who lay eggs in others nest. In cuckoo search, each egg in the nest indicates the candidate solution. In other words, we can say that each cuckoo bird can lay single egg only and having a unique shape too. This represents a unique set of solution. The key objective of cuckoo search is that it creates better solutions which replaces the worst from the current cluster [8].

Following steps define the process:

Step 1. *Initialization phase*: In this phase, random population of nest will be created. Assume S_i are the population of nest where i is 1, 2, 3, 4, 5, 6 and so on.
Step 2. *Generating new phase from step 1*: This phase will generate new cuckoos based on levy flights.
Step 3. *Fitness Evaluation*: After calculating the objectives, author has calculated the fitness solution based on the below mentioned formula

Fitness function = minimum objective functions (choose the best solution)

Step 4. *Updations phase*: After generating fitness evaluation, novel solution is developed based on the following equation

$$I_i^N = I_i^{(t+1)} = I_i^{(t)} + \alpha \text{ Levy } (\lambda)$$

Here, $\alpha > 0$ and levy flight is having random steps size with random walk.

$$\text{Levy } u = t^{-\lambda}$$

Here, λ is in the range from 1 and 3 that is $1 < \lambda < = 3$.

Step 5. *Reject all the worst solutions*: After step 4, that is, updations phase, we reject all the worst fitness values and consider all the optimal solutions.

Step 6. *Stopping Phase*: There will be maximum iterations accompanied this particular process and repeated until it will be maximized. The best possible solution will be selected for the further process.

Harmony Search Algorithm

Harmony search is known for the best searching technique of swarm intelligence. This algorithm is a meta-heuristic algorithm based on improvisation of musicians to determine and find the harmony. In past few years, due to its model, harmony search algorithm gains lot of attention. Harmony search algorithm is easy to implement and it finds optimal solutions very quickly and gives the best solution in stipulated time. The computational time of this algorithm is very less as compare to the other algorithms and it also solves the problems of various engineering applications and areas. The same harmony algorithm is explained in Fig. 1.

Step 1. *Initialization*: The best possible process in optimization is the process of finding the number of initialized solutions. In this process, first of all, we initialize the various parameters that are in the harmony search algorithm. The number of parameter that are used in this search algorithm is called harmony memory size and abbreviated as HMS. Second parameter here is Harmony Memory Considering Rate (HMCR) and lastly the Pitch Adjusting Rate (PAR).

Step 2. *Harmony Memory Initialization:* It is generated arbitrarily. Pseudo code for harmony memory initialization is described in Fig. 2.

The equation for harmony memory initialization is as below equivalent

$$\text{HMI} = \begin{bmatrix} \text{HMI}(1, 1) & \cdots & \text{HMI}(1, n) \\ \vdots & \ddots & \vdots \\ \text{HMI}(S, 1) & \cdots & \text{HMI}(S, n) \end{bmatrix}$$

Step 3. *Improvising harmony technique:* The harmony vectors are based on three rules. They are memory, random selection, and pitch change. Improvisation means creating a new harmony. When we consider memory, it is the choice of estimation in primary variable based on new vector condition. Harmony memory is basically like

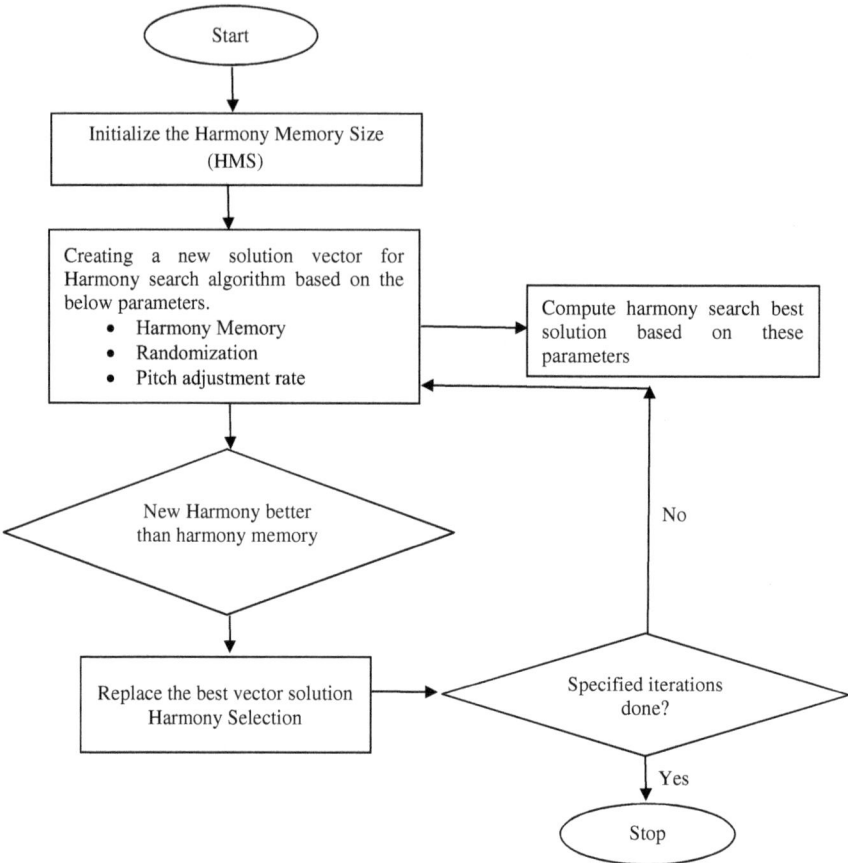

Fig. 1 Harmony selection process

a progression where we can say that musician can utilize the memory to get a new tune.

The HMCR $\in [0, 1]$ is the basic range for picking one initiative from the coronial values and the pitch adjustment is justified as the equation below.

$$HM_i^{New} = \{Pitch\ adjusted\ with\ y\ as\ probability\ (1 - PAR)$$

and after the execution of this pitch adjustment, if it comes to be YES for HM_i^{New} then it is replaced as follows.

$$HM_i^{New} = HM_i^{New} \pm rand \times b$$
$$(b\ is\ distance\ bandwidth\ and\ the\ range\ of\ rand\ is\ 0\ to\ 1)$$

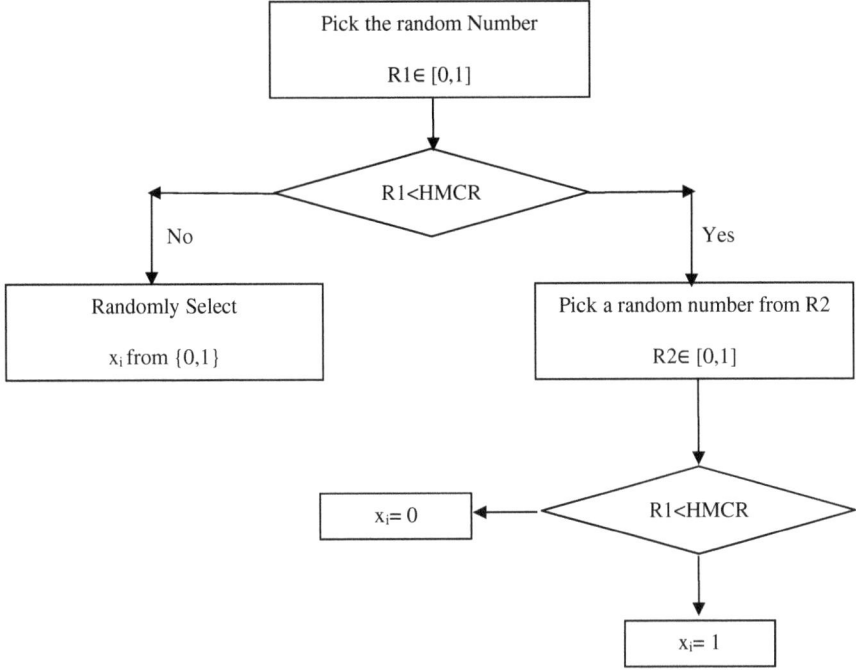

Fig. 2 Pseudo code for harmony memory initialization

Step 4. *Updating memory harmony*: After calculating harmony function, we calculate the harmony vector and that will be compared by previous one. If it is better than the previous one, then we will include the new harmony vector and discard the previous one.

Proposed Model of Cuckoo Harmony Search Algorithm

Major focus of the proposed model is to make an optimized algorithm for using the scheduled task using hybrid method. Here, the hybrid approach of optimized algorithm is comprising of the two main algorithms. They are cuckoo search algorithm and harmony search algorithm. For getting better accuracy, these proposed algorithms are combined with the optimized capabilities of both these algorithms. So, when hybridization of algorithms is done, it will remove all the shortcomings and give a better individual algorithm which is best in performance.

4   Problem Formulation and Proposed Framework

The key objective of task scheduling is allocation of task to other virtual machine those are minimal in usage with violating the service level agreement and also to have the minimal in fitness function. The proposed framework of task scheduling is explained in Fig. 3. In task scheduling algorithm, user sends request to the particular user. Then using the components of user interfaces, cloud consumer will send the application request and further request manager from the cloud computing receives generated task from particular user interface and also manage all the accepted requests. All the resource in the cloud pool are managed and monitored by resource monitor. Examples of cloud resource pools are CPU and RAM [9].

Cloud computing resources are pooled together and they are monitored by resource monitor. Then task has been allocated to scheduler. The scheduling process is based on the performance of various virtual machines. After this process, scheduler will collect data from resource manager and monitoring of process is done by

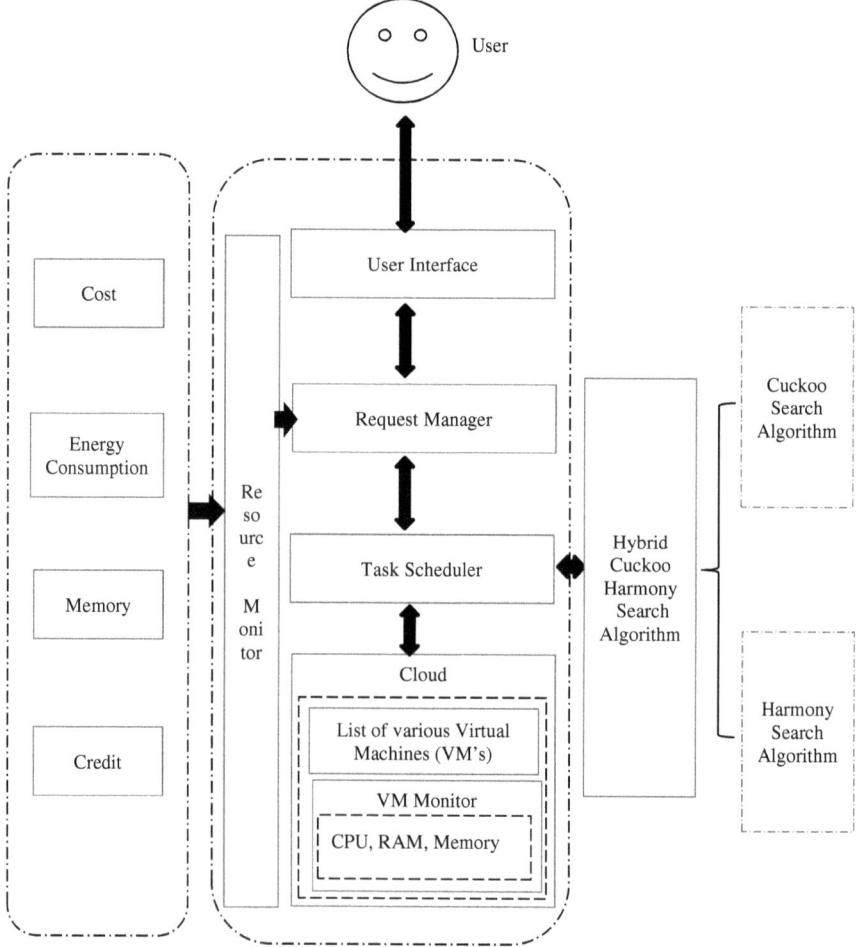

Fig. 3 Architecture of proposed task scheduling

resource monitor. The key work of task scheduler is to schedule the tasks in cloud computing and monitor the fitness function. The fitness function must be minimum. The scheduler assigns the tasks to corresponding virtual machines and obtains a routing problem. So to achieve this problem, author has proposed hybridization of task in this paper and find the minimum fitness function to solve the following equation [10].

$$\text{Minimum fitness function} = \text{Min}[\alpha_1(\text{cost of migrations})$$
$$+ \alpha_2(\text{consumption of energy per process})$$
$$+ \alpha_3(\text{usage of memory(CPU/RAM)}) + \alpha_4(1 - \text{credit})$$
$$+ (\text{penalty})]$$

α_1, α_2, α_3, and α_4 are the control parameters and they range from [11]. So, in this type of scheduling, Migration Cost (MC) must be the minimum. The total migration cost will be calculated using number of movements. The calculation of migration cost, movement factor, and cost factor are calculated using following equations.

$$\text{MC} = (M^F + C^F)/2 \quad \text{(Calculation of migration cost)}$$

$$\text{MF} = \frac{1}{\text{PM}} \left[\sum_{x=1}^{\text{VMi}} \frac{\text{number of movements}}{\text{total virtual machines}} \right] \quad \text{(Calculation of movement factor)}$$

$$\text{CF} = \sum_{x=1}^{\text{VMi}} \left(\frac{\text{memory of task} * \text{cost of run}}{\text{pm} * \text{vm}} \right) \quad \text{(Calculation of cost factor)}$$

Here PM denotes physical machines and VM denotes virtual machines.

From the above equations, we found the problem. The problem is to schedule the tasks and for solving this problem using mathematical formulas, we need large amount of computational programs and lot of efforts. So all these problems will be minimized by using the proposed algorithms.

5 Experimental Result Evaluation

The idea behind CHSA algorithm is virtual machine scheduling and that should be done at more efficient way. Here, author has done experiment considering physical machines and virtual machines based on N as number of tasks and R as number of resources. For task scheduling, this process is based in the various criteria that we had discussed in previous section of this paper. These criteria are cost, energy consumption, credit, memory usage, and penalty. In this experiment, performance analysis is done by using different number of physical machines and virtual machines.

Here, some different tasks were also used for scheduling the process and allocate to various virtual machines. Further, each task is categorized into 3 sub tasks and then the proposed result is compared with CS and HS algorithm [12].

Performance analysis and comparison of different algorithms

The pie charts shown below shows the performance of proposed methodology. This experiment is implemented with 4 physical machines and 13 virtual machines along with 5 tasks and 15 subtasks. Here, proposed algorithm is compared with the other algorithms, that is, harmony search and cuckoo search.

The figure above shows the comparison on various parameters. Figure 4 shows the performance of various algorithms based on memory usage. These algorithms must attain minimum fitness functions and the minimum fitness functions is only attained, if we have the proper attainment of all the parameters. These parameters must have a minimum in memory usage, minimum cost, and energy consumption but maximum is credit which are clearly mentioned in our results. Figures 5, 6, and 7 show the comparison of various parameters based on energy consumption, cost, and credits, respectively.

In the above comparisons of Fig. 4, proposed CHSA is having minimum memory usage, that is, 0.15 which is least as compared to other algorithms. Comparing the energy consumption with other algorithms, we have found that 0.21 is having the minimum energy consumption with performed on various iterations. Figure 6 shows the comparison based on cost and after various iterations we have found that proposed CHSA is having minimum cost that 0.008 when comparing with other algorithms. Figure 7 shows comparison based on credits and after various iterations, we have found that proposed CHSA is having maximum in credits. All these comparisons are implemented with 4 physical machines and 13 virtual machines along with 5 tasks and 15 subtasks.

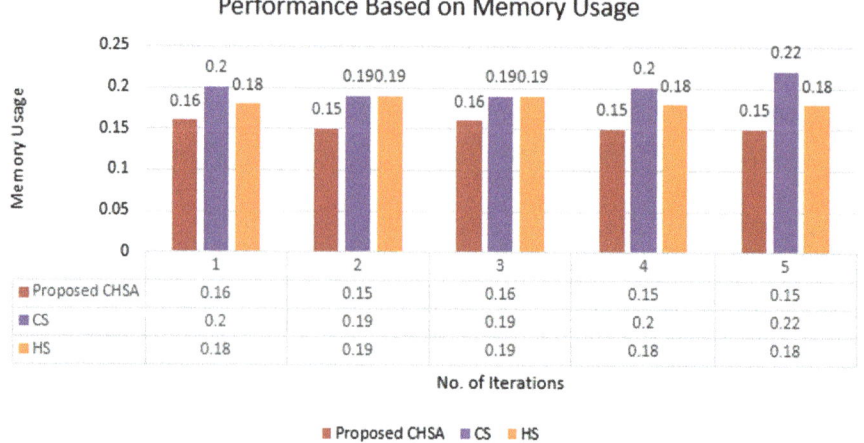

Fig. 4 Comparison based on memory usage

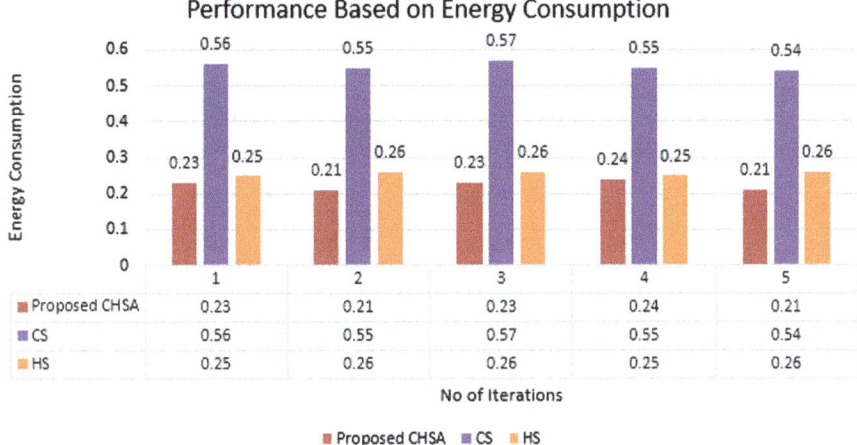

Fig. 5 Comparison based on energy consumption

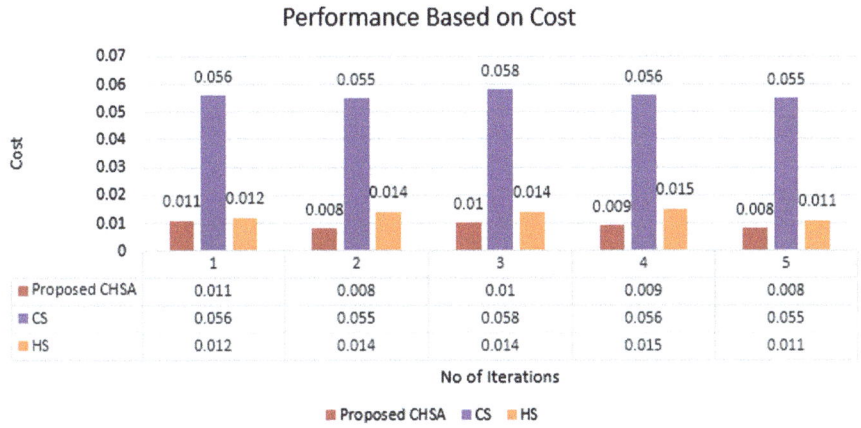

Fig. 6 Comparison based on cost

6 Conclusion and Future Scope

Multitask scheduling algorithm method was focused on the concept of Cuckoo Harmony Search Algorithm (CHSA). Basic feature of this optimization method is to improvise the scheduling performance when we are comparing with the basic function. Multitask scheduling function which we have discussed in the paper as well, like energy consumption, cost, memory usage, penalty, etc. are best performing when compared to other algorithms. Proposed algorithm provide dimensions to particular algorithm and compute the solution of various tasks. It also focuses on the basic task priorities. Harmony search algorithm will improve complete performance of cuckoo

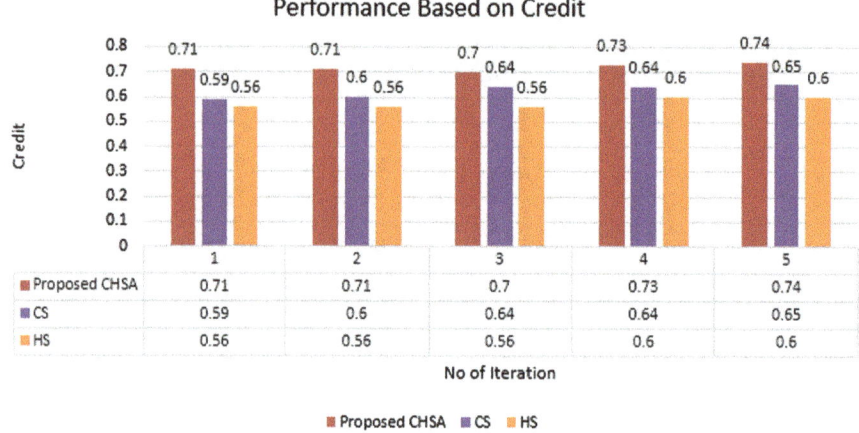

Fig. 7 Comparison based on credit

search algorithm while taking into consideration task scheduling in virtual machine. But comparing to hybridization of both these algorithms, this framework provides high quality of scheduling solution.

References

1. Nashaat H, Ashry N, Rizk R (2019) Smart elastic scheduling algorithm for virtual machine migration in cloud computing. J Supercomput 75(7):3842–3865
2. Krishna does P, Jacob P (2018) OCSA: task scheduling algorithm in the cloud computing environment. Int J Intell Eng Syst 11(3):271–279
3. Zhong Z, Chen K, Zhai X, Zhou S (2016) Virtual machine-based task scheduling algorithm in a cloud computing environment. Tsinghua Sci Technol 21(6):660–667
4. Sfrent A, Pop F (2015) Asymptotic scheduling for many task computing in big data platforms. Inf Sci 319:71–91
5. Latiff MSA, Madni SHH, Abdullahi M (2018) Fault tolerance aware scheduling technique for cloud computing environment using dynamic clustering algorithm. Neural Comput Appl 29(1):279–293
6. Gandomi AH, Yang X-S, Alavi AH (2013) Cuckoo search algorithm: a metaheuristic approach to solve structural optimization problems. Eng Comput 29(1):17–35
7. Pradeep K, Jacob TP (2018) A hybrid approaches for task scheduling using the cuckoo and harmony search in cloud computing environment. Wirel Pers Commun 101(4):2287–2311
8. Jiang H, Yi J, Chen S, Zhu X (2016) A multi-objective algorithm for task scheduling and resource allocation in cloud-based disassembly. J Manuf Syst 41:239–255
9. Zhu X, Chen C, Yang LT, Xiang Y (2015) ANGEL: agent-based scheduling for real-time tasks in virtualized clouds. IEEE Trans Comput 64(12):3389–3403

10. Tsai CW (2014) A hyper-heuristic scheduling algorithm for cloud. IEEE Trans Cloud Comput 2:236–250
11. Ali SA, Affan M, Alam M (2019) A study of efficient energy management techniques for cloud computing environment. In: 2019 9th international conference on cloud computing, data science & engineering (confluence), Jan 2019, pp 13–18
12. Wang C-M, Huang Y-F (2010) Self-adaptive harmony search algorithm for optimization. Expert Syst Appl 37(4):2826–2837

Accident Risk Prediction and Location Tracking of a Vehicle Using Real-Time Data Acquisition from Vehicle Sensors

C. Suhrit, Aarti Sharma, Deepa Kumari, and A. R. Kulkarni

Abstract A novel approach for vehicular Accident Risk Prediction is exhibited. Real-time sensor data is collected, fed to an embedded IoT application that computes the risk and alerts the user if the risk exceeds a threshold value. Location Tracking is performed to obtain the current location of the vehicle using a web application. In this paper, we present a low-cost but flexible internet-based location tracking and risk prediction system. The central core of the system is a real-time database. The real-time database updates data changes in real-time, which is visible across multiple devices.

Keywords Location tracking · Real-time model · Data Acquisition · Internet of things · Virtual objects · Risk prediction

1 Introduction

The rapid evolution in Networking and Communication would not have been possible without Internet of Things (IoT) systems. The evolution in IoT has paved the way for Real-Time IoT (RT-IoT). RT-IoT has more benefits and capabilities than a conventional IoT system [1]. A handy application of an RT-IoT system is an Intelligent Transportation System (ITS).

At this stage, automobile companies provide a certain amount of options to the customers regarding autonomy. However, these options are not yet fully autonomous. Even though autonomous systems aim to reduce risk due to human errors, the fully autonomous concept's actualization will require some time as even a single error in the whole process of manufacturing can produce disastrous accidents. Thorough research and precise manufacturing would solve this problem. Therefore, to reduce the time required to adopt autonomous vehicles, the RT-IoT systems will play a huge role.

C. Suhrit (✉) · A. Sharma · D. Kumari · A. R. Kulkarni
Delhi Technological University, New Delhi, New Delhi, India
e-mail: ashishkulkarni@dtu.ac.in

© The Author(s), under exclusive license to Springer Nature Singapore Pte Ltd. 2022   763
D. Gupta et al. (eds.), *Proceedings of Data Analytics and Management*,
Lecture Notes on Data Engineering and Communications Technologies 90,
https://doi.org/10.1007/978-981-16-6289-8_62

The paper attempts to provide an inexpensive RT-IoT-based solution to enhance the safety of the vehicle. Unlike expensive dedicated systems, we offer a multipurpose system that vehicle owners can use for both location tracking and real-time accident risk computation. The underlying machine learning model computes the risk index in real-time.

2 Motivation

The purpose of this section is to describe in detail the motivation to work on this paper. One of the reasons which lead to our work is the prevention of vehicle theft. Increasing vehicle theft cases lead to notable economic losses, challenging the quality of life in the communities [2]. Combating this problem is one purpose of the proposed system. In 2019 [3], reported vehicle theft cases in Delhi, India, were over 1233 cases per 100,000 inhabitants. Therefore, the goal of this paper is to make a system that prevents vehicle theft. Apart from the burdening issue of vehicle theft, another significant issue that plagues automobile drivers is road accidents. In 2017, India officially reported 464,910 road [4] accidents and claimed 147,913 deaths and 470,975 injured persons, that is, 405 deaths and 1290 injuries each day from 1274 accidents. Some accident cases also remain unreported, which further increases the rate of accidents. When compared to developed economies, this rate is relatively high. The present safety systems require a multi-pronged approach for making the road safer for the citizens. In this paper, we have considered environmental parameters that can make the drivers uncomfortable while driving the vehicle and lead to an accident. Machine learning techniques are used as a part of the novel approach to mitigate accidents.

3 Related Work

Some of the relevant work in the field of vehicle safety and risk prediction is as follows. M. H. Mohamad and team developed [5] a vehicle accident system embedded with an alcohol detector. The alert for the level of the drunkenness is displayed as a message and, a buzzer alarm is activated simultaneously with the ignition system's deactivation. Similarly, a portable device [6] based on water vapor and gas sensor unit as breath analyzer, a controller and communicator unit to cloud and health control center as the smartphone is projected as a remote alert system for the user.The evolution of IoT diminishes the contrast between the physical and natural world. Here, to bring forth customize information based on user priorities [7], the authors proposed a "Lightweight Context-Aware IoT Service Architecture." The authors [8] proposed an embedded system equipped with Arduino interfaced to a GSM module and GPS receiver along with a fingerprint device to offer a cheap and efficient solution for vehicle theft problems using Fingerprint Matching Algorithm for authentication purposes. If the deviation in the vehicle's location is detected without fingerprint verification, an alert message to the driver's cell phone is delivered. The advancements in the IoT system to offer a robust design for theft and simultaneously accident protection [9] are competently depicted with an IoT-based sensors embedded system in the paper. The authors of [10] presented an IoT system for tracking accident location using GPS and communicating the coordinates via a GSM module in critical situations. Data for the analysis was curated with Arduino as a data processing unit interfaced with sensors and, under emergency, an alert message with the location of the accident, sent via GSM module to the registered mobile numbers to ensure quick assistance to the victim. Implementation of machine learning algorithm with IoT is a robust solution with a practical approach as described further. Reference [11] implemented machine learning algorithms to predict risk and lower severity. They analyzed the four years of traffic accident data. The model focused on two result indicators: traffic violations and accident severity, and analyzed via stepwise logistic regression analysis.

The work of authors [12] is a stepping stone in our research where they analyzed vehicle safety data of the United Kingdom to target the major causes of accidents. Task modeling of a semi-autonomous vehicle based on basic safety parameters was done. Generated tasks were mapped and stored, serving as a source to the IoT system. For the development of an intelligent model, drawing on critical features in the data set, different prediction algorithms had been evaluated to predict the ferocity of accidents under a test input feature set. The outcomes of the model were compared. Random Forest delivered maximum accuracy.

4 Methodology

The interaction of different system components is exhibited in Fig. 1 as a UML diagram. The Proposed Real-time IoT system consists of three layers the physical, virtual, and the control layer. The entire system is displayed in Fig. 2. The physical layer consists of a combination of sensors and micro-controller boards attached to the vehicle. The efficacy of the physical layer depends upon the power consumed by the physical devices. The second layer is the Virtual Object Layer. In this layer, physical devices are exhibited as Virtual Objects (VOs) [13]. The virtual layer allows the client to interact with the physical system. The final layer is the supervisory layer that monitors any change and alerts the user when the vehicle's security or safety is compromised. The system performs two tasks, which are Location Tracking and Risk Prediction. The following subsections explain the approach used for these two tasks.

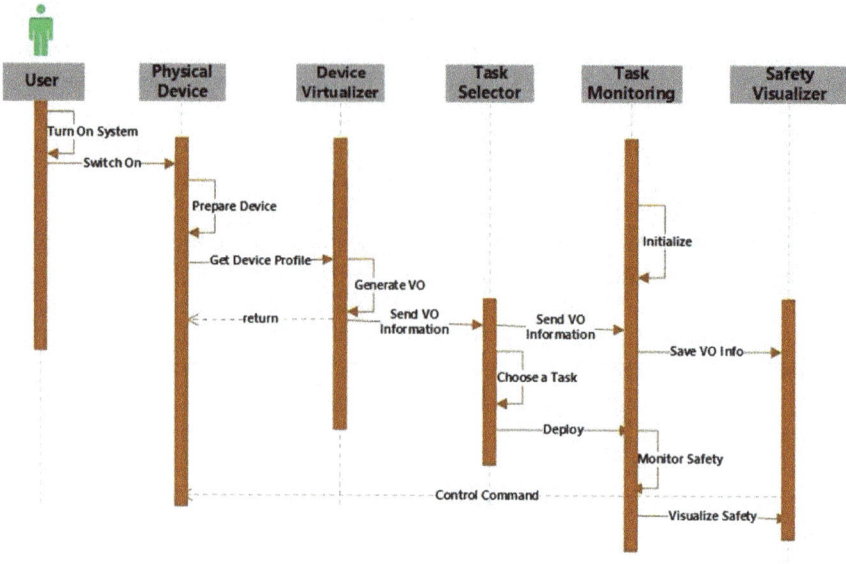

Fig. 1 UML sequence diagram

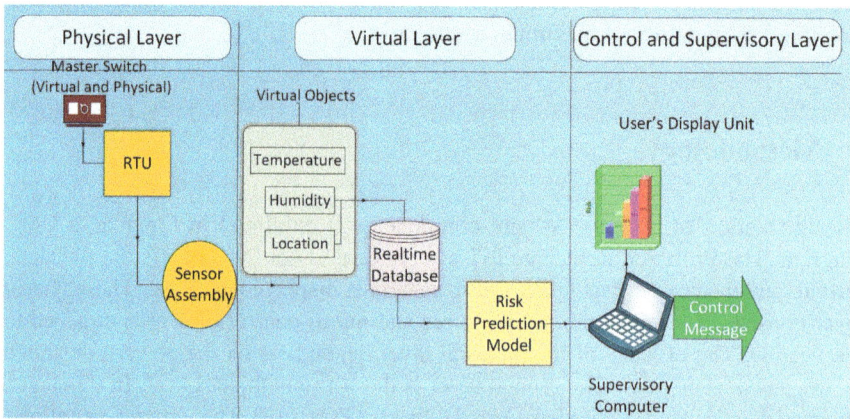

Fig. 2 Risk prediction flowchart

4.1 Location Tracking

Location Tracking system is proposed as a stand-alone system due to the innate advantages the individual system possesses for safety. The Location Tracking System is set to be continuously on, so that the owner of the vehicle may monitor his vehicle at all times. Hence, the location is tracked using the Web API method, which consumes significantly less power than the traditional hardware-based GPS

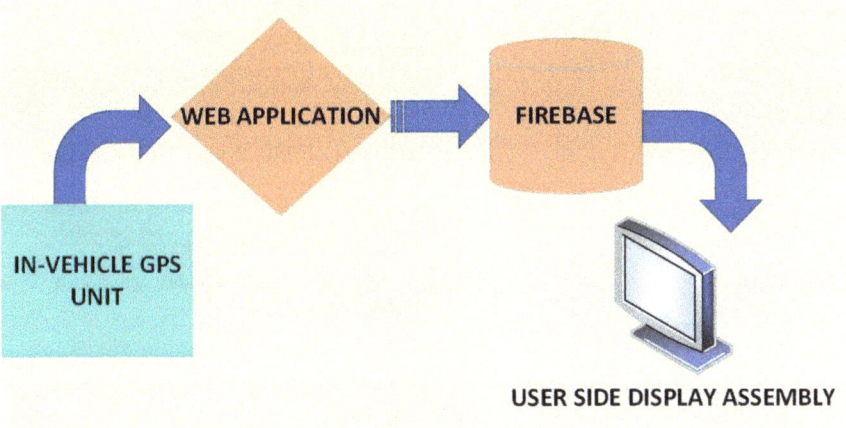

Fig. 3 Location tracking

unit. Figure 3 explains the methodology of location tracking. Further details about the implementation will be explained in section five. When the system is turned on either mechanically or virtually for the first time, a location request is sent to the Remote Terminal Unit (RTU) to obtain the vehicle's current location using the GPS unit connected to the RTU. The obtained coordinates of the vehicle are mapped to their respective VOs. The real-time VO representation of location increases the security of the overall system. When the vehicle is found to be changing position without the owner's permission, the owner is notified.

4.2 Accident Risk Prediction

The process of accident risk prediction is explained in Fig. 2. When the sensor information requests are sent to the RTU using the web-based IoT platform, the RTU collects the sensor data. Then, the sensor data is stored in its virtual Object and fed into the Python machine learning model, which predicts the risk index.

5 Proposed Architecture

The architecture of the system is explained in the following subsections. The technologies that have been used are illustrated in Table 1.

Table 1 Technologies used

Component	Summary
Hardware	NodeMCU, Arduino Uno, LCD Display 16X2
Sensors	Temperature, Humidity
Programming Languages	JavaScript, HTML, Python, C++ (Arduino), PHP
Realtime Database	Firebase
IDE	PyCharm,ArduinoIDE, Sublime Text
Python Libraries	Sci-kit Learn, Numpy, Pandas, Matplotlib, firebase admin

Fig. 4 Location tracking web application

5.1 Overview of the Location Tracking Web App-DTU Shuttle

Location tracking is implemented using a web application Fig. 4. The technologies used for the development are mentioned in the implementation stack table. The application is built for Driver Side Use, i.e., it is to be operated by the vehicle's driver. The App can be run on any smartphone/tablet device. The App uses the Web Geolocation API, which is available as an open-source API, to obtain the vehicle's location. The vehicle's location is then mapped to specific points of significance on the map so that users find it easy to visualize the vehicle's current location. The location is then stored in Firebase, which is a real-time database. The website was initially utilized for the tracking of the university shuttle to obtain testing data.

5.2 Real-Time Accident Risk Prediction Using Web IoT Platform

A screenshot of the web IoT platform is shown in Fig. 5. The real-time sensor data obtained from the RTU is fed into the Machine Learning model, which is integrated with the platform on the back-end. The machine learning model is built using Python

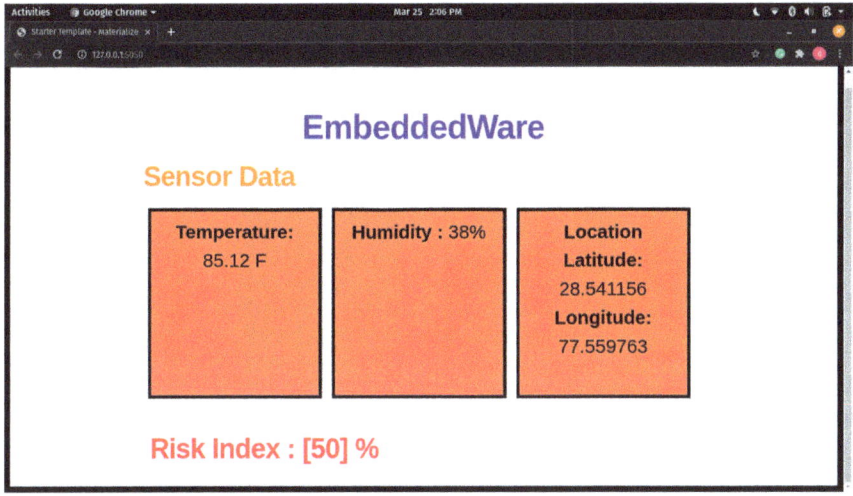

Fig. 5 IOT platform web application

and is run on a flask framework. The details of model evaluation are given in Sect. 6. Once the system is started, the accident risk value is calculated in real-time. Risk value is obtained in terms of percentage. Once the risk value exceeds the threshold value, the driver is alerted to stop the vehicle by a control message. The RTU sends the control message to the Driver's Display.

5.3 Model Implementation

The model used for the purpose of Accident Risk Prediction was built using Python and it's libraries and trained using the December 2020 USA Accidents data-set [14]. The approach used for Model Implementation has been exhibited in Fig. 6.

5.3.1 Data-Set Description

The December 2020 US Accidents Data-Set [14] has 47 features and 3 million records. Out of the 47 features, 17 features are string data, 13 features are boolean, 13 features are decimal, and the remaining four features are of other types, including Date-Time, etc. Many features describe the location of accidents, city, postal code, etc. In addition, some features provide information about the temperature, humidity, pressure, etc., features which were leveraged to build a more accurate real-time prediction system.

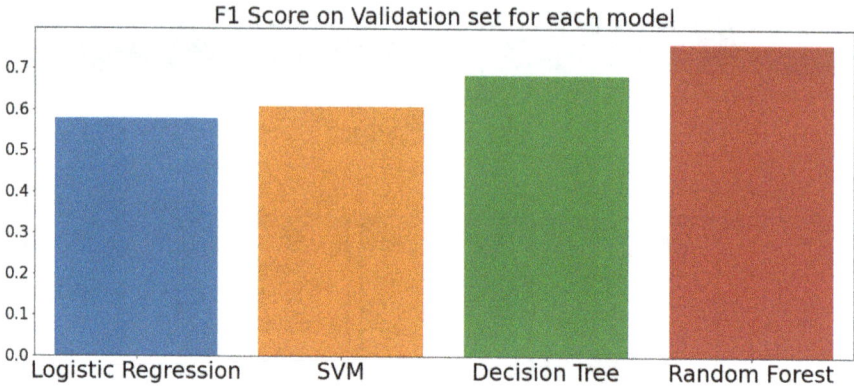

Fig. 6 F1 score on validation set of each model

5.3.2 Data Preprocessing

The dataset was obtained in a raw condition. The Python Libraries Pandas, Numpy, and Matplotlib were extensively used for pre-processing the data-set. Firstly, highly correlated fields were found out. For example, the Start Location and the End Location. Both of them were approximately equal. Next, one of the highly correlated fields was dropped to remove data redundancy. Further, the fields with missing values were replaced with mean values. Then outliers were removed from the data set. Finally, due to the highly unbalanced nature of the data-set, the data-set was balanced across all risk categories.

5.3.3 Data Normalization

In this paper, min-max normalization was used. The formula for min-max normalization is as follows:

$$(value - min)/(max - min) \qquad (1)$$

here, value is the value of current data point. The value of max is taken as 1, and min is taken as 0.

5.3.4 Feature Encoding

Categorical Features are features which have a discrete set of values which are of string data-type. The categorical features are converted into integer values as models cannot assign weights to string values. One-hot encoding technique is used to convert the categorical features.

5.3.5 Model

Sci-Kit Learn library of Python has been used to train the model. Firstly, the dataset was split into train set and validation set using n-fold cross-validation technique. Next, the training data is fitted to different models. We have used the Logistic Regression, Decision Tree, Support Vector Machine, and Random Forest classifier models for our analysis. Finally, the validation data-set is used to evaluate all the models. Further details on model evaluation are presented in Sect. 6.

6 Model Evaluation

The following metrics have been used to choose the best method out of the four applied methods.

6.1 Accuracy

Accuracy is a measure of performance that can be derived from the confusion matrix. While accuracy provides a better understanding of the performance of a model with respect to true predictions, it fails to provide the complete picture.

6.2 Recall

It is the percentage of True Positive Predictions that were identified correctly. Recall is always in contention with the precision of a model. However, high scores in both can be achieved by effective tweaking of the hyper-parameters. The precision of the Random Forest model was found out to be 0.757 and the recall 0.679.

6.3 F1 Score

The F1 score is computed using the following formula:

$$F1 = \text{Precision} * \text{Recall}/(\text{Precision} + \text{Recall}) \tag{2}$$

The comparison of the F1 score of the algorithms used is shown in Fig. 7.

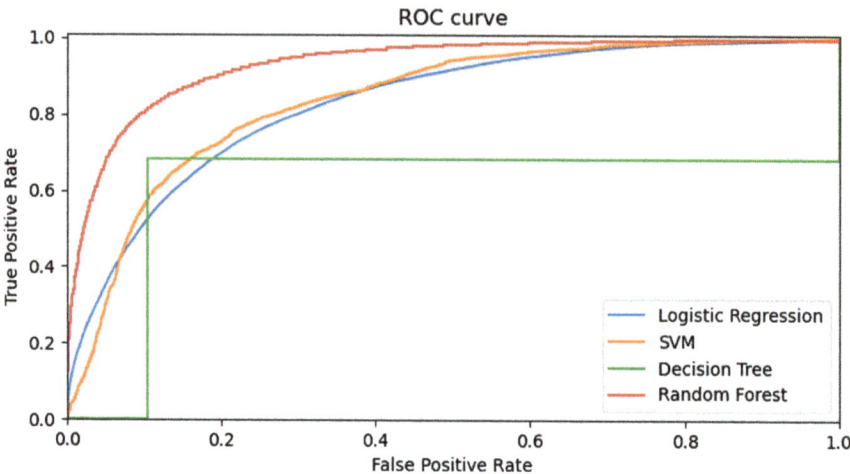

Fig. 7 ROC curve

6.4   ROC

Receiver Operating Characteristic curve. Area Under the Curve (AUC) gives an average measure of the performance. The AUC metric is chosen here because it does not vary with scale. The comparison of the AUC of the algorithms used is shown in Fig. 7.

As seen in Table 2, the Random Forest model performs significantly better than the remaining methods in all metrics. Accuracy obtained on the Random Forest Model is 81.4%.

Hence, Random Forest is the most suitable method for prediction of the Accident Risk.

7   Discussion

The paper proposes an Internet of Things system that performs two tasks to enhance the safety and security of driving in vehicles. The proposed system can be considered as a starting point of an economical alternative to the costly driver assistance systems that perform only a dedicated job. In this section, we present the comparative analysis of different methods.

Table 2 Comparison of algorithms based on different metrics

Algorithms Tested	Evaluation metrics			
	Accuracy	Precision	Recall	F1 score
Logistic Regression	0.594	0.4711	0.748	0.578
Support Vector Machine	0.618	0.4625	0.751	0.606
Decision Tree	0.683	0.654	0.592	0.683
Random Forest	0.814	0.757	0.679	0.76

7.1 Comparative Analysis

Four Different Techniques were used for training. Out of these four methods, the most suitable technique for the data-set was found out to be Random Forest. Table 2 provides the comparative study of different methods. Logistic Regression only provides an accuracy of 0.594. As we can see, the accuracy of the model improved by 37% (0.221) when the Random Forest Technique is used.

8 Conclusion

This Paper presents a Real-Time IoT system that utilizes IoT devices' interconnectivity with sensors and virtual devices. The first task performed by the system is Location Tracking of the Vehicle. Location tracking is facilitated by a Web Application that notifies the user about the possibility of any suspicious activity such as vehicle theft. The second task performed by the system is accident risk prediction. In this task, we have leveraged the US Accidents data-set to train the Random forest model with accuracy 81.4%. We have deployed various sensors to monitor temperature, humidity, and other parameters. These conditions are processed by the Embedded IoT application to compute the risk index of the vehicle. If the risk value exceeds a threshold value, the system will alert the user. Finally, an evaluation of the model is done along with the visualization of results. Model Evaluation shows the predicted accuracy of accident risk with the use of the trained Random Forest model.

9 Future Scope

The paper provides an entry-level system that can be utilized for accident risk prediction. Therefore, future work is required to enhance the system further. In future, we will train the model using different techniques like ensemble learning and deep neural networks to improve the precision and accuracy of the model, thereby improving the overall system. The location tracking system can be integrated with Google Maps to offer more excellent visualization. The Accident Risk Prediction system can be made more accurate by adding more sensors to the system, such as impact sensors, pressure sensors, etc. The Risk Prediction system can be integrated with the vehicle's internal On-Board Diagnostics system to offer seamless suggestions about any defective parts needed to be replaced.

References

1. Ahmad S, Malik S, Ullah I, Fayaz M, Park D-H, Kim K, Kim D (2018) An adaptive approach based on resource-awareness towards power-efficient real-time periodic task modeling on embedded IoT devices. Processes 6:7
2. Longman M (2006) Ch. 1 Forensic investigation of stolen-recovered & other crime-related vehicles. Academic, Arizona, USA, pp 1–21. https://doi.org/10.1016/B978-012088486-5/50034
3. Theft rate reported in India in 2019 by state (2021). https://www.statista.com/statistics/632912/reported-theft-rate-by-state-india. Accessed 25 Feb 2021
4. Road accidents in India claimed 405 lives, injured 1,290 each day in 2017 (2018). https://www.autocarpro.in/news-national/road-accidents-in-india-claimed-405-lives--injured-1-290-each-day-in-2017-41006. Accessed 20 Feb 2021
5. Mohamad M, Hasanuddin M, Ramli M (2018) Vehicle accident prevention system embedded with alcohol detector. Int J Rev Electron Commun Eng 1:4
6. Wakana H, Yamada M (2019) Portable alcohol detection system for driver monitoring. IEEE Sens 1:4. https://doi.org/10.1109/SENSORS43011.2019.8956885
7. Gochhayat SP, Kaliyar P, Conti M, Tiwari P, Prasath VBS, Gupta D, Khanna A (2019) LISA: lightweight context-aware IoT service architecture. J Clean Prod 212:1345. ISSN: 0959-6526. https://doi.org/10.1016/j.jclepro.2018.12.096
8. Dey M, Arif MA, Mahmud MA (2017) Anti-theft protection of vehicle by GSM and GPS with fingerprint verification. In: 2017 international conference on electrical, computer and communication engineering (ECCE), Cox's Bazar, Bangladesh. https://doi.org/10.1109/ECACE.2017.7913034
9. Sil S, Daw S, Deyasi A (2021) Smart intelligent system design for accident prevention and theft protection of vehicle. In: Nath V, Mandal J (eds) Nanoelectronics, circuits and communication systems. Lecture notes in electrical engineering, vol 692. Springer, Singapore. https://doi.org/10.1007/978-981-15-7486-3_47
10. Chourasia P, Choubey S, Verma R (2020) Vehicle accident detection, prevention and tracking system. Int Res J Eng Technol 7:8
11. Zhang G, Yau KKW, Chen G (2013) Risk factors associated with traffic violations and accident severity in China. Accid Anal Prev 59
12. Ahmad S, Jamil F, Khudoyberdiev A, Kim D (2020) Accident risk prediction and avoidance in intelligent semi-autonomous vehicles based on road safety data and driver biological behaviours. J Int Fuzzy 38(4):4591–4601. https://doi.org/10.3233/jifs-191375
13. Ullah I, Sohail Khan M, Kim D (2018) IoT services and virtual objects management in hyper-connected things network. Mob Inf Syst 2018. https://doi.org/10.1155/2018/2516972
14. US accidents data-set (2020). https://www.kaggle.com/sobhanmoosavi/us-accidents. Accessed 15 Feb 2021

An Hybrid Approach Based on Clustering and Synthetic Sample Generation for Imbalance Data Classification: ClustSyn

Gillala Rekha and Sailaja Madhu

Abstract One of the major problems to be investigated in various application domains in the real world is the skewed/imbalanced distribution of data. The class imbalance problem turns out when the number of samples from one class significantly exceeds the number samples from the other class and when such data is trained on traditional learning algorithms tends to be bias towards the majority class (i.e. class having more number of samples), resulting in a significant deterioration in the classification performance. Recent studies have shown that the presence of other features, such as small disjuncts, overlapping, and noise in data, can make classification much more difficult. In this paper, we propose an effective hybrid method based on clustering and synthetic sample generation for imbalance data classification called ClustSyn. It consists of clustering algorithm along with the synthetic data generation using Mahalanobis distance. The reason for using the Mahalanobis distance is that it can minimize the probability of overlap and maintain the structure of covariance when providing synthetic samples for the minority class. ClustSyn efficiency is compared with existing methods such as AdaBoost, RUSBoost, SMOTEBoost, based on ensemble learning. We have performed experiments with different Imbalance Ratios (IR) on 11 datasets. Results show ClustSyn outperformed existing methods for imbalanced and small disjunct datasets.

Keywords Ensemble learning · Class imbalance problem · Sampling · Imbalance ratio · Mahalanobis distance

G. Rekha (✉)
Department of Computer Science and Engineering, Koneru Lakshmaiah
Education Foundation, Hyderabad, India
e-mail: gillala.rekha@klh.edu.in

S. Madhu
Department of Computer Science and Engineering, Malla Reddy Engineering College
for Women (autonomous), Hyderabad, India

© The Author(s), under exclusive license to Springer Nature Singapore Pte Ltd. 2022
D. Gupta et al. (eds.), *Proceedings of Data Analytics and Management*,
Lecture Notes on Data Engineering and Communications Technologies 90,
https://doi.org/10.1007/978-981-16-6289-8_63

1   Introduction

Constructing an efficient learning model can be a challenging task in machine learning due to skewed data distribution. Training a conventional machine learning algorithm on skewed data would lead to decrease in classification accuracy. For example, real-world applications like fraud detection, oil spill detection [6, 12] and software defect prediction [11] suffer from skewed data distribution. Therefore, several
researchers have suggested many approaches in recent decades to resolve the issue
of class imbalance. According to [6, 16], the imbalanced data distribution problem
can be addressed broadly based on two categories, namely at the data preprocessing
level and at the algorithmic level. Data is pre-processed using sampling techniques
or by synthetic data generation to establish a balanced distribution between classes.
Whereas, the existing algorithms are updated or new algorithms are suggested to deal
with the problems of class imbalance in algorithmic-level techniques. In addition,
hybrid methods combine both preprocessing and algorithmic techniques together
in order to achieve better performance when training data has a skewed distribution. Recent literature work [6, 12] combines ensemble algorithms along with data
sampling techniques.

The main contribution of ClustSyn is to handle small disjunct that worsen the
performance of classification algorithms. The aim is of twofold: (1) improving the
sample selection process for over-sampling, and (2) improving the process for the
new sample generation.

Section 2 discusses literature work related to the issue of class imbalance. The
proposed ClustSyn approach is implemented in Sect. 3. Then, the experimental results
are presented in Sect. 4. Lastly, Sect. 5 concludes this work with a brief sketch of
possible future enhancements. Note that throughout this work, we use terms like
method, approach, and techniques interchangeably.

2   Literature Review

In class imbalance data, the large size of majority class samples leads to problem
when traditional classification algorithms are trained on such data. In the past decade,
several researchers have proposed many techniques to address the imbalanced data
problem [1]. Here, the literature review presents in two phases: (i) Generative Data
Sampling Methods and (ii) Ensemble Methods.

2.1   Generative Data Sampling Methods

To solve the problem of class imbalance distribution, data-level techniques may be
generally split into resampling and synthetic data generation methods. Under-

sampling, over-sampling, or both are used in resampling strategies to deal with class imbalance data. Synthetic data generation methods, on the other hand, create new instances from existing data. In general, the SMOTE technique (a synthetic data generation technique) is one of the useful and efficient techniques being used for skewed data distribution [4]. SMOTE generates new instances by randomly selecting minority data samples and interpolating between them and their nearest neighbours in the same class. Variations of methods have been suggested to enhance the selection process of data instances in SMOTE, such as Modified SMOTE (MSMOTE) [9], Borderline SMOTE [7], and many more. MSMOTE introduced a different strategy for selecting neighbours based on the existing samples to produces synthetic instances. In Borderline SMOTE, to construct synthetic data instances, only small instances near the boundary between classes are over-sampled. In [2], for dealing with unbalanced dataset challenges the author introduced the Majority Weighted Minority Over-sampling TEchnique (MWMOTE). The Euclidean distance (distance between two data points) from the nearest majority sample is used by MWMOTE to apply weights to minority cases. Bunkhumpornpat et al. [3] introduced the safe-level SMOTE technique, which samples minority occurrences along the same line with varying degrees of weight termed safe level. However, the application of clustering approaches in identifying the various classes has been the recent technique in addressing the class imbalance problem.

2.2 Ensemble Methods

A number of ensemble classifiers have been proposed [6, 15] for handling imbalance problems. By integrating different machine learning methods, ensemble algorithms increase the performance of a classifier. The most of ensemble learning algorithms were created specifically to deal with issues of class imbalance. The ensemble algorithms for class imbalance problems usually include preprocessing techniques or cost-sensitive approaches. Moreover, a combination of data-level approaches with ensemble learning are more robust in solving skewed data distribution. Ensemble approaches, such as SMOTEBagging [19], SMOTEBoosting [5], RUSBoost [18], and CUSBoost [14], are classified based on bagging or boosting techniques with data-level approaches. Such approaches integrated sampling methods and SMOTE using bagging and boosting algorithms. Aside from that, cluster-based methods for dealing with skewed distributions have been presented in the literature [13, 14]. Clustering-based Under-Sampling with Boosting (CUSBoost) was proposed in [14] to eliminate negative data in each cluster using random under-sampling. The entire dataset is initially partitioned into majority and minority classes, and k-means clustering algorithm is applied to the majority class. The resultant dataset along with minority class samples is trained using an ensemble method (AdaBoost algorithms).

CUSBoost performs best when the data is highly clusterable, and it gives best results according to the problem domain. The author [13] presented a new under-sampling method based on the cluster approach to tackle the problem of class imbalance.

3 ClustSyn: An Hybrid Approach Based on Clustering and Synthetic Sample Generation for Imbalance Data Classification—Methodology

This section discusses the proposed work. The proposed ClustSyn method combines the generation of synthetic instance using ensemble algorithm. In our technique, the clustering technique is implemented along with ensemble learning. It consists of the clustering phase, the synthetic data generation phase, and the classification phase.

3.1 Clustering Phase

In this phase, as shown in Fig. 1, instances are divided into negative class and positive class instances. Then, each class is grouped separately using clustering technique. This work applies Mahalanobis distance (MD) to preserve the diversity within the minority cluster to generate unique synthetic samples.

In this work, K-medoid clustering algorithm applied, which is more robust than the K-means clustering. The K-medoid algorithm creates K clusters, and the Partitioning Around Medoids (PAM) technique is the most widely used K-medoid clustering algorithm. The working of this algorithm can be described as follows:

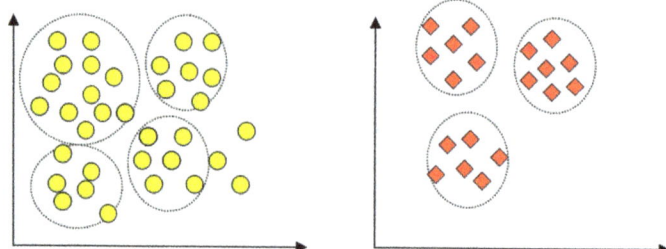

Fig. 1 Example of the clusters for majority classes and minority classes before synthetic data generation

1 **Initialization step**: Randomly choose K data points initially as cluster medoids (from datasets).
2 **Assignment step**: Assign each data point to the cluster that is closest to the medoid.
3 **Update step**: For each object 'i', recalculate the locations of the textit$K$ medoids using the distance measure $\sum_{j \epsilon C_i} d(i, j)$.
4 **Repeat** Steps 2 and 3 should be repeated until the medoids do not change any more.

3.2 Synthetic Data Generation Phase

The number of synthetic samples to be produced depends on the largest cluster size. To maintain the diversity within the minority cluster and generating unique synthetic samples, Mahalanobis distance (MD) is used. The MD measure is regarded as a unit-less measure and provides a relative measure of the distance of a dataset. Considering two data instances $x = (x_1, x_2, x_3, \ldots, x_n)^T$ and $y = (y_1, y_2, y_3, \ldots, y_n)^T$, the MD to be calculated is defined as follows:

$$d_M(x, y) = \sqrt{(x - y)^T S^{-1}(x - y)} \tag{1}$$

where 'S' is the matrix of sample covariance. The process of generating synthetic samples using MD is as follows:

- Select the cluster for which synthetic samples to be generated and also the number of samples to be generated as specified above.
- Compute MD of each data samples according to Eq. (2).
- Sort the samples based on MD in decreasing order and rank them.
- Now, generate the synthetic samples by averaging the highest rank data sample with the lowest rank data sample.
- Repeat the previous procedure until sufficient number of the synthetic samples are generated.

3.3 Ensemble Learning

To address the problem of class imbalance in the classification process, an ensemble learning technique has been developed. The Adaptive Boosting Method (AdaBoost) [6] is an iterative boosting algorithm for constructing a strong classifier from a linear combination of weak classifiers, as mentioned in the literature (Sect. 2). The working of ClustSyn algorithms is shown in Algorithm 1. It combines cluster-based over-sampling with AdaBoost.

Algorithm 1: ClustSyn Algorithm 1

    Input: Let Training Data

TD $=\{x_i, y_i\}, \ldots, \{x_m, y_m\}$

where $x_i \in X$ *and* $y_i \in Y = \{-1, +1\}$

C = Number of iterations; I = Weak Classifier

    Output: Final Classifier:

$H(X) = sign\left(\sum_{c=1}^{T} \alpha_c h_c(x)\right)$ *where* $h_C \epsilon = \{-1, +1\}$ and α_C = weight of respective

classifier.

1: Find the optimal 'K'(as shown in Eq. 2)

2: for i= 1 to K do

D_i using cluster-based over-sampling (as described in Sect. 4.2)

3: $D_1(i) = 1/m$

4: *for* $C = 1$ *to* C :

5: $h_C = I(TD, D_C)$

6: $\epsilon_C = \sum_{i, y_i \neq h(x_i)} D_C(i)$

7: *if* $\epsilon_C > 0.5$ then

8: C = C-1

9: return

10: endif

11: $\alpha_C = \frac{1}{2} ln\left(\frac{1-\epsilon_C}{\epsilon_C}\right)$

12: $D_{C+1}(i) = \frac{D_i exp(-\alpha_C y_i h_C(x_i))}{Z_i}$

where Z_C *is a normalization factor*

(*normalize* D_{C+1} *to be a proper distribution*)

13: end for

---

4   Experimental Results

The experimental analysis to evaluate the performance of our proposed ClustSyn algorithm is explained in this section. In this simulation, we have considered a total of 11 imbalanced datasets from KEEL data repository, having different imbalance ratio (IR) feature. The characteristics of the datasets are shown in Table 1 such as number of instances, number of attributes, and their imbalance ratio (IR).

4.1   Evaluation Parameters and Tools Used

The proposed ClustSyn method is compared with existing methods like AdaBoost [20], RUSBoost [17], SMOTEBoost [5], and CUSBoost [14].

Table 1 Dataset characteristics

Dataset name	No. of instance	No of attributes	Percentage IR (%)
Ecoli1	336	7	8.6
Glass1	214	9	2.06
Haberman	306	3	2.78
Iris	150	4	2.0
New-thyroid	215	5	5.14
Page-blocks	5472	10	8.79
Pima	768	8	1.87
Segment	2308	19	6.02
Vehicle	846	18	2.9
Wisconsin	683	9	1.86
Yeast	1484	8	2.46

4.2 Performance Measure

The efficiency of the proposed model is assessed using F-Score and Area Under the ROC Curve (AUC) [1, 7, 8, 10, 11, 13, 18]. Note that with imbalanced datasets, accuracy (acc) is a weak measure and not appropriately evaluate the performance of a classifier. Therefore, classification effectiveness needs to be calculated considering other metrics like AUC, F-score and geometric mean [6]. The formula for AUC and F-score is given below:

$$\text{AUC} = \frac{1}{2} \times \left(\frac{\text{TP}}{\text{TP} + \text{FN}} + \frac{\text{TN}}{\text{TN} + \text{FP}} \right) \tag{2}$$

where TP stands for True Positive, TN stands for True Negative, FP stands for False Positive, and FN for False Negative.

$$F\text{-Score} = \frac{2 * \text{precision} * \text{recall}}{\text{precision} + \text{recall}} \tag{3}$$

F-score consists of recall (previously defined) and precision (the proportion of samples classified as positive that are actually positive).

4.3 Results

The base classifier used is Decision Tree (C4.5) in boosting the algorithm to learn from the datasets. Using tenfold cross-validation, all results are assessed. The dataset for the training is split into ten partitions. To train the model, nine partitions are used while the one partition kept out is used to evaluate the model. To make each partition for test data, this process is repeated ten times. The experiments are carried out on eleven datasets taken from KEEL data repository with varying degree of imbalance ratio (IR) and data size. Figures 2 and 3 show the AUC and F-measure results obtained on the imbalanced datasets using AdaBoost, RUSBoost, SMOTEBoost, CUSBoost, and ClustSyn. From the figures, we can find that SMOTEBoost and ClustSyn methods show a better result than AdaBoost and RUSBoost. By incorporating improved sample generation technique, the proposed method over-sampled the right minority samples in the datasets and the same is been represented in results.

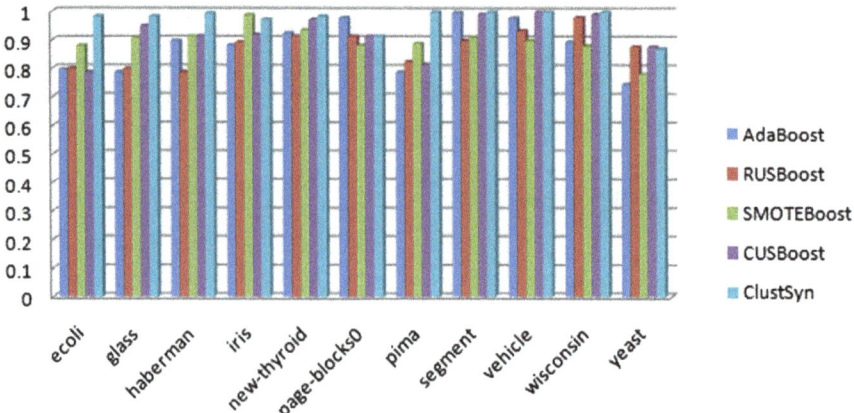

Fig. 2 AUC performance of boosting classifier with resampling method

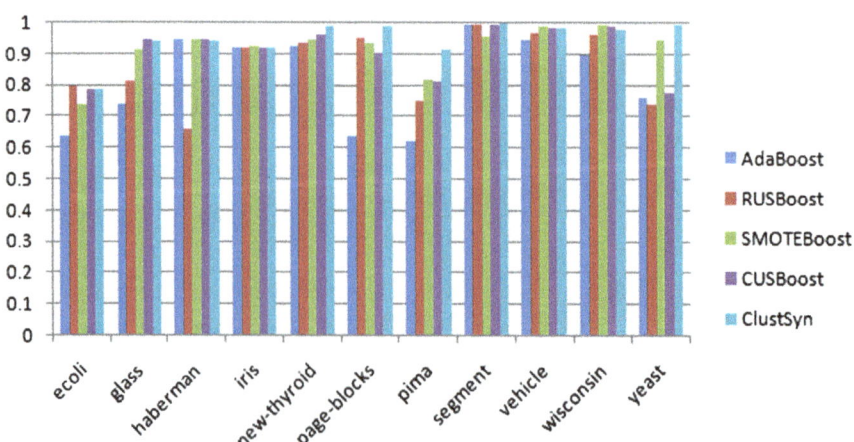

Fig. 3 F-measure performance of boosting classifier with resampling method

5   Conclusion

The proposed method presents a hybrid approach based on clustering and synthetic sample generation called ClustSyn for handling imbalanced datasets. This approach includes the use of clustering methods with ensemble classifiers in a novel synthetic data generation process. The ClustSyn approach efficiency has been compared with existing approaches such as AdaBoost, RUSBoost, CUSBoost, and SMOTEBoost. Based on the experimental results, we found that ClustSyn provided best performance against the current techniques. Remarkably, with datasets having higher imbalanced ratios, our proposed method yielded better results. Finally, as a future study, by using additional classifiers and their efficacy in a particular application domain, we expect to investigate the efficiency of the ClustSyn method. In addition, interested researchers are encouraged to expand our work on problems of multi-class imbalance.

References

1. Ali A, Shamsuddin SM, Ralescu AL (2015) Classification with class imbalance problem: a review. Int J Adv Soft Comput Appl 7(3):176–204
2. Barua S, Islam MM, Yao X, Murase K (2014) Mwmote-majority weighted minority oversampling technique for imbalanced data set learning. IEEE Trans Knowl Data Eng 26(2):405–425
3. Bunkhumpornpat C, Sinapiromsaran K, Lursinsap C (2009) Safe-level-smote: safe-level-synthetic minority over-sampling technique for handling the class imbalanced problem. In: Proceedings of the Pacific-Asia conference on knowledge discovery and data mining. Springer, pp 475–482
4. Chawla NV, Bowyer KW, Hall LO, Kegelmeyer WP (2002) Smote: synthetic minority over-sampling technique. J Artif Intell Res 16:321–357
5. Chawla NV, Lazarevic A, Hall LO, Bowyer KW (2003) Smoteboost: improving prediction of the minority class in boosting. In: Proceedings of the European conference on principles of data mining and knowledge discovery. Springer, pp 107–119
6. Galar M, Fernandez A, Barrenechea E, Bustince H, Herrera F (2012) A review on ensembles for the class imbalance problem: bagging-, boosting-, and hybrid-based approaches. IEEE Trans Syst Man Cybernet Part C Appl Rev 42(4):463–484
7. Han H, Wang WY, Mao BH (2005) Borderline-smote: a new over-sampling method in imbalanced data sets learning. In: Proceedings of the international conference on intelligent computing. Springer, pp 878–887
8. He H, Bai Y, Garcia EA, Li S (2008) Adasyn: adaptive synthetic sampling approach for imbalanced learning. In: Proceedings of the IEEE world congress on computational intelligence. IEEE, pp 1322–1328
9. Hu S, Liang Y, Ma L, He Y (2009) Msmote: improving classification performance when training data is imbalanced. In: Proceedings of the second international workshop on computer science and engineering, vol 2. IEEE, pp 13–17
10. Jo T, Japkowicz N (2004) Class imbalances versus small disjuncts. ACM Sigkdd Explor Newslett 6(1):40–49
11. Krstic M, Bjelica M (2015) Impact of class imbalance on personalized program guide performance. IEEE Trans Consum Electron 61(1):90–95
12. Lin M, Tang K, Yao X (2013) Dynamic sampling approach to training neural networks for multiclass imbalance classification. IEEE Trans Neural Netw Learn Syst 24(4):647–660

13. Lin WC, Tsai CF, Hu YH, Jhang JS (2017) Clustering-based undersampling in class-imbalanced data. Inf Sci 409:17–26
14. Rayhan F, Ahmed S, Mahbub A, Jani M, Shatabda S, Farid DM et al (2017) Cusboost: cluster-based under-sampling with boosting for imbalanced classification. arXiv preprint arXiv:1712.04356
15. Rekha G, Tyagi AK, Krishna Reddy V (2019) Solving class imbalance problem using bagging, boosting techniques, with and without using noise filtering method. Int J Hybrid Intell Syst 15(2):67–76
16. Rekha G, Tyagi AK, Krishna Reddy V (2019) A wide scale classification of class imbalance problem and its solutions: a systematic literature review. J. Comput. Sci. 15:886–929
17. Seiffert C, Khoshgoftaar TM, Van Hulse J, Napolitano A (2009) Rusboost: a hybrid approach to alleviating class imbalance. IEEE Trans Syst Man Cybernet Part A Syst Hum 40(1):185–197
18. Seiffert C, Khoshgoftaar TM, Van Hulse J, Napolitano A (2010) Rusboost: a hybrid approach to alleviating class imbalance. IEEE Trans Syst Man Cybernet Part A Syst Hum 40(1):185–197
19. Wang S, Yao X (2009) Diversity analysis on imbalanced data sets by using ensemble models. In: Proceedings of the IEEE symposium on computational intelligence and data mining. IEEE, pp 324–331
20. Ying C, Qi-Guang M, Jia-Chen L, Lin G (2013) Advance and prospects of adaboost algorithm. Acta Automat Sin 39(6):745–758

Multimodal Drowsiness Detection Using HM-LSTM Network

Siddharth Sharma, Ranit Dua, Aniket, and Vinod Kumar

Abstract A major factor identified in the ever increasing number of road accidents is drowsiness of the driver. With the advancements in today's technology, it is of utmost importance to come up with a viable and implementable solution to overcome this plight. In this paper, we analyze different methods and propose a hierarchical multi-scale long short-term memory (HM-LSTM) network for early drowsiness detection that captures the interplay between blinking and yawning modalities, the two main proposed attributes associated with fatigue and drowsiness. We use a combination of two modalities of visible drowsiness signs which are the blinking pattern and the event of a yawn of an individual. We use eye aspect ratio (EAR) and mouth aspect ratio (MAR) to carefully predict the state of the driver and eventually help to prevent a road accident. Our proposed multimodal approach achieves an accuracy of 78.3% which represents the viability of accurate timely detection of driver fatigue and drowsiness.

Keywords Drowsiness detection · Image processing · HM-LSTM · Machine learning

1 Introduction

Sleeping is one of the most basic needs of the human body alongside eating, drinking, and breathing [1]. Sleep deprivation can affect the reaction time and concentration

S. Sharma · R. Dua (✉) · Aniket · V. Kumar
Delhi Technological University, New Delhi, India
e-mail: ranitdua_2k17co262@dtu.ac.in

S. Sharma
e-mail: siddharthsharma_2k17co339@dtu.ac.in

Aniket
e-mail: aniket_2k17co55@dtu.ac.in

V. Kumar
e-mail: vinod_k@dtu.ac.in

of the driver which has been proven to be a major source of car accidents. A research conducted by the AAA Foundation for Traffic Safety approximated that the car accidents occurring due to fatigued driving are around 328,000 annually [2]. NHTSA estimates drowsiness-related accidents that impart injury of any kind costs around $109B annually, not including property damage [3]. All these unsettling facts communicate the dire need of a practical and cost-effective solution that can help expose fatigue at the earliest. As mutually deduced [4–6], features of drowsiness detection can be classified into three major categories: performance aspect, physiological aspect, and behavioral aspect. The performance measurement takes into consideration a combination of the speed of the vehicle, the path of the vehicle is following the general braking style of the driver and also but not limited to the pressure applied to the steering wheel, and the only drawback here being that these require extensive mechanisms to be installed in the vehicle leading to the cost increasing eminently. For instance, the famous vehicle producing company Mercedes installed such a feature into their top-models called attention assist system [7]. The physiological aspects suggest that drowsiness can also be detected using a permutation of BPM, heart rate, electrooculogram (EOG) [8], electroencephalogram (EEG) [8, 9], electromyogram (EMG), and electrocardiogram (ECG). These measurements can accurately filter out the cases of drowsiness but evidently require invasive methods, reducing their viability in vehicles. Although, with the advent of smart wearables like smart watches, this could be seen as a viable method in the near future [10]. The behavioral aspects on the other hand include features like facial expressions and gestures. These features use noninvasive sensors like cameras to detect signs which may advocate drowsiness [11]. We propose to use mobile phones of the respective drivers to detect drowsiness in order to overcome the problem of expensiveness and practicality.

2 Related Works

In the following section, few-related studies have been discussed and classified on the bases of (i) datasets and (ii) methods.

2.1 Datasets

As aforementioned, a tremendous amount of work has already been done revolving around this topic yet it is difficult to carry out a comparison due to the different techniques and datasets followed. Several studies conducted and used datasets [5, 12] having subjects staged to act drowsy which cannot lead to an accurate measure as subjects tend to overreact and lead to drastic difference in the results. Some were analyzed on a lesser number of cadets, and the visuals were also not available, and a few datasets [13] were created for the purpose of small expressions and subtle facial changes. The NTHU dataset [14] contains IR videos of participants professing to be drowsy leads us to the question of the validity of such a dataset. The DROZY dataset [9] is a dataset that brings forth numerous modalities of information to handle the plan of fatigue detecting frameworks and similar assessments. However, all videos in this particular dataset were captured in a regulated environment using the same camera position and background. As compared to our dataset, DROZY dataset also has fewer videos and provides only NIR video whereas the dataset used provides normal color videos. A few datasets like Friedrichs and Yang [15] having elaborate and genuine compiled driving time of 90 h with unfeigned drives are private and not available for public use.

2.2 Methods

Majority of the studies conducted for drowsiness detection use facial regions aiming to extract the relevant features. The study by McIntire et al. [16] brings to light the process by which blinking frequency, timing, and recurrence are generally proportionate with drowsiness. Svensson [8] manifests that the amplitude of how the eye blinks should also be considered as a prominent component. Other studies like Friedrichs and Yang [15] explore various blinking features which include blinking extent and timespan, eye opening velocity, and vitality of blinks besides head movement data. Nevertheless, all the features in [15] are withdrawn by the help of the seeing machines sensor [17] which utilize extensive indicators in order to detect fatigue as compared to the basic cell phone in our dataset. Ajay [18] observes the facial and eye movements from the Web camera to determine the drowsiness state of the subject. The study by Park et al. [19] tunes three convolutional neural networks and seeks and uses an SVM to the normalized features of the aforementioned three networks to distribute each frame into four categories.

3 The Real-Life Drowsiness Detection Dataset

We have used the real-life drowsiness detection (RLDD) dataset [20] which was developed by the students of University of Texas at Austin. This dataset provides authentic data which were formulated using 60 subjects of varied ethnicities and diverse facial features like beard, skin tone, and accessories like spectacles. Mode of video recording was majorly through mobile camera and Webcam, attempting to keep the position of the camera akin. Each subject was examined under three stages which were—alertness, low vigilance, and drowsiness. This dataset provides additional benefits of detecting even the slightest case of drowsiness in addition to the evident cases detected by other datasets.

4 Methodology

In our proposed strategy, we divide our approach into two parts. The first part explains our approach to detect blinks from the face of the driver using the technique of facial landmark detection using the Dlib library in Python. The eye aspect ratio of the subject at every instant in the input video is taken as the feature for the first part of our strategy. In the second part of our approach, along with the eye aspect ratio, we also included the mouth aspect ratio of the subject in the video input as an input to the HM-LSTM network. We took randomized 80% of the video input as our training data, and the remaining 20% of video input as our test data.

4.1 Blink Detection and Blink Feature Extraction

In our approach to detect blinks in the video, we will be calculating a metric called the eye aspect ratio (EAR), which was first utilized by Soukupová and Čech [21]. The eye aspect ratio is a fairly simple solution to detect blinks as it only takes into account the distance between the facial landmark points around the region of the eye. This approach is fast, uses less computing power, and it is easy to calculate and understand. In the first part, we will discuss the eye aspect ratio and how it can be used to determine if a person is blinking or not in a given video frame. The eye aspect ratio is mathematically calculating as in Eq. (1) (Fig. 1).

$$\text{EAR} = \frac{||p2 - p6|| + ||p3 - p5||}{2||p1 - p4||} \tag{1}$$

The following Eqs. (2), (3), (4), and (5) are the detected blink features:

$$\text{Duration}_i = \text{end}_i - \text{start}_i + 1 \tag{2}$$

$$\text{Amplitude}_i = \frac{\text{EAR}[\text{start}_i] - 2\text{EAR}[\text{bottom}_i] + \text{EAR}[\text{end}_i]}{2} \tag{3}$$

Fig. 1 EAR plotted against time. A sudden dip in the curve indicates a blink

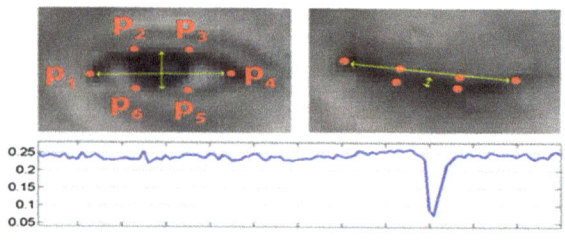

$$\text{Eye Opening Velocity}_i = \frac{\text{EAR}[\text{end}_i] - \text{EAR}[\text{bottom}_i]}{\text{end}_i - \text{bottom}_i} \tag{4}$$

$$\text{Frequency}_i = 100 \times \frac{\text{Number of blinks up to blink}_i}{\text{Nimber of frames upto end}_i} \tag{5}$$

Here, "i" is used to refer to the i-th blink in a blink sequence of the input video segment.

4.2 Yawn Detection and Yawn Feature Extraction

The second sign of early drowsiness in the driver is the occurrence of a yawn while behind the steering wheel. Yawning also can be picked up from the video input using the technique of facial feature extraction, where the mouth can be identified, and mouth aspect ratio can then be calculated.

Mouth Aspect Ratio
Just like the EAR, the mouth aspect ratio (MAR) is defined to be the ratio of the length of the mouth of a person to the width of the mouth at a given time (Fig. 2).

We can obtain the equation representing the mouth aspect ratio by mapping the relevant points on the mouth landmark. Let the points 49, 51, 52, 53, 55, 57, 58, and 59 be mapped as $B1$, $B2$, $B3$, $B4$, $B5$, $B6$, $B7$, and $B8$, respectively [22]. Then, the formula for mouth aspect ratio can be given in Eq. (6)

$$\text{MAR} = \frac{||B2 - B8|| + ||B3 - B7|| + ||B4 - B6||}{3||B1 - B5||} \tag{6}$$

MAR Over EAR Ratio
A yawn can be accurately predicted using MAR and EAR, as there exists an inverse relation between the two during the event of a yawn, that is, when we yawn, our EAR decreases (eyes almost close), whereas out MAR increases (mouth stretches open). Hence, we take the continuous ratio of MAR over EAR as a feature for our model to detect yawns in Eq. (7).

$$\text{MOE} = \frac{\text{MAR}}{\text{EAR}} \tag{7}$$

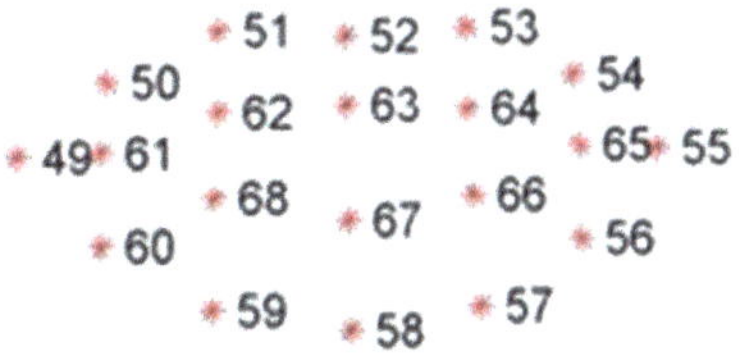

Fig. 2 Mouth region in facial landmarks

4.3 Preprocessing Features

Before the "blink" and "yawn" features are fed into the HM-LSTM algorithm, they have to be normalized because each person has a different blinking pattern, also they might have a different resting mouth aspect ratio. Whenever a person enters a new car, the normalization provides the calibration of the driver's blinking pattern while driving, for a certain initial amount of time. We try to not let outliers influence our modeling of the features. This comes with the added benefit of taking the whole training dataset at once to train the model. For normalization, we use Eq. (8).

$$\overline{\text{Feature}_{n,m}} = \frac{\text{Feature}_{n,m} - \mu_{n,m}}{\sigma_{n,m}} \tag{8}$$

where $\mu_{n,m}$ is the mean of the features and $\sigma_{n,m}$ is the standard deviation of the features observed about its mean.

4.4 The Model

In our attempt to detect drowsiness using blinking patterns, we need to identify long-term interdependencies between the blinking sequences of the person at all three stages, that is, alert, low vigilant, and drowsy. It was imperative to go for a deep learning solution so as to identify blink sequence dependencies, but the ability of LSTM networks to map long-term interdependencies and its proven success in the field of natural language processing made us explore the possibility of applying the same concepts of interword dependencies of information retrieval (IR) to our blink sequences in the video input. Long short-term memory organizations—normally called "LSTMs"—are an extraordinary sort of RNN, equipped for learning long haul conditions. They were presented by Hochreiter and Schmidhuber [23].

In our model, we utilized a hierarchical multiscale LSTM network to record the relationship between successive blinks over a period of time and detect a pattern in blinks corresponding to the event of the subject being drowsy. Also the mouth aspect ratio and the mouth over eye (MOE) have been fed into the HM-LSTM network along with the blinks. We expect the HM-LSTM to recognize the video part to be drowsy, with maximum accuracy coming when along with blinking pattern, mouth aspect ratio and MOE are included as input features.

5 Results

See Table 1.

Table 1 Normalized confusion matrix

Model	Category	Recall	Precision	F1 score	Average accuracy
Blinking pattern	Alert	0.74	0.75	0.74	0.6733
	Low vigilant	0.59	0.60	0.59	
	Drowsy	0.69	0.66	0.67	
Human judgement	Alert	0.76	0.77	0.76	0.6813
	Low vigilant	0.58	0.60	0.59	
	Drowsy	0.70	0.67	0.68	
Proposed model	Alert	0.79	0.85	0.82	0.7833
	Low Vigilant	0.73	0.74	0.73	
	Drowsy	0.83	0.77	0.80	

6 Discussion and Comparative Analysis

In the first part of our research where we trained the HM-LSTM network with blinking features from the videos, Table 1 shows the normalized confusion matrix obtained after running the trained model on the remaining 20% of the dataset. The average accuracy of this model in predicting either of the three states effectively comes out to be 67.33%.

The second evaluation to be done was of the human judgement baseline model, which too came as part of the RLDD dataset in which random individuals were selected and asked to predict the drowsiness state of the people in the videos. The normalized confusion matrix of the predictions of the people by looking at the video is given in Table 1, and the average accuracy observed across all the three drowsiness states was observed to be 68.13%.

Finally, the model we propose, which includes along with the blinking features—yawn detection and mouth over eye (MOE)—is evaluated against the test set of the video data, and the obtained confusion matrix is shown in Table 1. The average accuracy of prediction over all the three classes is observed to be 78.33%.

Our proposed approach in classifying real-time drivers video input as one of the three states of drowsiness shows a higher average accuracy than the model using only blinking features in identifying signs of early drowsiness as well as when the driver is visibly drowsy. It also outperforms the average human prediction of the driver who is showing early signs of drowsiness. We believe that the reason for the higher accuracy of our proposed model has been achieved as we aimed at using the ability of both—predicting early drowsiness from the blinking patterns as well as catching the visible signs as accurately as the human predictions. Therefore, upon taking the best out of both the approaches, our proposed model gives a much better overall result in terms of average accuracy, F1 score, higher precision, and recall across all the three classes.

7 Conclusion

In this project, we attempted to model a dataset that contains videos of genuinely drowsy subjects, showing subtle signs of drowsiness, and in some cases, no visible signs of drowsiness. Even, to the best of our intellect, our dataset which has been used is essentially larger compared to the data already present, with almost thirty hours of videography. We have likewise suggested a start to finish standard strategy utilizing the connection of blinking and yawning patterns for sleepiness recognition. The suggested approach has minimal processing and capacity requests. The outcomes exhibited that this approach outperforms human judgment on the data.

A potential direction for future work would be to design and include an appropriate deep learning network to learn different highlights of sleepiness other than blink sequences and yawn patterns in the video. In addition to these, we are hopeful of the prospect that we can model the distinctive head tilt angles in individuals who are drowsy, as compared to when they are not. Head tilt angle is also a desirable feature because head tilt angle would be free of dim illumination-related issues.

References

1. Sleep Deprivation and Deficiency | NHLBI, NIH (2021) [Online]. Available at: https://www.nhlbi.nih.gov/health-topics/sleep-deprivation-and-deciency#:~:text=Sleeing%20is-%20a%20basic%20human,well%2Dbeing%20throughout%20your%20lifetime
2. Owens JM, Dingus TA, Guo F, Fang Y, Perez M, McClafferty J, Tefft BC (2018) Prevalence of drowsy driving crashes: estimates from a large-scale naturalistic
3. Driving Study (Research Brief) (2021) AAA Foundation for Traffic Safety, Washington, DC Nsc.org. Fatigued Driver—National Safety Council [Online]. Available at: https://www.nsc.org/road-safety/safety-topics/fatigued-driving
4. Tadesse E, Sheng W, Liu M (2014) Driver drowsiness detection through hmm based dynamic modeling. In: 2014 IEEE international conference on robotics and automation (ICRA). IEEE, pp 4003–4008
5. Reddy B, Kim Y-H, Yun S, Seo C, Jang J (2017) Realtime driver drowsiness detection for embedded systems using model compression of deep neural networks. In: 2017 IEEE conference on computer vision and pattern recognition workshops (CVPRW). IEEE, pp 438–445
6. Ngxande M, Tapamo J-R, Burke M (2017) Driver drowsiness detection using behavioral measures and machine learning techniques: a review of state-of-art techniques. In: Pattern recognition association of South Africa and robotics and mechatronics (PRASA-RobMech). IEEE, pp 156–161
7. 'Attention Assist (2018) Available: https://www.mbusa.com/mercedes/technology/videos/detail/title-safety/videoId710835ab8d127410VgnVCM100000ccec1e35RCRD/
8. Svensson U (2004) Blink behavior based drowsiness detection. Technical report
9. Massoz Q, Langohr T, François C, Verly JG (2016). The ULg multimodality drowsiness database (called DROZY) and examples of use. In: 2016 IEEE winter conference on applications of computer vision (WACV). IEEE, pp 1–7
10. Lee BL, Lee BG, Chung WY (2016) Standalone wearable driver drowsiness detection system in a smartwatch. IEEE Sens J 2016(16):5444–5451
11. Johns M et al (2003) The amplitude-velocity ratio of blinks: a new method for monitoring drowsiness. Sleep 26(Suppl.)

12. Jo J, Lee SJ, Park KR, Kim I-J, Kim J (2014) Detecting driver drowsiness using feature-level fusion and user specific classification. Expert Syst Appl 41(4):1139–1152
13. Yan W-J, Li X, Wang S-J, Zhao G, Liu Y-J, Chen Y-H, Fu X (2014) Casme II: an improved spontaneous micro expression database and the baseline evaluation. PLoS ONE 9(1):e86041
14. Weng C-H, Lai Y-H, Lai S-H (2016) Driver drowsiness detection via a hierarchical temporal deep belief network. In: Asian conference on computer vision. Springer, pp 117–133
15. Friedrichs F, Yang B (2010) Camera-based drowsiness reference for driver state classification under real driving conditions. In: 2010 IEEE intelligent vehicles symposium (IV). IEEE, pp 101–106
16. McIntire LK, McKinley RA, Goodyear C, McIntire JP (2014) Detection of vigilance performance using eye blinks. Appl Ergon 45(2):354–362
17. Guardian (2018) [Online]. Available: http://www.seeingmachines.com/guardian/
18. Ajay S, Azariah JK, Subhashini R, Thomas J (2020) Drowsiness detection using eye blink and facial features image analysis. MLU 20(4):27–30
19. Park S, Pan F, Kang S, Yoo CD (2016) Driver drowsiness detection system based on feature representation learning using various deep networks. In: Asian conference on computer vision. Springer, pp 154–164
20. Ghoddoosian R, Galib M, Athitsos V (2019) A realistic dataset and baseline temporal model for early drowsiness detection. In: Computer vision and pattern recognition workshops
21. Soukupova T, Cech J (2016) Real-time eye blink detection using facial landmarks. In: 21st computer vision winter workshop (CVWW2016), pp 1–8
22. Chandra A (2018) Mouse cursor control using facial movements. Available at https://python awesome.com/mouse-cursor-control-using-facial-movements/
23. Hochreiter S, Schmidhuber J (1997) Long short-term memory. Neural Comput 9(8):1735–1780

Data Augmentation and Fine-Tuning the Radiography Images to Detect COVID-19 Patients with Pre-trained Network of Transfer Learning

Birjit Gope and Rachna Kohar

Abstract Around 26 million cases of COVID-19 have been acquired and currently are in inventory and still recorded worldwide as a consequence of the COVID-19 outbreak. For the automated identification of the coronavirus, a radiography imaging dataset of patients with common viral pneumonia, lungs opacity, reported COVID-19 patients, and regular patients was used in this study. The study's aim is to assess the efficiency of state-of-the-art VGG-16 architectures for medical image, i.e., CXR classification that has been proposed in recent years. Specifically, the technique of transfer learning was implemented. The identification of numerous abnormalities in specific medical image databases is an achievable goal with transfer learning, and the findings are often impressive. There are four groups of the datasets used in this experiment. In the first class, i.e., viral pneumonia, a collection of 1345 imaging images, in second class, i.e., lungs opacity having 6012 images, in third class, i.e., confirmed COVID-19 having 3616 images, and in fourth class, i.e., normal lungs having 10,192 images. The dataset was compiled using publicly accessible X-ray files from medical databases. The results show that transfer learning with X-ray visualization can extract biomarkers that are critical for the COVID-19 disease, whereas the best trainable accuracy and validation accuracy are achieved 97.57% and 93.48%, respectively. Because all diagnostic instruments already have failure rates that are cause for concern, the medical profession should determine the likelihood of integrating X-rays into disease detection based on the results, although further studies to examine the X-ray imaging method from various perspectives should be undertaken.

Keywords COVID-19 · CXR · Transfer learning · VGG-16 · RT-PCR

B. Gope (✉) · R. Kohar
School of Computer Science Engineering, Lovely Professional University, Phagwara, Punjab, India

© The Author(s), under exclusive license to Springer Nature Singapore Pte Ltd. 2022 795
D. Gupta et al. (eds.), *Proceedings of Data Analytics and Management*,
Lecture Notes on Data Engineering and Communications Technologies 90,
https://doi.org/10.1007/978-981-16-6289-8_65

1 Introduction

The pandemic of the coronavirus, which has an accelerated rate of infection, has reached an unsustainable level in health care systems. By March 2021, there had been over 21,000,000 active events registered worldwide, with over 26,000 fatalities [1]. Although, the most common method for diagnosing is reverse transcription polymerase chain reaction (RT-PCR) for COVID-19, which has poor precision, latency, and sensitivity [2–4]. Early diagnosis increases treating critically ill patients and reducing illness transmission to the population at large, as in the case of COVID-19. In order to classify the COVID-19 patients, the focus has been on the physiological and questionnaire inputs wearable devices for diagnostics and effective neural artificial networks [5]. Various machine-based prediction models have also been developed to forecast outbreak rates, the likelihood of second and third pandemic waves, and the possibility of travel expansion. The routine diagnostics of lung-related diseases such as pneumonia [6] or tubercles [7] are photographs of the chest X-ray (CXR) or CT, which may also be helpful to diagnose COVID-19 [8, 9]. A benefit of CXR is the ability to do them with compact X-ray devices that can be used to make CXR diagnostics faster and more precise [10–13]. CXRs have been shown to be potentially dangerous to the human body relative to CT with COVID-19 using artificial intelligence [11].

With the unexplained disease epidemic in China in late 2019, a number of people in the local market have become affected by the disease. The illness was at first totally unexplained, but the signs of coronavirus and influenza were diagnosed by specialists [1–4]. The exact cause of this common disease was originally unclear; however, after a laboratory examination and investigation for positive sputum by real-world polymerase chain reaction, viral infection was detected and finally dubbed the "COVID-19" (PCR) testing (WHO). The COVID-19 outbreak reached the regional frontiers for a brief time and has a crippling impact on the global population's well-being, environment and well-being [1–5]. By January 5, 2021, over 86 million citizens globally contracted COVID-19, with more than 1,870,000 of them reportedly dead as a result of the outbreak, on the basis of Worldometers (worldometers.info) figures. In addition to patient treatment, the early identification of COVID-19 is also essential in the public health sector by guaranteeing isolation and monitoring of the patient's pandemic [14]. The disease news meant that early on methods of combating it was not understood, but researchers found it to be necessary to detect and diagnose the affected patients and to separate them from the healthier population. COVID-19 has clinical characteristics including respiration, fever, toxins, dyspnea, and pneumonia. However, COVID-19 is not often seen, and sometimes, pneumonia results in medical difficulties in doctors [1, 8, 9]. These signs are not always seen in the prescription.

While the RT-PCR test is the gold standard for COVID-19 diagnosis, there are certain aspects which restrict the diagnosis of the illness, with certain features. RT-PCR is a long, complicated, expensive, and manual operation. One of the downsides is the requirement for a laboratory package, which in crisis and epidemics, many countries find problematic and sometimes impractical. This approach is not error-free and biased, as other diagnosis and experimental procedures in health systems. The testing of nasal and throat mucous membranes needs a specialist laboratory technician, and that is why, many people fail to perform nasal swap sampling [10–13]. More specifically, several tests showed that the RT-PCR test was of poor sensitivity; some studies have shown that it was 30–60% sensitive, which in several instances shows a reduction in the accuracy of COVID-19 diagnoses. Some reports have also shown its falsified negative rates and conflicting results [15, 16]. The radiological photographs like radiation and X-ray scanning are one of the most significant diagnoses of COVID-19. Chest imaging is prescribed by health and medical guidelines and is the first diagnostic technique of epidemics in some text messages [17, 18]. In the diagnosis and detection of cases with COVID-19, CT scans are highly sensitive compared to the RT-PCR; but, their accuracy is limited. This suggests that X-ray scans in COVID-19, but in non-viral pneumonia, are less effective. A diagnostic analysis in Wuhan, Germany, found that 14% of the conclusive cases of COVID-19 in X-ray scans were not reported to be fully stable dependent on their X-ray scans tests with consolidation and glass-bottomed opines (GGOs). Out of 18 COVID-19 patients with consolidated GGO, only 12 patients had GGO, which resulted in no consolidation or illness. In certain circumstances, it was challenging and practically impossible to identify COVID-19, despite the existence of consolidation without the introduction of GGO. The diagnosis of COVID-19 with CT scans showed an error [19–22] in these circumstances.

Although the chest X-ray scan has been effective in the detection of COVID-19-related lung injury, the usage of this diagnostic test presents some problems. In view of the recommendation from the WHO, chest X-ray results in certain patients have been common at the onset of the condition, which alone allows the usage of X-ray negative. The poor specificity of the X-ray scan will make non-COVID-19 detection issues. X-ray scanner rays can lead patients who need several X-ray scans during their illness to have complications. X-ray scans cannot be used as the first diagnostic channel; the American College of Radiology advises. Problems like the possibility of the disease being transmitted by the usage of an X-ray scan device and the high expense of this scan could pose serious problems for patients and health systems; thus, it is advised that CT scans be substituted by CXR [23], if medical imaging is necessary. In comparison with traditional diagnostics, X-ray imaging is significantly broader and more economically cost efficient. The transition of an X-ray visual image from the access point to the inspection point does not enable the diagnostic phase to be done very fast. Chest X-rays are easy and fast to triage patients medically. In contrast to X-ray scans, radiographic imaging needs fewer scarce and costly appliances, so considerable reductions in operating costs may be achieved. Furthermore, in enclosed spaces, portable CXR systems should be used to reduce the contamination risk in-patient hospital usage of these devices [6, 24]. Different experiments have shown CXR imaging deficiency when diagnosing and distinguishing COVID-19 from other pneumonia. In order to identify pleural effusion and the amount concerned, the radiologist cannot use X-rays. It has nevertheless several high points [21, 25] because of the poor diagnostic precision of COVID-19

X-rays. Several deep learning (DL) experiments were carried out in the radiology study for the resolution of vulnerabilities in radiological image diagnostic tests of COVID-19.

Recently, a series of studies with many COVID-19 detection strategies utilizing AI-based radiographs was carried out. The objective was to increase network capacity to distinguish between CB-19, common and other respiratory illnesses via various transfer learning processes, new network designs, and ensemble solutions. Given the relatively low amount of COVID-19 CXR images used on published papers, their analyzes cannot be widespread and cannot ensure that their observations are reproduced by analyzing them on a larger dataset. The various CXR methods of enhancement can be examined using an extensive range of regular (safe), non-COVID-19 (other pulmonary influences), and CXR contaminated images of COVID-19 patients. The following are specifically given for the contributions of this paper:

- This is the first research in which the effect on single is based on the best knowledge of the scholar, and segmented CXR image recognition of different image enhancement techniques was thoroughly analyzed.
- This is the biggest image dataset of CXR and lung segments, containing the COVID-19, the regular (healthy) images, and the non-COVID-19 images.
- This manuscript proposes an updated variant of the VGG-16 model, which outstrips the current CXR pneumonia U-Net design.
- The results of this analysis have been tested using the methodology of picture visualization in order to validate the deep networking results.
- Detection of COVID-19 has been enhanced by using CXR images with enhancement approaches of the transfer learning models.

The paper is organized as follows. Section 2 contains the specifics of the different pre-trained models, pulmonary segmentation models, improvement of different images, and strategies for visualization. Section 3 outlines the methods used and classification findings utilizing initial and segmented CXR images enriched using various approaches. The performance of the proposed model is discussed in Sect. 4. Section 5 explains the classification results. In Sects. 6 the discussion regarding the results is presented. The paper is concluded in Sect. 7.

2 Literature Review

References	Year	Findings
[26]	2020	The objective is to detect COVID-19 via 1500 images of X-ray using automatic feature extraction which is the feature engg. of CNN with transfer learning. VGG-16 SDD architecture is used, and validation results achieved accuracy is 94.92%, sensitivity is 94.92%, and specificity is 92%

(continued)

(continued)

References	Year	Findings
[27]	2020	The objectives of the study are to detect of 14,531 images of CT scan using prominent features selected which is the feature engg. of deep learning. And, the models used in the design of convolutional MVP-Net and 3D U-Net are used in order to obtain validation results; F1-score is 97% for lesions detecting, and sensibilities are 100%
[22]	2020	They use deep learning model, i.e., DenseNet121-FPN taking 5372 (two datasets) CT scan images to diagnosis and prognosis of patients having COVID-19. And, to achieve the validation results, AUC is 87% and 88%, sensitivity is 80.3% and 79.35%, and specificity is 76.3% and 81.1%
[28]	2020	Feature engineering (automatic feature extraction) is used to diagnose 11302 X-ray images (open source) with deep learning models, i.e., Xception and ResNet50_V2. Results accuracy is 95.5%, and overall average accuracy is 91.4%
[29]	2020	Finds the fast detection of X-ray (337) images (open source) using deep learning (nCOVnet)with pre-trained model VGG-16, and results are: sensitivity is 97.62%, specificity is 78.57%, and accuracy is 88.10%
[30]	2020	Detecting 194 CT scan images using deep learning via Alex_Net, VGG-16, VGG-19, Squeeze_Net, and Google_Net models. Sensitivity is 100%, specificity is 99.02%, and the best accuracy is 99.51%
[8, 9]	2020	Aim is to 4356 CT images diagnosis by automated function extraction (ResNet-50) to 3322 patients' exams as backbone of the major DL models with a 90% accuracy, and the specificity is 96%
[20]	2021	Used CNN with transfer learning model, that is, MobileNetV2 in 2914 X-ray with feature engineering of automatic feature extraction and got the accuracy of 96.78%
[31]	2020	Used 381 X-rays with the hybrid network (CNN and SVM with ResNet-50) and results: sensitivity is 95.33%
[32]	2020	Used in detection took 227 CT scan images with automatic feature extraction via deep learning (CoroNet) BigBiGANI, and results are: sensitivity is 85%; specificity is 88%
[33]	2020	Aim is to detection of 6523 X-ray images in deep learning model (CoroNet) and VGG-16, and accuracy is 97%
[34]	2020	Aim is to classification (diagnosis) of 618 CT images with CNN with ResNet-18, and results achieved: sensitivity is 98.2%, and specificity is 92.2%
[35]	2020	Among two datasets, automated detection is built with the deep learning (DarkNet) via X-ray images to detect COVID-19, and accuracy of binary case and multiclass case is 98.08% and 87.02%, respectively
[36]	2020	Using inception_ResNet_V2, diagnosis of6087 X-ray and CT scans is performed to achieved accuracy of 92.18%, and same for DenseNet201, accuracy is 88.09%
[37]	2020	Used the detection system of 295 CT scan images and apply DenseNet of deep learning, and accuracy achieved is 92%, sensitivities are 97%, and specificity is 87%

(continued)

(continued)

References	Year	Findings
[38]	2020	Used in detection 2492 (open source) CT scan images with automatic classification technique with deep transfer learning with DenseNet201, and the results are: precision is 96.29%, specificity is 96.21%, and accuracy is 96.25%
[39]	2020	Used in diagnosis of 5856 X-ray images with hybrid model CNN + CovXNet (deep learning), and accuracy of multiple classes is 90.2%
[40]	2020	Used in classification (diagnosis) CT images with automatic system using CNN, ANN, and ANFIS. The proposed model is compared and demonstrates high efficiency with the ANN, ANFIS, and CNN models
[41]	2020	Used in diagnosis (differentiate) 3993 patients with their CT scan images automatic extraction of functionality with deep learning (FCONet) and ResNet-50, and sensitivity is 99.58%, specificity is 100.00%, and accuracy is 99.87%
[42]	2020	Used for the screening (diagnostic) of 495 CT scanning images with automatic functional extraction feature engineering with the VGG-19 application of the deep lerning (CoroNet) model, and the findings are: accuracy is 76.0%, sensitivity is 81.1%, and specificity is 61.15%
[43]	2020	Used in 181 X-ray detections for automatic deep learning function extractions (CoroNet) models VGG-19, for COVID-19 X-ray detection and 96.3% accurate detection
[44]	2020	Used in classification (diagnosis) with feature engineering of automatic feature extraction with applied CNN of deep Bayes SqueezeNet model, and the results are: accuracy for overall class is 98.3%
[45]	2020	Used in diagnosis of two open sources ($n = 295$) which are X-ray data applied SqueezeNet and MobileNet for classification of COVID-19 with classification rate: 99.27%
[46]	2020	The objective is to detect and diagnose two ($n = 1300$) datasets with automatic feature extraction. Feature engineering Xception models are used for the specific deep learning (CoroNet). High accuracy is 89.6%
[7]	2020	Used in the classification (detection) of 88 X-ray images with CNN model with feature engineering of automatic feature extraction, and results are: sensitivity is 89%
[47]	2020	Screening (diagnosis)6845 X-ray with truncated inception network of CNN (deep learning), and the outcomes are 88% sensitivity and 100% specificity
[48]	2020	Diagnostic (classification) 321CT scan images for Q-deformed entropy extraction as a function engineering that is an LSTM neural network classifying deep transfer learning model and 99.68% of accuracy

We used VGG-16 for chest X-ray scans for classification of COVID-19 in our previous analysis in March 2020. But as the COVID-19 data was not sufficient, 50 typical cases and 50 optimistic COVID-19 cases have been trained [28]. In that research, we suggested an automated COVID-19 prediction-used pre-trained transfer learning model and chest x-ray images from a deep convolution neural network. To this end, we have used pre-trained models; VGG-16 model is used to gain higher predictive accuracy for four binary collections, namely regular (healthy) X-ray pictures, COVID-19, bacterial patients, and viral patients.

3 Methodology

In this paper, we have implemented transfer learning on radiology images and used in large datasets of about 28 k images of lung X-rays which are available on Kaggle, i.e., a chest X-ray database for COVID-19 positive cases along with normal and viral pneumonia images was established and also collecting all images of CXR by the team of researchers from the University of Qatar, Doha, Qatar, and Dhaka, Bangladesh in collaboration with their collaborators in Pakistan and Malaysia [49–51]. The architecture diagram followed technique of VGG-16 is shown in Fig. 1. Two separate bases of CXR images were used: (1) lung segmentation and (2) databases of classification. All the images are preprocessed according to the pre-trained neural network in which we are using them and then different augmentation techniques are

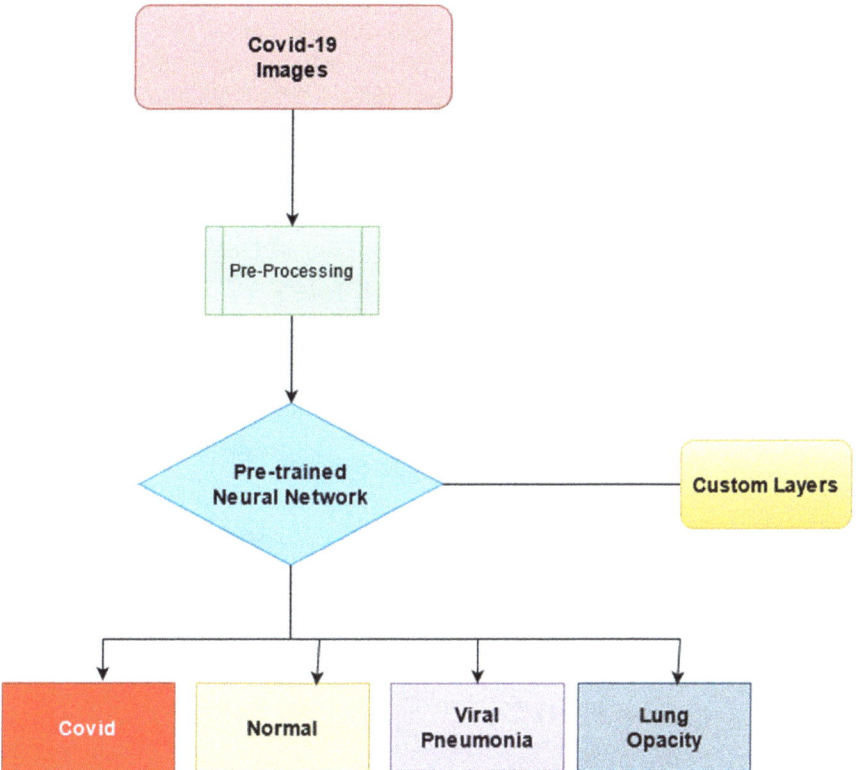

Fig. 1 Proposed methodology block diagram

applied. And, the fine-tuning is also performed in which we re-train some of the weights, and the classification accuracy increases. The pre-trained neural network architecture and the networks which we applied are VGG-16. The architecture of this pre-trained network is explained below. The following descriptions, i.e., details of the data collection, preprocessing and increase methods, output matrices used for the trial are addressed. The following is defined below.

4 Dataset

The dataset for this study contains lung X-ray images (CXR) is of four classes: viral pneumonia, which has 1345 images, lungs opacity, which has 6012 images, confirmed COVID-19, which has 3616 images, and normal lungs, which has 10,192 images. The data was obtained from the Kaggle analysis repository. A senior radiologist who has treated COVID-19 patients confirmed the dataset's effectiveness, according to the authors. The authors also recognize that radiography images during the outbreak were vital to the diagnosis and treatment of COVID-19 patients. For these factors, this dataset is used in this analysis for training and validation. Data increase on this dataset was used to increase the selection of images and boost the resilience of the training. Scaling, slicing, and adjustments in these augmentation techniques were used for horizontal and vertical flips, brightness, and contrast changes, as well as fine-tuning to achieve good results. To prevent learning bias and overfitting, make sure the dataset has a balanced distribution of all groups. Figure 2 depicts the dataset's final scale and shows balanced counts for all groups.

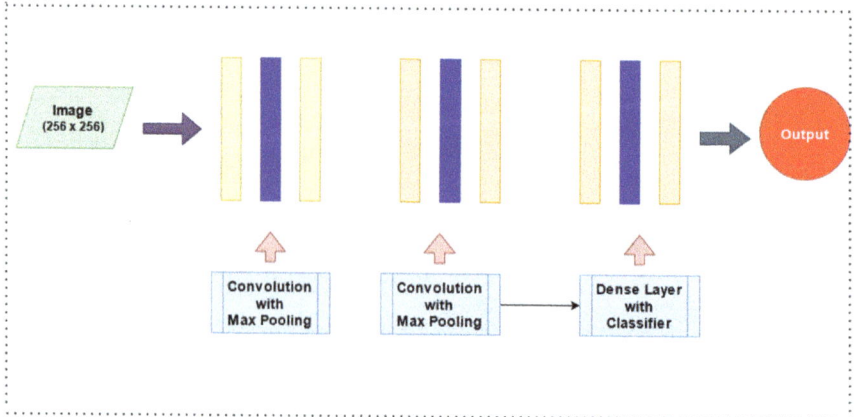

Fig. 2 Basic architecture of VGG-16

4.1 VGG-16

K. Simonyan and A. Zisserman first suggested VGG-16 [52] as a DCNN. It was one of the most powerful models in the 2015 large-scale visual recognition challenge (ILSVRC) [53], and the ImageNet dataset has been trained. This dataset contains 14,000,000 images and 1000 categories. This research together with the TL technology utilizes the pre-trained VGG-16 of the ImageNet dataset. Both pre-trained VGG-16 layers are frozen apart from the final, closely related layer and the output layer. The remaining unfrozen layers are trained using the dataset mentioned. The output layer generates either a COVID-19 or a belonging to other group shown in result. Figure 1 depicts the general design of VGG-16 architecture.

5 Results

In the implementation of datasets, the VGG-16 model performed better than other pre-trained models as per reviewed. VGG-16 shows promising results in detection of COVID-19 among the all CXR images. The performance of the models is summarized in the tabular form and is shown below in tables. In all the cases, we used 150 epochs. In Table 1, we got training and validation performance, i.e., accuracy and loss, respectively, and also find the kappa score; this performance is before fine-tuning with Adam optimizer means without weights. And also, plotting a graph of the particular (Fig. 3).

In Table 2, we got training and validation performance, i.e., accuracy and loss, respectively, and also find the kappa score; this performance is after fine-tuning with

Table 1 Before fine-tuning performance

Training loss	Training accuracy	Validation loss	Validation accuracy	Kappa score
0.031	90.15	0.345	85.45	0.88

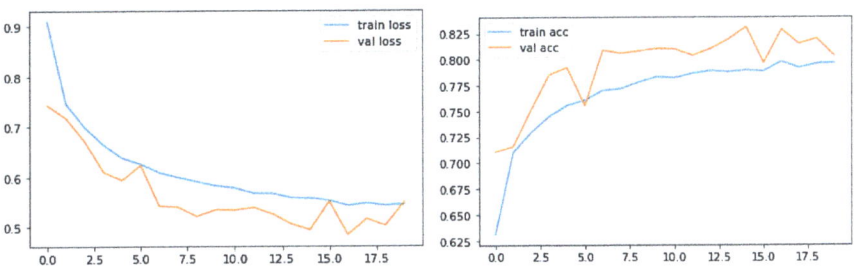

Fig. 3 Before fine-tuning, training and validation performance (accuracy and loss)

Table 2 After fine-tuning performance

Training loss	Training accuracy	Validation loss	Validation accuracy	Kappa score
0.0718	97.57	0.2276	93.48	0.9372

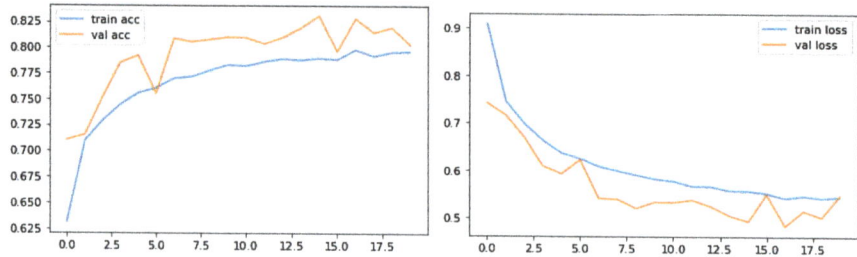

Fig. 4 After fine-tuning, training and validation performance (accuracy and loss)

Adam optimizer means with weights performance. And also, plotting a graph of the particular (Fig. 4).

There are not many research papers in which transfer learning is used for the chest X-ray images on the datasets which we used on this implementation. In the future, if more work is done, then our results could be improved more. One paper in which VGG-16 is used for the same dataset, but using a different algorithm [18] has 93.15% training accuracy and 89.25% testing accuracy. In our model, it has 97.57% training accuracy, and 93.48% validation accuracy is achieved.

6 Discussion

In this systemic study, we were reviewed to help researchers exploring and build artificial intelligence knowledge-based detection and diagnostic systems for COVID-19. The new research, which has analyzed a range of radiological image analysis approaches in order to assess radiological images, is one of the most detailed diagnostic and identification reviews. The latest study presented up-to-date material on the use of TL algorithms as an expression of COVID-19 X-ray imagery testing. Many studies found that TL algorithms would increase the rate of X-ray images metric features and improve radiographic images' sensitivity and accuracy compared to radiologists' diagnostics; therefore, as a reliable tool for COVID-19 diagnostics, this inexpensive and economic solution should be used. Through analyzing 23 articles on the use of X-rays in COVID-19 diagnosis utilizing TL techniques, science and the medical community will use the current early and fast diagnostic approach for this disease. The cheapest and safest imaging solution is possible to prevent COVID-19 transmission by designing imagery approaches based on artificial intelligence technologies. In some cases of pneumonia, in particular COVID-19, diagnosis of these diseases with TL algorithms was made under radiologists' surveillance. A study of published studies has shown. In the X-ray modality, the average diagnosis was >95%, the precision was >91%, and the diagnostics rate was better than in standard texts and methods.

The specificity of X-ray images is acquired by the TL system for COVID-19; the average scenario for the COVID-19 was more than 92%, with better specificity efficiencies in certain instances than previous texts. In X-ray COVID-19, the sensitivity of TL methods was in certain instances also higher or comparable to the normal methods for diagnosis. Given the uncontrollable and complex evaluation of these diseases by uncontrolled procedures, the over-arching results of COVID-19 on the pneumonia with various forms of bacterial and viral pneumonia are extremely challenging. The analysis of the algorithms and the TL architectures showed that almost all of the experiments were using the VGG-16 algorithm; other algorithms were naturally often used in other studies along with the VGG-16 algorithm. All the unique features of the VGG-16 architectures used in these studies include in it, and the ability to identify and diagnose the COVID-19 is not feasible without changing their parameters.

7 Conclusion

In order to avoid the transmission of virus, the rapid and precise identification of highly contagious COVID-19 is crucial. We used CXR images in this analysis because the images of X-rays are cheaper, easier to view, and quicker than widely used approaches such as RT-PCR and CT. CNN as a significant addition has been composed, and it is freely shared as a dataset benchmark by the largest CXR dataset comprising 3616 images of COVID-19, 6012 of non-COVID lung opacity, and 8851 of standard chest X-ray images. In addition, we examined the effects on the automated recognition of COVID-19 photographs by means of deep convolutional neural networks for the first time in our literature. As previously described, early identification and evaluation by DL strategies of COVID-19 and the least cost-effective and complicated are the simple measures to avoid the outbreak and pandemic progression. In the immediate future we introduced of TL algorithms in radiology center, equipment would require faster, cheaper, and better diagnosis of disease. This will help radiologists to minimize human errors and help them make decisions in vital and disease-conscious circumstances by using certain technologies in quick diagnostic decision-making on COVID-19. This investigation confirms the notion that TL algorithms are a promising way to optimize medical services and improve diagnostic and therapeutic outcomes. While TL is one of the most efficient pneumonia diagnostic calculation tools, particularly COVID-19, developers should ensure compliance and optimize the generalization and functionality of COVID-19 diagnostic TL models.

References

1. Huang C et al (2020) Clinical features of patients infected with 2019 novel coronavirus in Wuhan, China. Lancet 395(10223):497–506
2. Chen N et al (2020) Epidemiological and clinical characteristics of 99 cases of 2019 novel coronavirus pneumonia in Wuhan, China: a descriptive study. Lancet 395(10223):507–513
3. de Oliveira Lima CMA (2020) Information about the new coronavirus disease (COVID-19). Radiol Bras 53(2):V–VI
4. Struyf T et al (2020) Signs and symptoms to determine if a patient presenting in primary care or hospital outpatient settings has COVID-19 disease. Cochrane database of systematic reviews, vol 7. Wiley
5. Liao J et al (2020) Epidemiological and clinical characteristics of COVID-19 in adolescents and young adults. Innovation 1(1):100001
6. Baratella E et al (2020) Severity of lung involvement on chest X-rays in SARS-coronavirus-2 infected patients as a possible tool to predict clinical progression: an observational retrospective analysis of the relationship between radiological, clinical, and laboratory data. J Bras Pneumol 46(5)
7. Paul HY, Kim TK, Lin CT (2020) Generalizability of deep learning tuberculosis classifier to COVID-19 chest radiographs: new tricks for an old algorithm? J Thorac Imag 35(4):W102–W104
8. Li L et al (2020) Artificial intelligence distinguishes COVID-19 from community acquired pneumonia on chest CT. Radiology
9. Li Q et al (2020) Early transmission dynamics in Wuhan, China, of novel coronavirus–infected pneumonia. N Engl J Med
10. Lippi G, Plebani M (2018) A six-sigma approach for comparing diagnostic errors in healthcare—where does laboratory medicine stand? Ann Transl Med 6(10)
11. Lippi G, Simundic A-M, Plebani M (2020) 'Potential preanalytical and analytical vulnerabilities in the laboratory diagnosis of coronavirus disease 2019 (COVID-19). Clin Chem Lab Med (CCLM) 58(7):1070–1076
12. Sheridan C (2020) Coronavirus and the race to distribute reliable diagnostics. Nat Biotechnol 38(4):382
13. Wolach O, Stone RM (2015) How I treat mixed-phenotype acute leukemia. Blood J Am Soc Hematol 125(16):2477–2485
14. Wang L, Lin ZQ, Wong A (2020) Covid-net: a tailored deep convolutional neural network design for detection of covid-19 cases from chest X-ray images. Sci Rep 10(1):1–12
15. Ai T et al (2020) Correlation of chest CT and RT-PCR testing for coronavirus disease 2019 (COVID-19) in China: a report of 1014 cases. Radiology 296(2):E32–E40
16. Oliveira BA et al (2020) SARS-CoV-2 and the COVID-19 disease: a mini review on diagnostic methods. Rev Inst Med Trop Sao Paulo 62
17. Ng M-Y et al (2020) Imaging profile of the COVID-19 infection: radiologic findings and literature review. Radiol Cardiothorac Imag Radiol Soc North Am 2(1):e200034
18. Wang S et al (2021) A deep learning algorithm using CT images to screen for Corona Virus Disease (COVID-19). Eur Radiol 1–9
19. Gozes O et al (2020) Rapid ai development cycle for the coronavirus (covid-19) pandemic: initial results for automated detection and patient monitoring using deep learning ct image analysis. ArXiv preprint arXiv:2003.05037
20. Li J et al (2021) Radiology indispensable for tracking COVID-19. Diagn Interv Imag 102(2):69–75
21. Salehi S et al (2020) Coronavirus disease 2019 (COVID-19): a systematic review of imaging findings in 919 patients. Am J Roentgen Am Roentgen Ray Soc 215(1):87–93
22. Wang S et al (2020) A fully automatic deep learning system for COVID-19 diagnostic and prognostic analysis. Eur Respir J Eur Respir Soc 56(2)
23. Amis ES Jr et al (2007) American college of radiology white paper on radiation dose in medicine. J Am Coll Radiol 4(5):272–284

24. Rubin GD et al (2020) The role of chest imaging in patient management during the COVID-19 pandemic: a multinational consensus statement from the Fleischner society. Chest 158(1):106–116

25. Kallianos K et al (2019) How far have we come? artificial intelligence for chest radiograph interpretation. Clin Radiol 74(5):338–345

26. Saiz FA, Barandiaran I (2020) COVID-19 detection in chest X-ray images using a deep learning approach. Int J Interact Multimedia Artif Intell 1 (in press)

27. Ni Q et al (2020) A deep learning approach to characterize 2019 coronavirus disease (COVID-19) pneumonia in chest CT images. Eur Radiol 30(12):6517–6527

28. Rahimzadeh M, Attar A (2020) A modified deep convolutional neural network for detecting COVID-19 and pneumonia from chest X-ray images based on the concatenation of Xception and ResNet50V2. Inf Med Unlock 19:100360

29. Panwar H et al (2020) Application of deep learning for fast detection of COVID-19 in X-rays using nCOVnet. Chaos Solitons Fractals 138:109944

30. Ardakani AA et al (2020) Application of deep learning technique to manage COVID-19 in routine clinical practice using CT images: results of 10 convolutional neural networks. Comput Biol Med 121:103795

31. Sethy PK et al (2020) Detection of coronavirus disease (COVID-19) based on deep features and support vector machine. Preprints

32. Song J et al (2020) End-to-end automatic differentiation of the coronavirus disease 2019 (COVID-19) from viral pneumonia based on chest CT. Eur J Nucl Med Mol Imag 47(11):2516–2524

33. Brunese L et al (2020) Explainable deep learning for pulmonary disease and coronavirus COVID-19 detection from X-rays. Comput Methods Prog Biomed 196:105608

34. Xu X et al (2020) A deep learning system to screen novel coronavirus disease 2019 pneumonia. Engineering 6(10):1122–1129

35. Ozturk T et al (2020) Automated detection of COVID-19 cases using deep neural networks with X-ray images. Comput Biol Med 121:103792

36. El Asnaoui K, Chawki Y (2020) Using X-ray images and deep learning for automated detection of coronavirus disease. J Biomol Struct Dyn, 1–12

37. Yang S et al (2020) Deep learning for detecting corona virus disease 2019 (COVID-19) on high-resolution computed tomography: a pilot study. Ann Transl Med 8(7)

38. Jaiswal A et al (2020) Classification of the COVID-19 infected patients using DenseNet201 based deep transfer learning. J Biomol Struct Dyn, 1–8

39. Dey N et al (2020) Social group optimization–assisted Kapur's entropy and morphological segmentation for automated detection of COVID-19 infection from computed tomography images. Cogn Comput 12(5):1011–1023

40. Singh KK, Siddhartha M, Singh A (2020) Diagnosis of Coronavirus disease (COVID-19) from chest X-ray images using modified XceptionNet. Rom J Inf Sci Technol 23(657):91–115

41. Ko H et al (2020) COVID-19 pneumonia diagnosis using a simple 2D deep learning framework with a single chest CT image: model development and validation. J Med Int Res 22(6):e19569

42. Wu X et al (2020) Deep learning-based multi-view fusion model for screening 2019 novel coronavirus pneumonia: a multicentre study. Eur J Radiol 128:109041

43. Vaid S, Kalantar R, Bhandari M (2020) Deep learning COVID-19 detection bias: accuracy through artificial intelligence. Int Orthop 44:1539–1542

44. Ucar F, Korkmaz D (2020) COVIDiagnosis-net: deep bayes-SqueezeNet based diagnosis of the coronavirus disease 2019 (COVID-19) from X-ray images. Med Hypotheses 140:109761

45. Toğaçar M, Ergen B, Cömert Z (2020) COVID-19 detection using deep learning models to exploit social mimic optimization and structured chest X-ray images using fuzzy color and stacking approaches. Comput Biol Med 121:103805

46. Khan AI, Shah JL, Bhat MM (2020) CoroNet: a deep neural network for detection and diagnosis of COVID-19 from chest X-ray images. Comput Methods Program Biomed 196:105581

47. Das D, Santosh KC, Pal U (2020) Truncated inception net: COVID-19 outbreak screening using chest X-rays. Phys Eng Sci Med 43(3):915–925

48. Hasan AM et al (2020) Classification of covid-19 coronavirus, pneumonia and healthy lungs in CT scans using q-deformed entropy and deep learning features. Entropy 22(5):517

49. Chowdhury MEH et al (2020) Can AI help in screening viral and COVID-19 pneumonia? IEEE Access 8:132665–132676. https://doi.org/10.1109/ACCESS.2020.3010287

50. Rahman T et al (2021) Exploring the effect of image enhancement techniques on COVID-19 detection using chest X-ray images. Comput Biol Med 132:104319. https://doi.org/10.1016/j.compbiomed.2021.104319

51. Elaziz MA et al (2020) New machine learning method for image-based diagnosis of COVID-19. PLoS ONE 15(6):e0235187

52. Simonyan K, Vedaldi A, Zisserman A (2013) Deep inside convolutional networks: visualising image classification models and saliency maps. arXiv preprint arXiv:1312.6034

53. Russakovsky O et al (2015) Imagenet large scale visual recognition challenge. Int J Comput Vis 115(3):211–252

54. Apostolopoulos ID, Mpesiana TA (2020) Covid-19: automatic detection from X-ray images utilizing transfer learning with convolutional neural networks. Phys Eng Sci Med 43(2):635–640

55. Mei X et al (2020) Artificial intelligence–enabled rapid diagnosis of patients with COVID-19. Nat Med 26(8):1224–1228

56. Pereira RM et al (2020) COVID-19 identification in chest X-ray images on flat and hierarchical classification scenarios. Comput Methods Program Biomed 194:105532

57. Waheed A et al (2020) Covidgan: data augmentation using auxiliary classifier GAN for improved covid-19 detection. IEEE Access 8:91916–91923

Index Optimization Using Wavelet Tree and Compression

Sonam Gupta, Neha Katiyar, Arun Kumar Yadav, and Divakar Yadav

Abstract Retrieving of relevant information from a large corpus is a challenging task nowadays. Wavelet tree is an accomplished data structure to store and retrieve text, image, audio, and video data in efficient space and time. It has turn to be a leading tool in modernized full-text indexing or in its proficiency in compression. This study presents the contribution of wavelet trees to design indexing to retrieve information in the field of web, healthcare, agriculture, bioinformatics, and earthquake detection. In this paper, we proposed a technique of LZW compression on wavelet tree. It performs compression on the wavelet tree. This paper consists of several concepts of wavelets which show about the indexing procedure, empirical measures, wavelet packets, wavelet entropy, wavelet matrix, and complexity of wavelets. Further, the literature discusses the open issues where wavelet trees can be used to design indexing of other databases.

Keywords Indexing · Wavelet packets · Complexity · LZW compression · Tree structures

1 Introduction

In current sources of information, a huge quantity of data is stored either in the database or in the clouds. Indexing is one of the better approaches to store and retrieve information efficient time and space. In the past, various indexing methods have been proposed to retrieve quality information. It should be poor because surplus dead space lies between the indexed nodes and overlap between nodes. In the paper [1], authors explained that inverted indexing is associated with each term, point of a document, and its location of occurrences. The main issue of this type of indexing is that it takes a large amount of space to store it. The paper [2] proposed the concept

S. Gupta (✉) · N. Katiyar
Ajay Kumar Garg Engineering College, Ghaziabad, India

A. K. Yadav · D. Yadav
National Institute of Technology Hamirpur, Hamirpur, India

809

of self-indexing, i.e., more popular nowadays. This type of indexing required two times lesser space than traditional indexing. In the past, many data structures (*B*-tree, hash tree, etc.) have been proposed and implemented for the same. But wavelet is very popular due to its scanty property which was initially implemented in year 2003 paper [3] for text compression. Wavelet tree is also used for the representation of large volumes of the data because it takes less space to store data, due to this reason wavelet tree-based index can be stored in the RAM. Initially, it was formulated as a binary tree which was designed for text character and symbols [3].

Later on, it was extended with many applications by different researchers in the field of image processing, grid point, etc. Basic operations of wavelet trees are select, rank, and access queries.

Paper [4] given an overview on wavelet trees operations. Rank query counts the occurrences of a value x up to a certain index of i. The rank operation of wavelet tree includes rank $x(S, i) = |\{K\{1, ..., x\}S[i] = x\}|$ where S is the sequence, 'i' is the index, and x is the symbol. The total functioning time of rank operation is $O(\log)$. Quantile query of wavelet tree is having quantile $k(S, i, j)$ given return value the kth smallest symbol in range of (i, j); here i is the index, S is sequence, and j is another index. Quantile operation performed on the both side child nodes of the wavelet tree. The running time of quantile operation is $O(\log)$, but it is slightly bigger than the rank operation. Range query returned the no. of value which lies between the x, y and appears in range $[i, j]$3. The x is the x-axis of index, and y is the y-axis of actual value. The running time of range operation is $O(\log)$4. This value is higher than the rank operation or quantile operation.

Remaining part of the paper is classified in the following manner. Section 2 contains the contribution of wavelet trees and has its subsections. The first subsection includes indexing of wavelet tree. It consists of textual indexing, Huffman encoding, canonical Huffman code. It consists of the several wavelet techniques and its combinations with End Tagged Dense Code (ETDC) or Anatomy of a Tome witch Trials (ATWT) codes with wavelet tree. Second subsection includes indexing used by parallel wavelet tree. It consists of latest parallel algorithms and construction of parallel wavelet tree. Third subsection consists of indexing used in the healthcare and bioinformatics. It consists of MRI image sequencing, etc. Section 3 contains the proposed methodology, and Sect. 4 contains conclusion and discussion of the paper.

2   Contribution of Wavelet Tree in Indexing

As discussed in Sect. 1, wavelet tree is the versatile data structure which can be used in multiple dimensions, i.e., string processing, grid point, sequencing of alphabets, compression, and indexing. The contribution of wavelet tree technology is given in Table 2. In this section, we discuss the contribution of the wavelet tree to index different types of data and its further applications. It starts with terminology of the wavelet tree followed by its indexing application in text data, healthcare indexing, bioinformatics imaging, and earthquake indexing. Objective of this study is to a systematic study of indexing covering how indexing using wavelet trees can be used for different datasets, i.e., alphabets, text, and images. The popularity of wavelet trees is due to its discriminative features, i.e., range quantile, time quantile, wavelet matrix, rank query and wavelet packets, etc. Select and rank operations are the mechanism to permit the efficient implementation of operation on the set of non-binary dataset. It can be used to perform indexing for searching a document.

2.1 Indexing in Wavelet Tree

Textual indexing (alphabet, string) can be done using a wavelet tree. In the paper [1], authors proposed an idea of Huffman coding for searching the indices of a dataset. They focused on the self-indexing of wavelet trees and used the Huffman encoding for the compression of wavelet trees. In the same paper, they proposed new variants of indexing based on the word-based compressions and new improvement in the wavelet tree by joining it with char-based Huffman codes. Results are not accurate because the text indexing is used by words rather than the characters. It increased the size of dictionary. In the year "2007" [5], authors worked on binary sequencing to solve the complex structure of the wavelet tree by resolving complexity of select and rank operation for searching alphabets/text from index. The objective of this research was to reduce space utilization to design index. In paper [6], the authors have proposed implementation of the suffix array using a wavelet tree. This paper works on encoding the Burrow–Wheeler transform with canonical Huffman code using a wavelet tree. It has implemented the suffix automaton with space in an economical way. This research opens a new dimension of work to reduce search time and design index. In paper [7], author has reviewed the concept of indexing and concluded that theory of modified Bi-tree algorithms can work better for longer suffixes to text files. In paper [8], a research was conducted for sequencing, shorting, and portioning (left subtree and right subtree) of wavelet trees. This research is helpful for index construction using wavelet trees. In paper [9], authors proposed a feature which can identify discriminative features for texture data for texture indexing. In the family of trees, wavelet packets provide valid representations for indexing issues. Inverted indexing associated every term in a document and an index of pointers where location of its occurrences. In, inverted indexing, the indexing is performed using inverted files. The inverted files consist of three components: (1) vocabulary, (2) dictionary, and (3) the inverted lists. Rather than using only string indexing, researchers are also going into inverted indexing of documents for more details, readers can find in the paper of 2015 [10], and the technique of spatial, textual indexing is hybrid indexing, also in the paper of 2016 (IGI Global, Inverted indexing), also in the paper of 2019 [11], focused on the time complexity and efficiency of wavelet tree and also read the concept of dual indexing.

In the paper [12], authors proposed select/rank and range quantile operation algorithms for inverted indexing using wavelet trees. But results are not faster for basic operations. They also focused on self-indexing using wavelet trees and designed for fast searching using the suffix tree and suffix arrays. Main aim of this research was to lower the space complexity of indexes, but they do not discuss the search time complexity. In the paper [2], the author works on the full-text indexing of wavelet trees. It supported the fast full-text pattern matching technique using wavelet trees, which required a limited amount of space. It studied the various codes of wavelet trees like AWWT codes, Huffman codes and performed some experiments with that. It is also done a study of wavelet tree size, wavelet tree construction time, and wavelet query time. In the paper [13], the author has examined the various issues involved in the characteristics of data structures. The wavelet tree has the feature of range searching in the data structure. It acts as a self-index compression on the suffix arrays. They provide an algorithm of compression on the mechanism of the wavelet tree. Again in paper [3], the authors show the space for lower-order index terms and achieve the asymptotically empirical entropy of texts. If the text is highly compressible and the alphabet proportional is small, its entropy is $H_h = 0\,(1)$. On the bit arrays, time enhancement is needed, and more research should be done on this side. In the year 2015, a new version of wavelet tree was formed; it is called parallel wavelet tree.

2.2 Indexing Using Parallel Wavelet Tree

In the paper [10], the author worked on the complexity of parallel wavelet trees. It provides two algorithms on parallel wavelet trees having complexity $O(n \log)$. The first algorithm is positioned on the polylogarithmic depth, and the second algorithm is based on the depth. These algorithms also work on the arbitrarily shaped binary trees as well as multinary trees and wavelet matrices which are having lower work complexity. In the paper [14], the author represents the parallel wavelet tree construction algorithm. It has done improvement upon the linear depth of the latest parallel algorithms. It provided the two parallel algorithms for each bitmap and each stage of the parallel wavelet tree. The first algorithm consists of construction of all the levels of the wavelet tree in parallel with $O(n)$ time and working space of $(On \log + \log n)$ where n is the size of input sequence and is the alphabet. The second algorithm consists of wavelet decomposition of domain for each segment; the reaching time is $O(\log n)$ time and $O(n \log + p \log n/\log)$ bits of extra space where p is the number of available cores. The parallel construction wavelet tree would need more space in memory, and for reducing the space, more enhancements are needed for tree construction. Again in paper [15], author proposed index construction using parallel wavelet trees. It provided a sorted algorithm named sort WT and two theorems to prove work and depth complexity. According to the first theorem, the level requires $O(n \log)$ work and $O(\log n \log)$ depth and according to the second theorem; it requires $O(\text{sort}\,(n)\,\log)$ work and $O\,(l\,\text{sort}\,(n) + \log\,(n))$ depth.

2.3 Indexing in Healthcare and Bioinformatics

In paper [16], author also used the wavelet tree in healthcare. It used the wavelet technique for magnetic resonance imaging (MRI). The wavelet tree is used for compressive sensing for reconstruction of MRI image. It has used two main methods for static MRI sequencing of wavelets. So dynamic MRI indexing is also needed for imaging in healthcare. Bioinformatics is used in wavelet trees. In the paper [17], authors have used the wavelets for detecting the earthquakes. It worked on the seismic signals and applied wavelet technology over it. Wavelet packets are used for searching. It provided a linear sequence of time frequency atoms which are collected from the dilation of analyzing functions. It can detect the earthquake, but faster techniques are needed for detection. The transformation of signals is required for detection of seismic vibrations. In the year "2018" [18], author proposed a new dimension of application of wavelet trees to index bioinformatics data for searching sensitive data on demand. This work proposed a secured algorithm named secure wavelet matrix [SWM]. The string search time of mentioned work is $|\Sigma|$. But, it can be executed on homographic encryption of a wavelet tree. In the paper [19], authors created a new concept of wavelet tree that can be connected with machine learning procedures for weather forecasting. It has combined traditional quantile mapping approach with WSVM wavelet support vector machine and wavelet random forest techniques. It has used the pattern reading methods of wavelet forecasting the summer or winter. But new improvements are needed in forecasting, quantile mapping.

3 Proposed Methodology

In the proposed methodology, firstly take the normal string, which consists of characters and having a set of alphabets. Suppose we take the set of alphabets like "A BAAAABAAA" and apply the LZW compression over it. The LZW compression used a code table 0.255 always assigned to represent a single byte from the input file. After applying the LZW compression, we could get the result in the form of tables. The whole set of alphabets solved in ten tables. The start table or end table is shown in Fig. 1.

At the end of table, we get the string "AABBAAAAAAAAABBAAAAA" the string is in the sorted form, but it is greater than the original string. The more compression of this string is required. Then, the construction of wavelet tree using LZW string is done as shown in Fig. 2. The flowchart of proposed methodology is shown in Fig. 3.

In this paper, we represent the two character set, one is alphabet set, and another is compressed alphabet set. The complexity of alphabet set is 0.6 s, and space complexity is 320 bits.

1

Encoder Output		String Table	
Output Code	Representing	Codeword	String
66	A	256	A

2

Encoder Output		String Table	
Output code	Representing	Codeword	String
66	A	256	A
67	B	257	AB

9

Encoder Output		String Table	
Output code	Representing	Codeword	String
66	A	256	A
67	B	257	AB
66	A	258	BA
66	A	259	AA
66	A	260	AA
66	A	261	AA
67	B	262	AB
66	A	263	BA
66	A	264	AA

10

Encoder Output		String Table	
Output code	Representing	Codeword	String
66	A	256	A
67	B	257	AB
66	A	258	BA
66	A	259	AA
66	A	260	AA
66	A	261	AA
67	B	262	AB
66	A	263	BA
66	A	264	AA
66	A	265	AA

Fig. 1 Construction of wavelet tree using LZW string

The time complexity of compressed alphabet set is 0.6 s, and space complexity is 608 bits as given in Table 1. The LZW compression takes more space as comparison of without compressed string. LZW compression mainly depends upon dictionaries. Its compression mainly depends upon the no. of repetitions between the old data and former data. It is well-suited algorithm for the characters. It is having

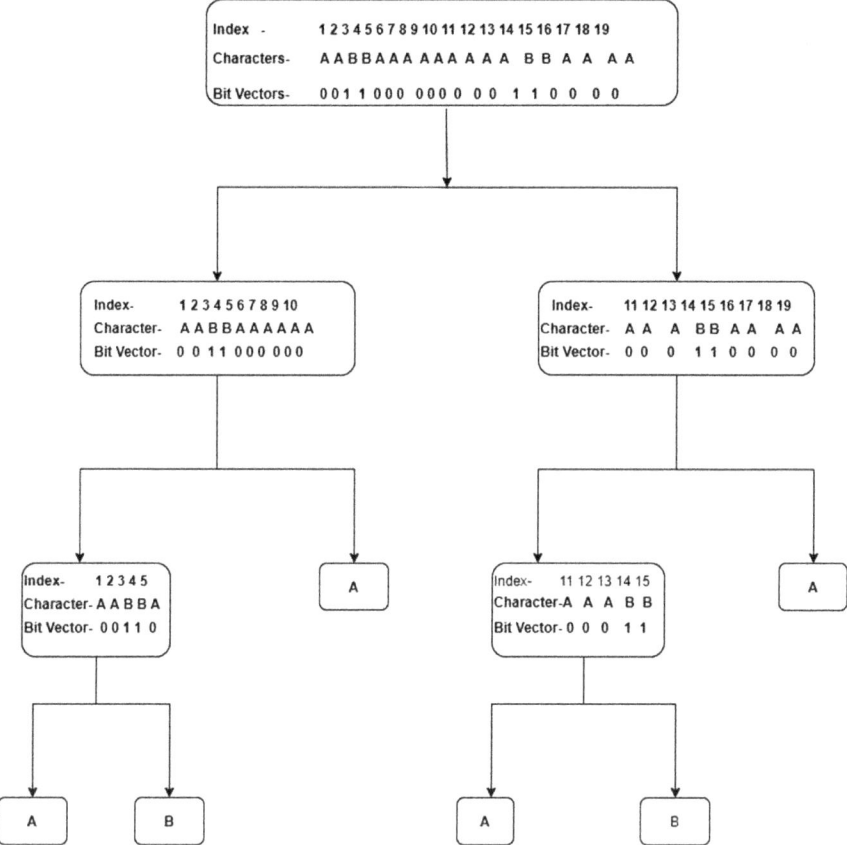

Fig. 2 Construction of wavelet tree with LZW string

two main features that are (1) latency tolerance and (2) real-time compression. It reduced the data amount of storage and transmissions. Compression of redundant data reduced the amount of data transferred from one system to another system. LZW compression is utilized and improves the reliability of non-volatile storage system. The LZW compression is the integral part of many data compression systems, for image compression and storage. In LZW compression, the process is started with empty dictionary when the next symbol is encoded in the dictionary. Then, it start the process of its compression. It encodes symbols and applies compression constantly and reached the end of dictionary with highest encoded symbol.

It used the 8-bit ASCII code for encoding the symbols in the dictionary. It starts encoding the symbols from the value of 0–255. Then, decoding process is also started by initializing the dictionary to all 256 symbols in the alphabet. Then, it reads its input stream (which consists of pointers to the dictionary) and uses each pointer to retrieve uncompressed symbols from its dictionary and write them on its output file/stream. It does not need to receive the dictionary from compressor. It builds its

Fig. 3 Flowchart of proposed methodology

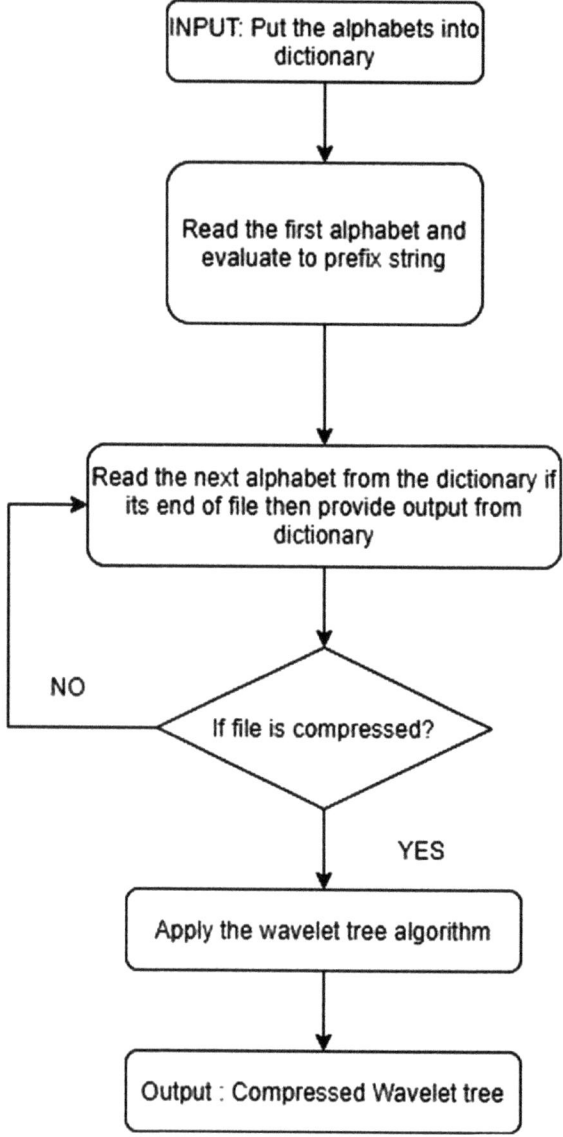

INPUT: Put the alphabets into dictionary

Read the first alphabet and evaluate to prefix string

Read the next alphabet from the dictionary if its end of file then provide output from dictionary

NO

If file is compressed?

YES

Apply the wavelet tree algorithm

Output : Compressed Wavelet tree

Table 1 Character set complexity

S. No.	Space and time complexity of different character set		
	Character set	Time complexity (s)	Space complexity (bits)
1	Alphabet set	0.60	320
2	Compressed alphabet set	0.60	608

dictionary in the same way as the encoder. In LZW compression technique, encoder and decoder are synchronized (Table 2).

Table 2 Contribution of wavelet tree technology

References	Year	Contribution in wavelet tree technology
[1]	2006	It contributed to the self-indexing of wavelet trees and organized as a binary tree. It performs indexing as an indexed natural language text. Joining the Huffman tree or char-based Huffman codes, word-based Huffman codes, wavelet tree build with ETDC codes
[5]	2007	Contributed in the select and rank queries of the wavelet tree. It mainly denotes a structure by Rank $c(s, q)$, the no. of times c appears to select queries, and representation of the no. of binary sequences and select can be seen for $c(s, q)$. The point in the QTH occurrence of bit c. Size of a function which is having empirical entropy zero order is $Ho(s) = -\Sigma(nc/n)$ $\log(nc/n)$ where nc is the number of occurrences of symbol in S, $n = \Sigma c$, $nc = |s|$ and discussed the concept of compression boosting which shows about the "best possible" contexts
[17]	2009	In this paper, the author has done the analysis on wavelet trees used for the ground seismic records sequencing motion records of grounds. Wavelet transforms used for sequential data in time axis earthquake accelerations. In this, wavelet packet having function Ψ, and i is the parameter of modulation with dilation parameter k. $\Psi i j, k(t) = 2 - 1/2\,\Psi i\,(2j\,t - k)$, and corresponding signal to wavelet tree is $Cij, k = -\infty \square \infty f(t)\,\Psi i j, k(t)\,dt$
[6]	2010	In this paper, the contribution mainly depends upon the space of wavelets and given the canonical Huffman code. It reduces the space of wavelet trees with Huffman code. It implemented the wavelet tree which is having height $[\log |\Sigma|]$, and the vector of a node is $i + 1$th bits, where 'i' is the binary prefix of a node. It consists of a comparative study of canonical Huffman code, Huffman tree-indexed-based code, and Wavelet tree-based indexed code
[12]	2011	Wavelet trees contributed for solving the fundamental issues such as range next value queries, range insertion queries, and range quantile queries. The new algorithm is implemented on the range of wavelets. It has range next value, range quantile, and range intersection for range to achieve adaptive complexity. It has matched with one achieved for sorted ranges. Total space included the sequence of $\eta Ho(s) + \Theta(\eta + \sigma)$ where $[1, \sigma]$ alphabet of S and $Ho(s)$ is its zero-order empirical entropy
[2]	2011	Contribution in the wavelet tree of this paper is that it works on the full-text pattern matching technique of wavelet tree which requires less space. It shows study of different codes of wavelet trees like Huffman codes, AWTT codes, etc.
[13]	2012	Contribution in the wavelet tree of this paper contains wavelet tree data structure and text compression applications. It handles the various applications of wavelet trees. It focused on the self-indexing method for wavelet trees and how the self-index data structure manages itself and how it moves to the compression point from the origin text to indexing module. It grants more functionality to a wavelet tree indexing procedure

(continued)

Table 2 (continued)

References	Year	Contribution in wavelet tree technology
[7]	2014	In this paper, they showed the indexing of text data and their incarnations or applications. It also provided specifications to the ubiquitous tools of string matching and also studied various algorithms. They discussed how searching of the text takes place and how text differs in patterns and a single text character insertion, deletion, and substitution take place. It is also focused on the issue and solved it by finding the longest substring, which appears twice
[3]	2014	New implementations of compressed suffix arrays that exhibit several trade-offs compressed suffix array requires self-indexing $n\,H_h + O(m\,\lg |\Sigma|+\text{Poly}\,\log{(n)})$ time. To reduce the query time, they desire $\lg + k_x \leq Z_k\,\lg |\Sigma|$ encoding space $n\,H_h + 0\,(n_k)$ bits per level
[18]	2015	Wavelet tree represents novel algorithms of string in logarithmic time $|\Sigma|$. The algorithm for wavelet matrix is (SWM) which shows the additively homomorphic encryption built and efficient data structure wavelet matrix. It could search any type of string, and protecting its privacy is an important approach The first iteration $f_0 + O_0 \rightarrow F_1 = v[f_0 + O_0]$ $G_0 + O_0 \leftarrow g_1 = v[g_0 + O_0]$ The Kth iteration $f_{K-1} + O_{k-1} \rightarrow g_k = v[g_{k-1} + O_{K-1}]$ $G_{k-1} + O_{k-1} \leftarrow g_k = v[g_{K-1} + O_{k-1}]$
[10]	2015	Contributed in the indexing of wavelet trees. It provides a technique on spatial-textual hybrid indexing. It searches the textual keywords and designs an inverted index. It can propose two indexing algorithms in the hybrid indexing. The first one works on the spatial hybrid index. The second one works on searching documents on wavelet trees from hybrid index
[16]	2015	The wavelet contributed to the expansion and standardization of the wavelet tree and the experiments on three parallel algorithms for constructions which includes the number of threads, range length, and alphabet size on it
[14]	2016	Wavelet trees contributed to the document retrieval. The wavelet tree does not use Huffman coding and replaces the bitmap with an integer array. For depth first search traversal (DFT), σ be length of alphabet and $|A|$ be length of array, we can say that we have six documents have total $|A|$ characters
[19]	2016	In this paper, comparative study of inverted files in data structure is to be performed. They focused on indexing resolving methods. Indexing is used for re solving the issues of textual query, geographical query, and inversion of list. Indexing is based on the tree or in the hashing. It can simplify the process by reducing text size and stop words, and reduction of spaces
[8]	2016	Wavelet trees contributed to invert indexing. Indexing technique proposed on wavelet trees for fast text retrieval. It focused on the inverted indexing which can consist of three components: terminology, dictionary, and inverted list. There are several different structures in which an inverted list is implemented like a hash tree, B-tree, and sorted arrays

(continued)

Table 2 (continued)

References	Year	Contribution in wavelet tree technology
[9]	2017	Algorithm that reduced the time complexity of wavelet trees by using ubiquitous multicore machines. It focused on the extension of the wavelet matrix. It focused on the general representation of w tree, and also the variants of w tree
[15]	2018	Wavelet trees contributed in weather forecasting. The weather forecast was done in the local areas of china. The wavelet SVM and random forest algorithms are used for examining the local areas. It also provides information regarding seasonal pattern rainfall focused in china
[4]	2019	In this, Julian mainly focused on the construction of wavelet tree and its traversal of the tree, and its queries solved the rank query of wavelet and provided an algorithm on it
[20]	2019	Wavelet packet bases are a computational efficient algorithm explored for the tree structure of wavelet packet collection. The adaptive quality of the proposed solution demonstrates constructed and real data structures. In this, we check the wavelet packet base. Wavelet packets used for the minimum cost tree pruning is CART. Wavelet tree indexing is used for sufficient balance between feature discrimination and determining (over-fitting) complexity
[11]	2019	In this paper, they contributed to the efficiency and time complexity of wavelet trees. They show how time complexity reduces the searching time. It also performed indexing on the R-tree or $R*$-tree. It also show efficiently work on B-tree, R-tree, R^+ tree, $R*$ tree, and Hilbert R-tree
[21]	2020	In this paper, they focused on the IoT applications used in the health care. IoT applications has affected the human life personally as well as professionally. The IoT applications used for healthcare monitoring. It records the patient's data. Doctors can easily check out patient status and prescribed medicines according to it
[22]	2020	Improved the work complexity of wavelet tree construction. The wavelet parallel tree algorithm with $O(n \log \sigma / \sqrt{\log n})$ works on polylogarithmic depths for the any value of σ

4   Conclusion and Discussion

Wavelet tree technology is more significant and perfect technique for indexing. It is used for the indexing of all the sufficient data structures and programming. This technology is used in the multidimensional areas of indexing. This paper consists of systematic study on the contribution of indexing in wavelet trees. Indexing techniques are also called by researcher space efficient techniques. Indexing is mainly used by trees like H-tree, B-tree, R-tree, $R*$-tree, etc. Wavelet technology is used for the self-indexing methods. It is used for range quantile problems, range insertion queries, and perfect use of space sequences. Wavelet trees are having new indexing methods regarding the time and space complexities. It uses multicore architecture for optimizing the data. Wavelets packet formation for indexing of data and parallel wavelet tree, which can reduce the time taken for indexing of the data with indexing procedures from the dataset. This paper consists of a contribution of wavelet tree and a proposed technique of LZW compression. This technique required the characters' indexing method and its compressed construction. In the future, it may be used to perform the exhaustive experiments over the different texts, and it may work efficiently implementations on byte-oriented select rank operations.

Acknowledgements This research is supported by Council of Science and Technology, Lucknow, Uttar Pradesh via Project Sanction letter number CST/D-3330.

References

1. Brisaboa NR, Cillero Y, Fariña A, Ladra S, Pedreira O (2007) A new approach for document indexing using wavelet trees. https://doi.org/10.1109/DEXA.2007.118
2. Grossi R, Vitter JS, Xu B (2011) Wavelet trees: from theory to practice. In: Proceedings—1st international conference on data compression, communication, and processing, CCP 2011, pp 210–221. https://doi.org/10.1109/CCP.2011.16
3. Grossi R, Gupta A, Vitter JS (2003) High-order entropy-compressed text indexes
4. Grossi R (2014) Wavelet trees. Encyclopedia algorithms. Published online 2014, pp 1–6. https://doi.org/10.1007/978-3-642-27848-8_642-1
5. Ferragina P, Manzini G, Mäkinen V, Navarro G (2007) Compressed representations of sequences and full-text indexes. ACM Trans Alg 3(2). https://doi.org/10.1145/1240233.124 0243
6. Yang W et al (2013) Compressed format index based on suffix arrays and it's implementing in bioinformatics. J Bionanosci 7(1):110–113
7. Apostolico A, Crochemore M, Farach-Colton M, Galil Z, Muthukrishnan S (2013) Forty years of text indexing. In: Lecture notes in computer science (including subseries lecture notes in artificial intelligence and lecture notes in bioinformatics). LNCS, vol 7922, pp 1–10. https://doi.org/10.1007/978-3-642-38905-4_1
8. Yadav AK, Yadav D, Prasad R (2016) Efficient textual web retrieval using wavelet tree. Int J Inf Retr Res 6(4):16–29. https://doi.org/10.4018/ijirr.2016100102
9. Fuentes-Sepúlveda J, Elejalde E, Ferres L, Seco D (2017) Parallel construction of wavelet trees on multicore architectures. Knowl Inf Syst 51(3):1043–1066. https://doi.org/10.1007/s10115-016-1000-6
10. Yadav A, Yadav D (2015) Wavelet tree based hybrid geo-textual indexing technique for geographical search. Indian J Sci Technol 8(33):1–7. https://doi.org/10.17485/ijst/2015/v8i33/72962
11. Yadav AK, Yadav D (2019) Wavelet tree based dual indexing technique for geographical search. Int Arab J Inf Technol 16(4):624–632
12. Gagie T, Navarro G, Puglisi SJ (2012) New algorithms on wavelet trees and applications to information retrieval. Theor Comput Sci 426–427:25–41. https://doi.org/10.1016/j.tcs.2011.12.002
13. Makris C (2012) Wavelet trees: a survey, vol 9. https://doi.org/10.2298/CSIS110606004M
14. Institute of Electrical and Electronics Engineers, IEEE Signal Processing Society (2016) IEEE international conference on image processing: proceedings. Phoenix Convention Center, Phoenix, Arizona, USA, 25–28 Sept 2016
15. Xu L, Chen N, Zhang X, Chen Z, Hu C, Wang C (2019) Improving the North American multi-model ensemble (NMME) precipitation forecasts at local areas using wavelet and machine learning. Clim Dyn 53(1–2):601–615. https://doi.org/10.1007/s00382-018-04605-z

16. Shun J (2015) Parallel wavelet tree construction. In: Data compression conference proceedings, vol 2015. Institute of Electrical and Electronics Engineers Inc., July 2015, pp 63–72. https://doi.org/10.1109/DCC.2015.7
17. Ghodrati Amiri G, Asadi A (2009) Comparison of different methods of wavelet and wavelet packet transform in processing ground motion records. Int J Civ Eng 7(4):248–257
18. Sudo H, Jimbo M, Nuida K, Shimizu K (2019) Secure wavelet matrix: Alphabet-friendly privacy-preserving string search for bioinformatics. IEEE/ACM Trans Comput Biol Bioinf 16(5):1675–1684. https://doi.org/10.1109/TCBB.2018.2814039
19. Yadav AK, Yadav D, Rai D (2016) Efficient methods to generate inverted indexes for IR. Adv Intell Syst Comput 435:431–440. https://doi.org/10.1007/978-81-322-2757-1_43
20. Vidal A, Silva JF, Busso C (2019) Discriminative features for texture retrieval using wavelet packets. IEEE Access 7:148882–148896. https://doi.org/10.1109/ACCESS.2019.2947006
21. Gupta S, Goel L, Agarwal AK (2020) Technologies in health care domain: a systematic review. Int J E-Collab 16(1):33–44. https://doi.org/10.4018/IJeC.2020010103
22. Shun J (2020) Improved parallel construction of wavelet trees and rank/select structures. Inf Comput 273. https://doi.org/10.1016/j.ic.2020.104516

Author Index

Lightning Source UK Ltd.
Milton Keynes UK
UKHW020829110123
415103UK00002B/3